INVERSE SPECTRA

# North-Holland Mathematical Library



VOLUME 53

ELSEVIER
Amsterdam – Lausanne – New York – Oxford – Shannon – Tokyo

# Inverse Spectra

A. CHIGOGIDZE
*Department of Mathematics and Statistics*
*University of Saskatchewan*
*McLean Hall*
*Saskatoon, Canada*

1996
ELSEVIER
Amsterdam – Lausanne – New York – Oxford – Shannon – Tokyo

ELSEVIER SCIENCE B.V.
Sara Burgerhartstraat 25
P.O. Box 211, 1000 AE Amsterdam, The Netherlands

ISBN: 0 444 82225 9



This book is printed on acid-free paper

Transferred to digital printing 2005

# Preface

The notion of an inverse sequence and its limit first appeared in the well-known memoir by Alexandrov [**3**], where a special case of inverse spectra – the so called projective spectra – was considered. The concept of an inverse spectrum in its present form was first introduced by Lefschetz in [**209**] and later, in full generality, in [**210**]. In the meantime Freudenthal [**152**] introduced the notion of a morphism of inverse spectra. The foundations of the entire method of inverse spectra were laid down in these basic works. The book of Eilenberg and Steenrod [**143**] contains a detailed discussion of (algebraic) inverse spectra and must be mentioned as well. Subsequently, inverse spectra began to be widely studied and applied not only in the various major branches of Topology, but also in Functional Analysis and Algebra. This is not surprising because of the categorical nature of inverse spectra and of the extraordinary power of the related techniques.

In Topology inverse spectra have been used both for the construction of objects with special properties (synthetic applications) and for investigation of complicated objects by means of their spectral approximation by simpler ones (analytic applications). For instance, using projective spectra Alexandrov approximated compacta by complexes and this method allowed him to extend the basic results of the combinatorial topology of polyhedra to arbitrary compacta. The next significant success of analytic applications of inverse spectra was the investigation by Pontrjagin and A.Weil of the structure of compact groups using approximation by Lie groups. Pontrjagin also considered continuous transfinite spectra – the so called Lie series.

Since their creation, inverse spectra have been extensively used for constructing various exotic examples of spaces and maps. We mention here the example of the open map of a one-dimensional compactum onto a two-dimensional one constructed by Kolmogorov [**196**]. Several counterexamples in dimension theory, as well as in the theory of cardinal invariants, have been produced using this approach (see, for example, [**147**]). It should be emphasized that all the special dimension-raising maps presented in Chapter 4 are also constructed using the techniques of inverse spectra.

Mardešić made the next step in analytic applications of inverse spectra in

[**213**]. His principal result states that every (non-metrizable) compactum admits a representation as the limit space of an inverse spectrum, consisting of metrizable compacta of the same dimension. The significance of this statement becomes clear after remarking that there exist [**213**] one-dimensional non-metrizable compacta which do not have such representations, consisting of polyhedra. Further work in this direction made possible the development of the dimension theory (including both Lebesgue and cohomological) of non-metrizable compact spaces.

The work of Haydon [**164**] was a landmark in the development of the method of inverse spectra. Haydon proved that the class of compact absolute extensors in dimension 0 coincides with the class of Dugundji compacta. But undoubtedly the most significant ingredient of his work was the discovery of what are now called Haydon spectra. It should be emphasized especially that in the work of Alexandrov on the homological theory of compacta and in the works of Pontrjagin and A.Weil on compact groups all of the results are natural extensions of the results in the metric case. Mardešić's theorem, mentioned above, also has, to some extent, a "countable" counterpart – the theorem of Freudenthal [**152**] which states that every metrizable compactum is the limit of an inverse sequence, consisting of polyhedra of the same dimension. In this sense Haydon's method has an "uncountable" nature. It is based on a special spectral construction whose roots are in a simple set-theoretical principle known as "the effect of uncountability." This principle, extracted in its present form by *Ščepin* (see [**278**]), serves as the foundation for the Spectral Theorem. This theorem is perhaps the most powerful tool of the whole theory. It states the following: if the limit spaces of two uncountable inverse spectra (with some additional non-restrictive properties) are homeomorphic, then these spectra contain isomorphic cofinal subspectra. The Spectral Theorem has no analog for inverse sequences: for example, two inverse sequences, consisting of even- and odd-dimensional cubes, produce the same limit space – the Hilbert cube – but, of course, they contain no isomorphic subspectra whatsoever. Theorems stating that "limit spaces of spectra, consisting of "good" spaces are "good"" were typical in spectral applications. The Spectral Theorem allows us to obtain theorems stating that "limit spaces of spectra, consisting of "bad" spaces or "bad" projections are "bad"."

Detailed discussion of various versions (including compact and non-compact cases) of the Spectral Theorem is presented in Chapter 1.

The remainder of the book illustrates the extraordinary strength of the spectral method. It shows exactly how this method works on a range of examples and what kinds of difficulties it helps to overcome. We present various problems from different areas of Topology where the Spectral Theorem is effective. The list of applications presented in Chapter 8 includes: topological characterizations of uncountable powers of the Cantor cube and of the separable Baire space; spectral representations of topological groups (including a simple proof of the existence of Lie series); topological characterization of locally convex linear topological spaces that are homeomorphic to the powers of the real line (generalization of the classical result of Anderson-Kadec); the Spectral Theorem in Shape Category; actions of non-metrizable topological groups and the structure of fixed point sets in non-metrizable manifolds; the Spectral Theorem for Baire maps (isomorphisms); connections with direct spectra and $\Sigma$-products. The reader can find several other particular applications of the spectral technique in other parts of the book as well.

Manifold theory, or, more generally, the theory of absolute extensors, is one of the major areas of modern topology where the spectral approach has already shown its full strength. We discuss this theory in detail. We start with metrizable infinite-dimensional manifolds. This subject is almost completely covered in three excellent books written by C. Bessaga and A. Pelczyński [**32**], by T.A.Chapman [**69**] and by J. van Mill [**228**]. For this reason, in Chapter 2 we present an updated survey of this theory. The reader can find here proofs of several statements of $Q$-manifold and $l_2$-manifold theories. We begin with the introduction of the concept of (strong) $Z$-sets and discuss all major results of these theories including Toruńczyk's characterization theorems of the model spaces: countable infinite powers of the closed unit interval [**298**] and the real line [**299**].

Chapter 3 contains a solution of the well-known and long standing problem concerning coincidence of Lebesgue and integral cohomological dimensions.

Chapter 4 can be considered as an independent introduction to Menger manifold theory. We present a detailed discussion of this theory including the basic geometric constructions. We also present examples of $n$-soft maps of $(n+1)$-dimensional Menger compacta onto the Hilbert cube. These maps, constructions of which, as was mentioned above, are performed using the spectral techniques, are used both in Menger manifold theory and later in the general theory of absolute extensors.

Unlike Menger compacta very little is known about Nöbeling universal spaces. In Chapter 5 we illustrate the close ties between Menger and Nöbeling spaces by showing how Nöbeling spaces can be identified with the so-called pseudo-interiors of Menger compacta. Using the spectral approach we construct an $n$-soft map of $n$-dimensional Nöbeling space onto the separable Hilbert space.

In Chapters 6 and 7 we develop the general theory of absolute extensors in dimension $n$, $n \in \omega$. The Spectral Theorem plays a crucial role in almost every statement here. As the culmination of this theory we present its two major elements – the topological characterizations of uncountable powers of the closed unit interval and of the real line (originally done by Ščepin [**277**] and the author [**84**] respectively). Coupled with the corresponding results of Toruńczyk we therefore obtain topological characterizations of all infinite products of the closed unit interval and the real line.

Each section is ended with Historical and Bibliographical notes. Throughout the text the reader can find discussions of several unsolved problems.

This book significantly extends and updates the previously published (in Russian) book written by the author [**102**] (with the help of V. V. Fedorchuk). Responding to a suggestion to translate [**102**] I instead decided to rewrite book completely in order to include several new topics and recent developments. The result of this work is before the reader.

I wish to express my thanks to Dr. J. van Mill for his support. I also thank Dr. R. B. Sher who read the entire manuscript and made many valuable suggestions.

Alex Chigogidze

Saskatoon, CANADA

The author was supported in part by NSERC Grant # OGP0155552.

# Contents

CHAPTER 1

# Inverse Spectra

In most cases all necessary notions are defined in the text. Nevertheless we assume that the reader is familiar with basics of general topology as it is presented in [**145**] or in any other standard textbook. All spaces under consideration are assumed to be completely regular (i.e. Tychonov) and maps are continuous. $D$, $N$, $I$ and $R$ stand for the two-point (discrete) space, natural numbers, the closed unit segment and the real line respectively. $\omega$ and $\omega_1$ denote the first infinite and the first uncountable cardinals respectively.

## 1.1. Preliminary information

**1.1.1. Completeness and compactness.** We begin with the key notion of *completeness* of a metrizable space. A sequence $\{x_n : n \in \omega\}$ of points in a metric space $(X, d)$ is called a *Cauchy sequence* if for every $\epsilon > 0$, there exists a natural number $k = k(\epsilon)$, such that $d(x_n, x_m) < \epsilon$ whenever $n, m \geq k$. As can be easily seen every convergent sequence of points in any metric space is a Cauchy sequence, but not vice versa.

DEFINITION 1.1.1. *A metric space $(X, d)$ is* complete *if every Cauchy sequence in $(X, d)$ is convergent to a point of $X$.*

Not every metric, generating the topology of a given complete metrizable space, is complete. However, the very property of existence of at least one complete metric generating the given topology of a metrizable space is a topological invariant. In what follows, we call separable completely metrizable spaces the *Polish spaces.*

It is well-known that the product of countably many Polish spaces is a Polish space.

PROPOSITION 1.1.2. *The following conditions are equivalent for any subspace $Y$ of any Polish space $X$:*

(i) *$Y$ is a Polish space.*

(ii) *$Y$ is a $G_\delta$-subspace of $X$.*

Since the real line $R$ is a Polish space we conclude by Proposition 1.1.2 that the space of rational numbers (considered as a subspace of $R$) is not Polish. At the same time irrational numbers form a Polish space. The validity of the last fact can also be seen as follows. Since the space of irrationals is homeomorphic to the countable infinite power $N^\omega$ of the space $N$ of natural numbers, it suffices to note that $N$ itself, as a closed subset of $R$, is a Polish space and to observe once again that the operation of taking countable products maintains the property of being a Polish space. Similarly, we see that the countable infinite power $R^\omega$ of the real line $R$ is also a Polish space. The spaces $N^\omega$ and $R^\omega$ will play key roles below.

The following statement is one of the most convenient definitions of a Polish space.

PROPOSITION 1.1.3. *The following conditions are equivalent for any space $X$:*

(i) *$X$ is a Polish space.*

(ii) *$X$ is homeomorphic to a closed subspace of the space $R^\omega$.*

A space $X$ has the *Baire property* if an intersection of any countable collection of dense open subsets of $X$ is dense. This property of the real line $R$ was shown by Baire (which is why the property bears his name). A much more general statement is true.

PROPOSITION 1.1.4. *Every Polish space has the Baire property.*

Let us discuss the properties of the space $N^\omega$ in more detail (by the way, this space appears in the literature under the name of the *separable Baire space* as well). First of all, we present a classical result of Alexandrov-Urysohn , which gives a topological characterization of this space.

THEOREM 1.1.5. *The following conditions are equivalent for any space $X$:*

(i) *$X$ is homeomorphic to $N^\omega$.*

(ii) *$X$ is a zero-dimensional Polish space without compact open subspaces.*

It is very interesting that the space $N^\omega$ plays the same role with respect to the class of zero-dimensional Polish spaces, as the space $R^\omega$ does with respect to the class of all Polish spaces. The following statement can be viewed as an illustration of this fact.

PROPOSITION 1.1.6. *The following conditions are equivalent for any space $X$:*

(i) *$X$ is a zero-dimensional Polish space.*

(ii) *$X$ is homeomorphic to a closed subspace of the space $N^\omega$.*

It is worth noting that all Polish spaces can be described by means of the space $N^\omega$. Indeed, let us observe, first of all, that the real line $R$ can be presented as an open image of $N^\omega$ (the reader can easily check this fact by analyzing the proof of Lemma 6.4.1). Further, taking the countable infinite power of the such an open mapping we see that there exists an open mapping $f : N^\omega \to R^\omega$. Consider now an arbitrary Polish space $X$, which, by Proposition 1.1.3, can be identified with a closed subspace of $R^\omega$. Clearly, the inverse image $f^{-1}(X)$ is a zero-dimensional Polish space and consequently, the product $f^{-1}(X) \times N^\omega$ is homeomorphic to $N^\omega$ (apply Theorem 1.1.5). Clearly, the composition of the projection $f^{-1}(X) \times N^\omega \to f^{-1}(X)$ and of the open mapping $f^{-1}(X) \to X$ is open. Therefore $X$ is an open image of $N^\omega$. The converse statement is also well-known. Thus we have proved the following theorem.

THEOREM 1.1.7. *The following conditions are equivalent for any space $X$:*
(i) *$X$ is a Polish space.*
(ii) *$X$ is an open image of the space $N^\omega$.*

An important subclass of the class of Polish spaces is formed by the compact (metrizable) spaces. The theory of these spaces is one of the most developed areas of general topology, and for this reason we restrict ourselves to a very brief discussion.

The *Cantor cube* $D^\omega$ and the *Hilbert cube* $I^\omega$ are basic objects for the entire class of (zero-dimensional) metrizable compacta. Their roles are emphasized by the following statements (compare with Propositions 1.1.3, 1.1.6).

PROPOSITION 1.1.8. *The following conditions are equivalent for any space $X$:*
(i) *$X$ is a metrizable compactum.*
(ii) *$X$ is homeomorphic to a closed subspace of the Hilbert cube $I^\omega$.*

PROPOSITION 1.1.9. *The following conditions are equivalent for any space $X$:*
(i) *$X$ is a zero-dimensional metrizable compactum.*
(ii) *$X$ is homeomorphic to a closed subspace of the Cantor cube $D^\omega$.*

Here is a topological characterization of $D^\omega$.

THEOREM 1.1.10. *The following conditions are equivalent for any space $X$:*
(i) *$X$ is homeomorphic to $D^\omega$.*
(ii) *$X$ is a zero-dimensional metrizable compactum without isolated (i.e. open) points.*

We finish our brief discussion with the following surjective characterization of metrizable compacta.

THEOREM 1.1.11. *The following conditions are equivalent for any space $X$:*
(i) *$X$ is a metrizable compactum.*
(ii) *$X$ is a surjective image of $D^\omega$.*

**1.1.2. Realcompact spaces.** We begin with the following important definition.

DEFINITION 1.1.12. *A space $X$ is called* realcompact *if it is homeomorphic to a closed subspace of some power of the real line.*

Realcompact spaces have been introduced by Hewitt [**170**] (under a somewhat different name). The same class of spaces has been independently defined in terms of uniformities by Nachbin [**239**]. It is necessary to note that realcompact spaces appear in the literature under different names, such as: realcomplete spaces, Hewitt spaces, $Q$-spaces, $R$-spaces. This class of spaces plays a very important role in the sequel. Therefore, for the reader's convenience, we discuss some of their properties here.

The following statement is the straightforward consequence of the above definition.

PROPOSITION 1.1.13. *Every closed subspace of a realcompact space is realcompact, as is the product of an arbitrary family of realcompact spaces.*

COROLLARY 1.1.14. *Let $\{X_t : t \in T\}$ be a family of realcompact subspaces of a space $X$. Then the intersection $\cap\{X_t : t \in T\}$ is realcompact.*

PROOF. Obviously, the intersection $\cap\{X_t : t \in T\}$ is homeomorphic to the intersection of the subspaces $\prod\{X_t : t \in T\}$ and $\triangle_X$ (the diagonal of the product $X^T$). Consequently, the indicated intersection is closed in the product $\prod\{X_t : t \in T\}$. It only remains to apply Proposition 1.1.13 $\square$

COROLLARY 1.1.15. *A functionally open subspace of a realcompact space is realcompact.*

PROOF. Let $G$ be a functionally open subspaces of a realcompact space $X$. Consider the embedding $i = \triangle\{\varphi : \varphi \in C(X)\} : X \to R^{C(X)}$ and identify the space $X$ with its homeomorphic image $i(X)$ in $R^{C(X)}$. It follows from the definition of the embedding $i$ that $X$ is $C$-embedded in $R^{C(X)}$. Consequently, there exists a functionally open subset $V$ of $R^{C(X)}$ such that $G = V \cap X$. Further, by Corollary 1.1.14, it only remains to show that any functionally open subspace $V$ of an arbitrary power $R^A$ of the real line is realcompact. If $\mid A \mid\leq \omega$, then $V$, as an open subspace of a Polish space, is realcompact. Suppose now that $\mid A \mid> \omega$. Since any continuous real-valued function defined on $R^A$ depends on countably many coordinates (see, for example, [**145**]), we can easily conclude that $V = V_B \times R^{A-B}$, where $B$ is a countable subset of $A$ and $V_B$ is an open subspace of the product $R^B$. It only remains to apply Proposition 1.1.13. $\square$

PROPOSITION 1.1.16. *A realcompact and $C$-embedded subspace of a space is closed.*

PROOF. Let $X$ be a realcompact and $C$-embedded subspace of a space $Y$. By Definition 1.1.12, there exist a set $A$ and a homeomorphism $f : X \to X'$ onto the closed subspace $X'$ of the product $R^A$. Denote by $\pi_\alpha : R^A \to R_\alpha = R$, $\alpha \in A$, the projection onto the $\alpha$-th coordinate. Since $X$ is $C$-embedded in $Y$, for each $\alpha \in A$, there is a function $\varphi_\alpha \in C(Y)$ such that $\varphi_\alpha \mid X = \pi_\alpha \cdot f$. Let $\varphi = \Delta\{\varphi_\alpha : \alpha \in A\}$. Clearly $\varphi$ maps the space $Y$ into $R^A$ and $\varphi \mid X = f$. It follows from the continuity of $\varphi$ that

$$\varphi(cl_Y X) \subseteq cl_{R^A}\varphi(X) = cl_{R^A} f(X) = cl_{R^A} X' = X'.$$

In other words, the restriction $g = \varphi \mid cl_Y X$ maps the closure $cl_Y X$ onto $X'$. In this situation the composition $r = f^{-1} \cdot g : cl_Y X \to X$ is a retraction. Hence, $cl_Y X = X$. □

DEFINITION 1.1.17. *A subspace $X$ of a space $Y$ is said to be* $z$-embedded *in $Y$ if every functionally open (closed) subset $G$ of $X$ can be written as $G = X \cap \tilde{G}$, where $\tilde{G}$ is a functionally open (closed) subset of $Y$.*

$C$-embedded subspaces (closed subspaces of normal spaces, for example) are $z$-embedded. In a perfectly normal space every subspace is $z$-embedded.

LEMMA 1.1.18. *A functionally open subspace of any space is $z$-embedded.*

PROOF. Let $X$ be a functionally open subspace of a space $Y$. Fix a non-negative function $\varphi \in C(Y)$ such that $Y - X = \{y \in Y : \varphi(y) = 0\}$ and consider a functionally closed subspace $Z$ in $X$. Our goal is to find a functionally closed subspace $\tilde{Z}$ of $Y$ such that $Z = X \cap \tilde{Z}$. Fix a non-negative function $\psi \in C(X)$ such that $Z = \{x \in X : \psi(x) = 0\}$. Define the function $h : Y \to R$ as follows:

$$h(y) = \begin{cases} 0, & \text{if } y \in Y - X, \\ \min\{\varphi(y), \psi(y)\}, & \text{if } y \in X. \end{cases}$$

Let $\tilde{Z} = \{y \in Y : h(y) = 0\}$. Obviously, $Z = X \cap \tilde{Z}$. Therefore, it only remains to show continuity of the function $h$. The continuity of $h$ at points of $X$ follows from our definitions. If $y \in Y - X$ and $\epsilon > 0$, then

$$\{y \in Y : h(y) < \epsilon\} = \{y \in Y : \varphi(y) < \epsilon\} \cup \{x \in X : \psi(x) < \epsilon\}.$$

The continuity of $\varphi$ implies the openness of the first term on the right side of this equation. The second term is also open in $Y$. This immediately follows from the continuity of $\psi$ and the openness of $X$ in $Y$ and completes the verification of the continuity of $h$. □

LEMMA 1.1.19. *A Lindelöf subspace of an arbitrary space is $z$-embedded.*

PROOF. Let $X$ be a Lindelöf subspace of a space $Y$ and $G$ be a functionally open subspace of $X$. For each point $x \in G$ take a functionally open subspace $V_x$ such that $x \in V_x \cap X \subseteq G$. From the open cover $\{V_x \cap X : x \in G\}$ of the Lindelöf space $G$ select a countable subcover $\{V_n \cap X : n \in \omega\}$. Clearly, $G = X \cap \tilde{G}$, where $\tilde{G} = \cap\{V_n : n \in \omega\}$. It only remains to note that the set $\tilde{G}$, as a countable union of functionally open subspaces, is itself functionally open in $Y$. □

It should be noted that $z$-embeddability of certain subspaces completely determines the class of so-called perfectly $\kappa$-normal spaces. First, we give the corresponding definition.

DEFINITION 1.1.20. *A space is called* perfectly $\kappa$-normal *if the closure of each open subset is functionally closed.*

Obviously, perfectly normal spaces are perfectly $\kappa$-normal. It can be shown [**274**] that the product of an arbitrary family of metrizable spaces is also perfectly $\kappa$-normal.

PROPOSITION 1.1.21. *The following conditions are equivalent for any space $X$:*

(i) *$X$ is perfectly $\kappa$-normal.*
(ii) *All dense subsets of $X$ are $z$-embedded.*
(ii) *All open subsets of $X$ are $z$-embedded.*

PROOF. (*i*) $\Rightarrow$ (*ii*). Let $Y$ be a dense subset of $X$. It is enough to show that for every pair $(Z_1, Z_2)$ of disjoint functionally closed subsets of $Y$ there are functionally closed subsets $F_1$ and $F_2$ of $X$ such that $F_1 \cap F_2 \cap Y = \emptyset$ and $Z_i \subseteq F_i, i = 1,2$. Consider open (in $Y$) neighborhoods $G_1$ and $G_2$, of $Z_1$ and $Z_2$ respectively, with disjoint closures. Since $Y$ is dense in $X$ we conclude that the sets $F_i = cl_X G_i$, $i = 1,2$ are canonically closed in $X$. Therefore, by (*i*), these sets are functionally closed in $X$. Clearly, $F_1 \cap F_2 \cap Y = \emptyset$ and $Z_i \subseteq F_i$, $i = 1,2$. (*ii*) $\Rightarrow$ (*ii*). Let $Y$ be an open subset of $X$. Clearly, the subset $A = Y \cup (Y - cl_X Y)$ is dense in $X$. It can be easily seen that each functionally closed subset $Z$ of $Y$ is functionally closed in $A$. By (*ii*), there is a functionally closed subset $F$ in $X$ such that $F \cap A = Z$. But then $F \cap Y = Z$.

(*ii*) $\Rightarrow$ (*i*). Let $Y$ be an open subset of $X$. As above, consider the set $A = Y \cup (Y - cl_X Y)$. $A$ is open in $X$ and consequently, by(*ii*), is $z$-embedded in $X$. The set $Y$ is functionally closed in $A$. Therefore, there is a functionally closed subset $Z$ of $X$ such that $Y = Z \cap A$. Clearly, $Z = cl_X Y$. □

PROPOSITION 1.1.22. *Let $Y$ be a $z$-embedded subset of a space $X$. Then, for each function $\varphi \in C(Y)$ there exist a countable family $\{G_n : n \in \omega\}$ of functionally open subsets in $X$ and a function $\psi \in C(\cap\{G_n : n \in \omega\})$ such that $Y \subseteq \cap\{G_n : n \in \omega\}$ and $\psi \mid Y = \varphi$.*

PROOF. Since every space can be identified with a $C$-embedded subspace of $R^\tau$ for some $\tau \geq \omega$ (see the proof of Corollary 1.1.15) we can assume without loss of generality that $X = R^\tau$.

First we consider the case of a bounded function $\varphi$, i.e. $|\varphi| \leq a$, where $a$ is a non-negative real number. We follow the proof of the well-known theorem of Brouwer-Tietze-Urysohn (see, for example, [7]). Let $\varphi_0 = \varphi$ and consider functionally closed subsets $A_0 = \{y \in Y : \varphi_0(y) \leq -\frac{a}{3}\}$ and $B_0 = \{y \in Y : \varphi_0(y) \geq \frac{a}{3}\}$ of $Y$. Since, by our assumption, $Y$ is $z$-embedded in $X$, there exist functionally closed subsets $A_0'$ and $B_0'$ of $X$ such that $A_0 = A_0' \cap Y$ and $B_0 = B_0' \cap Y$. Let $G_0 = X - (A_0' \cap B_0')$. Obviously, $G_0$ is functionally open in $X$ and $Y \subseteq G_0$. Note also that the sets $\bar{A}_0 = A_0' \cap G_0$ and $\bar{B}_0 = B_0' \cap G_0$ are disjoint and functionally closed in $G_0$. Consequently, there exists a function $g_0 \in C(G_0)$ such that $|g_0| \leq \frac{a}{3}$, $\bar{A}_0 = g_0^{-1}(-\frac{a}{3})$ and $\bar{B}_0 = g_0^{-1}(\frac{a}{3})$. Let $\varphi_1 = \varphi_0 - g_0/Y$. It can be easily seen that $|\varphi_1| \leq \frac{2a}{3}$. Continuing this process we obtain a sequence of functions $\varphi_n \in C(Y)$ satisfying the inequalities $|\varphi_n| \leq (\frac{2}{3})^n a$, $n \in \omega$. We also have a sequence $g_n \in C(G_n), n \in \omega$, (where $G_n$ is a functionally open subset of $X$ containing $Y$ and $G_n \subseteq G_{n+1}$ for each $n \in \omega$) of functions satisfying the following conditions: $|g_n| \leq (\frac{2}{3})^n \frac{a}{3}$ and $\varphi_{n+1} = \varphi_n - g_n/Y$. Let $G = \cap\{G_n : n \in \omega\}$. Straightforward verification shows that the (bounded on $Y$) functions $\psi_n = g_0 + \cdots + g_n$, $n \in \omega$ uniformly converge to some continuous (on $Y$) function $\psi$. It only remains to note that $\psi$ extends the function $\varphi$ and is bounded by the same number $a$.

Suppose now that the function $\varphi$ in unbounded. Then the composition $arctan \cdot \varphi \in C(Y)$ is bounded. By the case, considered above, there is a function $f$ extending $arctan \cdot \varphi$, bounded by the same number as $arctan \cdot \varphi$ and defined on some set $G = \cap\{G_n : n \in \omega\}$, where each $G_n$ is a functionally open subset of $X$ containing $Y$. Let $Z = \{x \in G : |f(x)| \geq \frac{\pi}{2}\}$. Obviously, $Z$ is functionally closed in $G$ and $Y \subseteq G - Z$. Therefore the function $tan \cdot f$ is defined on $G - Z$ and coincides with $\varphi$ on $Y$. It only remains to note that the set $G - Z$ can also be represented as the intersection of countably many functionally open subsets of $X$. □

Using a similar argument one can prove the following statement.

PROPOSITION 1.1.23. *If a function $\varphi \in C(Z)$, defined on a functionally closed subset $Z$ of a space $X$, can be extended to a functionally open neighborhood of $Z$ in $X$, then $\varphi$ has an extension defined on $X$.*

PROPOSITION 1.1.24. *Let $X$ be a $z$-embedded subset of a space $Y$. Then the following conditions are equivalent:*

(i) *$X$ is realcompact.*

(ii) *$X$ coincides with the intersection of all functionally open subsets of $Y$ containing $X$.*

PROOF. (*i*) $\Rightarrow$ (*ii*). Let $G$ denote the intersection of all functionally open subsets of $Y$ containing $X$. Let us show that $X$ is dense in $G$. Suppose that this assumption is not true. Then there exists an open subset $V$ of $Y$ such that $V \cap G \neq \emptyset$ and $V \cap X = \emptyset$. Take a point $y_0 \in V \cap G$. Choose also a functionally closed subset $Z$ of $Y$ such that $y_0 \in Z \subseteq V$. Obviously, $X \subseteq Y - Z$ and, consequently, by definition of $G$, $G \subseteq Y - Z$. On the other hand, the point $y_0$, as can be easily seen, does not belong to $Y - Z$. This shows that $X$ is dense in $G$. By Proposition 1.1.22, $X$ is $C$-embedded in $G$. In this situation the realcompactness of $X$ guarantees, by Proposition 1.1.16, that $X = G$.

(*ii*) $\Rightarrow$ (*i*). Apply Corollaries 1.1.14 and 1.1.15. □

Lemma 1.1.19 and Proposition 1.1.24 imply the following statement.

COROLLARY 1.1.25. *Every Lindelöf space is realcompact.*

PROPOSITION 1.1.26. *Suppose that a space $X$ can be represented as the union of countable many $z$-embedded realcompact subspaces. Then $X$ is realcompact.*

PROOF. Let $X = \cup\{X_n : n \in \omega\}$, where $X_n$ is a $z$-embedded and realcompact subspace of $X$, $n \in \omega$. Consider the Stone-Čech compactification $\beta X$ of the space $X$ and take a point $p \in \beta X - X$. By Proposition 1.1.24, it suffices to construct a functionally closed subset $Z$ of $\beta X$ such that $p \in Z$ and $Z \cap X = \emptyset$. If $p \notin cl_{\beta X}(X_n)$, then there is a functionally closed subset $Z_n$ in $\beta X$ such that $p \in Z_n$ and $Z_n \cap cl_{\beta X}(X_n) = \emptyset$. If $p \in cl_{\beta X}(X_n)$, then, applying Proposition 1.1.24 (replacing in it $X$ by $X_n$ and $Y$ by $cl_{\beta X}(X_n)$), we conclude that there exists a functionally closed subset $F_n$ of $cl_{\beta X}(X_n)$ such that $p \in F_n$ and $F_n \cap X_n = \emptyset$. Finally, let $Z_n$ denote a functionally closed subset of $\beta X$ such that $Z_n \cap cl_{\beta X}(X_n) = F_n$. It only remains to note that $Z = \cap\{Z_n \colon n \in \omega\}$ is the desired functionally closed subset of $\beta X$. □

To each space $X$ we can assign a completely determined realcompact space $\nu X$, the so called *Hewitt realcompactification* of $X$. Let us describe the construction of $\nu X$. Consider the canonical embedding $i_X = \triangle\{\varphi \colon \varphi \in C(X)\} \colon X \to R^{C(X)}$ and identify $X$ with its image $i_X(X)$. The closure of $X$ in $R^{C(X)}$ is a realcompact space which contains $X$ as a dense $C$-embedded subspace. This closure is the Hewitt realcompactification of $X$. It can be shown that $\nu X$ is completely characterized (up to a homeomorphism) by these properties. It follows from Proposition 1.1.16 that realcompact spaces can be defined as spaces satisfying the equation $X = \nu X$. Another, sometimes more convenient description of $\nu X$ can be obtained as follows. Embed $X$ as a $z$-embedded subspace into a realcompact space $Y$ (for example, the Stone-Čech compactification $\beta X$) and consider the intersection of all functionally open subsets of $Y$ containing $X$. As indicated in the proof of Proposition 1.1.24, $X$ is $C$-embedded in this intersection, which of course is realcompact (apply Corollary 1.1.14). Using this description, we may assume that $X \subseteq \nu X \subset \beta X$ and, consequently, $\dim X = \dim \nu X$. Further, one can show that:

(a) $\nu X$ is the smallest realcompact subspace of $\beta X$ containing $X$.

(b) $\nu X$ is the largest subspace of $\beta X$ containing $X$ as a $C$-embedded subspace.

The correspondence $X \mapsto \nu X$ is even functorial. For each map $f\colon X \to Y$ we can assign a map $\nu f\colon \nu X \to \nu Y$ which coincides with $f$ on $X$. One can get this conclusion, by analyzing the proof of Proposition 1.1.16.

**1.1.3. Directed sets.** Let $A$ be a partially ordered *directed set* (i.e. for every two elements $a, b \in A$ there exists an element $c \in A$ such that $c \geq a$ and $c \geq b$). We say that a subset $A_1 \subseteq A$ of $A$ *majorates* another subset $A_2 \subseteq A$ of $A$ if for each element $a_2 \in A_2$ there exists an element $a_1 \in A_1$ such that $a_1 \geq a_2$. A subset which majorates $A$ is called *cofinal* in $A$. A subset of $A$ is said to be a *chain* if every two elements of it are comparable. The symbol $\sup B$ , where $B \subseteq A$ , denotes the lower upper bound of $B$ (if such an element exists in $A$ ). Let $\tau \geq \omega$ be a cardinal number. A subset $B$ of $A$ is said to be $\tau$-*closed* in $A$ if for each chain $C \subseteq B$, with $|C| \leq \tau$, we have $\sup C \in B$, whenever the element $\sup C$ exists in $A$ . Finally, a directed set $A$ is said to be $\tau$-*complete* if for each chain $B$ of elements of $A$ with $|C| \leq \tau$, there exists an element $\sup C$ in $A$ . A standard example of a $\tau$-complete set is the set $\exp_\tau A$ of all subsets of cardinality $\leq \tau$ of any set $A$ with $| A | > \tau$ .

The standard example of a $\tau$-complete set can be obtained as follows. For an arbitrary set $A$ let $expA$ denote, as usual, the collection of all subsets of $A$. There is a natural partial order on $expA$: $A_1 \geq A_2$ if and only if $A_1 \supseteq A_2$. With this partial order $expA$ becomes a directed set. If we consider only those subsets of the set $A$ which have cardinality $\leq \tau$, then the corresponding subcollection of $expA$, denoted by $exp_\tau A$, serves as a basic example of a $\tau$-complete set.

Let us consider a second way of constructing $\tau$-complete sets. Let $A$ be an arbitrary directed set. We call two countable chains in $A$ *equivalent* if each of them majorates the other. The collection of all equivalence classes of countable chains in $A$ is denoted by $[A]_\omega$ and is called the $\omega$-completion of $A$. The order relation on $[A]_\omega$ is defined by the relation of "majorantness": one class majorates another if one of the chains in the first class majorates a chain belonging to the second class. It is easy to see that with respect to the described order the set $[A]_\omega$ is $\omega$-complete. Assigning to each element $a \in A$ the countable stationary chain (consisting of one element $\{a\}$) we obtain a natural embedding of the set $A$ into its $\omega$-completion. If, for all $\lambda < \tau$, the $\lambda$-completions $[A]_\lambda$ of $A$ have already been defined, then the $\tau$-completion $[A]_\tau$ can be defined as the set of equivalence classes of chains of cardinality $\leq \tau$ from the union $\cup\{[A]_\lambda\colon \lambda < \tau\}$.An order on $[A]_\tau$, as above, is generated by the relation of majorantness. Straightforward verification shows that the $\tau$-completion of an arbitrary directed set is $\tau$-complete for each infinite cardinal $\tau$.

PROPOSITION 1.1.27. *Let $\{A_t : t \in T\}$ be a collection of $\tau$-closed and cofinal subsets of a $\tau$-complete set $A$. Then the intersection $\cap\{A_t : t \in T\}$ is also cofinal and $\tau$-closed in $A$ .*

PROOF. Without loss of generality we can assume that the set $T$ is well-ordered. Consider the product $\omega \times T$ endowed with the lexicographic order. Let $a$ be an arbitrary element in $A$. In an obvious way, by transfinite induction, we define a chain $B = \{a_{(n,t)}: (n,t) \in \omega \times T\}$ in $A$ so that the following conditions are satisfied:

(i) $a_{(n,t)} \geq a$ for each $(n,t) \in \omega \times T$.
(ii) $a_{(n,t)} \leq a_{(m,t')}$ whenever $(n,t) \leq (m,t')$.
(iii) $a_{(n,t)} \in A$ for each $(n,t) \in \omega \times T$.

The $\tau$-completeness of $A$ guarantees the existence of an element $b = \sup\{a_{(n,t)} : n \in \omega\}$. By closedness of $A_t$, $b \in A_t$ for each $t \in T$. Therefore, $b \in \cap\{A_t : t \in T\}$. By the construction, $b \geq a$. This shows the cofinality of this intersection in $A$. Its $\tau$-closedness is obvious. □

COROLLARY 1.1.28. *For each subset $B$, with $|B| \leq \tau$, of a $\tau$-complete set $A$ there exists an element $c \in A$ such that $c \geq b$ for each $b \in B$ .*

PROOF. Apply Proposition 1.1.27 to the collection $\{A_b : b \in B\}$, where $A_b = \{a \in A : a \geq b\}$, $b \in A$. □

PROPOSITION 1.1.29. *Let $A$ be a $\tau$-complete set, $L \subseteq A^2$, and suppose the following three conditions are satisfied:*

**Existence:** *For each $a \in A$ there exists $b \in A$ such that $(a,b) \in L$ .*
**Majorantness:** *If $(a,b) \in L$ and $c \geq b$, then $(a,c) \in L$ .*
**$\tau$-closedness:** *Let $\{a_t : t \in T\}$ be a chain in $A$ with $a = \sup\{a_t : t \in T\}$. If $(a_t, c) \in L$ for some $c \in A$ and each $t \in T$, then $(a,c) \in L$ .*

*Then the set of all $L$-reflexive elements of $A$ (an element $a \in A$ is said to be $L$-reflexive if $(a,a) \in L$) is cofinal and $\tau$-closed in $A$ .*

PROOF. First of all let us show that the set of all $L$-reflexive elements is cofinal in $A$. Let $a \in A$. We are going to construct a chain $B = \{b_\alpha : \alpha < \tau\}$, indexed by all ordinals strictly less than $\tau$ and satisfying the following conditions:

(a) $b_0 = a$.
(b) $(b_\beta, b_\alpha) \in L$, whenever $\beta < \alpha < \tau$.
(c) $b_\beta \leq b_\alpha$, whenever $\beta < \alpha < \tau$.

Suppose that the elements $b_\beta$ have already been constructed for all $\beta$ smaller than some $\alpha$, where $\alpha < \tau$. Let us construct the element $b_\alpha$.

If $\alpha$ is a limit ordinal then let $b_\alpha = \sup\{b_\beta : \beta < \alpha\}$. Clearly, $b_\alpha \geq b_\beta$ for each $\beta < \alpha$. Note that $b_\alpha \geq b_{\beta+1} \geq b_\beta$ and $(b_\beta, b_{\beta+1}) \in L$. Therefore, by the majorantness condition, $(b_\beta, b_\alpha) \in L$ for each $\beta < \alpha$.

If $\alpha$ is non-limit ordinal, then $\alpha = \beta + 1$. By the existence condition, there exists an element $c \in A$ such that $(b_\beta, c) \in L$. Since $A$ is directed we can find an element $b_\alpha$ which majorates both $b_\beta$ and $c$, i.e. $b_\alpha \geq b_\beta$ and $b_\alpha \geq c$. The majorantness condition implies that $(b_\beta, b_\alpha) \in L$. This finishes the construction of the chain $B$.

Let $b = \sup\{b_\alpha\colon \alpha < \tau\}$. Let us show that $b$ is a $L$-reflexive element. Indeed, by the majorantness condition, $(b_\alpha, b) \in L$ for each $\alpha < \tau$. Therefore, by the closeness condition, $(b, b) \in L$. This shows cofinality of the set of all $L$-reflexive elements in $A$.

If $C = \{a_t\colon t \in T\}$ is a chain, consisting of $L$-reflexive elements, $\mid T \mid \leq \tau$ and $a = supC$, then, by the majorantness condition, we have $(a_t, a) \in L$ whenever $t \in T$. Finally, by the closeness condition, $(a, a) \in L$. $\square$

*Historical and bibliographical notes* 1.1. The results of subsection 1.1.1 are widely known. For example, Proposition 1.1.2 has been proven by Alexandrov in [**2**] (see also [**208**]). The characterization theorems 1.1.5 and 1.1.10 have been obtained by Alexandrov and Urysohn [**8**] and Brouwer [**56**] respectively. Theorems 1.1.7 and 1.1.11 belong respectively to Hausdorff and Alexandrov. Propositions 1.1.13, 1.1.16, as well as Corollary 1.1.14, have been proved by Shirota [**281**] (see also [**185**]). $z$-embedded subspaces were systematically studied in [**40, 41, 42, 10**]. In particular, these references contain proofs of Lemmas 1.1.18 and 1.1.19. The property included in the definition of perfectly $\kappa$-normal spaces was apparently known to Bokshtein (see also [**144, 261**]). The concept of perfect $\kappa$-normality itself was independently introduced (under different names) by Blair [**40**] and Terada [**292**]. Shortly after the same notion was also defined by Ščepin [**274**]. Propositions 1.1.21 and 1.1.22 appear in [**40**] and [**42**] respectively. Proposition 1.1.24 was in fact proved in [**236**]. Corollary 1.1.25 belongs to Hewitt [**170**]. Originally Proposition 1.1.26 was obtained in [**40**] and later rediscovered in [**74**]. Earlier, in the presence of normality of the given space coupled with closeness of summands, Proposition 1.1.26 was proved in [**236**]. It should be noted that the assumption of $z$-embeddedness of the summands in Proposition 1.1.26 is essential [**237, 238**]. The results of Subsection 1.1.3 were originally obtained in [**278**]; see also [**107**].

## 1.2. Definitions and elementary properties of inverse spectra

If $\prod\{X_\alpha : \alpha \in A\}$ is a Cartesian product of topological spaces $X_\alpha$, $\alpha \in A$, and $B \subseteq A$ then $\pi_B : \prod\{X_\alpha : \alpha \in A\} \to \prod\{X_\alpha : \alpha \in B\}$ denotes the standard projection onto the corresponding subproduct. The same meaning is given to the projection $\pi^B_C : \prod\{X_\alpha : \alpha \in B\} \to \prod\{X_\alpha : \alpha \in C\}$, $C \subseteq B \subseteq A$. For simplicity, sometimes we also use the following notation: $X_B = \prod\{X_\alpha : \alpha \in B\}$, where $B \subseteq A$. We will maintain these notations throughout the whole text.

**1.2.1. Definition of inverse spectra.** Let $A$ be a directed set and suppose that to each $\alpha \in A$ there corresponds a topological space $X_\alpha$ in such a way that whenever $\beta \succeq \alpha$, a continuous map $p_\alpha^\beta : X_\beta \to X_\alpha$ is also given. Suppose further that:

- $p_\beta^\gamma \cdot p_\alpha^\beta = p_\alpha^\gamma$ for each triple $(\alpha, \beta, \gamma)$ of indexes with $\alpha \succeq \beta \succeq \gamma$.
- $p_\alpha^\alpha = id_{X_\alpha}$ for each $\alpha \in A$.

In this situation we say that the *inverse spectrum* (*inverse system*) or simply *spectrum* $\mathcal{S}_X = \{X_\alpha, p_\alpha^\beta, A\}$ is given. We call the spaces $X_\alpha$ *elements* of the spectrum. The mappings $p_\alpha^\beta : X_\beta \to X_\alpha$ are called *projections* (or *bonding mappings*) of the spectrum.

An *inverse sequence* is a spectrum whose indexing set coincides with $\omega$ (the first infinite cardinal directed by its natural order).

A point $\{x_\alpha : \alpha \in A\}$ of the Cartesian product $\prod\{X_\alpha : \alpha \in A\}$ of spaces $X_\alpha$ is called a *thread* of the spectrum $\mathcal{S}_X = \{X_\alpha, p_\alpha^\beta, A\}$ if $p_\alpha^\beta(x_\beta) = x_\alpha$ for any $\alpha, \beta \in A$ such that $\alpha \preceq \beta$. The subspace of the product $\prod\{X_\alpha : \alpha \in A\}$ consisting of all threads is called the *inverse limit* (or, simply, limit) of the spectrum $\mathcal{S}_X = \{X_\alpha, p_\alpha^\beta, A\}$ and is denoted by $\lim \mathcal{S}_X$ or, more explicitly, by $\lim \{X_\alpha, p_\alpha^\beta, A\}$ . By $p_\alpha \colon \lim \mathcal{S}_X \to X_\alpha$, $\alpha \in A$, we denote the restriction of the projection $\pi_{\{\alpha\}} \colon \prod\{X_\alpha : \alpha \in A\} \to X_\alpha$, and we call this the $\alpha$-th *limit projection* of the spectrum $\mathcal{S}_X$.

By definition of the product topology, the sets of the form $\lim \mathcal{S}_X \cap \{\pi_\beta^{-1}(G_\beta) \colon \beta \in B\}$, where $B$ is a finite subset of the indexing set $A$ and $G_\beta$ is an open set in $X_\beta$, constitute a base of the limit space $\lim \mathcal{S}_X$. Let us show that in fact we can be more specific.

PROPOSITION 1.2.1. *The collection of all sets $p_\alpha^{-1}(G_\alpha)$, where $G_\alpha$ is open in $X_\alpha$ and $\alpha \in A$, forms a base for the topology of the limit space of the spectrum $\lim \mathcal{S}_X$.*

PROOF. Let $G = \lim \mathcal{S}_X \cap \{\pi_\beta^{-1}(G_\beta) \colon \beta \in B\}$, where $B$ is a finite subset of the indexing set $A$ and $G_\beta$ is open in $X_\beta$ Take an index $\alpha \in A$ such that $\alpha \succeq \beta$ for each $\beta \in B$. Then the set $G_\alpha = \cap\{(p_\beta^\alpha)^{-1}(G_\beta) \colon \beta \in B\}$ is open in $X_\alpha$, and it easy to see that $G = p_\alpha^{-1}(G_\alpha)$. □

PROPOSITION 1.2.2. *The limit space of a spectrum $\mathcal{S}_X = \{X_\alpha, p_\alpha^\beta, A\}$ is a closed subspace of the product $\prod\{X_\alpha : \alpha \in A\}$.*

PROOF. Let $x = \{x_\alpha\} \in \prod\{X_\alpha : \alpha \in A\} - \lim \mathcal{S}_X$. Then there exist indexes $\alpha, \beta \in A$ such that $\beta > \alpha$ and $x'_\alpha = p_\alpha^\beta(x_\beta) \neq x_\alpha$. Consider disjoint open subsets $V$ and $V'$ in $X_\alpha$ such that $x'_\alpha \in V'$ and $x_\alpha \in V$. Clearly, the set $U = (p_\alpha^\beta)^{-1}(V')$ is an open neighborhood of the point $x_\beta$ in $X_\beta$. Therefore, the set $W = \pi_\alpha^{-1}(V) \cap \pi_\beta^{-1}(U)$ is an open neighborhood of $x$ in $\prod\{X_\alpha : \alpha \in A\}$. It only remains to note that $W \cap \lim \mathcal{S}_X = \emptyset$. □

COROLLARY 1.2.3. *The limit space of an inverse spectrum consisting of completely regular spaces is completely regular.*

Consider now a spectrum $\mathcal{S}_X = \{X_\alpha, p_\alpha^\beta, A\}$ and a subspace $Y$ of its limit space lim $\mathcal{S}_X$. The subspace $Y$ defines the inverse spectrum $\mathcal{S}_Y = \{Y_\alpha, q_\alpha^\beta, A\}$, where $Y_\alpha = p_\alpha(Y)$, $\alpha \in A$, and $q_\alpha^\beta$ denotes the restriction $p_\alpha^\beta / X_\beta$, $\beta \succeq \alpha$. In this situation we say that the spectrum $\mathcal{S}_Y = \{Y_\alpha, q_\alpha^\beta, A\}$ is *induced by* $Y$. Obviously, the product $\prod\{Y_\alpha : \alpha \in A\}$ is a subspace of the product $\prod\{X_\alpha : \alpha \in A\}$ and the threads of the spectrum $\mathcal{S}_Y$ are threads of the spectrum $\mathcal{S}_X$. Therefore the limit space $\tilde{Y} = \lim\ \mathcal{S}_Y$ is naturally embedded in the limit space $X = \lim\ \mathcal{S}_X$. We claim that

$$\tilde{Y} = \cap\{p_\alpha^{-1}(Y_\alpha) : \alpha \in A\}.$$

Indeed, if $x = \{x_\alpha : \alpha \in A\} \in \cap\{p_\alpha^{-1}(Y_\alpha) : \alpha \in A\}$, then $x_\alpha \in Y_\alpha$ for each $\alpha \in A$. This shows that the thread $x$ of the spectrum $\mathcal{S}_X$ is in fact a thread of the spectrum $\mathcal{S}_Y$. Therefore, $\cap\{p_\alpha^{-1}(Y_\alpha) : \alpha \in A\} \subseteq \tilde{Y}$. Conversely, if $x \in \tilde{Y}$, then $p_\alpha(x) \in Y_\alpha$ for each $\alpha \in A$. Therefore, $\tilde{Y} \subseteq \cap\{p_\alpha^{-1}(Y_\alpha) : \alpha \in A\}$ and we have proved the following statement.

PROPOSITION 1.2.4. *Let $Y$ be a subspace of the limit space of a spectrum $\mathcal{S}_X = \{X_\alpha, p_\alpha^\beta, A\}$. Then the limit space* lim $\mathcal{S}_Y$ *of the induced (by $Y$) spectrum $\mathcal{S}_Y = \{Y_\alpha, q_\alpha^\beta, A\}$ contains $Y$. The equality $Y = \lim\ \mathcal{S}_Y$ holds if and only if*

$$Y = \cap\{p_\alpha^{-1}(Y_\alpha) : \alpha \in A\}.$$

COROLLARY 1.2.5. *Let $Y$ be a subspace of the limit space $X$ of a spectrum $\mathcal{S}_X = \{X_\alpha, p_\alpha^\beta, A\}$. Then the limit space of the spectrum $\{cl_{X_\alpha}(p_\alpha(Y)), p_\alpha^\beta / cl_{X_\beta}(p_\beta(Y), A\}$ coincides with $cl_X Y$.*

PROOF. Apply Propositions 1.2.1 and 1.2.4 □

COROLLARY 1.2.6. *Let $Y$ be a closed subspace of the limit space of a spectrum $\mathcal{S}_X = \{X_\alpha, p_\alpha^\beta, A\}$. Then*

$$Y = \lim\ \{p_\alpha(Y), p_\alpha^\beta / p_\beta(Y), A\}$$

*and*

$$Y = \lim\ \{cl_{X_\alpha}(p_\alpha(Y)), p_\alpha^\beta / cl_{X_\beta}(p_\beta(Y)), A\}.$$

Suppose now that a spectrum $\mathcal{S}_X = \{X_\alpha, p_\alpha^\beta, A\}$ is given and $A'$ is a cofinal subset of the indexing set $A$. In this case we say that the spectrum $\mathcal{S}'_X = \{X_\alpha, p_\alpha^\beta, A'\}$ is a *cofinal subspectrum* of $\mathcal{S}_X$.

PROPOSITION 1.2.7. *Let $\mathcal{S}'_X = \{X_\alpha, p_\alpha^\beta, A'\}$ be a cofinal subspectrum of the spectrum $\mathcal{S}_X = \{X_\alpha, p_\alpha^\beta, A\}$. Then the map $f\colon \lim\ \mathcal{S}_X \to \lim\ \mathcal{S}'_X$ consisting of restricting all threads from $X = \lim\ \mathcal{S}_X$ to those from $X' = \lim\ \mathcal{S}'_X$ is a homeomorphism of $X$ onto $X'$.*

PROOF. We construct a map $f'\colon X' \to X$ inverse to $f$. Let $x' = \{x_\alpha\colon \alpha \in A'\} \in X'$. For each $\alpha \in A$ let $x_\alpha = p_\alpha^\beta(x_\beta)$, where $\beta \succeq \alpha$ and $\beta \in A'$. We claim that the point $x_\alpha$ does not depend on the choice of index $\beta \in A'$. Indeed, consider a third index $\gamma \in A'$ such that $\gamma \succeq \alpha$. Since $A$ is directed and $A'$ is cofinal in $A$, there is another index $\delta \in A'$ such that $\delta \succeq \beta$ and $\delta \succeq \gamma$. The fact that $x'$ is a thread of the spectrum $\mathcal{S}'_X$ implies that

$$p_\alpha^\beta(x_\beta) = p_\alpha^\beta p_\beta^\delta(x_\delta) = p_\alpha^\delta(x_\delta) = p_\alpha^\gamma p_\gamma^\delta(x_\delta) = p_\alpha^\gamma(x_\gamma).$$

This shows that $x = \{x_\alpha\colon \alpha \in A\}$ is a well-defined point of the product $\prod \{X_\alpha\colon \alpha \in A\}$. In fact, $x$ is a thread of the spectrum $\mathcal{S}_X$. Indeed,

$$x_\alpha = p_\alpha^\gamma(x_\gamma) = p_\alpha^\beta p_\beta^\gamma(x_\gamma) = p_\alpha^\beta(x_\beta).$$

Thus, to each point $x' \in X'$ we have assigned a point $x = f'(x') \in X$. It follows immediately from the above definition that $ff'(x') = x'$ for each $x' \in X'$ and $f'f(x) = x$ for each $x \in X$. Therefore, $f$ is a bijection and $f'$ is the inverse of $f$.

Next we show that $f$ is an open map (the continuity of $f$ is obvious). Let $G_\alpha$ be an open subset of $X_\alpha$, $\alpha \in A$, and $\beta \succeq \alpha$, where $\beta \in A'$. Then

$$p_\alpha^{-1}(G_\alpha) = (p_\alpha^\beta)^{-1}(G_\alpha) = p_\beta^{-1}\left(\left(p_\alpha^\beta\right)^{-1}(G_\alpha)\right).$$

Therefore, the sets of the form $p_\beta^{-1}(G_\beta)$, where $\beta \in A'$ and $G_\beta$ is open in $X_\beta$, constitute a base for the topology of $X$ (compare with Proposition 1.2.1). Denote by $q_\beta$ the $\beta$-th limit projection of the spectrum $\mathcal{S}'_X$. Then

$$p_\beta^{-1}(G_\beta) = \{x = \{x_\alpha\colon \alpha \in A\}\colon x_\beta \in G_\beta\}$$

and

$$q_\beta^{-1}(G_\beta) = \{x' = \{x_\alpha\colon \alpha \in A'\}\colon x_\beta \in G_\beta\}.$$

Consequently,

$$fp_\beta^{-1}(G_\beta) = q_\beta^{-1}(G_\beta).$$

This shows that $f$ is a homeomorphism. □

COROLLARY 1.2.8. *Let $\mathcal{S}_X = \{X_\alpha, p_\alpha^\beta, A\}$ be an inverse spectrum whose indexing set $A$ has a maximal element $\alpha$. Then the inverse limit* lim $\mathcal{S}_X$ *of $\mathcal{S}_X$ is homeomorphic to the space $X_\alpha$.*

EXAMPLE 1.2.9. *Suppose we are given a family $\{X_t\colon t \in T\}$ of spaces with $|T| \geq \omega$. Obviously the set $exp_{<\omega}T$ of all finite subsets of $T$ is directed by inclusion, i.e. by letting $T_1 \preceq T_2$ if and only if $T_1 \subseteq T_2$. For each $S \in exp_{<\omega}T$, let $X_S = \prod\{X_t\colon t \in S\}$. Also, in cases when $S \preceq R$, $S, R \in exp_{<\omega}T$, denote by $\pi_S^R\colon X_R \to X_S$ the natural projection onto the corresponding subproduct. Then we obtain an inverse spectrum $\mathcal{S} = \{X_S, \pi_S^R, exp_{<\omega}T\}$. One can readily verify that the limit space of this spectrum is homeomorphic to the product $\prod\{X_t\colon t \in T\}$.*

*If* $|T| > \omega$, *then in the same way we can define the spectrum consisting of countable subproducts of the product* $\prod\{X_t : t \in T\}$, *of natural projections between them and with* $exp_\omega T$ *as the indexing set. The limit space of the latter spectrum will still be homeomorphic to the whole product* $\prod\{X_t : t \in T\}$.

EXAMPLE 1.2.10. *Let* $\{X_t : t \in T\}$ *be a family of subspaces of a space* $X$, *indexed by a directed set* $T$. *Assume also that* $X_{t'} \subset X_t$, *whenever* $t \preceq t'$. *Denote by* $i_t^{t'} : X_{t'} \hookrightarrow X_t$, $t \preceq t'$, *the inclusion map. One can easily verify that the limit space of the spectrum* $\mathcal{S} = \{X_t, i_t^{t'}, T\}$ *is homeomorphic to the intersection* $\cap\{X_t : t \in T\}$ *of the given subspaces.*

Using the general observation made in Example 1.2.10 as a guide, the reader can construct examples of inverse spectra consisting of non-empty spaces and projections with dense images (even surjections), the limit spaces of which are empty.

On the other hand we have the following statement, the proof of which follows from the compactness of arbitrary products of compact spaces (Tychonov's theorem) coupled with the fact that an intersection of any centered collection (i.e. a collection, all finite subcollections of which have non-empty intersection) of compact subspaces of any compact space is non-empty.

PROPOSITION 1.2.11. *The limit space of any inverse spectrum consisting of non-empty compacta is a non-empty compactum.*

The following technical statement is useful in various situations.

PROPOSITION 1.2.12. *Let* $Y_1$ *and* $Y_2$ *be disjoint closed subsets of the limit space* $\lim \mathcal{S}_X$ *of a spectrum* $\mathcal{S}_X = \{X_\alpha, p_\alpha^\beta, A\}$ . *If* $Y_1$ *is compact, then there exist an index* $\alpha \in A$ *and an open subset* $G_\alpha \subseteq X_\alpha$ *such that* $Y_1 \subseteq p_\alpha^{-1}(G_\alpha)$ *and* $Y_2 \cap p_\alpha^{-1}(G_\alpha) = \emptyset$. *In particular,* $p_\alpha(Y_1) \cap p_\alpha(Y_2) = \emptyset$.

PROOF. For each point $y \in Y_1$ consider an index $\alpha_y \in A$ and an open subset $G_{\alpha_y} \subseteq X_{\alpha_y}$ such that $y \in p_{\alpha_y}^{-1}(G_{\alpha_y}) \subseteq X - Y_2$, where $X = \lim \mathcal{S}_X$. Since $Y_1$ is compact, there is a finite subcollection $\{p_{\alpha_i}^{-1}(G_{\alpha_i}) : i = 1, ..., k\}$ of the collection $\{p_{\alpha_y}^{-1}(G_{\alpha_y}) : y \in Y_1\}$ which still covers $Y_1$. Take an index $\alpha \in A$ such that $\alpha \succeq \alpha_i$ for each $i = 1, ..., k$. One can easily verify that the set $G_\alpha = \cup\{(p_{\alpha_i}^\alpha)^{-1}(G_{\alpha_i}) : i = 1, ..., k\}$ has the desired property. □

**1.2.2. Morphisms of inverse spectra.** Suppose we are given two inverse spectra $\mathcal{S}_X = \{X_\alpha, p_\alpha^\beta, A\}$ and $\mathcal{S}_Y = \{Y_\gamma, q_\gamma^\delta, B\}$ . A *morphism of the spectrum* $\mathcal{S}_X$ *into the spectrum* $\mathcal{S}_Y$ is a family $\{\varphi, \{f_\gamma : \gamma \in B\}\}$ consisting of a nondecreasing function $\varphi : B \to A$ such that the set $\varphi(B)$ is cofinal in $A$, and of continuous maps $f_\gamma : X_{\varphi(\gamma)} \to Y_\gamma$ defined for all $\gamma \in B$ such that

$$q_\gamma^\delta f_\delta = f_\gamma p_{\varphi(\gamma)}^{\varphi(\delta)},$$

whenever $\gamma, \delta \in B$ and $\gamma \preceq \delta$. In other words, we require (in the above situation) the commutativity of the following diagram

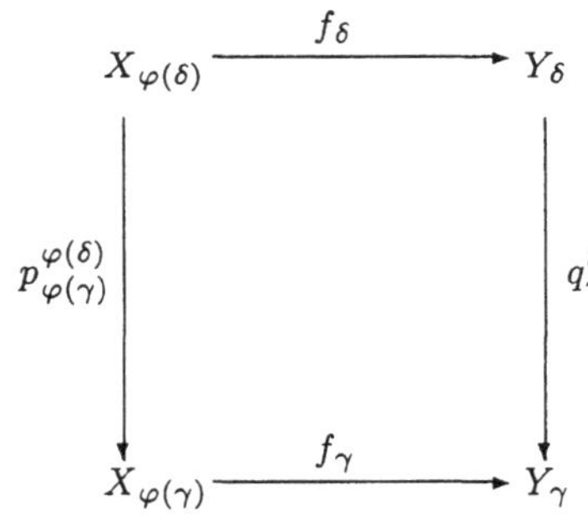


Any morphism

$$\{\varphi, \{f_\gamma \colon \gamma \in B\}\} \colon \mathcal{S}_X \to \mathcal{S}_Y$$

induces a (continuous) map, called *the limit map of the morphism*

$$\lim \{\varphi, \{f_\gamma \colon \gamma \in B\}\} \colon \lim \mathcal{S}_X \to \lim \mathcal{S}_Y.$$

To see this, assign to each thread $x = \{x_\alpha \colon \alpha \in A\}$ of the spectrum $\mathcal{S}_X$ the point $y = \{y_\gamma \colon \gamma \in B\}$ of the product $\prod\{Y_\gamma \colon \gamma \in B\}$ by letting

$$y_\gamma = f_\gamma(x_{\varphi(\gamma)}), \gamma \in B.$$

It is easily seen that the point $y = \{y_\gamma \colon \gamma \in B\}$ is in fact a thread of the spectrum $\mathcal{S}_Y$. Therefore, assigning to $x = \{x_\alpha \colon \alpha \in A\} \in \lim \mathcal{S}_X$ the point $y = \{y_\gamma \colon \gamma \in B\} \in \lim \mathcal{S}_Y$, we define a map $\lim \{\varphi, \{f_\gamma \colon \gamma \in B\}\} \colon \lim \mathcal{S}_X \to \lim \mathcal{S}_Y$. Straightforward verification shows that this map is continuous.

The morphisms of inverse spectra which arise most frequently in practice are those defined on the same indexing set. In this case, the map $\varphi \colon A \to A$ of the definition of morphism is taken to be the identity. Below we shall mostly deal with such situations and use the following notation:

$$\{f_\alpha \colon X_\alpha \to Y_\alpha; \alpha \in A\} \colon \mathcal{S}_X \to \mathcal{S}_Y$$

or sometimes the even shorter form

$$\{f_\alpha\} \colon \mathcal{S}_X \to \mathcal{S}_Y.$$

PROPOSITION 1.2.13. *Let $\mathcal{S}_Y = \{Y_\alpha, q^\beta_\alpha, A\}$ be an inverse spectrum and $X$ be a space. Suppose that for each $\alpha \in A$ a map $f_\alpha \colon X \to Y_\alpha$ is given in such a way that*

$$f_\alpha = q^\beta_\alpha f_\beta$$

*whenever $\alpha, \beta \in A$ and $\alpha \preceq \beta$. Then there exists a natural map $f \colon X \to \lim \mathcal{S}_Y$ (diagonal product $\triangle\{f_\alpha \colon \alpha \in A\}$) satisfying, for each $\alpha \in A$, the condition*

$$f_\alpha = q_\alpha f.$$

PROOF. Indeed, we only have to note that a space $X$, together with its identity map $id_X$, forms the inverse spectrum $\mathcal{S}$. So the collection $\{f_\alpha : \alpha \in A\}$ is in fact a morphism $\mathcal{S} \to \mathcal{S}_Y$. The rest follows from the definitions given above. □

The proof of the following statement is an easy exercise.

PROPOSITION 1.2.14. *Let* $\{f_\alpha\}: \mathcal{S}_X \to \mathcal{S}_Y$ *be a morphism consisting of homeomorphisms. Then the limit map* $f = \lim \{f_\alpha\}: \lim\ \mathcal{S}_X \to \lim\ \mathcal{S}_Y$ *is a homeomorphism.*

*Historical and bibliographical notes* 1.2. The basic definitions and introductory statements presented in this Section are widely known and can be found in many textbooks.

## 1.3. Factorizing spectra and the spectral theorem

**1.3.1. Factorizing spectra.** We say that a spectrum $\mathcal{S}_X = \{X_\alpha, p_\alpha^\beta, A\}$ is *factorizing* is for each real-valued continuous function $\varphi$: lim $\mathcal{S}_X \to R$ there exist an index $\alpha \in A$ and a function $\varphi_\alpha : X_\alpha \to R$ such that $\varphi = \varphi_\alpha p_\alpha$.

PROPOSITION 1.3.1. *Let* $\mathcal{S}_X = \{X_\alpha, p_\alpha^\beta, A\}$ *be an inverse spectrum whose indexing set* $A$ *is* $\omega$*-complete. If all limit projections* $p_\alpha$, $\alpha \in A$, *of* $\mathcal{S}_X$ *are surjective, then the following conditions are equivalent:*

(i) *The spectrum* $\mathcal{S}_X$ *is factorizing.*
(ii) *Each functionally open subset* $G$ *of* lim $\mathcal{S}_X$ *is* $\alpha$*-cylindrical for some* $\alpha \in A$, *i.e.* $G = p_\alpha^{-1}(G_\alpha)$, *where* $\alpha \in A$ *and* $G_\alpha$ *is a functionally open subset of* $X_\alpha$.

PROOF. If $G$ is a functionally open subset of $X$ = lim $\mathcal{S}_X$, then there is a function $\varphi \in C(X)$ such that $G = \varphi^{-1}(R - \{0\})$. By (i), there exist an index $\alpha \in A$ and a function $\varphi_\alpha \in C(X_\alpha)$ such that $\varphi = \varphi_\alpha p_\alpha$. Obviously, the set $G_\alpha = \varphi^{-1}(R - \{0\})$ is functionally open in $X_\alpha$. It only remains to note that $G = p_\alpha^{-1}(G_\alpha)$.

Conversely, let $\varphi \in C(X)$. Fix a countable open base $\mathcal{U} = \{U_n : n \in \omega\}$ of the real line $R$. Obviously, the sets $G_n = \varphi^{-1}(U_n)$, $n \in \omega$, are functionally open subsets of $X$. Consequently, by (ii), there exist indices $\alpha_n \in A$, $n \in \omega$ and functionally open subsets $V_{\alpha_n} \subseteq X_{\alpha_n}$ such that $G_n = p_{\alpha_n}^{-1}(V_{\alpha_n})$ for each $n \in \omega$. By Corollary 1.1.28, there exists an index $\alpha \in A$ which majorates each of the indices $\alpha_n$, $n \in \omega$. Let $V_n = (p_{\alpha_n}^\alpha)^{-1}(V_{\alpha_n})$, $n \in \omega$. Define a function $\varphi_\alpha \in C(X_\alpha)$ by letting $\varphi_\alpha(x_\alpha) = \varphi(p_\alpha^{-1}(x_\alpha))$ for each point $x_\alpha \in X_\alpha$. First let us show that the function $\varphi_\alpha$ is well-defined. Surjectivity of the limit projection $p_\alpha$ guarantees that the set $\varphi(p_\alpha^{-1}(x_\alpha))$ is non-empty. Suppose that $\varphi(p_\alpha^{-1}(x_\alpha))$ contains two different points. Choose an element $U_n$ of the base $\mathcal{U}$ that contains exactly one of these points. Then it is easy to see that $G_n \cap p_\alpha^{-1}(x_\alpha) \neq \emptyset$ and $p_\alpha^{-1}(x_\alpha) \not\subseteq G_n$. These two relations, coupled with the choice of the index $\alpha$, lead to the following two contradictoty relations: $x_\alpha \in V_n$ and $x_\alpha \notin V_n$. Therefore, the map $\varphi_\alpha \colon X_\alpha \to R$ is well-defined and satisfies the equality $\varphi = \varphi_\alpha p_\alpha$. By the construction, the inverse images $\varphi_\alpha^{-1}(U_n)$, $n \in \omega$, of elements of the base $\mathcal{U}$ are open in $X_\alpha$ (because they coincide with $V_n$). This shows the continuity of $\varphi_\alpha$. □

COROLLARY 1.3.2. *Let $\mathcal{S}_X = \{X_\alpha, p_\alpha^\beta, A\}$ be an inverse spectrum with surjective limit projections and $\omega$-complete indexing set. If the limit space* lim $\mathcal{S}_X$ *is Lindelöf, then the spectrum $\mathcal{S}_X$ is factorizing.*

PROOF. Let $X = \lim \mathcal{S}_X$ and $G$ be a functionally open subset of $X$. By Proposition 1.3.1, it suffices to show that $G$ is cylindrical, i.e. $\alpha$-cylindrical for some $\alpha \in A$. Since functionally open cylindrical sets from a base of the limit space of any inverse spectrum (see Proposition 1.2.1), we conclude that $G = \cup\{p_{\alpha_t}^{-1}(G_{\alpha_t}) \colon t \in T\}$, where $G_{\alpha_t}$ is a functionally open subset of $X_{\alpha_t}$ for each $t \in T$. Obviously, a functionally open subspace of a Lindelöf space is Lindelöf itself. Therefore, $G$ is Lindelöf and there exists a countable subcollection $\{p_{\alpha_{t_n}}^{-1}(G_{\alpha_{t_n}}) \colon n \in \omega\}$ of the collection $\{p_{\alpha_t}^{-1}(G_{\alpha_t} \colon t \in T\}$. Corollary 1.1.28 provides an index $\alpha \in A$ majorating each of the indices $\alpha_{t_n}$, $n \in \omega$. Let $V_n = (p_{\alpha_{t_n}}^\alpha)^{-1}(G_{\alpha_{t_n}})$, $n \in \omega$. Obviously, the set $V = \cup\{V_n \colon n \in \omega\}$ is functionally open in $X_\alpha$ and $G = p_\alpha^{-1}(V)$. Proposition 1.3.1 finishes the proof. □

PROPOSITION 1.3.3. *Let $\tau \geq \omega$ and $\mathcal{S}_X = \{X_\alpha, p_\alpha^\beta, A\}$ be an inverse spectrum whose limit projections $p_\alpha$ are open, whose spaces $X_\alpha$ have Suslin number $\leq \tau$ and whose indexing set $A$ is $\tau$-complete. Then the spectrum $\mathcal{S}_X$ is factorizing.*

PROOF. First let us show that for each open subset $G \subseteq X = \lim \mathcal{S}_X$ there exists an index $\alpha \in A$ such that $cl_X G = p_\alpha^{-1}(cl_{X_\alpha} p_\alpha(G))$. To see this, represent the set $G$ as the union $\{p_{\alpha_t}^{-1}(V_{\alpha_t} \colon t \in T\}$, where $V_{\alpha_t}$ is open in $X_{\alpha_t}$ for each $t \in T$. It is known [**278**] that the Suslin number of the limit space $X$ does not exceed $\tau$. Therefore there is a subset $T' \subseteq T$ of cardinality $\leq \tau$ such that

$$cl_X G = cl_X(\cup\{p_{\alpha_t}^{-1} \colon t \in T'\}).$$

By Corollary 1.1.28, there exists an index $\alpha \in A$ such that $\alpha \succeq \alpha_t$ for each $t \in T'$. Let $V = \cup\{V_t \colon t \in T'\}$, where $V_t = (p_{\alpha_t}^\alpha)^{-1}(V_{\alpha_t})$, $t \in T'$. Obviously, by the construction, $cl_X G = cl_X p_\alpha^{-1}(V)$. Consequently, by openness of the limit projection $p_\alpha$, we have $cl_X G = p_\alpha^{-1}(cl_{X_\alpha} V)$ and, hence, $cl_X G = p_\alpha^{-1}(cl_{X_\alpha} p_\alpha(G))$.

The rest of the proof only differs slightly from the proof of Proposition 1.3.1. Consider a function $\varphi \in C(X)$ and fix a countable open base $\mathcal{U} = \{U_n \colon n \in \omega\}$ of the real line $R$. By the above established fact, coupled with Corollary 1.1.28, we may assume that

$$cl_X \varphi^{-1}(U_n) = p_\alpha^{-1}(cl_{X_\alpha} p_\alpha(\varphi^{-1}(U_n)))$$

for each $n \in \omega$. We claim that the function $\varphi_\alpha\colon X_\alpha \to R$, defined by letting $\varphi_\alpha(x_\alpha) = \varphi(p_\alpha^{-1}(x_\alpha)$, $x_\alpha \in X_\alpha$, is a well-defined single-valued function. Assume the contrary. Then, for some point $x_\alpha \in X_\alpha$, there are points $x_1$ and $x_2$ in the fiber $p_\alpha^{-1}(x_\alpha)$, such that $\varphi(x_1) \neq \varphi(x_2)$. Take elements $U_{n_1}$ and $U_{n_2}$ of the base $\mathcal{U}$ such that $cl_R U_{n_1} \cap cl_R U_{n_2} = \emptyset$ and $\varphi(x_i) \in U_{n_i}$, $i = 1, 2$. Since $\varphi$ is continuous, we have $cl_X \varphi^{-1}(U_{n_1}) \cap cl_X \varphi^{-1}(U_{n_2}) = \emptyset$. By the above construction and the choice of the index $\alpha$, both $cl_X \varphi^{-1}(U_{n_1})$ and $cl_X \varphi^{-1}(U_{n_2})$ are $\alpha$-cylindrical. Consequently, we have $p_\alpha(\varphi^{-1}(U_{n_1})) \cap p_\alpha(\varphi^{-1}(U_{n_2})) = \emptyset$. On the other hand, $x_\alpha \in p_\alpha(\varphi^{-1}(U_{n_1})) \cap p_\alpha(\varphi^{-1}(U_{n_2}))$. This contradiction shows that $\varphi_\alpha \in C(X_\alpha)$. Obviously, $\varphi = \varphi_\alpha p_\alpha$. The continuity of $\varphi_\alpha$ follows immediately from the openness of the limit projection $p_\alpha$. □

**1.3.2. The Spectral Theorem.** In Proposition 1.2.14 we have already seen that any isomorphism (i.e. morphism, consisting of homeomorphisms) of inverse spectra induces a homeomorphism of their limit spaces. Of course, the converse statement is not true. Indeed, consider two inverse sequences $\mathcal{S}_{odd} = \{I^{2k+1}, \pi_k^{k+1}, \omega\}$ and $\mathcal{S}_{even} = \{I^{2k+2}, \lambda_k^{k+1}, \omega\}$, where $\pi_k^{k+1}\colon I^{2k+3} \to I^{2k+1}$ and $\lambda_k^{k+1}\colon I^{2k+4} \to I^{2k+2}$ denote the natural projections onto the corresponding subproducts. Obviously, both lim $\mathcal{S}_{odd}$ and lim $\mathcal{S}_{even}$ are homeomorphic to the Hilbert cube $Q$. At the same time, as one can easily see, there is no isomorphism between the given inverse sequences whose limit map is a homeomorphism.

Our next goal is to show that for certain types of inverse spectra, all maps between their limits are induced by morphisms of these spectra (or their cofinal subspectra). First we need some definitions.

The spectrum $\mathcal{S}_X = \{X_\alpha, p_\alpha^\beta, A\}$ is said to be $\tau$-*continuous* if:

- For each chain $B$ in $A$ with $|B| < \tau$ and sup$B = \beta$, the diagonal product $\triangle\{p_\alpha^\beta\colon \alpha \in B\}$ maps the space $X_\beta$ homeomorphically *into* the space lim $\{X_\alpha, p_\alpha^\beta, B\}$ .

The spectrum $\mathcal{S}_X = \{X_\alpha, p_\alpha^\beta, A\}$ is said to be a $\tau$-*spectrum* if:

- $w(X_\alpha) \leq \tau$ for each $\alpha \in A$.
- The spectrum $\mathcal{S}_X$ is $\tau$-continuous.
- The indexing set $A$ of $\mathcal{S}_X$ is $\tau$-complete.

We call a spectrum *transfinite* if its indexing set is an uncountable cardinal $\tau$ and the spectrum is $\kappa$-continuous (in the above sense) for each $\kappa$ with $\omega \leq \kappa < \tau$.

THEOREM 1.3.4. *If a $\tau$-spectrum $\mathcal{S}_X = \{X_\alpha, p_\alpha^\beta, A\}$ with surjective limit projections is factorizing, then each map of its limit space into the limit space of another $\tau$-spectrum $\mathcal{S}_Y = \{Y_\alpha, q_\alpha^\beta, A\}$ is induced by a morphism of cofinal and $\tau$-closed subspectra. If two factorizing $\tau$-spectra with surjective limit projections and the same indexing set have homeomorphic limit spaces, then they contain isomorphic cofinal and $\tau$-closed subspectra.*

PROOF. Let $f\colon \lim \mathcal{S}_X \to \lim \mathcal{S}_Y$ be a given map. We perform the spectral search by means of the following relation

$$L = \{(\alpha, \beta) \in A^2 \colon \text{there exists a map } f_\alpha^\beta \colon X_\beta \to Y_\alpha \text{ such that } f_\alpha^\beta p_\beta = q_\alpha f\}.$$

Let us verify the conditions of Proposition 1.1.29.

**Existence Condition.** Since the weight of the space $Y_\alpha$ does not exceed $\tau$, there exists a subset $\{\varphi_t \colon t \in T\} \subseteq C(Y_\alpha)$ of cardinality $|T| \leq \tau$ such that the diagonal product $\triangle\{\varphi_t \colon t \in T\}$ is a topological embedding of $Y_\alpha$ into $R^T$. Since the spectrum $\mathcal{S}_X$ is factorizing, we conclude that for each $t \in T$ there exist an index $\beta_t \in A$ and a function $\psi_\alpha^{\beta_t} \in C(X_{\beta_t})$ such that $\psi_\alpha^{\beta_t} p_{\beta_t} = \varphi_t q_\alpha f$. Corollary 1.1.28, coupled with the inequality $|T| \leq \tau$, guarantees the existence of an index $\beta \in A$ which majorates each of the indices $\beta_t$, $t \in T$. Consider the diagonal product $f_\alpha^\beta = \triangle\{\psi_\alpha^{\beta_t} p_{\beta_t}^\beta \colon t \in T\}$. Obviously, $f_\alpha^\beta$ maps the space $X_\beta$ into the product $R^T$ and $f_\alpha^\beta p_\beta = q_\alpha f$. The last equality, coupled with the surjectivity of the limit projection $p_\beta$ of the spectrum $\mathcal{S}_X$, shows that $f_\alpha^\beta(X_\beta) \subseteq Y_\alpha$.

**Majorantness Condition.** The verification of this condition is trivial. Indeed, it is enough to consider the composition $f_\alpha^\gamma = f_\alpha^\beta p_\beta^\gamma$.

**$\tau$-closeness Condition.** Suppose that for some chain $C = \{\alpha_t \colon t \in T\}$ in $A$ with $|T| \leq \tau$ and $\alpha = \sup C$, the maps $f_{\alpha_t}^\beta \colon X_\beta \to Y_{\alpha_t}$ are already defined is such a way that $f_{\alpha_t}^\beta p_\beta = q_{\alpha_t} f$ for each $t \in T$ (in other words, $(\alpha_t, \beta) \in L$ for each $t \in T$). These equalities and surjectivity of the limit projection $p_\beta$ of the spectrum $\mathcal{S}_X$ imply the equalities $f_{\alpha_t}^\beta = q_{\alpha_t}^{\alpha_{t'}} f_{\alpha_{t'}}^\beta$ whenever $\alpha_t \preceq \alpha_{t'}$. Since the spectrum $\mathcal{S}_Y$ is $\tau$-continuous, $Y_\alpha$ is canonically homeomorphic to the limit space of the subspectrum $\mathcal{S}_Y/C$. Therefore the diagonal product $f_\alpha^\beta = \triangle\{f_{\alpha_t}^\beta \colon t \in T\}$ maps the space $X_\beta$ into the space $Y_\alpha$ and satisfies the equalities $f_\alpha^\beta p_\beta = q_\alpha f$. Consequently, $(\alpha, \beta) \in L$.

Now denote by $A'$ the set of all $L$-reflexive elements in $A$. By Proposition 1.1.29, $A'$ is a cofinal and $\tau$-closed subset of $A$. One can easily see that the $L$-reflexivity of an element $\alpha \in A$ is equivalent to the existence of a map $f_\alpha = f_\alpha^\alpha \colon X_\alpha \to Y_\alpha$ satisfying the equality $f_\alpha p_\alpha = q_\alpha f$. Consequently, the collection $\{f_\alpha \colon \alpha \in A'\}$ is a morphism of the cofinal and $\tau$-closed subspectrum $\mathcal{S}_X/A'$ of the spectrum $\mathcal{S}_X$ into the cofinal and $\tau$-closed subspectrum $\mathcal{S}_Y/A'$ of the spectrum $\mathcal{S}_Y$. It only remains to remark that the original map $f$ is induced by the constructed morphism. This finishes the proof of the first part of our Theorem.

The second part of the Theorem can be obtained from the first as follows. Let $f\colon \lim \mathcal{S}_X \to \lim \mathcal{S}_Y$ be a homeomorphism. Denote by $f^{-1}\colon \lim \mathcal{S}_Y \to \lim \mathcal{S}_X$ the inverse of $f$. By the first part proved above, there exist a cofinal and $\tau$-closed subset $A_1$ of $A$ and a morphism

$$\{f_\alpha \colon X_\alpha \to Y_\alpha \colon \alpha \in A_1\} \colon \mathcal{S}_X/A_1 \to \mathcal{S}_Y/A_1$$

such that $f = \lim\{f_\alpha \colon \alpha \in A_1\}$. Similarly, there exist a cofinal and $\tau$-closed subset $A_2$ of $A$ and a morphism

$$\{g_\alpha : Y_\alpha \to X_\alpha : \alpha \in A_2\} : \mathcal{S}_Y/A_2 \to \mathcal{S}_X/A_2$$

such that $f^{-1} = \lim\{g_\alpha : \alpha \in A_2\}$.

By Proposition 1.1.27, the set $B = A_1 \cap A_2$ is still cofinal and $\tau$-closed in $A$. Therefore, in order to complete the proof, it suffices to show that for each $\alpha \in B$ the map $f_\alpha : X_\alpha \to Y_\alpha$ is a homeomorphism. Indeed, take a point $x_\alpha \in X_\alpha$ and, using surjectivity of the limit projection $p_\alpha$ of the spectrum $\mathcal{S}_X$, fix a point $x \in \lim\ \mathcal{S}_X$ such that $x_\alpha = p_\alpha(x)$. Then

$$g_\alpha f_\alpha(x_\alpha) = g_\alpha f_\alpha p_\alpha(x) = g_\alpha q_\alpha f(x) = p_\alpha f^{-1} f(x) = p_\alpha(x) = x_\alpha.$$

Similarly, we see that $f_\alpha g_\alpha = id_{Y_\alpha}$ for each $\alpha \in B$. It follows easily that in this situation each $f_\alpha$, $\alpha \in B$, is a homeomorphism. □

The above Theorem provides a topological sense of the so-called "uncountability effect," the set-theoretical justification of which is contained in Proposition 1.1.27. As was remarked above, the Spectral Theorem has no counterpart in the case of inverse sequences. This means that two inverse sequences with homeomorphic limit spaces do not have to contain isomorphic cofinal subsequences. The Spectral Theorem allows us to give a practical sense to the method of "spectral search," the roots of which, in fact, are still set-theoretical (see Proposition 1.1.29). More formally, if, for example, we need to show that a given (non-metrizable) compactum can be represented as the limit space of some $\tau$-spectrum, consisting of open limit projections, and if we already have a $\tau$-spectrum representing the given compactum, then (if the problem is solvable) there is a cofinal and $\tau$-closed subspectrum of the given one whose limit projections are open and, therefore, the entire problem reduces to the problem of finding this subspectrum.

Let us remark also that the requirement of surjectivity of limit projections of spectra, hypothesized in Theorem 1.3.4, is indeed essential. First observe that any separable metrizable space $X$ can be represented as the limit space of some factorizing $\omega$-spectrum with surjective projections (it is enough to consider the trivial spectrum all elements of which coincide with $X$ and all projections of which are identity maps $id_X$). Moreover, one can easily see that every separable metrizable space, $X$ in particular, can be represented as the limit space of factorizing $\omega$-spectrum consisting of Polish spaces with, generally, non-surjective limit projections (to see this embed $X$ into, say, the Hilbert cube $Q$ and consider the collection of all $G_\delta$-subsets of $Q$ containing $X$, together with inclusion maps; compare with Example 1.2.10). Let us now assume that $X$ is a non-complete separable metrizable space and represent it as the limit space of the spectrum $\mathcal{S}_X = \{X_\alpha, p_\alpha^\beta, A\}$ consisting of Polish spaces. Assuming that Theorem 1.3.4 is valid in this situation, we easily conclude that there exist an index $\alpha \in A$ and a map $f_\alpha \colon X_\alpha \to X$ such that $f_\alpha p_\alpha = id_X$. The latter immediately implies the completeness of $X$, contradicting our choice of $X$.

This example shows that spectra arising in different situations do not necessarily have surjective limit projections and, therefore, that we are somewhat restricted in using such a powerful tool of spectral analysis as the Spectral Theorem. If we wish to understand what sort of generalizations of Theorem 1.3.4 we might expect in order to cover wider classes of inverse spectra, let us first characterize the class of spaces which can be represented as the limit spaces of factorizing $\omega$-spectra (a more general problem, of characterizing the limit spaces of factorizing $\tau$-spectra, has been considered in [**75**]). This is done in the next statement.

PROPOSITION 1.3.5. *The following conditions are equivalent for any space $X$:*
(i) *$X$ is realcompact.*
(ii) *$X$ is homeomorphic to the limit space of a factorizing $\omega$-spectrum, consisting of Polish spaces.*
(iii) *$X$ is homeomorphic to the limit space of a factorizing $\omega$-spectrum.*

PROOF. **(i) ⇒ (ii).** Embed $X$ as a $C$-embedded subspace into $R^A$ for sufficiently large $A$ (one can take $A = C(X)$). By $(i)$ and Proposition 1.1.16, $X$ is closed in $R^A$. Consider the standard $\omega$-spectrum $\mathcal{S} = \{R^B, \pi^B_C, exp_\omega A\}$, consisting of countable subproducts of $R^A$ and natural projections between them. Obviously, $\lim \mathcal{S} = R^A$ (see Example 1.2.9). Note also that since every continuous real-valued function, defined on $R^A$, depends on countably many coordinates, the spectrum $\mathcal{S}$ is factorizing (the same conclusion follows, for example, from Proposition 1.3.3). Closeness of $X$ in $R^A$ and Corollary 1.2.6 show that the limit space of the spectrum $\mathcal{S}_X = \{X_B, p^B_C, exp_\omega A\}$ coincides with $X$. Here $X_B = cl_{R^B}\pi_B(X)$, $B \in exp_\omega A$, and $p^B_C = \pi^B_C/X_B$ whenever $C, B \in exp_\omega A$ and $C \subseteq B$. Obviously, the spectrum $\mathcal{S}_X$ is $\omega$-continuous and consists of Polish spaces. Since $X$ in $C$-embedded in $R^A$ we see that $\mathcal{S}_X$ is a factorizing spectrum.

**(ii) ⇒ (iii).** Trivial.

**(iii) ⇒ (i).** Apply Propositions 1.2.2, 1.1.13 and Corollary 1.1.25. □

Thus, we now have a result characterizing the class of realcompact spaces as the maximal class of spaces which can be studied using the concept of factorizing $\omega$-spectra. But here it is necessary to make an important remark. Namely, although the factorizing $\omega$-spectra, constructed in Proposition 1.3.5, have a very promising additional property - all spaces in this spectra are Polish - nevertheless, these spectra have an essential deficiency since their limit projections are not surjective. And as we have seen above, under the latter circumstances we cannot use the fundamental result of the whole spectral analysis - the Spectral Theorem. Of course, we can take spaces $\pi_B(X)$ as elements $X_B$ of the spectra $\mathcal{S}_X$ (instead of their closures in $R^B$; see the proof of Proposition 1.3.5). Doing so we lose completeness in favor of surjectivity of the limit projections. But, unfortunately, the inverse spectra obtained in this way would not be continuous!

A simple observation saves the situation. First of all, let us note that according to Corollary 1.2.5 we may assume, without loss of generality, that

**all limit projections of inverse spectra, considered below, have dense images.**

Now we prove another version of the Spectral Theorem.

THEOREM 1.3.6. *Each map* $f\colon \lim \mathcal{S}_X \to \lim \mathcal{S}_Y$ *between the limit spaces of two factorizing* $\omega$*-spectra* $\mathcal{S}_X = \{X_\alpha, p_\alpha^\beta, A\}$ *and* $\mathcal{S}_Y = \{Y_\alpha, q_\alpha^\beta, A\}$ *, consisting of Polish spaces, is induced by a morphism of cofinal and* $\omega$*-closed subspectra. If* $f$ *is a homeomorphism, we may assume that the above morphism is an isomorphism.*

PROOF. The proof of this statement differs just slightly from the proof of Theorem 1.3.4. As in that proof we perform the spectral search with respect to the following relation:

$$L = \{(\alpha, \beta) \in A^2\colon \text{there exists a map } f_\alpha^\beta\colon X_\beta \to Y_\alpha \text{ such that } f_\alpha^\beta p_\beta = q_\alpha f\}.$$

The majorantness and $\omega$-closedness conditions of Proposition 1.1.29 can be verified just as in the proof of Theorem 1.3.4.

Let us verify the conditions of Proposition 1.1.29. Let $\alpha \in A$. Since, by our assumption, the space $Y_\alpha$ is Polish, we can, by Proposition 1.1.3, identify $Y_\alpha$ with a closed subspace of $R^\omega$ . Therefore there exists a countable subset $\{\varphi_n\colon n \in \omega\} \subseteq C(Y_\alpha)$ such that the diagonal product $\triangle\{\varphi_n\colon n \in \omega\}$ is a closed embedding of $Y_\alpha$ into $R^\omega$ . As in the proof of Theorem 1.3.4, we can find an index $\beta \in A$ and a map $f_\alpha^\beta\colon X_\beta \to R^\omega$ such that $q_\alpha f = f_\alpha^\beta p_\beta$. The latter implies that

$$f_\alpha^\beta(p_\beta(\lim \mathcal{S}_X)) = q_\alpha(f(\lim \mathcal{S}_X)) \subseteq q_\alpha(\lim \mathcal{S}_Y) \subseteq Y_\alpha.$$

Now recall that the image of the limit projection $p_\beta$ is dense in $X_\beta$. Therefore, by continuity of $f_\alpha^\beta$ and closeness of $Y_\alpha$ in $R^\omega$ , we have

$$f_\alpha^\beta(X_\beta) = f_\alpha^\beta(cl_{X_\beta} p_\beta(\lim \mathcal{S}_X)) \subseteq cl_{R^\omega} f_\alpha^\beta(p_\beta(\lim \mathcal{S}_X)) \subseteq cl_{R^\omega} Y_\alpha = Y_\alpha.$$

This finishes the verification of the existence condition and the proof of the Theorem. □

Let us now consider some curious examples indicating that Theorem 1.3.6, unlike Theorem 1.3.4, works even for separable metrizable spaces.

Indeed, let $X$ be a dense subspace of a Polish space $X'$ and consider the collection of all $G_\delta$-subspaces of $X'$ containing $X$. Obviously (compare with Example 1.2.10), this collection, together with the corresponding inclusion maps, forms an $\omega$-spectrum $\mathcal{S}_{X,X'}$ consisting of Polish spaces. Lemma 1.1.19 and Proposition 1.1.22 show that this spectrum is factorizing. Obviously, $\lim \mathcal{S}_{X,X'} = X$. Suppose now that $Y$ is a dense subspace of another Polish space $Y'$ and consider a factorizing $\omega$-spectrum $\mathcal{S}_{Y,Y'}$ constructed in a similar way. Finally assume that there is a homeomorphism $f\colon X \to Y$. Theorem 1.3.6, applied to spectra $\mathcal{S}_{X,X'}$ and $\mathcal{S}_{Y,Y'}$, guarantees that there exist $G_\delta$-subspaces $\tilde{X} \subset X'$ and $\tilde{Y} \subset Y'$ of $X'$ and $Y'$ respectively and a map $\tilde{f}\colon \tilde{X} \to \tilde{Y}$ such that $X \subseteq \tilde{X}$, $Y \subseteq \tilde{Y}$ and $f = \tilde{f}/X$. In other words we have obtained the classical result of Lavrentieff [**208**].

In the next subsection we present another application of Theorem 1.3.6 dealing with a well-known result of Tumarkin [**301**].

**1.3.3. Immediate consequences.** We begin by establishing some dimensional properties of realcompact spaces. We start with the following simple lemma, the straightforward proof of which is left to the reader.

LEMMA 1.3.7. *Let $\mathcal{S}_X = \{X_\alpha, p_\alpha^\beta, A\}$ be a factorizing spectrum and $X = \lim \mathcal{S}_X$. If $\dim\, p_\alpha(X) \leq n$ for each $\alpha \in A$, then $\dim\, X \leq n$.*

PROPOSITION 1.3.8. *Let $f\colon X \to Y$ be a map of at most $n$-dimensional compactum $X$ into a metrizable compactum $Y$. Then there exist at most $n$-dimensional metrizable compactum $Z$ and two maps $g\colon X \to Z$ and $h\colon Z \to Y$ such that $f = h \cdot g$.*

PROOF. By Proposition 1.3.5, we can represent $X$ as the limit space of some factorizing $\omega$-spectrum $\mathcal{S}_X = \{X_\alpha, p_\alpha^\beta, A\}$ . We shall show that almost all spaces $X_\alpha$ in the spectrum $\mathcal{S}_X$ are at most n-dimensional, and that as the compactum $Z$ one can take an element of our spectrum.

We perform the spectral search with respect to the following relation:

$$L = \{(\alpha,\beta) \in A^2\colon \alpha \leq \beta \text{and for each open cover } \mathcal{U} \text{ of } X_\alpha \text{ there is an open}$$

$$\text{cover } \mathcal{V} \text{ of } X_\beta \text{ of order} \leq n \text{ such that } (p_\alpha^\beta)^{-1}(\mathcal{U}) \text{ refines } \mathcal{V}\}.$$

Let us the check conditions of Proposition 1.1.29.

**Existence condition.** Let $\{\mathcal{U}_k\colon k \in \omega\}$ be a fundamental sequence of finite open covers of $X_\alpha$. This means that for each open cover $\mathcal{U}$ of $X_\alpha$ there is an index $k \in \omega$ such that $\mathcal{U}_k$ refines $\mathcal{U}$. For each $k \in \omega$ take a finite functionally open cover $\mathcal{W}_k$ of $X$ of order $\leq n$ such that $\mathcal{W}_k$ refines $p_\alpha^{-1}(\mathcal{U}_k)$. Since our spectrum is factorizing, we conclude (see Proposition 1.3.1) that there exist an index $\beta_k \in A$ such that $\beta_k \geq \alpha$, and an open cover $\mathcal{V}_k^{\beta_k}$ of the compactum $X_{\beta_k}$ such that $\mathcal{W}_k = p_{\beta_k}^{-1}(\mathcal{V}_k^{\beta_k})$. Observe that the order of $\mathcal{V}_k^{\beta_k}$ is at most $n$. By Corollary 1.1.28, there is an index $\beta \in A$ which majorates each of the indices $\beta_k$, $k \in \omega$. Let $\mathcal{V}_k = (p_{\beta_k}^{\beta})^{-1}(\mathcal{V}_k^{\beta_k})$. Clearly $\mathcal{V}_k$ is a finite open cover of $X_\beta$ of order at most $n$, $k \in \omega$. It only remains to note that $(\alpha,\beta) \in L$.

**Majorantness condition.** Let $(\alpha,\beta) \in L$ and $\gamma \succ \beta$. If $\mathcal{U}$ is an open cover of $X_\alpha$, then there is an open cover $\mathcal{V}'$ of $X_\beta$ of order at most $n$ which refines

$(p^{\beta}_{\alpha})^{-1}(\mathcal{U})$. Then it is easy to see that the open cover $\mathcal{V} = (p^{\gamma}_{\beta})^{-1}(\mathcal{V}')$ verifies the fact that $(\alpha, \gamma) \in L$.

**$\omega$-closeness condition.** Let $\{\alpha_i : i \in \omega\}$ be a countable chain of elements in $A$ such that $(\alpha_i, \beta) \in L$ for each $i \in \omega$ and $\alpha = \sup\{\alpha_i : i \in \omega\}$. Let $\mathcal{U}$ be an open cover of the compactum $X_\alpha$. Since our spectrum is $\omega$-continuous, the space $X_\alpha$ is naturally homeomorphic to the limit space of the inverse sequence $\{X_{\alpha_i}, p^{\alpha_{i+1}}_{\alpha_i}, \omega\}$. Consequently there exist an index $i \in \omega$ and open cover $\mathcal{U}_i$ of $X_i$ such that $(p^{\alpha}_{\alpha_i})^{-1}(\mathcal{U}_i)$ is refinement of $\mathcal{U}$. Further, by assumption, $(\alpha_i, \beta) \in L$. Therefore, there is an open cover $\mathcal{V}$ of $X_\beta$ of order $\leq n$ such that $\mathcal{V}$ refines $(p{\alpha_i}^{\beta})^{-1}(\mathcal{U}_i)$. Clearly, in this situation, $\mathcal{V}$ refines $(p^{\beta}_{\alpha})^{-1}(\mathcal{U})$ as well. This finishes the verification of all three conditions.

Therefore, by Proposition 1.1.29, the set $A'$ of all $L$-reflexive indices is cofinal and $\omega$-closed in $A$. Observe that if an index $\alpha$ is $L$-reflexive (i.e. $(\alpha, \alpha) \in L$), then $\dim X_\alpha \leq n$.

Since $Y$ is metrizable and the spectrum $\mathcal{S}_X/A'$ is still factorizing (see Corollary 1.3.2) there exist an index $\alpha \in A'$ and a map $h : X_\alpha \to Y$ such that $f = h \cdot p_\alpha$. This finishes the proof. □

LEMMA 1.3.9. *For any countable collection $\{p_k : X \to X_k; k \in \omega\}$ of maps of an at most $n$-dimensional space $X$ into Polish spaces $X_k$, there exist an at most $n$-dimensional Polish space $P$ and maps $g : X \to P$, $g_k : P \to P_k$, $k \in \omega$, such that $p_k = g_k \cdot g$ for each $k \in \omega$.*

PROOF. Let us first consider the situation in which the given collection contains only one element, a map $p_0 : X \to P_0$. Let $\bar{P}_0$ denotes a metrizable compactification of $P_0$. Then there is an extension $\bar{p}_0 : \beta X \to \bar{P}_0$ of the map $p_0$, where $\beta X$ denotes the Stone-Čech compactification of $X$. Since $\dim X = \dim \beta X$, we conclude, using Proposition 1.3.8, that there exist an at most $n$-dimensional metrizable compactum $K$ and two maps $\bar{g} : \beta X \to K$ and $\bar{g}_0 : K \to \bar{P}_0$ such that $\bar{p}_0 = \bar{g}_0 \cdot \bar{g}$. Obviously the Polish space $P = (\bar{g}_0)^{-1}(P_0)$ and maps $g = \bar{g}/X : X \to P$ and $g_0 = \bar{g}_0/P : P \to P_0$ satisfy the desired conditions.

Next we consider the general case. The diagonal product $\triangle\{p_k : k \in \omega\}$ maps the space $X$ into the Polish space $P' = \prod\{P_k : k \in \omega\}$. By the above considered case, there exist an at most $n$-dimensional Polish space $P$ and two maps $g : X \to P$ and $h : P \to P'$ such that $p = h \cdot g$. Then the space $P$ and maps $g$ and $g_k = \pi_k \cdot h$, $k \in \omega$, where $pi_k : P' \to P_k$ denotes the projection onto the $k$-th coordinate, satisfy the desired conditions. □

THEOREM 1.3.10. *The following conditions are equivalent for any realcompact space $X$:*

(i) $\dim X \leq n$.

(ii) *$X$ can be represented as the limit space of a factorizing $\omega$-spectrum, consisting of at most $n$-dimensional Polish spaces.*

PROOF. One part of our statement follows from Lemma 1.3.7. Now we consider an at most $n$-dimensional realcompact space $X$, and we try to represent it as the limit space of a factorizing $\omega$-spectrum consisting of at most $n$-dimensional Polish spaces. If $X$ itself is a Polish space, then the above conclusion is trivially true. Therefore, assuming that $X$ is $C$-embedded (and, by Proposition 1.1.16, closed) in $R^A$ , we only have to consider the case in which $A$ is uncountable. Denote by $\pi_B\colon R^A \to R^B$, $B \subseteq A$, the natural projection onto the corresponding subproduct and let $X_B = cl_{R^B}\pi_B(X)$. Next, denote by $exp_{<\omega}A$ the directed (by inclusion) set of all finite subsets of $A$. Closeness of $X$ in $R^A$ implies that the limit space of the spectrum $\mathcal{S} = \{X_B, p^B_C, exp_{<\omega}A\}$ coincides with $X$ (see Example 1.2.9). Let us now construct another spectrum $\mathcal{S}'$, consisting of at most $n$-dimensional Polish spaces, the limit space of which also coincides with $X$. Simultaneously, there will be constructed a morphism $\{h_B\}\colon \mathcal{S}' \to \mathcal{S}$.

The set $exp_{<\omega}A$ can be represented as the union of pairwise disjoint sets $T_k$, $k \in \omega$, where $T_k$ stands for the collection of all subsets of $A$ whose cardinality is exactly $k$. Let $k \in T_1$. Apply Lemma 1.3.9 to the map $p_B\colon X \to X_B$. We clearly get an at most $n$-dimensional Polish space $P_B$ and two maps $g_B\colon X \to P_B$ and $h_B\colon P_B \to X_B$ such that $p_B = h_B \cdot g_B$. Suppose now that for each $k \leq m-1$, at most $n$-dimensional Polish spaces $P_B$, $B \in T_1 \cup ... \cup T_{m-1}$, and maps $g_B\colon X \to P_B$, $h_B\colon P_B \to X_B$ and $g^B_C\colon P_B \to P_C$, $C \subseteq B$, have already been constructed in such a way that $g_C = g^B_C \cdot g_B$ and $p_B = h_B \cdot g_B$. Let $B_0 \in T_m$. Take all sets $B$ contained in $B_0$. There are only finitely many of those. By Lemma 1.3.9, applied to the maps $g_B\colon X \to P_B$, $B \subset B_0$, and $p_{B_0}\colon X \to X_{B_0}$, there exist an at most $n$-dimensional Polish space $P_{B_0}$ and maps $g_{B_0}\colon X \to P_{B_0}$, $g^{B_0}_B\colon P_{B_0} \to P_B$ and $h_{B_0}\colon P_{B_0} \to X_{B_0}$ such that $g_B = g^{B_0}_B \cdot g_{B_0}$ and $p_{B_0} = h_{B_0} \cdot g_{B_0}$. This finishes the construction of the spectrum $\mathcal{S}' = \{P_B, g^B_C, exp_{<\omega}\}$ and of the morphism $\{h_B\}\colon \mathcal{S}' \to \mathcal{S}$. It immediately follows from the construction that the limit space of the spectrum $\mathcal{S}'$ is homeomorphic to $X$ and that the limit map of the morphism coincides with the identity map $id_X$. Consider now an $\omega$-completion (see subsection 1.1.3) of the indexing set $exp_{<\omega}A$ of our spectra. Clearly this $\omega$-completion coincides with $exp_\omega A$. Further, add to the spectra $\mathcal{S}'$ and $\mathcal{S}$ the limit spaces of all inverse sequences of the form $\mathcal{S}'/C$ and $\mathcal{S}/C$ respectively, where $C \in exp_\omega A$. In this way we obtain two $\omega$-spectra $[\mathcal{S}']_\omega$ and $[\mathcal{S}]_\omega$, consisting of Polish spaces. Additionally, as is easily seen, elements of the spectrum $[\mathcal{S}']_\omega$ will be at most $n$-dimensional. The above constructed morphism $\{h_B\}\colon \mathcal{S}' \to \mathcal{S}$ generates (in a canonical way) a certain morphism, with the identity map $id_X$ as its limit, between the spectra $[\mathcal{S}']_\omega$ and $[\mathcal{S}]_\omega$. Now observe that the spectrum $[\mathcal{S}]_\omega$ coincides with the spectrum constructed in the proof of Proposition 1.3.5 and, consequently, is factorizing. This immediately implies that the spectrum $[\mathcal{S}']_\omega$ is also factorizing. □

Now we are ready to explain how Theorem 1.3.6 can be used to obtain a simple proof of Tumarkin's "$G_\delta$-enlargement theorem."

Let $X$ be a dense subspace of a Polish space $X'$. Assume, additionally, that $\dim X = n$. Let $\mathcal{S}_{X,X'}$ be a factorizing $\omega$-spectrum consisting of all $G_\delta$-subspaces of $X'$, containing $X$, together with their inclusion maps (see the end of Subsection 1.3.2). By Theorem 1.3.10 and Corollary 1.1.25, the space $X$ is homeomorphic to the limit space of a factorizing $\omega$-spectrum $\mathcal{S}$ consisting of $n$-dimensional Polish spaces. Now, applying Theorem 1.3.6, we see that the spectra $\mathcal{S}_{X,X'}$ and $\mathcal{S}$ contain isomorphic cofinal and $\omega$-closed subspectra. In particular, there is an $n$-dimensional $G_\delta$-subspace of $X'$ containing $X$.

The following statement will be needed later in Chapter 5.

PROPOSITION 1.3.11. *Let $f\colon P \to Y$ be a map of a Polish space $P$ into a metrizable compactum $Y$. Then $f$ admits an extension $F\colon \tilde{P} \to Y$ to a metrizable compactification $\tilde{P}$ of $P$ such that* $\dim \tilde{P} = \dim P$.

PROOF. Let $\beta P$ denote the Stone-Čech compactification of $P$. Obviously, $\dim \beta P = \dim P \equiv n$. By Theorem 1.3.10, $\beta P = \lim \mathcal{S}$, where $\mathcal{S} = \{X_\alpha, p_\alpha^\beta, A\}$ is a factorizing $\omega$-spectrum consisting of $n$-dimensional metrizable compacta. Since $P$ is Polish, we see that $P$ is the intersection of countably many functionally open subsets of $\beta P$. Therefore there exists an index $\alpha_0 \in A$ such that $P = p_{\alpha_0}^{-1}(p_{\alpha_0}(P))$. Consider now the spectrum $\mathcal{S}_P = \{P_\alpha, q_\alpha^\beta, A_0\}$, where $A_0 = \{\alpha \in A\colon \alpha \geq \alpha_0\}$, $P_\alpha = p_\alpha(P)$ and $q_\alpha{}^\beta = p_\alpha^\beta/P_\beta$ for each $\alpha, \beta \in A_0$ with $\alpha \leq \beta$. By Corollary 1.3.2, the spectrum $\mathcal{S}_P$ is factorizing. Therefore, by Theorem 1.3.4, there is a cofinal and $\omega$-closed subset $A_1 \subseteq A_0$ such that $q_\alpha\colon P \to P_\alpha$ is a homeomorphism for each $\alpha \in A_1$. Now consider the extension $\beta f\colon \beta P \to Y$. Again, by Theorem 1.3.4, we can find an index $\alpha \in A_1$ and a map $f_\alpha\colon X_\alpha \to Y$ such that $\beta f = f_\alpha p_\alpha$. It only remains to take $\tilde{P} = X_\alpha$ and $F = f_\alpha$. □

We conclude this subsection with some generalizations of the Spectral Theorem.

PROPOSITION 1.3.12. *Let $Z_0$ be a subspace of a space $Z$ and $f\colon Z_0 \to X$ be a map into a realcompact space $X$. If $C(f)(C(X)) \subseteq C(Z)/Z_0$, then $f$ admits an extension $g\colon cl_Z Z_0 \to X$ such that $C(g)(C(X)) \subseteq C(Z)/cl_Z Z_0$.*

PROOF. We assume that $X$ is $C$-embedded in $R^{C(X)} = \prod\{R_\varphi\colon \varphi \in C(X)\}$, where $R_\varphi = R$ for each function $\varphi \in C(X)$. By Proposition 1.1.16, $X$ is closed in $R^{C(X)}$. Denote by $\pi_\varphi\colon R^{C(X)} \to R_\varphi$ projection onto the corresponding coordinate and by $p_\varphi$ its restriction on $X$. Clearly the diagonal product $\triangle\{p_\varphi\colon \varphi \in C(X)\}$ coincides with the inclusion map of $X$ into $R^{C(X)}$. Since $C(f)(C(X)) \subseteq C(Z)/Z_0$, each of the maps $p_\varphi \cdot f$ has an extension $g_\varphi$ to the space $Z$. The diagonal product $\bar{g} = \triangle\{g_\varphi\colon \varphi \in C(X)\}$ maps the space $Z$ into $R^{C(X)}$ and coincides with $f$ on $Z_0$. By continuity of $\bar{g}$ and closedness of $X$ in $R^{C(X)}$ we see that $\bar{g}(cl_Z Z_0) \subseteq X$. It only remains to let $g = \bar{g}/cl_Z Z_0$. □

PROPOSITION 1.3.13. *Let $\mathcal{S}_X = \{X_\alpha, p_\alpha^\beta, A\}$ and $\mathcal{S}_Y = \{Y_\alpha, q_\alpha^\beta, A\}$ be two factorizing $\omega$-spectra consisting of Polish spaces, $X = \lim\ \mathcal{S}_X$, $Y = \lim\ \mathcal{S}_Y$ and*

*$Z$ is closed in $X$. Suppose also that $f\colon Z \to Y$ is a map such that $C(f)(C(Y)) \subseteq C(X)/Z$. Then $f$ is induced by a morphism of cofinal and $\omega$-closed subspectra of the spectra $\mathcal{S}_Z$ and $\mathcal{S}_Y$, where $\mathcal{S}_Z = \{cl_{X_\alpha} p_\alpha(Z), p_\alpha^\beta, A\}$.*

PROOF. We perform the spectral search with respect to the following relation:

$$L = \{(\alpha,\beta) \in A^2\colon \alpha \preceq \beta \text{and there exists a map } f_\alpha^\beta\colon cl_{X_\beta} p_\beta(Z) \to Y_\alpha$$

$$\text{such that } q_\alpha \cdot f = f_\alpha^\beta \cdot p_\beta/Z\}.$$

Let us first verify the Existence Condition of Proposition 1.1.29, i.e. let us show that for each $\alpha \in A$ there is $\beta \in A$ such that $(\alpha,\beta) \in L$. Since $Y_\alpha$ is a Polish space, we can assume, by Proposition 1.1.3, that $Y_\alpha$ is closed in $R^\omega$. Denote by $\pi_n$ the restriction to $Y_\alpha$ of the natural projection of $R^\omega$ onto the $n$-th coordinate. Clearly, the diagonal product $\triangle\{\pi_n\colon n \in \omega\}$ coincides with the inclusion map of $Y_\alpha$ into $R^\omega$. The function $\pi_n \cdot q_\alpha$, $n \in \omega$ is an element of $C(Y_\alpha)$ and, consequently, by our assumption, the function $\pi_n \cdot q_\alpha \cdot f$ has an extension $\varphi_n$ to the space $X$. The spectrum $\mathcal{S}_X$ is factorizing. Therefore, for each $n \in \omega$, there exist an index $\beta_n \in A$ and a function $\varphi_n^{\beta_n} \in C(X_{\beta_n})$ such that $\varphi_n = \varphi_n^{\beta_n} \cdot p_{\beta_n}$. Since the indexing set $A$ is $\omega$-complete, there is, by Corollary 1.1.28, an index $\beta \in A$ which majorates each of the indices $\beta_n$, $n \in \omega$. Let $\varphi_n^\beta = \varphi_n^{\beta_n} \cdot p_{\beta_n}^\beta$, $n \in \omega$. The diagonal product $g_\alpha^\beta = \triangle\{\varphi_n^\beta\colon n \in \omega\}$ maps the space $X_\beta$ into $R^\omega$. It follows from the construction that $q_\alpha \cdot f = g_\alpha^\beta \cdot p_\beta/Z$. Consequently, $g_\alpha^\beta(p_\beta(Z)) \subseteq Y_\alpha$. By continuity of $g_\alpha^\beta$ and closedness of $Y_\alpha$ in $R^\omega$, we have $g_\alpha^\beta(cl_{X_\beta} p_\beta(Z)) \subseteq Y_\alpha$. It only remains to note that the map $f_\alpha^\beta = g_\alpha^\beta / cl_{X_\beta} p_\beta(Z)$ satisfies the desired properties. This finishes the verification of the Existence Condition. The verification of the Majorantness and $\omega$-closedness Conditions of Proposition 1.1.29 is trivial. Therefore, by that Proposition, the set of $L$-reflexive indices is cofinal and $\omega$-closed in $A$. □

Later we will find it helpful to have a cardinal invariant distinguishing the Polish spaces among all separable metrizable spaces.

DEFINITION 1.3.14. *The* $R$-weight *of a space $X$ is the minimal infinite cardinal $\tau$ such that $X$ admits a $C$-embedding into $R^\tau$ (notation: $R - w(X) = \tau$).*

Obviously, the usual topological weight of any space $X$ does not exceed its $R$-weight. The converse is true, for example, for any infinite compactum. The reader can easily verify that:

- $R - w(X) \leq \mid C(X) \mid$ for any space $X$.
- $R - w(X) = R - w(\nu X)$ for any space $X$.

Countability of the $R$-–weight characterizes Polish spaces (see Propositions 1.1.3 and 1.1.16). Monotonicity of the $R$-weight with respect to $C$-embedded subspaces also follows immediately from the definition.

LEMMA 1.3.15. *Let $\mathcal{S}_X = \{X_\alpha, p_\alpha^\beta, A\}$ be a factorizing spectrum consisting of Polish spaces. Then $R - w(\lim\ \mathcal{S}_X) \leq \max\{\omega, |A|\}$.*

PROOF. If the indexing set $A$ of the spectrum $\mathcal{S}_X$ is countable, then the limit space $\lim\ \mathcal{S}_X$ is Polish and, consequently, in this case the conclusion of the Lemma is true. Suppose that $A$ is uncountable. Since $\mathcal{S}_X$ is a factorizing spectrum, we can easily see that the limit space $\lim\ \mathcal{S}_X$ is $C$-embedded in the product $\prod\{X_\alpha \colon \alpha \in A\}$ (compare with Proposition 1.2.2). But as noted before, the $R$-weight in monotone with respect to $C$-embedded subspaces. Therefore, it suffices to show that the $R$-weight of the product $\prod\{X_\alpha \colon \alpha \in A\}$ of uncountable family of Polish spaces does not exceed $|A|$. Without loss of generality we may assume that $X_\alpha$ is closed in $R_\alpha^\omega = R^\omega$ (Proposition 1.1.3). Consequently,

$$X = \prod\{X_\alpha \colon \alpha \in A\} \subseteq \prod\{R_\alpha^\omega \colon \alpha \in A\} = R^A$$

and we only have to show that $X$ is $C$-embedded in $R^A$. Consider a function $\varphi \in C(X)$. By Proposition 1.3.3, there exist a countable subset $B$ of $A$ and a function $\psi \in C(X_B)$, where $X_B = \prod\{X_\alpha \colon \alpha \in B\}$, such that $\varphi = \psi \cdot p_B$ ($p_B$ denotes the natural projection of $X$ onto $X_B$). The countability of $B$ implies that $R^B$ is Polish, and since $X_B$ is closed in $R^B$ there is an extension $\bar{\psi} \in C(R^B)$ of $\psi$. Then the function $\bar{\varphi} = \bar{\psi} \cdot \pi_B$, where $\pi_B \colon R^A \to R^B$ denotes the projection, is the desired extension of the original function $\varphi$ to the whole $R^A$. □

Finally we present one more version of the Spectral Theorem.

THEOREM 1.3.16. *Let $f \colon Z \to F$ be a map between closed subspaces $Z$ and $F$ of an uncountable product $\prod\{X_\alpha \colon \alpha \in A\}$ of Polish spaces. Suppose that $B \subseteq A$ and there exists a map $f_B \colon Z_B \to F_B$ such that $\pi_B \cdot f = f_B \cdot \pi_B/Z$. Then the collection*

$$\mathcal{K}_B = \{C \in exp_\omega(A - B) \colon \textit{there is a map } f_{B\cup C} \colon Z_{B\cup C} \to F_{B\cup C}$$

$$\textit{such that } \pi_{B\cup C} \cdot f = f_{B\cup C} \cdot \pi_{B\cup C}/Z\}$$

*is cofinal and $\omega$-closed in $exp_\omega(A - B)$.*

PROOF. Consider the following relation $L_B$ on the set $(exp_\omega(A - B))^2$:

$$L_B = \{(C, D) \in (exp_\omega(A - B))^2 \colon C \subseteq D \text{ and there is a map}$$

$$f_{C,D} \colon Z_{B\cup D} \to F_{B\cup C} \text{ such that } \pi_{B\cup C} \cdot f = f_{C,D} \cdot \pi_{B\cup D}/Z\}.$$

Now we are going to perform the spectral search with respect to $L_B$.

**Existence Condition.** We have to show that for each $C \in exp_\omega(A - B)$ there exists $D \in exp_\omega(A - B)$ such that $(C, D) \in L_B$.

Using the standard factorization argument we can find a countable set $R \in exp_\omega A$ and a map $f_R \colon Z_R \to F_C$ such that $C \subset R$ and $f_R \cdot \pi_R/Z = \pi_C \cdot f$. Let

$D = R - B$. Clearly $C \subseteq D$. Define the map $f_{C,D}$ to be the diagonal product $f_{C,D} = f_B \cdot \pi_B^{B\cup D} \triangle f_R \cdot \pi_R^{B\cup D}$ (note that $B \cup D = B \cup R$). It only remains to note that $\pi_{B\cup C} \cdot f = f_{C,D} \cdot \pi_{B\cup D}/Z$. Therefore $(C, D) \in L_B$.

**Majorantness Condition.** Let $(C, D) \in L_B$, $E \in exp_\omega(A - B)$ and $D \subseteq E$. We need to show that $(C, E) \in L_B$.

Consider the map $f_{C,E} = \pi_{B\cup D}^{B\cup E} \cdot f_{C,D}$.

**$\omega$-closedness Condition.** Let $(C_i, D) \in L_B$, $i \in \omega$, where $\{C_i : i \in \omega\}$ is a countable chain in $exp_\omega(A - B)$. Then we need to show that $(C, D) \in L_B$, where $C = \cup\{C_i : i \in \omega\}$.

Clearly $F_C$ coincides with the limit space of an inverse sequence

$$\{F_{B\cup C_i}, \pi_{B\cup C_i}^{B\cup C_{i+1}}/F_{B\cup C_{i+1}}, i \in \omega\}.$$

A straightforward verification shows that $f_{C_i,D} = \pi_{B\cup C_i}^{B\cup C_{i+1}} \cdot f_{C_{i+1},D}$ for each $i \in \omega$. It only remains to note that the map $f_{C,D} = \triangle\{f_{C_i,D} : i \in \omega\}$ confirms the fact that $(C, D) \in L_B$.

By Proposition 1.1.29, the set $\mathcal{K}_B$ is cofinal and $\omega$-closed in $exp_\omega(A - B)$. □

Observe that the case $B = \emptyset$ yields, in fact, the above discussed version (Theorem 1.3.4) of the Spectral Theorem for $\omega$-spectra consisting of compacta.

COROLLARY 1.3.17. *If in Theorem 1.3.16 the maps $f$ and $f_B$ are homeomorphisms, then the collection $\mathcal{K}_B$ contains a cofinal and $\omega$-closed subcollection $\mathcal{K}'_B$ such that $f_{B\cup C}$ is a homeomorphism for each $C \in \mathcal{K}'_B$.*

We conclude this Chapter with the following statement.

PROPOSITION 1.3.18. *Dimension* dim *is monotone with respect to $z$-embedded subspaces.*

PROOF. Let $Y$ be a $z$-embedded subspace of a space $X$ with $\dim X = n$. Since $\dim X = \dim \nu X$ and since $X$ is $z$-embedded (even $C$-embedded) in $\nu X$ we may assume, without loss of generality, that $X$ is realcompact. Consider a finite functionally open cover $\mathcal{U} = \{U_1, \dots, U_k\}$ of the subspace $Y$. Since $Y$ is $z$-embedded in $X$ there is a functionally open subset $V_i$ of $X$ such that $U_i = V_i \cap Y$, $i = 1, \dots, k$. Let $V = \cup_{i=1}^k V_i$. By Theorem 1.3.10, $X$ can be represented as the limit space of a factorizing $\omega$-spectrum $\mathcal{S}_X = \{X_\alpha, p_\alpha^\beta, A\}$ consisting of at most $n$-dimensional Polish spaces. Since $V_i$ is functionally open in $X$ we conclude, by Proposition 1.3.1, that there exist an index $\beta \in A$ and an open subset $W_i$, $i = 1, \dots, k$, of $X_\beta$ such that $V_i = p_\beta^{-1}(W_i)$ for each $i = 1, \dots, k$. Now consider the open subspace $W = \cup_{i=1}^k W_i$ of $X_\beta$ and its open cover $\mathcal{W} = \{W_1, \dots, W_k\}$. Since dim is monotone with respect to any subspaces within the class of Polish spaces, we conclude that there exists a finite open cover $\mathcal{W}'$ of $W$ of order $\leq n$ which refines $\mathcal{W}$. It only remains to remark that $p_\beta^{-1}(\mathcal{W}')/Y$ is a functionally open cover of $Y$ of order $\leq n$ and which refines $\mathcal{U}$. Therefore $\dim Y \leq n$. □

*Historical and bibliographical notes* 1.3. The notion of factorizing $\tau$-spectrum first appeared in [**278**]. The Spectral Theorem for such spectra with surjective limit projections (Theorem 1.3.4) was established by Ščepin [**278**]. Its version (Theorem 1.3.6) for spectra consisting of Polish spaces was proved in [**84**]. Theorem 1.3.10 is taken from [**81**]. In the case of compact spaces this result contains Mardešić's theorem mentioned in the Preface. Proposition 1.3.11 was originally obtained in [**153**]. Theorem 1.3.16 is taken from [**107**]. Proposition 1.3.18 first appeared in [**72, 77, 80**] and was later rediscovered (in a more concrete situation) in [**250**] and [**240**]. All other statements from this Section are due to the author.

CHAPTER 2

# Infinite-Dimensional Manifolds

## 2.1. Absolute extensors and absolute retracts

**2.1.1. Spaces of maps.** The collection of all open (countable) covers of a Polish space $X$ is denoted by $cov(X)$. If another Polish space $Y$ is given then $C(Y,X)$ denotes the set of all continuous maps from $Y$ into $X$. Let $f, g\colon Y \to X$ be two maps and $\mathcal{U} \in cov(X)$. We call these maps *$\mathcal{U}$-close* if for each point $y \in Y$ there exists an element $U \in \mathcal{U}$ such that $f(y), g(y) \in U$. Denote by $B(f,\mathcal{U})$ the collection of all those maps $g \in C(Y,X)$ which are $\mathcal{U}$-close to $f$. We define the topology on the set $C(Y,X)$ as follows. A set $T \subseteq C(Y,X)$ is open if for each $f \in T$ there is an open cover $\mathcal{U} \in cov(X)$ such that $B(f,\mathcal{U}) \subseteq T$. This topology is precisely the *limitation topology* on $C(Y,X)$. *Everywhere below, unless otherwise stated, we consider only the above defined topology*[1].

Suppose that a map $f\colon Y \to X$ is given and $\mathcal{U} \in cov(Y)$. We say that $f$ is a *$\mathcal{U}$-map* if there exists an open cover $\mathcal{V} \in cov(X)$ such that the cover $f^{-1}(\mathcal{V}) = \{f^{-1}(V)\colon V \in \mathcal{V}\}$ refines $\mathcal{U}$.

LEMMA 2.1.1. *Let $\mathcal{U} \in cov(Y)$. Then the set of all $\mathcal{U}$-maps is open in $C(Y,X)$.*

PROOF. Without loss of generality we may assume that the set of all $\mathcal{U}$-maps is not empty. Let $f\colon Y \to X$ be an $\mathcal{U}$-map. Our goal is to find an open cover $\mathcal{W} \in cov(X)$ such that the set $B(f,\mathcal{W})$ consists entirely of $\mathcal{U}$-maps. Since $f$ is a $\mathcal{U}$-map, there is an open cover $\mathcal{V} \in cov(X)$ such that $f^{-1}(\mathcal{V})$ refines $\mathcal{U}$. Take an open cover $\mathcal{W} \in cov(X)$ star of which refines $\mathcal{V}$. This means that the star $St(W,\mathcal{W}) = \cup\{W' \in \mathcal{W}\colon W' \cap W \neq \emptyset\}$ of each element $W \in \mathcal{W}$ is contained in an element of the cover $\mathcal{V}$ (paracompactness of the space $X$ implies that such refinements always exist). Let us show that the set $B(f,\mathcal{W})$ consists

[1]It should be emphasized that the sets $B(f,\mathcal{U})$ are not necessarily open (although their interiors are non-empty); see **[53]** for additional details.

of $\mathcal{U}$-maps. First observe that, by the choice of $\mathcal{W}$, it suffices to prove that $g^{-1}(W) \subseteq f^{-1}(St(W,\mathcal{W}))$ for each $g \in B(f,\mathcal{W})$ and each $W \in \mathcal{W}$. Indeed, suppose that $y \in g^{-1}(W)$. Since the maps $f$ and $g$ are $\mathcal{W}$-close, there is an element $W_y \in \mathcal{W}$ such that $g(y), f(y) \in W_y$. Since $g(y) \in W$, we conclude that $W \cap W_y \neq \emptyset$. Therefore $f(y) \in St(W,\mathcal{W})$. Consequently, $g^{-1}(W) \subseteq f^{-1}(St(W,\mathcal{W}))$ as desired. □

LEMMA 2.1.2. *Let $d$ be a complete metric on a Polish space $Y$ and suppose a sequence $\{\mathcal{U}_n : n \in N\} \subseteq cov(Y)$ is chosen so that $diam_d U < \frac{1}{n}$ for each $U \in \mathcal{U}_n$ and each $n \in N$. Then a map $f : Y \to X$ is a closed embedding (i.e. an embedding with closed image) if and only if $f$ is a $\mathcal{U}_n$-map for each $n \in N$. In particular, the set of all closed embeddings of $Y$ into $X$ is a $G_\delta$-subset of $C(Y,X)$.*

PROOF. Let $f : Y \to X$ be a closed embedding and $\mathcal{U} \in cov(Y)$. Then $f$ is a $\mathcal{U}$-map. Indeed, consider an open cover $f(\mathcal{U}) = \{f(U) : U \in \mathcal{U}\}$ of the subspace $f(Y) \subseteq X$. For each $U \in \mathcal{U}$ select an open subset $U'$ of $X$ so that $f(U) = U' \cap f(Y)$. Obviously, $\mathcal{V} = \{U' : U \in \mathcal{U}\} \cup \{X - f(Y)\}$ is an open cover of $X$. It only remains to note that $f^{-1}(\mathcal{V})$ refines $\mathcal{U}$.

Now suppose that a map $f : Y \to X$ has the properties indicated in our Lemma. Let us show that $f$ is a closed embedding. First we show that $f$ is a one-to-one map. Let $y_1, y_2 \in Y$ and $y_1 \neq y_2$. Take an integer $n$ so large that $\frac{1}{n} < d(y_1,y_2)$. Since $f$ is a $\mathcal{U}_n$-map, there is an open cover $\mathcal{V}_n \in cov(X)$ such that $f^{-1}(\mathcal{V}_n)$ refines $\mathcal{U}_n$. Suppose that $f(y_1) = x = f(y_2)$ and consider an element $V \in \mathcal{V}_n$, containing $x$. Then there exists an element $U \in \mathcal{U}_n$ such that $f^{-1}(V) \subseteq U$. In particular, $y_1, y_2 \in U$. This contradicts the choice of $n$. Thus, $f$ is a one-to-one map.

Consider now an arbitrary sequence $\{x_k \; k \in N\} \subseteq f(Y)$ which converges to a point $x \in X$. Let $y_k = f^{-1}(x_k)$, $k \in N$, $\epsilon > 0$, $\frac{1}{n} < \epsilon$ and $\mathcal{V}_n \in cov(X)$ be an open cover of $X$ such that $f^{-1}(\mathcal{V}_n)$ refines $\mathcal{U}_n$. Take an element $V \in \mathcal{V}_n$ containing the point $x$. Since $\lim x_k = x$, there is an integer $m$ such that $x_k \in V$ for each $k \geq m$. Let $U \in \mathcal{U}_n$ such that $f^{-1}(V) \subseteq U$. Obviously, $y_k \in U$ for each $k \geq m$. But $diam_d(U) < \epsilon$. Therefore $d(y_k, y_{k'}) < \epsilon$ whenever $k, k' \geq m$. This shows that the sequence $\{y_k : k \in N\}$ is a Cauchy sequence. Let $y = \lim y_k$. By the continuity of $f$, $x = f(y) \in f(Y)$. Therefore the set $f(Y)$ is closed in $X$.

The remaining part of our statement follows from Lemma 2.1.1. □

The proof of the following statement is trivial.

LEMMA 2.1.3. *Let $A$ be a closed subset of a Polish space $Y$, $\mathcal{U} \in cov(Y)$ and $f \in C(Y,X)$. If the restriction $f/A : A \to X$ is a $\mathcal{U}/A$-map (here $\mathcal{U}/A = \{U \cap A : U \in \mathcal{U}\} \in cov(A)$), then there exists an open neighborhood $G$ of $A$ in $Y$ such that the restriction $f/G : G \to X$ is a $\mathcal{U}/G$-map.*

In order to obtain other, not so trivial, properties of the space $C(Y, X)$ we describe the topology introduced above in a somewhat different way.

It is easy to see that every bounded metric $d$ on the space $X$ generates the sup-metric $\check{d}$ on $C(Y, X)$. The definition is as follows:

$$\check{d}(f, g) = \sup\{d(f(y), g(y)) \colon y \in Y\}, f, g \in C(Y, X).$$

The topology generated by the metric $\check{d}$ on the set $C(Y, X)$ is called the *topology of d-uniform convergence.*

By $Metr(X)$ we denote the collection of all bounded metrics on a space $X$ generating the topology of $X$. The collection $\{\check{d} \colon d \in Metr(X)\}$ generates a topology on $C(Y, X)$ which is precisely the *topology of uniform convergence with respect to all metrics.* Obviously, this topology is stronger than the topology of $d$-uniform convergence for each particular metric $d$.

LEMMA 2.1.4. *The limitation topology coincides with the topology of uniform convergence with respect to all metrics.*

PROOF. First, suppose that $G$ is a neighborhood of $f \in C(Y, X)$ with respect to the topology of uniform convergence. This means that there exist a number $\epsilon > 0$ and a bounded metric $d \in Metr(X)$ such that

$$f \in \{g \in C(Y, X) \colon \check{d}(f, g) < \epsilon\} \subseteq G.$$

It is easy to check that $B(f, \mathcal{U}) \subseteq \{g \in C(Y, X) \colon \check{d}(f, g) < \epsilon\}$, where $\mathcal{U} = \{B_d(x, \frac{\epsilon}{2} \colon x \in X\}$ (here $B_d(x, r)$ denotes the open $d$-ball of radius $r$ with center at $x$).

For the converse, consider a neighborhood $B, f, \mathcal{U})$ of the map $f$ in the limitation topology. Take a metric $d \in Metr(X)$ such that $\{B_d(x, 1) \colon x \in X\}$ refines $\mathcal{U}$ (see, for example, [**32**, Theorem 4.1]). Obviously, $\{g \in C(Y, X) \colon \check{d}(f, g) < 1\} \subseteq B(f, \mathcal{U})$. □

*Remark* 2.1.5. If the space $X$ in the second half of the proof of Lemma 2.1.4 is Polish, then the metric $d$, corresponding to a cover $\mathcal{U}$, may be assumed to be complete.

The following statement gives us another description of the limitation topology.

LEMMA 2.1.6. *Let $d \in Metr(X)$ and $f \in C(Y, X)$. The collection $\{B_d(f, \alpha) \colon \alpha \in C(X, (0, \infty))\}$, where $B_d(f, \alpha) = \{g \in C(Y, X) \colon \check{d}(f(y), g(y)) \leq \alpha(f(y))$ for each $y \in Y\}$, forms a local basis at $f$ in $C(Y, X)$.*

PROOF. For a given open cover $\mathcal{U} \in cov(X)$, define a map $\alpha_{\mathcal{U}} \colon X \to (0, \infty)$ as follows:

$$\alpha_{\mathcal{U}}(x) = \frac{1}{2} \sup\{dist_d(x, X - U) \colon U \in \mathcal{U}\}, x \in X.$$

Observe that $B_d(f, \alpha_{\mathcal{U}}) \subseteq B(f, \mathcal{U})$.

Conversely, if $\alpha \in C(X, (0, \infty))$, then we define an open cover $\mathcal{U}_\alpha \in cov(X)$ as $\mathcal{U}_\alpha = \{U_x : x \in X\}$, where

$$U_x = B_d(x, \frac{\alpha(x)}{4}) \cap \alpha^{-1}((\frac{\alpha(x)}{2}, \infty)).$$

Note that $B(f, \mathcal{U}_\alpha) \subseteq B_d(f, \alpha)$. □

Note that if the spaces $X$ and $Y$ are compact (and metrizable), then the limitation topology coincides with the compact-open topology.

The following statement expresses a very important and frequently used property of the limitation topology.

PROPOSITION 2.1.7. *Let $X$ be a Polish space, $F$ be a subspace of the space $C(Y, X)$ and the set $G_n$ bee open in $C(Y, X)$ for each $n \in N$. If the intersection $G_n \cap F$ is dense in $F$ for each $n \in N$, then $F \subseteq cl_{C(Y,X)}((\cap G_n) \cap F_d)$, where $F_d$ denotes the closure of $F$ in the topology of $d$-uniform convergence and $d$ is any bounded metric on $X$. In particular, $C(Y, X)$ has the Baire property.*

We use Proposition 2.1.7 to characterize maps between Polish spaces which are approximable by homeomorphisms. Such maps are called near-homeomorphisms. More formally, a map $f : Y \to X$ is a *near-homeomorphism* if, given $\mathcal{U} \in cov(X)$, there is a homeomorphism $h_{\mathcal{U}} : Y \to X$ which is $\mathcal{U}$-close to $f$.

THEOREM 2.1.8. *A map $f : Y \to X$ between Polish spaces is a near-homeomorphism if and only if $f(Y)$ is dense in $X$ and the following condition is satisfied:*

- ($\star$) *For each $\mathcal{U} \in cov(Y)$ and $\mathcal{V} \in cov(X)$ there exist an open cover $\mathcal{W} \in cov(X)$ and a homeomorphism $h : Y \to Y$ such that $fh \in B(f, \mathcal{V})$ and $hf^{-1}(\mathcal{W}) \prec \mathcal{U}$.*

PROOF. Let $f : Y \to X$ be a near-homeomorphism. Obviously, $f(Y)$ is dense in $X$. Let us show that the condition ($\star$) is satisfied. Let $\mathcal{U} \in cov(Y)$ and $\mathcal{V} \in cov(X)$. Choose a star-refinement $\mathcal{V}_0 \in cov(X)$ of $\mathcal{V}$ and let $p : Y \to X$ be a homeomorphism which is $\mathcal{V}_0$-close to $f$. Now consider a star-refinement $\mathcal{W} \in cov(X)$ of the open cover $p(\mathcal{U}) \cap \mathcal{V}_0 = \{p(U) \cap V_0 : U \in \mathcal{U}, V_0 \in \mathcal{V}_0\} \in cov(X)$. Take a homeomorphism $q : Y \to X$ which is $\mathcal{W}$-close to $f$. Let us show that the composition $h = p^{-1}q : Y \to Y$ satisfies the condition ($\star$).

First we show that $fh$ and $f$ are $\mathcal{V}$-close. Indeed, let $y \in Y$. Since $f$ and $q$ are $\mathcal{V}_0$-close, there is an element $V_0^y \in \mathcal{V}_0$ such that $f(y), q(y) \in V_0^y$. Consider the point $p^{-1}q(y) \in Y$. Since $f$ and $p$ are $\mathcal{V}_0$-close, there is an element $V_1^y \in \mathcal{V}_0$ such that $q(y), fp^{-1}q(y) \in V_1^y$. Consequently, $V_0^y \cap V_1^y \neq \emptyset$. This implies that both points $fp^{-1}q(y)$ and $f(y)$ belong to the set $St(V_0^y, \mathcal{V}_0)$. Since, by our choice, $\mathcal{V}_0$ is a star-refinement of $\mathcal{V}$, we see that there is an element $V_y \in \mathcal{V}$ such that $fp^{-1}q(y), f(y) \in V_y$. Therefore, the maps $fh$ and $f$ are $\mathcal{V}$-close.

Next we show that $hf^{-1}(\mathcal{W})$ refines $\mathcal{U}$. Let $W \in \mathcal{W}$ and $y \in hf^{-1}(W)$. Then $fq^{-1}p(y) \in W$. Since the maps $f$ and $q$ are $\mathcal{W}$-close, there is an element $W_y \in \mathcal{W}$ such that $fq^{-1}p(y), p(y) \in W_y$. Consequently, $W \cap W_y \neq \emptyset$ and $p(y) \in St(W, \mathcal{W})$. By the construction, there is an element $U \in \mathcal{U}$ such that $St(W, \mathcal{U}) \subseteq p(U)$. Hence $p(y) \in p(U)$ and $y \in U$. This proves the inclusion $hf^{-1}(W) \subseteq U$ as desired.

Now we assume that a map $f \colon Y \to X$, satisfying our conditions, is given. Let us show that $f$ is a near-homeomorphism. Fix a complete metric $d_Y$ on $Y$ and denote by $\mathcal{U}_n$, $n \in N$, an open cover of $Y$, the diameters (with respect to the metric $d_Y$) of whose elements are less than $\frac{1}{n}$. Let $G_n$ denote the set of all $\mathcal{U}_n$-maps in $C(Y, X)$. Consider the following set:

$$F = \{fh \colon h \text{ is an autohomeomorphism of the space } Y\}.$$

We claim that $G_n \cap F$ is dense in $F$. To see this consider an arbitrary element in $F$. Such an element has the form $fh$, where $h$ is an autohomeomorphism of $Y$. Consider also an element $\mathcal{V} \in cov(X)$. Our goal is to show that $B(fh, \mathcal{V}) \cap G_n \cap F \neq \emptyset$. By the condition $(\star)$, applied to the open covers $h(\mathcal{U}_n) \in cov(Y)$ and $\mathcal{V} \in cov(X)$, there exist an open cover $\mathcal{W} \in cov(X)$ and a homeomorphism $g \colon Y \to Y$ such that the composition $fg$ is $\mathcal{V}$-close to $f$ and $gf^{-1}(\mathcal{W})$ refines $h(\mathcal{U}_n)$. Then $h^{-1}gf^{-1}(\mathcal{W})$ refines $\mathcal{U}_n$. Therefore, $fg^{-1}h$ is a $\mathcal{U}_n$-map. Obviously, $fg^{-1}h \in F$. Thus, it only remains to show that $fg^{-1}h$ and $fh$ are $\mathcal{V}$-close maps. In turn, it suffices to prove that $fg^{-1}$ and $f$ are $\mathcal{V}$-close. Indeed, consider a point $y \in Y$. Since the maps $fg$ and $f$ are $\mathcal{V}$-close, there is an element $V \in \mathcal{V}$ such that $f(y) = fgg^{-1}(y), fg^{-1}(y) \in V$. Consequently, $fg^{-1} \in B(f, \mathcal{V})$. This implies that $fg^{-1}h(y), fh(y) \in V$ and $fg^{-1}h \in B(fh, \mathcal{V})$. Therefore $B(fh, \mathcal{V}) \cap G_n \cap F \neq \emptyset$ as desired. This shows that the set $G_n \cap F$ is dense in $F$ for each $n \in N$. By Lemma 2.1.1, $G_n$ is open in $C(Y, X)$. Therefore, by Proposition 2.1.7, the set $F$ is contained in the closure (in the space $C(Y, X)$) of the intersection $\cap G_n \cap F_{d_X}$, where $F_{d_X}$ denotes the closure of $F$ in the topology of $d_X$-uniform convergence and $d_X$ stands for an arbitrary bounded complete metric on $X$. Since $f \in F$, we conclude that $f \in cl(\cap G_n \cap F_{d_X})$. Consequently, for each open cover $\mathcal{V} \in cov(X)$ the intersection $B(f, \mathcal{V}) \cap (\cap G_n \cap F_{d_X})$ is non-empty. Let $p$ be an element of this intersection. We claim that $p$ is a homeomorphism. Obviously this is all we need to show. Since $p \in \cap G_n$, we conclude, by Lemma 2.1.2, that $p$ is a closed embedding. On the other hand, since $p \in F_{d_X}$, $p$ is the $d_X$-uniform limit of a sequence of maps, belonging to $F$. But maps in $F$ have dense images in $X$. This is enough to conclude that $p(Y)$ is also dense in $X$. Therefore, $p$ is a homeomorphism. □

COROLLARY 2.1.9. *A closed surjection $f \colon Y \to X$ between Polish spaces is a near-homeomorphism if and only if the following condition is satisfied:*

$(\star\star)$ *For each $\mathcal{U} \in cov(Y)$ and each $\mathcal{V} \in cov(X)$ there exists a homeomorphism $h \colon Y \to Y$ such that $fh \in B(f, \mathcal{V})$ and the collection $\{hf^{-1}(x) \colon$*

$x \in X\}$ *refines* $\mathcal{U}$.

PROOF. Clearly, $(\star) \Rightarrow (\star\star)$. Let us prove the reverse implication. The closedness of $f$ implies that each of the sets $X - f(Y - h^{-1}(U))$, $U \in \mathcal{U}$, is open in $X$. The collection $\mathcal{W} = \{X - f(Y - h^{-1}(U)) : U \in \mathcal{U}\}$ covers $X$. Indeed, let $x \in X$. Then, by $(\star\star)$, we can find an element $U \in \mathcal{U}$ such that $hf^{-1}(x) \subseteq U$. Therefore $f^{-1}(x) \cap (Y - h^{-1}(U)) = \emptyset$ and hence $x \in X - f(Y - h^{-1}(U))$. It only remains to note that $hf^{-1}(\mathcal{W})$ refines $\mathcal{U}$. □

**2.1.2. Main definitions.** Each map $f : A \to [0, 1]$, defined on a closed subspace $A$ of a normal space $B$, can be extended to $B$. This classical result, known as the Brouwer-Tietze-Urysohn extension theorem, serves as a prototype of the entire theory, which studies, roughly speaking, the possibility of extending maps into given spaces. The important property of the unit segment (or the real line) expressed in this theorem can be formalized as follows.

DEFINITION 2.1.10. *A Polish space $X$ is called an* absolute (neighborhood) extensor, *or shortly, an $A(N)E$-space, if any map $f : A \to X$, defined on a closed subspace $A$ of an arbitrary Polish space $B$ can be extended to a map of the space $B$ (respectively, of a neighborhood of $A$ in $B$) into $X$.*

Obviously, the notion of absolute neighborhood extensor can be introduced for different classes of topological spaces. In each particular situation, the choice of the appropriate class depends, as a rule, on a variety of different reasons. In Chapter 6 the general definition of this notion, suitable for the class of completely regular spaces, will be given. But for now we restrict ourselves to the class of Polish spaces.

The following notion is also very important.

DEFINITION 2.1.11. *Let $n \in \omega$. A Polish space $X$ is called an* absolute (neighborhood) extensor in dimension $n$, *or shortly, an $A(N)E(n)$-space, if any map $f : A \to X$, defined on a closed subspace of a Polish space $B$ with $\dim B \leq n$, can be extended to a map of the space $B$ (respectively of a neighborhood of $A$ in $B$) into $X$.*

It is not difficult to verify that a metrizable compactum $X$ is an $A(N)E(n)$-space (respectively, $A(N)E$-space) if the condition, formulated in Definition 2.1.11 (respectively, in Definition 2.1.10), is satisfied only for compact metrizable spaces $B$. At the same time, it should be especially emphasized that there exists a Polish space, which is not an absolute extensor, such that the condition of Definition 2.1.10 is satisfied for each metrizable compactum $B$. This follows, as has been noted by van Mill (see [**22**]), from Corollary 3.2.16. The following characterizations of the classes of spaces introduced above are well-known. Their proofs can be found in several topological textbooks (see, for example, [**203, 176**]).

THEOREM 2.1.12. *Let $n \in \omega$. The following conditions are equivalent for any Polish space $X$;*

(i) *$X$ is an $ANE(n)$-space.*

(ii) *$X$ is an $\mathcal{LC}^{n-1}$-space. In other words, for each point $x \in X$ and any neighborhood $U$ of $x$ in $X$ there exists a smaller neighborhood $V$ of $x$ such that any map $f\colon S^k \to V$ of the $k$-dimensional sphere $S^k$, where $k \leq n-1$, can be extended to a map $g\colon B^{k+1} \to U$ of the $(k+1)$-dimensional disk $B^{k+1}$ into $U$.*

(iii) *Any map $f\colon A \to X$, defined on a closed subspace $A$ of a Polish space $B$ with $\dim(B - A) \leq n$, can be extended to a map of a neighborhood of $A$ (in $B$) into $X$.*

(iv) *For any Polish space $B$, containing $X$ as a closed subspace so that $\dim(B - X) \leq n$, there exists a retraction $r\colon G \to X$, defined on a neighborhood $G$ of $X$ in $B$.*

(v) *For each point $x \in X$ and for any neighborhood $U$ of $x$ in $X$ there exists a smaller neighborhood $V$ such that any map $f\colon A \to V$, defined on a closed subspace $A$ of an at most $n$-dimensional Polish space $B$, can be extended to a map $g\colon B \to U$ of $B$ into $X$.*

(vi) *For each point $x \in X$ and for any neighborhood $U$ of $x$ in $X$ there exists a smaller neighborhood $V$ such that any map $f\colon B \to V$ of an at most $(n-1)$-dimensional Polish space $B$ into $V$ is homotopic in $U$ to a constant map of $B$ into $U$.*

(vii) *For each open cover $\mathcal{U} \in cov(X)$ there is an open cover $\mathcal{V} \in cov(X)$, refining $\mathcal{U}$, such that any two $\mathcal{V}$-close maps $f, g\colon B \to X$, defined on an at most $(n-1)$-dimensional Polish space $B$, are $\mathcal{U}$-homotopic (let us recall that two maps $f$ and $g$ are $\mathcal{U}$-*homotopic *if there is a homotopy $H: B \times [0,1] \to X$, connecting $f$ and $g$ such that the collection $\{H(\{b\} \times [0,1]) : b \in B\}$ refines $\mathcal{U}$).*

(viii) *For each open cover $\mathcal{U} \in cov(X)$ there exist a countable locally finite simplicial complex $K$ and a map $p\colon |K| \to X$ such that for any map $f\colon B \to X$, defined on an at most $(n-1)$-dimensional Polish space $B$, there is a map $g\colon B \to |K|$ such that the maps $f$ and $pg$ are $\mathcal{U}$-homotopic.*

Note that an $ANE(n)$-space is an $AE(n)$-space if and only if it is $(n-1)$-connected, i.e. belongs to the class $C^{n-1}$. The latter, as usual, means that any map of the $k$-dimensional sphere, $k \leq n-1$, into the given space is null-homotopic or, equivalently, admits an extension onto the $(k+1)$-dimensional disk.

PROPOSITION 2.1.13. *Each Polish space is an $AE(0)$-space.*

One can prove this statement in several ways. For example, it suffices to note that any closed subspace of any zero-dimensional Polish space is a retract of it. Alternatively, we can use one of the results dealing with the existence of single-valued continuous selections of *many-valued maps*. First, we need some definitions.

By a many-valued map (between Polish spaces) $F\colon X \to Y$ we understand a map which for each point $x \in X$ assigns a closed subset $F(x)$ of a space $Y$. We call such a map a *lower semi-continuous* if the set

$$F^{-1}(U) = \{x \in X : F(x) \cap U \neq \emptyset\}$$

is open in $X$ whenever $U$ is open in $Y$. In the case when the sets of the form

$$F^{\#}(U) = \{x \in X : F(x) \subseteq U\}$$

are open in $X$ (whenever $U$ is an open subset of $Y$) we say that $F$ is *upper semi-continuous*. A single-valued map $f\colon X \to Y$ is a *selection* of $F$ if $f(x) \in F(x)$ for each point $x \in X$. Lower semi-continuous maps occur quite often in the theory of Absolute Extensors. Consider, for example, the following situation.

Suppose that $f\colon X \to Y$ is an open surjection between Polish spaces. Suppose, in addition, that $A$ is a closed subspace of a Polish space $B$ and that two maps $g\colon A \to X$ and $h\colon B \to Y$, satisfying the condition $fg = h/A$, are given. Let us define a many-valued map $F\colon B \to X$ generated by the diagram

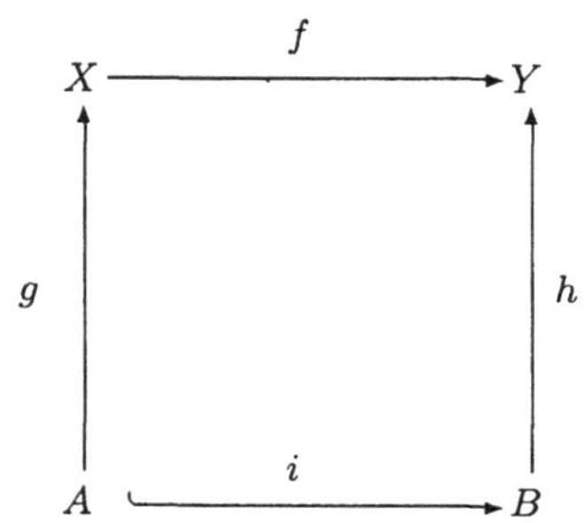


by the following rule:

$$F(b) = \begin{cases} g(b), & \text{if } b \in A \\ f^{-1}(h(b)), & \text{if } b \in B - A \end{cases}$$

Obviously the openness of the map $f$ implies the lower semi-continuity of $F$.

The following statement was proved in [**225**]. Note that Proposition 2.1.13 is an easy consequence of it.

THEOREM 2.1.14. *Let $F\colon X \to Y$ be a lower semi-continuous many-valued map between Polish spaces. If* $\dim X = 0$*, then $F$ has a selection.*

In order to formulate a selection theorem for positive dimensional cases we need additional definitions.

Let $n \in \omega$ and $\mathcal{F} = \{X_t : t \in T\}$ be a collection of closed subsets of a Polish space $X$. We say that $\mathcal{F}$ is *uniformly locally connected in dimensions* $< n$

(shortly, $\mathcal{F}$ is an *equi* $-\mathcal{LC}^{n-1}$-collection) if for each open cover $\mathcal{U} \in cov(X)$ there is an open cover $\mathcal{V} \in cov(X)$, refining $\mathcal{U}$, such that the following condition is satisfied:

- For each index $t \in T$ and for each element $U \in \mathcal{U}$ there is an element $V \in \mathcal{V}$ such that $V \subseteq U$ and any map $f\colon S^k \to X_t \cap V$, with $k \leq n-1$, has an extension $g\colon B^{k+1} \to X_t \cap U$.

We also say that $\mathcal{F}$ is connected in dimensions $< n$ (shortly, $\mathcal{F}$ is a $\mathcal{C}^{n-1}$-collection) if $X_t \in \mathcal{C}^{n-1}$ for each $t \in T$.

The following important theorem holds.

THEOREM 2.1.15. *Let $n \in \omega$ and $f\colon X \to Y$ be a lower semi-continuous many-valued map between Polish spaces. If $\dim X \leq n$ and $F$ is an $\mathcal{LC}^{n-1} \cap \mathcal{C}^{n-1}$-collection, then $F$ has a selection.*

Suppose now that a lower semi-continuous many-valued map $F\colon X \to E$ of a Polish space $X$ into a locally convex Polish space $E$ (i.e. $E$ is a separable *Fréchet space*) is given. We say that the map $F$ is *convex* if the image $F(x)$ of any point $x \in X$ is a convex (and closed) subset of $E$. In this situation we have the following statement.

THEOREM 2.1.16. *Every convex lower semi-continuous many-valued map of any Polish space into any separable Fréchet space has a selection.*

This theorem has several important corollaries. Here are two of them.

THEOREM 2.1.17. *Every closed and convex subspace of a separable Fréchet space is an absolute extensor.*

It was shown by R.Cauty [**62**] that the requirement of local convexity in the above statement is essential.

THEOREM 2.1.18. *A continuous linear surjection $f\colon E \to L$ between separable Fréchet spaces is topologically a trivial bundle with fiber $f^{-1}(0)$.*

PROOF. By the classical theorem of Banach, $f$ is an open map. Therefore the many-valued map $f^{-1}\colon L \to E$ is lower semi-continuous. Linearity of $f$ implies that the sets $f^{-1}(l)$, $l \in L$ are convex (and obviously closed). Theorem 2.1.17 guarantees the existence of a selection $g\colon L \to E$ of $f^{-1}$. Then the desired trivialization of $f$ (i.e. a homeomorphism $h\colon E \to L \times f^{-1}(0)$ such that $f = \pi_L h$) can be given according to the following formula: $h(e) = (e - gf(e), f(e))$, $e \in E$. $\square$

Theorem 2.1.17 provides an important class of Polish absolute extensors. In particular, all Euclidean spaces $R^n$, as well as the countable infinite power $R^\omega$ of the real line $R$ (which is homeomorphic to the separable Hilbert space $l_2$) are absolute extensors. Note also that the class of Polish absolute (neighborhood) extensors coincides with the class of retracts of $R^\omega$ (respectively, retracts of open subspaces of $R^\omega$). The latter class can easily be seen to coincide with the class of

Polish *absolute (neighborhood) retracts* (shortly, $A(N)R$-spaces). In particular, every locally compact polyhedron is an $ANR$-space.

The following criterion helps us to recognize $ANR$-spaces.

THEOREM 2.1.19. *A Polish space $X$ is an $ANR$-space if and only if for any open cover $\mathcal{U} \in cov(X)$ there exist a countable locally finite simplicial complex $K$ (finite-dimensional in case $X$ is finite-dimensional) and maps $p\colon X \to |K|$ and $q\colon |K| \to X$ such that the composition $qp$ is $\mathcal{U}$-homotopic to the identity map $id_X$. In addition, we may assume that the cover $p^{-1}(\mathcal{U}_0)$ refines $\mathcal{U}$, where $\mathcal{U}_0$ denotes the cover of $|K|$ consisting of open stars (with respect to the triangulation of $K$) of vertices of $K$.*

Each Polish $ANR$-space has the homotopy type of a locally finite polyhedron [**232**]. Since an open subspace of a Polish $ANR$-space is itself a Polish $ANR$-space (see [**161**]), we see that any open subspace of a Polish $ANR$-space has the homotopy type of a locally finite polyhedron. The converse statement (the validity of which was a long standing unsolved problem) has been recently proved by R. Cauty [**63**]. Thus, we have the following characterization of Polish $ANR$-spaces.

THEOREM 2.1.20. *The following conditions are equivalent for any Polish space $X$:*

(i) *$X$ is an $ANR$-space.*
(ii) *each open subspace of $X$ has the homotopy type of a locally finite polyhedron.*

The following two statements are also well-known.

PROPOSITION 2.1.21. *Let $X$ be a Polish $ANR$-space. Then for each open cover $\mathcal{U} \in cov(X)$ there is an open cover $\mathcal{V} \in cov(X)$, refining $\mathcal{U}$, such that any two $\mathcal{V}$-close maps $f, g\colon B \to X$, defined an a Polish space $B$, are $\mathcal{U}$-homotopic in $X$.*

PROPOSITION 2.1.22. *Let $X$ be a Polish $ANR$-space. Then for each open cover $\mathcal{U} \in cov(X)$ there is an open cover $\mathcal{V} \in cov(X)$, refining $\mathcal{U}$, such that the following condition is satisfied:*

- *If one of two $\mathcal{V}$-close maps $f, g\colon A \to X$, defined on a closed subspace of a Polish space $B$ (say $f$) has an extension $F\colon B \to X$ to $B$, then the other (the map $g$) also has an extension $G\colon B \to X$. Moreover, we may assume that the extensions $F$ and $G$ are $\mathcal{U}$-homotopic.*

COROLLARY 2.1.23. *Let $A$ be a closed subspace of a Polish space $X$. Then the "restriction" map $C(B, X) \to C(A, X)$ is open.*

**2.1.3. Maps of special type.** Fine homotopy equivalences are very important in the theory of $ANR$-spaces. In this Subsection we present some of their properties.

DEFINITION 2.1.24. *A map $f\colon X \to Y$ is a* fine homotopy equivalence *if for each open cover $\mathcal{U} \in cov(Y)$ there is a map $g_{\mathcal{U}}\colon Y \to X$ such that the composition $fg$ is $\mathcal{U}$-homotopic to $id_Y$ and the composition $gf$ is $f^{-1}(\mathcal{U})$-homotopic to $id_X$.*

PROPOSITION 2.1.25. *Let $f\colon X \to Y$ be a fine homotopy equivalence and $X$ be an $ANR$-space. Then $Y$ is also an $ANR$-space.*

PROOF. Let $\mathcal{U} \in cov(Y)$. Take an open cover $\mathcal{V} \in cov(Y)$ which star refines $\mathcal{U}$. Take a map $g\colon Y \to X$ such that $fg \overset{\mathcal{V}}{\simeq} id_Y$ (and $gf \overset{f^{-1}(\mathcal{V})}{\simeq} id_X$). By Theorem 2.1.19, there exist a locally compact polyhedron $K$ and maps $p\colon X \to K$ and $q\colon K \to X$ such that the composition $qp$ is $f^{-1}(\mathcal{V})$-homotopic to $id_X$. Consider the maps $\alpha = pg\colon Y \to K$ and $\beta = fq\colon K \to Y$. By the same Theorem 2.1.19, it suffices to show that the composition $\beta\alpha$ is $\mathcal{U}$-homotopic to $id_Y$.

Since $qp \overset{f^{-1}(\mathcal{V})}{\simeq} id_X$, we see that $qpg \overset{f^{-1}(\mathcal{V})}{\simeq} g$. Then $fqpg \overset{\mathcal{V}}{\simeq} fg$. By construction, $fg \overset{\mathcal{V}}{\simeq} id_Y$. Therefore, $\beta\alpha = fqpg \overset{\mathcal{U}}{\simeq} id_Y$. □

Maps introduced in the following Definition are closely related to fine homotopy equivalences.

DEFINITION 2.1.26. *A map $f\colon X \to Y$ between Polish spaces is* approximately soft *if for each open cover $\mathcal{U} \in cov(Y)$, each Polish space $B$, each closed subspace $A$ of $B$, and any two maps $g\colon A \to X$ and $h\colon B \to Y$ satisfying the condition $fg = h/A$, there exists a map $k\colon B \to X$ such that $g = k/A$ and the composition $fk$ is $\mathcal{U}$-close to $h$.*

PROPOSITION 2.1.27. *Let $f\colon X \to Y$ be an approximately soft map and $Y$ be a Polish $ANR$-space. Then $f$ is a fine homotopy equivalence.*

PROOF. Take an open cover $\mathcal{U} \in cov(Y)$ and let $\mathcal{U}_1 \in cov(Y)$ be its star-refinement. Since $Y$ is an $ANR$-space, there exists, by Proposition 2.1.21, an open refinement $\mathcal{V} \in cov(Y)$ of $\mathcal{U}_1$ such that any two $\mathcal{V}$-close maps into $Y$ (defined on an arbitrary space) are $\mathcal{U}_1$-homotopic. Since $f$ is approximately soft, there exists a map $f\colon Y \to X$ such that the composition $fg$ is $\mathcal{V}$-close to $id_Y$. Obviously, by the choice of $\mathcal{V}$, we have that

$$fg \overset{\mathcal{U}_1}{\simeq} id_Y \text{ and, consequently, } fg \overset{\mathcal{U}_1}{\simeq} id_Y.$$

Let us now show that the composition $gf$ is $f^{-1}(\mathcal{U})$-homotopic to $id_X$. Since the maps $fg$ and $id_Y$ are $\mathcal{U}_1$-homotopic, we see that the maps $fgf$ and $f$ are also $\mathcal{U}_1$-homotopic. Consider the corresponding homotopy $h\colon X \times [0,1] \to Y$ connecting the maps $fgf$ and $f$. Define a map $H'\colon X \times \{0,1\} \to X$ as follows:

$$H'(x,t) = \begin{cases} gf(x), & \text{if } t = 0 \\ x, & \text{if } t = 1. \end{cases}$$

Obviously, $fH' = h/(X \times \{0,1\})$. Since $f$ is approximately soft, there exists a map $H\colon X \times [0,1] \to X$ such that $H/(X \times \{0,1\}) = H'$ and the composition $fH$ is $\mathcal{U}_1$-close to $h$. It is not hard to see that $H$ is a homotopy connecting the maps $fg$ and $id_X$. Next we show that $H$ is a $f^{-1}(\mathcal{U})$-homotopy. Consider a point $x \in X$. It suffices to verify that the set $H(\{x\} \times [0,1])$ is contained in an element of the cover $f^{-1}(\mathcal{U})$. Since the homotopy $h$ is limited by the cover $\mathcal{U}_1$, there exists an element $U_1 \in \mathcal{U}_1$ such that $h(\{x\} \times [0,1]) \subseteq U_1$. Let $t \in [0,1]$. Since the maps $fH$ and $h$ are $\mathcal{U}_1$-close, we can find an element $U_1^t \in \mathcal{U}_1$ such that $fH(x,t), h(x,t) \in U_1^t$. Consequently, $U_1 \cap U_1^t \neq \emptyset$ for each $t \in [0,1]$. But then

$$fH(\{x\} \times [0,1]) = \cup\{fH(x,t)\colon t \in [0,1]\} \subseteq \cup\{U_1^t\colon t \in [0,1]\} \subseteq st(U_1, \mathcal{U}_1).$$

Recall that $\mathcal{U}_1$ star refines $\mathcal{U}$. Therefore, there is an element $U \in \mathcal{U}$ such that $fH(\{x\} \times [0,1]) \subseteq U$. This obviously implies that $H(\{x\} \times [0,1]) \subseteq f^{-1}(U)$. □

In should be noted that even polyhedrally soft maps between Polish $ANR$-spaces are fine homotopy equivalences (we obtain the definition of such maps by considering only polyhedral pairs $(A,B)$ in Definition 2.1.26). This fact has been established in [**163**].

PROPOSITION 2.1.28. *Let $f\colon X \to Y$ be a near-homeomorphism and $Y$ be a Polish $ANR$-space. Then $f$ is approximately soft.*

PROOF. Let $A$ be a closed subset of a Polish space $B$ and let maps $\varphi\colon A \to X$ and $\psi\colon B \to Y$ such that $f\varphi = \psi/A$ be given. Consider an arbitrary open cover $\mathcal{U} \in cov(Y)$ and suppose that $\mathcal{W} \in cov(Y)$ is a star-refibement of $\mathcal{U}$. By Proposition 2.1.22, there is an open refinement $\mathcal{V} \in cov(Y)$ of $\mathcal{W}$ such that the following condition is satisfied:

- If one of the two arbitrarily given $\mathcal{V}$-close maps into $Y$, defined on $A$, has an extension to $B$, then the second also has an extension to $B$. Moreover, we may assume that these extensions are $\mathcal{W}$-close.

Take a homeomorphism $g\colon X \to Y$ which is $\mathcal{V}$-close to $f$ and consider the composition $g\varphi\colon A \to Y$. Obviously, $g\varphi$ is $\mathcal{V}$-close to $f\varphi$. But the map $f\varphi$ has the extension $\psi$ onto the space $B$. Consequently, by the choice of the cover $\mathcal{V}$, there is a $\mathcal{W}$-close to $\psi$ extension $h\colon B \to Y$ of the map $g\varphi$. Define the map $k\colon B \to X$ as the composition $k = g^{-1}h$. Obviously, $k/A = \varphi$. It now suffices to show that the composition $fk$ is $\mathcal{U}$-close to $\psi$. Let $b \in B$. Since the maps $h$ and $\psi$ are $\mathcal{W}$-close, there exists an element $W \in \mathcal{W}$ such that $h(b), \psi(b) \in W$. Consider the point $g^{-1}h(b) \in X$. $\mathcal{W}$-closeness of the maps $g$ and $f$ guarantees the existence of an element $W^b \in \mathcal{W}$ such that $h(b) = gg^{-1}h(b), fg^{-1}h(b) \in W^b$.

Therefore, $W \cap W^b \neq \emptyset$. Then $fg^{-1}h(b), \psi(b) \in st(W, \mathcal{W})$. Since $st(\mathcal{W})$ refines $\mathcal{U}$, there is an element $U \in \mathcal{U}$ such that $fk(b) = fg^{-1}h(b), \psi(b) \in U$. This verifies that the maps $fk$ and $\psi$ are $\mathcal{U}$ close. □

Now we introduce some notions from Shape Theory (the reader can find a comprehensive introduction to Shape Theory in several textbooks; for instance, in [**47**], [**135**] or [**215**]). We say that a compactum $X$ has *trivial shape* (and we write $Sh(X) = *$) if for some (any) embedding of $X$ into $I^\omega$ (or into any other $AR$-compactum) the following condition is satisfied:

- for each open neighborhood $U$ of $X$ in $I^\omega$ there exists an open neighborhood $V$ of $X$ in $I^\omega$ such that $V \subseteq U$ and $V$ contracts to a point in $U$.

Several other equivalent definitions of compacta with trivial shape are known. They appear under the name of *cell-like* compacta as well as under the name of $UV^\infty$*-compacta*. Proper maps all fibers of which are cell-like are known as *cell-like maps*. The proof of the following statement involves only elementary arguments.

PROPOSITION 2.1.29. *Let $f\colon X \to Y$ be a proper map between Polish $ANR$-spaces. Then the following conditions are equivalent:*

(i) *$f$ is a fine homotopy equivalence.*
(ii) *$f$ is cell-like.*

DEFINITION 2.1.30. *Let $n \in \omega$. We say that a compactum $X$ is an $UV^n$-*compactum *if for some (any) embedding of $X$ into $I^\omega$ the following condition is satisfied:*

- *for each open neighborhood $U$ of $X$ in $I^\omega$ there exists an open neighborhood $V$ of $X$ in $I^\omega$ such that $V \subseteq U$ and any map $f\colon S^k \to V$, $k \leq n$, has an extension $g\colon B^{k+1} \to U$.*

It is perhaps of some interest to remark that the $UV^n$-compacta are exactly those compacta with trivial $n$-shape (see Subsection 4.4.3).

$UV^n$*-maps* are proper maps all fibers of which are $UV^n$-compacta. The following statement is well known.

PROPOSITION 2.1.31. *Let $n \in N$ and $f\colon X \to Y$ be a proper map between locally compact $LC^{n-1}$-spaces. Then the following conditions are equivalent:*

(i) *$f$ is a $UV^{n-1}$-map.*
(ii) *For each at most $n$-dimensional locally compact space $B$, each closed subset $A$ of $B$, any two proper maps $g\colon A \to X$ and $h\colon B \to Y$ with $fg = h/A$, and any open cover $\mathcal{U} \in cov(Y)$, there exists a proper map $k\colon B \to X$ such that $k/A = g$ and the composition $fk$ is $\mathcal{U}$-close to $h$ (in other words, $f$ is* approximately $n$-soft*).*
(iii) *The same as condition (ii), but for polyhedral pairs $(A, B)$ (i.e. $f$ is* polyhedrally approximately $n$-soft*).*

The following proposition holds.

PROPOSITION 2.1.32. *Let $n \in \omega$ and let $f: X \to Y$ be a $UV^n$-map between compacta. Then the following conditions are satisfied:*

(i) *$X \in UV^n$ if and only if $Y \in UV^n$.*

(ii) *If $X \in LC^n$, then $Y \in LC^n$.*

The following concept is related to Definition 2.1.26.

DEFINITION 2.1.33. *Let $n = 0, 1, \dots, \infty$. We say that a map $f: X \to Y$ between Polish spaces is* $n$-soft *if for each at most $n$-dimensional Polish space $B$, each closed subset $A$ of $B$, and any two maps $g: A \to X$ and $h: B \to Y$ with $fg = h/A$, there exists a map $k: B \to X$ such that $g = k/A$ and $fk = h$. The $\infty$-soft maps are also called* soft.

Obviously, every $n$-soft map between compacta is a $UV^{n-1}$-map and every soft map is a fine homotopy equivalence.

PROPOSITION 2.1.34. *Let $f: X \to Y$ be a map between Polish spaces. Then the following conditions are equivalent:*

(i) *$f$ is an open map.*

(ii) *$f$ is a 0-soft map.*

PROOF. (i)$\Longrightarrow$(ii). Let $A$ be a closed subset of a zero-dimensional Polish space $B$. Suppose also that two maps $g: A \to X$ and $h: B \to Y$ are given so that $fg = h/A$. Consider the many-valued map $F: B \to X$ generated by the diagram

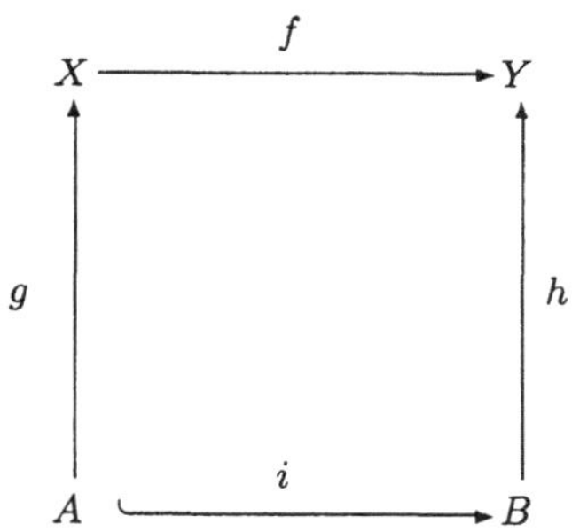


by the following rule:

$$F(b) = \begin{cases} g(b), & \text{if } b \in A \\ f^{-1}(h(b)), & \text{if } b \in B - A \end{cases}$$

Obviously, by(i), $F$ is lower semi-continuous. Since $\dim B = 0$ we conclude, by Theorem 2.1.14, that $F$ has a (continuous) selection $k: B \to X$. It only remains to note that $g = k/A$ and $fk = h$.

(ii)$\Longrightarrow$(i). First observe that every 0-soft map is surjective. Let $U$ be an open subset of $X$. Our goal is to show that $f(U)$ is open in $Y$. Consider an arbitrary converging sequence $\{y_n : n \in N\}$ of points of the complement $Y - f(U)$. Suppose that $y$ is the limit point of this sequence. We will be done if we show that $y \in Y - f(U)$. Assume the contrary and take a point $x \in U$ such that $f(x) = y$. Consider the homeomorphism

$$\psi \colon \alpha N \to \cup\{y_n : n \in N\} \cup \{y\},$$

where $\psi(\alpha) = y$ and $\psi(n) = y_n$ for each $n \in N$ (here $\alpha N$ denotes the one-point compactification of $N$). Let $\varphi(\alpha) = x$. By the zero-dimensionality of $\alpha N$ and the 0-softness of $f$, there is a map $k \colon \alpha N \to X$ such that $k(\alpha) = x$ and $fk = \psi$. The last equality implies that each of the points $k(n)$, $n \in N$, is contained in the complement $X - U$. The continuity of $k$ implies that the sequence $\{k(n) : n \in N\}$ converges in $X$ to the point $x$. This contradicts the openess of $U$. Consequently, $y \in Y - f(U)$. □

*Historical and bibliographical notes* 2.1. The majority of the results of Subsection 2.1.1 are taken from [**299**]. Corollary 2.1.9 was proven by Bing [**38**] and is known as Bing's shrinking criterion. The same name is applied to Theorem 2.1.8 (see [**299**, Theorem 1.5]). Selection Theorems 2.1.14, 2.1.15 and 2.1.16 were proved by Michael [**224**], [**225**], [**226**]. Theorem 2.1.17 was in fact proved in [**45**] and [**133**]. Theorem 2.1.18 appears in [**27**]. Theorem 2.1.19 is due to Hanner [**161**] (see also [**299**]). Propositions 2.1.21 and 2.1.22 are widely known (see, for example, [**176**]). Lacher's survey [**207**] is an excellent introduction to the theory of $UV^n$-compacta and $UV^n$-maps. Proposition 2.1.31 is taken from this work (see also [**33**]).

## 2.2. $Z$-sets in $ANR$-spaces

**2.2.1. General properties of $Z$-sets.** If a subset $A$ is closed and nowhere dense in a Polish space $X$, then it easy to see that the set $\{f \in C(I^0, X) : f(I^0) \cap A = \emptyset\}$ is dense in the space $C(I^0, X)$ ($I^0$ denotes the zero-dimensional cube, i.e. the one-point space). Of course, the converse is also true. This simple observation allows us to define higher degrees of nowhere density of subsets. This can be formalized as follows.

DEFINITION 2.2.1. *Let $n \in \omega$. We say that a closed subset $A$ of a Polish space $X$ is a $Z_n$-set in $X$ if the set $\{f \in C(I^n, X) : f(I^n) \cap A = \emptyset\}$ is dense in the space $C(I^n, X)$.*

DEFINITION 2.2.2. *We say that a closed subset $A$ of a Polish space $X$ is a $Z$-set in $X$ if the set $\{f \in C(I^\omega, X)\colon f(I^\omega) \cap A = \emptyset\}$ is dense in the space $C(I^\omega, X)$.*

Thus, having in mind the remark made above, we see that the $Z_0$-sets are precisely the closed and nowhere dense sets.

$Z$-sets in $ANR$-spaces admit some useful characterizations.

PROPOSITION 2.2.3. *Let $A$ be a closed subset of a Polish $ANR$-space $X$. Then the following conditions are equivalent:*

(i) *$A$ is a $Z$-set in $X$.*
(ii) *For each locally compact polyhedron $P$, the set $\{f \in C(P, X)\colon f(P) \cap A = \emptyset\}$ is dense in the space $C(P, X)$.*
(iii) *The set $\{f \in C(X, X)\colon f(X) \cap A = \emptyset\}$ is dense in the space $C(X, X)$.*
(iv) *For each Polish space $Y$, the set $\{f \in C(Y, X)\colon f(Y) \cap A = \emptyset\}$ is dense in the space $C(Y, X)$.*

PROOF. **(i) $\Rightarrow$ (ii).** First let us note that if $T$ is an $AR$-compactum, then the set $\{f \in C(T, X)\colon f(T) \cap A = \emptyset\}$ is dense in the space $C(T, X)$. Indeed, embed $T$ into $I^\omega$ and fix a retraction $r\colon I^\omega \to T$. Consider also an arbitrary map $f\colon T \to X$ and an open cover $\mathcal{U} \in cov(X)$. By (i), there exists a map $g'\colon I^\omega \to X$ such that $g'(I^\omega) \cap A = \emptyset$ and the composition $fr$ is $\mathcal{U}$-close to $g'$. Obviously $g = g'/t$ is $\mathcal{U}$-close to $f$ and $g(T) \cap A = \emptyset$.

Suppose now that $K$ is a countable locally finite simplicial complex such that $| T | = P$. Let us show that for each simplex $\sigma \in K$ the set $B_\sigma = \{f \in C(P, X)\colon f(| \sigma |) \cap A = \emptyset\}$ is dense in the space $C(P, X)$. Consider any map $f\colon P \to X$ and an open cover $\mathcal{U} \in cov(X)$. Consider an open cover $\mathcal{V} \in cov(X)$ satisfying the condition of Proposition 2.1.22 relative to $\mathcal{U}$. By the case considered above, there is a map $g'\colon |\sigma| \to X$ such that $g'$ is $\mathcal{V}$-close to $f/|\sigma|$ and $g'(|\sigma|) \cap A = \emptyset$. Consequently, by the choice of $\mathcal{V}$, the map $g'$ has an extension $g\colon P \to X$ which is $\mathcal{U}$-close to $f$. Clearly, $g(| \sigma |) \cap A = \emptyset$.

Next we show that the set $B_\sigma$ is open in the space $C(P, X)$ for each $\sigma \in K$. Take an element $f \in B_\sigma$. Since $f(| \sigma |)$ and $A$ are disjoint closed sets of the space $X$, we can find an open neighborhood $U$ of $f(| \sigma |)$ which does not intersect $A$. It only remains to note that the neighborhood $B(f, \{U, X - f(| \sigma |)\})$ of $f$ in $C(P, X)$ is completely contained in the set $B_\sigma$.

Observe now that

$$\{f \in C(P, X)\colon f(P) \cap A = \emptyset\} = \cap\{B_\sigma\colon \sigma \in K\}$$

and note that, by Proposition 2.1.7, the set written in the right side of the above equality is dense in the space $C(P, X)$.

**(ii) $\Rightarrow$ (ii).** Consider a map $f\colon X \to X$ and an open cover $\mathcal{U} \in cov(X)$. Let $\mathcal{V} \in cov(X)$ be a star-refinement of $\mathcal{U}$. For an open cover $f^{-1}(\mathcal{V}) \in cov(X)$, by Theorem 2.1.19 there exist a locally compact polyhedron $P$ and maps $p\colon X \to P$

and $q: P \to X$ such that the composition $qp$ if $f^{-1}(\mathcal{V})$-close to $id_X$. By (ii), there is a map $g': P \to X$ which is $\mathcal{V}$-close to the composition $fq$ and such that $g'(P) \cap A = \emptyset$. Let $g = g'p$. It is easy to check that $g$ and $f$ are $\mathcal{U}$-close and $g(X) \cap A = \emptyset$.

Implications (iii) $\Rightarrow$ (iv) and (iv) $\Rightarrow$ (i) are obvious. □

If $X$ happens to be locally compact, then we have a stronger conclusion.

COROLLARY 2.2.4. *Let $A$ be a $Z$-set in a locally compact $ANR$-space $X$. Then for each space $Y$ the set $\{f \in C(Y,X): cl_X f(Y) \cap A = \emptyset\}$ is dense in the space $C(Y,X)$.*

PROOF. Obviously, it suffices to show that each neighborhood of the point $id_X$ in the space $C(X,X)$ contains a map $f: X \to X$ such that $cl f(X) \cap A = \emptyset$. Consider an open cover $\mathcal{U} \in cov(X)$. Local compactness of $X$ guarantees the existence of an open cover $\mathcal{V} \in cov(X)$ such that every map $f: X \to X$ which is $\mathcal{V}$-close to $id_X$ has a closed image in $X$. By Proposition 2.2.3, the neighborhood $B(id_X, \mathcal{U} \wedge \mathcal{V})$ contains a map $f: X \to X$ such that $f(X) \cap A = \emptyset$. It only remains to note that, by the choice of $\mathcal{V}$, $f(X)$ is closed in $X$. □

We reserve the term *strong $Z$-set* for those closed sets $A$ of a space $X$ satisfying the property of Corollary 2.2.4

DEFINITION 2.2.5. *We say that a closed subset $A$ of a Polish space $X$ is a strong $Z$-set in $X$ if the set $\{f \in C(X,X): cl f(X) \cap A = \emptyset\}$ is dense in the space $C(X,X)$.*

Thus Corollary 2.2.4 can be reformulated as follows: each $Z$-set in a locally compact $ANR$-space is a strong $Z$-set. The following simple example shows that the converse need not be true. Consider the subset

$$X = ([0,1] \times \{0\}) \cup (\bigcup \{\{\frac{1}{n}\} \times [0,1]\})$$

of the plane:

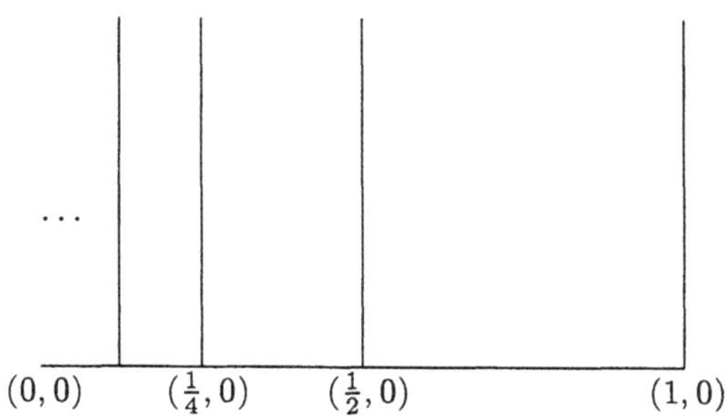

It is not hard to see that $X$ is a Polish $AR$-space. Moreover, $X$ is even locally compact at each point other than the origin $(0,0)$. For each $t \in [0,1)$ consider a map $h_t: X \to X$ determined by the following properties:

- $h_t(1,0) = (1,0)$.
- $h_t(\frac{1}{n},1) = (\frac{1}{n},1)$.
- $h_t$ linearly shrinks the segment $[0,1] \times \{0\}$ onto the segment $[t,1] \times \{0\}$.
- $h_t$ linearly expands the segment $\{\frac{1}{n}\} \times [t,1]$ onto the segment $\{\frac{1}{n}\} \times [0,1]$.
- $h_t$ linearly maps the segment $\{\frac{1}{n}\} \times [0,t]$ onto the segment $[\frac{1}{n}, t + \frac{1-t}{n}] \times \{0\}$ (sending $(\frac{1}{n},t)$ onto $(\frac{1}{n},0)$ and $(\frac{1}{n},0)$ onto $(t+\frac{1-t}{n},0)$ ).

Clearly, the identity map $id_X$ is in the closure (in the space $C(X,X)$) of the set $\{h_t: t \in (0,1)\}$, and since $h_t(X) \subseteq X - (0,0)$ it follows that the origin $(0,0)$ is a $Z$-set in $X$. At the same time $(0,0)$ is not a strong $Z$-set in $X$. To see this, observe that $X - G$ is disconnected for sufficiently small neighborhoods $G$ of the point $(0,0)$.

The following statement provides a characterization of strong $Z$-sets in Polish $ANR$-spaces. Its proof is similar to the proof of Proposition 2.2.3 and is therefore omitted.

PROPOSITION 2.2.6. *Let $A$ be a closed subset of a Polish $ANR$-space $X$. Then the following conditions are equivalent:*

(i) *$A$ is a strong $Z$-set in $X$.*

(ii) *The set $\{f \in C(N \times I^\omega, X): cl_X f(N \times I^\omega) = \emptyset\}$ is dense in the space $C(N \times I^\omega, X)$.*

(iii) *For each locally compact polyhedron $P$, the set $\{f \in C(P,X): cl_X f(P) \cap A = \emptyset\}$ is dense in the space $C(P,X)\}$.*

(iv) *The set $\{f \in C(X,X): cl_X f(X) \cap A = \emptyset\}$ is dense in the space $C(X,X)$.*

(v) *For each Polish space $Y$, the set $\{f \in C(Y,X): cl_X f(Y) \cap A = \emptyset\}$ is dense in the space $C(Y,X)$.*

Condition (v) demonstrates the main advantage of the concept of strong $Z$-set over the concept of $Z$-set (in the locally compact case see Corollary 2.2.4). The proof of the following statement is a useful exercise for the reader.

PROPOSITION 2.2.7. *Let $X$ and $Y$ be Polish spaces.*

(i) *If $A$ is a (strong) $Z$-set in $X$, then $A \times Y$ is a (strong) $Z$-set in $X \times Y$.*

(ii) *Let $A_n$, $n \in \omega$ be a $Z$-set in $X$ and suppose the union $A = \cup\{A_n: n \in \omega\}$ is closed in $X$. Then $A$ is a $Z$-set in $X$.*

PROPOSITION 2.2.8. *Let $A$ be a closed subset of the product $X = \prod\{X_n: n \in \omega\}$ of Polish spaces $X_n$. If $\pi_n(A) \neq X_n$ for infinitely many indices $n$, then $A$ is a $Z$-set in $X$.*

PROOF. Take an increasing sequence $n_0 < n_1 < \cdots$, satisfying the relation $\pi_{n_k}(A) \neq X_{n_k}$ for each $k \in \omega$. Take a point $x_k \in X_{n_k} - \pi_{n_k}(A)$. Now, for each map $f \in C(I^\omega, X)$ define the map $f_k\colon I^\omega \to X$, $k \in \omega$ as follows:

$$\pi_j \cdot f_k(x) = \begin{cases} \pi_j \cdot f(x), & \text{if } j \neq n_k \\ x_k, & \text{if } j = n_k. \end{cases}$$

It only remains to observe that the sequence $\{f_k\colon k \in \omega\}$ converges to the map $f$ (in the space $C(I^\omega, X)$) and that $f_k(I^\omega) \cap A = \emptyset$ for each $k \in \omega$. □

PROPOSITION 2.2.9. *Let $A$ be a closed subset of a Polish $ANR$-space $X$ and $G$ be an open neighborhood of $A$ in $X$. Then the following conditions are equivalent:*

(i) *$A$ is a (strong) $Z$-set in $X$.*

(ii) *$A$ is a (strong) $Z$-set in $G$.*

PROOF. Since the proofs of the two cases are similar, we consider only the case of strong $Z$-sets. First, suppose that $A$ is a strong $Z$-set in $G$ and let $\mathcal{U} \in cov(X)$. Our goal is to find a map $h\colon X \to X$ which is $\mathcal{U}$-close to the identity $id_X$ and satisfies $cl_X h(X) \cap A = \emptyset$. Consider open subsets $G_1$ and $G_2$ of $X$ such that

$$A \subseteq G_1 \subseteq cl_X G_1 \subseteq G_2 \subseteq cl_X G_2 \subseteq G.$$

Let $\mathcal{V} = \{G_2, X - cl_X G_1\}$ and denote by $\mathcal{W} \in cov(X)$ an open refinement of $\mathcal{U} \wedge \mathcal{V}$ such that the pair $(\mathcal{W}, \mathcal{U} \wedge \mathcal{V})$ satisfies the condition of Proposition 2.1.22. Consider now the open cover $\mathcal{W}/G \in cov(G)$. Since $A$ is a strong $Z$-set in $G$, there exists a map $g'\colon G \to G$ such that $g'$ is $\mathcal{W}/G$-close to $id_G$ and $cl_G g'(G) \cap A = \emptyset$. Let $g = g'/cl_X G_2$. Then, by the choice of $\mathcal{W}$, there exists a map $h\colon X \to X$, $\mathcal{U} \wedge \mathcal{V}$-close to $id_X$, such that $h/cl_X G_2 = g$. The latter implies that $cl_X(h(cl_X G_2)) \cap A = \emptyset$. Let $x \in X - cl_X G_2$. The $\mathcal{V}$-closeness of $h$ and $id_X$ guarantees that $h(x) \in X - cl_X G_1$ and, consequently, $cl_X h(X - cl_X G_2) \cap A = \emptyset$. This shows that $cl_X h(X) \cap A = \emptyset$. Therefore $A$ is a strong $Z$-set in $X$.

Let us show the converse. Let $A$ be a strong $Z$-set in $X$. As above, take open subsets $G_1$ and $G_2$ of $X$ such that

$$A \subseteq G_1 \subseteq cl_X G_1 \subseteq G_2 \subseteq cl_X G_2 \subseteq G.$$

Let $\mathcal{U} \in cov(G)$ and $\mathcal{V} = \{G_2, G - cl_X G_1\}$. Since $G$ is also a Polish $ANR$-space, we can, by Proposition 2.1.22, choose an open refinement $\mathcal{W} \in cov(G)$ of $\mathcal{U} \wedge \mathcal{V}$ such that the pair $(\mathcal{W}, \mathcal{U} \wedge \mathcal{V})$ satisfies the condition of that Proposition. Now let $\mathcal{W}_1 = \mathcal{W} \cup \{X - cl_X G_2\}$. Clearly, $\mathcal{W}_1 \in cov(X)$. Since, by our assumption, $A$ is a strong $Z$-set in $X$, there exists a map $g'\colon X \to X$, $\mathcal{W}_1$-close to $id_X$, such that $cl_X g'(X) \cap A = \emptyset$. Let $g = g'/cl_X G_2$. Then $cl_X g(cl_X G_2) \cap A = \emptyset$. Moreover, the $\mathcal{W}_1$-closeness of $g$ and $id_{cl_X G_2}$ implies the inclusion $g(cl_X G_2) \subseteq G$. In addition, it is easy to see that $g$ and $id_{cl_X G_2}$ are $\mathcal{W}$-close and, consequently, by the choice of $\mathcal{W}$, there exists a map $h\colon G \to G$, $\mathcal{U} \wedge \mathcal{V}$-close to $id_G$, such that $h/cl_X G_2 = g$. In particular, $cl_G h(cl_X G_2) \cap A = \emptyset$. Consider now a point $x \in G - cl_X G_2$. The $\mathcal{V}$-closedness of the maps $h$ and $id_G$ guarantees that $h(x) \in G - cl_X G_1$. Therefore, $cl_G h(G - cl_X G_2) \cap A = \emptyset$. But then $cl_G h(G) \cap A = \emptyset$ and the proof is complete. □

COROLLARY 2.2.10. *If $A$ is a (strong) $Z$-set in a Polish $ANR$-space $X$, then $A \cap U$ is a (strong) $Z$-set in $U$ for each open subspace $U$ of $X$.*

PROOF. $A \cap U$ can be represented as the union $A \cap U = \cup\{A_n : n \in \omega\}$, where each of the sets $A_n$, $n \in \omega$, is closed in $X$. Obviously, $A_n$ is a strong $Z$-set in $X$. Proposition 2.2.9 allows us to conclude that each of these sets $A_n$, $n \in \omega$, is then a strong $Z$-set in $U$. Therefore, by Proposition 2.2.7, the set $A \cap U$, as a countable union of strong $Z$-sets of $U$, is itself a strong $Z$-set in $U$. □

COROLLARY 2.2.11. *Let $A$ be a closed set in a Polish $ANR$-space $X$. If there exists an open cover $\mathcal{U} \in cov(X)$ such that $A \cap U$ is a strong $Z$-set in $U$ for each $U \in \mathcal{U}$, then $A$ is a strong $Z$-set in $X$.*

PROOF. Without loss of generality we can assume that $\mathcal{U} = \{U_n : n \in \omega\}$ is countable. Each of the intersections $A \cap U_n$, $n \in \omega$, can be represented as $A \cap U_n = \cup\{A_n^m : m \in \omega\}$, where $A_n^m$ is a strong $Z$-set in $X$, $m \in \omega$. Consequently, by Proposition 2.2.9, the sets $A_n^m$ are strong $Z$-sets in $U_n$, $n, m \in \omega$. Since $A = \cup\{A_n^m : n, m \in \omega\}$, Proposition 2.2.7 finishes the proof. □

**2.2.2. Skeletoids and absorbers in Polish spaces.** In this Subsection $X$ denotes a Polish space and $Auth(X)$ the set of all autohomeomorphisms of $X$, endowed with the topology induced by $C(X, X)$. If $d$ is a complete bounded metric on $X$, then this topology is generated by the following metric

$$D(f, g) = \check{d}(f, f) + \check{d}(f^{-1}, g^{-1}).$$

It is well known that $Auth(X)$ is a topological group. Further assume that $\Gamma$ is a closed subgroup of $Auth(X)$ and that $\mathcal{K}$ is an *additive* collection of $\Gamma$-*invariant* closed subsets of $X$. By $\mathcal{K}_\sigma$ we denote the collection consisting of the countable unions of elements of $\mathcal{K}$. The subgroup $\{f \in \Gamma : f/(X - M) = id_{X-M}\}$ of $Auth(X)$, where $M$ be a subspace of $X$, is denoted by $\Gamma_M$. Finally, $\Gamma/M$ denotes the subgroup $\{g/M : g \in \Gamma_M\}$ of the group $Auth(M)$.

The key idea exploited in the following definition is the notion of compact absorption set, due to R. D. Anderson. This notion has been subsequently generalized by different authors (see [**35**], [**31**], [**31**], [**158**], [**294**], [**295**], [**313**]).

DEFINITION 2.2.12. *We call a set $M \subseteq X$ a $(\Gamma, \mathcal{K})$*-absorbing set *(or, shortly, a $(\Gamma, \mathcal{K})$-absorber), if $M \in \mathcal{K}_\sigma$ and the following condition is satisfied:*

(∗)

$$id_X \in cl(\{g \in \Gamma_U : g(A \cap U) \subseteq M\}),$$

*whenever $A \in \mathcal{K}$ and $U$ is an open set in $X$.*

The proof of the following statement is a direct consequence of the above definition.

PROPOSITION 2.2.13. *Let $M$ be a $(\Gamma, \mathcal{K})$-absorbing set in $X$.*

(i) *If* $A \in \mathcal{K}_\sigma$, *then* $M \cup A$ *is a* $(\Gamma, \mathcal{K})$*-absorbing set in* $X$.
(ii) *If* $g \in \Gamma$, *then the set* $g(M)$ *is a* $(\Gamma, \mathcal{K})$*-absorbing set in* $X$.
(iii) *If* $U$ *is an open set in* $X$, *then* $M \cap U$ *is a* $(\Gamma_U, \mathcal{K}_U)$*-absorbing set in* $X$, *where* $\mathcal{K}_U$ *denotes the collection of all closed subsets of* $X$ *contained in the sets of the form* $U \cap A$, $A \in \mathcal{K}$.
(iv) *If* $U$ *is an open subset of* $X$, *then* $M \cap U$ *is a* $(\Gamma/U, \mathcal{K}_U)$*-absorbing set in* $U$.

The following statement also can be easily proved.

LEMMA 2.2.14. *Let* $U$ *be an open subset of* $X$, $g \in \Gamma$ *and* $A \in \mathcal{K}$. *Then*

$$g \in cl(\{f \in \Gamma \colon f/(X - U) = g/(X - U) \text{ and } f(A \cap U) \subseteq M\}).$$

THEOREM 2.2.15. *Let* $M_1$ *and* $M_2$ *be two* $(\Gamma, \mathcal{K})$*-absorbing sets in* $X$. *Then the set* $\{g \in \Gamma \colon g(M_1) = M_2\}$ *is dense in* $\Gamma$.

PROOF. Obviously, it suffices to show that

$$id_X \in cl_\Gamma(\{g \in \Gamma \colon g(M_1) = M_2\}).$$

In other words, for each $\alpha \in C(X, (0,1))$ we need to show the existence of an element $g \in \Gamma \cap B_d(id_X, \alpha)$ such that $g(M_1) = M_2$ (compare with Subsection 2.1.1). Let $A_0 = K_1 = L_1 = \emptyset$ and $g_0 = id_X$. Represent the sets $M_1$ and $M_2$ as the unions

$$M_1 = \cup\{K_n \colon n = 2, 3, ...\} \text{ and } M_2 = \cup\{L_n \colon n = 2, 3, ...\},$$

where $K_n, L_n \in \mathcal{K}$. Let us now inductively construct a sequence of homeomorphisms $\{g_n\} \subseteq \Gamma$ and a sequence $\{A_n\}$ of closed sets of $X$ satisfying the following conditions:

$(a)_n$ $M_1 \supseteq A_n \supseteq \cup\{K_i \colon i \leq n\}$ and $M_2 \supseteq g_n(A_n) \supseteq \cup\{L_i \colon i \leq n\}$.
$(b)_n$ $g_n/\cup\{A_i \colon i < n\} = g_{n-1}/\cup\{A_i \colon i < n\}$.
$(c)_n$ $d(g_{n-1}(x), g_n(x)) + d(g_{n-1}^{-1}(x), g_n^{-1}(x)) \leq 2^{-(n+1)}\alpha(x)$ for each $x \in X$.

It can easily be seen that the homeomorphism $g = \lim_{n\to\infty} g_n$ is the desired one.

Let $g_1 = id_X$ and $A_1 = \emptyset$.

Suppose now that the homeomorphisms $g_0, g_1, ..., g_{n-1}$ and the closed sets $A_0, A_1, ..., A_{n-1}$, satisfying conditions $(a)_k$, $(b)_k$ and $(c)_k$, $0 \leq k \leq n-1$, have already been constructed. We are going to define a homeomorphism $g_n$ and a closed set $A_n$. Applying Lemma 2.2.14 to the objects $M = M_1$, $g = g_{n-1}^{-1}$, $A = L_n$ and $U = X - g_{n-1}(A_{n-1})$, we can conclude that there exists a homeomorphism $f \in \Gamma$ such that

$$f/g_{n-1}(A_{n-1}) = g_{n-1}^{-1}/g_{n-1}(A_{n-1}),\ f(L_n) \subseteq M_1$$

and

$$d(g_{n-1}^{-1}(x), f(x)) + d(g_{n-1}(x), f^{-1}(x)) \leq 2^{-(n+1)}\alpha(x) \text{ for each } x \in X.$$

Applying Lemma 2.2.14 to the objects $M = M_2$, $g = f^{-1}$, $A = K_n$ and $U = X - (f^{-1}(L_n)) \cup A_{n-1})$ we find a homeomorphism $g_n \in \Gamma$ such that

$$g_n/(A_{n-1} \cup f^{-1}(L_n)) = f/(A_{n-1} \cup f^{-1}(L_n)),\ g_n(K_n) \subseteq M_2$$

and

$$d(f^{-1}(x), g_n(x)) + d(f(x), g_n^{-1}(x)) \leq 2^{-(n+1)}\alpha(x) \text{ for each } x \in X.$$

Straightforward verifications show that the conditions $(a)_n$, $(b)_n$ and $(c)_n$ are satisfied. This completes the inductive step of our construction and finishes the proof. □

COROLLARY 2.2.16. *Let $M$ be a $(\Gamma, \mathcal{K})$-absorbing set in $X$. If $A \in \mathcal{K}_\sigma$ and $U$ is an open subset of $X$, then*

$$id_U \in cl_{\Gamma/U}(\{g \in \Gamma/U \colon g((M \cap U) \cup (A \cap U)) = M \cap U\}).$$

PROOF. By Proposition 2.2.13, both $M \cap U$ and $(M \cap A) \cup (A \cap U)$ are $(\Gamma/U, \mathcal{K}_U)$-absorbing sets in $U$. It only remains to apply Theorem 2.2.15. □

COROLLARY 2.2.17. *Let $M_1$ and $M_2$ be $(\Gamma, \mathcal{K})$-absorbing sets in $X$. If $g \in \Gamma$ and $A$ is a closed subset of $X$ such that $g(A) \cap M_2 = g(A \cap M_1)$, then*

$$g \in cl(\{h \in \Gamma \colon h/A = g/A \text{ and } h(M_1) = h(M_2).$$

PROOF. Consider an open cover $\mathcal{U} \in cov(X)$ and let $M_3 = g(M_1)$, $U = X - A$. By Proposition 2.2.13, both $M_2 \cap U$ and $M_3 \cap U$ are $(\Gamma_U, \mathcal{K}_U)$-absorbing sets in $X$. Therefore, by Theorem 2.2.15, there exists a homeomorphism $f \in \Gamma_U$, $\mathcal{U}$-close to $id_X$, such that $f(M_3 \cap U) = M_2 \cap U$. Then the homeomorphism $h = f \cdot g \in \Gamma$ is $\mathcal{U}$-close to $id_X$, $h/A = g/A$ and $h(M_1) = h(M_2)$. □

COROLLARY 2.2.18. *Let $A$ be a closed subset of $X$ and suppose there exists a $(\Gamma, \mathcal{K})$-absorbing set $M_0$ in $X$ such that $M_0 \cap A = \emptyset$. Then for each $(\Gamma, \mathcal{K})$-absorbing set $M$ in $X$, the complement $M - A$ is a $(\Gamma, \mathcal{K})$-absorbing set in $X$.*

PROOF. Consider the open set $U = X - A$ of $X$. By Proposition 2.2.13, both $M \cap A = M - A$ and $M_0 \cap U = M_0$ are $(\Gamma_U, \mathcal{K}_U)$-absorbing sets in $X$ and, consequently, by Theorem 2.2.15, there exists a homeomorphism $h \in \Gamma_U$ such that $h(M_0 \cap U) = M \cap U$. Then, by Proposition 2.2.13, the set $M - A = h(M_0)$ is $(\Gamma, \mathcal{K})$-absorbing in $X$. □

Let us now consider the closely related concept of *skeleton*.

DEFINITION 2.2.19. *By a* $\Gamma - \mathcal{K}$-skeleton, *we mean any increasing sequence $\{A_i \colon i \in N\}$ of members of $\mathcal{K}$ such that, for every $i \in N$, for every $A, B \in \mathcal{K}$ and for every $F \in \Gamma$ with $F(B) \subseteq A_i$, we have*

$$F \in cl(\cup_j\{H \in \Gamma \colon H/B = F/B \text{ and } H(A) \subseteq A_j\}).$$

*A subset of $X$ which can be expressed as the union of elements of a* $\Gamma-\mathcal{K}$-skeleton *is called a* $\Gamma-\mathcal{K}$-skeletoid. *When $\Gamma = Auth(X)$, it is simply called a* $\mathcal{K}$-skeletoid.

It is easy to see that any $\Gamma-\mathcal{K}$-skeletoid is a $\Gamma-\mathcal{K}$-absorber. Any two $\Gamma-\mathcal{K}$-skeletoids are equivalent in the sense of the next Theorem. In particular, if there exists a $\Gamma-\mathcal{K}$-skeletoid, then it is equivalent to any $\Gamma-\mathcal{K}$-absorber.

THEOREM 2.2.20. *If $A$ and $B$ are $\Gamma-\mathcal{K}$-skeletoids in $X$, then the set $\{H \in \Gamma\colon H(A) = B\}$ is dense in $\Gamma$.*

A closed subset $A$ of a complete metric space $X$ is called a *thin set*, if for each open neighborhood $V$ of $A$ and for each open cover $\mathcal{U}$ of $X$, there is an element $F \in Auth(X)$ such that $F$ is $\mathcal{U}$-close to $id_X$, $F/(X-V) = id$ and $F(A)\cap A = \emptyset$.

DEFINITION 2.2.21. *An additive $Auth(X)$-invariant collection $\mathcal{K}$ consisting of closed subsets of a Polish space $X$ is called a* perfect collection *if $\mathcal{K}$ satisfies the following conditions:*

(i) *Each member of $\mathcal{K}$ is a compact thin set in $X$.*
(ii) *If $B$ is a compact subset of $A \in \mathcal{K}$, then $B \in \mathcal{K}$.*
(iii) *(*Estimated Extension Property*) For each $A \in \mathcal{K}$, for each neighborhood $V$ of $A$ and for each open cover $\mathcal{U}$ of $X$, there exists an open cover $\mathcal{V}$ of $X$ refining $\mathcal{U}$ such that*
   - *for each $B \in \mathcal{K}$ with $B \subseteq V$, and each homeomorphism $f\colon A \to B$ which is $\mathcal{V}$-close to id, there is an extension $F \in Auth(X)$ of $f$ which is $\mathcal{U}$-close to id and $F/(X-V) = id$.*

When the collection $\mathcal{K}$ is perfect, $\Gamma-\mathcal{K}$-skeletoids are characterized by the following theorem.

THEOREM 2.2.22. *Suppose that the collection $\mathcal{K}$ is perfect and let $\{A_i\colon i \in N\}$ be an increasing sequence of members of $\mathcal{K}$. Then $\{A_i\colon i\in N\}$ is a $\Gamma-\mathcal{K}$-skeleton if and only if, for every $K \in \mathcal{K}$, for every $i \in N$ and for every $\epsilon > 0$, there exist an index $j > i$ and an embedding $f\colon K \to X$ such that*

(*) $d(f, id_K) < \epsilon$, $f/(K\cap A_i) = id$ *and* $f(K) \subseteq A_j$.

Suppose that $U$ is an open set in $X$ and let $\mathcal{K}(U) = \{K \in \mathcal{K}\colon K \subseteq U\}$, $Auth(X||X-U) = \{H \in Auth(X)\colon H/(X-U) = id\}$ and $\Gamma(U) = \{H/U\colon H \in Auth(X||X-U)\}$. $Auth(X||X-U)$ is a closed subgroup of $Auth(X)$ and $\Gamma(U)$ can be regarded as a closed subgroup of $Auth(U)$.

THEOREM 2.2.23. *Suppose that $A$ is a $\mathcal{K}$-skeletoid in $X$ and $U$ is an open subset of $X$. Then $A\cap U$ is a $\Gamma(U)-\mathcal{K}(U)$-skeletoid in $U$ and is also an $Auth(X||X-U)-\mathcal{K}(U)$-skeletoid in $X$.*

PROPOSITION 2.2.24. *Let $A$ be a $\mathcal{K}$-skeletoid in $X$ and $f\colon K_1 \to K_2$ be a homeomorphism between members of $\mathcal{K}$ such that $f(K_1\cap A) = K_2 \cap A$. Then there exists $F \in Auth(X)$ such that $F/K_1 = f$ and $F(A) = A$.*

PROPOSITION 2.2.25. *If $A$ is a $\mathcal{K}$-skeletoid in $X$ and $B$ is a countable union of members of $\mathcal{K}$, then $A \cup B$ is also a $\mathcal{K}$-skeletoid.*

### 2.2.3. Properties of $Z$-sets in $I^\omega$ and $R^\omega$.

In this Subsection the Hilbert cube $I^\omega$ is denoted by $Q$ and is represented as the countable infinite power of the closed segment $[-1,1]$. Obviously the countable infinite power $P = (-1,1)^\omega$ is a copy of $R^\omega$. We call $P$ the *pseudo-interior of the cube $Q$* and the complement $Q - P$ the *pseudo-boundary of $Q$*. Evidently, $Q$ can be considered as a subspace of $R^\omega$ and, since the latter space has the structure of a topological linear space, we are in position to add elements of $Q$ as well as multiply them by real numbers. Of course, the results of these operations are not necessarily elements of $Q$. We consider the following metric on $Q$:

$$d(\{x_n\},\{y_n\}) = \sum_{n=1}^{\infty} \frac{|x_n - y_n|}{2^n},$$

where $x_n$ is an $n$-th coordinate of a point $\{x_n\} \in Q$ (i.e. $x_n = \pi_n(\{x_n\})$, where $\pi_n \colon Q \to [-1,1]_n$ denotes the corresponding projection). We will also denote by $\operatorname{Auth}(X,A)$, where $A \subseteq X$, the subgroup $\{f \in \operatorname{Auth}(X) \colon f(A) = A\}$ of the group $\operatorname{Auth}(X)$.

LEMMA 2.2.26. *For each compactum $A$ in $P$ there exists a homeomorphism $h \in \operatorname{Auth}(Q,P)$ such that $\pi_n h(A) = \{0\}$ for each odd index $n$.*

PROOF. Without loss of generality we may assume that the compactum $A$ is contained in the countable infinite product of the segment $[-\frac{1}{2},\frac{1}{2}]$. First we show that there is a homeomorphism $g \in \operatorname{Auth}(Q,P)$ such that $\pi_1 g(A) = \{0\}$. Consider the homeomorphism $f \in \operatorname{Auth}(Q,P)$ defined by letting $f(\{x_n\}) = \{y_n\}$, where

$$y_1 = x_1, \quad \text{and} \quad y_n = \frac{1}{n}x_n + \frac{1}{4}x_1 + \lambda(x_n)\left(\frac{n-1}{n}x_n - \frac{1}{4}x_1\right), \quad n \geq 2,$$

and where

$$\lambda(t) = \frac{1}{2}(|2t-1| + |2t+1| - 2) \text{ for each } t \in [-1,1].$$

Obviously, $f(A) \subseteq P$.

Now represent the cube $Q$ as the product $[-1,1]_1 \times Q_1$, where

$$Q_1 = \prod\{[-1,1]_n \colon n \geq 2\}.$$

Denote by

$$s \colon [-1,1]_1 \times Q_1 \to \{0\} \times Q_1$$

the map defined by letting $s(t,x) = (0,x)$ for each $(t,x) \in [-1,1]_1 \times Q_1$. It follows from the construction of the homeomorphism $f$ that the restriction $f_1 = s/f(A) \colon f(A) \to sf(A)$ is a homeomorphism. Straightforward verification shows that $f_1$ is the restriction (to $f(A)$) of some homeomorphism $g_0 \in \mathrm{Auth}(Q,P)$. Let $g = g_0 f$. It only remains to observe that $g \in \mathrm{Auth}(Q,P)$ and $\pi_1 g(A) = \{0\}$.

Let us now prove our statement in full generality. Represent the set $N$ of all natural numbers as the union $\cup\{N_n : n \in N\}$, where $N_n$'s are pairwise disjoint infinite subsets of $N$ such that $2n-1 \in N_n$ for each $n \in N$. Let

$$\pi_{N_n} : Q \to Q_{N_n} = \prod\{[-1,1]_k : k \in N_n\}$$

denote the corresponding projection and $A_n = \pi_{N_n}(A)$. By the case considered above, there exists a homeomorphism $g_n \in \mathrm{Auth}(Q_{N_n}, P_{N_n})$ such that $\pi_{2n-1} g_n(A_n) = \{0\}$, $n \in N$. Then the desired homeomorphism $g \in \mathrm{Auth}(Q,P)$ is defined to be the product $g \times \{g_n : n \in N\}$. □

The following statement implies, in particular, the topological homogeneity of $R^\omega$.

LEMMA 2.2.27. *Let $f : K \to L$ be a homeomorphism between compacta contained in $P$. Then $f$ has an extension $h \in \mathrm{Auth}(Q,P)$.*

PROOF. Let

$$Q_i = \prod\{[-1,1]_n : n \in N_i\} \text{ and } P_i = \prod\{(-1,1)_n : n \in N_i\}, i = 1,2,$$

where $N_1$ denotes the set of all odd indices and $N_2 = N - N_1$. First consider the special case when $K \subseteq P_1 \times \{0\}$ and $L \subseteq \{0\} \times P_2$. Extend the map $\pi_{N_2} f : K \to P_2$ to a map $\lambda : Q_1 \times \{0\} \to P_2$. For each $\epsilon \in (-1,1)$ denote by $g_\epsilon \in \mathrm{Auth}([-1,1])$ a non-decreasing function, linear on the intervals $[-1,0]$ and $[0,1]$ and such that $g_\epsilon(0) = \epsilon$. For each point $y \in P_2$ denote by $r_y \in \mathrm{auth}(Q_2, P_2)$ the homeomorphism defined by letting $r_y(x) = \{g_{y_n}(x_n) : n \in N_2\}$. Now define the map $f_1 : Q_1 \times Q_2 \to Q_1 \times Q_2$ as follows:

$$f_1(x,y) = (x, r_{\lambda(x)}(y)), \text{ whenever } (x,y) \in Q_1 \times Q_2.$$

Obviously,

$$f_1 \in \mathrm{Auth}(Q_1 \times Q_2, P_1 \times P_2) \text{ and } f_1(x,y) = (x, \pi_{N_2} f(x,y)), \ (x,y) \in K.$$

Similarly, we can construct a homeomorphism $f_2 \in \mathrm{Auth}(Q_1 \times Q_2, P_1 \times P_2)$ such that $f_2(x,y) = (\pi_{N_1} f^{-1}(x,y), y)$ whenever $(x,y) \in L$. It is easy to see that the composition $h = f_2 f_1^{-1}$ is the desired extension of the homeomorphism $f$.

Let us now consider the general case. By Lemma 2.2.26, there exist homeomorphisms $f_i \in \mathrm{Auth}(Q,P)$, $i = 1,2$, such that $f_1(K) \subseteq P_1 \times \{0\}$ and $f_2(L) \subseteq \{0\} \times P_2$. By the case considered above, the homeomorphism $g = f_2 f f_1^{-1} : f_1(K) \to f_2(L)$ has an extension $g' \in \mathrm{auth}(Q,P)$. Then the composition $h = f_2^{-1} g' f_1$ is the desired extension of $f$ □

Similar arguments prove the following statement.

PROPOSITION 2.2.28. *Let $K$ be a closed subset of $Q$. Then $K$ is a $Z$-set in $Q$ if and only if $K \cap P$ is a $Z$-set in $P$. In particular, this is the case if $K \subseteq P$ or $K \subseteq Q - P$.*

LEMMA 2.2.29. *Let $K$ be a $Z$-set in $Q$. Then there is a homeomorphism $f \in \mathrm{Auth}(Q)$ such that $f(K) \subseteq P$.*

PROOF. Let

$$W_{2k-1} = \{x \in Q : x_k = -1\} \text{ and } W_{2k} = \{x \in Q : x_k = 1\}.$$

Then $Q - P = \cup\{W_n : n \in N\}$. Therefore, it suffices to show that the set

$$B_n = \{f \in \mathrm{Auth}(Q) : f(K) \cap W_n = \emptyset\}$$

is dense in $\mathrm{Auth}(Q)$. In turn, by symmetry, it suffices to establish the denseness of $B_1$ in $\mathrm{Auth}(Q)$.

Consider an arbitrary homeomorphism $g \in \mathrm{Auth}(Q)$ and take a homeomorphism $f \in \mathrm{Auth}(Q)$ such that $f(W_1) \subseteq P$. Since the set $fg(K)$ is a $Z$-set in $f(W_1) \approx Q$ we can conclude, by Proposition 2.2.3, that the map $id_{f(W_1)}$ can be arbitrarily closely approximated by embeddings $h : f(W_1) \to P$ such that $hf(W_1) \cap fg(K) = \emptyset$. Then the identity map $id_Q$ can be approximated by homeomorphisms $h \in \mathrm{Auth}(Q, P)$ such that $hf(W_1) \cap fg(K) = \emptyset$ (we use Lemma 2.2.27). Therefore, as close as we wish to the map $id_Q$ there is a homeomorphism $h \in \mathrm{Auth}(Q, P)$ such that $f^{-1}h^{-1}fg(K) \cap W_1 = \emptyset$. □

THEOREM 2.2.30. *Every homeomorphism $f : A_1 \to A_2$ between $Z$-sets of the Hilbert cube $Q$ can be extended to a homeomorphism $h \in \mathrm{Auth}(Q)$.*

PROOF. By Lemma 2.2.29, there are two homeomorphisms $g_1, g_2 \in \mathrm{Auth}(Q)$ such that $g_i(A_i) \subseteq P$, $i = 1, 2$. Then, by Lemma 2.2.27, the homeomorphism $g = g_2 f(g_1/A_1)^{-1} : g_1(A_1) \to g_2(A_2)$ has an extension $g' \in \mathrm{Auth}(Q, P)$. It only remains to remark that the composition $h = g_2^{-1} g' g_1$ is the desired extension of $f$. □

By Proposition 2.2.28, each point in $Q$ is a $Z$-set. Consequently, applying Theorem 2.2.30, we obtain the following result.

COROLLARY 2.2.31. *The Hilbert cube is topologically homogeneous.*

More detailed consideration of the above situation allows us to obtain the following "controlled" version of Theorem 2.2.30.

THEOREM 2.2.32. *For each open cover $\mathcal{U} \in cov(Q)$ there exists an open cover $\mathcal{V} \in cov(Q)$ such that the following condition is satisfied:*

- *if a homeomorphism $f : Z_1 \to Z_2$ between two $Z$-sets of $Q$ is $\mathcal{V}$-close to the inclusion map $i_1 : Z_1 \hookrightarrow Q$, then $h$ can be extended to a homeomorphism $h : Q \to Q$ which is $\mathcal{U}$-close to the identity map $id_Q$.*

Let us now consider some properties of $Z$-sets in the space $R^\omega$.

THEOREM 2.2.33. *Every homeomorphism* $f\colon A_1 \to A_2$ *between* $Z$*-sets of* $P$ *can be extended to a homeomorphism* $h \in \mathrm{Auth}(P)$. *Moreover, for each open cover* $\mathcal{U} cov(P)$ *there exists an open cover* $\mathcal{V} \in cov(P)$ *such that the following condition is satisfied:*

- *if a homeomorphism* $f\colon A_1 \to A_2$ *between two* $Z$*-sets of* $P$ *is* $\mathcal{V}$*-close to the inclusion map* $i_1\colon A_1 \hookrightarrow P$, *then* $f$ *can be extended to a homeomorphism* $h\colon P \to P$ *which is* $\mathcal{U}$*-close to the identity map* $id_P$.

PROOF. We prove only the first part of the theorem. Let $f\colon A_1 \to A_2$ be a homeomorphism between $Z$-sets of $R^\omega$ (we identify $P$ with $R^\omega$ in order to use the linear structure). First we consider the case when there exist homeomorphisms $g_1, g_2\colon R^\omega \times R^\omega \to R^\omega$ such that $g_1(A_1) \subseteq R^\omega \times \{0\}$ and $g_2(A_2) \subseteq \{0\} \times R^\omega$. Let $K_i = g_i(A_i)$, $i = 1, 2$ and $g = g_2 f(g_1/A_1)^{-1}\colon K_1 \to K_2$. Let us show that there is a homeomorphism $g' \in \mathrm{Auth}(R^\omega \times R^\omega)$ such that $g'/K_1 = g$. Since $K_1$ is closed in $R^\omega \times \{0\}$ and since $R^\omega$ is an absolute extensor, we easily see that there is a map $\varphi\colon R^\omega \times \{0\} \to R^\omega$ such that $\varphi(x) = \pi_2 g(x)$ for each $x \in K_1$. Define a homeomorphism $h_1 \in \mathrm{Auth}(R^\omega \times R^\omega)$ by letting

$$h_1(x_1, x_2) = (x_1, x_2 + \varphi(x_1, 0)) \quad \text{whenever} \quad (x_1, x_2) \in R^\omega \times R^\omega.$$

It is easy to see that

$$h_1(x) = (\pi_1(x), \pi_2 g(x)) \quad \text{for each} \quad x \in K_1.$$

Similarly, there exists a homeomorphism $h_2 \in \mathrm{Auth}(R^\omega \times R^\omega)$ satisfying the equality

$$h_2(x) = (\pi_1 g^{-1}(x), \pi_2(x)) \quad \text{for each} \quad x \in K_2.$$

Then $h_2 g(x) = h_1(x)$ for each $x \in K_1$ and, consequently, the homeomorphism $g' = h_2^{-1} h_1 \in \mathrm{Auth}(R^\omega \times R^\omega)$ extends the homeomorphism $g$. The desired extension $h$ can now be defined by letting $h = g_2^{-1} g' g_1$.

Thus, in order to complete the proof, it suffices to show that for each $Z$-set $A$ in $P$ there is a homeomorphism $g\colon P \to P \times P$ such that $g(A) \subseteq P \times \{0\}$. Denote by $A'$ the closure of $A$ in $Q$. By Proposition 2.2.28, $A'$ is a $Z$-set in $Q$. Observe also that, by Proposition 2.2.7, $A' \times \{0\}$ is a $Z$-set in the product $Q \times Q$. By Theorem 2.2.30, there is a homeomorphism $h\colon Q \to Q \times Q$ such that $h(x) = (x, 0)$ for each $x \in A'$. Now consider the sets $M_1 = Q - P$ and $Q \times Q - (P \times P)$. It is not hard to see that both these sets are absorbing (with respect to the collection of all $Z$-sets in $Q$ and $Q \times Q$ and groups $\mathrm{Auth}(Q)$ and $\mathrm{Auth}(Q \times Q)$ respectively). Therefore the set $M_2 = h^{-1}(Q \times Q - (P \times P))$ is also an absorbing set in $Q$. By Corollary 2.2.17, there is a homeomorphism $h' \in \mathrm{Auth}(Q)$ such that $h'/A' = id$ and $h'(M_1) = M_2$. It only remains to remark that the composition $g = hh'/P$ is a homeomorphism of $P$ onto $P \times P$ and $g(A) \subseteq P \times \{0\}$. □

We also have the following statement.

PROPOSITION 2.2.34. *Let $A$ be a $Z_\sigma$-set in $R^\omega$. Then the inclusion $R^\omega - A \to R^\omega$ is a near-homeomorphism.*

PROOF. We identify $R^\omega$ with the *pseudo-interior* $P$ of $Q$. Let $A = \cup_{n=1}^{\infty} A_n$, where each of the sets $A_n$ is a $Z$-set in $P$. By Proposition 2.2.28, each of the sets $B_n = cl_Q A_n$ is a $Z$-set in $Q$. As observed in the proof of Theorem 2.2.33, the pseudo-boundary $Q - P$ of the cube $Q$ is an absorbing set in $Q$. Therefore, by Proposition 2.2.13, the union $(Q - P) \cup B$, where $B = \cup_{n=1}^{\infty} B_n$, is also an absorbing set in $Q$. In this situation Theorem 2.2.15 guarantees the existence of a homeomorphism (as close to the identity as we wish) $g \in \mathrm{Auth}(Q)$ such that $g((Q - P) \cup B) = Q - P$. Then $g(P - B) = P$. It only remains to remark that $P - B = P - A$. □

COROLLARY 2.2.35. *let $A$ be a $Z_\sigma$-set in an $R^\omega$-manifold $X$. Then $X - A$ is also an $R^\omega$-manifold.*

PROOF. Represent $X$ as the countable union $X = \cup_{n=1}^{\infty} X_n$ of open subspaces $X_n$ each of which is homeomorphic to $R^\omega$. Let $A = \cup_{k=1}^{\infty} A_k$, where $A_k$ is a $Z$-set in $X$ for each $k$. By Corollary 2.2.10, each of the sets $X_n \cap A_k$ is a $Z$-set in $X_n$, $k = 1, 2, \ldots$. Therefore, $X_n \cap A = X_n \cap \cup_{k=1}^{\infty} A_k$ is a $Z_\sigma$-set in $X_n$ for each $n$. Applying Proposition 2.2.34, we conclude that $X_n - A$ is homeomorphic to $R^\omega$. Now observe that $X - A = \cup_{n=1}^{\infty}(X_n - A)$ and that each of the summands $X_n - A$ is open in $X - A$. Consequently, $X - A$ is an $R^\omega$-manifold. □

*Historical and bibliographical notes* 2.2. The notion of $Z$-set was first introduced by Anderson in [**19**] (see also [**142**]) in order to generalize the concept of infinite codimension. He used the following definition: a closed subset $A$ of a space $X$ has a property $Z$ in $X$ if for each non-empty contractible open set $G \subseteq X$, the complement $G - A$ is contractible in $X - A$. In all cases considered in this book, this definition is equivalent to Definition 2.2.2 (compare with [**293**]; see also [**111**]). The concept of a stong $Z$-set was introduced by Henderson [**167**]. There are a number of papers devoted to these two notions (see, for example, [**20**], [**297**], [**35**], [**34**]). Corollary 2.2.4, Proposition 2.2.6 and an example of $Z$-set which is not a strong $Z$-set appeared in [**34**]. The key idea exploited in the definition of the notion of compact absorption set (skeletoid) belongs to Anderson. This notion was subsequently generalized by various authors (see [**35**], [**31**], [**32**], [**158**], [**294**], [**295**], [**313**], [**64**]). The results of Subsection 2.2.3 are mainly obtained in [**19**], [**20**], [**24**], [**26**] (see also [**69**] and [**32**]). Corollary 2.2.31 was proved in [**192**].

## 2.3. $R^\omega$ - and $I^\omega$ -manifolds

Traditionally, by a *Y-manifold* (i.e. the manifold, modeled over the space $Y$) we understand a space, each point of which has an open neighborhood homeomorphic to an open subspace of the space $Y$. Below we consider the two important cases in which $Y = I^\omega$ and $Y = R^\omega$.

**2.3.1. Stability of $I^\omega$ - and $R^\omega$ -manifolds.** In this Subsection we show that if $Y \in \{ I^\omega, R^\omega \}$, then every $Y$-manifold $X$ is $Y$-stable, meaning that the product $X \times Y$ is homeomorphic to $X$. For the proof of these important statements we need some preliminary results.

First we recall the following topological construction. For a continuous map $f\colon X \to Y$ and a closed subset $A$ of in $Y$, the *adjunction space* $X \cup_f Y$ is defined to be the disjoint union of $(X - f^{-1}(A))$ and $A$ with the topology consisting of the usual open subsets $X - f^{-1}(A)$ together with sets of the form $f^{-1}(U - A) \cup (U \cap A)$ for open subsets $U \subseteq Y$. It is easy to see that in this situation we have two naturally arising maps $f_A\colon X \to X \cup_f A$ and $p_A\colon X \cup_f A \to Y$. Their definitions are:

$$p_A(x) = \begin{cases} x, & \text{if } x \in X - f^{-1}(A) \\ f(x), & \text{if } x \in f^{-1}(A) \end{cases}$$

and

$$f_A(x) = \begin{cases} f(x), & \text{if } x \in X - f^{-1}(A) \\ x, & \text{if } x \in A. \end{cases}$$

Obviously, $f = p_A f_A$. It is left as a manageable exercise for the reader to verify that if $f\colon X \to Y$ is a fine homotopy equivalence between Polish $ANR$-spaces and $A$ is a (strong) $Z$-set in $Y$, then $X \cup_f A$ is also a Polish $ANR$-space and $A$ is a (strong) $Z$-set in $X \cup_f A$. Additionally, in this situation, the maps $p_A$ and $f_A$ also are fine homotopy equivalences.

Consider now an important particular case of the above construction. Suppose that two spaces $X$ and $Y$ are given and $A$ is closed in $X$. Consider the projection $\pi\colon X \times Y \to X$ onto the first coordinate and form the adjunction space $(X \times Y) \cup_\pi A$. This space, for simplicity, is denoted by $(X \times Y)_A$ and is called the *reduced (or partial) product* of spaces $X$ and $Y$ over the set $A$. The map $p\colon (X \times Y)_A \to X$ is denoted in this particular case by $p_X$. Clearly, if $A = \emptyset$, then the space $(X \times Y)_A$ coincides with the usual product $X \times Y$. The reader can easily observe that by enlarging set $A$ this construction "gradually" turns the product $X \times Y$ into the space $X$.

DEFINITION 2.3.1. *The space $X$ is* strictly $Y$-stable *if for each closed subset $A \subseteq X$, open neighborhood $G$ of $A$, and neighborhood $W$ of the projection $\pi_X\colon X \times Y \to X$ (in the space $C(X \times Y, X)$), there exists a homeomorphism $f\colon X \times Y \to (X \times Y)_A$ such that $p_X f \in W$ and $f(z) = z$ for each point $z \in (X - G) \times Y$.*

PROPOSITION 2.3.2. *Let $X$ be a $Y$-manifold. If each open subspace of $Y$ is strictly $Y$-stable, then the space $X$ is also strictly $Y$-stable.*

PROOF. Consider the collection $\mathcal{U}$ of open subspaces of the space $X$ satisfying the following condition:

(*) If $U \in \mathcal{U}$, $V$ is open in $X$ and $V \subseteq U$, then $V$ is strictly $Y$-stable.

Obviously, the collection $\mathcal{U}$ is hereditary with respect to open subspaces. It is also easy to see that a discrete countable union of elements of $\mathcal{U}$ is an element of $\mathcal{U}$. Therefore, by the well-known localization principle (see, for example, [**32**]), it only remains to show that if $V_1, V_2 \in \mathcal{U}$, then $V = V_1 \cup V_2$ is strictly $Y$-stable.

Let $A$ be a closed subset of $V$ and $G$ be an open neighborhood of $A$ in $V$. Take any metric $d$ on $V$ generating the topology of $V$ and let $\alpha \in C(V,(0,\infty))$. Let $A_k = A \cap V_k$, $k = 1, 2$. Take open neighborhoods $G_1$ and $G_2$ of the sets $A_1$ and $A_2$ respectively (in $X$) so that

$$A_k \subseteq A \cap G_k \subseteq \operatorname{cl} G_k \subseteq V_k, \quad k = 1, 2.$$

Since $V_1$ is strictly $Y$-stable, there exists a homeomorphism

$$g\colon V_1 \times Y \to (V_1 \times Y)_{A_1}$$

such that $g(z) = z$ for each point $Z \in (V_1 - G_1) \times Y$ and

$$d(p_{V_1} g(z), \pi_{V_1}(z)) < \frac{1}{2}\alpha\pi_{V_1}(z) \text{ for each } z \in V_1 \times Y.$$

Completing the homeomorphism $g$ by the identity map on $(V_2 - V_1) \times Y$, we obtain the homeomorphism $g'\colon V \times Y \to (V \times Y)_{A_1}$.

Since $V_2 - A_1$ is strictly $Y$-stable, we have a homeomorphism

$$h\colon (V_2 - A_1) \times Y \to ((V_2 - A_1) \times Y)_{A_2 - A_1}$$

such that $h(z) = z$ for each $z \in (V_2 - G_2 - A_1) \times Y$ and

$$d(p_{V_2 - A_1} h(z), \pi_{V_2 - A_1}(z)) < \beta(\pi_{V_2 - A_1}(z)) \text{ for each } z \in (V_2 - A_1) \times Y,$$

where

$$\beta(x) = \min\{\frac{1}{2}\alpha(x), d(x, A_1)\}.$$

Completing the homeomorphism $h$ by the identity map on $((V_1 - A - V_2) \times Y) \cup A_1$, we obtain the homeomorphism $h'\colon (V \times Y)_{A_1} \to (V \times Y)_A$. It only remains to note that the composition $f = h'g'$ is a homeomorphism of the product $V \times Y$ onto the space $(V \times Y)_A$, coinciding with the identity on $(V - G) \times Y$ and satisfying the desired inequality $d(p_V f(z), \pi_V(z)) < \alpha\pi_V(z)$ for each $z \in V \times Y$. $\square$

DEFINITION 2.3.3. *A space $Y$ is said to have the* reflective isotopy property *(or, briefly, RIP) if there exists a homotopy $g\colon Y \times Y \times [0,1] \to Y \times Y$ (called a* reflection isotopy*) such that the following conditions are satisfied:*

(i) $g(y_1, y_2, 0) = (y_1, y_2)$ *for each* $(y_1, y_2) \in Y \times Y$.

(ii) $g(y_1, y_2, 1) = (y_2, y_1)$ *for each* $(y_1, y_2) \in Y \times Y$.

(iii) *The map* $\bar{g}\colon Y \times Y \times [0,1] \to Y \times [0,1]$ *defined by letting* $\bar{g}(y_1, y_2, t) = (g(y_1, y_2), t)$ *is a homeomorphism.*

LEMMA 2.3.4. *The spaces* $R^\tau$ *and* $I^\tau$ *have the reflective isotopy property for each* $\tau \geq \omega$.

PROOF. Consider a complex linear topological space $E$ and define a map $g\colon E \times E \times [0,1] \to E \times E$ as follows:

$$g(x_1, x_2, t) = \begin{cases} (e^{2\pi i t}x_1, x_2), & \text{if } 0 \leq t \leq \frac{1}{2} \\ (-x_1 \sin \pi t - x_2 \cos \pi t, -x_1 \cos \pi t + x_2 \sin \pi t), & \text{if } \frac{1}{2} \leq t \leq 1. \end{cases}$$

Straightforward verification shows that $g$ is a reflection isotopy.

We also remark that for each linear topological space over the field $R$ of real numbers, its square can be considered as a complex linear topological space. Therefore, the plane $R^2$ has the reflective isotopy property.

Let us now show that the square $[-1,1]^2$ also has the reflective isotopy property. Consider the reflection isotopy $g\colon R^2 \times R^2 \times [0,1] \to R^2 \times R^2$ constructed above. Let

$$K = \{(z_1, z_2, z_3, z_4) \in R^4\colon z_1^2 + z_2^2 + z_3^2 + z_4^2 \leq 1\}.$$

Obviously, the 4-dimensional ball $K$ is invariant with respect to $g/(R^2 \times R^2 \times \{t\})$ for each $t \in [0,1]$. Consider now the radial homeomorphism

$$f\colon [-1,1]^2 \times [-1,1]^2 \to K,$$

shrinking the 4-dimensional cube $[-1,1]^2 \times [-1,1]^2$ onto $K$ along the lines passing through the origin. Finally, observe that the map

$$h\colon [-1,1]^2 \times [-1,1]^2 \times [0,1] \to [-1,1]^2 \times [-1,1]^2,$$

defined by letting

$$h(x_1, x_2, t) = f^{-1}g(f(x_1, x_2), t) \quad \text{for each} \quad (x_1, x_2, t) \in [-1,1]^2 \times [-1,1]^2 \times [0,1]$$

is the reflection isotopy for $[-1,1]^2$.

Thus we have shown that the spaces $R^2$ and $[-1,1]^2$ both have the reflective isotopy property. Taking products of the corresponding reflection isotopies we can easily conclude that the spaces $R^\tau$ and $I^\tau$, $\tau \geq \omega$ also have reflective isotopy property. $\square$

DEFINITION 2.3.5. *A homotopy* $Y^N \times Y \times [1,\infty) \to Y^N$ *is called a* displacement isotopy *if the following two conditions are satisfied:*

(i) $s_n h(x, y, t) = s_n(x)$ *for each* $n \leq t$ *and* $(x, y, t) \in Y^N \times Y \times [1, \infty)$, *where* $S_n\colon Y^N \to Y^n$ *denotes the projection onto the corresponding subproduct.*

(ii) *the map* $\bar{h}\colon Y^N \times Y \times [1,\infty) \to Y^N \times [1,\infty)$ *defined by letting* $\bar{h}(x, y, t) = (h((x, y, t, ), t)$ *is a homeomorphism.*

LEMMA 2.3.6. *If a space has the reflective isotopy property, then it admits a displacement isotopy.*

PROOF. Let $g: Y \times Y \times [0,1] \to Y$ be a reflection isotopy for the space $Y$. Then the corresponding displacement homotopy can now be defined as follows. For each $(\{y_k\}, y) \in Y^N \times Y$ and for each $n \in N$ and $t \in [0,1]$, let $h(\{y_k\}, y, n+t) = \{z_k\}$, where $z_k = y_k$ if $k < n$, $z_k = y_{k-1}$ if $k \geq n+2$, and $(z_n, z_{n+1}) = g(y, y_n t)$. □

Let $V$ be an open subset in $Y^N$. We call a function $\lambda \in C(V, (0,\infty))$ a *stearing function* if for each $n \in N$ and for $x, x' \in V$, the conditions

$$\lambda(x) \leq n+1 \text{ and } s_n(x) = s_n(x')$$

imply the equality $\lambda(x) = \lambda(x')$.

LEMMA 2.3.7. *The following conditions are satisfied:*

(a) *If $\lambda_1$ and $\lambda_2$ are stearing functions on $V$, then $\lambda(x) = \max\{\lambda_1(x), \lambda_2(x)\}$ is also a stearing function on $V$.*

(b) *Let $V$ be an open subset in $Y^N$ and $\alpha \in C(V, (0,\infty))$. Then there is a stearing function $\lambda$ on $V$ such that $\lambda > \alpha$.*

(c) *Let $V_1$ and $V_2$ be open subsets of $Y^N$, $B$ be a closed subset of $V_1$ and $B \subseteq V_2 \subseteq V_1$. Then for any two functions $\alpha_1 \in C(V_1, (0,\infty))$ and $\alpha_2 \in C(V_2, (0,\infty))$, there exists a stearing function $\lambda_k$ on $V_k$, $k = 1,2$, such that $\lambda_1/B = \lambda_2/B$ and $\lambda_k > \alpha_k$, $k = 1,2$.*

(d) *Let $\lambda$ be a stearing function on an open subset $V \subseteq Y^N$. Suppose that $h: Y^N \times Y \times [1,\infty) \to Y^N$ is a displacement isotopy for $Y$. Suppose also that on $Y^N$ we have the metric*

$$d(\{y_n\}, \{y'_n\}) = \sum_{n=1}^{\infty} \min\{\frac{1}{2^n}, \rho(y_n, y'_n)\},$$

*where $\rho$ is a metric on $Y$. If $2^{-\lambda(x)} < \frac{1}{2} d(x, Y^N - V)$ for each $x \in V$, then the map $f: V \times Y \to Y^N$, defined by letting*

$$f(x,y) = h(x, y, \lambda(x)), \quad (x,y) \in V \times Y$$

*is a homeomorphism of the product $V \times Y$ onto $V$. In addition,*

$$d(f(x,y), x) < 2^{-\lambda(x)-1} \text{ for each } (x,y) \in V \times Y.$$

PROOF. Conditions (a)-(c) are obvious. Let us verify condition (d). For a point $(x,y) \in V \times Y$ take a number $n$ such that $\lambda(x) \in [n, n+1]$. Then, by the definition of displacement isotopy, we have the following two conditions:

(i) $s_n f(x,y) = s_n(x)$.

(ii) $\lambda f(x,y) = \lambda(x)$.

The first condition implies that $g(f(x,y),x) < 2^{-n} < 2^{-\lambda(x)+1}$ for all points $(x,y) \in V \times Y$. In particular, $f(V \times Y) \subseteq V$. The second condition implies that the map $g(x) = (h/(Y^N \times Y \times \{\lambda(x)\}))^{-1}$, $x \in V$, sends the set $V$ into the product $V \times Y$. Moreover, one can see that $g$ is an inverse of $f$. □

PROPOSITION 2.3.8. *If a space $Y$ admits a displacement isotopy, then every open subspace of $Y^N$ is strictly $Y$-stable.*

PROOF. Let $U$ be an open subspace of $Y^N$ and $A$ and $B$ be disjoint closed subsets of $U$. Also let $\alpha \in C(U,(0,\infty))$. Our goal is to construct a homeomorphism $f: U \times Y \to (U \times Y)_A$ such that $f(z) = z$ for each $z \in B \times Y$, and $d(p_U f(x,y),x) < \alpha(x)$ for each $(x,y) \in Y \times Y$ (here $d$ denotes the metric from condition (d) of Lemma 2.3.7). Let $U_1 = U$, $U_2 = U - A$ and let $\beta_k(x) = -\log_2 \alpha_k(x)$ for each $x \in U_k$, where

$$\alpha_k(x) = \frac{1}{4}\min\{\alpha(x), d(x, Y^N - U_k)\}, \quad k = 1,2.$$

By condition (c) of Lemma 2.3.7, there exist two stearing functions $\lambda_k$ on $U_k$ such that $\lambda_1/B = \lambda_2/B$ and $\lambda_k > \beta_k$, $k = 1,2$. We define maps $f_k: U_k \times Y \to Y^N$ by letting

$$f_k(x,y) = h(x,y,\lambda_k(x)), \quad k = 1,2$$

(here $h$ denotes the displacement isotopy for $Y$). It follows from condition (d) of Lemma 2.3.7 that $f_k$ is a homeomorphism of the product $U_k \times Y$ onto $U_k$ and $d(f_k(x,y),x) < 2\alpha_k(x)$ for each $(x,y) \in U_k \times Y$, $k = 1,2$.. Now consider the map $f_3: (U \times Y)_A \to U$ coinciding with $f_2$ on $U_2 \times Y$ and with the identity map on $A$. Obviously, $f_3$ is a homeomorphism. Finally, let $f = f_3^{-1} f_1$. It is not difficult to see that $f$ sends the product $U \times Y$ homeomorphically onto the space $(U \times Y)_A$. It only remains to remark that $f(z) = z$ for each $z \in B \times Y$ and that $d(p_U f(x,y),x) < 2\alpha_1(x) + 2\alpha_2(x) \leq \alpha(x)$ for each $(x,y) \in U \times Y$. □

The statements proved so far in this section imply the following two important stability theorems for $R^\omega$ - and $I^\omega$ -manifolds.

THEOREM 2.3.9. *If $X$ is an $R^\omega$ -manifold, then $X \times R^\omega \approx X$. Moreover, the projection $\pi_X: X \times R^\omega \to X$ is a near-homeomorphism.*

THEOREM 2.3.10. *If $X$ is an $R^\omega$ -manifold, then $X \times I^\omega \approx X$. Moreover, the projection $\pi_X: X \times I^\omega \to X$ is a near-homeomorphism.*

COROLLARY 2.3.11. *If $X$ is an $R^\omega$ - or $I^\omega$ -manifold, then the projection $\pi: X \times I \to X$ is a near-homeomorphism.*

PROOF. Consider only the case of $R^\omega$ -manifolds (the remaining case is completely analogous to this one). Denote by $\pi_X: X \times R^\omega \to X$ and $\pi_1: R^\omega \times I \to R^\omega$ the projections onto the first coordinates. By Theorem 2.3.9, $\pi_X$ is a near-homeomorphism. The projection $\pi_1$ is also a near-homeomorphism (see **[25]** for details). Now consider the following commutative diagram:

$$\begin{array}{ccc} X \times R^{\omega} \times I & \xrightarrow{\pi_X \times id} & X \times I \\ \Big\downarrow{\scriptstyle id \times \pi_1} & & \Big\downarrow{\scriptstyle \pi} \\ X \times R^{\omega} & \xrightarrow{\pi_X} & X \end{array}$$

As observed above, both horizontal arrows, as well as the left vertical arrow, are near-homeomorphisms. In this situation the projection $\pi$ is also a near-homeomorphism. □

**2.3.2. The strong universality of $I^{\omega}$- and $R^{\omega}$-manifolds.** In Subsection 1.1.1 we have already seen (Propositions 1.1.3 and 1.1.8) that the spaces $R^{\omega}$ and $I^{\omega}$ are inuversal with respect to the classes of all Polish spaces and all compact metrizable spaces respectively. Furthermore, in the first case, it is essential that each Polish space admits a *closed* embedding into $R^{\omega}$ . This property of the space $R^{\omega}$ is non-trivial, because it is easy to see every Polish (and, in general, every separable metrizable) space can be embedded into the Hilbert cube $I^{\omega}$ as well. Below, we denote by $\mathcal{A}_{\omega,\infty}$ and $\mathcal{B}_{\omega,\infty}$ the class of all Polish spaces and the class of all metrizable compacta respectively. The sense of these notations will become clear in Section 6.5.

DEFINITION 2.3.12. *Let $\mathcal{P} \in \{\mathcal{A}_{\omega,\infty}, \mathcal{B}_{\omega,\infty}\}$. A Polish space $X$ is said to be* strongly $\mathcal{P}$-universal *if for each space $Y$ from the class $\mathcal{P}$, the set of all closed embeddings of $Y$ in $X$ is dense in the space $C(Y, X)$.*

PROPOSITION 2.3.13. *Each $I^{\omega}$ -manifold is strongly $\mathcal{B}_{\omega,\infty}$-universal.*

PROOF. We consider only the compact case. Let $X$ be a compact $I^{\omega}$-manifold. By Theorem 2.3.10, it suffices to establish the strong $\mathcal{B}_{\omega,\infty}$-universality of the product $X \times I^{\omega}$ . Let $f: Y \to X \times I^{\omega}$ , where $Y$ is a metrizable compactum and $\mathcal{U} \in cov(X \times I^{\omega})$. Without loss of generality we can assume that the cover $\mathcal{U}$ has the form $\mathcal{V} \times \mathcal{W}$, where $\mathcal{V} \in cov(X)$ and $\mathcal{W} \in cov(I^{\omega})$. Take a finite subset $A \subseteq \omega$ and an open cover $\mathcal{W}_A \in cov(I^A)$ such that the cover $\pi_A^{-1}(\mathcal{W}_A)$ refines $\mathcal{W}$ (here $\pi_A: I^{\omega} \to I^A$ denotes the projection onto the corresponding subproduct). Since the space $I^{\omega - A}$ is homeomorphic to $I^{\omega}$ , we can see, by Proposition 1.1.8, that there is an (closed) embedding $g: Y \to I^{\omega - A}$. It only remains to note that the diagonal product

$$\pi_1 f \triangle \pi_A \pi_2 \triangle g: Y \to X \times I^{\omega}$$

is an embedding $\mathcal{U}$-close to $f$ (here $\pi_1$ and $\pi_2$ denote the projection of $X \times I^\omega$ onto the first and the second coordinate respectively). □

PROPOSITION 2.3.14. *Each $R^\omega$ -manifold is strongly $\mathcal{A}_{\omega,\infty}$-universal.*

PROOF. By Theorem 2.3.9, the general case can be reduced to the case $X = R^\omega$. The space $R^\omega$ is identified with its countable power $X = \prod\{R^\omega_n : n \in \omega\}$, where $R^\omega_n = R^\omega$ for each $n \in \omega$. Let $\pi_n : X \to R^\omega_n$ denote the projection onto the corresponding coordinate. Take a metric $d_n$ on the space $R^\omega_n$, bounded by $\frac{1}{2^n}$, and consider the following metric on the product $X$:

$$d(x, (x') = \max\{d_n(\pi_n(x), \pi_n(x')) : n \in \omega\}, \quad x, x' \in X.$$

Let $Y$ be a Polish space. Suppose also that maps $f : Y \to X$ and $\alpha : X \to (0,1)$ are given. For our purposes it suffices to construct a closed embedding $g : Y \to X$ such that $d(f(y), g(y)) \leq \alpha f(y)$ for each point $y \in Y$ (see Subsection 2.1.1).

Since $R^\omega_n$ is a Polish absolute retract, there exists a continuous map

$$h_n : R^\omega_n \times R^\omega_n \times [0, \infty) \to R^\omega_n$$

such that

$$h_n(x_1, x_2, t) = x_1 \text{ for each } t \leq 1$$

and

$$h_n(x_1, x_2, t) = x_2 \text{ for each } t \geq 2.$$

By Proposition 1.1.3, there exists a closed embedding $\varphi_n : Y \to R^\omega_n$, $n \in \omega$. Let the map $g_n : Y \to R^\omega_n$, $n \in \omega$, be defined by letting

$$g_n = h_n(\pi_n f \triangle \varphi_n \triangle 2^n \alpha f).$$

We now define the map $g : Y \to X$ to be the diagonal product

$$g = \triangle\{g_n : n \in \omega\}.$$

If $y \in Y$, then there exists $n \in \omega$ such that $2^{-(n+1)} \leq \alpha f(y) \leq 2^{-n}$. Therefore, for $k \leq n$ we have $g_k(y) = \pi_k f(y)$. This implies that

$$d(f(y), g(y)) = \max\{d_k(p_k f(y), p_k g(y)) : k = n+1, \dots\} \leq 2^{-(n+1)} \leq \alpha f(y).$$

The injectivity of $g$ is obvious. Let us show that $g$ is a closed embedding. Consider an arbitrary sequence $\{y_k\}$ of points in $Y$ such that the sequence $\{g(y_k)\}$ converges in $X$ to a point $x = \lim g(y_k)$. Obviously, it suffices to show that the sequence $\{y_k\}$ also converges in $Y$ to some point $y \in Y$ (indeed, this will imply, first of all, that the map $g : Y \to g(Y)$ is open and, secondly, closedness of the set $g(Y)$ in $X$).

First, assume that $0 < \epsilon = \inf\{\alpha f(y_i)\}$. Choose a number $n \in \omega$ so that $\epsilon \geq 2^{-n+1}$. Then we have $\pi_n g(y_k) = g_n(y_k) = \varphi_n(y_k)$ for each $k$. Continuity of the projection $\pi_n$ guarantees that $\pi_n(x) = \lim \varphi_n(y_k)$. But $\varphi_n$ is a closed embedding. Therefore, the sequence $\{y_k\}$ converges in $Y$ to the point $y = \varphi_n^{-1}(\pi_n(x))$.

Thus it only remains to show that $\epsilon > 0$. Assume to the contrary that $\epsilon = 0$. Then there exists a subsequence $\{y_{k_i}\}$ of the sequence $\{y_k\}$ such that $\lim \alpha f(y_{k_i}) = x$. But this equality, coupled with the continuity of $\alpha$, leads us to $\alpha(x) = 0$ and this is a contradiction. This completes the proof. □

LEMMA 2.3.15. *Every $Z$-set in a strongly $\mathcal{A}_{\omega,\infty}$-universal Polish $ANR$-space is a strong $Z$-set.*

PROOF. Let $A$ be a $Z$-set in a strongly $\mathcal{A}_{\omega,\infty}$-universal Polish $ANR$-space $X$. According to Proposition 2.2.6, it suffices to show that if $P$ is a locally compact polyhedron then the set of those maps $f\colon P \to X$ such that $cl_X f(P) \cap A = \emptyset$ is dense in the space $C(P,X)$. Represent $P$ as the countable union of its compact subspaces (simplexes) $P_n$, $n \in \omega$. Denote by $B_n$ the set of those maps $f\colon P \to X$ for which $f(P_n) \cap A = \emptyset$. Obviously, $B_n$ is an open and dense subset of $C(P,X)$ for each $n \in \omega$. Recall that the space $C(P,X)$ contains a dense (and obviously $G_\delta-$) subspace $F$, consisting of all closed embeddings (apply Lemma 2.1.2). By the Baire property in $C(P,X)$ (see Proposition 2.1.7), the set

$$F \cap \cap_{n=1}^{\infty} B_n$$

is also dense in $C(P,X)$. □

LEMMA 2.3.16. *Each compact set in a strongly $\mathcal{A}_{\omega,\infty}$-universal Polish $ANR$-space is a strong $Z$-set.*

PROOF. Let $K$ be a compact subset of a strongly $\mathcal{A}_{\omega,\infty}$-universal Polish $ANR$-space $X$. By Lemma 2.3.15, it suffices to show that $K$ is a $Z$-set in $X$. Let $\mathcal{U} \in cov(X)$ and $f\colon I^\omega \to X$ be a map. Denote by $\pi\colon N \times X \to X$ the natural projection. The strong $\mathcal{A}_{\omega,\infty}$-universality of $X$ guarantees the existence of a closed embedding $h\colon N \times X \to X$ which is $\mathcal{U}$-close to the composition $\pi f$. Obviously, the collection $\{h(\{k\} \times I^\omega)\colon k \in N\}$ is discrete in $X$. Therefore, by compactness of $K$, there exists an index $k_0 \in N$ such that $h(\{k_0\} \times I^\omega) \cap K = \emptyset$. Now note that the map $g\colon I^\omega \to X$ defined by letting $g(a) = h(k_0, a)$, $a \in I^\omega$, is $\mathcal{U}$-close to $f$ and $g(I^\omega) \cap K = \emptyset$. □

The following two statements show that, in fact, the spaces $R^\omega$ and $I^\omega$ both satisfy a stronger version of the universality property discussed above.

THEOREM 2.3.17. *Let $X$ be a $R^\omega$-manifold and $A$ be a closed subset of a Polish space $B$. Then each map $f\colon B \to X$, such that $f/A$ is a $Z$-embedding, can be arbitrarily closely approximated by $Z$-embeddings coinciding with $f$ on $A$. In particular, the set of all $Z$-embeddings $B \to X$ is dense (and a $G_\delta$) in the space $C(B,X)$.*

THEOREM 2.3.18. *Let $X$ be an $I^\omega$-manifold and $A$ be a closed subset of a locally compact space $B$. Then each proper map $f\colon B \to X$, such that $f/A$ is a $Z$-embedding, can be arbitrarily closely approximated by $Z$-embeddings coinciding with $f$ on $A$.*

The following two statements also belong to the same category of results.

THEOREM 2.3.19. *Let $X$ be an $I^\omega$-manifold, $A$ be a locally compact space and $F\colon A\times I\to X$ be a proper map such that $F_0$ and $F_1$ are $Z$-embeddings. Then there exists an isotopy $H\colon X\times I\to X$ such that $H_0 = id$ and $H_1F_0 = F_1$. Moreover, if $F$ is a $\mathcal{U}$-homotopy ($\mathcal{U}\in cov(X)$), then we may choose $H$ to also be a $\mathcal{U}$-isotopy.*

THEOREM 2.3.20. *Let $X$ be an $R^\omega$-manifold, $A$ be a Polish space, and $F\colon A\times I\to X$ be a map such that $F_0$ and $F_1$ are $Z$-embeddings. Then there exists an isotopy $H\colon X\times I\to X$ such that $H_0 = id$ and $H_1F_0 = F_1$. Moreover, if $F$ is an $\mathcal{U}$-homotopy ($\mathcal{U}\in cov(X)$), then we may choose $H$ to also be a $\mathcal{U}$-isotopy.*

**2.3.3. Further properties of $I^\omega$- and $R^\omega$-manifolds.** The following two statements, known as $ANR$-theorems, are among the main results of infinite-dimensional manifold theory.

THEOREM 2.3.21. *The following conditions are equivalent for each locally compact space $X$:*

(i) *$X\times I^\omega$ is a $I^\omega$-manifold.*
(ii) *$X$ is an $ANR$-space.*

A complete proof of this theorem is contained in Chapman's Lectures [**69**], and for this reason a proof is not presented here. A similar approach proves the following statement.

THEOREM 2.3.22. *The following conditions are equivalent for each Polish space $X$:*

(i) *$X\times R^\omega$ is homeomorphic to $R^\omega$ (respectively, $X\times R^\omega$ is a $R^\omega$-manifold).*
(ii) *$X$ is an $A(N)R$-space.*

The fact that every compact contractible $I^\omega$-manifold is homeomorphic to $I^\omega$ (see [**65**]), coupled with Theorem 2.3.21, implies the following result.

COROLLARY 2.3.23. *The following conditions are equivalent for each compact space $X$:*

(i) *$X\times I^\omega$ is homeomorphic to $I^\omega$.*
(ii) *$X$ is an $AR$-space.*

Open embedding theorems are also very important in infinite-dimensional manifold theory. In the case of $R^\omega$-manifolds, the following shows that the class of $R^\omega$-manifolds coincides with the class of open subspaces of the model space $R^\omega$.

THEOREM 2.3.24. *The following conditions are equivalent for any space $X$:*

(i) *$X$ is an $R^\omega$-manifold.*
(ii) *$X$ admits an open embedding into $R^\omega$.*

PROOF. Obviously, every open subspace of $R^\omega$ is an $R^\omega$-manifold.

Consider now an arbitrary $R^\omega$-manifold $X$, and denote by $Y$ the open cone of $X$, i.e. the reduced product $([0,1) \times X)_{\{0\}}$. It is not hard to see that $Y$ is a Polish absolute retract. Consequently, by Theorem 2.3.22, the product $Y \times R^\omega$ can be identified with $R^\omega$. Observe also that the product $(0,1) \times Y$ is open in $Y$ (its complement in $Y$ is one point). Consequently, the product $(0,1) \times X \times R^\omega$ is open in $R^\omega$. It only remains to note that, by Theorem 2.3.9, the indicated product is homeomorphic to $X$. □

Of course, not every $I^\omega$-manifold can be identified with an open subspace of $I^\omega$. Compactness is an obvious obstruction. Nevertheless, in this case we have the following result.

THEOREM 2.3.25. *Let $X$ be an $I^\omega$-manifold. Then the product $X \times [0,1)$ admits an open embedding into $I^\omega$.*

PROOF. First assume that $X$ is compact. Then its cone, $\mathrm{Con}(X)$, is an absolute retract. By Corollary 2.3.23, the product $\mathrm{Con}(X) \times I^\omega$ is homeomorphic to $I^\omega$. Since $X \times [0,1)$ is open in $\mathrm{Con}(X)$, Theorem 2.3.10 finishes the proof in this case.

Now suppose that $X$ is a non-compact $I^\omega$-manifold. Denote by $Y$ the one-point compactification of the product $X \times [0,1)$. It follows from [**178**] that $Y$ is an absolute retract. Corollary 2.3.23 implies that the product $Y \times I^\omega$ is homeomorphic to $I^\omega$. Consequently, the product $X \times [0,1) \times I^\omega$ is homeomorphic to an open subspace of $I^\omega$. Theorem 2.3.10 completes the proof. □

A detailed proof of the following statement can be found in [**69**].

THEOREM 2.3.26. *$I^\omega$ manifolds $X$ and $Y$ are homotopy equivalent if and only if the products $X \times [0,1)$ and $Y \times [0,1)$ are homeomorphic.*

A similar result is valid for $R^\omega$-manifolds as well. Observing, additionally, that every $R^\omega$-manifold is $[0,1)$-stable (this follows from Theorems 2.3.9 and 2.3.22) we obtain the following homotopy classification of $R^\omega$-manifolds.

THEOREM 2.3.27. *Homotopy equivalent $R^\omega$-manifolds are homeomorphic.*

We conclude this section with triangulation theorems for $I^\omega$- and $R^\omega$-manifolds.

THEOREM 2.3.28. *For each $I^\omega$-manifold $X$ there exists a locally compact polyhedron $K$ such that $X \approx K \times I^\omega$.*

THEOREM 2.3.29. *For each $R^\omega$-manifold $X$ there exists a locally compact polyhedron $K$ such that $X \approx K \times R^\omega$.*

Applying Theorems 2.3.21 and 2.3.28, we obtain the following result (compare with Theorem 2.1.20).

COROLLARY 2.3.30. *Each metrizable $ANR$-compactum has the homotopy type of a compact polyhedron.*

Finally we mention relative versions of Theorems 2.3.28 and 2.3.29.

THEOREM 2.3.31. *Let the $I^\omega$ -manifold $X_0$ be a $Z$-set in the $I^\omega$ -manifold $X$. Suppose that $h_0\colon X_0 \to K_0 \times I^\omega$ is a triangulation of $X_0$ (i.e. $h_0$ is a homeomorphism and $K_0$ is a locally compact polyhedron). Then there exists a triangulation $h\colon X \to K \times I^\omega$ such that $K_0$ is a subpolyhedron of $K$ and $h$ extends $h_0$.*

THEOREM 2.3.32. *Let the $R^\omega$ -manifold $X_0$ be a $Z$-set in the $R^\omega$ -manifold $X$. Suppose that $h_0\colon X_0 \to K_0 \times R^\omega$ is a triangulation of $X_0$ (i.e. $h_0$ is a homeomorphism and $K_0$ is a locally compact polyhedron). Then there exists a triangulation $h\colon X \to K \times R^\omega$ such that $K_0$ is a subpolyhedron of $K$ and $h$ extends $h_0$.*

*Historical and bibliographical notes* 2.3. The Stability Theorems 2.3.9 and 2.3.10 for $R^\omega$ - and $I^\omega$ -manifolds were obtained by Anderson and Schori [**25**]. Lemma 2.3.4 was proved by West in [**311**]. The strong $\mathcal{B}_{\omega,\infty}$-universality of $I^\omega$ -manifolds was observed in [**298**]. For locally compact $ANR$-spaces, the property of strong $\mathcal{B}_{\omega,\infty}$-universality is equivalent to the existence of the so-called *Disjoint n-Disks Property* ($DD^nP$) for every $n \in \omega$. The last property can be obtained from Definition 2.3.12 by replacing in it the space $Y$ by the disjoint sum $\{0,1\} \times I^n$ of two $n$-dimensional cubes ($n$-disks). The Disjoint $n$-Disks Property plays a very important role in modern geometric topology. This property was formally introduced by Cannon [**60**], [**61**], although it was used earlier in [**37**]. The strong $\mathcal{A}_{\omega,\infty}$-universality of $R^\omega$-manifolds was observed by Toruńczyk [**299**]. Theorems 2.3.17 and 2.3.20 were proved in [**24**] and Theorems 2.3.18 and 2.3.19 in [**23**] (see also [**69**]). Theorems 2.3.21 and 2.3.22 are due to Edwards [**139**] and Toruńczyk [**296**] respectively. Theorems 2.3.24, 2.3.27 and 2.3.29 were proved by Henderson [**165**], [**166**]. Theorems 2.3.25, 2.3.26 and 2.3.28 are due to Chapman [**65**], [**66**]. The complete classification of $I^\omega$-manifolds in terms of (infinite) simple homotopy types (see Whitehead [**317**]) was obtained by Chapman [**68**] (see also [**283**]). Corollary 2.3.30 was obtained by West [**314**]. It solved the long standing problem of Borsuk concerning the finiteness of homotopy types of $ANR$-compacta.

## 2.4. Topology of $R^\omega$- and $I^\omega$-manifolds

**2.4.1. Topology of $R^\omega$-manifolds.** In the previous section we have seen that each $R^\omega$-manifold is strongly $\mathcal{A}_{\omega,\infty}$-universal. Our goal in this Subsection is to outline the proof of the converse statement, which in fact gives us a topological characterization of $R^\omega$-manifolds.

THEOREM 2.4.1. *Let $X$ be a Polish $ANR$-space. Then the following conditions are equivalent:*

(i) *$X$ is homeomorphic to $R^\omega$ (respectively, $X$ is an $R^\omega$-manifold).*

(ii) *$X$ is strongly $\mathcal{A}_{\omega,\infty}$-universal.*

The strategy of the proof of this fundamental result consists of two major steps. The first proves the existence of a "good" resolution for a given Polish $ANR$-space $X$. This means that there exists a fine homotopy equivalence $f: M \to X$, where $M$ is an $R^\omega$-manifold. Of course, formally, such a resolution can be obtained by applying Theorem 2.3.22. Indeed, by this theorem, the product $X \times R^\omega$ is a $R^\omega$-manifold and the projection $\pi_X : X \times R^\omega \to X$, being a soft map is, by Proposition 2.1.27, a fine homotopy equivalence. Since the proof of Theorem 2.3.22 was not presented, we refer the reader to [**231**] and [**139**] (see also discussion in [**34**]). The second step of this strategy proves the following fact: if $M \to X$ is a fine homotopy equivalence, $M$ is an $R^\omega$-manifold and $X$ is a strongly $\mathcal{A}_{\omega,n}$-universal Polish $ANR$-space, then $f$ is a near-homeomorphism. Obviously these two steps complete the proof of Theorem 2.4.1.

First let us introduce some additional definitions. For a map $f: X \to Y$ between Polish spaces, a point $y \in Y$ is called a *nondegenerate value of $f$* provided, for a complete metric $d$ on $X$, there is an $\epsilon > 0$ such that the $d$-diameter of $f^{-1}(U)$ is greater than $\epsilon$ for each neighborhood $U$ of $y$. It is easily seen that different choices of complete metrics for $X$ yield the same non degenerate values. Indeed, if $f(X)$ is dense in $Y$, a purely topological (in other words metric independent) description of a nondegenerate value $y \in Y$ is that either

- $f^{-1}(y) = \emptyset$,
- $f^{-1}$ contains at least two points,

or

- $f^{-1}(y) = \{x\}$ but $f^{-1}(\mathcal{B})$ is not a neighborhood basis for $x$, where $\mathcal{B}$ is a neighborhood basis for $y$.

The set of all non degenerate values of $f$ is denoted by $N_f$. A straightforward verification shows that $N_f$ is an $F_\sigma$-subset of $Y$. Observe also that the restriction of $f$ is a homeomorphism from $f^{-1}(Y - N_f)$ to $Y - N_f$. Of course, if $f$ is a proper map, then $N_f = \{y \in Y : |f^{-1}(y)| > 1\}$. In general, fine homotopy equivalences are not surjective, though they have dense images, and points not in the image are necessarily non degenerate values. We begin with the following lemma.

LEMMA 2.4.2. *Let $X$ be a strongly $\mathcal{A}_{\omega,\infty}$-universal Polish $ANR$-space. Then the space $C(I^\omega, X)$ contains a dense set consisting of $Z$-embeddings with pairwise disjoint images.*

PROOF. Let $d$ be a complete metric on $X$, and let $\mathcal{U}_k \in cov(X)$ denote an open cover of $X$ whose elements have diameter less than $\frac{1}{k}$, $k \in N$. Let $\{\alpha_n : n \in N\}$ be a countable dense subset of the space $C(I^\omega, X)$. Consider the map $\alpha\colon N \times I^\omega \to X$ defined by letting $\alpha(n,a) = \alpha_n(a)$ for each $(n,a) \in N \times I^\omega$. By Lemma 2.3.16, each compact subset of $X$ is a strong $Z$-set in $X$. Therefore there exists a map $\beta_1\colon N \times I^\omega \to X$ such that $beta_1$ is $\mathcal{U}_1$-close to $\alpha$ and $\beta_1(N \times I^\omega) \cap \alpha(N \times I^\omega) = \emptyset$. If maps $\beta_i\colon N \times I^\omega \to X$, each $\mathcal{U}_i$-close to $\alpha$ and satisfying

$$\beta_i(N \times I^\omega) \cap (\alpha(N \times I^\omega) \cup \cup_{j<i}\beta_j(N \times I^\omega)) = \emptyset,$$

have already been constructed for each $i$ with $i \le k$, then the strong $\mathcal{A}_{\omega,\infty}$-universality of $X$ implies the existence of a $Z$-embedding $\beta_{k+1}\colon N \times I^\omega \to X$, $\mathcal{U}_{k+1}$-close to $\alpha$ such that

$$\beta_{k+1}(N \times I^\omega) \cap (\alpha(N \times I^\omega) \cup \cup_{j\le k}\beta_j(N \times I^\omega)) = \emptyset.$$

It only remains to note that $\{\beta_k^n : n, k \in N\}$, where $\beta_k^N(a) = \beta_k(n,a)$, is the desired countable dense subset of $C(I^\omega, X)$ consisting of $Z$-embeddings. □

PROPOSITION 2.4.3. *If a Polish $ANR$-space $X$ is expressed as the union $X = M \cup A$ of an $R^\omega$-manifold $M$ and a strong $Z$-set $A$, then the inclusion $i\colon M \hookrightarrow X$ is a near-homeomorphism.*

PROOF. Let $\mathcal{U} \in cov(X)$ be an open cover of $X$ and $\mathcal{V} \in cov(M)$ an open cover of $M$. Since $A$ is a strong $Z$-set in $X$, there is a map $g\colon X \to M$ that is $\mathcal{U}$-homotopic to $id_X$. Take open covers $\mathcal{U}_0 \in cov(X)$ and $\mathcal{V}_0 \in cov(M)$ such that $\mathcal{U}_0$ refines both $\mathcal{U}$ and $g^{-1}(\mathcal{V})$ and $\mathcal{V}_0$ refines both $\mathcal{V}$ and $\mathcal{U}_0/M = \{U \cap M : U \in \mathcal{U}_0\}$. Also consider a star-refinement $\mathcal{V}_1 \in cov(M)$ of $\mathcal{V}_0$ and a homeomorphism $H\colon M \times I \to M$ that is $\mathcal{V}_1$-close to the projection $\pi\colon M \times I \to M$ (see Corollary C:1.4.9). Note that any map of the form $HhH^{-1}\colon M \to M$, where $h\colon M \times I \to M \times I$ preserves the first coordinate, is $\mathcal{V}_0$-close to $id_M$.

In order to prove our statement we need, by Theorem 2.1.8, a shrinking homeomorphism $h\colon M \to M$. We construct such a homeomorphism as a composition $h = h_2h_3h_1^{-1}$ of homeomorphisms of $M$.

*Construction of $h_1$.* Let $\mathcal{U}_1 \in cov(X)$ be an open cover of $X$ which star refines $\mathcal{U}_0$. Take an $\mathcal{U}_1$-homotopy $\alpha\colon X \times I \to X$ such that $\mathrm{cl}_X(\alpha_1(X)) \cap A = \emptyset$. Choose a $Z$-embedding $e\colon H(M \times \{0\}) \to M$ such that $e$ is $\mathcal{U}_1/M$-homotopic to $\alpha_1/H(M \times \{0\})$ and such that the image of $e$ is disjoint from $A$ (observe that this image is closed in $M$). Since $e$ is $\mathcal{U}_0/M$-homotopic to the inclusion $H(M) \times \{0\}) \hookrightarrow M$, an application of Theorem 2.3.20 gives a homeomorphism $h_1\colon M \to M$ extending $e$ that is $\mathcal{U}_0/M$-homotopic to $id_M$.

*Construction of* $h_2$. Choose a $Z$-embedding $e_0: H(M \times \{1\}) \to M$ which is $\mathcal{V}_0$-close to the restriction $gih_1/H(M \times \{1\})$, and extend it to a homeomorphism $h_2: M \to M$. Since the composition $gih_1$ is $st(\mathcal{U}_0/M, \mathcal{U}/M)$-homotopic to the identity $id_M$, we may assume that $h_2$ is $st^2(\mathcal{U}/M)$-closed to $id_M$.

Observe that for each $a \in A$, if $W$ is a neighborhood of $a$ in $X$ that is contained in an element of $\mathcal{U}_0$ and $\pi: M \times I \to M \times \{1\}$ is the projection, then $h_2 H\pi H^{-1} h_1^{-1}(W - A)$ is contained in an element of $st^4(\mathcal{V})$.

In order to verify this, notice that, as $h_2$ is $\mathcal{V}_0$-close to $gih_1$ on $H(M \times \{1\})$, it suffices to show that $gih_1 H\pi H^{-1} h_1^{-1}(W - A)$ is contained in an element of $st^3(\mathcal{V})$. Since $\mathcal{U}_0$ refines $g^{-1}(\mathcal{V})$, it suffices to show that $ih_1 H\pi H^{-1} h_1^{-1}(W - A)$is contained in an element of $st^3(\mathcal{U}_0)$. The last containment is clear since $W - A$ is contained in an element of $\mathcal{U}_0$, $h_1$ is $\mathcal{U}_0/M$-close to $id_M$, $\mathcal{V}_0$ refines $\mathcal{U}_0/M$ and $H\pi H^{-1}$ is $\mathcal{V}_0$-close to $id_M$.

*Construction of* $h_3$. Since $A$ and $h_1(H(M \times \{0\}))$ are disjoint closed subsets of $X$, $A$ has a closed neighborhood $N$ disjoint from $h_1(H(M \times \{0\}))$. Consequently, $H^{-1}h_1^{-1}(N - A)$ is a closed subset of $M \times I$ and is contained in $M \times (0, 1]$. Let $h_0: M \times I \to M \times I$ be a homeomorphism that preserves the first coordinates and "pushes" $H^{-1}h_1^{-1}(N - A)$ near $M \times \{1\}$. For an appropriately chosen $h_0$, the homeomorphism $h_3 = Hh_0H^{-1}$ is "close enough" to $H\pi H^{-1}$ on neighborhoods $h_1^{-1}(W - A)$ (where $W \subseteq N$) as in the above observation that $h_2 h_3 h_1^{-1}(W - A)$ is contained in an element of $st^4(\mathcal{V})$. This finishes the verification of the conditions of Theorem 2.1.8 and completes the proof of the proposition. □

PROPOSITION 2.4.4. *If* $f: M \to X$ *is a fine homotopy equivalence from the* $R^\omega$*-manifold into the Polish* $ANR$*-space* $X$, *and* $\mathrm{cl}(N_f)$ *is a strong* $Z$*-set in* $X$, *then* $f$ *is a near-homeomorphism.*

PROOF. The idea of the proof is to approximate $f$ by a fine homotopy equivalence $g: M \to X$ such that$N_g$ is contained in $\mathrm{cl}(N_f)$ and $g^{-1}(\mathrm{cl}(N_g))$ is a (possibly empty) $Z$-set. Since $M$ is an $R^\omega$-manifold, we conclude, by Proposition 2.3.14 and Lemma 2.3.15, that $g^{-1}(\mathrm{cl}(N_g))$ is a strong $Z$-set in $M$. Applying Proposition 2.4.3, we see that both inclusions $i: g^{-1}(X - \mathrm{cl}(N_g)) \hookrightarrow M$ and $j: (X - \mathrm{cl}(N_g)) \hookrightarrow X$ are near-homeomorphisms. Therefore, $g$ is approximable by a homeomorphism of the form $h_2 g_0 h_1^{-1}$, where $h_1$ approximates $i$, $h_2$ approximates $j$ and $g_0 = g/g^{-1}(X - \mathrm{cl}(N_g))$. The fine homotopy equivalence $g$ approximating $f$ is the limit of a sequence $f_0, f_1, f_2, \ldots$ of fine homotopy equivalences constructed inductively. First, applying Lemma 2.4.2, we take a countable dense subset $\{e_n: n \in N\} \subseteq C(I^\omega, M)$ consisting of $Z$-embeddings with pairwise disjoint images. Let $f_0 = f$. The map $f_0 e_1: I^\omega \to X$ is homotopic, by a "small" homotopy, to a $Z$-embedding $j_1: I^\omega \to X - \mathrm{cl}(N_{f_0})$. Using an approximate lift of the homotopy as a guide, Theorem 2.3.20 produces a homeomorphism $h_1: M \to M$, fixed outside a small neighborhood of $f_0^{-1} f_0(e_1(I^\omega))$, such that $h_1 e_1 = f_0^{-1} j_1$ and $f_0 h_1$ is "sufficiently close" to $f_0$. Let $f_1 = f_0 h_1$. Observe that $f_1$ is one-to-one over the set $e_1(I^\omega)$. This means that $f_1^{-1} f_1(m) = m$ for each

$m \in e_1(I^\omega)$. Continuing this process we obtain a sequence $f_0, f_1, f_2, \ldots$ such that $f_{i+1} = f_i h_{i+1}$, where $h_{i+1}\colon M \to M$ is a homeomorphism fixed outside a small neighborhood of $f_i^{-1} f_i(e_{i+1}(I^\omega))$ missing $\cup_{k=1}^{i} e_k(I^\omega)$. Further, letting $N_{i+1}$ denote the non degeneracy set of $f_{i+1}$, we want $h_{i+1}e_{i+1} = f_i^{-1} j_{i+1}$, where $j_{i+1}\colon I^\omega \to X - \mathrm{cl}(N_i)$ is a $Z$-embedding approximating $f_i e_{i+1}$, and we also want $f_i h_{i+1}$ to be "close" to $f_i$. Note that

$$\cup_{k=1}^{i+1} f_{i+1}(e_k(I^\omega)) \subseteq X - \mathrm{cl}(N_{i+1}).$$

Since $N_i = N_f$, we can verify directly that if $f_{i+1}$ was chosen sufficiently close to $f_i$, then the limit map $g = \lim\{f_i\}$ is a fine homotopy equivalence approximating $f$ and satisfying the following two conditions

$$N_g \subseteq \mathrm{cl}(N_f) \text{ and } g(e_i(I^\omega)) \subseteq X - \mathrm{cl}(N_f) \text{ for each } i.$$

It only remains to remark that $g^{-1}(\mathrm{cl}(N_g))$ is a $Z$-set in $M$. □

PROPOSITION 2.4.5. *Let $f\colon M \to X$ be a fine homotopy equivalence from an $R^\omega$ -manifold to a strongly $\mathcal{A}_{\omega,\infty}$-universal Polish ANR-space. If $N_f$ and $f^{-1}(N_f)$ are $Z_\sigma$-sets in $X$ and $M$ respectively, then $f$ is a near-homeomorphism.*

PROOF. Let $\mathcal{U} \in cov(X)$ be an open cover of $X$ and $\mathcal{V} \in cov(M)$ an open cover of $M$. Take a map $g\colon X \to M$ such that $gf$ is $f^{-1}(\mathcal{U})$-homotopic to $id_M$. Take open covers $\mathcal{U}_0 \in cov(X)$ and $\mathcal{V}_0 \in cov(M)$ such that $\mathcal{U}_0$ refines both $\mathcal{U}$ and $g^{-1}(\mathcal{V})$ and $\mathcal{V}_0$ refines both $\mathcal{V}$ and $f^{-1}(\mathcal{U}_0)$. Take a star-refinement $\mathcal{V}_1 \in cov(M)$ of $\mathcal{V}_0$ and a homeomorphism $H\colon M \times I \to M$ that is $\mathcal{V}_1$-close to the projection $\pi\colon M \times I \to M$ (see Corollary C:1.4.9). Note that any map of the form $HhH^{-1}\colon M \to M$, where $h\colon M \times I \to M \times I$ preserves the first coordinate, is $\mathcal{V}_0$-close to $id_M$.

As in the proof of Proposition 2.4.3, the shrinking homeomorphism will be constructed as the composition $h = h_2 h_3 h_1^{-1}$ of homeomorphisms of $M$.

With minor changes homeomorphisms $h_1$ and $h_2$ can be constructed as the corresponding homeomorphisms in the proof of Proposition 2.4.3.

*Construction of $h_3$.* This homeomorphism is constructed as the composition $h_3 = Hh_0H^{-1}$, where $h_0$ preserves the first coordinates. Since $M \times \{0\}$ misses the inverse image under $(h_1H)^{-1}$ of an $f$-saturated neighborhood[2] of $f^{-1}(N_f)$, for any given point $a \in N_f$, we could choose a neighborhood $W_a$ of $a$ contained in an element of $\mathcal{U}_0$ with $\mathrm{cl}(f^{-1}(W_a)) \cap h_1H(M \times \{0\}) = \emptyset$ and specify that $h_0$ "push" $(h_1H)^{-1}(f^{-1}(W_a))$ so near $M \times \{1\}$ that $h_2Hh_0H^{-1}h_1^{-1}$ "shrinks" $f^{-1}(W_a)$. The problem is that, presumably, any inverse image under $(h_1H)^{-1}$ of an $f$-saturated neighborhood of $f^{-1}(N_f)$ is dense, thereby making it impossible to simultaneously do this for every point of $N_f$. This situation can be rectified by observing that for any point $a \in N_f$ sufficiently close to $fh_1H(M \times \{0\})$, the inverse image $f^{-1}(a)$ has an $f$-saturated neighborhood having diameter close to

[2]By an $f$-saturated neighborhood of a set $f^{-1}(A)$, for $A \subseteq X$, we mean a neighborhood of the form $f^{-1}(N)$ for some open neighborhood $N$ of $A$.

zero. We now proceed with the construction of $h_0$. As above, $h_0$ is the identity on $M \times \{0\}$ and moves only the first coordinates.

For a pair of maps $\epsilon, \delta\colon M \to [0,1]$ with $\epsilon(m) \le \delta(m)$ for each $m \in M$, let

$$\Gamma(\epsilon, \delta) = \{(m,t)\colon \epsilon(m) \le t \le \delta(m)\}$$

and let $p_{\epsilon,\delta}\colon M \times I \to \Gamma(\epsilon,\delta)$ be the retraction sending $\{m\} \times [0, \epsilon(m)]$ to $(m, \epsilon(m))$ and $\{m\} \times [\delta(m), 1]$ to $(m, \delta(m))$.

Take a map $\delta\colon M \to (0,1)$ such that

(i) each $a \in N_f$ has a neighborhood $W_a$ such that

$$h_2 H \pi_{1-\delta,1} H^{-1} h_1^{-1}(f^{-1}(W_a))$$

is $st^4(\mathcal{V})$-small, and

(ii) for each $(m,t) \in M \times I$, $h_2 H(B_{2\delta(m)}(m) \times B_{2\delta(m)}(t))$ is $st^4(\mathcal{V})$-small.

Next we inductively specify maps $\epsilon(i)\colon M \to (0,1]$, $i = 0,1,2,\ldots$, with $\epsilon(0)(m) = 1$ and $\epsilon(i)(m) > \epsilon(i+1)(m)$ for each $m \in M$ so that every point $a \in N_f$ has a neighborhood $W_a$ such that either

$(iii)_1$

$$H^{-1}h_1^{-1}(f^{-1}(W_a)) \subseteq \Gamma_{\epsilon(i+2),\epsilon(i)}$$

from some $i \ge 0$ and $\pi H^{-1} h_1^{-1}(f^{-1}(W_a))$ is contained in $B_{\delta(m)}$ for each $m \in M$, or

$(iii)_2$ $H^{-1}h_1^{-1}(f^{-1}(W_a)) \subseteq \Gamma_{\epsilon(1),\epsilon(0)}$.

*Choosing* $\epsilon(1)$. Consider the set $Z_0$ consisting of those points $a \in X$ that do not have a neighborhood $W_a$ such that $\pi H^{-1} h_1^{-1}(f^{-1}(W_a))$ is contained in $B_{\delta(m)}(m)$ for some $m \in M$. This set is closed in $X$ and is contained in $N_f$. Thus, there is a closed $f$-saturated neighborhood $N_0$ of $f^{-1}(Z_0)$ disjoint from $h_1 H(M \times \{0\})$. Choose $\epsilon(1)$ so that $\Gamma(0, \epsilon_1)$ is disjoint from $H^{-1}h_1^{-1}(\operatorname{int} N_0)$[3]. Observe that points in $Z_0$ have neighborhoods satisfying $(iii)_2$.

*Choosing* $\epsilon(i+1)$. Consider the set $Z_i$ consisting of those points $a \in X$ that do not have a neighborhood $W_a$ such that $H^{-1}h_1^{-1}(f^{-1}(W_a)) \subseteq \Gamma(0, \epsilon(i))$. This set is closed in $X$, and $f^{-1}(Z_i)$ is disjoint from $h_1 H(M \times \{0\})$. Choose any neighborhood $N$ of $h_1 H(M \times \{0\})$ missing an $f$-saturated neighborhood of $f^{-1}(Z_i)$, and any $\epsilon(i+1)$ with $\epsilon(i+1)(m) < \epsilon(i)(m)$ for each $m \in M$ and with $\Gamma(0, \epsilon(i+1)) \subseteq H^{-1}h_1^{-1}(\operatorname{int} N)$.

Making the further restriction that $\epsilon(i)(m) < \frac{1}{i}$, $i = 1,2,\ldots$, we specify a homeomorphism $h_0\colon M \times I \to M \times I$ that is the identity on $M \times \{0,1\}$ and sends the graph of $\epsilon(i)$ onto the graph of $(1-\delta)^i$, $i = 1,2,\ldots$[4].

Setting $h_3 = h_0 H^{-1}$ and $h = h_2 h_3 h_1^{-1}$, we claim that $h$ is the desired shrinking homeomorphism. For, if $a \in N_f$ and $(iii)_1$ is satisfied, then

$$h_0 H^{-1} h_1^{-1}(f^{-1}(W_a)) \subseteq \Gamma((1-\delta)^{i+2}, (1-\delta)^i)$$

---

[3] here "0" denotes the constant map $M \to \{0\}$.

[4] $(1-\delta)^i(m) = (1-\delta(m))^i$.

and so $h_0H^{-1}h_1^{-1}(f^{-1}(W_a))$ is contained in $B_{2\delta(m)}(m) \times B_{2\delta(m)}(t)$ for some $(m.t) \in M \times I$ and condition (ii) implies that $h(f^{-1}(W_a))$ is $st^4(\mathcal{V})$-small. If $a \in N_f$ and $(iii)_2$ is satisfied, then

$$h_0H^{-1}h_1^{-1}(f^{-1}(W_a)) \subseteq \Gamma(1-\delta, 1)$$

and condition (i) implies that $h(f^{-1}(W_a))$ is $st^4(\mathcal{V})$-small. □

PROPOSITION 2.4.6. *If $f\colon M \to X$ is a fine homotopy equivalence from an $R^\omega$-manifold to a strongly $\mathcal{A}_{\omega,\infty}$-universal Polish $ANR$-space and $N_f$ is a $Z_\sigma$-set in $X$, then $f$ is a near-homeomorphism.*

PROOF. The proof parallels closely that of Proposition 2.4.4. An approximating fine homotopy equivalence $g$ is produced using Proposition 2.4.5 in place of Proposition 2.4.3. □

PROPOSITION 2.4.7. *A fine homotopy equivalence $f\colon M \to X$ from an $R^\omega$-manifold to a strongly $\mathcal{A}_{\omega,\infty}$-universal Polish $ANR$-space is a near-homeomorphism.*

PROOF. By lemma 2.4.2, there exists a countable dense subset $\{e_n : n \in N\} \subseteq C(I^\omega, X)$, consisting of $Z$-embeddings with pairwise disjoint images. The map $f\colon M \to X$ factors through the adjunction space as $f = p_1q$, where the maps

$$q\colon M \to M \cup_f e_1(I^\omega) \text{ and } p_1\colon M \cup_f e_1(I^\omega) \to X$$

are described in Subsection 2.3.1 on page 61. Obviously, $N_q \subseteq e_1(I^\omega)$ and $e_1(I^\omega)$ is a strong $Z$-set in $M \cup_f e_1(I^\omega)$. By Proposition 2.4.4, the map $q$, being clearly a fine homotopy equivalence, is a near-homeomorphism. Take a homeomorphism $h$ approximating $q$ and let $f_1 = p_1h\colon M \to X$. Observe that $f_1$ approximates $f$ and is one-to-one over the set $e_1(I^\omega)$. Continuing this process, we obtain a sequence $f = f_0, f_1, f_2, \ldots$ consisting of fine homotopy equivalences (from $M$ to $X$) such that $f_{n+1}$approximates $f_n$ and is one-to one over the set $\cup_{k=1}^{n+1} e_k(I^\omega)$. If $f_{n+1}$ is sufficiently close to $f_n$ (for each $n \in N$), then the map $g = \lim f_n$ is also a fine homotopy equivalence approximating $f$. It is not hard to see that $g$ is one-to -one over the set $\cup_{n=1}^{\infty} e_n(I^\omega)$. This fact, coupled with the choice of the collection $\{e_n\}$, guarantees that $N_g$ is a $Z_\sigma$-set in $X$. It only remains to apply Proposition 2.4.6. □

We are now ready to prove Theorem 2.4.1.

*Proof of Theorem* 2.4.1. One part of the statement follows from Proposition 2.3.14. Consider now a strongly $\mathcal{A}_{\omega,\infty}$-universal Polish $ANR$-space $X$. By Theorem 2.3.22, there exists a fine homotopy equivalence $f\colon M \to X$ where $M$ is an $R^\omega$-manifold. Proposition 2.4.7 implies that $f$ is a near-homeomorphism and, consequently, that $X$ is an $R^\omega$-manifold. If, in addition, $X$ is an absolute retract, then $M$ is contractible (recall that $f$ is a fine homotopy equivalence). It only remains to note that in this case, according to Theorem 2.3.27, $M$ (as well as $X$) is homeomorphic to $R^\omega$. This completes the proof of Theorem 2.4.1.

Sometimes it is much more convenient to use the following version of Theorem 2.4.1.

PROPOSITION 2.4.8. *Let $X$ be a Polish $A(N)R$-space. Then the following conditions are equivalent:*

(i) *$X$ is homeomorphic to $R^\omega$ (respectively, $X$ is an $R^\omega$-manifold).*

(ii) *The set*

$$\{f \in C(N \times I^\omega, X)\colon \text{the collection } \{f(\{n\} \times I^\omega)\colon n \in N\} \text{ is discrete}\}$$

*is dense in the space $C(N \times I^\omega, X)$.*

(iii) *The set*

$$\{f \in C(\oplus\{I^n\colon n \in \omega\}, X)\colon \text{the collection } \{f(I^n)\colon n \in \omega\} \text{ is discrete}\}$$

*is dense in the space $C(\oplus\{I^n\colon n \in \omega\}, X)$.*

In order to prove this statement we need some preliminary results. The following lemma is geometrically obvious.

LEMMA 2.4.9. *Let $K$ be an $n$-dimensional countable locally finite simplicial complex and let $G$ be an open neighborhood of $|K^{(n-1)}|$ in $|K|$. Then the convex halls $C_\sigma$ of the sets $|\sigma| - G$, $\sigma \in K$, forms a discrete collection in $|K|$.*

PROOF. Assuming the contrary, there is a point $x \in \operatorname{cl} C - C$, where $C = \cup\{C_\sigma\colon \sigma \in K\}$. Then $x \in |\tau|$ for some $\tau \in K^{(n-1)}$. By an obvious geometrical argument, $\operatorname{dist}(|\sigma| - G, |\tau|) = \operatorname{dist}(C_\sigma, |\tau|)$ for all $\sigma \in st(\tau, K)$. Since $|\tau| \subseteq G$, we have $\operatorname{dist}(C \cap |st(\tau, K), |\tau|) > 0$, contrary to the fact that $x \in \operatorname{cl} C \cap |\tau|$. □

LEMMA 2.4.10. *Let $X$ be a Polish $ANR$-space satisfying condition (iii) of Proposition 2.4.8, $K$ be a finite-dimensional countable locally finite simplicial complex, and $\mathcal{U} \in cov(|K|)$ be the cover of $|K|$ by open stars of the vertices of $K$. Then the set of all $\mathcal{U}$-maps of the polyhedron $|K|$ into the product $X \times I^\omega$ is open and dense in the space $C(|K|, X \times I^\omega)$.*

PROOF. That the set of all $\mathcal{U}$-maps from $|K|$ to $X \times I^\omega$ is open follows from Lemma 2.1.1. Let us show that this set is dense. If $\dim K = 0$, then this follows immediately from condition (iii) of Proposition 2.4.8. Suppose that $\dim K = n$ and that the lemma has already been proved for complexes of dimension $\leq n - 1$. Suppose also that a map $f\colon |K| \to X \times I^\omega$ and an open cover $\mathcal{V} \in cov(X \times I^\omega)$ are given. Our goal is to show the existence of a $\mathcal{U}$-map $g\colon |K| \to X \times I^\omega$ that is $\mathcal{V}$-close to $f$.

By our inductive assumption and Corollary 2.1.23, we may assume without loss of generality that the restriction $f/|K^{(n-1)}|$ is an $\mathcal{U}$-map. By Lemma 2.1.3, there is an open neighborhood $G$ of $|K^{(n-1)}|$ in $|K|$ such that the restriction $f/G$ is still a $\mathcal{U}$-map. Denote by $\mathcal{W} \in cov(X \times I^\omega)$ a double star-refinement of $\mathcal{V}$ such that for each map $g\colon |K| \to X \times I^\omega$ that is $\mathcal{W}$-close to $f$, the restriction $g/G$ is an $\mathcal{U}$-map (here we use Lemma 2.1.1 and Corollary 2.1.23). By Lemma 2.4.9, the complement $|K|-G$ can be covered by a discrete (in $|K|$) collection $\{C_\sigma : \sigma \in K\}$ such that $C_\sigma \approx I^n$ and $C_\sigma \subseteq \operatorname{int}|\sigma|$ for each $\sigma \in K$. By condition (iii) of Proposition 2.4.8 and Corollary 2.1.23, there exists a map $g_0 : |K| \to X \times I^\omega$ such that $g_0$ is $\mathcal{W}$-close to $f$ and the collection $\{g_0(C_\sigma : \sigma \in K\}$ is discrete in $X \times I^\omega$. Let $\{D_\sigma : \sigma \in K\}$ be a discrete collection of open subsets of $|K|$ such that $C_\sigma \subseteq D_\sigma \subseteq \operatorname{int}|\sigma|$ for each $\sigma \in K$. Take a function $\varphi : |K| \to [0,1]$ such that

$$\varphi(\cup\{C_\sigma : \sigma \in K\}) = 1 \text{ and } \varphi(|K| - \cup\{D_\sigma : \sigma \in K\}) = 0.$$

A straightforward verification shows that the diagonal product

$$g' = g_0 \triangle \varphi : |K| \to X \times I^\omega \times [0,1]$$

is a $\mathcal{U}$-map. Since the projection $\pi : I^\omega \times [0,1] \to I^\omega$ is a near-homeomorphism (Corollary 2.3.11), there exists a homeomorphism

$$h : X \times I^\omega \times [0,1] \to X \times I^\omega$$

that is $\mathcal{V}$-close to $id \times \pi$. Then the desired map $g$ can be defined as the composition $g = hg'$. □

LEMMA 2.4.11. *For each $n \in \omega$ and for each $n$-dimensional Polish space $X$, there exists an $(n+1)$-dimensional Polish absolute retract containing $X$ as a closed subspace.*

PROOF. Represent $X$ as the limit space of some inverse sequence $\mathcal{S}_X = \{X_i, p_i^{i+1}\}$ consisting of $n$-dimensional locally compact polyhedra and surjective limit projections (see [**179**]). We construct a new inverse sequence $\mathcal{S}_Y = \{Y_i, q_i^{i+1}\}$ as follows. As the first element $Y_1$ of this spectrum we take $con(X_1)$ the cone over the polyhedron $X_1$. Obviously, $Y_1$ is an $(n+1)$-dimensional Polish absolute retract. We identify $X_1$ with its naturally embedded copy in $Y_1$. Suppose that for each $i$, with $i \leq k$, an $(n+1)$-dimensional Polish absolute retract $Y_i$, containing $X_i$ as a closed subspace, and a fine homotopy equivalence $q_{i-1}^i : Y_i \to Y_{i-1}$, extending $p_{i-1}^i$, have already been constructed. As the space $Y_{k+1}$ we take the mapping cylinder $Map(p_k^{k+1})$ of the map $p_k^{k+1}$, considered as a map from $X_{k+1}$ into $Y_k$ (recall that $p_k^{k+1}(X_{k+1}) = X_k$ is closed in $Y_k$). Obviously, $Y_{k+1}$ is an $(n+1)$-dimensional Polish absolute retract (see [**161**] for details). The closed embedding of the polyhedron $X_{k+1}$ into $Y_{k+1}$ and the fine homotopy equivalence $q_k^{k+1} : Y_{k+1} \to Y_k$ are defined canonically. It is also clear that $q_k^{k+1}(X_{k+1}) = X_k$ and $q_k^{k+1}/X_{k+1} = p_k^{k+1}$. Let $Y$ denote the limit space of the inverse sequence $\mathcal{S}_Y$. It follows from the construction that $Y$ is an $(n+1)$-dimensional Polish space containing $X$ as a closed subspace. Since all projections of this spectrum are fine homotopy equivalences and since $Y_1$ is a Polish absolute retract we easily conclude that $Y$ is also an absolute retract (see [**112**] for details). □

LEMMA 2.4.12. *Let $X$ be a Polish space satisfying condition (iii) of Proposition 2.4.8. Then for each finite-dimensional Polish space $Y$, the set of closed embeddings of $Y$ into $X \times I^\omega$ is dense in the space $C(Y, X \times I^\omega)$.*

PROOF. By Lemmas 2.1.1, 2.1.2 and Proposition 2.1.7, it suffices to show that for each open cover $\mathcal{U} \in cov(Y)$, the set of all $\mathcal{U}$-maps is dense in the space $C(Y, X \times I^\omega)$. By Lemma 2.4.11 we may additionally assume that $Y$ is a finite-dimensional Polish absolute retract. But then, by Theorem 2.1.19, we may assume, without loss of generality, that $Y = |K|$, where $K$ is a finite-dimensional countable locally finite simplicial complex and $\mathcal{U}$ is the cover of the polyhedron $|K|$ by open stars of its vertices. In this situation Lemma 2.4.10 completes the proof. □

LEMMA 2.4.13. *Let $X$ be a Polish $ANR$-space satisfying condition (iii) of Proposition 2.4.8, $K$ be a countable locally finite simplicial complex, and $\mathcal{U}$ be is the cover of the polyhedron $|K|$ by open stars of its vertices. Then the set of maps $f\colon |K| \to X \times I^\omega$ such that the collection of fibers of the composition $\pi_X f$, where $\pi_x\colon X \times I^\omega \to X$ is the projection, refines $\mathcal{U}$ and is a dense and $G_\delta$-subset of the space $C(|K|, X \times I^\omega)$.*

PROOF. Given integers $m$ and $n$, consider for each $\sigma \in K^{(n)} - K^{(n-1)}$ the image of $|\sigma|$ under the $(1 - \frac{1}{m})$-homothery with respect to the barycenter of $|\sigma|$. Obviously, the collection $\mathcal{A}_{n,m}$ of all such images is discrete in $|K|$. By condition (iii) of Proposition 2.4.8 and the compactness of $I^\omega$, the set

$$F_{n,m} = \{f \in C(|K|, X \times I^\omega)\colon \text{ the collection } \pi_X f(\mathcal{A}_{n.m}) \text{ is discrete in } X\}$$

is open and dense in $C(|K|, X \times I^\omega)$. It only remains to remark that the intersection $\cap F_{n,m}$ is the desired set. □

LEMMA 2.4.14. *Let $X$ be a Polish $ANR$-space satisfying condition (iii) of Proposition 2.4.8. Then for each Polish space $Y$, the set of closed embeddings of $X$ into the product $X \times I^\omega$ is dense in the space $C(Y, X \times I^\omega)$.*

PROOF. As in the proof of Lemma 2.4.12 it suffices to show that if $K$ is a locally compact polyhedron and $\mathcal{U} \in cov(K)$, then the set of all $\mathcal{U}$-maps is dense in the space $C(Y, X \times I^\omega)$. For simplicity we assume that $X$ is a Polish absolute retract (this simplification has a purely technical nature and does not affect the general argument). Represent $K$ as the union $K = \cup\{K_n\colon n \in \omega\}$ of its finite-dimensional subcompacta in such a way that

(a) $K_n \subseteq \operatorname{int} K_{n+1}, n \in \omega$, and
(b) The collections $\{K_{2n+1} - \operatorname{int} K_{2n}\colon n \in \omega\}$ and $\{K_{2n} - \operatorname{int} K_{2n-1}\colon n \in N\}$ are discrete in $K$.

Next, take an arbitrary map $f\colon K \to X \times I^\omega$ and introduce the notation $A = \cup\{K_{2n+1} - \operatorname{int} K_{2n}\colon n \in \omega\}$ and $B = \cup\{K_{2n} - \operatorname{int} K_{2n-1}\colon n \in N\}$. Obviously, the sets $A$ and $B$ are closed in $K$. By Proposition 2.1.7, Corollary 2.1.23 and Lemma 2.4.12, we may assume without loss of generality that the restrictions

$f/K_n$ are closed embeddings (we use the finite-dimensionality of $K_n$ also). Next, identifying each of the "rings" $K_{2n+1} - \operatorname{int} K_{2n}$, $n \in \omega$, with a subspace of a finite-dimensional cube $I^{n_m}$, we may also assume that the subspace $A$ is identified with a closed subspace of the discrete sum $\oplus\{I^{n_m} : m \in \omega\}$. The restriction $f/A$ can be extended to a map

$$f' \colon \oplus \{I^{n_m} : m \in \omega\} \to X \times I^\omega .$$

By condition (iii) of Proposition 2.4.8, we may assume without loss of generality that the collection

$$\{f'(I^{n_m}) \colon m \in \omega\}$$

is discrete in the product $X \times I^\omega$. Therefore, the collection

$$\{f'(K_{2n+1} - \operatorname{int} K_{2n}) \colon n \in \omega\}$$

is also discrete in $X \times I^\omega$. As above, we can now conclude that the restriction $f/A$ is a closed embedding of $A$ into $X \times I^\omega$. A completely analogous argument shows that the restriction $f/B$ is also a closed embedding of $B$ into $X \times I^\omega$. Since every closed embedding is a $\mathcal{U}$-map, there exists an open neighborhood $G$ of $A$ in $K$ such that the map $f/G$ is still a $\mathcal{U}$-map (here we apply Lemma 2.1.3; observe also that, by Lemma 2.1.1 and Corollary 2.1.23, the restriction $g/G$ of a map $g \colon K \to X \times I^\omega$ that is sufficiently close to $f$ is also a $\mathcal{U}$-map). In addition, we assume that $G \cap B = \emptyset$. Now take a function $\varphi \colon K \to [0,1]$ such that $\varphi(A) = 0$ and $\varphi(K - G) = 1$. A straightforward verification shows that the diagonal product

$$g' = f \triangle \varphi \colon K \to X \times I^\omega \times [0,1]$$

is a $\mathcal{U}$-map. Take a homeomorphism $h \colon I^\omega \times [0,1] \to I^\omega$ sufficiently close to the projection $\pi \colon I^\omega \times [0,1] \to I^\omega$. Then the composition

$$(id_X \times h)g' \colon K \to X \times I^\omega$$

is a $\mathcal{U}$-map, sufficiently close to $f$. □

We are now ready to prove Proposition 2.4.8.

*Proof of Proposition* 2.4.8. The implications (i)$\Longrightarrow$(ii) and (ii)$\Longrightarrow$(iii) are obvious. Let us prove the validity of the implication (iii)$\Longrightarrow$(i). By Theorem 2.4.1, it suffices to show that for each Polish space $Y$, the set of closed embedding of $Y$ into $X$ is dense in the space $C(Y, X)$. As above, in order to show this last fact, it suffices to prove that for each countable locally finite simplicial complex $K$ the set of all $\mathcal{U}$-maps of $|K|$ into $X$, where $\mathcal{U}$ is the cover of $|K|$ by open stars (with respect to the triangulation $K$) of its vertices, is dense in the space $C(|K|, X)$. By Lemmas 2.4.13, 2.4.14 and Proposition 2.1.7, the space $C|K|, X \times I^\omega)$ contains a dense set $F$, consisting of closed embeddings such that for each $f \in F$ the collection of fibers of the composition $\pi_X f$, where $\pi_X \colon X \times I^\omega \to X$ denotes the projection, refines $\mathcal{U}$. Each such composition is a proper map. Therefore, as can be seen in the proof of Corollary 2.1.9, each such composition is a $\mathcal{U}$-map. It only remains to remark that the set of maps $g\colon |K| \to X$, that can be represented as a composition $g = \pi_X f$, $f \in F$ is dense in the space $C(|K|, X)$. The proof of Proposition 2.4.8 is complete.

**2.4.2. Topology of $I^\omega$-manifolds.** The strategy of the proof of the main result of this Subsection (Theorem 2.4.18 and Proposition 2.4.19) is the same as that of Subsection 2.4.1. As a very useful (and sufficiently difficult, but manageable) exercise we recommend that the reader supply the omitted pieces of the corresponding proofs.

PROPOSITION 2.4.15. *Let $f\colon M \to X$ be a proper $CE$-map of an $I^\omega$-manifold onto a strongly $\mathcal{B}_{\omega,\infty}$-universal locally compact $ANR$-space $X$. If the set $\mathrm{cl}\, N_f$ is a $Z$-set in $X$, then $f$ is a near-homeomorphism.*

PROPOSITION 2.4.16. *Let $f\colon M \to X$ be a proper $CE$-map of an $I^\omega$-manifold onto a strongly $\mathcal{B}_{\omega,\infty}$-universal locally compact $ANR$-space $X$. If the set $N_f$ is a $Z_\sigma$-set in $X$, then $f$ is a near-homeomorphism.*

As in Subsection 2.4.1 these two propositions imply the following statement.

PROPOSITION 2.4.17. *A proper $CE$-map of an $I^\omega$-manifold onto a strongly $\mathcal{B}_{\omega,\infty}$-universal locally compact $ANR$-space is a near-homeomorphism.*

As in the case of $R^\omega$-manifolds, the problem of the existence of a resolution for a given locally compact $ANR$-space $X$ (i.e. the existence of a proper $CE$-map $f\colon M \to X$, where $M$ is an $I^\omega$-manifold) arises here as well. The existence of such a resolution follows, for example, from Theorem 2.3.21. Alternatively we can apply Miller's approach (see, for instance, [**231**] or [**139**]). Therefore, Proposition 2.4.17 implies the following characterization theorem.

THEOREM 2.4.18. *The following conditions are equivalent for each (locally) compact $A(N)R$-space $X$:*

(i) *$X$ is homeomorphic to $I^\omega$ (respectively, $X$ is an $I^\omega$-manifold).*
(ii) *$X$ is strongly $\mathcal{B}_{\omega,\infty}$-universal.*

Sometimes it is more convenient to use the following version of Theorem 2.4.18.

PROPOSITION 2.4.19. *The following conditions are equivalent for each locally compact $ANR$-space $X$:*

(i) *$X$ is an $I^\omega$-manifold.*
(ii) *The set*

$$\{f \in C(D \times I^\omega, X)\colon f(\{0\} \times I^\omega) \cap f(\{1\} \times I^\omega) = \emptyset\}^5$$

*is dense in the space $C(D \times I^\omega, X)$.*

[5] Recall that $D = \{0, 1\}$ denotes the two-point discrete space

(iii) *For each* $n \in \omega$, *the set*

$$\{f \in C(D \times I^n, X)\colon f(\{0\} \times I^n) \cap f(\{1\} \times I^n) = \emptyset\}$$

*is dense in the space* $C(D \times I^n, X)$.

PROOF. The implications (i)$\Longrightarrow$(ii) and (ii)$\Longrightarrow$(iii) are obvious. Let us prove the validity of the implication (iii)$\Longrightarrow$(i). According to Theorem 2.4.18 we only have to show the strong $\mathcal{B}_{\omega,\infty}$-universality of $X$.

First of all let us establish the following fact:

- For metrizable compacta $T_1$ and $T_2$ the set

$$\{f \in C(T_1 \oplus T_2, X)\colon f(T_1) \cap f(T_2) = \emptyset\}$$

is dense in the space $C(T_1 \oplus T_2, X)$.

Indeed, by Theorem 2.1.19, we may assume without loss of generality that $T_1 = |K_i|$, where $K_i$ is a finite simplicial complex, $i = 1, 2$. For each pair of simplexes $\sigma_1 \in K_1$ and $\sigma_2 \in K_2$, let

$$C(\sigma_1, \sigma_2) = \{f \in C(|K_1| \oplus |K_2|, X)\colon f(|\sigma_1|) \cap f(|\sigma_2|) = \emptyset\}.$$

By condition (iii) and Corollary 2.1.23, the set $C(\sigma_1, \sigma_2)$ is dense (and obviously open) in the space $C(|K_1| \oplus |K_2|, X)$. Consequently, according to Proposition 2.1.7, we conclude that the intersection

$$\cap\{C(\sigma_1, \sigma_2)\colon \sigma_1 \in K_1, \sigma_2 \in K_2\}$$

is also a dense subset of the space $C(|K_1| \oplus |K_2|, X)$. It only remains to observe that for each map $f$ from this intersection, we obviously have $f(|K_1|) \cap f(|K_2|) = \emptyset$.

We proceed now to the direct proof of the strong $\mathcal{B}_{\omega,\infty}$-universality of the space $X$. Consider a metrizable compactum $Y$ and for each pair of disjoint subcompacta $T_1$ and $T_2$ of $Y$ denote by $C(T_1, T_2)$ the set

$$\{f \in C(Y, X)\colon f(T_1) \cap f(T_2) = \emptyset\}.$$

By the fact established above, the set $C(T_1, T_2)$ is dense (and obviously open) in the space $C(Y, X)$. Take now a countable open basis $\mathcal{V}$ for the space $Y$. Then for each pair $(V_1, V_2)$ of elements of $\mathcal{V}$, with $\operatorname{cl} V_1 \cap \operatorname{cl} V_2 = \emptyset$, the set $C(\operatorname{cl} V_1, \operatorname{cl} V_2)$ is an open and dense subset of $C(Y, X)$. Consequently, by Proposition 2.1.7, the intersection of all sets of the form $C(\operatorname{cl} V_1, \operatorname{cl} V_2)$ is dense in the space $C(Y, X)$. It only remains to observe that each map from this intersection embeds the compactum $Y$ into the space $X$. $\square$

**2.4.3. Applications of characterization theorems.** Theorems 2.4.1 and 2.4.18 (as well as their alternate versions — Propositions 2.4.8 and 2.4.19) can be successfully used in many different situations. In this Subsection we present some of them.

PROPOSITION 2.4.20. *Let $f$ be a map between $R^\omega$-manifolds. Then the following conditions are equivalent:*

(i) *$f$ is approximately soft.*
(ii) *$f$ is a fine homotopy equivalence.*
(iii) *$f$ is a near-homeomorphism.*

PROOF. The implications (i)$\Longrightarrow$(ii) and (ii)$\Longrightarrow$(iii) follow from Propositions 2.1.27 and 2.1.28 respectively. The implication (iii)$\Longrightarrow$(i) follows from Propositions 2.3.14 and 2.4.7. □

PROPOSITION 2.4.21. *Let $f$ be a proper map between $I^\omega$-manifolds. Then the following conditions are equivalent:*

(i) *$f$ is approximately soft.*
(ii) *$f$ is a fine homotopy equivalence.*
(iii) *$f$ is a $CE$-map.*
(iv) *$f$ is a near-homeomorphism.*

Corollary 2.2.35 states that if $A$ is a $Z_\sigma$-set in an $R^\omega$-manifold $X$, then the complement $X-A$ is also an $R^\omega$-manifold. Moreover, since the inclusion $X-A \hookrightarrow X$ is a (fine) homotopy equivalence, we can, by virtue of Theorem 2.3.27, conclude that the $R^\omega$-manifolds $X-A$ and $X$ are homeomorphic. Proposition 2.4.7 implies an even stronger result.

PROPOSITION 2.4.22. *Let $A$ be a $Z_\sigma$-set in an $R^\omega$-manifold $X$. Then the inclusion $X - A \hookrightarrow X$ is a near-homeomorphism.*

PROPOSITION 2.4.23. *A countable infinite product of non-compact Polish absolute retracts is homeomorphic to $R^\omega$.*

PROOF. Let $X = \prod_{n=1}^{\infty} X_n$, where each $X_n$ is a non-compact Polish absolute retract. Obviously, $X$ itself is a Polish absolute retract. Consequently, by Theorem 2.4.1, it suffices to show the strong $\mathcal{A}_{\omega,\infty}$-universality of $X$. It follows from the proof of Proposition 2.3.14 that we will be done if each of the spaces $X_n$, $n \in N$, contains a closed copy of $R^\omega$. Represent the set $N$ of natural numbers as a countable infinite disjoint union of infinite subsets $N_k$, $k \in N$. Then

$$X = \prod_{k=1}^{\infty} Y_k, \text{ where } Y_k = \prod\{X_k \colon n \in N_k\}, k \in N.$$

Therefore it only remains to show that each of the spaces $Y_k$, $k \in N$, contains a closed copy of $R^\omega$. This last condition will be satisfied if we show that the real line $R$ admits a closed embedding into the product $A \times B$ of any two non-compact Polish absolute retracts. Take countable infinite and discrete subsets $\{a_n : n \in Z\}$ and $\{b_n : n \in Z\}$ of the spaces $A$ and $B$ respectively (use non-compactness of these spaces). Let $[a_n, a_{n+1}]$ denote the copy of the closed unit segment in $A$ with end points coinciding with $a_n$ and $a_{n+1}$. The same meaning shall be given to the symbol $[b_n, b_{n+1}]$, $n \in Z$. Let $c_n = (a_n, b_n) \in A \times B$ and $c_{n+1} = (a_{n+1}, b_{n+1}) \in A \times B$. Obviously the segment

$$[c_n, c_{n+1}] = ([a_n, a_{n+1}] \times \{b_n\}) \cup (\{a_{n+1} \times [b_n, b_{n+1}])$$

connects the points $c_n$ and $c_{n+1}$ in $A \times B$. Moreover, different segments of the form $[c_n, c_{n+1}]$ are either disjoint or have exactly one common point. It only remains to observe that the union $\cup\{[c_n, c_{n+1}] : n \in Z\}$ is closed in $A \times B$ and is homeomorphic to the real line $R$. □

PROPOSITION 2.4.24. *A proper retract of an $R^\omega$-manifold is an $R^\omega$-manifold.*

PROOF. Let $r: M \to X$ be a proper retraction of an $R^\omega$-manifold $M$ into a subspace $X$. First of all we show that each compact subset $K \subseteq X$ is a $Z$-set in $X$. Let $\mathcal{U} \in cov(X)$ be an open cover of $X$ and suppose a map $f: I^\omega \to X$ is given. Define a map $g: N \times I^\omega \to M$ by letting $g(n, a) = f(a)$ for each $(n, a) \in N \times I^\omega$. The strong $\mathcal{A}_{\omega,\infty}$-universality of the $R^\omega$-manifold $M$ guarantees the existence of a map $h: N \times I^\omega \to M$ which is $r^{-1}(\mathcal{U})$-close to $g$ and such that the collection $\{h(\{n\} \times I^\omega) : n \in N\}$ is discrete in $M$. Consider the collection $\{rh(\{n\} \times I^\omega) : n \in N\}$. Properness of the retraction $r$ implies that this collection is locally finite in $X$. Consequently, for each point $x \in K$ there is an open neighborhood $G_x$ of $x$ in $X$ intersecting only finitely many sets of the form $rh(\{n\} \times I^\omega)$. Selecting a finite subcover of the cover $\{G_x : x \in K\}$ of $K$, we obtain an open neighborhood $G$ of the compactum $K$ in $X$ which intersects only finitely many sets having the indicated form. Therefore there exists an index $n \in N$ such that $rh(\{n\} \times I^\omega) \cap K = \emptyset$. Obviously, the map $f': I^\omega \to X$, defined by letting $f'(a) = rh(n, a)$ for each $a \in I^\omega$, is $\mathcal{U}$-close to $f$ and $f'(I^\omega) \cap K = \emptyset$. Thus, every compact subset of $X$ is a $Z$-set in $X$.

This fact, coupled with Proposition 2.1.7 and Corollary 2.1.23, shows that the set

$$A_1 = \{f \in C(N \times I^\omega, X) : f(\{n\} \times I^\omega) \cap f(\{m\} \times I^\omega) = \emptyset;\ n \neq m\}$$

is a dense $G_\delta$-subset of the space $C(N \times I^\omega, X)$. On the other hand, the set

$$A_2 = \{f \in C(N \times I^\omega, X) : \text{ the collection } \{f(\{n\} \times I^\omega) : n \in N\} \text{ is locally finite}\}$$

is open and dense in the space $C(N \times I^\omega, X)$.

By Proposition 2.1.7, the intersection $A_1 \cap A_2$ is also dense in the space $C(N \times I^\omega, X)$. Finally, observe that if $f \in A_1 \cap A_2$, then the collection $\{f(\{n\} \times I^\omega) : n \in N\}$ is discrete in $X$. Proposition 2.4.8 completes the proof. □

The following result (compare with Proposition 2.4.23) follows directly from Proposition 2.4.19.

PROPOSITION 2.4.25. *A countable infinite product of non-trivial*[6] *compact absolute retracts is homeomorphic to* $I^\omega$.

Let $X$ be a Polish space. As usual, by $\exp X$ we denote the hyperspace of all non-empty compact subsets of $X$ endowed with the Hausdorff metric (or, equivalently, having the Vietoris topology; see, for example, [**145**]). The correspondence $X \longmapsto \exp X$ can be naturally extended to maps, i.e. for each map $f: X \to Y$ we can assign a map $\exp f: \exp X \to \exp Y$. Moreover, it is easy to see that this correspondence is functorial. Obviously the functor exp preserves the class of compacta. The following statement shows that *exp* significantly improves the properties of spaces.

PROPOSITION 2.4.26. *If* $X$ *is a Peano continuum, i.e.* $X \in AE(1)$, *then* $\exp X$ *is homeomorphic to* $I^\omega$.

PROOF. Take a *convex metric* $\rho$ on $X$ (see [**36**]). This means that each pair of points in $X$ is contained in a subspace isometric (with respect to metric $\rho$) to a closed segment on the real line. Consider the following metric on $\exp X$:

$$d(K_1, K_2) = \max\{\rho(x, K_i) \colon i = 1, 2, x \in K_1 \cup K_2\}.$$

By the result of Wojdyslawski [**321**], $\exp X$ is an $AR$-compactum.

For each $\epsilon > 0$, the map $f_\epsilon$ defined by letting

$$f_\epsilon(K) = \{x \in X : \rho(x, K) \leq \epsilon\} \quad K \in \exp X$$

is $\epsilon$-close to the identity map $id_{\exp X}$. It is not hard to see that the image $f_\epsilon(\exp X)$ is a $Z$-set in $\exp X$ (see [**116**] for details). Proposition 2.4.19 completes the proof. □

The following is a non-compact counter-part of the previous result.

PROPOSITION 2.4.27. *Let* $X$ *be a nowhere locally compact Polish* $AE(1)$*-space. Then* $\exp X$ *is homeomorphic to* $R^\omega$.

Throughout the remainder of this Section, $X$ stands for a *separable Fréchet space.* It is well known (see, for example, [**262**, Theorem 1.24]) that every Fréchet space admits an *invariant metric* $d$ such that all maps of the half-line $(0, \infty)$ into itself of the form $t \to d(tx, 0)$, $x \in X$, are non-decreasing. For simplicity we use the notation $\|x\| = d(x, 0)$, $x \in X$ (*warning:* $\|\cdot\|$ *is not a norm*).

The proofs of the following statements are left to the reader.

LEMMA 2.4.28. *Let* $\epsilon > 0$ *and let* $f: I^n \to X$ *be a map. Then there exists a map* $g: I^n \to X$ *such that* $g(I^n)$ *is contained in the convex hall of a finite subset of* $f(I^n)$ *and* $\|f(q) - g(q)\| < \epsilon$ *for each* $q \in I^n$.

[6] i.e. containing at least two points

LEMMA 2.4.29. *Let $E$ be a proper closed linear subspace of $X$ and $0 < a < b < \sup\{\|x - e\| : (x, e) \in X \times E\}$. Then there exists a map $u: [a, b] \to X$ such that the image $u([a, b])$ is contained in a finite-dimensional linear subspace of $X$ and the following conditions are also satisfied:*

$$\|u(t)\| \leq 5t \text{ and } \|u(t) - e\| \geq \frac{t}{2} \text{ for each } t \in [a, b] \text{ and } e \in E.$$

PROPOSITION 2.4.30. *Each infinite-dimensional separable Fréchet space is homeomorphic to $R^\omega$.*

PROOF. By Theorem 2.1.17, $X$ is a Polish absolute retract. Therefore, it suffices to verify condition (iii) of Proposition 2.4.8. If $X$ contains a linear subspace $X_0$ isomorphic to $R^\omega$, then, by Theorem 2.1.18, $X$ is homeomorphic to the product $R^\omega \times Y$, where $Y$ is a separable Fréchet space (namely, the factor space $X/X_0$). By Theorems 2.1.17 and 2.3.22, $X$ is homeomorphic to $R^\omega$.

Next, consider the case when $X$ does not contain a linear subspace isomorphic to $R^\omega$. In this case, there exists a number $\epsilon > 0$ such that no finite-dimensional linear subspace of $X$ is an $\epsilon$-net in $(X, d)$. Let us verify that condition (iii) of Proposition 2.4.8 holds. Take two maps $f: \oplus\{I^n : n \in N\} \to X$ and $\alpha: X \to (0, \infty)$. Without loss of generality we may assume that $\alpha < \epsilon$. In addition, by Lemma 2.4.28, we may assume that for each $n \in N$, the compactum $f(I^n)$ is contained in a finite-dimensional linear subspace of $X$. We construct a map $g: \oplus\{I^n : n \in N\} \to X$ by induction. Let $g/I^1 = f/I^1$ and assume that $g$ has already been defined on the union $\oplus\{I^k : k < n\}$ in such a way that $g(I^k)$, $k < n$, is contained in a finite-dimensional linear subspace $E$ of $X$ and the following condition is satisfied:

(*)

$$\|g(q) - f(q)\| \leq 5\alpha f(q) \text{ and } \|g(q) - g(q_1)\| \geq \frac{1}{2}\alpha f(q)$$

for each $q, q_1 \in I^k$, where $k < m < n$.

Apply Lemma 2.4.29 to the segment $[a, b] = \alpha f(I^n)$ and proper (recall that $X$ is infinite-dimensional) linear subspace $E$. Let $u: [a, b] \to X$ be the map whose existence is guaranteed by that Lemma. Let

$$g(q) = f(q) + u\alpha f(q), \quad q \in I^n.$$

This completes the inductive step of our construction.

Let us show that the map $g: \oplus\{I^n : n \in N\} \to X$ constructed above sends the collection $\{I^n : n \in N\}$ into a discrete (in $X$) collection. Assuming the contrary, we can find sequences of indices $k_1 < k_2 < \ldots$ and points $q_m \in I^{k_m}$ such that $\{g(q_m)\}$ converges in $X$. By (*), $\lim \alpha f(q_m) = 0$ and the sequence $\{f(q_m)\}$ also converges in $X$. This, in turn, implies the false equality $\alpha(\lim f(q_m)) = 0$. □

PROPOSITION 2.4.31. *Each infinite-dimensional compact convex subset of a separable Fréchet space is homeomorphic to $I^\omega$.*

PROOF. Let $K$ be an infinite-dimensional compact convex subset of a separable Fréchet space. By Theorem 2.1.17, $K$ is an absolute retract. Let us verify that condition (iii) of Proposition 2.4.19 holds. Let $\epsilon > 0$ and let $f, g\colon I^n \to K$ be maps. By Lemma 2.4.28, we may assume that $f(I^n) \cup g(I^n)$ is contained in the convex hull of a finite set $\{k_1, k_2, \cdots, k_m\} \subseteq K$. Since $\dim K = \infty$, there exists a point $k \in K$ which does not belong to the indicated convex hull. The compactness of $K$ guarantees that there is a $\delta > 0$ such that $\delta(K - K)$ is contained in the ball of radius $\epsilon$ with center at the origin. Let $f_\epsilon(q) = (1-\delta)f(q) + \delta k$, $q \in I^n$. Then

$$d(f_\epsilon(q), f(q)) = d(\delta(k - f(q)), 0) < \epsilon, \quad q \in I^n.$$

Let $g_\epsilon = g$. It only remains to note that $f_\epsilon(I^n) \cap g_\epsilon(I^n) = \emptyset$. □

### 2.4.4. Trivial Bundles.

In this Subsection we present parameterized versions of Theorems 2.4.1 and 2.4.18. We also establish some corollaries which will be essential in Chapter 6.

First of all let us introduce some notations. Let a map $f\colon X \to Y$ be given. For any space $Z$ and any map $g\colon Z \to X$, $C^g(Z, X)$ shall denote the subspace of the space $C(Z, X)$ consisting of all maps $h\colon Z \to X$ satisfying the equality $fh = fg$. The following two theorems characterize trivial $R^\omega$- and $I^\omega$-bundles.

THEOREM 2.4.32. *Let $f\colon X \to Y$ be a soft map between Polish ANR-spaces. Then the following conditions are equivalent:*

(i) *$f$ is a trivial bundle with fiber $R^\omega$.*
(ii) *For each Polish space $Z$ and any map $g\colon Z \to X$, the set of closed embeddings is dense in the space $C^g(Z, X)$.*

THEOREM 2.4.33. *Let $f\colon X \to Y$ be a proper soft map between locally compact ANR-spaces. Then the following conditions are equivalent:*

(i) *$f$ is a trivial bundle with fiber $I^\omega$.*
(ii) *For each locally compact space $Z$ and any proper map $g\colon Z \to X$, the set of closed embeddings is dense in the space $C^g(Z, X)$.*

The following statements are direct consequences of Theorems 2.4.32 and 2.4.33.

PROPOSITION 2.4.34. *Let $f\colon X \to Y$ be a soft map between Polish ANR-spaces. Then the composition $f\pi_X$, where $\pi_X\colon X \times R^\omega \to X$ is the projection, is a trivial bundle with fiber $R^\omega$.*

PROPOSITION 2.4.35. *Let $f\colon X \to Y$ be a proper soft map between locally compact ANR-spaces. Then the composition $f\pi_X$, where $\pi_X\colon X \times I^\omega \to X$ is the projection, is a trivial bundle with fiber $I^\omega$.*

PROPOSITION 2.4.36. *Let $f\colon X \to Y$ be a proper soft map between locally compact $ANR$-spaces. If for each open cover $\mathcal{U} \in cov(X)$ there are maps $f_1, f_2\colon X \to X$, $\mathcal{U}$-close to $id_X$, such that $f_1(X) \cap f_2(X) = \emptyset$ and $ff_i = f$, $i = 1,2$, then $f$ is a trivial bundle with fiber $I^\omega$.*

PROOF. By Theorem 2.4.33, it suffices to show that if $g\colon Z \to X$ is a proper map of a locally compact space $Z$ into $X$, then the set of closed embeddings of $Z$ into $X$ is dense in the space $C^g(Z,X)$. Take any two disjoint compact subsets $K_1$ and $K_2$ of $Z$. Obviously the set

$$L(K_1,K_2) = \{h \in C^g(Z,X)\colon h(K_1) \cap h(K_2) = \emptyset\}$$

is open in the space $C^g(Z,X)$. Let us show that these sets are dense in $C^g(Z,X)$. Take any $p \in C^g(Z,X)$ and any $\mathcal{U} \in cov(X)$. Softness of the map $f$ guarantees the existence of a refinement $\mathcal{V} \in cov(X)$ of $\mathcal{U}$ such that the following condition is satisfied:

(*) If a map $h'\colon K_1 \cup K_2 \to X$ is $\mathcal{V}$-close to the restriction $p/(K_1 \cup K_2)$ and $fh' = fp/(K_1 \cup K_2)$, then there exists a map $h\colon Z \to X$ that is $\mathcal{U}$-close to $p$ and satisfies $h/(K_1 \cup K_2) = h'$ and $fh = fp$.

By our assumption, there are maps $f_1, f_2\colon X \to X$, $\mathcal{V}$-close to $id_X$, such that $f_1(X) \cap f_2(X) = \emptyset$ and $ff_i = f$, $i = 1,2$. Define the map $h'\colon K_1 \cup K_2 \to X$ by letting it be equal to $f_ip/K_i$ on $K_i$, $i = 1,2$. Condition (*) guarantees the existence of a map $h\colon Z \to X$ such that $h$ is $\mathcal{U}$-close to $p$, $h/(K_1 \cup K_2) = h'$, and $fh = fp$. It only remains to remark that $h \in L(K_1,K_2)$. Therefore, the set $L(K_1,K_2)$ is open and dense in the space $C^g(Z,X)$.

Next, consider (in the space $Z$) a countable open basis $\mathcal{B} = \{V_i\colon i \in N\}$, the elements of which have compact closures. Let $L$ denote the intersection of all sets of the form $L(\mathrm{cl}\,V_i, \mathrm{cl}\,V_j)$, where $V_i, V_j \in \mathcal{B}$ and $\mathrm{cl}\,V_i \cap \mathrm{cl}\,V_j = \emptyset$. By Proposition 2.1.7, $L$ is dense in the space $C^g(Z,X)$. It only remains to remark that each map $h \in L$ is a closed embedding. □

The proof of the following statement is similar to that of Proposition 2.4.19.

PROPOSITION 2.4.37. *Let $f\colon X \to Y$ be a proper soft map between locally compact $ANR$-spaces. Then the following conditions are equivalent:*

(i) *$f$ is a trivial bundle with fiber $I^\omega$.*

(ii) *For each map $g\colon D \times I^\omega \to X$, the set*

$$\{h \in C^g(D \times I^\omega, X)\colon h(\{0\} \times I^\omega) \cap h(\{0\} \times I^\omega) = \emptyset\}$$

*is dense in the space $C^g(D \times I^\omega, X)$.*

As an immediate consequence of Proposition 2.4.37 we have the following statement.

PROPOSITION 2.4.38. *Let $\mathcal{S} = \{X_n, p_n^{n+1}, \omega\}$ be an inverse sequence consisting of locally compact $ANR$-spaces and proper soft short projections, each* *of which has two sections with disjoint images. Then the limit projection* $p_0\colon \lim \mathcal{S} \to X_0$ *is a trivial bundle with fiber* $I^\omega$.

The reader can provide the proof of the following useful statement by analyzing the proof of Proposition 2.4.8 (keeping in mind Theorem 2.4.32).

PROPOSITION 2.4.39. *Let* $f\colon X \to Y$ *be a soft map between Polish* $ANR$*-spaces. Then the following conditions are equivalent:*

(i) $f$ *is a trivial bundle with fiber* $R^\omega$.

(ii) *For each map* $g\colon N \times I^\omega \to X$, *the set*

$$\{h\colon C^g(N \times I^\omega, X)\colon \text{collection } \{h(\{n\} \times I^\omega)\colon n \in N\} \text{ is discrete in } X\}$$

*is dense in the space* $C^g(N \times I^\omega, X)$.

PROPOSITION 2.4.40. *Let* $\mathcal{S} = \{X_n, p_n^{n+1}, \omega\}$ *be an inverse sequence consisting of Polish* $ANR$*-spaces and soft short projections, each of which has a countable (infinite) family of sections, the images of which form discrete collections. Then the limit projection* $p_0\colon \lim \mathcal{S} \to X_0$ *is a trivial bundle with fiber* $R^\omega$.

PROOF. Obviously the limit projection $p_0$ is a soft map and the limit space $X = \lim \mathcal{S}$ is a Polish $ANR$-space. Consequently, we only have to verify condition (ii) of Proposition 2.4.39. Take a metric $d_n$ on the space $X_n$, $n \in \omega$, bounded by $\frac{1}{2^n}$. On $X$ we consider the metric $d$, defined as follows:

$$d(\{x_n\}, \{x'_n\}) = \max\{d_N(x_n, x'_n)\colon n \in \omega\}.$$

Consider two maps $f\colon N \times I^\omega \to X$ and $\alpha\colon X \to (0,1)$. Our goal is to construct a map $g\colon N \times I^\omega \to X$ satisfying the following conditions:

(a) $p_0 g = p_0 f$.

(b) $d(f(y), g(y)) \leq \alpha f(y)$ for each $y \in N \times I^\omega$.

(c) The collection $\{g(\{k\} \times I^\omega)\colon k \in \omega\}$ is discrete in $X$.

For each $n \in \omega$ take a closed embedding $i_{n+1}\colon X_{n+1} \to R^\omega$. Since $R^\omega$ is a Polish absolute retract there exists a map

$$h\colon R^\omega \times R^\omega \times [0,\infty) \to R^\omega$$

such that

$$h(a,b,t) = a \quad \text{for each } t \leq 1$$

and

$$h(a,b,t) = b \quad \text{for each } t \geq 2.$$

Since $i_{n+1}$ is a closed embedding, the diagonal product

$$p_n^{n+1} \triangle i_{n+1}\colon X_{n+1} \to X_n \times R^\omega$$

is also a closed embedding. Consequently, softness of the short projection

$$p_n^{n+1}\colon X_{n+1} \to X_n$$

guarantees the existence of a map

$$r_{n+1}\colon X_n \times R^\omega \to X_{n+1}$$

such that $p_n^{n+1} r_{n+1} = \pi_{X_n}$ and $r_{n+1}/A_{n+1} = (p_n^{n+1} \triangle i_{n+1})^{-1}$, where

$$\pi_{X_n}\colon X_n \times R^\omega \to X_n$$

is the projection onto the first coordinate and $A_{n+1} = (p_n^{n+1} \triangle i_{n+1})(X_{n+1})$.

Let $g_0 = p_0 f$.

By our assumptions, the projection $p_0^1\colon X_1 \to X_0$ admits a countable infinite family of sections, the images of which form a discrete collection in $X_1$. This means that the space $X_1$ contains a closed subspace homeomorphic to the product $N \times X_0$ in such a way that $p_0^1/(N \times X_0) = \lambda_{X_0}$, where $\lambda_{X_0}\colon N \times X_0 \to X_0$ is the projection onto the first coordinate. Consider the map $\varphi_1\colon N \times I^\omega \to X_1$ defined by letting $\varphi_1 = \lambda \triangle p_0 f$, where $\lambda\colon N \times I^\omega \to N$ is the projection onto the first coordinate. Obviously, $p_0^1 \varphi_1 = g_0$ and the collection $\{\varphi_1(\{k\} \times I^\omega)\colon k \in N\}$ is discrete in $X_1$. Now define a map $g_1\colon N \times I^\omega \to X_1$ as follows:

$$g_1 = r_1(g_0 \triangle h(i_1 p_1 f \triangle i_1 \varphi_1 \triangle 2\alpha f)).$$

It is easy to see that $p_0^1 g_1 = g_0$. If $y \in N \times I^\omega$ and $\alpha f(y) \le \frac{1}{2}$, then, by the properties of $h$, we have $g_1(y) = p_1 f(y)$.

Let us describe the construction of the map $g_2\colon N \times I^\omega \to X_2$. As above we take a map $\varphi_2\colon N \times I^\omega \to X_2$ such that $p_1^2 \varphi_2 = g_1$ and for which the collection $\{\varphi_2(\{k\} \times I^\omega)\colon k \in N\}$ is discrete in $X_2$. Let

$$g_2 = r_2(g_1 \triangle h(i_2 p_2 f \triangle i_2 \varphi_2 \triangle 2^2 \alpha f)).$$

Clearly, $p_1^2 g_2 = g_1$. As above, if $\alpha f(y) \le \frac{1}{2^2}$, then $g_2(y) = p_2 f(y)$. If $\alpha f(y) \ge \frac{1}{2}$, then it follows from the construction that $g_2(y) = \varphi(y)$.

Suppose now that for each $i$, with $2 \le i \le n$, we have already constructed maps $g_i\colon N \times I^\omega \to X_i$ and $\varphi_i\colon N \times I^\omega \to X_i$ satisfying the following conditions:

$(1)_i$ $p_{i-1}^i g_i = g_{i-1}$.

$(2)_i$ $p_{i-1}^i \varphi_i = g_{i-1}$.

$(3)_i$ If $\alpha f(y) \le \frac{1}{2^i}$, then $g_i(y) = p_i f(y)$.

$(4)_i$ If $\alpha f(y) \ge \frac{1}{2^{i-1}}$, then $g_i(y) = \varphi_i(y)$.

$(5)_i$ The collection $\{\varphi_i(\{k\} \times I^\omega)\colon k \in N\}$ is discrete in $X_i$.

We now construct the map

$$g_{n+1}\colon N \times I^\omega \to X_{n+1} \text{ and } \varphi_{n+1}\colon N \times I^\omega \to X_{n+1}.$$

Following the above description, we obtain a map $\varphi_{n+1}\colon N \times I^\omega \to X_{n+1}$ such that $p_n^{n+1} \varphi_{n+1} = g_n$ and the collection $\{\varphi_{n+1}(\{k\} \times I^\omega)\colon k \in N\}$ is discrete in $X_{n+1}$. Then

$$g_{n+1} = r_{n+1}(g_n \triangle h(i_{n+1} p_{n+1} f \triangle i_{n+1} \varphi_{n+1} \triangle 2^{n+1} \alpha f)).$$

Straightforward verification shows that the maps $g_{n+1}$ and $\varphi_{n+1}$ satisfy conditions $(1)_{n+1} - (5)_{n+1}$.

It follows from the conditions $(1)_n$ that the diagonal product

$$g = \triangle\{g_n : n \in \omega\} : N \times I^\omega \to X$$

satisfies the equalities $p_n g = g_n$ for each $n \in \omega$. In particular, we have $p_0 g = g_0 = p_0 f$.

If $y \in N \times I^\omega$, then there is an index $n \in \omega$ such that

$$\frac{1}{2^{n+1}} \leq \alpha f(y) \leq \frac{1}{2^n}.$$

Therefore, for each $i \leq n$, we conclude, by conditions $(3)_i$, that $g_i(y) = p_i f(y)$. This shows that

$$d(f(y), g(y)) = \max\{d_k(p_k f(y), p_k g(y)) : k \geq n+1\} \leq \frac{1}{2^{n+1}} \alpha f(y).$$

This establishes the required closeness of the maps $f$ and $g$.

Next we show that

$$g(\{k\} \times I^\omega) \cap g(\{k'\} \times I^\omega) = \emptyset \text{ whenever } k \neq k'.$$

Indeed, let $y_1 \in \{k\} \times I^\omega$ and $y_2 \in \{k'\} \times I^\omega$. Take a sufficiently large $n$ to ensure that $\alpha f(y_i) \leq \frac{1}{2^{i-1}}$, $i = 1, 2$. Then, by condition $(4)_n$, $p_n g(y_i) = g_n(y_i) = \varphi_n(y_i)$. Consequently, by condition $(5)_n$, $y_1 \neq y_2$.

Finally, let us show that the collection $\{g(\{k\} \times I^\omega) : k \in\}$ is discrete in $X$. Since this collection consists (as was just shown) of pairwise disjoint compacta, it suffices to show that for each (infinite) subset $N'$ of $N$ the set $g(N' \times I^\omega)$ is closed in $X$. Consider an arbitrary sequence $\{y_i\}$ of points in $N' \times I^\omega$ so that the sequence $\{g(y_i)\}$ converges to a point $x \in X$. Assume that $0 < \epsilon = \inf\{\alpha f(y_i)\}$. Take $n \in \omega$ so large that $\epsilon \geq \frac{1}{2^{n-1}}$. Then, by condition $(4)_n$, $p_n g(y_i) = g_n(y_i) = \varphi_n(y_i)$ for each $i$. Continuity of the projection $p_n$ implies that $p_n(x) = \lim \varphi_n(y_i)$. By condition $(5)_n$, the collection $\{\varphi_n(\{k\} \times I^\omega) : k \in N'\}$ is discrete in $X_n$. Therefore, without loss of generality, we may assume that the convergent (in $X_n$) sequence $\{\varphi_n(y_i)\}$ is completely contained in the compactum $\{\varphi_n(\{k\} \times I^\omega)$ for some $k \in N'$. But then $x \in g(\{k\} \times I^\omega)$.

Thus, it only remains to see that $\epsilon > 0$. Assume the contrary. Then there exists a subsequence $\{y_{i_j}\}$ of $\{y_i\}$ such that $\lim \alpha f(y_{i_j}) = 0$. Then, keeping in mind the closedness of the maps $f$ and $g$, we conclude that $\lim f(y_{i_j}) = x$. This equality implies, by continuity of $\alpha$, the contradictory statement $\alpha(x) = 0$. Therefore $\epsilon > 0$. □

*Historical and bibliographical notes* 2.4. The characterization Theorems 2.4.1 and 2.4.18 were obtained by Toruńczyk in [**299**] and [**298**] respectively. The proof presented here is taken from [**34**] (see also [**139**] and [**306**]). Proposition 2.4.8 is due to Toruńczyk [**299**]. Lemma 2.4.11 is taken from [**86**] (in the compact case this result had been obtained earlier in [**48**]). Proposition 2.4.20 appears in [**299**] and [**149**]. Proposition 2.4.21 is in fact contained in [**139**]. Propositions 2.4.23 and 2.4.24 were also proved in [**299**]. Proposition 2.4.25 is due to West [**312**]. Propositions 2.4.26 and 2.4.27 can be found in [**115**], [**117**] and [**113**]. Proposition 2.4.30 is the well known theorem of Anderson-Kadec. Proposition 2.4.31 were proved in [**192**] (see also [**195**]). The proofs of the last two statements are taken from [**124**]. Theorems 2.4.32 and 2.4.33 — parametric versions of Theorems 2.4.1 and 2.4.18 — are due to Toruńczyk and West [**300**]. Formally this work contains only the proof of Theorem 2.4.33, and the proof of Theorem 2.4.32 is still unpublished. Nevertheless, this result is known to experts and has often been used. [**300**] contains an example of an $n$-soft map of the Hilbert cube onto itself, all fibers of which are also copies of the Hilbert cube, but which does not admit two sections with disjoint images. Therefore, a "fibered" version of Theorem 2.4.18 is not generally true. Propositions 2.4.38 and 2.4.40 were proved by Ščepin and the author respectively.

## 2.5. Incomplete manifolds

In Subsection 2.2.3, on page 56, we have already encountered the pseudo-boundary $\Sigma$ (and the pseudo-interior $P$ of the Hilbert cube $I^\omega$). This space can also be described as

$$\Sigma = \left\{ \{x_i\} \in l_2 \colon \sum_{i=1}^{\infty} (ix_i)^2 < \infty \right\}.$$

Consider also the space

$$\sigma = \{\{x_i\} \in l_2 \colon \text{ all but finitely many } x_i = 0\}.$$

Alternatively, $\Sigma$ can be thought as the span of the standard Hilbert cube in the Hilbert space $l_2$ and $\sigma$ as the span of the usual orthonormal basis of $l_2$. These two spaces, $\sigma$ and $\Sigma$, represent the minimal and maximal topological types of infinite-dimensional $\sigma$-compact locally convex metrizable linear spaces in the following sense: every infinite-dimensional $\sigma$-compact locally convex linear metrizable space contains a copy of $\sigma$ and is contained in $\Sigma$. It is not hard to see that both spaces $\sigma$ and $\Sigma$ are $\sigma$-compact absolute retracts. Each of these spaces can be represented as an increasing (countable) union of strong $Z$-sets. Incidently, every compact subset in either of these spaces is a strong $Z$-set. The topological characterizations of these spaces also involve a version of the universality property. Namely we have the following two results.

THEOREM 2.5.1. *Let $X$ be a metrizable $A(N)R$-space. Then $X$ is homeomorphic to $\Sigma$ (respectively, is a $\Sigma$-manifold) if and only if the following conditions are satisfied:*

(i) *X is σ-compact.*
(ii) *X is a countable union of strong Z-sets.*
(iii) *For each compactum B and closed subspace A of B, every map $f\colon B \to \sigma$, such that $f/A$ is an embedding, can be arbitrarily closely approximated by embeddings coinciding with f on A.*

THEOREM 2.5.2. *Let X be a metrizable A(N)R-space. Then X is homeomorphic to σ (respectively, is a σ-manifold) if and only if the following conditions are satisfied:*
(i) *X is a countable union of finite-dimensional compacta.*
(ii) *X is a countable union of strong Z-sets.*
(iii) *For each finite-dimensional compactum B and closed subspace A of B, every map $f\colon B \to \sigma$, such that $f/A$ is an embedding, can be arbitrarily closely approximated by embeddings coinciding with f on A.*

One of the main ingredient of the proofs of the above theorems is the following version of Bing's shrinking criterion (Theorem 2.1.8) for incomplete σ-compact spaces.

PROPOSITION 2.5.3. *Let X and Y be metrizable ANR-spaces which are countable unions of (finite-dimensional) compacta. Suppose that X has the estimated extension property*[7] *for compacta and Y has the universality property formulated in condition (iii) of Theorem 2.5.1 (of Theorem 2.5.2, respectively). Let $f\colon X \to Y$ be a map with the property that for every compactum A in Y and closed subset B of A, the map*[8]

$$f_A/(X - f^{-1}(B))\colon X - f^{-1}(B) \to (X \cup_f A) - B$$

*is a near-homeomorphism. Then f is a near-homeomorphism.*

PROOF. We consider only the parenthetical case, the other case being similar. Let $X = \cup_{n=1}^{\infty} A_n$ and $Y = \cup_{n=1}^{\infty} B_n$, where $A_n$ and $B_n$ are finite-dimensional compacta for each $n = 1, 2, \ldots$. Let $d$ be any metric on $Y$. It is sufficient (see Subsection 2.1.1) to show that there is a homeomorphism $h\colon X \to Y$ such that $d(h(x), f(x)) < 1$ for each $x \in X$. We inductively construct a sequence $\{C_n\}_{n=0}^{\infty}$ of compact subsets of $Y$ and a sequence $\{h_n\}_{n=0}^{\infty}$ of homeomorphisms of $X$ onto $X \cup_f C_n = X_n$ such that, for each $n = 1, 2, \ldots$:
$(a)_n$ $C_n \supseteq B_n \cup C_{n-1}$.
$(b)_n$ $h_n(A_n) \subseteq C_n$.
$(c)_n$ $h_n/h_{n-1}^{-1}(C_{n-1}) = h_{n-1}/h_{n-1}^{-1}(C_{n-1})$.
$(d)_n$ $p_n h_n/(X - h_{n-1}^{-1}(C_{n-1})$ is $\alpha_n$-close to $p_{n-1}h_{n-1}/(X - h_{n-1}^{-1}(C_{n-1})$, where $\alpha_n\colon Y - C_{n-1} \to (0,1)$ is defined by

$$\alpha_n(y) = \frac{1}{2^n} \min\{1, d(y, C_{n-1})\}$$

[7]See page 55.
[8]See definition of an adjunction space and induced maps on page 61

and $p_n\colon X_n \to Y$ is the map defined by $P_n f_{C_n} = f$.
We let $C_0 = \emptyset$ and $h_0 = id$. Assume that $h_i\colon X \to X_i$ and $X_i$ satisfying conditions $(a)_i$, $(b)_i$, $(c)_i$ and $(d)_i$ for each $i$ with $0 \le i \le n$ have already been constructed. Observe that $p_n(X_n - C_n) \subseteq Y - C_n$. Let $\mathcal{U} \in cov(Y - C_n)$ be an open cover of $Y - C_n$ such that

$$B(p_n/(X_n - C_n, st^3\mathcal{U}) \subseteq B(p_n/(X_n - C_n, \alpha_{n+1}).$$

By the properties of $Y$, we can find an embedding

$$v\colon D_{n+1} = h_n(A_{n+1}) \cup C_n \to Y$$

such that $v/C_n = id_{C_n}$ and $v/(D_{n+1} - C_n)$ is $\mathcal{U}$-homotopic to $p_n/(D_{n+1} - C_n)$. Take

$$C_{n+1} = B_{n+1} \cup v(D_{n+1}) \cup H((D_{n+1} - C_n) \times [0,1]),$$

where $H\colon (D_{n+1} - C_n) \times [0,1] \to Y$ is a $\mathcal{U}$-homotopy with $H(x,0) = p_n(x)$ and $H(x,1) = v(x)$ for each $x \in D_{n+1} - C_n$. Since the restriction $f_{C_n}/(X - f^{-1}(C_n))$ is a homeomorphism of $X - f^{-1}(C_n)$ onto $X_n - C_n$, by the assumption about the map $f$, there exists a homeomorphism $g_{n+1}\colon X_n \to X_{n+1}$ such that $g_{n+1}/C_n = id$ and $g_{n+1}/(X_n - C_n)$ is $\overset{-1}{p_{n+1}}(\mathcal{U})$-homotopic to the map $f_{C_{n+1},C_n)}/(X_n - C_n)$, where $f_{C_{n+1},C_n)}$ is defined by the equality $f_{C_{n+1},C_n)}f_{C_n} = f_{C_{n+1}}$. The embeddings $g_{n+1}/D_{n+1}$ and $(p_{n+1}/C_{n+1})^{-1}v$ are $st(\overset{-1}{p_{n+1}}(\mathcal{U}))$-homotopic. Since $X_{n+1}$, being homeomorphic to $X$, has the estimated extension property for compacta, there exists a homeomorphism $u_{n+1}\colon X_{n+1} \to X_{n+1}$ which is $st^2(\overset{-1}{p_{n+1}}(\mathcal{U}))$-close to the identity map and such that

$$u_{n+1}g_{n+1}/D_{n+1} = (p_{n+1}/C_{n+1})^{-1}v.$$

Let $h_{n+1} = u_{n+1}g_{n+1}h_n$. Then $p_{n+1}h_{n+1}$ is $st^2(\mathcal{U})$-close to $p_{n+1}g_{n+1}h_n$ and, consequently, $p_{n+1}h_{n+1}$ is $st^3(\mathcal{U})$-close to $p_nh_n$. It is not hard to see that the conditions $(a)_{n+1}$, $(b)_{n+1}$, $(c)_{n+1}$ and $(d)_{n+1}$ are all satisfied. We leave it as a manageable exercise for the reader, to verify the fact that the map $h = \lim p_nh_n$ is a homeomorphism of $X$ onto $Y$ such that $d(f(x), h(x)) < 1$ for all $x \in X$. □

Theorems 2.5.1 and 2.5.2 have several corollaries.

PROPOSITION 2.5.4. *The following conditions are equivalent for each space $X$:*

(i) *$X \times \Sigma$ is homeomorphic to $\Sigma$ (respectively, is a $\Sigma$-manifold).*
(ii) *$X$ is a retract of $\Sigma$ (respectively, of an open subspace of $\Sigma$).*

Similarly, we have the following.

PROPOSITION 2.5.5. *The following conditions are equivalent for each space $X$:*

(i) *$X \times \sigma$ is homeomorphic to $\sigma$ (respectively, is a $\sigma$-manifold).*
(ii) *$X$ is a retract of $\sigma$ (respectively, of an open subspace of $\sigma$).*

Triangulation theorems also hold.

PROPOSITION 2.5.6. *Each $\Sigma$-manifold is homeomorphic to the product $K \times \Sigma$, where $K$ is a locally compact polyhedron.*

PROPOSITION 2.5.7. *Each $\sigma$-manifold is homeomorphic to the product $K \times \sigma$, where $K$ is a locally compact polyhedron.*

We also have open embedding and homotopy classification results.

PROPOSITION 2.5.8. *Each $\Sigma$-manifold is homeomorphic to an open subspace of $\Sigma$.*

PROPOSITION 2.5.9. *Each $\sigma$-manifold is homeomorphic to an open subspace of $\sigma$.*

PROPOSITION 2.5.10. *Homotopy equivalent $\Sigma$-manifolds are homeomorphic.*

PROPOSITION 2.5.11. *Homotopy equivalent $\sigma$-manifolds are homeomorphic.*

*Historical and bibliographical notes* 2.5. The characterization Theorems 2.5.1 and 2.5.2 were obtained by Mogilski [**234**]. The triangulation theorems (Propositions 2.5.6 and 2.5.7), open embedding theorems (Propositions 2.5.8 and 2.5.9), homotopy classification theorems (Propositions 2.5.10 and 2.5.11) were proved by Chapman [**64**]. The spaces $\Sigma$ and $\sigma$ are, in some sense, basic $\sigma$-compact (incomplete) $ANR$-spaces. In [**35**] the reader can find constructions of similar objects for higher Borel classes (additive or multiplicative). See also Section 5.6.

CHAPTER 3

# Cohomological Dimension

## 3.1. Cohomological Dimension

For each commutative group $G$ and any nonnegative integer $n$, we denote by $K(G,n)$ the corresponding *Eilenberg-MacLane complex*, i.e. a $\mathcal{CW}$-complex, satisfying the following conditions:

$$\pi_i(K(G,n)) = \begin{cases} 0, & \text{if } i \neq n \\ G, & \text{if } i = n. \end{cases}$$

We are especially interested in the spaces $K(\mathbb{Z},n)$. The complex $K(\mathbb{Z},n)$ can be thought as a $\mathcal{CW}$-complex obtained from the $n$-dimensional sphere $S^n$ by attaching cells of dimension $\geq n+2$. In this interpretation, the $n$-dimensional and $(n+1)$-dimensional skeleta of $K(\mathbb{Z},n)$ both coincide with $S^n$. Obviously, every compact subset of $K(\mathbb{Z},n)$ is contained in the union of finitely many cells (recall that $\mathcal{CW}$-complexes are endowed with the weak topology). Therefore, if a map $f\colon A \to K(\mathbb{Z},n)$ is defined on a closed subspace $A$ of a compactum $B$, then $f$ can be extended over an open neighborhood of $A$ in $B$. In other words, the complexes $K(\mathbb{Z},n)$ are absolute extensors with respect to compact spaces.

Denote by $H^i(X,G)$ the $i$-dimensional *Čech* cohomology group of the compactum $X$ with coefficients in $G$. It is well known [**291**], [**287**] that

$$H^i(X,G) = [X, K(G,i)],$$

where the square brackets denotes the set of homotopy classes of maps of the indicated spaces.

Now we recall the definition of the cohomological dimension $\dim_G$ with respect to a group $G$.

DEFINITION 3.1.1. *The cohomological dimension of a compactum $X$ with respect to a group $G$ does not exceed $n$, i.e.* $\dim_G X \leq n$, *if for each closed subspace $A$ of $X$ we have $H^m(X,A;G) = 0$ whenever $m > n$. If* $\dim_G X \leq n$ *and* $\dim_G X \not\leq n-1$, *then* $\dim_G X = n$. *If for every $n$, we have* $\dim_G X \not\leq n$, *then* $\dim_G X = \infty$.

We begin with the following statement.

THEOREM 3.1.2. *The following conditions are equivalent for each compactum* $X$:

$(1)_n$ $\dim_G X \leq n$.

$(2)_n$ $H^{n+1}(X,A;G)=0$ *for each closed subspace* $A$ *of* $X$.

$(3)_n$ *For each closed subspace* $A$ *of* $X$, *any map* $f\colon A \to K(G,n)$ *has an extension* $F\colon X \to K(G,n)$.

PROOF. The implication $(1)_n \Longrightarrow (2)_n$ is obvious. Condition $(3)_n$ simply means that for each closed subspaces $A$ of $X$, the inclusion homomorphism $H^n(X;G) \to H^n(A;G)$ is an epimorphism. Consequently, by the exact cohomology sequence of the pair $(X,A)$, condition $(2)_n$ implies condition $(3)_n$.

Therefore it suffices to establish the implication $(3)_n \Longrightarrow (1)_n$. Consider an additional condition:

$(4)_n$ *For each closed subspace* $A$ *of* $X$ *we have* $H^{n+1}(A;G)=0$.

Observe that condition $(4)_n$ implies condition $(3)_{n+1}$, and conditions $(3)_n$ and $(4)_n$ imply condition $(2)_n$. In turn, condition $(1)_n$ is equivalent to the collection of conditions $(2)_k$ with $k \geq n$. Therefore, it suffices to show the implication $(3)_k \Longrightarrow (4)_k$ for $k \geq n$.

Consider an arbitrary map $f\colon A \to K(G,k+1)$, defined on a closed subspace $A$ of $X$. Take a point $k_0 \in K(G,k+1)$, and by $E = P(K(G,k+1)),k_0)$ denote the space of paths $\omega\colon I \to K(G,k+1)$, such that $\omega(0)=k_0$, endowed with the compact open topology. Consider the natural map $\alpha\colon E \to K(G,k+1)$ assigning to a path $\omega \in E$ the point $\omega(1) \in K(G,k+1)$. The map $\alpha$ is a locally trivial bundle with fiber $K(G,k)$. It is also important to remark that the space $E$ is contractible. Consequently, in order to show that the map $f$ is homotopic to a constant map, it suffices to construct a map $g\colon A \to E$ such that $\alpha g = f$. By compactness of the space $A$, there exists a compact polyhedron $P \subseteq K(G,k+1)$ containing the image $f(A)$. Let $P^{(i)}$denote the $i$-dimensional skeleton of $P$ (with respect to some triangulation of $P$). We construct the map $g$ by induction, first on $f^{-1}(P^{(0)})$, then on $f^{-1}(P^{(1)})$, and so forth until we reach $f^{-1}(P)=A$. Choose a section $s\colon P^{(0)} \to E$ of the map $\alpha$ over $P^{(0)}$ and let $g_0 = sf/f^{-1}(P^{(0)})$. Suppose that $g_i\colon f^{-1}(P^{(i)}) \to E$ is a lifting of the map $f/f^{-1}(P^{(i)})$. We are going to construct a lifting $g_{i+1}\colon f^{-1}(P^{(i+1)}) \to E$ of the map $f/f^{-1}(P^{(i+1)})$. Consider an $(i+1)$-dimensional simplex $\sigma$ in $P$. The contractibility of $\sigma$ implies that the space $\alpha^{-1}(\sigma)$ is homeomorphic to the product $\sigma \times K(G,k)$. Denote by

$$h\colon \alpha^{-1}(\sigma) \to \sigma \times K(G,k)$$

the corresponding homeomorphism and by

$$\pi\colon \sigma \times K(G,k) \to K(G,k)$$

the projection onto the second coordinate. By condition $(3)_k$, the map

$$\pi h g_i / f^{-1}(\partial\sigma) \colon f^{-1}(\partial\sigma) \to K(G,k)$$

has an extension $\psi \colon f^{-1}(\sigma) \to K(G,k)$. But then the map

$$h^{-1}(f/f^{-1}(\sigma)\Delta\psi) \colon f^{-1}(\sigma) \to \alpha^{-1}(\sigma)$$

is an extension of the lifting $g_i$ onto the whole set $f^{-1}(\sigma)$. Performing the same procedure with all of the $(i+1)$-dimensional simplexes, we obtain a lifting $g_{i+1}$ defined on the set $f^{-1}(P^{(i+1)})$. □

Thus, each of the conditions from Theorem 3.1.2 can be taken as the definition of cohomological dimension. Below, in most cases, we use the definition given by condition $(3)_n$. In this connection, recall that if we replace the complex $K(G,n)$ by the sphere $S^n$ in condition $(3)_n$, then we obtain the definition of the usual Lebesgue dimension dim. More formally,we have the following statement.

THEOREM 3.1.3. *For each compactum $X$, the inequality* $\dim X \leq n$ *is equivalent to the condition:*

- *For each closed subspace $A$ of $X$, any map $f \colon A \to S^n$ into the $n$-dimensional sphere $S^n$ has an extension $F \colon X \to S^n$.*

Theorems 3.1.2 and 3.1.3 imply the following statement.

PROPOSITION 3.1.4. *If $X$ is a compactum, then*

$$\dim_{\mathbb{Z}} X \leq \dim X.$$

PROOF. If $\dim X = \infty$, then the indicated inequality is trivially true. Assume that $\dim X \leq n$ and consider a map $f \colon A \to K(\mathbb{Z},n)$, defined on a closed subspace $A$ of the compactum $X$. Since $\dim X \leq n$, the map $f$ is homotopic to a map $g \colon A \to K(\mathbb{Z},n)^{(n)} = S^n$. By Theorem 3.1.3, there is an extension $G \colon X \to S^n$ of $g$. Since the complex $K(\mathbb{Z},n)$ is an absolute extensor with respect to compact spaces, we conclude that the original map $f$ also has an extension $F \colon X \to K(\mathbb{Z},n)$ (homotopic to $G$). □

THEOREM 3.1.5. *If $X$ is a finite-dimensional (in the sense of* dim*) compactum, then*

$$\dim X = \dim_{\mathbb{Z}} X.$$

PROOF. By Proposition 3.1.4, it suffices to show that $\dim X \leq \dim_{\mathbb{Z}} X$. Assume the contrary. Then there exists a compactum $X$ such that for some integer $m$ we have

$$\dim X = m + 1 \text{ and } \dim_{\mathbb{Z}} X \leq m.$$

By the first equality, there exist a closed subspace $A$ of $X$ and a map $f \colon A \to S^m$ which cannot be extended to the space $X$. Identify the sphere $S^m$ with the $m$-dimensional skeleton of the complex $K(\mathbb{Z},m)$. Then, by the inequality $c - dim_{\mathbb{Z}} X \leq m$, there exists an extension $G \colon X \to K(\mathbb{Z},m)$ of $f$. Again, by the first equality and Theorem 3.1.3, there is a map $F\colon X \to K(\mathbb{Z}, m)$ homotopic to $G$ and such that

$$F/G^{-1}(K(\mathbb{Z}, m)^{(m+1)}) = G/G^{-1}(K(\mathbb{Z}, m)^{(m+1)}).$$

In this situation, it can easily be seen that $F$ is an extension of the map $f$ into the sphere $S^m$. This contradiction the choice of $f$. □

It follows from the universal coefficients formula (see, for example, [**287**]) that for each group $G$ and compactum $X$ we have $\dim_G X \le \dim_{\mathbb{Z}} X$. This fact, coupled with Proposition 3.1.4, shows that the dimension $\dim_{\mathbb{Z}}$ is "closer" to the dimension dim than is the dimension $\dim_G$ for any other group $G$. Below we consider only the cohomological dimension $\dim_{\mathbb{Z}}$ with integral coefficients. The main problem of determining whether the dimensions dim and $\dim_{\mathbb{Z}}$ coincide is known as *Alexandrov's problem* [**5**]. Keeping in mind Theorem 3.1.5, this problem can be formulated as follows:

- *Does there exist an infinite-dimensional compactum having finite cohomological dimension* $\dim_{\mathbb{Z}}$ *?*

In this section we show that the above problem is equivalent to the so-called $CE$-problem:

- *Do cell-like maps of compacta raise dimension?*

First we present a spectral characterization of cohomological dimension. For this we need to introduce the following two notions.

DEFINITION 3.1.6. *The* $n$-dimensional diameter of a map $f\colon X \to Y$ *is defined as the following number:*

$$a_n(f) = \inf\{dist(f, \varphi)\colon \varphi \in C(X, Y) \text{ and } \dim \varphi(X) \le n\}.$$

DEFINITION 3.1.7. *The* binary $(n, k)$-dimensional diameter of a map $f\colon X \to Y$ *is defined as the following number:*

$$a_k^n(f) = \sup\{a_k(f/A)\colon A \text{ is closed in } X \text{ and } \dim A \le n\}.$$

THEOREM 3.1.8. *Let* $X$ *be the limit space of an inverse sequence* $\mathcal{S} = \{X_i, p_i^{i+1}\}$, *consisting of compact polyhedra. Then the following conditions are equivalent:*

(i) $\dim_{\mathbb{Z}} X \le n$.
(ii) $\lim_{i\to\infty} a_n^{n+1}(p_j^i) = 0$ *for each index* $j$.

PROOF. *(ii)* $\Longrightarrow$ *(i)*. We assume that the sequence $\mathcal{S} = \{X_i, p_i^{i+1}\}$ is realized in the Hilbert cube $I^\omega$, with a given metric on it. Let $f\colon A \to K(\mathbb{Z}, n)$ be a map defined on a closed subspaces $A$ of $X$. Since the complex $K(\mathbb{Z}, n)$ is an absolute extensor with respect to compact spaces, we can extend the map $f$ to a map

$\bar{f}\colon OA \to K(\mathbb{Z}, n)$ defined on a compact neighborhood $OA$ of the set $A$ in $X$. Choose a number $\epsilon > 0$ so small that any two $\epsilon$-close maps of any compactum into $OA$ have homotopic compositions with $\bar{f}$.

Take an index $k$ such that for each $i \geq k$ the projection $p_i$ moves points not more than $\epsilon$ and, additionally, the $\epsilon$-neighborhood $O_\epsilon(p_i(A))$ of the set $p_i(A)$ in $I^\omega$ is contained in $OA$.

By $(ii)$, there is an index $i(k)$ such that $a_n^{n+1}(p_k^{i(k)} < \frac{\epsilon}{2}$. Take a triangulation of mesh $< \frac{\epsilon}{2}$ of the polyhedron $X_{i(k)}$ and denote by $A_{i(k)}$ the union of all those simplexes of this triangulation that intersect $p_{i(k)}(A)$. Obviously,

$$p_{i(k)}(A) \subseteq A_{i(k)} \subseteq OA.$$

By the choice of the index $i(k)$, there exists a map $\psi\colon X_{i(k)}^{(n+1)} \to X_k$ of the $(n+1)$-dimensional skeleton of the polyhedron $X_{i(k)}$ (with respect to the above indicated triangulation) with at most $n$-dimensional image such that $dist(\psi, p_k^{i(k)}) < \frac{\epsilon}{2}$.

Obviously, $\psi(A_{i(k)}^{(n+1)}) \subseteq OA$ and the restrictions of the maps $\bar{f}\psi$ and $\bar{f}$ onto $A_{i(k)}^{(n+1)}$ (i.e. $n$-dimensional skeleton of $A_{i(k)}$ are homotopic.

Since $\dim \psi(X_{i(k)}^{(n+1)}) \leq n$, the map $\bar{f}$ admits an extension defined on the union $OA \cup \psi(X_{i(k)}^{(n+1)})$.

Consequently, the map $\bar{f}\psi/A_{i(k)}^{(n+1)}$ can be extended to $X_{i(k)}^{(n+1)}$. By the remark made in the beginning, $f/A_{i(k)}^{(n+1)}$ also has an extension to $X_{i(k)}^{(n+1)}$. Now recall that there are no obstructions to extending a map from the $m$-dimensional skeleton onto the $(m+1)$-dimensional one if the range has trivial $m$-dimensional homotopy group. Therefore there is a map $g\colon X_{i(k)} \to K(\mathbb{Z}, n)$ extending $\bar{f}/A_{i(k)}^{(n+1)}$. Since the maps $g/A_{i(k)}$ and $\bar{f}/A_{i(k)}$ coincide on the $(n+1)$-skeleton, they are homotopic. This shows that $\bar{f}/A_{i(k)}$ has an extension to the space $X_{i(k)}$. It only remains to remark that $f$ is homotopic to $\bar{f}p_{i(k)}/A$ and, consequently, has an extension to $X$. By Theorem 3.1.2, $\dim_{\mathbb{Z}} X \leq n$.

$(i) \Longrightarrow (ii)$. It is not hard to see that the inequality $\dim_{\mathbb{Z}} X \leq 1$ implies the inequality $\dim X \leq 1$. This observation shows that the implication $(i) \Longrightarrow (ii)$ is true for $n = 1$. Thus, we may assume below that $n \geq 2$.

Consider a finite simplicial complex $L$. We now present a description of a connected $\mathcal{CW}$-complex $\hat{L}$ associated with $L$. First, represent $L$ as the union $L = L^{(n)} \cup \sigma_1 \cup \cdots \cup \sigma_s$, where the simplexes $\sigma_i$ are indexed so that

$$n + 1 \leq \dim \sigma_1 \leq \dim \sigma_2 \leq \cdots \leq \sigma_s.$$

We construct the complex $\hat{L}$ by induction on $s$. In the meantime the following conditions will be satisfied:

(a) $\hat{L} = L^{(n)} \cup K(\sigma_1) \cup \cdots \cup K(\sigma_s)$, where $K(\sigma_i)$ is an Eilenberg-MacLane complex of type $K(\pi_n(\sigma_i^{(n)}, n)$.

(b) $\hat{L}^{(n)} = \hat{L}^{(n+1)} = L^{(n)}$ and $L^{(n)} \cap K(\sigma_i) = \sigma_i^{(n)}$ for each $i \leq s$.

(c)

$$K(\sigma_i) \cap K(\sigma_j) = \begin{cases} \sigma_i \cap \sigma_j, & \text{if } \dim(\sigma_i \cap \sigma_j) \leq n \\ K(\sigma_i \cap \sigma_j), & \text{if } \dim(\sigma_i \cap \sigma_j) \geq n+1. \end{cases}$$

If $s = 0$, then we let $\hat{L} = l^{(n)}$. If a $\mathcal{CW}$-complex $L^{(n)} \cup K(\sigma_1) \cup \cdots \cup K(\sigma_k)$, associated to the complex $L^{(n)} \cup \sigma_1 \cup \cdots \cup \sigma_k$ and satisfying the above conditions, has already been constructed, then the complex $K(\sigma_{k+1})$ can be obtained by killing all homotopy groups in dimensions larger than $n$ of the complex

$$K(\partial\sigma_{k+1}) = \sigma_{k+1}^{(n)} \cup \bigcup\{K(\sigma_i)\colon \sigma_i \subseteq \partial\sigma_{k+1}\}.$$

Since the embedding $\partial\sigma_{k+1}^{(n)} \hookrightarrow K(\partial\sigma_{k+1})$ induces isomorphisms of the homotopy groups of dimension $\leq n$, we conclude that condition (a) is satisfied. The remaining conditions are obvious.

We now proceed to the direct proof of the implication $(i) \Longrightarrow (ii)$. Let $\epsilon > 0$ and an index $j$ be given. Consider a triangulation of the polyhedron $X_j$ of *mesh* $< \frac{\epsilon}{2}$. Consider the complex

$$\hat{X}_j = X_j^{(n)} \cup K(\sigma_1) \cup \cdots \cup K(\sigma_s)$$

associated with this triangulation. Define a map $\hat{p}_j\colon X \to \hat{X}_j$ so that the restrictions of $p_j$ and $\hat{p}_j$ onto $p_j^{-1}(X_j^{(n)})$ coincide and such that for each $k \leq s$ we have

$$\hat{p}_j(p_j^{-1}(\sigma_k)) \subseteq K(\sigma_k).$$

We construct the desired map $\hat{p}_j$ by induction, defining it on the sets of the form $p_j^{-1}(X_j^{(n)} \cup \sigma_1\sigma \cdots \cup \sigma_k)$, where $0 \leq k \leq s$. If the map $\hat{p}_j$ has already been defined on the set $p_j^{-1}(X_j^{(n)} \cup \sigma_1\sigma \cdots \cup \sigma_k)$, then, by the inductive assumption,

$$\hat{p}_j(p_j^{-1}(\partial\sigma_{k+1})) \subseteq K(\sigma_{k+1}).$$

Since $\dim_{\mathbb{Z}} X \leq n$, we conclude that the map $\hat{p}_j/p_j^{-1}(\partial\sigma_{k+1})$ admits an extension to a map of the whole inverse image $p_j^{-1}(\sigma_{k+1})$ into

$$K(\sigma_{k+1}) \simeq K(\oplus_1^r \mathbb{Z}, n) \simeq \prod_1^r K(\mathbb{Z}, n).$$

This completes the inductive step and shows the existence of the map $\hat{p}_j$.

Extend the map $\hat{p}_j$ onto a neighborhood $OX$ of the limit space $X$ in $I^\omega$ and denote by $\hat{p}_j^i$ the restriction of this extension on $X_i$. For sufficiently large $i$ the following condition will be satisfied:

(∗) If $p_j^i(x) \in \sigma \subseteq X$ and $\hat{p}_j^i(x) \in \sigma' \subseteq X_j^{(n)}$, then $\sigma \cap \sigma' \neq \emptyset$; if $\hat{p}_j^i(x) \in K(\sigma_m)$, then $\sigma \cap \sigma_m \neq \emptyset$.

Next consider an at most $(n+1)$-dimensional compactum $Y \subseteq X_i$. Since, by the construction, $\hat{X}_j^{(n+1)} = X_j^{(n)}$, there exists a map $g_j^i\colon Y \to X_j^{(n)}$ such that

$$\hat{p}_j^i/(\hat{p}_j^i)^{-1}(X_j^{(n)}) = g_j^i/(\hat{p}_j^i)^{-1}(X_j^{(n)})$$

and, moreover, $g_j^i(y) \in \sigma$ whenever $\hat{p}_j^i(y)\sigma$, $y \in Y$ and $\sigma \in X_j$. It only remains to observe that, by $(*)$, the maps $p_j^i$ and $g_j^i$ are $\epsilon$-close. The proof is complete. □

Now we are ready to prove the equivalence of Alexandrov's problem and the $CE$-problem.

First notice that every inverse sequence $\mathcal{S} = \{X_i, p_i^{i+1}\}$, consisting of metrizable compact, can be realized in the Hilbert cube $I^\omega$ (see above). Take any metric $d$ on $I^\omega$ .Without loss of generality we can assume that all projections of the sequence $\mathcal{S}$ are *non-stretching*[1] maps with respect to the metric $d$.

Next, suppose that two inverse sequences $\mathcal{S} = \{X_i, p_i^{i+1}\}$ and $\mathcal{S}' = \{Y_i, q_i^{i+1}\}$ are realized in the Hilbert cube $I^\omega$ . We say that a sequence $\{f_i \colon X_i \to Y_i\}$ of maps converges to a map $f \colon X = \lim \mathcal{S} \to \lim \mathcal{S}' = Y$ if the map $|f| \colon |\mathcal{S}| \to |\mathcal{S}'|$, defined by relations $|f|/X = f$ and $|f|/X_i = f_i$ for each $i$, is continuous. Here $|\mathcal{S}|$ denotes the union of all elements of $\mathcal{S}$ and the limit space $\lim \mathcal{S}$ (recall that all these spaces are considered as subspaces of $I^\omega$ ).

LEMMA 3.1.9. *Let $\mathcal{S}' = \{Y_i, q_i^{i+1}\}$ be an inverse sequence realized in the Hilbert cube $I^\omega$ with a given metric d. If all projections of $\mathcal{S}'$ are non-stretching, then the convergence of the series $\sum_{i=1}^{\infty} \epsilon_i$, where $\epsilon_i = dist(f_i p_i^{i+1}, q_i^{i+1} f_{i+1})$, implies the convergence of the sequence $\{f_i\}$. Furthermore, the limit map $f \colon X \to Y$ satisfied the following condition for each $i$:*

$$dist(f_i p_i, q_i f) \leq \sum_{j=i}^{\infty} \epsilon_j.$$

PROOF. One can easily see, using induction on $k$, that

$$dist(f_i p_i^{i+k}, q^{i+k} f_{i+k}) \leq \sum_{j=0}^{k-1} \epsilon_{i+j}.$$

Then, for each $j \leq k$ we have

$$dist(q_i^j f_j p_j, q_i^k f_k p_k) \leq \sum_{n=j}^{k-1} \epsilon_n.$$

Therefore the sequence $\{q_i^j f_j p_j\}_{j=1}^{\infty}$ converges uniformly. It only remains to observe that the limit map $f \colon X \to Y$ can be uniquely determined from the equalities $q_i f = \lim_{j \to \infty} q_i^j f_j p_j$. □

THEOREM 3.1.10. *Let $X$ be a metrizable compactum. Then the following conditions are equivalent:*

(i) $\dim_{\mathbb{Z}} X \leq n$.
(ii) *There exists a cell-like map $f \colon Y \to X$ such that* $\dim Y \leq n$.

[1]A map $f \colon X \to Y$ is non-stretching if $dist(f(x), f(y)) \leq dist(x, y)$ whenever $x, y \in X$.

PROOF. $(ii) \Longrightarrow (i)$. Let $f\colon Y \to X$ be a cell-like map and $\dim Y \leq n$. Consider a closed subspace $A$ of the compactum $X$ and denote by $f_A$ the restriction of the map $f$ to the inverse image $f^{-1}(A)$. The natural inclusion maps of $A$ into $X$ and of $f^{-1}(A)$ into $Y$ are denoted by $i$ and $j$ respectively. By Proposition 3.1.4, we have $\dim Y \leq n$. Consequently, by Theorem 3.1.2, the map induced by the inclusion $j$, i.e. the map

$$j^*\colon [Y, K(\mathbb{Z}, n)] \to [f^{-1}(A), K(\mathbb{Z}, n)]$$

is a surjection (recall that the symbol $[A, B]$ denotes the set of homotopy classes of maps from $A$ into $B$). Since both maps $f\colon Y \to X$ and $f_A\colon f^{-1}(A) \to A$ are cell-like, we conclude, using the Vietoris-Beagle Theorem [**287**], that the induced maps

$$f^*\colon [X, K(\mathbb{Z}, n)] \to [Y, K(\mathbb{Z}, n)]$$

and

$$f_A^*\colon [A, K(\mathbb{Z}, n)] \to [f^{-1}(A), K(\mathbb{Z}, n)]$$

are bijections. The equality $fj = if_A$ implies the commutativity of the following diagram:

$$\begin{array}{ccc} [X, K(\mathbb{Z}, n)] & \xrightarrow{\quad i^* \quad} & [A, K(\mathbb{Z}, n)] \\ \Big\downarrow f^* & & \Big\downarrow f_A^* \\ [Y, K(\mathbb{Z}, n)] & \xrightarrow{\quad j^* \quad} & [f^{-1}(A), K(\mathbb{Z}, n)] \end{array}$$

It is now easy to see that the map

$$i^*\colon [X, K(\mathbb{Z}, n)] \to [A, K(\mathbb{Z}, n)]$$

is a surjection. Consequently, by Theorem 3.1.2, $\dim_{\mathbb{Z}} X \leq n$.

$(i) \Longrightarrow (ii)$. Represent $X$ as the limit space of an inverse sequence $\mathcal{S} = \{X_i, p_i^{i+1}\}$ consisting of compact polyhedra and surjective projections. As above, we may assume without loss of generality that the sequence $\mathcal{S}$ is realized in the Hilbert cube $I^\omega$. We may assume also that all projections of our spectra are non-stretching. If a sequence $\{\epsilon_i\}$ of positive numbers is given, then, by Theorem 3.1.8, we can define an increasing sequence $\{k(i)\}$ of natural numbers such that $a_n^{n+1}(p_{k(i)}^{k(i+1)}) < \epsilon_i$ for each $i$.

On each polyhedron $X_{k(i)}$, consider a triangulation of *mesh* $< \epsilon_i$ and denote by $X_{k(i)}^{(m)}$ an $m$-dimensional skeleton of this triangulation. Since $a_n^{n+1}(p_{k(i)}^{k(i+1)}) < \epsilon_i$, there exists an $\epsilon_i$-close to $p_{k(i)}^{k(i+1)}/X_{k(i+1)}^{(n+1)}$ map of $X_{k(i+1)}^{(n+1)}$ into $X_{k(i)}$ with at most $n$-dimensional image. Pushing this image into $X^{(n)}_{k(i)}$, we obtain a map $g^{i+1}_i \colon X^{(n+1)}_{k(i+1)} \to X^{(n)}_{k(i)}$ which is $2\epsilon_i$-close to the restriction $p^{k(i+1)}_{k(i)}/X^{(n+1)}_{k(i+1)}$.

Let $q^{i+1}_i = g^{i+1}_i/X^{(n)}_{k(i+1)}$, $q^j_i = q^{i+1}_i \cdot q^{i+2}_{i+1} \cdot \dots \cdot q^j_{j-1}$ and consider an inverse sequence $\mathcal{S}' = \{X^{(n)}_{k(i)}, q^{i+1}_i\}$. Let $e_i \colon X^{(n)}_{k(i)} \hookrightarrow X_{k(i)}$ denote the inclusion map. If the series $\sum \epsilon_i$ converges, then, by Lemma 3.1.9, the limit map $f \colon Y \to X$ is well defined (here $Y = \lim \mathcal{S}'$). Now we show that if we choose the sequence $\{\epsilon_i\}$ in a certain way then the map $f$ will be cell-like.

Let $\bar{\epsilon}_i = \sum_{j=i}^{\infty} 2\epsilon_i$. Then, by Lemma 3.1.9, we have the following inequality for each $i$:

$(*)$ $dist(p_{k(i)}f, q_i) \leq \bar{\epsilon}_i$.

The required condition on the sequence $\{\epsilon_i\}$ is the following:

*For each point $x \in X_{k(i+1)}$, its $2\bar{\epsilon}_{i+1}$-neighborhood can be contracted to a point within its $\epsilon_i$-neighborhood.*

The construction of such a sequence is trivial, and therefore we assume below that this condition for the sequence $\{\epsilon_i\}$ is satisfied.

Let $x$ be an arbitrary point in $X$. We are going to show that the inverse image $f^{-1}(x)$ is a cell-like set. Denote by $\bar{O}_\epsilon y$ the closed $\epsilon$-neighborhood of the point $y$. Also let $x_i = p_{k(i)}(x)$.

The map $q^{i+1}_i$ sends the set $O_{2\bar{\epsilon}_{i+1}} x_{i+1}$ into the set $O_{2\bar{\epsilon}_i} x_i$. Denote by $\bar{q}^j_i$ the restriction of the map $q^j_i$ to the set $O_{2\bar{\epsilon}_j} x_j \cap X^{(n)}_{k(j)}$. Observe that the indicated intersection is non-empty (recall that the *mesh* of the triangulation of $X_{k(j)}$ is less than $\epsilon_j$ and $\epsilon_j < 2\bar{\epsilon}_j$). Now we show that the limit space of the inverse sequence

$$\mathcal{S}_x = \{\bar{O}_{2\bar{\epsilon}_i} x_i \cap X^{(n)}_{k(i)}, \bar{q}^{i+1}_i\}$$

coincides with the inverse image $f^{-1}(x)$. If a point $y \in Y$ does not belong to the limit space $\lim \mathcal{S}_x$, then $q_i(y) \notin O_{2\bar{\epsilon}_i} x_i$ for some index $i$. It follows from condition $(*)$ that $dist(q_I(y), p_{k(i)}f(y)) \leq \bar{\epsilon}_i$. Therefore $p_{k(i)}f(y) \neq x_i$ and $y \notin f^{-1}(x)$. If $y \in \lim \mathcal{S}_x$, then $\lim_{i\to\infty} dist(q_i(y), x_i) = 0$. By condition $(*)$, we then have

$$\lim_{i\to\infty} dist(q_i(y), p_{k(i)}f(y)) = 0.$$

Consequently, $\lim_{i\to\infty} dist(x_i, p_{k(i)}f(y)) = 0$ and hence $y \in f^{-1}(x)$.

Since all spaces of the inverse sequence $\mathcal{S}_x$ are non-empty and compact, so is its limit. This shows that $f$ is a surjective map.

Further, $im(\bar{q}^{i+1}_i) \subseteq \bar{O}_{2\epsilon_i} x_i$ and, consequently, contracts to a point within $O_{\epsilon_{i+1}} x_i$. Pushing the corresponding homotopy into $X^{(n+1)}_{k(i)}$ we obtain a homotopy in $O_{\epsilon_{i-1}+\epsilon_i}(x) \cap X^{(n+1)}_{k(i)}$ (recall that the *mesh* of the triangulation of $X_{k(i)}$ is less than $\epsilon_i$). Consider the composition of the final homotopy and the map $g^i_{i-1}$. The image of this composition is contained in the neighborhood of the point $x_{i-1}$ of radius not greater than $\epsilon_{i-1} + \epsilon_i + 2\epsilon_{i-1} < 4\epsilon_{i-1} < 2\bar{\epsilon}_{i-1}$. Therefore, $\bar{q}^{i+1}_{i-1}$ is homotopic to a constant map.

Finally, observe that all projections of the subspectrum of the spectrum $\mathcal{S}_x$, indexed by even numbers, are non essential and, consequently, its limit space, i.e. the fiber $f^{-1}(x)$, is cell-like. This completes the proof of the Theorem. □

*Remark* 3.1.11. The proof of the implication $(ii) \Longrightarrow (i)$ in Theorem 3.1.10 shows that cell-like maps cannot raise the cohomological dimension $\dim_{\mathbb{Z}}$.

COROLLARY 3.1.12. *Alexandrov's problem is equivalent to CE-problem.*

*Historical and bibliographical notes* 3.1. Originally Theorem 3.1.2 was proved in [**110**], although the proof presented in the text is due to Ferry (see [**305**], [**130**]). Theorems 3.1.3 and 3.1.5 were obtained by Alexandrov [**4**]. Cohomological dimension theory (in particular, the $CE$-problem) is an important and popular area of investigation; see, for instance, [**5**], [**6**], [**43**], [**44**], [**118**], [**126**], [**128**], [**130**], [**132**], [**136**], [**205**], [**206**], [**198**], [**61**], [**207**], [**254**], [**262**], [**263**], [**138**]. Theorems 3.1.8 and 3.1.10 are due to Edwards (see [**305**]).

## 3.2. Cell-like mappings raising dimension

In this Section we present a positive solution of Alexandrov's problem (see p.100).

We begin with the following definition.

DEFINITION 3.2.1. *Let $f\colon X \to Y$ be a map. We write* $\dim_G f \leq n$ *if the following condition is satisfied:*

- *For each closed subspace $A$ of $Y$ and map $\varphi\colon A \to K(G,n)$, there is an extension $\psi\colon X \to K(G,n)$ of the composition $\varphi f/f^{-1}(A)$.*

Obviously, $\dim_G id_X = \dim_G X$, where $id_X$ denotes the identity map of $X$. If $f\colon X \to P$ is a map into a polyhedron $P$ with triangulation $\tau$, then the inequality $\dim_G(f,\tau)$ stands for the above condition, with $A$ a subpolyhedron of $P$ with respect to the triangulation $\tau$.

LEMMA 3.2.2. *Suppose that an inverse sequence $\mathcal{S} = \{X_i, p_i^{i+1}\}$, consisting of compact polyhedra, is realized in the Hilbert cube $I^\omega$ with a metric $d$. Assume, in addition, that $\tau_i$ is a triangulation of $X_i$ such that $\lim_{i\to\infty} mesh(\tau_i) = 0$. If $\dim_G(p_i^{i+1},\tau_i) \leq n$ for almost all indexes, then $\dim_G \lim \mathcal{S} \leq n$.*

PROOF. Let $A$ be a closed subspace of the limit space $X = \lim \mathcal{S}$ and $\varphi\colon A \to K(G,n)$ be an arbitrary map. Compactness of $A$ guarantees that there is a finite subcomplex $K \subseteq K(G,n)$ such that $\varphi(A) \subseteq K$. Since $K$ is an $ANR$-space, there exists an extension $\varphi'\colon U \to K$ of $\varphi$ to some neighborhood $U$ of the set $A$ in the Hilbert cube $I^\omega$ . Obviously, if $i$ is sufficiently large, then we have $p_i(A) \subseteq U$. Denote by $A_i$ the star of the set $p_I(A)$ in $X_i$ with respect to the triangulation

$\tau_i$, i.e. $A_i$ is the union of those simplexes of $\tau_i$ that intersect the set $p_i(A)$. Since $mesh(\tau_i) \to 0$, we conclude that for almost all indexes $i$ we have $A_i \subseteq U$. It is also easy to see that the sequence $\{\varphi' p_i/A\colon A \to K\}$ converges to the map $\varphi$. Since sufficiently close maps into an $ANR$-compactum are homotopic, there exists an index $i$ such that $\varphi' p_i/A \simeq \varphi$. Consequently, it suffices to construct an extension of the map $\varphi' p_i/A$. By the assumptions, for sufficiently large indexes $i$ we have $\dim_G(p_i^{i+1}, \tau_i) \leq n$. This guarantees the existence of an extension $\psi\colon X_{i+1} \to K(G,n)$ of the map $\varphi' p_i^{i+1}/(p_i^{i+1})^{-1}(A_i)$. It only remains to observe that the composition $\psi p_{i+1}$ extends the map $\varphi' p_i/A$. □

LEMMA 3.2.3. *Let a compactum $X$ be the limit space of an inverse sequence $\mathcal{S} = \{P_i, p_i^{i+1}\}$, consisting of compact polyhedra, $\tau$ be a triangulation of the polyhedron $P_0$ and $\dim_G(p_0, \tau) \leq n$, where $G$ is a finitely generated group ($G = \mathbb{Z}$ for instance). Then there exists an index $i$ such that $\dim_G(p_0^i, \tau) \leq n$.*

PROOF. Let $A_0$ be a subpolyhedron of the polyhedron $P_0$ with respect to the triangulation $\tau$, and $\varphi\colon A_0 \to K(G,n)$ be a map defining one of the generators of the group $H^n(A_0; G)$. As before, we may assume that the sequence $\mathcal{S}$ is realized in the Hilbert cube $I^\omega$. By the condition $\dim_G(p_0, \tau) \leq n$, there exists an extension $\bar{\varphi}\colon X \to K(G,n)$ of the map $\varphi p_0/p_0^{-1}(A_0)$. Consider the closed subset

$$B = X \cup \left(\cup_{i=0}^{\infty}(p_0^i)^{-1}(A_0)\right)$$

of the Hilbert cube $I^\omega$. The collection of maps

$$\{\bar{\varphi}\} \cup \{\varphi p_0^i/(p_0^i)^{-1}(A_0)\}_{i=0}^{\infty}$$

defines a continuous map $\psi\colon B \to K(G,n)$. By the compactness of $B$, there is a finite complex $K$ such that $\psi(B) \subseteq K \subseteq K(G,n)$. Since $K$ is an $ANR$-space, there is an extension $\bar{\psi}\colon U \to K$ of the map $\psi$ to a neighborhood $U$ of the set $B$ in $I^\omega$. Obviously, $id_X = \lim p_i$. This shows that there is an index $k$ such that $P_i \subseteq U$ for each $i \geq k$. Also observe that the map $\bar{\psi}/P_i$ extends the map $\varphi p_0^i/(p_0^i)^{-1}(A_0)$ whenever $i \geq k$. Since there are only finitely many different subpolyhedra $A$ of the polyhedron $P_0$, and since the corresponding cohomologies $H^n(A;G)$ are finitely generated, we see that there exists an index $k$ such that for each subpolyhedron $A \subseteq P_0$ and map $\varphi\colon A \to K(G,n)$, corresponding to one of the generators of the group $H^n(A;G)$, there exists an extension $\psi_{A,\varphi}\colon P_k \to K(G,n)$ of the map $\varphi p_0^k/(p_0^k)^{-1}(A)$. In terms of cohomology, this just means that for each subpolyhedron $A \subseteq P_0$, and for each generator $\alpha \in H^n(A;G)$, there is an element $\beta(\alpha) \in H^n(P_k;G)$ such that $j^*(\beta(\alpha)) = (p_0^k/\ldots)^*(\alpha)$, where $j^*$ denotes the homomorphism induced by the inclusion $j\colon (p_0^k)^{-1}(A) \subseteq P_k$. If $\gamma \in H^n(A;G)$ is not a generator, then $\gamma$ can be expressed in terms of the generators, $\gamma = \sum \lambda_i \alpha_i$, and, consequently,

$$(p_0^k/\ldots)^*(\gamma) = j^*(\sum \lambda_i \beta(\alpha_i)).$$

Translating this fact back into the language of maps into $K(G,n)$, we obtain the desired conclusion. □

LEMMA 3.2.4. *Let $L$ be a compact polyhedron with a triangulation $\tau$. Then there exists a map $\psi_\tau\colon E(\tau) \to L$ satisfying the following conditions:*

(i) *For each simplex $\sigma \in \tau$, the inverse image $\psi_\tau^{-1}(\sigma)$ has the homotopy type of the complex $K(\oplus_1^{r_\sigma}\mathbb{Z}, 3)$, where $r_\sigma = \operatorname{rank}\pi_3(\sigma^{(3)})$.*
(ii) *The space $E(\tau)$ is a $\mathcal{CW}$-complex admitting a triangulation and such that its $m$-dimensional skeleton $E(\tau)^{(m)}$ is a finite subcomplex for each $m$.*
(iii) *The map $\psi_\tau$ is combinatorial with respect to the triangulation $\tau$ and the $\mathcal{CW}$-complex structure of $E(\tau)$.*
(iv) $\dim_{\mathbb{Z}}(\psi_\tau, \tau) \leq 3$.

PROOF. We use induction on the dimension of $L$. First let us consider one additional condition:

(v) For each 3-connected subpolyhedron $K \subseteq L$, the space $\psi_\tau^{-1}(K)$ is 2-connected and
$$H_3(\psi_\tau^{-1}(K)) = \oplus_1^{r_K}\mathbb{Z},$$
where $r_K = \operatorname{rank}\pi_3(K^{(3)})$.

If $\dim L \leq 3$, then $E(\tau) = L$ and $\psi_\tau = id_L$. Next we consider the case $\dim L = 4$. For each 4-dimensional simplex $\sigma$, consider the inclusion $j_\sigma\colon \sigma^{(3)} \hookrightarrow K(\mathbb{Z},3)$ as the 3-dimensional skeleton. We may assume that the finite-dimensional skeleta of the $\mathcal{CW}$-complex $K(\mathbb{Z},3)$ are finite complexes. In addition, we may assume that $K(\mathbb{Z},3)^{(4)} = \sigma^{(3)}$. For each $\sigma$ we attach the mapping cylinder $M_{j_\sigma}$ of the inclusion $j_\sigma$ to $L^{(3)}$ along the subset $\sigma^{(3)}$. Denote by $c_\sigma$ the center of the simplex $\sigma$. Now we define $\psi_\tau(K(\mathbb{Z},3)) = c_\sigma$ for each $\sigma$, and $\psi_\tau(L^{(3)}) = id_{L^{(3)}}$. Next we extend $\psi_\tau$ onto $L^{(3)} \cup (\cup_\sigma M_{j_\sigma})$ by linearity. Finally we let
$$E(\tau) = L^{(3)} \cup (\cup_\sigma M_{j_\sigma}).$$

Conditions (i)-(v) are satisfied by the construction.

Suppose that the lemma has already been proved for polyhedra $M$ with $\dim M \leq m$, $m > 3$, and consider an $(m+1)$-dimensional polyhedron $L$ with triangulation $\tau$. By the inductive assumption, there exists a map
$$\psi_{\tau^{(m)}}\colon E(\tau^{(m)}) \to L^{(m)}$$
satisfying conditions (i)-(v). Take any $(m+1)$-dimensional simplex $\sigma \in \tau$ and consider the space $\psi_{\tau^{(m)}}^{-1}(\sigma^{(m)})$. By condition (v), we have
$$H_3(\psi_{\tau^{(m)}}^{-1}(\sigma^{(m)})) = \oplus_1^{r_\sigma}\mathbb{Z}.$$

Let $Y_3 = \psi_{\tau^{(m)}}^{-1}(\sigma^{(m)})$. By condition (v) and the Hurewicz theorem, we have $\pi_3(Y_3) = \oplus_1^{r_\sigma}\mathbb{Z}$. By condition (ii), $Y_3^{(5)}$ is a finite complex. Since $\pi_4(Y_3) = \pi_4(Y_3^{(5)})$, and since $Y_3^{(5)}$ is one-connected, we conclude that $\pi_4(Y_3)$ is a finitely generated group (see [**270**]). Attaching 5-dimensional cells to $Y_3$, we construct a complex $Y_4$ with $\pi_4(Y_4) = 0$. Attaching 5-dimensional cells to $Y_4$, we obtain a complex $Y_5$ with $\pi_3(Y_5) = \oplus_1^{r_\sigma}\mathbb{Z}$ and $\pi_i(Y_5) = 0$ for each $i < 3$ and $i = 4, 5$. Continuing in this manner we obtain an increasing sequence

$$Y_3 \subseteq Y_4 \subseteq Y_5 \subseteq \cdots \subseteq Y_{3+i} \subseteq \cdots .$$

Let

$$Y_\infty = \varinjlim\{Y_{3+i}\}.$$

Obviously the complex $Y_\infty$ is homotopy equivalent to the space $K(\oplus_1^{r_\sigma}\mathbb{Z}, 3)$.

Denote by $j_\sigma$ the inclusion of the space $\psi^{-1}_{\tau^{(m)}}(\sigma^{(m)})$ into $Y_\infty$ and by $M_{j_\sigma}$ the mapping cylinder of this inclusion. For each $(m+1)$-dimensional simplex $\sigma$, we attach the space $M_{j_\sigma}$ to the space $E(\tau^{(m)})$ along the subspace $\psi^{-1}_{\tau^{(m)}}(\sigma^{(m)})$. The complex obtained in this way is the required space $E(\tau)$. Extend the map $\psi_{\tau^{(m)}}\colon E(\tau^{(m)}) \to L^{(m)}$ to the map $\psi_\tau\colon E(\tau) \to L$ as follows. For each simplex $\sigma$, let $\psi_\tau(Y_\infty) = c_\sigma$ be the center of the simplex $\sigma$, and further extend $\psi_\tau$ by linearity. In other words, shrinking $Y_\infty$ into a point, we obtain from $M_{j_\sigma}$ the cone over the space $\psi^{-1}_{\tau^{(m)}}(\sigma^{(m)})$, and since $\sigma$ is the cone over $\sigma^{(m)}$ we see that the map

$$\psi_{\tau^{(m)}}/\ldots\colon \psi^{-1}_{\tau^{(m)}}(\sigma^{(m)}) \to \sigma^{(m)}$$

between the bases of these cones can be naturally extended to the map between these cones.

It only remains to verify conditions (i)-(v). Conditions (i)-(iii) are satisfied by the construction. Let $K \subseteq L$ be a 3-connected subpolyhedron of the polyhedron $L$. Consider an arbitrary $(m+1)$-dimensional simplex $\sigma \subseteq K$. If there is no such $\sigma$, then condition (v) is satisfied by the inductive hypothesis. In the mapping cylinder $M_{j_\sigma}$, shrink each segment connecting a point $x \in \psi^{-1}_{\tau^{(m)}}(\sigma^{(m)})$ with the point $j_\sigma(x)$, to a point. This defines a cell-like map $\xi\colon \psi_\tau^{-1}(K) \to Z_\sigma$. Obviously, $\xi(M_{j_\sigma}) = Y_\infty$. Observe also that

$$H_3(\psi_\tau^{-1}(K)) = H_3(\psi^{-1}_{\tau^{(m)}}(K^{(m)}) \cup \left(\cup_{\sigma'\neq\sigma} M_{j'_\sigma}\right) \cup M_{j_\sigma} =$$

$$H_3(\psi^{-1}_{\tau^{(m)}}(K^{(m)}) \cup \left(\cup_{\sigma'\neq\sigma} M_{j'_\sigma}\right) \cup Y_3) = H_3(\psi^{-1}_{\tau^{(m)}}(K^{(m)}) \cup \left(\cup_{\sigma'\neq\sigma} M_{j'_\sigma}\right)).$$

Consequently, removing single simplexes, we get

$$H_3(\psi_\tau^{-1}(K)) = H_3(\psi^{-1}_{\tau^{(m)}}(K^{(m)})) = \oplus_1^{r_K}\mathbb{Z}.$$

Let us show now that $\pi_i(\psi_\tau^{-1}(K)) = 0$ for each $i < 3$. Obviously, the cells of dimension more than 3 do not affect the $i$-dimensional homotopy groups with $i < 3$. Therefore, for each simplex $\sigma \subseteq K$ we can remove the cells of dimension larger than 3 from the space $M_{j_\sigma} \subseteq \psi_\tau^{-1}(\sigma)$. As a result we obtain a space, deformable onto $\psi_\tau^{-1}(K^{(m)})$. Observing that $\pi_i(\psi_\tau^{-1}(K^{(m)})) = 0$ for each $i < 3$, we finish the verification of condition (v).

Let us verify the remaining condition (iv). Let $A \subseteq L$ be a subpolyhedron of the polyhedron $L$ and $\varphi\colon A \to K(\mathbb{Z}, 3)$ be an arbitrary map. By the inductive hypothesis there is a map

$$\varphi'\colon \psi_\tau^{-1}(L^{(m)}) \cup \psi_\tau^{-1}(A) \to K(\mathbb{Z}, 3),$$

extending the map $\varphi\psi_\tau$. All we now have to do is extend the map $\varphi'/\psi_\tau^{-1}(\sigma^{(m)})$ to a map $\varphi_\sigma\colon M_{j_\sigma} \to K(\mathbb{Z}, 3)$ for each simplex $\sigma \subseteq L$ with $\sigma \not\subseteq A$. But by the construction, the inclusion $j_\sigma$ induces an epimorphism (even an isomorphism) of 3-dimensional cohomologies. Therefore the indicated extension exists for each $\sigma$. The proof is complete. □

*Remark* 3.2.5. The 3-dimensional skeleton $E(\tau)^{(3)}$ of the complex $E(\tau)$ is, by construction, homeomorphic to the 3-dimensional skeleton $L^{(3)}$ of the polyhedron $L$. Moreover, the restriction $\psi_\tau/E(\tau)^{(3)}$ serves as the corresponding homeomorphism.

Let us now recall some facts from the *generalized cohomology theories.* A comprehensive introduction to the corresponding results can found in several textbooks (for instance, see [**291**]). Recall that the *reduced generalized cohomology theory* $h^*$ on the category $\mathcal{PT}'$ of pointed spaces with morphisms homotopy classes of maps, preserving base points, is a collection $\{h^n\colon \mathcal{PT}' \to \mathcal{A}\}$ of contravariant functors into the category of Abelian groups and natural equivalences $\sigma^n\colon h^n \to h^{n+1}\Sigma^*$, where $\Sigma^*$ - is the functor of taking the *reduced suspension* and $n \in \mathbb{Z}$. These objects satisfy the following *exactness axiom*:

- For each pair $(X, A, x_0)$ the following sequence is exact

$$\cdots \leftarrow h^n(A) \xleftarrow{i^*} h^n(X) \xleftarrow{j^*} h^n(X \cup C^*A) \xleftarrow{\delta_n} h^{n-1}(A) \leftarrow \cdots$$

Here $i\colon (A, x_0) \hookrightarrow (X, x_0)$ and $j\colon (X, x_0) \hookrightarrow (X \cup C^*A, *)$ denote inclusions, $C^*$ stands for the *reduced cone*, $i^* = h^n([i])$ and $j^* = h^n([j])$. The homomorphism $\delta_n$ coincides with the composition $k^* \cdot \nu^* \cdot \sigma^{n-1}$, where $\nu\colon \Sigma^*A \to \Sigma^*$ is the inverse in the *H-cogroup* and $k\colon X \cup C^*A \to \Sigma^*A$ shrinks $X$ into a point.

Applying the above exact sequence to $(\{x\}, \{x\}, x)$ we see that $h^n(\{x\}) = 0$ for each $n$. The following lemma is also obvious.

LEMMA 3.2.6. *If a map $f\colon X \to Y$ is homotopic to a constant map, then the homomorphism $f^*\colon h^*(Y) \to h^*(X)$ is trivial.*

One of the major ways of defining the generalized cohomology theories is based on the so called *spectra*. The spectrum $E$ is a collection $\{(E_n, *)\colon n \in \mathbb{Z}\}$ of $\mathcal{CW}$-complexes such that $\Sigma^*E_n \subseteq E_{n+1}$. The cofunctor $h^*$ can then be defined by letting

$$h^n(\cdot) = \varinjlim [\Sigma^k \cdot; e_{n+k}].$$

It is known that under some restrictions, all generalized cohomology theories can be obtained as indicated above[2].

[2]This fundamental fact was proved by Brown, see [**291**].

The *Unreduced generalized cohomology theory* $k^*$ on the category $\mathcal{T}^2$ of pairs of topological spaces is a sequence $\{k^n\colon \mathcal{T}^2 \to \mathcal{A}, n \in \mathbb{Z}\}$ of contravariant functors and natural equivalences $\{\delta^n\colon k^{n-1} \cdot R \to k^n, n \in \mathbb{Z}\}$ (here $R(X, A) = (A, \emptyset)$) satisfying the following axioms:

(1) Values of $k^*$ are the same for homotopic maps.

(2) For each pair $(X, A)$, the following sequence is exact

$$\cdots \leftarrow k^{n+1}(X, A) \overset{\delta^{n+1}}{\longleftarrow} k^n(A, \emptyset) \overset{i^*}{\longleftarrow} k^n(X, \emptyset) \overset{j^*}{\longleftarrow} k^n(X, A) \leftarrow \cdots .$$

(3) For each pair $(X, A)$, and for any set $U$ with $U \subseteq \operatorname{cl} U \subseteq \operatorname{int} A$, the homomorphism induced by the inclusion $(X - U, A - U) \hookrightarrow (X, A)$ is an isomorphism.

Conditions (1)-(3) are the well known Eilenberg-Steenrod axioms 4-6 (see [**290**]). Axioms 1-3 of Eilenberg-Steenrod are contained in the above definition. Generally speaking, it is not required that axiom 7 of Eilenberg-Steenrod be satisfied. This is exactly why these cohomology theories are called generalized. Below we use the notation $k^*(X) = k^*(X, \emptyset)$.

The following statement is well known and can be found in [**291**].

LEMMA 3.2.7. *For a cellular triad $(X; A, B)$ we have the following* Mayer-Vietoris exact sequence

$$\cdots \leftarrow k^n(A \cap B) \leftarrow k^n(A) \oplus k^n(B) \leftarrow k^n(X) \leftarrow k^{n-1}(A \cap B) \leftarrow \cdots$$

*and this sequence is functorial.*

As usual $X = A \cup B$ and $(A, B)$ is an ordered pair.

Each reduced cohomology theory $h^*$ defines a corresponding unreduced cohomology $\hat{h}^*$ simply by letting

$$\hat{h}^n(X, A) = h^n(X \cup C^*A \cup \mathrm{pt}).$$

Here an isolated point pt is taken as the base point. If $A = \emptyset$, then $\hat{h}^n(X) = h^n(X \cup \mathrm{pt})$.

LEMMA 3.2.8. *Let $f\colon (X, x) \to (Y, y)$ be a continuous map. Then the following conditions are equivalent:*

(i) *$f^*\colon h^*(Y) \to h^*(X)$ is an isomorphism.*

(ii) *$\hat{f}^*\colon \hat{h}^*(Y) \to \hat{h}^*(X)$ is an isomorphism.*

PROOF. Our statement follows from the five lemma, applied to the diagram generated by the exact sequences of pairs $(X \cup \mathrm{pt}, \{x\} \cup \mathrm{pt})$, $(Y \cup \mathrm{pt}, \{y\} \cup \mathrm{pt})$ and their transformation $f \cup \mathrm{id}_{\mathrm{pt}}$. $\square$

Let $\mathbf{U}$ be an infinite *unitary group* and $\mathbf{BU}$ its *classifying space.* It is known [**291**] that $\mathbf{BU}$ is a $\mathcal{CW}$-complex. Define a spectrum $E$ by letting $E_{2n} = \mathbf{BU}$ and $E_{2n+1} = \mathbf{U}$ for each $n \in \mathbb{Z}$. *Bott periodicity* implies that $E$ is a spectrum.

The corresponding reduced generalized cohomology theory is known to be the reduced complex $K$-theory $\tilde{K}^*_{\mathbf{C}}$ (see [**291**] for details).

Let $B^2_2$ be the *Moore space*, that is $B^2_2 = S^1 \cup_2 B^2$ is the circle with an attached (via a map of degree 2) 2-dimensional cell. By $\wedge$ we denote the *smash product.* Recall that

$$X \wedge Y = (X \times Y)/(X \times \{y_0\}) \cup (\{x_0\} \times Y),$$

where $x_0$ and $y_0$ are base points in the spaces $X$ and $Y$ respectively and $(X \times \{y_0\}) \cup (\{x_0\} \times Y)$ defines the base point in $X \wedge Y$. The $K$-functor with coefficients in $\mathbb{Z}_2$ is defined as follows:

$$\tilde{K}^*_{\mathbf{C}}(X; \mathbb{Z}_2) = \tilde{K}^*_{\mathbf{C}}(X \wedge B^2_2).$$

Observe that $\tilde{K}^*_{\mathbf{C}}(\cdot; \mathbb{Z}_2)$ is the reduced generalized cohomology theory.

The following two statements play a key role.

PROPOSITION 3.2.9. *Let $X = \cup X^n$ be an arbitrary* filtration *of a $\mathcal{CW}$-complex $X$ by its subcomplexes. Then*

$$\tilde{K}^*_{\mathbf{C}}(X; \mathbb{Z}_2) = \varprojlim \tilde{K}^*_{\mathbf{C}}(X^n; \mathbb{Z}_2).$$

PROPOSITION 3.2.10. *If $r$ is a natural number, then*

$$\tilde{K}^*_{\mathbf{C}}\left(K\left(\oplus_1^r \mathbb{Z}, 3\right); \mathbb{Z}_2\right) = 0.$$

LEMMA 3.2.11. *The map $\psi_\tau \colon E(\tau) \to L$ from Lemma 3.2.4 induces an isomorphism*

$$(\psi_\tau)^* \colon \tilde{K}^*_{\mathbf{C}}(L; \mathbb{Z}_2) \to \tilde{K}^*_{\mathbf{C}}(E(\tau); \mathbb{Z}_2).$$

PROOF. Let us show that each continuous map $f \colon E(\tau) \to L$, satisfying conditions (i) and (iii) of Lemma 3.2.4, has the required property. We use induction on the dimension of $L$.

If $\dim L = 0$, then $L$ consists of finitely many points, and for each of these points $x$, we have $f^{-1}(x) \simeq K(\oplus \mathbb{Z}, 3)$. Therefore the required conclusion follows from Proposition 3.2.10.

Suppose that our statement has already been proved for polyhedra $M$ with $\dim M \leq n$. Consider an $(n+1)$-dimensional compact polyhedron $L$ with triangulation $\tau$. We now proceed by induction on the number of $(n+1)$-dimensional simplexes of $\tau$. If there is no $(n+1)$-dimensional simplex, then $\dim L \leq n$. Let $L$ contains $m$ simplexes of dimension $n+1$. Take one of these simplexes $\sigma$, and denote by $M$ the polyhedron $L - \operatorname{int} \sigma$. Consider the restriction $g$ of the map $f \colon E(\tau) \to L$ to the subspace $f^{-1}(M)$ and the restriction $q$ of the map $f$ to the subspace $f^{-1}(\partial \sigma)$. By the inductive assumption, the homomorphisms $g^*$ and $q^*$ induce isomorphisms of the corresponding $\tilde{K}^*_{\mathbf{C}}(\cdot; \mathbb{Z}_2)$ groups. Since $f^{-1}(\sigma) \simeq K\left(\oplus_1^r \mathbb{Z}, 3\right)$, we conclude, by Proposition 3.2.10, that the restriction $s$ of the map $f$ to the inverse image $f^{-1}(\sigma)$ induces a trivial isomorphism

$$s^* \colon \tilde{K}^*_{\mathbf{C}}(\sigma; \mathbb{Z}_2) \to \tilde{K}^*_{\mathbf{C}}(f^{-1}(\sigma); \mathbb{Z}_2).$$

Lemma 3.2.8 guarantees that the homomorphisms $\hat{g}^*$, $\hat{q}^*$ and $\hat{s}^*$ induce isomorphisms of the corresponding reduced $K$-groups. Applying Lemma 3.2.7 to the triads $(L;\sigma,M)$ and $(E(\tau);f^{-1}(\sigma),f^{-1}(M))$ we obtain the following commutative diagram:

$$\begin{array}{ccccccccc}
\cdots\leftarrow h^i(f^{-1}(\partial\sigma)) & \leftarrow & h^i(f^{-1}(\sigma))\oplus h^i(f^{-1}(M)) & \leftarrow & h^i(E(\tau)) & \leftarrow & h^{i-1}(f^{-1}(\partial\sigma)) & \leftarrow \\
\uparrow & & \uparrow & & \uparrow & & \uparrow & \\
\cdots\leftarrow h^i(\partial\sigma) & \longleftarrow & h^i(\sigma)\oplus h^i(M) & \longleftarrow & h^i(L) & \longleftarrow & h^{i-1}(\partial\sigma) & \leftarrow
\end{array}$$

Here $h^*(\cdot)=\hat{K}^*_{\mathbb{C}}(\cdot;\mathbb{Z}_2)$. The five lemma implies that $\hat{f}^*$ is an isomorphism. It only remains to remark that, by Lemma 3.2.8, $f^*$ is also an isomorphism. □

The following statement follows immediately from Definition 3.2.1.

LEMMA 3.2.12. *Let $f\colon X\to L$ be a map, $\dim_{\mathbb{Z}}(f,\tau)\le n$, and $Y\subseteq X$. Then $\dim_{\mathbb{Z}}(f/Y,\tau)\le n$.*

LEMMA 3.2.13. *Let $L$ be a compact polyhedron with triangulation $\tau$. If $\alpha$ is a non-trivial element of the group $\tilde{K}^*_{\mathbb{C}}(L;\mathbb{Z}_2)$, then there exists a compact polyhedron $x_\alpha\subseteq E(\tau)$ such that $(\psi_\tau/X_\alpha)^*(\alpha)\neq 0$.*

PROOF. By condition (ii) of Lemma 3.2.4, there is a filtration $X^1\subseteq X^2\subseteq X^3\subseteq\cdots$ of the complex $E(\tau)$, consisting of finite subcomplexes. By Lemma 3.2.11, $\psi^*_\tau(\alpha)\neq 0$. By Proposition 3.2.9, there exists an index $i$ such that $\gamma^*_i(\psi^*_\tau(\alpha))\neq 0$, where $\gamma^*_i$ denotes the homomorphism induced by the inclusion $\gamma_i\colon X^i\hookrightarrow E(\tau)$. Let $X_\alpha=X^i$. It only remains to remark that $(\psi_\tau/X_\alpha)^*(\alpha)=\gamma^*_i\psi^*_\tau(\alpha)\neq 0$. □

Now we are ready to prove the main result of this Section.

THEOREM 3.2.14. *There exists an infinite-dimensional compactum $X$ such that $\dim_{\mathbb{Z}}X=3$.*

PROOF. We construct the compactum $X$ as the limit space of an inverse sequence $\mathcal{S}=\{X_i,p_i^{i+1}\}$ consisting of compact polyhedra. The spectrum $\mathcal{S}$ itself is constructed by induction and realized in the Hilbert cube $I^\omega$. Represent the Hilbert cube $I^\omega$ as an infinite product

$$I_0^\omega\times I_1^\omega\times\cdots\times I_n^\omega\times\cdots,$$

where $I_n^\omega = I^\omega$ for each $n \in \omega$. The polyhedron $X_n$ will be constructed as a subspace of the subproduct

$$I_0^\omega \times I_1^\omega \times \cdots \times I_n^\omega \times \{0\} \times \cdots \times \{0\} \times \cdots,$$

where $0 \in I^\omega$ is the base point. In addition, the embedding of $X_n$ into $I^\omega$ will be chosen so that the restriction of the projection

$$\pi_n^{n+1} \colon \prod_{i=0}^{n+1} I_i^\omega \to \prod_{i=0}^{n} I_i^\omega$$

to $X_{n+1}$ coincides with $p_n^{n+1}$. Clearly, in this situation, $X = \lim \mathcal{S} \subseteq I^\omega$ and $\lim_{i\to\infty} p_i = id_X$, i.e. the sequence $\mathcal{S}$ is realized in the Hilbert cube $I^\omega$.

Take any metric $d$ on $I^\omega$, and define a sequence $\{\tau_i\}$ of triangulations of the polyhedra $X_i$ so that $\lim_{i\to\infty} mesh(\tau_i) = 0$. Further, by induction, we define a sequence $\{\alpha_i\}$ of non-trivial elements $\alpha_i \in \tilde{K}^*_{\mathbf{C}}(X_i;\mathbb{Z}_2)$. As the polyhedron $X_0$ we take the 4-dimensional sphere $S^4$. Embed $X_0$ into $I_0^\omega \times \{0\} \cdots \times \{0\} \times \cdots$, and take a triangulation $\tau_0$ of $X_0$ with $mesh(\tau_0) < 1$. As the element $\alpha_0$ take a generator of the group $\tilde{K}^*_{\mathbf{C}}(S^4;\mathbb{Z}_2) = \mathbb{Z}_2$.

Suppose that the polyhedra $X_i$, projections $p_{i-1}^i$, triangulations $\tau_i$ of $X_i$ and non-trivial elements $\alpha_i \in \tilde{K}^*_{\mathbf{C}}(X_i;\mathbb{Z}_2)$, satisfying the above conditions, have already been constructed for each $i \le n$. By Lemma 3.2.13, there exists a compact polyhedron $X_{\alpha_n} \subseteq E(\tau_n)$ such that $(\psi_{\tau_n}/X_{\alpha_n})^*(\alpha_n) \neq 0$. Let $X_{n+1} = X_{\alpha_n}$, $p_n^{n+1} = \psi\tau_n/X_{\alpha_n}$ and $\alpha_{n+1} = (p_n^{n+1})^*(\alpha_n)$. Embed $X_{n+1}$ into $I_{n+1}^\omega$ and consider the graph of $p_n^{n+1}$ in the product

$$I_0^\omega \times \cdots \times I_{n+1}^\omega \times \{0\} \times \cdots \times \{0\} \times \cdots.$$

This graph is the desired embedding of $X_{n+1}$ into $I^\omega$. Finally, we take any triangulation of $X_{n+1}$ with $mesh(\tau_{n+1}) < \frac{1}{n+1}$.

By condition (iv) of Lemma 3.2.4 and by Lemma 3.2.12, $\dim_{\mathbb{Z}}(p_n^{n+1}, \tau_n) \le 3$ for each $n \in \omega$. Lemma 3.2.2 guarantees that $\dim_{\mathbb{Z}} X \le 3$. It is also obvious (see Remark 3.2.5) that $\dim_{\mathbb{Z}} \ge 3$. Therefore, $\dim_{\mathbb{Z}} = 3$.

Let us verify that $X$ is infinite-dimensional. Assume the contrary. Then, by Theorem 3.1.5, $\dim X = 3$. Consequently, the limit projection $p_o \colon X \to X_0 = S^4$ is homotopic to a constant map. Then we can find an index $i$ such that the short projection $p_0^i \colon X_i \to X_0$ of the spectrum $\mathcal{S}$ is also homotopic to a constant map (see, for example, [**290**]). By Lemma 3.2.6, the homomorphism

$$(p_0^i)^* \colon \tilde{K}^*_{\mathbf{C}}(X_0;\mathbb{Z}_2) \to \tilde{K}^*_{\mathbf{C}}(X_i;\mathbb{Z}_2)$$

is trivial. On the other hand, by the above construction, $(p_0^i)^*(\alpha_0) = \alpha_i \neq 0$. This contradiction finishes the proof. □

The specific nature of $K$-theory plays no direct role in the above construction. The essential feature is that it is a generalized cohomology theory for which the complex $K(\mathbb{Z}, 3)$ behaves as a point. The absence of a generalized cohomology theory for which the complex $K(\mathbb{Z}, 2)$ is known to behave as a point requires an alternate approach in order to produce an infinite-dimensional compactum having integral cohomological dimension equal to 2. Such an approach, based on the validity (see [**230**]) of the following *Sullivan Conjecture*, was carried out in [**136**]:

- If $G$ is a locally finite group (i.e., each finitely generated subgroup is finite) and $L$ is a finite-dimensional $\mathcal{CW}$-complex, then the space of pointed maps from $K(G,1) \to L$ has trivial homotopy groups.

This leads to the following result.

THEOREM 3.2.15. *There exists an infinite-dimensional compactum $X$ such that* $\dim_{\mathbb{Z}} X = 2$.

Theorems 3.1.10 and 3.2.15 imply the following.

COROLLARY 3.2.16. *There exists a cell-like map $f: X \to Y$ between compacta $X$ and $Y$ for which* $\dim X = 2$ *and* $\dim Y = \infty$.

COROLLARY 3.2.17. *There exists a cell-like map $f: I^5 \to Y$, where* $\dim Y = \infty$.

PROOF. Embed the compactum $X$ from Corollary 3.2.16 into the cube $I^5$. Then consider the quotient space of $I^5$ with respect to the decomposition, whose non-trivial elements are the fibers $f^{-1}(y)$, where $f: X \to Y$ denotes the map from Corollary 3.2.16. □

It is known (see [**307**]) that cell-like maps defined on the 3-dimensional cube cannot raise dimension. Thus, the following problem remains open.

PROBLEM 3.2.18. *Can cell-like maps defined on the 4-dimensional cube raise dimension?*

We conclude this Chapter by showing that cell-like images of $ANR$-compacta are not necessarily $ANR$-spaces. First we need the following concept.

DEFINITION 3.2.19. *We say that a map $f: X \to Y$ is* approximately invertible *if for each open cover $\mathcal{U} \in cov(Y)$ there is a map $g_{\mathcal{U}}: Y \to X$ such that the composition $fg_{\mathcal{U}}$ is $\mathcal{U}$-close to the identity map $id_Y$.*

LEMMA 3.2.20. *If $f: X \to Y$ is approximately invertible, then* $\dim X \geq \dim Y$.

PROOF. Consider an arbitrary open cover $\mathcal{U} \in cov(Y)$. Denote by $\mathcal{V}$ a star-refinement of $Y$, and take a map $g: Y \to X$ such that the composition $fg$ is $\mathcal{V}$-close to $id_Y$. Consider the open cover

$$f^{-1}(\mathcal{V}) = \{f^{-1}(V): V \in \mathcal{V}\}$$

of the space $X$. Take an open refinement $\mathcal{W} \in cov(X)$ of $f^{-1}(\mathcal{V})$ of order at most $\dim X + 1$. Obviously, the order of the cover

$$g^{-1}(\mathcal{W}) = \{g^{-1}(W) \colon W \in \mathcal{W}\}$$

is also at most $\dim X + 1$. Consequently, it suffices to show that $g^{-1}(\mathcal{W})$ refines $\mathcal{U}$. In turn, in order to show this fact it is sufficient to show that $g^{-1}f^{-1}(\mathcal{V})$ refines $\mathcal{U}$. Thus, let $V \in \mathcal{V}$ and $y \in g^{-1}f^{-1}(V)$. Since the maps $fg$ and $id_Y$ are $\mathcal{V}$-close, there exists an element $V_y \in \mathcal{V}$ such that both points $fg(y)$ and $y$ are contained in $V_y$. Since $y \in g^{-1}f^{-1}(V)$, we see that $fg(y) \in V$ and, consequently, $V_y \cap V \neq \emptyset$. This implies that $g^{-1}f^{-1}(V) \subseteq St(V, \mathcal{V})$. □

COROLLARY 3.2.21. *There are non cell-like images of the* 5*-dimensional cube that are not ANR's.*

PROOF. Let $f \colon I^5 \to Y$ be the map from Corollary 3.2.17. Assuming that $X$ is an $ANR$-compactum, by Lemma 3.2.20 and Proposition 2.1.29 we conclude that $\dim Y \leq 5$. This is a contradiction. □

*Historical and bibliographical notes* 3.2. This Section contains the solution of the $CE$-problem given by Dranishnikov [**130**]. Propositions 3.2.9 and 3.2.10 appeared in [**59**]. Theorem 3.2.15, as well as Corollaries 3.2.16, 3.2.17, were proved by Dydak and Walsh [**136**].

## 3.3. Universal space for cohomological dimension

In this Section we show that there exist universal spaces with respect to integral cohomological dimension.

### 3.3.1. $(\mathbb{Z}, n)$-*invertible and* $(\mathbb{Z}, n)$-*soft maps.*

THEOREM 3.3.1. *There exists a Polish subspace* $Y \subseteq D^\omega \times I^\omega$ *satisfying the following conditions:*

(i) $\dim_{\mathbb{Z}} Y = n$;

(ii) *For every* $G_\delta$*-subset* $Y' \subseteq I^\omega$ *with* $\dim_{\mathbb{Z}} Y \leq n$, *there exists* $c \in D^\omega$ *such that* $(I^\omega \times \{c\}) \cap Y = Y' \times \{c\}$.

PROOF. Let $\alpha \colon M \to I^\omega$ be an $n$-invertible map of an $n$-dimensional compactum $M$ onto the Hilbert cube $I^\omega$. The results of Chapter 4 show that $\alpha$ has the following property:

- For any map $g \colon Z \to I^\omega$ of an at most $n$-dimensional separable metrizable space $Z$, there is an embedding $i \colon Z \to M$ such that $\alpha i = g$.

Let $\beta\colon D^\omega \to expM$ be a surjection, where $expM$ denotes the hyperspace of $M$. Consider the subspace

$$A = \bigcup\{\{c\} \times \beta(c)\colon c \in D^\omega\}$$

of the product $D^\omega \times M$ and define the map $\mu\colon A \to D^\omega \times I^\omega$ by letting

$$\mu(c,x) = (c, \alpha(x)) \text{ for each } (c,x) \in A.$$

Let

$$Y = \{y \in D^\omega \times I^\omega : \mathrm{Sh}(\mu^{-1}(y)) = *\}.$$

It can easily be seen that $Y$ is a $G_\delta$-subspace of the product $D^\omega \times I^\omega$. Clearly

$$\dim \mu^{-1}(Y) \leq \dim A \leq \dim(D^\omega \times M) = n.$$

Therefore, by the non-compact version of Proposition 3.1.4, $\dim_{\mathbb{Z}} Y = n$ (observe that by the definition of $Y$, the restriction $\mu/\mu^{-1}(Y)\colon \mu^{-1}(Y) \to Y$ is a proper $CE$-map).

Let us now show that if $Y'$ is a $G_\delta$-subset of $I^\omega$ then there is a point $c \in D^\omega$ such that $(I^\omega \times \{c\}) \cap Y = y' \times \{c\}$. Using once again the non-compact version of Proposition 3.1.4, we have an at most $n$-dimensional Polish space $X'$ admitting a proper $CE$-map $f'\colon X' \to Y'$ onto $Y'$. Choose an embedding $i'\colon X' \to M$ such that $\alpha i' = f'$. Without loss of generality, we may assume that $i' = id$, i.e. $X' \subseteq M$. Finally, choose $c \in D^\omega$ with $\operatorname{cl} X' = \beta(c)$. □

A map $f : X \to Y$ is called *$(\mathbb{Z}, n)$-invertible* provided that for each map $g : Z \to Y$ with $\dim_{\mathbb{Z}} Z \leq n$, there exists a map $h : Z \to X$ such that $f = gh$.

COROLLARY 3.3.2. *There exists a $(\mathbb{Z}, n)$-invertible map $f : Y \to I^\omega$, where $Y$ is a Polish space with $\dim_{\mathbb{Z}} Y \leq n$.*

PROOF. Let $Y \subseteq I^\omega \times D^\omega$ be the space from Theorem 3.3.1 and $\tilde{f}\colon Y \to I^\omega$ be the restriction of the projection map $\pi_1 : I^\omega \times D^\omega \to I^\omega$. Split $I^\omega$ into the product, $I^\omega \cong I^\omega \times I^\omega$, and let $F$ be the composition

$$Y \xrightarrow{\tilde{f}} I^\omega \xrightarrow{\cong} I^\omega \times I^\omega \xrightarrow{\pi_1} I^\omega. \quad \square$$

The following definition is a natural counterpart of that of $n$-soft map (Definition 2.1.33).

DEFINITION 3.3.3. *A map $f : X \to Y$ is called $(\mathbb{Z}, n)$-*soft *provided for each commutative diagram*

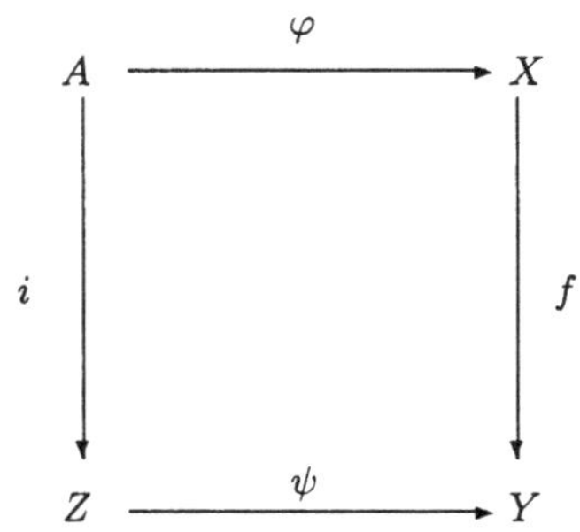

*where $A$ is a closed subset of a space $Z$, $\dim_{\mathbb{Z}} \leq n$, there exists a map $\Phi : Z \to X$ such that $f\Phi = \psi$ and $\Phi|A = \phi$.*

DEFINITION 3.3.4. *A space $X$ is called an* absolute extensor in integral cohomological dimension $n$ *(briefly, $X \in AE(Z, n)$) provided for every map $f : A \to X$ defined on a closed subset of a space $Y$, $\dim_Z \leq n$, there exists an extension of $f$ to the space $Y$.*

Obviously, $X \in AE(\mathbb{Z}, n)$ if the constant map $X \to \{x\}$ is $(\mathbb{Z}, n)$-soft.

PROPOSITION 3.3.5. *Let $f : X \to Y$ be a $(\mathbb{Z}, n)$-soft map, $\dim_{\mathbb{Z}} X \leq n$ and $Y \in AE$. Then $X \in AE(\mathbb{Z}, n)$.*

LEMMA 3.3.6. *For any closed subset $A$, $\dim_{\mathbb{Z}} A \leq n$, of a Polish space $B$ there exists a $(\mathbb{Z}, n)$-invertible map $g : K \to B$ of a Polish space $K$, $\dim_{\mathbb{Z}} K \leq n$, such that $g/g^{-1}(A) : g-1(A) \to A$ is an embedding.*

PROOF. Let $f : X \to B$ be a $(\mathbb{Z}, n)$-invertible map of a Polish space $X$ with $\dim_{\mathbb{Z}} X \leq n$. Embed $f$ into a map $\tilde{f} : \tilde{X} \to \tilde{B}$ of compact spaces, and let $\mathcal{G}$ be the decomposition of $\tilde{X}$ into singletons and sets of the form $\tilde{f}^{-1}(a), a \in \bar{A}$ (= the closure of $A$ in $\tilde{B}$). Let $q : \tilde{X} \to \tilde{X}/\mathcal{G}$ be the quotient map and $\tilde{f} = g'q$ a factorization of $\tilde{f}$. We can take $K = q(X)$ and $g = g'/K$. □

PROPOSITION 3.3.7. *Let $n \in \omega$. There exists a $(\mathbb{Z}, n)$-soft map $f : X \to I^{\omega}$, where $\dim_{\mathbb{Z}} X \leq n$.*

PROOF. Let $s$ denote the pseudo-interior of the Hilbert cube $I^{\omega}$.

Let $q_1 : \tilde{X}_1 \to I^{\omega} = Z_0$ be a $(\mathbb{Z}, n)$-invertible map of a Polish space $\tilde{X}_1$ with $\dim_{\mathbb{Z}} \tilde{X}_1 = n$ (apply Corollary 3.3.2). There exists a closed embedding $\tilde{X}_1 \hookrightarrow I^{\omega} \times s = Z_1$ such that $q_1 = \pi_0^1/\tilde{X}_1$ ($\pi_0^1$ denotes the projection of $I^{\omega} \times s$ onto the first factor).

By Lemma 3.3.6 there exists a $(\mathbb{Z}, n)$-invertible map $q_2 : \tilde{X}_2 \to Z_1$ of a Polish space $\tilde{X}_1$ with $\dim_{\mathbb{Z}} \tilde{X}_2 = n$ such that the restriction

$$q_2/q_2^{-1}(\tilde{X}_1) : q_2^{-1}(\tilde{X}_1) \to \tilde{X}_1$$

is a homeomorphism. Let $e_1 : \tilde{X}_1 \to \tilde{X}_2$ be a map such that $q_2 e_1 = id_{\tilde{X}_1}$.

Proceeding as above we obtain a commutative diagram

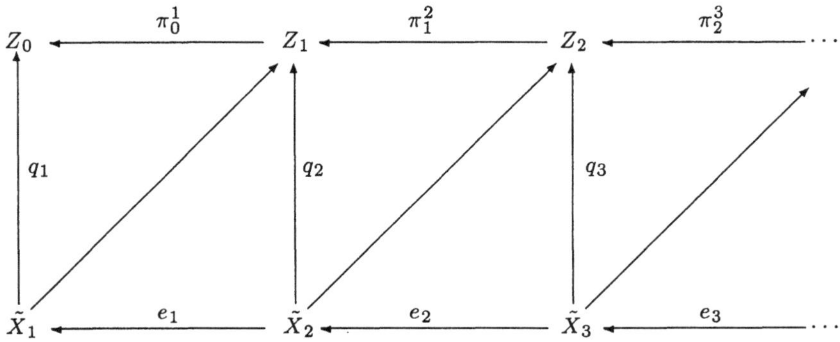


where $Z_i = I^\omega \times s \times \cdots \times s$ ($i+1$ factors), $\pi_i^{i+1}\colon Z_{i+1} \to Z_i$ are the projections onto the initial $i$ factors, $\tilde{X}_i$ is a closed subspace of $Z_i$ with $\dim_{\mathbb{Z}} \tilde{X}_i = n$, and $q_i$ are $(\mathbb{Z}, n)$-invertible maps such that the restrictions

$$q_i/q_i^{-1}(\tilde{X}_{i-1})\colon q_i^{-1}(\tilde{X}_{i-1}) \to \tilde{X}_{i-1}$$

are maps such that $q_{i+1}e_{i+1} = id_{\tilde{X}_i}$.

Consider the inverse system $\mathcal{S} = \{Z_i, \pi_i^{i+1}\}$ and let $\pi_0 : \lim \mathcal{S} \to Z_0 = I^\omega$ be the limit projection. Let

$$X_i = \{(x_j)_{j=1}^\infty\colon x_i \in \tilde{X}_j \text{ and } x_{j+1} = e_i(x_j) \text{ for every } j \geq i\} \subseteq \lim \mathcal{S}.$$

Also let

$$X = \cup_{i=1}^\infty X_i \text{ and } f = \pi_0/X : X \to Z_0 = I^\omega .$$

Obviously, the $X_i$ are closed subspaces of $X$, $\dim_{\mathbb{Z}} X_i = n$, and hence, by the Countable Sum Theorem, $\dim_{\mathbb{Z}} X = n$.

To show that $f$ is a $(\mathbb{Z}, n)$-soft map, consider a map $\gamma : D \to Z_0$, with $\dim_Z D \leq n$, and a map $\alpha\colon C \to X$ defined on a closed subset $C$ of $D$ such that $f\alpha = \gamma/D$. Let $\alpha = (\alpha_i)_{i=1}^\infty$, where $\alpha_i\colon C \to Z_i$ are the coordinate maps. Let $C_i = \alpha^{-1}(X_i)$ and let $D_1 \subseteq D_2 \subseteq \ldots$ be a sequence of closed subsets of $D$ such that $D_i \cap C = C_i$ and $D = \cup_{i=1}^\infty D_i$.

We are going to construct the map $\beta\colon D \to X$ in the form $\beta = (\beta_j)_{j=0}^\infty$, where $\beta_j : D \to Z_j$ are the coordinate maps. Let $\beta_0 = \gamma$. Since $s \in AE$, there exists a map $\beta_1\colon D \to Z_1$ such that $\beta_1/C = \alpha_1$ and $\pi_0^1\beta_1 = \beta_0 = \gamma$.

Assume that for every $j \leq k$, a map $\beta_j\colon D \to Z_j$ is defined in such a way that $\beta_j/C = \alpha_j$ and $\pi_{j-1}^j\beta_j = \beta_{j-1}$. Since the map $q_{k+1}$ is $(\mathbb{Z}, n)$-invertible, there exists a map $g_k\colon D_k \to \tilde{X}_{k+1}$ such that $q_{k+1}g_k = \beta_k/D_k$. Note that $g_k/C_k = e_k/C_k = \alpha_{k+1}/C_k$ and, therefore, the map

$$\alpha_{k+1} \cup g_k : C \cup D_k \to \tilde{X}_{k+1}$$

is well-defined. Since the map $\pi_k^{k+1}$ is soft, there exists a map $\beta_{k+1}\colon D \to Z_{k+1}$ such that $\pi_k^{k+1}\beta_{k+1} = \beta_k$ and $\beta_{k+1}/(C \cup D_k) = \alpha_{k+1} \cup g_k$. The map $\beta = (\beta_j)_{j=0}^{\infty}\colon D \to X$ is well-defined, $\beta/C = \alpha$ and $f\beta = \gamma$. □

We need a slightly stronger version of the above statement.

PROPOSITION 3.3.8. *For any closed subset $A$ of a separable metrizable space $Y$ such that $\dim_{\mathbb{Z}} A \le n$, there exists a $(\mathbb{Z}, n)$-soft map $f\colon X \to Y$ such that $\dim_{\mathbb{Z}} X \le n$ and the restriction $f/f^{-1}(A)\colon f^{-1}(A) \to A$ is a homeomorphism.*

PROOF. Let $f'\colon X' \to Y$ be a $(\mathbb{Z}, n)$-soft map with $\dim X' \le n$. Embed $f'$ into a map $\tilde{f}\colon \tilde{X} \to \tilde{Y}$ of compacta and let $\mathcal{G}$ be a decomposition of $\tilde{X}$ into singletons and the inverse images of the form $\tilde{f}^{-1}(a)$, $a \in \bar{A}$ (here $\bar{A}$ denotes the closure of $A$ in $\tilde{Y}$). Let $q\colon \tilde{X} \to \tilde{X}/\mathcal{G}$ denote the quotient map and $\tilde{f} = rq$ the factorization of $f$. Set $X = q(X')$ and $f = r/X : X \to Y$.

Note that $\dim_{\mathbb{Z}} X \le n$ and $f/f^{-1}(A)\colon f^{-1}(A) \to A$ is a homeomorphism.

Consider a commutative diagram

$$\begin{array}{ccc} B & \xrightarrow{\ \varphi\ } & X \\ {\scriptstyle i}\downarrow & & \downarrow{\scriptstyle f} \\ Z & \xrightarrow{\ \psi\ } & Y \end{array}$$

where $B$ is a closed subset of a separable metrizable space $Z$ with $\dim_{\mathbb{Z}} Z \le n$. We add the set $\psi^{-1}(A)$ and its derivatives to the topology of $Z$. The space thus obtaoned is denoted by $Z_\tau$ and its underlying set is naturally identified with $Z$.

Since $f'$ is $(\mathbb{Z}, n)$-soft, there exists a map

$$\varphi\colon B \cap \psi^{-1}(A) \to X'$$

such that $f'\varphi = f\varphi/(B \cap \psi^{-1}(A))$. Obviously, there exists a map

$$\bar{\varphi}\colon B - \psi^{-1}(A) \to X'$$

such that $q\bar{\varphi} = \varphi/(B - \psi^{-1}(A))$. The maps $\varphi$ and $\bar{\varphi}$ uniquely determine the map

$$\hat{\varphi} = \varphi \cup \bar{\varphi}\colon B_\tau \to X'$$

(here $B_\tau$ denotes the set $B$ with the topology inherited from $Z_\tau$).

Since $f'\hat{\varphi} = \psi/B_\tau$, by the $(\mathbb{Z}, n)$-softness of $f'$ there exists a map $\hat{\Phi}\colon Z_\tau \to X'$ such that $\hat{\Phi}/B_\tau = \hat{\varphi}$ and $f'\hat{\Phi} = \psi$. Then the map $\Phi = q\hat{\Phi}\colon Z \to X$ is continuous and

$$\Phi/B = q\hat{\Phi}/B = q\hat{\varphi} = \varphi \quad \text{and} \quad f\Phi = fq\hat{\Phi} = f'\hat{\Phi} = \psi.$$

Hence, the map $f$ is $(\mathbb{Z}, n)$-soft. □

PROPOSITION 3.3.9. *Let $f\colon X \to I^\omega$ be a $(\mathbb{Z}, n)$-soft map, where $\dim_{\mathbb{Z}} X \leq n$. There exists a complete space $\tilde{X} \supseteq X$, $\dim_{\mathbb{Z}} X \leq n$ and a $(\mathbb{Z}, n)$-soft map $\tilde{f}\colon \tilde{X} \to I^\omega$ such that $\tilde{f}/X = f$.*

PROOF. We may assume that $X \subset I^\omega \times I^\omega$ and $f = \pi_1/X : X \to I^\omega$ where $\pi_1\colon I^\omega \times I^\omega \to I^\omega$ denotes the projection onto the first coordinate. We identity $I^\omega \times I^\omega$ with the subset $I^\omega \times I^\omega \times \{0\} \subset I^\omega \times I^\omega \times I^\omega$.

*Claim. Let $K \subset I^\omega \times I^\omega \subset I^\omega \times I^\omega \times I^\omega$ and $L_K = (I^\omega \times I^\omega \times I^\omega) - (I^\omega \times I^\omega \times \{0\} - K)$. Then the map $\pi_1/L_K\colon L_K \to I^\omega$ is soft.*

Proof of Claim. Consider a commutative diagram

$$\begin{array}{ccc} A & \xrightarrow{\ \varphi\ } & L_K \\ {\scriptstyle i}\downarrow & & \downarrow{\scriptstyle \pi_1/L_K} \\ B & \xrightarrow{\ \psi\ } & I^\omega . \end{array}$$

Let $\alpha = (\alpha_i)_{i=1}^\infty : B \to I^\omega$ be a map such that $\alpha^{-1}(\{0\}) = A$ (here, $I^\omega = \prod_{i=1}^\infty I_i$, $I_i = [0,1]$, and $\alpha_i\colon B \to I_i$ is a coordinate map, $i \in N$). Let

$$\Phi' = (\Phi_1', \Phi_2', \Phi_3')\colon B \to I^\omega \times I^\omega \times I^\omega$$

be a map such that $\Phi'/A = \varphi$ and $\pi_1\Phi' = \psi$.

Let $\Phi_3' = (\Phi_{3i}')_{i=1}^\infty$, where $\Phi_{3i}'\colon B \to I_i$ is a coordinate map for each $i \in N$. Put $\Phi_3 = (\Phi_{3i})_{i=1}^\infty$, where

$$\Phi_{3i}(b) = \min\{\Phi_{3i}'(b) + \alpha(b), 1\}.$$

Now we can take $\Phi = (\Phi_1', \Phi_2', \Phi_3)$. The claim has been established.

Let $X \subset X' \subset I^\omega \times I^\omega \subset I^\omega \times I^\omega \times I^\omega$, where $X'$ is a Polish space, $\dim_{\mathbb{Z}} X' = n$, and $\bar{X} \supset X'$ (the closure $\bar{X}$ is taken in $I^\omega \times I^\omega \times I^\omega$).

Let $g\colon Y' \to L_{X'}$ be a $(\mathbb{Z}, n)$-soft map, where $\dim_Z Y' = n$ and

$$g|g^{-1}(X')\colon G^{-1}(X') \to X'$$

is a homeomorphism.

Let $Y = g^{-1}(L_X)$. Consider the commutative diagram

$$\begin{array}{ccc} g^{-1}(X) & \xrightarrow{g/g^{-1}(X)} & X \\ {\scriptstyle i}\downarrow & & \downarrow{\scriptstyle f} \\ Y & \xrightarrow{(\pi_1/L_X)(g/Y)} & I^\omega \end{array}$$

By the $(\mathbb{Z}, n)$-softness of $f$, there exists a map $r: Y \to X$ such that $r/g^{-1}(X) = g/g^{-1}(X)$ and $fr = (\pi_1/L_X)(g/Y)$.

By the Lavrentieff theorem, there exist $G_\delta$-subsets $\tilde{X} \subset X'$ and $\tilde{Y} \subset Y'$ and a map $\tilde{r}: \tilde{Y} \to \tilde{X}$ such that $\tilde{r}/Y = r$. Obviously, we may assume that $\tilde{Y} \supset g^{-1}(\tilde{X})$ (if not, take $\tilde{Y} = Y \cup g^{-1}(\tilde{X})$).

We shall show that the map $\tilde{f} = \pi_1/\tilde{X} : \tilde{X} \to I^\omega$ is $(\mathbb{Z}, n)$-soft. Let

$$\begin{array}{ccc} A & \xrightarrow{\varphi} & \tilde{X} \\ {\scriptstyle i}\downarrow & & \downarrow{\scriptstyle \tilde{f}} \\ Z & \xrightarrow{\psi} & Y \end{array}$$

be a commutative diagram, where $A$ is a closed subset of a space $B$ and $\dim_{\mathbb{Z}} B \leq n$. By the Claim, there exists a map $\beta: B \to L_{\tilde{X}}$ such that $\beta/A = \varphi$ and $\tilde{f}\beta = \psi$. Further, by the $(\mathbb{Z}, n)$-softness of $g$ there exists a map $\gamma: B \to Y'$ such that $\gamma/A = g^{-1}\varphi$ and $g\gamma = \beta$. Obviously, $\gamma(A) \subset \tilde{Y}$ and the map $\Phi = \tilde{r}\gamma$ satisfies the conditions $\Phi/A = \varphi$ and $\tilde{f}\Phi = \psi$. □

COROLLARY 3.3.10. *For every Polish space $Y$, there exist a Polish space $K$ with* $\dim_{\mathbb{Z}} K \leq n$ *and a $(Z, n)$-soft map $g: K \to Z$.*

PROOF. Consider $Z$ as a subset of $I^\omega$ and let $K = \tilde{f}^{-1}(Z)$ and $g = \tilde{f}/K : K \to Z$. □

**3.3.2. Universal space for cohomological dimension.** Let $\mathcal{M}_1(\mathbb{Z}, n)$ denote the class of Polish spaces $X$ with $\dim_{\mathbb{Z}} X \leq n$.

LEMMA 3.3.11. *Let* $X = \lim\{X_i, p_i^{i+1}\}$, *where the maps* $p_i^{i+1}$ *are surjections and* $\dim_{\mathbb{Z}} X_i \leq n$. *Then* $\dim_{\mathbb{Z}} X \leq n$.

PROOF. Let $\sim$ be an equivalence relation on $X \times [0,1]$ defined as follows: $(x,t) \sim (y,t')$ if $t = t'$ and either $x = y$ or $\frac{1}{i} < t = t' \leq \frac{1}{i+1}$ and $p_i^{i+1}(x) = p_i^{i+1}(y)$. Let

$$q\colon X \times [0,1] \to (X \times [0,1])/_{\sim} = Y$$

denote the quotient map. We identify $X$ with the subset $q(X \times \{0\})$ and $X_i$ with $q(X_i \times \{\frac{1}{i}\})$.

Let $f\colon A \to K(\mathbb{Z}, n)$ be map defined on a closed subset $A$ of $X$. Assuming $K(\mathbb{Z}, n)$ to be an absolute neighborhood extensor, we can obtain an extension $\bar{f}\colon U \to K(\mathbb{Z}, n)$ of $f$ to a neighborhood $U$ of $A$ in $Y$ (recall that $X \subset Y$). Let $V$ be a neighborhood of $A$ with $\operatorname{cl} V \subset U$. There exists a continuous function $\varphi\colon A \to (0,1]$ such that

$$d(q(\varphi(a)), Y - V) < \frac{1}{2} d(a, Y - V)$$

for every $a \in A$ ($d$ denotes a compatible metric on $Y$) and such that the map $g = \bar{f}q(id_A \times \varphi)$ is homotopic to $f$.

Let $\bar{\varphi}\colon X \to (0,1]$ be an extension of $\varphi$. Then the space $q(id_X \times \bar{\varphi})(X)$ is a countable union of closed subspaces which are homeomorphic to closed subsets of spaces $X_i$, $i \in N$. Thus, $\dim_{\mathbb{Z}} q(id_X \times \bar{\varphi})(X) \leq n$ and the map

$$\bar{f}/(id_A \times \bar{\varphi})(A)\colon (id_A \times \bar{\varphi})(A) \to K(\mathbb{Z}, n)$$

has an extension $\bar{g}\colon (id_X \times \bar{\varphi})(X) \to K(\mathbb{Z}, n)$.

Therefore, the map

$$\bar{f}q(id_A \times \varphi)\colon A \to K(\mathbb{Z}, n)$$

has an extension, namely,

$$\bar{g}q(id_X \times \varphi)\colon X \to K(\mathbb{Z}, n).$$

By the Homotopy Extension Theorem, the map $f\colon A \to K(\mathbb{Z}, n)$, which is homotopic to $\bar{f}q(id_A \times \varphi)$, can be extended over $X$. □

THEOREM 3.3.12. *There exists a space* $Y(\mathbb{Z}, n)$ *satisfying the following properties:*

(i) $Y(\mathbb{Z}, n) \in \mathcal{M}_1(\mathbb{Z}, n)$.

(ii) $Y(\mathbb{Z}, n) \in AE(\mathbb{Z}, n)$.

(iii) $Y(\mathbb{Z}, n)$ *is strongly* $\mathcal{M}_1(\mathbb{Z}, n)$*-universal, i.e. for any Polish space* $C$ *with* $\dim_Z C \leq n$, *the set of closed (even* $Z$*-) embeddings is dense in the space* $C(C, Y(\mathbb{Z}, n))$.

PROOF. We shall define an inverse sequence $\mathcal{S} = \{X_i, p_i^{i+1}\}$ by induction. Let $X_0$ be a singleton. Assuming that the spaces $X_i$ and maps $p_{i-1}^i$ have already been defined, let $f : X_{i+1} \to X_i \times s$ (here $s$ denotes the pseudo-interior of $I^\omega$ ) be a $(\mathbb{Z}, n)$-soft map of a space $X_{i+1} \in \mathcal{M}_1(\mathbb{Z}, n)$ (see Corollary 3.3.10) and let $p_i^{i+1} = \pi_1 f$.

Now let $Y(\mathbb{Z}, n) = lim \mathcal{S}$. By Lemma 3.3.11, $\dim Y(\mathbb{Z}, n) = n$. Since all the bonding projections are $(\mathbb{Z}, n)$-soft maps, we easily conclude that $Y(\mathbb{Z}, n) \in AE(\mathbb{Z}, n)_Z$ (compare with Lemma 6.2.6).

Note that the bonding projections satisfy the following property:

- For each $Z \in \mathcal{M}_1(\mathbb{Z}, n)$ and each map $g\colon Z \to X_{i+1}$ there exists a closed embedding $g'\colon Z \to X_{i+1}$ such that $p_i^{i+1} g = p_i^{i+1} g'$.

Following the argument of Proposition 5.1.11, it can now be shown that the limit projection $p_1 : Y(\mathbb{Z}, n)$ satisfies the following fiberwise universality property:

(∗) For every space $Z \in \mathcal{M}_1(\mathbb{Z}, n)$, every map $g\colon Z \to Y(\mathbb{Z}, n)$ and every open cover $\mathcal{U} \in cov(Y(\mathbb{Z}, n))$ there exists a closed embedding $g'\colon Z \to Y(\mathbb{Z}, n)$ which is $\mathcal{U}$-close to $g$ and such that $p_1 g = p_1 g'$.

In fact, slightly modifying the argument we may assume that $g'$ is even a $Z$-embedding. □

*Remark* 3.3.13. Comparing the properties of the space $Y(\mathbb{Z}, n)$ from Theorem 3.3.12 with the properties of the universal $n$-dimensional Nöbeling space $N_n^{2n+1}$ we can see that the space $Y(\mathbb{Z}, n)$ plays the similar role with respect to the cohomological dimension as that of $N_n^{2n+1}$ with respect to the usual dimension.

It is still unknown whether there exists a compact space universal in the class of spaces with $\dim_{\mathbb{Z}} \leq n$.

The space $Y$ in Theorem 3.3.2 cannot be made compact. Indeed, we have the following statement.

PROPOSITION 3.3.14. *There is no $(\mathbb{Z}, n)$-invertible map $f\colon Y \to I^\omega$ where $Y$ is a compactum with $\dim_{\mathbb{Z}} Y \leq n$. In other words, for any map $f\colon Y \to I^\omega$ of a compactum $Y$ to the Hilbert cube $I^\omega$ such that $\dim_{\mathbb{Z}} Y \leq n$, there is a compactum $X \subseteq I^\omega$ such that $\dim_{\mathbb{Z}} X \leq n$ and the inclusion $X \hookrightarrow I^\omega$ does not lift to $Y$.*

PROOF. Choose a $\mathcal{CW}$-complex $K$ which is a $K(\mathbb{Z}, n)$ and has infinite skeleta. Let $X_k \subseteq I^\omega$ be compacta with the following properties:

(a) $\dim_{\mathbb{Z}} X_k \leq n$ for each $n \in \omega$.

(b) Each $X_k$ contains the same copy of the $n$-dimensional sphere $S^n$.

(c) The inclusion $S^n \hookrightarrow K$ cannot be extended over $X_k$ so that the image of the extension is contained in the $k$-dimensional skeleton $K^{(k)}$ of $K$.

Suppose now that there is a $(\mathbb{Z}, n)$-invertible map $f\colon Y \to I^\omega$ where $Y$ is a compactum with $\dim_{\mathbb{Z}} Y \leq n$. Choose an extension $g\colon Y \to K$ of $f^{-1}(S^n) \to S^n \hookrightarrow K$. Then there exists an integer $k$ such that $g(Y) \subseteq K^{(k)}$, which demonstrates that the inclusion $X_k \hookrightarrow I^\omega$ cannot be lifted to $Y$. This contradicts with the choice of $f$. □

*Historical and bibliographical notes* 3.3. Theorem 3.3.1 was in fact obtained in [**134**]. All of the other results of this Section appear in [**326**]. The existence of a Polish space universal in the class of Polish spaces with $\dim_{\mathbb{Z}} \leq n$ (a weaker version of Theorem 3.3.12) was shown earlier in [**134**] (see also [**96**]). The compacta $X_k$ used in the proof of Proposition 3.3.14 were constructed in [**137**].

CHAPTER 4

# Menger Manifolds

In this Chapter we present a survey of Menger manifold theory.

## 4.1. General Theory

Before we start a detailed discussion, let us outline the construction of the $k$-dimensional universal Menger compactum $\mu^k$, $k \geq 0$. We partition the standard unit cube $I^{2k+1}$, lying in $(2k+1)$-dimensional Euclidean space $R^{2k+1}$, into $3^{2k+1}$ congruent cubes of the "first rank" by hyperplanes drawn perpendicular to the edges of the cube $I^{2k+1}$ at points dividing the edges into three equal parts, and we choose from these $3^{2k+1}$ cubes those which intersect the $k$-dimensional skeleton of the cube $I^{2k+1}$. The union of the selected cubes of the "first rank" is denoted by $I(k,1)$. In an analogous way, we divide every cube entering as a term in $I(k,1)$ into $3^{2k+1}$ congruent cubes of the "second rank" and the union of all analogously selected cubes of the "second rank" is denoted by $I(k,2)$. If we continue the process we get a decreasing sequence of compacta

$$I(k,1) \supseteq I(k,2) \supseteq \cdots .$$

The compactum $\mu^k = \cap\{I(k,i) \colon i \in N\}$ is called the *k-dimensional universal Menger compactum*. Note that $\mu^k = M_k^{2k+1}$, where $M_k^n$, $0 \leq k < n$, denotes the "$k$-dimensional Menger compactum constructed in the $n$-dimensional cube $I^n$" (a precise definition of $M_k^n$ is given below).

Obviously, $\mu^0$ coincides with the Cantor discontinuum $D^\omega$and, consequently, is the only zero-dimensional compactum with no isolated points (Theorem 1.1.10). It was shown by Sierpinski [**285**] that $\mu^0$ is a universal space for the class of all zero-dimensional metrizable compacta.

Positive dimensional Menger compacta $M_k^n$ were originally defined within the classical dimension theory by Menger in [**221**]. They are generalizations of the Cantor discontinuum and Sierpinski's universal curve $M_1^2$ [**284**]. Let us recall that the space $M_1^2$ is universal for the class of all at most 1-dimensional planar compacta [**284**]. Further, it was shown by Menger [**221**] that the 1-dimensional Menger compactum $\mu^1 = M_1^3$ is universal for the class of all at most 1-dimensional compacta. Generally, it was conjectured [**221**] that $M_k^n$ is a universal space for the class of all at most $k$-dimensional compacta embeddable in $R^n$ (Menger's problem). As was already mentioned, this problem was known to have a positive solution for $(k, n) = (1, 2)$ and $(k, n) = (1, 3)$. A positive solution in the case $k = n - 1$ was also given by Menger [**221**]. Results of Lefschetz [**209**] and Bothe [**48**] produced a positive solution in the case $n = 2k + 1$. The ultimate affirmative solution of Menger's was obtained by Štanko [**289**].

The topological characterization of $\mu^0$ has already been mentioned. Note also that $\mu^0$, being a topological group, is homogeneous. The 1-dimensional Menger compacta $M_1^2$ and $\mu^1 = M_1^3$ were also characterized topologically by Whyburn [**318**] and Anderson [**14**] (see also [**15**] ), respectively. Anderson's theorem characterizes $\mu^1$ as a 1-dimensional locally connected continuum with no local separating points and with no non-empty open subspaces embeddable into the plane.

Comparing the characterizations of $\mu^0$ and $\mu^1$, it is very hard to see what is at the root of these results, what is common between them and, finally, how they can be generalized to higher dimensions in order to (at least) make a reasonable conjecture concerning the characteristics of $M_k^n$. In full generality, some of these questions still remain open. To the best of our knowledge, there are no conjectures concerning the characteristic properties of the compacta $M_k^n$ when $1 < k < n \leq 2k$.

### 4.1.1. Construction of Menger compacta.

In this Subsection we describe the constructions of Menger compacta given by Menger, Lefschetz and Bestvina. Although the Menger construction has already been described above, we restate it in a slightly different (but equivalent) form for later use. Throughout this section we fix integers $0 \leq k \leq n$.

#### I. Menger's construction [221]

As a metric on $R^n$, we use the maximum metric, i.e.

$$d(\{x_i\}, \{y_i\}) = \max\{| x_i - y_i |: 1 \leq i \leq n\}$$

for each $\{x_i\}, \{y_i\} \in R^n$. If $A$ is a subset of $R^n$, then $N(A, \epsilon)$ denotes the $\varepsilon$-neighborhood of $A$ with respect to the above metric.

Let $I^n$ be the $n$-cell in $R^n$ with the standard linear structure. For each integer $i \geq 0, L_i$ denotes the cell complex structure of $I^n$ whose $n$-cells are of the form

$$\prod\left\{\left[\frac{m_t}{3^i}, \frac{m_t+1}{3^i}\right] : m_t = 0, 1, \ldots, 3^i - 1\right\}.$$

We define the Menger compactum $M_k^n$ as follows: Let $M_0 = I^n$ and (by induction) for each integer $i \geq 1$ let

$$M_{i+1} = st(|L_i^{(k)}|, L_{i+1}) \cap M_i = st(|L_i^{(k)}|, L_{i+1}|M_i) = st(|L_i^{(k)}|M_i|, L_{i+1}|M_i).$$

Clearly, $\{M_i\}$ is a decreasing sequence of compacta and $M_k^n = \cap M_i$ is called the *Menger compactum of type* $(k,n)$. When $n = 2k+1$, we use the symbol $\mu^k = M_k^{2k+1}$ to denote the *k-dimensional universal Menger compactum.*

We will also use another description of $M_k^n$ [**120**, Chap.2].

Let $V_i = \{(2t+1)/2 \cdot 3^i : t = 0, 1, \ldots, 3^i - 1\}$ and $V = \cup V_i$. Note that $B_i = V_i \times \cdots \times V_i$ ($n$ factors) is the set of centers of $n$-cells of $L_i$. Let $\mathcal{P}$ be the finite collection of homeomorphisms of $R^n$ defined by permutations of coordinates of $R^n$. For each $i$ we define

$$D_i = \cap\{\alpha(\{c\} \times I^{n-k-1}) : \alpha \in \mathcal{P} \text{ and } c \in V_i^{k+1}\}$$

and $N_i = N(D_i, 1(2 \cdot 3^i))$. Then $D_i$ can be regarded as the "dual $(n-k-1)$-skeleton"of $L_i$ and $N_i$ as the regular neighborhood of $D_i$. It is easy to see that $M_k^n = I^n - \cup\{\text{int}(N_i) : i = 1, \ldots, n\}, I^n = M_i \cup N_i$ and $\partial M_i = M_i \cap (N_i \cup \partial I^n)$.

If we perform the above construction starting with $R^n$ (instead of $I^n$) we get a closed subspace $U_k^n$ of $R^n$ which is a countable union of copies of $M_k^n$ (in this case $L_0$ is the partition of $R^n$ into unit cubes).

## II. Lefschetz's construction [209]

Replacing the cell complexes in (I) by simplicial complexes, we obtain Lefschetz's construction. We describe it in slightly general form.

Let $M$ be a PL $n$-manifold with a (combinatorial) triangulation $L$. Inductively, we define a sequence $\{M_i\}$ of PL $n$-manifolds and their triangulations $L_i$ as follows.

Let $M_0 = M$ and $L_0 = L$. Let $M_1 = st(L^{(k)}, \beta^2 L_0), L_1 = \beta^2 L_0 | M_1$ and suppose that $M_i$ and $L_i$ have already been defined. Consider $\beta^2 L_i$ and let $M_{i+1} = st(L_i^{(k)}, \beta^2 L_i)$ and $L_{i+1} = \beta^2 L_i | M_{i+1}$. Then $\{M_i\}$ is a decreasing sequence and $\cap M_i \neq \emptyset$.

If $M$ is the $n$-simplex with the standard simplicial complex structure, then the resulting compactum $\cap M_i$ is denoted by $L_k^n$. In particular, $L_k^{2k+1} = \mu^k$ (We use the same symbol as in (I). This notation is justified by the Characterization Theorem 2.4.1). Notice that $M_{i+1}$ may be regarded as a regular neighborhood of the k-skeleton of $M_i$ (with respect to $L_i$).

## III. Bestvina's construction [33]

In Bestvina's construction, the $k$-skeleta in (II) are replaced by the dual $k$-skeleta. Suppose that $M$ is a PL $n$-manifold with a (combinatorial) triangulation $L$. As in (II), we define a sequence $\{M_i\}$ of PL $n$-manifolds and their triangulations $\{L_i\}$ as follows:

Let $M_0 = M, L_0 = L$ and suppose that we already have defined $M_i$ and $L_i$. Then $M_{i+1} = \cup\{st(b_\sigma, \beta^2 L_i) : b_\sigma$ is the barycenter of $\sigma \in L_i$ with $\dim \sigma \geq n-k\}$ and $L_{i+1} = \beta^2 L_i | M_{i+1}$.

If $M$ is the $n$-simplex with the standard simplicial complex structure, the resulting compactum $\cap M_i$ is denoted by $T_k^n$. In particular, $T_k^{2k+1}$ is denoted by $\mu^k$. (Again this is justified by the Characterization Theorem). Observe that $M_{i+1}$ is regarded as a regular neighborhood of the dual $k$-skeleton of $M_i$ (with respect to $L_i$).

It might be worth noting the differences among these constructions. Consider the properties of the partitions which are naturally induced by each of the above constructions. For simplicity, we formulate these properties only for $M_k^n, L_k^n$ and $T_k^n$.

PROPOSITION 4.1.1. *There are sequences $\{P_i\}, \{Q_i\}$ and $\{R_i\}$ of partitions of $M_k^n, L_k^n$ and $T_k^n$, respectively, satisfying the following conditions:*

(a) *$P_{i+1}, Q_{i+1}$ and $R_{i+1}$ are refinements of $P_i, Q_i$ and $R_i$, respectively.*
(b) $\lim \operatorname{mesh} P_i = \lim \operatorname{mesh} Q_i = \lim \operatorname{mesh} R_i = 0$.
(c) $\operatorname{ord} P_i = n+1, \lim \operatorname{ord} Q_i = \infty$ *and* $\operatorname{ord} R_i = k+1$.
(d) *For each $p_1, p_2, \dots, p_t \in P_i$, $\cap\{p_j : j = 1, \dots, t\}$ is an at most $k$-dimensional $LC^{k-1} \cap C^{k-1}$-compactum or an at most $k$-dimensional cell.*
(e) *For each $q_1, q_2, \dots, q_t \in Q_i$, $\cap\{q_j : j = 1, \dots, t\}$ is an at most $k$-dimensional $LC^{k-1} \cap C^{k-1}$-compactum or an at most $k$-dimensional simplex.*
(f) *For each $r_1, r_2, \dots, r_t \in R_i$, $\cap\{r_j : j = 1, \dots, t\}$ is a $(k-t+1)$-dimensional $LC^{k-t} \cap C^{k-t}$ -compactum.*

PROOF. The partitions defined below satisfy the desired conditions:

$P_i = \{e \cap M_k^n : e \in L_i\}, Q_i = \{s \cap L_k^n : s \in L_i\}$ and $R_i = \{s \cap T_k^n : s \in L_i\}$.

□

*Remark* 4.1.2. (1) In the last case, if $(k, n) = (1, 3)$ we have a partition of $\mu^1$ with 0-dimensional intersections of all adjacent elements. In this sense, the partition determined by Bestvina's construction can be regarded as a generalization of the partition of the Menger curve considered in [**14, 15**] and [**235**].

(2) We may obtain characterizations of (compact) Menger manifolds as well as (compact) $Q$-manifolds in terms of the existence of certain types of partitions [**188**].

**4.1.2. $n$-homotopy.** We are going to describe an adequate homotopy language for $\mu^{n+1}$-manifold theory. This is the so called $n$-homotopy theory. The related notion of $\mu^{n+1}$-homotopy was first exploited in [**33**].

DEFINITION 4.1.3. *Two maps $f, g : X \to Y$ are said to be $n$-homotopic (written $f \overset{n}{\simeq} g$) if the compositions $f\alpha$ and $g\alpha$ are homotopic in the usual sense for any map $\alpha : Z \to X$ of an at most $n$-dimensional space $Z$ into $X$.*

It can easily be seen [**87**, Proposition 2.3] that if $\dim X \leq n+1$ and $Y \in LC^n$, then maps $f, g : X \to Y$ are $\mu^{n+1}$-homotopic in the sense of Bestvina [**33**, Definition 2.1.7] if and only if they are $n$-homotopic. Note also that if, in the above definition, we consider, instead of compact, only polyhedral $Z$, then we get Fox's definition of $n$-homotopy [**154**].

In practice it is convenient to use the following statement.

PROPOSITION 4.1.4. *Maps $f, g : X \to Y$ are $n$-homotopic if and only if for some (or, equivalently, any) $n$-invertible map $\alpha : Z \to X$ with* $\dim Z \leq n$, *the compositions $f\alpha$ and $g\alpha$ are homotopic.*

DEFINITION 4.1.5. *A map $\alpha : A \to X$ is said to be $n$-invertible if for any map $\beta : B \to X$ with* $\dim B \leq n$, *there is a map $\gamma : B \to A$ such that $\gamma\alpha = \beta$.*

Note that 0-invertible maps between metrizable compacta are precisely surjections with a regular averaging operator (see Definition 6.1.24). Note also that each compactum is an $n$-invertible image of an $n$-dimensional compactum.

Of course, homotopic maps are $n$-homotopic for each $n \geq 0$, but not conversely. Indeed, consider the identity map and the constant map of an arbitrary non-contractible $LC^\infty \cap C^\infty$-compactum. Nevertheless, $n$-homotopic maps have several useful properties.

PROPOSITION 4.1.6. *For each $Y \in LC^n$, there exists an open cover $\mathcal{U} \in cov(Y)$ such that any two $\mathcal{U}$-close maps of any space into $Y$ are $n$-homotopic.*

PROPOSITION 4.1.7. *($n$-Homotopy Extension Theorem). Let $Y \in LC^n$. Then for each $\mathcal{U} \in cov(Y)$, there exists $\mathcal{V} \in cov(Y)$ refining $\mathcal{U}$ such that the following condition holds:*

$(*)_n$ *For any at most $(n+1)$-dimensional space $B$, any closed subspace $A$ of $B$, and any two $\mathcal{V}$-close maps $f, g : A \to Y$ such that $f$ has an extension $F : B \to Y$, it follows that $g$ also has an extension $G : B \to Y$ which is $\mathcal{U}$-close to $F$.*

PROPOSITION 4.1.8. *Let $Y \in LC^n$. Suppose that $A$ is closed in $B$ and* $\dim B \leq n+1$. *If maps $f, g : A \to Y$ are $n$-homotopic and $f$ admits an extension $F : B \to Y$, then $g$ also admits an extension $G : B \to Y$, and it may be assumed that $F \overset{n}{\simeq} G$.*

A map $f : X \to Y$ is an *$n$-homotopy equivalence* if there is a map $g : Y \to X$ such that $gf \overset{n}{\simeq} id_X$ and $fg \overset{n}{\simeq} id_Y$ [**87**]. The spaces $X$ and $Y$ in this case are said to be *$n$-homotopy equivalent.* For example, any map between arcwise connected spaces is a 0-homotopy equivalence. Note also that the $(n+1)$-dimensional sphere $S^{n+1}$ is $n$-homotopy equivalent to the one-point space.

In general, we have the following algebraic characterization of $n$-homotopy equivalences [**316**, Theorem 2].

PROPOSITION 4.1.9. *A map $f : X \to Y$ between at most $(n+1)$-dimensional locally finite polyhedra is an $n$-homotopy equivalence if and only if it induces isomorphisms of homotopy groups of dimension $\leq n$, i.e., $f$ induces a bijection between the components of $X$ and $Y$ and the homomorphism $\pi_k(f') : \pi_k(C_X) \to \pi_k(C_Y)$ is an isomorphism for each $k \leq n$ and each pair of components $C_X \subseteq X$ and $C_Y \subseteq Y$ with $f(C_X) \subseteq C_Y$, where $f' : C_X \to C_Y$ denotes the restriction of $f$.*

Recall that each $ANR$-compactum is homotopy equivalent to a finite polyhedron (Corollary 2.3.30). The following statement is an "$n$-homotopy version" of West's result.

PROPOSITION 4.1.10. *Every at most $(n+1)$-dimensional locally compact $LC^n$-space is properly $n$-homotopy equivalent to an at most $(n-1)$-dimensional locally finite polyhedron.*

Therefore, 4.1.9 holds even for maps between at most $(n+1)$-dimensional locally compact $LC^n$-spaces.

Proper $n$-homotopies, and all associated notions, are defined in the natural way and we do not repeat them here. In order to state an algebraic characterization of proper $n$-homotopy equivalences similar to Proposition 4.1.9, we need some preliminary definitions. We say that a proper map $f : X \to Y$ between locally compact spaces induces an *epimorphism of $i$-th homotopy groups of ends* ($i \geq 0$) if for every compactum $C \subseteq Y$ there exists a compactum $K \subseteq Y$ such that for each point $x \in X - f^{-1}(K)$ and every map $\alpha : (S^i, *) \to (Y - K, f(x))$ there exists a map $\beta : (S^i, *) \to (X - f^{-1}(C), x)$ and a homotopy $f\beta \simeq \alpha(rel\, *)$ in $Y - C$. We say that $f : X \to Y$ induces a *monomorphism of $i$-th homotopy groups of ends* if for every compactum $C \subseteq Y$ there exists a compactum $K \subseteq Y$ such that for every map $\alpha : S^i \to X - f^{-1}(K)$ with the property that $f\alpha$ is null-homotopic in $Y - K$ it follows that $\alpha$ is null-homotopic in $X - f^{-1}(C)$. As usual, $f$ is said to induce an *isomorphism* of $i$-th homotopy groups of ends if it simultaneously induces an epimorphism and a monomorphism.

PROPOSITION 4.1.11. *A proper map $f : X \to Y$ between at most $(n+1)$-dimensional locally compact $LC^n$-spaces is a proper $n$-homotopy equivalence if and only if it induces isomorphisms of homotopy groups of dimension $\leq n$ and isomorphisms of homotopy groups of ends of dimension $\leq n$.*

Note that proper $n$-homotopies have also been studied from the categorical point of view [**169**].

The following proposition will be used below and indicates a difference between the $n$-homotopy and usual homotopy theories (compare with [**303**]).

PROPOSITION 4.1.12. *Let $M$ be an at most $(n+1)$-dimensional locally finite polyhedron. Suppose that there exists an at most $(n+1)$-dimensional finite polyhedron $K$ and two maps $f : M \to K$ and $g : K \to M$ such that $gf \overset{n}{\simeq} id_M$ (i.e.*

*$g$ is an $n$-*homotopy domination*). Then there exist an $(n+1)$-dimensional finite polyhedron $T$, containing $K$ as a subpolyhedron, and an $n$-homotopy equivalence $h : T \to M$ extending $g$ such that $f$ is a $n$-homotopy inverse of $h$.*

PROOF. Obviously, it suffices to consider only connected polyhedra. Consequently, the case $n = 0$ is trivial. If $n = 1$, then, by the assumption, $\pi_1(g)\colon \pi_1(K) \to \pi_1(M)$ is an epimorphism and $Ker(\pi_1(g))$ is a finitely generated group. Select finitely many generators of $Ker(\pi_1(g))$ and use them to attach 2-cells to $K$ and to extend $g$ over these cells. In this way we obtain a 2-dimensional finite polyhedron $T$, containing $K$ as a subpolyhedron, and a map $h\colon T \to M$, extending $g$, which induces an isomorphism of fundamental groups. By Proposition 4.1.9, $h$ is a 1-homotopy equivalence.

Assume, by way of induction, that the proposition is already proved in the cases $n \leq m$, $m \geq 1$, and consider the case $n = m+1$. Without loss of generality we can suppose that $f(M^{(i)}) \subseteq K^{(i)}$ and $g(K^{(i)}) \subseteq M^{(i)}$ for each $i \leq m+1$. Since $gf \overset{m+1}{\simeq} id_M$ it follows easily that $gf/M^{(m+1)} \overset{m}{\simeq} id_{M^{(m+1)}}$. By the inductive hypothesis, there are an $(m+1)$-dimensional finite polyhedron $R$, containing $K^{(m+1)}$ as a subpolyhedron, and an $m$-homotopy equivalence $r\colon R \to M^{(m+1)}$ extending $g/K^{(m+1)}$. Sewing together the polyhedra $K$ and $R$ along naturally embedded copies of $K^{(m+1)}$, we obtain the $(m+2)$-dimensional finite polyhedron $L$, containing $K$ and $R$ as subpolyhedra, and the map $s\colon L \to M$ which coincides with $g$ on $K$ and with $r$ on $R$, whence $sf = gf \overset{m+1}{\simeq} id_M$ and $fs/L^{(m+1)} = fr \overset{m}{\simeq} id_{L^{(m+1)}}$. By these conditions, we conclude that $\pi_i(s)\colon \pi_i(L) \to \pi_i(M)$ is an isomorphism for each $i \leq m$ and an epimorphism for $i = m+1$. One can easily verify that in this situation $Ker(\pi_{m+1}(s))$ is a finitely generated $Z(\pi_1(L))$-module. Select $Z(\pi_1(L))$-generators for $Ker(\pi_{m+1}(s))$ and use them to attach $(m+2)$-cells to $L$ and to extend $s$ over these cells. Let $T$ denote the resulting $(m+2)$-dimensional finite polyhedron, containing $L$ as a subpolyhedron, and $h\colon T \to M$ the corresponding extension of $s$. Then $\pi_i(h)$ is an isomorphism for each $i \leq m+1$. Again, by Proposition 4.1.9, $h$ is an $(m+1)$-homotopy equivalence. This completes the inductive step and finishes the proof. □

The analogous statement for proper $n$-homotopy dominations (near $\infty$) will be discussed in Subsection 4.4.4

### 4.1.3. *Z-set unknotting and topological homogeneity.*

PROPOSITION 4.1.13. *Let $A$ be a closed subset of a Polish $ANE(n)$-space $X$. Then the following conditions are equivalent:*

(i) *$A$ is a $Z_n$-set .*
(ii) *For each at most $n$-dimensional locally finite polyhedron $P$, the set $\{f \in C(P,X) : f(P) \cap A = \emptyset\}$ is dense in $C(P,X)$.*
(iii) *For each at most $n$-dimensional Polish space $Y$, the set $\{f \in C(Y,X) : f(Y) \cap A = \emptyset\}$ is dense in $C(Y,X)$.*

*Note that each $Z_n$-set in any at most $n$-dimensional $LC^{n-1}$-space is a $Z$-set. If, in Proposition 4.1.13, $X$ is locally compact, then the listed conditions are equivalent to the following:*

(iv) *For each at most $n$-dimensional Polish space $Y$, the set $\{f \in C(Y,X) : cl(f(Y)) \cap A = \emptyset\}$ is dense in $C(Y,X)$.*

Closed subsets satisfying the property (4), as in the case of infinite-dimensional manifolds, are called *strong $Z_n$-sets.* These sets are especially important in the non-locally compact setting.

PROPOSITION 4.1.14. *One-point subsets of Menger manifolds are $Z$-sets.*

The following statements are versions of the powerful $Z$-set unknotting theorem.

THEOREM 4.1.15. *Let $Z_1$ and $Z_2$ be two $Z$-sets in a $\mu^{n+1}$-manifold $M$, and let $h : Z_1 \to Z_2$ be a homeomorphism. Denote by $i_j : Z_j \to M$ the inclusion map ($j = 1,2$). If $i_1$ and $i_2 h$ are properly $n$-homotopic, then $h$ extends to a homeomorphism $H : M \to M$ which is properly $n$-homotopic to $id_M$.*

COROLLARY 4.1.16. *Every homeomorphism between $Z$-sets of $\mu^n$ can be extended to an autohomeomorphism of $\mu^n$.*

For $n = 0$ this result is well-known. A closed subset of $\mu^1$ is a $Z$-set if and only if it does not locally separate $\mu^1$. The result for $n = 1$ originally appeared in [**235**].

A compactum $X$ is called *strongly locally homogeneous* if for each point $x \in X$ and each neighborhood $U$ of $x$ , there is a neighborhood $V$ of $x$ contained in $U$ such that the following condition holds: for each point $y \in V$, there is a homeomorphism $h : X \to X$ such that $h(x) = y$ and $h/(X - U) = id$.

COROLLARY 4.1.17. *$\mu^n$ is topologically homogeneous. Moreover, it is strongly locally homogeneous.*

THEOREM 4.1.18. *Let $M$ be a $\mu^n$-manifold. For each open cover $\mathcal{U} \in cov(M)$, there is an open cover $\mathcal{V} \in cov(M)$ with the following property:*

(*) *if a homeomorphism $h : Z_1 \to Z_2$ between two $Z$-sets of $M$ is $\mathcal{V}$-close to $i_1$ (see notations in 2.3.5), then $h$ can be extended to a homeomorphism $H : M \to M$ which is $\mathcal{U}$-close to $id_M$.*

**4.1.4. Topological characterization.** The following characterization theorem [**33**, Theorem 5.2.1] is central to the whole theory. We recall that a space $X$ has $DD^nP$ (Disjoint $n$-Disks Property) if for each open cover $\mathcal{U} \in cov(X)$ and any two maps $\alpha : I^n \to X$ and $\beta : I^n \to X$, there are maps $\alpha_1 : I^n \to X$ and $\beta_1 : I^n \to X$ such that $\alpha_1$ is $\mathcal{U}$-close to $\alpha$, $\beta_1$ is $\mathcal{U}$-close to $\beta$ and $\alpha_1(I^n) \cap \beta_1(I^n) = \emptyset$.

THEOREM 4.1.19. *The following conditions are equivalent for any $n$-dimensional locally compact $ANE(n)$-space $X$ :*

(i) *$X$ is a $\mu^n$-manifold.*
(ii) *$X$ is strongly $\mathcal{B}_{\omega,n}$-universal, i.e. has $DD^nP$.*
(iii) *Each map of the discrete union $I^n \oplus I^n$ into $X$ can be approximated arbitrarily closely by embeddings.*
(iv) *Each proper map of any at most $n$-dimensional locally compact space into $X$ can be approximated arbitrarily closely by closed embeddings.*
(v) *Each proper map of any at most $n$-dimensional locally compact space into $X$ can be approximated arbitrarily closely by $Z$-embeddings.*
(vi) *Each proper map $f : Y \to X$ of any at most $n$-dimensional locally compact space $Y$ into $X$ such that the restriction $f/_{Y_0}$ onto a closed subset $Y_0$ is a $Z$-embedding can be arbitrarily closely approximated by $Z$-embeddings coinciding with $f$ on $Y_0$.*

Additionally, if $X$ is compact and $(n-1)$-connected (i.e. $X \in AE(n)$), then conditions (ii)-(vi) give a topological characterization of the compactum $\mu^n$.

Note that the theorem remains true even in the case $n = \infty$ (see Chapter 2).

If $n = 0$ and $X$ is compact, condition (ii) trivially implies that $X$ has no isolated points. Therefore, in this case, $X$ is homeomorphic to the Cantor discontinuum as has already been noted above.

Applying the above characterization in the case $k = 1$, we see that a compactum is homeomorphic to $\mu^1$ if and only if it is a 1-dimensional, locally connected continuum with $DD^1P$. It is known (see [**15**] or [**235**]) that a locally connected continuum has $DD^1P$ if and only if it has no local separating points and has no open subspaces embeddable in the plane. In this sense, Bestvina's characterization of $\mu^1$ reduces to Anderson's.

In Subsection 4.1.1, three major geometric constructions of the universal Menger compactum have been presented. Let us indicate another, spectral construction, given in [**248**]. Let $\{G_i\}$ be a basis of open sets of $S^1$, such that $G_i$ is an open cell with the property that $diam G_i \to 0$. Let us construct an inverse sequence $\{X_i, p_i^{i+1}\}$ as follows. We set $X_0 = S^1$, and we get $X_{i+1}$ from $X_i$ by "bubbling over $G_i$" , i.e. $X_{i+1}$ is the quotient space obtained from the disjoint union $X_i \oplus X_i$ by identifying the two copies of $x \in X_i$ precisely when $p_0^i(x) \notin G_i$ (here $p_0^i : X_i \to X_0$ denotes the corresponding projection). The projection $p_i^{i+1} : X_{i+1} \to X_i$ is defined in the obvious way. Note that $p_i^{i+1}$is a retraction. One can check directly that $X = \lim\{X_i, p_i^{i+1}\}$ is a 1-dimensional

$AE(1)$-compactum with $DD^1P$ and hence, by 4.1.19, $X$ is a copy of $\mu^1$. More careful consideration shows [**33**] that we get $\mu^n$ if we start with $S^n = X_0$ (and proceed as above).

There are several other constructions of $\mu^n$. To the best of our knowledge, all of them are defined as the limit spaces of inverse sequences(see, for example, [**127**], [**97**], [**155**], [**186**]).

#### 4.1.5. *Approximation by Homeomorphisms.*

THEOREM 4.1.20. *Proper $UV^n$-maps between $\mu^{n+1}$-manifolds are near-homeomorphisms.*

Another important result of $I^\omega$-manifold theory states that an infinite simple homotopy equivalence between $I^\omega$-manifolds is homotopic to a homeomorphism. Let us note that this is not the case for homotopy equivalences (i.e. there exist non-homeomorphic but homotopy equivalent compact $I^\omega$ -manifolds). In $\mu^n$-manifold theory we do not have simple homotopy obstructions and this significantly simplifies the corresponding result.

THEOREM 4.1.21. *Each proper $n$-homotopy equivalence between $\mu^{n+1}$-manifolds is properly $n$-homotopic to a homeomorphism.*

The following result, due to Ferry [**150**, Proposition 1.7] and improved slightly in [**188**], also illustrates this situation.

PROPOSITION 4.1.22. *Let $f : P \to L$ be a map between compact polyhedra which induces an isomorphism between the i-th homotopy groups for each $i \leq n$. Then there is a compact polyhedron $Z$ and $UV^n$-maps $\alpha : Z \to P$ and $\beta : Z \to L$ such that $f\alpha \overset{n}{\simeq} \beta$.*

*Historical and bibliographical notes* 4.1. Most of the results of this Section were obtained by Bestvina in his fundamental work [**33**]. The notion of $n$-homotopy is due to the author [**87**]. Proposition 4.1.12 was proved by the author [**100**].

### 4.2. $n$-soft mappings of compacta, raising dimension

The following statement is the first resolution theorem for locally compact $LC^{n-1}$-spaces, in particular for $\mu^n$-manifolds.

THEOREM 4.2.1. *Every locally compact $LC^{n-1}$-space is an (n-invertible) proper $UV^{n-1}$-image of a $\mu^n$-manifold.*

We also will be using the following result, which states that the $UV^n$-resolutions from Theorem 4.2.1 can be improved over $Z$-sets.

PROPOSITION 4.2.2. *Let $f: M \to X$ be a proper $UV^n$-surjection of a $\mu^{n+1}$-manifold onto a locally compact $LC^n$-space and let $Z$ be a $Z$-set in $M$. If $g: Z \to X$ is any proper map properly $n$-homotopic to $f/Z$, then there exists a proper $UV^n$-surjection $h: M \to X$ such that $h/Z = g$.*

Below, in this subsection, we prove much stronger results. We begin by fixing a natural number $n$. Let $B^{n+1} = \{x \in R^n : \|x\| \leq 1\}$ denote the unit closed ball in $R^{n+1}$. Consider a many-valued retraction $R_n: B^{n+1} \to S^n$ of the ball $B^{n+1}$ onto its boundary, the sphere $S^n = \partial B^{n+1}$, defined as follows:

$$R_n(x) = \{y \in S^n: (x, y) \geq 4\|x\|^2 - 3\|x\|\} \text{ for each point } x \in B^{n+1}.$$

Here $(x, y)$ denotes the standard scalar product in the Euclidean space $R^{n+1}$ and $\|x\| = \sqrt{(x, x)}$ denotes the induced norm. The above many-valued map assigns to each point of the ball $B_0^{n+1} = \{x \in R^{n+1}: \|x\| \leq \frac{1}{2}\}$ the whole sphere $S^n$. If $x \in B^{n+1} - (B_0^{n+1} \cup S^n)$, then $R_n(x)$ is a ball on the sphere $S^n$. Finally, if $x \in S^n$, then $R_n(x) = x$, i.e. $R_n$ is a retraction. It is easy to see that $R_n$ is continuous. Consider the graph $\Gamma_{R_n}$ of this map, i.e.

$$\Gamma_{R_n} = \{((x, y) \in B^{n+1} \times S^n: y \in R_n(x)\}.$$

By $p_n: \Gamma_{R_n} \to B^{n+1}$ and $q_n: \Gamma_{R_n} \to S^n$ we denote the restrictions to $\Gamma_{R_n}$ of the natural projections of the product $B^{n+1} \times S^n$ onto its coordinates.

LEMMA 4.2.3. *The map $p_n: \Gamma_{R_n} \to B^{n+1}$ is $(n-1)$-soft.*

PROOF. Continuity of the retraction $R_n$ implies openness of $p_n$. Therefore, by Theorem 2.1.15, it suffices to show that the collection $\{R_n(x): x \in B^{n+1}\}$ is connected and uniformly locally connected in all dimensions less than $n - 1$. As was mentioned above, topologically there are only three types of elements of the collection $\{R_n(x): x \in B^{n+1}\}$. Indeed, $R_n(x)$ is the sphere $S^n$ (for $x \in B_0^{n+1}$), a closed ball $B^n$ with center at $\frac{x}{\|x\|}$ in the sphere $S^n$ (for $x \in B^{n+1} - B_0^{n+1}$), or a point (for $x \in S^n$). Consequently the indicated collection is connected in all dimensions not exceeding $n - 1$. Let us show the uniform local connectivity of this collection in all dimensions less than $n - 1$. Let $k \leq n - 2$, $\epsilon > 0$ and $\delta = \frac{\epsilon}{2}$. Consider a point $x \in B^{n+1}$ and an arbitrary map $\alpha: S^k \to R_n(x)$ such that the diameter of its image $\alpha(S^k)$ is less than $\delta$. Evidently there is a ball $B^n$ on the sphere $S^n$, containing this image $\alpha(S^k)$. Let $\bar{\alpha}: B^{k+1} \to B^n$ be an extension of $\alpha$ from the boundary $S^k = \partial B^{k+1}$ onto the whole $B^{k+1}$. Additionally, we can assume that if $R_n(x) = x$ (which occurs when $x \in S^n$), then $\bar{\alpha} = \alpha$. Obviously, in this case as well as in the case when $R_n(x) = S^n$, $\bar{\alpha}$ is the desired extension. Suppose now that $x \in B^{n+1} - (B^{n+1} \cup S^n)$. Then $R_n(x)$ is a ball on $S^n$ of non-zero diameter. If $S^n - R_n(x) \not\subseteq B^n$, then the intersection $B^n \cap R_n(x)$ is homeomorphic to the $n$-dimensional ball and, consequently, there exists a retraction $r: B^n \to B^n \cap R_n(x)$. Therefore, in this case, the composition $\bar{\alpha}r$ is the desired extension of $\alpha$ with diameter of the image less than $\epsilon$. In the remaining case, when $S^n - R_n(x) \subseteq B^n$, we first observe that the boundary

$\partial R_n(x)$ of the ball $R_n(x)$ is contained in $B^n$. Since $k+1 < n$, there is a homotopy (in $B^n$) connecting the map $\bar{\alpha}$ with a map $\beta\colon B^{k+1} \to B^n \cap R_n(x)$ fixed on $\bar{\alpha}^{-1}(R_n(x))$. Then $\beta$ does not move points of $\partial B^{k+1} = S^k$ as well. Clearly, $\beta$ is the required extension of $\alpha$. □

LEMMA 4.2.4. *The map $p_n\colon \Gamma_{R_n} \to B^{n+1}$ is $n$-invertible.*

PROOF. Let us consider a lower semi-continuous many-valued retraction $F_n\colon B^{n+1} \to S^n$, defined as follows:

$$F_n(x) = \begin{cases} \frac{x}{\|x\|}, & \text{if } \|x\| \geq \frac{1}{2}, \\ S^n, & \text{if } \|x\| < \frac{1}{2}. \end{cases}$$

Obviously, $F_(x) \subseteq R_(x)$ for each point $x \in B^{n+1}$. It is also easy to see that the collection $\{F_n(x)\colon x \in B^{n+1}\}$ is connected and uniformly locally connected in all dimensions less than $n$. Let $\Gamma_{F_n} \subseteq B^{n+1} \times S^n$ denote the graph of the map $F_n$ and let $p_{F_n} = \pi_{B^{n+1}}/\Gamma_{F_n}$, where $\pi_{B^{n+1}}\colon B^{n+1} \times S^n \to B^{n+1}$ is the projection onto the first coordinate. It follows from Theorem 2.1.15 that $p_{F_n}$ is $n$-soft. In particular, $p_{F_n}$ is $n$-invertible. Now observe that $\Gamma_{F_n} \subseteq \Gamma_{R_n}$ and $p_{F_n} = p_n/\Gamma_{F_n}$. This is obviously enough to conclude that $p_n$ is also $n$-invertible. □

LEMMA 4.2.5. *Let $Y$ be a closed subset of the sphere $S^n$ and $f\colon Y \to B^{n+1}$ be a map into the ball $B^{n+1}$. Then there exists a homotopy $H\colon Y \times [0,1] \to B^{n+1}$, fixed on $f^{-1}(\partial B^{n+1})$, connecting $f$ with a map whose image is contained in $\partial B^{n+1}$. Moreover, $H(Y \times [0,1])$ does not contain the center of $B^{n+1}$.*

PROOF. The case when $0 \notin f(Y)$ is trivial. Consider the case when $0 \in f(Y)$ and let $U = f^{-1}(B_0^{n+1} - \{0\})$. Obviously, $U$ is a proper subset of the sphere $S^n$ and, consequently, has trivial $n$-dimensional cohomology. Let us show that $f/U\colon U \to B_0^{n+1} - \{0\}$ is null-homotopic. Obviously, $B_0^{n+1} - \{0\}$ is homeomorphic to the product $S^n \times (0, \frac{1}{2}]$. Therefore we can fix a canonical homeomorphism sending $S^n \times \{t\}$ onto the set $\{x \in R^{n+1}\colon \|x\| = t\}$, $t \in (0, \frac{1}{2}]$. Suppose now that the map $f/U$ is not null-homotopic. Since the projection $\pi_1\colon S^n \times (0, \frac{1}{2}] \to S^n$ is a homotopy equivalence (notice that the half-interval is an $AR$-space) we conclude that the composition $\pi f/U\colon U \to S^n$ is also not null-homotopic. In this situation, this composition generates a non-trivial element in the $n$-dimensional cohomology, contrary to the remark above.

Let $\pi_2\colon S^n \times (0, \frac{1}{2}] \to (0, \frac{1}{2}]$ denote the projection onto the second coordinate. We assume that the half-interval $(0, \frac{1}{2}]$ is isometrically embedded into $B_0^{n+1}$ as a subspace $\{x \in R^{n+1}\colon x = te, t \in (0, \frac{1}{2}]\}$, where $e = (1, 0, \ldots, o) \in R^{n+1}$. Since $(0, \frac{1}{2}]$ is contractible, the projection $\pi_2$ is null-homotopic. Consequently, the composition $\pi_2 f/U$ is also null-homotopic. Thus, keeping in mind the above facts, we can conclude that the maps $f/U$ and $\pi_2 f/U$, considered as maps into $B_0^{n+1} - \{0\}$, are homotopic. Take the corresponding homotopy $G\colon U \times [0,1] \to B_0^{n+1} - \{0\}$ such that $G_0 = f/U$ and $G_1 = \pi_2 f/U$). Observe that $\pi_2 G_0(y) = \pi_2 G_1(y)$ for each $y \in U$. Therefore we may assume without loss of generality that $\pi_2 G_0(y) = \pi_2 G_1(y)$ for each $y \in U$ and each $t \in [0,1]$. The homotopy $H$ can be extended to a homotopy $F\colon (U \cup f^{-1}(0)) \times [0,1] \to B_0^{n+1}$. Further, the homotopy $F$ can be extended to a homotopy $H'\colon Y \times [0,1]$ so that

$$H'(f^{-1}(B^{n+1} - B_0^{n+1}) \times [0,1]) \subseteq B^{n+1} - IntB_0^{n+1}.$$

We also can assume that $H'$ is fixed on the set $f^{-1}(\partial B^{n+1})$. The intersection of the image of $H_1'$ with $B_0^{n+1}$ is a segment, in fact a radius of the ball $B_0^{n+1}$. Let us now, after applying the homotopy $H'$, shrink the indicated segment into its end which belongs to the boundary $\partial B_0^{n+1}$. Next push the ring $B^{n+1} - IntB_0^{n+1}$ onto the sphere $\partial B^{n+1}$. As a result, we obtain a homotopy $\bar{H}: Y \times [0,1] \to B^{n+1}$. Observe that $\bar{H}^{-1}(0) = \bar{H}_0^{-1}(0) \times [0,a]$, where $a < 1$. The decomposition of the product $Y \times [0,1]$ into the intervals $\{\{y\} \times [0,a]\colon y \in \bar{H}_0^{-1}(0)\}$ and single points generates a $CE$-map $\mu\colon Y \times [0,1] \to Y \times [0,1]$ which is divisor of the homotopy $\bar{H}$. In other words, there is a map $H: Y \times [0,1] \to B^{n+1}$ such that $\bar{H} = H\mu$. The homotopy $H$ satisfies our requirements. □

LEMMA 4.2.6. *There exists $\delta > 0$ such that for each map $\psi\colon B^n \to B^{n+1}$, with $diam(im(\psi)) < \delta$, and for each map $\varphi\colon \partial B^n\colon \Gamma_{R_n}$, satisfying the equality $p_n\varphi = \psi/\partial B^n$, there is a map $\bar{\varphi}\colon B^n \to \Gamma_{R_n}$ such that $p_n\bar{\varphi} = \psi$ and $\varphi = \bar{\varphi}/\partial B^n$.*

PROOF. Let $\delta = \frac{1}{6}$ and consider a maps $\psi\colon B^n \to B^{n+1}$ and $\varphi\colon \partial B^n \to \Gamma_{R_n}$, satisfying the following two conditions:

$$diam\psi(B^n) < \frac{1}{6} \text{ and } p_n\varphi = \psi/\partial B^n.$$

First assume that $\psi(B^n) \cap \partial B_0^{n+1} = \emptyset$. Then, by the connectedness of $\psi(B^n)$, either $\psi(B^n) \subseteq B_0^{n+1}$ or $\psi(B^n) \subseteq B^{n+1} - B_0^{n+1}$. It follows from the construction of the many-valued map $R_n$ that the map

$$p_n' = p_n/p_n^{-1}(B_0^{n+1})\colon p_n^{-1}(B_0^{n+1}) \to B_0^{n+1}$$

is a locally trivial bundle with fiber the sphere $S^n$ and, consequently, $p_n'$ is $n$-soft. Thus, if $\psi(B^n) \subseteq B_0^{n+1}$, then the conclusion of our lemma is true. Observe that the map

$$\bar{p}_n = p_n/p_n^{-1}(B^{n+1} - B_0^{n+1})\colon p_n^{-1}(B^{n+1} - B_0^{n+1}) \to B^{n+1} - B_0^{n+1}$$

is also $n$-soft (it is even soft). Therefore the lemma is true in the case $\psi(B^n) \subseteq B^{n+1} - B_0^{n+1}$. Next we consider the situation when $\psi(B^n) \cap \partial B_0^{n+1} \neq \emptyset$. Since $diam(im(\psi)) < \frac{1}{6}$, we see that the set $\psi(B^n)$ does not contain the center of the ball $B^{n+1}$. Denote by $J\colon B^{n+1} - \{0\} \to \partial B^{n+1}$ the central projection of $B^{n+1} - \{0\}$ onto the boundary $\partial B^{n+1}$. Then the set $J(\psi(B^n))$ is contained in the ball (on the sphere $S^n$) of an angular radius $\frac{\pi}{6}$. Let $z$ denote the center of this ball. Straightforward calculations show that for each point $x \in \psi(B^n)$, the set $R_n(x)$ is a ball on the sphere $S^n$ with center at $\frac{x}{\|x\|}$ and of radius at least $\frac{\pi}{2}$. Define a section $\eta\colon B^n \to B^n \times S^n$ of the projection $B^n \times S^n \to B^n$ by letting

$$\eta(x) = (x, -\frac{\psi(x)}{\|\psi(x)\|}) \text{ for each point } x \in B^n.$$

Since for each $x \in B^n$ angle between $-z$ and $-\frac{\psi(x)}{\|\psi(x)\|}$ is less than $\frac{\pi}{6}$, we see that there is an isometry $h\colon B^n \times S^n \to B^n \times S^n$, with $\pi_{B^n} = \pi_{B^n} h$, transforming $\eta$ into the trivial section $\theta\colon B^n \to B^n \times S^n$ of the projection $\pi_{B^n}$ (i.e. $h\eta = \theta$) determined by $-z$: $\theta(x) = (x, -z)$ for each $x \in B^n$. Denote by $D^n$ the ball (on the sphere $S^n$) with center at $-z$ and of radius $\frac{\pi}{2}$. The map $\varphi\colon \partial B^n \to \Gamma_{R_n}$ induces a section $\alpha\colon \partial B^n \to \partial B^n \times S^n$ of the projection $\partial B^n \times S^n \to \partial B^n$. One can see that $\alpha(x) \in R_n(\psi(x))$ whenever $x \in \partial B^n = S^{n-1}$. By Lemma 4.2.5, we can connect the composition $\pi_{S^n} h\alpha\colon \partial B^n \to S^n$ with a map $f\colon S^{n-1} \to S^n - IntD^n$ via a homotopy $H_t$ such that $H_t(x) \neq -z$ whenever $t > 0$. Further, let $G_t\colon S^{n-1} \times [0,1] \to S^n$ be a homotopy connecting $\pi_{S^n} h\alpha$ with a constant map ($G_1(S^{n-1}) = d$), which can be obtained from $H$ by adding the contraction of the set $S^n - IntD^n$. For each $x \in B^n$ denote by $r_x$ the push (fixed on $R_n(x)$) in $S^n - R_n(\psi(x))$ with the center at $-\frac{\psi(x)}{\|\psi(x)\|}$, onto the sphere

$$\Sigma_x = \{y \in S^n\colon (\psi(x), y) = 4\|\psi(x)\|^2 - 3\|\psi(x)\|\}$$

if the sphere $\Sigma_x$ is defined. If not, then $r_x$ denotes the identity map of $S^n$. Observe that if $\psi(x) \in \partial B_0^{n+1}$, then $\Sigma_x$ consists of the single point, and if $\psi(x) \in IntB_0^{n+1}$, then $\Sigma_x = \emptyset$.

Define an extension $\bar{\alpha}\colon B^n \to \partial B^n \times S^n$ of $\alpha$ by letting

$$\bar{\alpha}(x) = \begin{cases} r_x h^{-1}\left(x, G_{1-\|x\|}\left(\frac{x}{\|x\|}\right)\right), & \text{if } x \neq 0 \\ r_x h^{-1}(x, d), & \text{if } x = 0. \end{cases}$$

Note that the map $\bar{\alpha}$ is well-defined and that $\bar{\alpha}(x) \in R_n(\psi(x))$ for each point $x \in B^n$. It only remains to note that the desired map $\bar{\varphi}$ can now be defined by letting $\bar{\varphi} = (\psi \times id_{S^n})\bar{\alpha}$. □

DEFINITION 4.2.7. *A map $f\colon X \to Y$ is called* polyhedrally $n$-soft *if for each at most $n$-dimensional polyhedron $B$, subpolyhedron $A$ of $B$, and any two maps $g\colon A \to X$ and $h\colon B \to Y$ with $fg = h/A$, there exists a map $k\colon B \to X$ such that $fk = h$ and $k/A = g$.*

The following statement expresses one of the most important properties of $p_n$.

LEMMA 4.2.8. *The map $p_n\colon \Gamma_{R_n} \to B^{n+1}$ is polyhedrally $n$-soft.*

PROOF. It suffices to show that for every pair of maps $h\colon B^n \to B^{n+1}$ and $g\colon \partial B^n \to \Gamma_{R_n}$ with $p_n g = h/\partial B^n$, there exists a map $k\colon B^n \to \Gamma_{R_n}$ such that $p_n k = h$ and $k/\partial B^n = g$. Take a triangulation of the ball $B^n$ small enough to ensure that the diameters of images (under $h$) of simplexes of this triangulation are less than $\frac{1}{6}$. Denote by $B^n_{(n-1)}$ the $(n-1)$-dimensional skeleton of this triangulation. Since, by Lemma 4.2.3, the map $p_n$ is $(n-1)$-soft, we conclude that there exists a map $k'\colon B_{(n-1)} \to \Gamma_{R_n}$ such that $p_n k' = h/B^n_{(n-1)}$ and $k'/\partial B^n = g$. Apply Lemma 4.2.6 to each of the $n$-dimensional simplexes of the indicated triangulation $B^n$ to obtain the desired extension $k$. □

LEMMA 4.2.9. *The map $q_n\colon \Gamma_{R_n} \to S^n$ is a trivial bundle with fiber homeomorphic to the ball $B^{n+1}$.*

PROOF. It follows from the definition of the retraction $R_n$ that the fiber $q_n^{-1}(y)$ of each point $y \in S^n$ can be written as

$$q_n^{-1}(y) = \{x \in B^{n+1}\colon (x,y) \geq 4\|x\|^2 - 3\|x\|\}.$$

Let $B_0^{n+1} = \{z\colon \|z\| \leq \frac{1}{2}\}$ and define a map $f_y\colon q_n^{-1}(y) \to B_0^{n+1}$ as follows

$$f_y(x) = \frac{4x}{3 + \sqrt{9 + 16(x,y)}}.$$

Straightforward verification shows that the map $f_y$ is a homeomorphism. Additionally, the collection of homeomorphisms $\{f_y\colon y \in S^n\}$ continuously depends on $y$, i.e. $q_n$ is homeomorphic to the trivial bundle $B_0^{n+1} \times S^n \to S^n$. □

Let $K$ be a finite simplicial complex. By $K^{(n)}$ we denote the $n$-dimensional skeleton of $K$ and by $|K|$ the underlying polyhedron of $K$. If for each at most $(n+1)$-dimensional simplex $\sigma \in K$ we take a many-valued retraction $R_n\colon |\sigma| \to |\sigma^{(n+1)}|$, homeomorphic to the retraction $R_n\colon B^{n+1} \to S^n$ constructed above, and then consider the union of these retractions, we obtain a retraction

$$R_{K^{(n)}}\colon |K^{(n+1)}| \to |K^{(n)}|.$$

Generally there is no canonical homeomorphism of the ball $B^{n+1}$ onto the $(n+1)$-dimensional simplex. Therefore the retractions $R_{K^{(n)}}$ are not uniquely defined. For this reason $R_{K^{(n)}}$ denotes any retraction constructed in the above indicated way.

Further, by $\Gamma_{R_{K^{(n)}}} \subseteq |K^{(n+1)}| \times |K^{(n)}|$ we denote the graph of the many-valued retraction $R_{K^{(n)}}\colon |K^{(n+1)}| \times |K^{(n)}|$. As above, $p_{R_{K^{(n)}}}\colon \Gamma_{R_{K^{(n)}}} \to |K^{(n+1)}|$ and $q_{R_{K^{(n)}}}\colon \Gamma_{R_{K^{(n)}}} \to |K^{(n)}|$ shall denote the restrictions of the projections of the product $|K^{(n+1)}| \times |K^{(n)}|$ onto the first and second coordinates respectively.

LEMMA 4.2.10. *The map $p_{R_{K^{(n)}}}\colon \Gamma_{R_{K^{(n)}}} \to |K^{(n+1)}|$ is $n$-invertible, $(n-1)$-soft and polyhedrally $n$-soft.*

PROOF. Apply Lemmas 4.2.3, 4.2.4 and 4.2.8. □

For each pair $(n,k)$ of natural numbers $(n > k)$ and for each finite simplicial complex $K$, let us now assign a commutative diagram $\mathcal{E}^{k,n}_{i,j}(K)$ as follows. Consider the many-valued retractions $R_{K^{(n-1)}}\colon |K^{(n-i+1)}| \to |K^{(n-i)}|$ constructed above, $i = 1, \ldots, k$. First, by induction on $i+j$ (beginning with $i+j=k$), we define spaces $E^{k,n}_{i,j}(K)$ and maps

$$f^{k,n}_{i,j}(K)\colon E^{k,n}_{i,j}(K) \to E^{k,n}_{i-1,j}(K),$$

$$g^{k,n}_{i,j}(K)\colon E^{k,n}_{i,j}(K) \to E^{k,n}_{i,j-1}(K),$$

that will be part of the diagram. For each non-negative integer $i \leq k$, we define our space as $E^{k,n}_{i,k-i}(K) = | K^{(n-1)} |$. For each non-negative integer $i \leq k-1$, let

$$E^{k,n}_{i+1,k-i}(K) = \Gamma_{R_{K^{(n-i-1)}}},$$

$$f^{k,n}_{i+1,k-i}(K) = p_{R_{K^{(n-i-1)}}}$$

and

$$g^{k,n}_{i+1,k-i}(K) = q_{R_{K^{(n-i-1)}}}.$$

Suppose now that the spaces and maps, required for these diagrams have already been constructed for all $i, j$ with $i+j<m$. Let us define the space $E^{k,n}_{i,m-i}(K)$ as a fibered product (i.e. pullback) of spaces $E^{k,n}_{i-1,m-i}(K)$ and $E^{k,n}_{i,m-i-1}(K)$ with respect to the maps

$$f^{k,n}_{i,m-i-1}(K)\colon E^{k,n}_{i,m-i-1}(K) \to E^{k,n}_{i-1,m-i-1}(K)$$

and

$$g^{k,n}_{i-1,m-i-1}(K)\colon E^{k,n}_{i-1,m-i}(K) \to E^{k,n}_{i-1,m-i-1}(K).$$

The maps

$$f^{k,n}_{i,m-i}(K)\colon E^{k,n}_{i,m-i}(K) \to E^{k,n}_{i-1,m-i}(K)$$

and

$$g^{k,n}_{i,m-i}(K)\colon E^{k,n}_{i,m-i}(K) \to E^{k,n}_{i,m-i-1}(K)$$

are defined to be the canonical projections of the corresponding fibered product. In these notations, we have a commutative square diagram

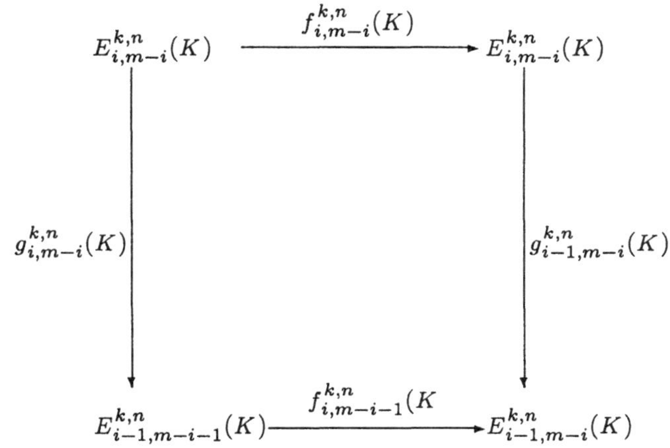


After performing $k$ steps we get a single space $E^{k,n}_{k,k}(K)$ and two maps $f^{k,n}_{k,k}(K)$ and $g^{k,n}_{k,k}(K)$. This finishes the construction of the diagram $\mathcal{E}^{k,n}_{i,j}(K)$. Below, when there is no confusion, we omit upper indices in the notations of spaces and maps of the diagram $\mathcal{E}^{k,n}_{i,j}(K)$.

We formulate some elementary properties of these diagrams.

- The map $f = f_{1,k}f_{2,k}\cdots f_{k,k}\colon E_{k,k} \to E_{0,k}$ is $(n-k-1)$-soft, $(n-k)$-invertible and polyhedrally $(n-k)$-soft.
- Let $g$ be a projection of the space $E_{k,k}$ onto the space $E_{k,0}$. Then
$$gf^{-1} = R_{K^{(n-k)}}R_{K^{(n-k+1)}}\cdots R_{K^{(n-1)}}.$$
- The part of the diagram $\mathcal{E}^{k,n}_{i,j}(K)$ consisting of spaces $E^{k,n}_{i,j}(K)$ with indices $i \geq m$ is naturally isomorphic to the diagram $\mathcal{E}^{k-m,n-m}_{i,j}(K)$. In particular, $E^{k,n}_{i,j}(K) \approx E^{k-m,n-m}_{i-m,j}(K)$.
- Let $K_1$ be a subcomplex of $K$. Then the diagram $E^{k,n}_{i,j}(K_1)$ can be naturally embedded into the diagram $E^{k,n}_{i,j}(K)$ in the sense that for all indices $i,j$ there exist embeddings $\varphi_{i,j}\colon E^{k,n}_{i,j}(K_1) \hookrightarrow E^{k,n}_{i,j}(K)$ such that the maps $f^{k,n}_{i,j}(K_1)$ and $g^{k,n}_{i,j}(K_1)$ coincide with the restrictions of the maps $f^{k,n}_{i,j}(K)$ and $g^{k,n}_{i,j}(K)$ respectively. Observe also that for each $i > 0$ and $j \geq 0$, we have
$$E^{k,n}_{i,j}(K_1) = (f^{k,n}_{i,j}(K))^{-1}(E^{k,n}_{i,j}(K_1)).$$

LEMMA 4.2.11. *If $K$ is a finite simplicial complex, then $E^{k,n}_{i,j}(K)$ is an ANR-compactum.*

PROOF. We prove our statement by induction on $k$. First consider the case $k=1$. Represent the space $E^{1,n}_{0,1}(K) = \mid K^{(n)} \mid$ as the union $\sigma_1 \cup \sigma_2 \cup \cdots \cup \sigma_r$ of its simplexes. Then

$$E_{1,1}^{1,n}(K) = f_{1,1}^{-1}(\sigma_1) \cup \cdots \cup f_{1,1}^{-1}(\sigma_r).$$

Since the map $f_{1,1}$ is a projection of the graph of the many-valued retraction $R_{K^{(n-1)}}$, it can easily be seen that for each $i \leq r$ we have

$$f_{1,1}^{-1}(\sigma_i) \cap \cup_{j<i} f_{1,1}^{-1}(\sigma_j) \approx \sigma_i \cap \cup_{j<i}\sigma_j \in ANR.$$

We also remark that for each $n$-dimensional simplex $\sigma_i$, the inverse image $f_{1,1}^{-1}(\sigma_i)$ is homeomorphic to the graph $\Gamma_{R_{n-1}}$ and that for each simplex $\sigma_j$ of dimensiona less than $n$, its inverse image $f_{1,1}^{-1}(\sigma_j)$ is homeomorphic to $\sigma_j$. By Lemma 4.2.9, the space $\Gamma_{R_{n-1}}$ is a (space of) trivial fibration over the sphere $S^{n-1}$ with fiber $B^n$. Therefore $\Gamma_{R_{n-1}}$ is an $ANR$-compactum. Now by induction, it easily follows that $E_{1,1}^{1,n}$ in an $ANR$-compactum.

Suppose now that for each finite simplicial complex $K$ and for each natural number $n$, the space $E_{k,k}^{k,n}(K)$ is an $ANR$-space. Let us show that in this situation the space $E_{k+1,k+1}^{k+1,n}(K)$ is also an $ANR$-compactum.

Consider the following part of the diagram $\mathcal{E}_{i,j}^{k+1,n}(K)$:

$$\begin{array}{ccccccc}
E_{k+1,k+1} & \xrightarrow{f_{k+1,k+1}} & E_{k,k+1} & \longrightarrow \cdots \longrightarrow & E_{1,k+1} & \xrightarrow{f_{1,k+1}} & E_{0,k+1} \\
\downarrow {\scriptstyle g_{k+1,k+1}} & & \downarrow {\scriptstyle g_{k,k+1}} & & \downarrow {\scriptstyle g_{1,k+1}} & & \\
E_{k+1,k} & \xrightarrow{f_{k+1,k}} & E_{k,k} & \longrightarrow \cdots \longrightarrow & E_{1,k} & &
\end{array}$$

Denote by $\lambda\colon K^{(n)} \to \omega$ the map assigning to each simplex $\sigma \in K^{(n)}$ the number of $n$-dimensional simplexes of $K$ containing $\sigma$ as a subcomplex. Applying Lemma 4.2.9, we can see that for each simplex $\sigma \in K^{(n)}$ the map $g_{1,k+1}$ restricted to the inverse image $g_{1,k+1}^{-1}(Int \mid \sigma \mid)$, is a locally trivial bundle with base $Int \mid \sigma \mid$ and fiber $\vee_{\lambda(\sigma)} B^n$, where $\vee_{\lambda(\sigma)} B^n$ is a wedge of $\lambda(\sigma)$ $n$-dimensional spheres.. Contractibility of $Int \mid \sigma \mid$ guarantees that this bundle is trivial. Similarly, it can be seen that the space $cl(g_{1,k+1}^{-1}(Int \mid \sigma \mid)) \subseteq g_{1,k+1}^{-1}(\mid \sigma \mid)$ (closure is taken in $E_{1,k+1}$) is also the space of a trivial bundle over $\mid \sigma \mid$ with the same fiber. Let $a_1, \ldots, a_{m_0}$ be the vertices of $K$, and in general denote by $a_{m_{i-1}}, \ldots, a_{m_i}$ $i$-dimensional simplexes of $K$. Let $M_l = \cup_{i \leq l} \mid a_i \mid$. Denote by $f_0$ the composition $f_{1,k+1} \cdots f_{k+1,k+1}$ and by $f_1$ the composition $f_{2,k} \cdots f_{k+1,k}$. Suppose now that

$M$ is a subcomplex of $K^{(n-1)}$. It follows from the above listed properties of the diagrams $\mathcal{E}_{i,j}^{k,n}(K)$ that

$$f_1^{-1}(\mid M \mid) = E_{k+1,k}^{k+1,n}(M) \approx E_{k,k}^{k,n-1}(M).$$

Therefore, by our inductive assumption, $f_1^{-1}(\mid M \mid) \in ANR$. Observe that

$$g_{1,k+1}^{-1}(\mid a \mid) \approx \vee_{\lambda(a)} B^n$$

for each vertex $a \in K^{(0)}$. It is not hard to see that $g_{k+1,k+1}^{-1} f_1^{-1}(a) \approx f_1^{-1}(a) \times \vee_{\lambda(a)} B^n$ and, consequently, $g_{k+1,k+1}^{-1} f_1^{-1}(a)$ is an $ANR$-space for each vertex $a \in K^{(0)}$.

Suppose now that the inverse image $g_{k+1,k+1}^{-1} f_1^{-1}(M_l)$ is an $ANR$-space for some $l$ ($l \geq m_0$). Observe that $\mid a_{l+1} \mid - M_l = Int \mid a_{l+1} \mid$. Therefore

$$g_{k+1,k+1}^{-1} f_1^{-1}(M_{l+1}) = g_{k+1,k+1}^{-1} f_1^{-1}(Int \mid a_{l+1} \mid) \cup g_{k+1,k+1}^{-1} f_1^{-1}(M_l) =$$

$$= f_0^{-1}(cl(g_{1,k+1}^{-1}(Int \mid a_{l+1} \mid)) \cup g_{k+1,k+1}^{-1} f_1^{-1}(M_l).$$

Notice also that

$$f_0^{-1}(cl(g_{1,k+1}^{-1}(Int \mid a_{l+1} \mid)) \approx f_1^{-1}(\mid a_{l+1} \mid) \times \vee_{\lambda(a_{l+1})} B^n.$$

In this situation it suffices to show that

$$f_0^{-1}(cl(g_{1,k+1}^{-1}(Int \mid a_{l+1} \mid)) \cap g_{k+1,k+1}^{-1} f_1^{-1}(M_l)$$

is an $ANR$-space. Note that

$$f_0^{-1}(cl(g_{1,k+1}^{-1}(Int \mid a_{l+1} \mid)) \cap g_{k+1,k+1}^{-1} f_1^{-1}(M_l) =$$

$$f_0^{-1}(cl(g_{1,k+1}^{-1}(Int \mid a_{l+1} \mid)) \cap g_{1,k+1}^{-1} f_1^{-1}(M_l) =$$

$$f_0^{-1}(cl(g_{1,k+1}^{-1}(Int \mid a_{l+1} \mid)) \cap g_{1,k+1}^{-1} f_1^{-1}(\partial \mid a_{l+1} \mid).$$

Consider the projection

$$\pi\colon f_1^{-1}(\partial|a_{l+1}|) \times \vee_{\lambda a_{l+1})} B^n \to E_{k+1,k}$$

onto $f_1^{-1}(\partial|a_{l+1}|)$ which, as shown above, is an $ANR$-space. Let

$$\psi\colon f_1^{-1}(\partial|a_{l+1}|) \times \vee_{\lambda a_{l+1})} B^n \to E_{1,k+1}$$

be the composition of $f_1 \times id_{\vee B^n}$ and the trivialization

$$\eta\colon \partial|a_{l+1}| \times \vee B^n \to g_{1,k+1}^{-1}(\partial|a_{l+1}|)$$

of the bundle $g_{1,k+1}/cl(g_{1,k+1}^{-1}(Int \mid a_{l+1} \mid)) \cap g_{1,k+1}^{-1}(\partial \mid a_{l+1} \mid)$. Notice that $f_1\pi = g_{1,k+1}\psi$. Consequently there exists a map

$$\zeta\colon f_1^{-1}(\partial|a_{l+1}|) \times \vee_{\lambda(a_{l+1})} B^n \to E_{k+1,k+1},$$

lifting both $\pi$ and $\psi$. Since the collection $\{\pi, \psi\}$ separates points of the compactum $f_1^{-1}(\partial|a_{l+1}|) \times \vee_{\lambda(a_{l+1})} B^n$, we conclude that $\zeta$ is injective. On the other hand,

$$\zeta(f_1^{-1}(\partial|a_{l+1}|) \times \vee_{\lambda(a_{l+1})} B^n) = f_0^{-1}(cl(g_{1,k+1}^{-1}(Int|a_{l+1}|)) \cap g_{1,k+1}^{-1} f_1^{-1}(M_l).$$

Therefore

$$f_0^{-1}(cl(g_{1,k+1}^{-1}(Int|a_{l+1}|)) \cap g_{k+1,k+1}^{-1} f_1^{-1}(M_l)$$

is an $ANR$-space.

Thus $g_{k+1,k+1}^{-1} f_1^{-1}(M_l)$ is an $ANR$-space for each $l$. It only suffices to note that $E_{k+1,k+1} = g_{k+1,k+1}^{-1} f_1^{-1}(E_{1,k})$. This finishes the proof. □

LEMMA 4.2.12. *For each finite simplicial complex $K$ and each natural number $n$, there exist an $ANR$-compactum $X$ and two maps $f\colon X \to |K|$ and $g\colon X \to |K^{(n)}|$ satisfying the following conditions:*

(a) *The map $f$ is $n$-invertible, polyhedrally $n$-soft and $(n-1)$-soft.*
(b) $g/f^{-1}(|K^{(n)}|) = f/f^{-1}(|K^{(n)}|)$.
(c) $gf^{-1}(|\sigma|) \subseteq |\sigma^{(n)}|$.

PROOF. Let $dim K = n + k$. Consider the diagram $\mathcal{E}_{i,j}^{k,n+k}(K)$ and let

$$X = E_{k,k}^{k,n+k}(K),$$
$$f = f_{1,k} \cdots f_{k,k}\colon E_{k,k}^{k,n+k}(K) \to E_{0,k}^{k,n+k}(K) = |K|,$$
$$g = g_{1,k} \cdots g_{k,k}\colon E_{k,k}^{k,n+k}(K) \to E_{k,0}^{k,n+k}(K) = |K^{(n)} \quad \text{for each } \sigma \in K.$$

By Lemma 4.2.11, $X$ is an $ANR$-compactum. By the properties of the diagrams $\mathcal{E}_{i,j}^{k,n}(K)$, the map $f$ is $n$-invertible, polyhedrally $n$-soft and $(n-1)$-soft. These properties also imply that $gf^{-1} = R_{K^n} \cdots R_{K^{(n+k-1)}}$, and consequently $f$ and $g$ coincide on the inverse image $f^{-1}(|K^{(n)}|)$. For the same reason, $gf^{-1}(|\sigma|) \subseteq |\sigma^{(n)}|$. □

*Remark* 4.2.13. All statements proved in this section so far are valid for countable locally finite simplicial complexes as well. In such cases, the resulting spaces are locally compact $ANR$-spaces and the resulting maps are proper.

*Remark* 4.2.14. More careful consideration shows that the compactum $X$ from Lemma 4.2.12 is a finite polyhedron. Respectively, if $K$ is countable and locally finite then the polyhedron $X$ is also countable and locally finite.

We are ready to prove the main result of this section.

THEOREM 4.2.15. *For each natural number $n$, there exists a map $f_n\colon \mu^n \to I^\omega$ of the universal $n$-dimensional Menger compactum onto the Hilbert cube satisfying the following properties:*

(i) *The map $f_n$ is $n$-invertible, polyhedrally $n$-soft and $(n-1)$-soft.*

(ii) *For each at most $n$-dimensional compactum $Y$, each map $f: Y \to \mu^n$ and each open cover $\mathcal{U} \in cov(\mu^n)$, there is an embedding $g: Y \to \mu^n$ that is $\mathcal{U}$-close to $f$ and such that $f_n g = f_n f$.*

PROOF. We construct the map $f_n$ as the limit projection of an inverse sequence $\mathcal{S} = \{M_i, \alpha_i^{i+1}\}$ consisting of the Hilbert cube manifolds (in particular, $M_1 = I^\omega$ ). The construction is carried out by induction. The limit of this inverse sequence is contained in the product $I_1^\omega \times I_2^\omega \times \cdots \times I_i^\omega \times \cdots$ which has the metric

$$d(x,y) = \Sigma_{i=1}^{\infty} \frac{d_i(x_i, y_i)}{2^{(i+1)}},$$

where $d_i$ is a (bounded by 1) metric on the Hilbert cube $I_i^\omega$, $x = (x_1, \dots, x_i, \dots)$, $y = (y_1, \dots, y_i, \dots)$ and $x_i, y_i \in I_i^\omega$ for each $i$. Let $a_i \in I^\omega$ for each $i$. Then each of the products $I_1^\omega \times \cdots \times I_m^\omega$ can be naturally identified with the subspace $I_1^\omega \times \cdots \times I_m^\omega \times \{a_{m+1}\} \times \cdots \times \{a_i\} \times \cdots$ of the product $\prod_{i=1}^{\infty} I_i^\omega$. We construct an $I^\omega$-manifold as a subspace of the product $I_1^\omega \times \cdots \times I_m^\omega$ and consider the restriction (to $M_{i+1}$) of the projection $\pi_i^{i+1}: \prod_{j=1}^{i} I_j^\omega \times I_{i+1}^\omega \to \prod_{j=1}^{i} I_j^\omega$ as a projection $\alpha_i^{i+1}: M_{i+1} \to M_i$ of the spectrum $\mathcal{S}$. In this situation, as can easily be seen, the limit space $X = \lim \mathcal{S}$ is naturally embedded into $I^\omega = \prod_{i=1}^{\infty} I_i^\omega$.

Let $M_1 = I_1^\omega$ and suppose that the $I^\omega$-manifold $M_j \subseteq \prod_{i=1} j I_i^\omega \subseteq I^\omega$ has already been constructed. By Theorem 2.3.28, the $I^\omega$-manifold $M_j$ is homeomorphic to the product of the Hilbert cube ${}_jI^\omega$ and a finite polyhedron $K_j$. Obviously, we may assume that the composition of the projection $w_j: K_j \times_j I^\omega \to K_j$ with a certain section $s_j: K_j: K_j \times_j I^\omega$ is a $2^{-j}$-move, i.e. for each point $x \in K_j \times_j I^\omega$, we have $d(x, s_j w_j(x)) < 2^{-j}$. We also assume that $K_j$ is given together with a triangulation such that $diam s_j(\sigma) < 2^{-j}$ for each simplex $\sigma$ of this triangulation. By Lemma 4.2.12, there exist an $ANR$-compactum $X_{j+1}$ and maps $f_{j+1}: X_{j+1} \to K_j$ and $g_{j+1}: X_{j+1} \to K_j^{(n)}$ such that $f_{j+1}$ is $n$-invertible, polyhedrally $n$-soft, $(n-1)$-soft and, in addition, $g_{j+1} f_{j+1}^{-1}(\sigma) \subseteq \sigma^{(n)}$ for each simplex $\sigma$ of the triangulation given on $K_j$. Let $M_{j+1} = X_{j+1} \times_{j+1} I^\omega$. By Theorem 2.3.21, $M_{j+1}$ is a $I^\omega$-manifold. Next, represent the cube ${}_{j+1}I^\omega$ as the product ${}_jI^\omega \times_{j+1} I_1^\omega$ of two Hilbert cubes, and denote by $\eta_{j+1}: X_{j+1} \times_j I^\omega \times_{j+1} I_1^\omega \to X_{j+1} \times jI^\omega$ the product of the identity map $id_{X_{j+1}}$ and the projection $\pi_1^j: {}_jI^\omega \times_{j+1} I_1^\omega \to_j I^\omega$. Define the map $\beta_j^{j+1}: M_{j+1} \to M_j$ by letting

$$\beta_j^{j+1} = (f_{j+1} \times id_{{}_jI^\omega}) \eta_{j+1}.$$

In this situation we have the following diagram:

$$\begin{array}{ccc}
M_{j+1} = X_{j+1} \times_j I^\omega \times_{j+1} I_1^\omega & & \\
\Big\downarrow {\scriptstyle \eta_{j+1} = id \times \pi_1^j} & & \\
X_{j+1} \times_j I^\omega & \xrightarrow{f_{j+1} \times id} & K_j \times_j I^\omega = M_j \\
\Big\downarrow & & \Big\downarrow \\
X_{j+1} & \xrightarrow{f_{j+1}} & K_j
\end{array}$$

Let $\lambda\colon M_{j+1} \hookrightarrow I_{j+1}^\omega$ be an embedding. Then the desired embedding of the $I^\omega$-manifold $M_{j+1}$ into the product $I_1^\omega \times \cdots \times I_{j+1}^\omega$ can be defined as the diagonal product of the map $\beta_j^{j+1}$ and the embedding $\lambda$. Let $\alpha_j^{j+1} = \pi_j^{j+1}/M_{j+1}$, and observe that this map $\alpha_j^{j+1}\colon M_{j+1} \to M_j$ is $n$-invertible, polyhedrally $n$-soft and $(n-1)$-soft (because both $f_{j+1}$ and $\pi_1^j$ have the corresponding properties).

Thus the construction of spectrum $\mathcal{S} = \{M_i, \alpha_i^{i+1}\}$ is complete. Let $\alpha_i\colon X = \lim \mathcal{S} \to M_i$ denote the $i$-th limit projection of this spectrum.

As the desired map $f_n\colon \mu^n \to I^\omega$ we take the first limit projection $\alpha_1\colon X \to M_1 = I^\omega$. Of course, we still have to show that $X \approx \mu^n$.

Let us investigate the properties of the compactum $X$ and the map $f_n$. Since each of the short projections $\alpha_i^{i+1}$ of the spectrum $\mathcal{S}$ is $n$-invertible, polyhedrally $n$-soft and $(n-1)$-soft, we easily see that $f_n$ (as well as all other limit projections of the spectrum $\mathcal{S}$, see Lemma 6.2.6) has the same properties. This proves one part of the theorem.

Let us now show that $\dim X = n$. Since $M_1 = I^\omega$, $M_1$ contains a topological copy of the $n$-dimensional cube $I^n$. The $n$-invertibility of $f_n$ then guarantees that $X$ also contains a copy of $I^n$. Therefore $\dim X \geq n$. In order to show that $\dim X \leq n$ we proceed as follows. Obviously, all we need is an existence, for each $i$, of a $\frac{3}{2^i}$-move of $X$ into an $n$-dimensional polyhedron. Such a move is indicated below:

$$s_i w_i (g_{i+1} \times id_{I^\omega}) \eta_{i+1} \alpha_{i+1}\colon X \to K_i^{(n)}.$$

Consequently, $\dim X = n$.

Next we show that $f_n$ has the second property formulated in the theorem. Let $Y$ be an at most $n$-dimensional compactum, $f\colon Y \to X$ be a map and

$\mathcal{U} \in cov(X)$. Choose an index $i$ and an open cover $\mathcal{V} \in cov(M_i)$ such that $\alpha_i^{-1}(\mathcal{V})$ refines $\mathcal{U}$. It follows from the construction of the map $\alpha_i^{i+1}$ that $\alpha_i^{i+1}$ can represented as the composition $\gamma_{i+1}\eta_{i+1}$, where $\eta_{i+1}$ is a trivial bundle with fiber the Hilbert cube. Consequently, there exists an embedding $h_{i+1}\colon Y \to M_{i+1}$ such that $\alpha_i^{i+1}h_{i+1} = \alpha_i f$. Since $dim Y \leq n$, and since the limit projection $\alpha_{i+1}\colon X \to M_{i+1}$ in $n$-invertible, we can find a map $h\colon Y \to X$ such that $\alpha_{i+1}h = h_{i+1}$. It is not hard to see that $h$ is an embedding. Moreover, $h$ is $\mathcal{U}$-close to $f$ and $f_n h = f_n f$.

Finally, let us show that $X$ is homeomorphic to $\mu^n$. By the property of $f_n$ established above, $X$ satisfies the condition from Theorem 4.1.19., i.e. every map of an at most $n$-dimensional compactum into $X$ can be arbitrarily closely approximated by embeddings. The equality $\dim X = n$ was also established above. Therefore, by Theorem 4.1.19, it suffices to show that $X \in LC^{n-1} \cap C^{n-1}$.

First we show that $X \in LC^{n-1}$. Let $x$ be an arbitrary point of $X$ and $U$ be a neighborhood of $x$. Take an index $i$ and a neighborhood $U_i$ of $\alpha_i(x)$ in $M_i$ such that $\alpha_i^{-1}(U_i) \subseteq U$. Since, by our construction, $M_i$ is an $ANR$-compactum (even an $I^\omega$ manifold), $M_i \in LC^{n-1}$. Consequently, there exists a neighborhood $V_i$ of $\alpha_i(x)$ in $M_i$ such that $V_i \subseteq U_i$ and the following condition is satisfied:

- for each $k \leq n-1$ and each map $\varphi_i\colon S^k \to V_i$, there is a map $\psi_i\colon B^{k+1} \to U_i$ such that $\varphi_i = \psi_i/S^k$.

Let $V\alpha_i^{-1}(V_i)$. Clearly, $V$ is a neighborhood of $x$ contained in $U$. Take any map $\varphi\colon S^k \to V$, $k \leq n-1$. By the choice of the neighborhood $V_i$, there is a map $\psi_i\colon B^{k+1} \to U_i$ such that $\alpha_i\varphi = \psi_i/S^k$. Since $(S^k, B^{k+1})$ is a polyhedral pair and the map $\alpha_i$ is polyhedrally $n$-soft (recall that $k \leq n-1$), we conclude that there is a map $\psi\colon B^{k+1} \to X$ such that the following diagram commutes:

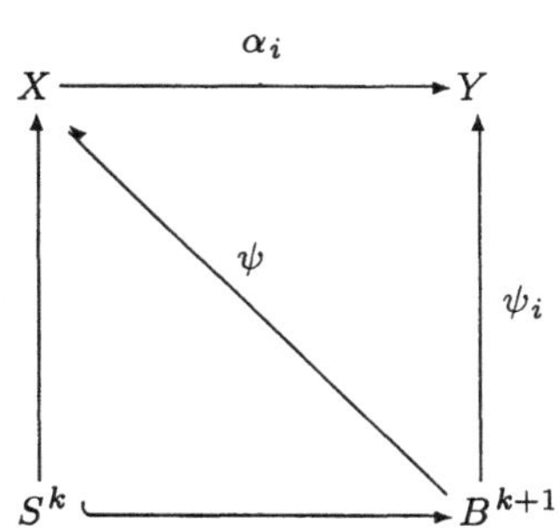


In other words, $\psi/S^k = \varphi$ and $\alpha_i\psi = \psi_i$. This obviously implies that $\psi(B^{k+1}) \subseteq U$. Thus, $X \in LC^{n-1}$.

A similar (but simpler) argument shows that $X \in C^{n-1}$. Therefore, $X \approx \mu^n$. The proof is finished. □

In order to obtain other important properties of the map $f_n$ we need some preliminary statements. The proof of the following one is, in fact, contained in the proof of Lemma 4.2.6.

LEMMA 4.2.16. *For each $\epsilon > 0$, there is a $\delta > 0$ such that for any map $\varphi\colon S^{n-1} \to \Gamma_{R_n}$, with $diam(im(\varphi)) < \delta$, and any $\delta$-homotopy $H : S^{n-1} \times [0,1] \to B^{n+1}$, with $H_0 = p_n\varphi$, there exists a homotopy $G\colon S^{n-1} \times [0,1] \to \Gamma_{R_n}$ such that $G_0 = \varphi$, $p_n G = H$ and $diam G(S^{n-1} \times [0,1]) < \epsilon$.*

LEMMA 4.2.17. *Let $Y$ be an $LC^{n-1}$-compactum, $K$ be a finite simplicial complex and $\varphi\colon Y \to |K^{(n+1)}|$ be a map. Then the fibered product (pullback) $X$ of the spaces $\Gamma_{R_{K^{(n)}}} = \Gamma_{n+1}$ and $Y$ with respect to the maps $p_{n+1} = p_{R_{K^{(n)}}} : \Gamma_{n+1} \to |K^{(n+1)}|$ and $\varphi\colon Y \to |K^{(n+1)}|$ is an $LC^{n-1}$-compactum.*

PROOF. Let $\alpha\left(\cup_{i=1}^{\infty} B_i^n\right)$ be the one-point compactification of a discrete collection of $n$-dimensional disks $B_i^n$, and let $x$ be the compactifying point. Let $S_i^{n-1}$ denote the boundary of the disk $B_i^n$.

Assume the contrary. Then there exists a map $\psi\colon \alpha\left(\cup_{i=1}^{\infty} S_i^{n-1}\right) \to X$ such that for each $i$ and for any extension of $\psi/S_i^{n-1}$ (to the disk $B_i^n$) the diameter of its image is more than some positive number $a$. Let $\sigma$ be an $(n+1)$-dimensional simplex in $K^{(n+1)}$ containing the point $y = p_{n+1}\varphi'\psi(x)$. Here $\varphi'\colon X \to \Gamma_{n+1}$ denotes the canonical projection (parallel to $\varphi$) of the fibered product $X$. The map $q\colon X \to Y$ has similar meaning (see the following diagram):

$$\begin{array}{ccccc}
\alpha\left(\cup_{i=1}^{\infty} S_i^{n-1}\right) & \xrightarrow{\ \psi\ } & X & \xrightarrow{\ \varphi'\ } & \Gamma_{n+1} \\
 & & \Big\downarrow q & & \Big\downarrow p_{n+1} \\
 & & Y & \xrightarrow{\ \varphi\ } & |K^{(n+1)}|
\end{array}$$

We identify the simplex $\sigma$ with the unit $(n+1)$-dimensional ball $B^{n+1}$. Also, we identify the restriction of the many-valued retraction $R_{K^{(n)}} : |K^{(n+1)}| \to |K^{(n)}|$ to $|\sigma|$ with the many-valued retraction $R_n\colon B^{n+1} \to \partial B^{n+1}$ (see the beginning of this section).

We consider two cases.

**Case 1.** Assume first that $y \notin \partial B_0^{n+1}$, and take a neighborhood $V$ of $y$ in $|K^{(n+1)}|$ disjoint from $\partial B_0^{n+1}$. The restriction of $p_{n+1}$ to this neighborhood is $n$-soft by construction. But then the restriction of $q$ to $q^{-1}(\varphi^{-1}(V))$ is also $n$-soft. It only remains to observe that the latter fact, coupled with the condition $Y \in LC^{n-1}$, contradicts the assumption made in the beginning.

**Case 2.** Now assume that $y \in \partial B_0^{n+1}$. Since $Y \in LC^{n-1}$ we can conclude that there is an extension $\zeta\colon \alpha\left(\cup_{i=1}^{\infty} B_i^n\right) \to Y$ of the composition $q\psi$. Take an $\epsilon > 0$ such that for each set $F \subseteq X$, the inequalities $diam(\varphi'(F)) < \epsilon$ and $diam(q(F)) < \epsilon$ imply the inequality $diam F < a$. Next, choose $\delta$ as in Lemma 4.2.16. Let $b_i$ be the center of the ball $B_i^n$. Connect the points $\zeta(b_I)$ and $q\psi(x)$ by paths $\eta_i\colon [0,1] \to Y$ so that $\lim_{i\to\infty} diam \eta_i([0,1]) = 0$. Then there is a number $k$ such that

$$\max\{diam(\eta_k([0,1]) \cup \zeta(B_k^n)), diam(\varphi\eta_k([0,1]) \cup \varphi\zeta(B_k^n))\} < \delta.$$

Let $H_t\colon S_k^{n-1} \times [0,1] \to Y$ be a homotopy which connects $\zeta/S_k^{n-1}$ with the constant map to the point $\zeta(b_k)$ inside of the image $\zeta(B_k^n)$, and then sends this constant map along the path $\eta_k$ to another constant map to the point $q\psi(x)$. By Lemma 4.2.16, the homotopy $\varphi H_t$ can be lifted to a homotopy $G_t\colon S_k^{n-1} \times [0,1] \to \Gamma_{n+1}$ so that $G_0 = \varphi'\psi$, $G_1(S_k^{n-1}) \subseteq p_{n+1}^{-1}(y)$ and $diam(G_t(S_k^{n-1} \times [0,1]) < \epsilon$. Since $p_{n+1}^{-1}(y) \approx S^n$ and the convex hall of $G_1(S_k^{n-1})$ in the sphere $S^n$ has diameter less than $\epsilon$, we may conclude that, also shrinking $conv(G_1(S_k^{n-1})$ to a point, the homotopy $G_t$ can be extended to an $\epsilon$-homotopy $L_t\colon S_k^{n-1} \times [0,1] \to \Gamma_{n+1}$. Extending the homotopy $H_t$ by adding the identity map, we obtain a homotopy $N_t\colon S_k^{n-1} \times [0,1] \to Y$ such that $\varphi N_t = p_{n+1}L_t$. But then (recall that $X$ is the fibered product in the above indicated diagram) we get a well-defined $a$-homotopy $M_t\colon S_k^{n-1} \times [0,1] \to X$ such that $M_0 = \psi/S_k^{n-1}$ and $M_1(S_k^{n-1}) = \psi(x)$. This contradiction finishes the proof. □

LEMMA 4.2.18. *Let $K$ be an $m$-dimensional finite simplicial complex and $f\colon X \to |K|$ be a map, constructed in Lemma 4.2.12. Then for each $LC^{n-1}$-compactum $Y$, $m > n$, and for each map $\varphi\colon Y \to |K|$, the fibered product $Z$ of $X$ and $Y$ with respect to $f$ and $\varphi$ is an $LC^{n-1}$-compactum.*

PROOF. Denote by $X_1$ the inverse limit of the following diagram

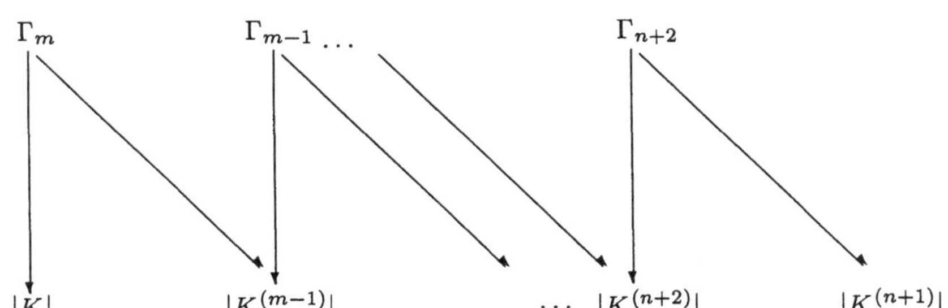


and assume that $p_i'\colon X_1 \to |K^{(i)}|$ and $q_i'\colon X_1 \to \Gamma_i$ denote the corresponding projections. $n$-softness of the projection $p_i'$ implies that the fibered product $Y'$ of $X_1$ and $Y$ with respect to $p_m'$ and $\varphi$ is an $LC^{n-1}$-compactum. Let $\varphi'\colon Y' \to X_1$ be the projection of this fibered product, parallel to $\varphi$. It is easy to see that $X$ is also the fibered product of spaces $X_1$ and $\Gamma_{n+1}$ with respect to maps $p_{n+1}'\colon X_1 \to |K^{(n+1)}|$ and $p_{n+1}\colon \Gamma_{n+1} \to |K^{(n+1)}|$. Then $Z$ itself is the fibered product of spaces $\Gamma_{n+1}$ and $Y'$ with respect to $p_{n+1}\colon \Gamma_{n+1} \to |K^{(n+1)}|$ and

$p'_{n+1}\varphi'\colon Y' \to |K^{(n+1)}|$. By Lemma 4.2.17, $Z$ is an $LC^{n-1}$-compactum. The lemma is proved. □

The following statement expresses one of the most important properties of the map $f_n\colon \mu^n \to I^\omega$ constructed above.

THEOREM 4.2.19. *For each $LC^{n-1}$-compactum $Y$ contained in the Hilbert cube $I^\omega$, the inverse image $f_n^{-1}(Y)$ is a $\mu^n$-manifold. Additionally, if $Y \in C^{n-1}$, then $f_n^{-1}(Y)$ is homeomorphic to $\mu^n$.*

PROOF. We use the same notations as in the proof of Theorem 4.2.15. Recall that the map $f_n\colon \mu^n \to I^\omega$ was constructed as the limit projection of the inverse sequence $\mathcal{S} = \{M_i, \alpha_i^{i+1}\}$ consisting of $I^\omega$-manifolds ($M_1 = I^\omega$) and $n$-invertible, polyhedrally $n$-soft and $(n-1)$-soft short projections. If $Y$ is an $LC^{n-1}$-compactum contained in $I^\omega$, then the inverse image $f_n^{-1}(Y)$ is the limit space of the induced inverse sequence $\mathcal{S}' = \{Y_i, r_i^{i+1}\}$, where $Y_1 = Y$, $Y_{i+1} = (\alpha_i^{i+1})^{-1}(Y_i)$ and $r_i^{i+1} = \alpha_i^{i+1}/Y_{i+1}$. Since the restriction of an $n$-invertible, polyhedrally $n$-soft and $(n-1)$-soft map onto the inverse image still has all these properties, we see that all short projections of the spectrum $\mathcal{S}' = \{Y_i, r_i^{i+1}\}$ are $n$-invertible, polyhedrally $n$-soft and $(n-1)$-soft. While proving Theorem 4.2.15 we have already seen that the inverse limit of an inverse sequence consisting of $LC^{n-1}$-compacta and polyhedrally $n$-soft short projections is an $LC^{n-1}$-compactum. Therefore it suffices to show that $Y_i \in LC^{n-1}$ for each $i$. We prove this fact by induction. By assumption, $Y_1 \in LC^{n-1}$. Assume that $Y_i \in LC^{n-1}$. The projection $\alpha_i^{i+1}$ can be represented as the composition $\gamma_{i+1}\eta_{i+1}$, where $\eta_{i+1}$ is a trivial bundle with fiber the Hilbert cube and $\gamma_{i+1}$ is homeomorphic to the map $f_{i+1} \times id_{I^\omega}$. Here $f_{i+1}\colon X_{i+1} \to K_i$ is the map from Lemma 4.2.12 (see the diagram in the proof of Lemma 4.2.12). Consequently, by Lemma 4.2.18, $\eta_{i+1}(Y_{i+1})$ is an $LC^{n-1}$-compactum. But then $Y_{i+1} \approx \eta_{i+1}(Y_{i+1}) \times I^\omega$ is also an $LC^{n-1}$-compactum. Thus $f_n^{-1}(Y) \in LC^{n-1}$. Polyhedral $n$-softness of the short projections guarantees that if, in addition, $Y \in C^{n-1}$, then $f_n^{-1}(Y) \in C^{n-1}$.

Obviously, $\dim f_n^{-1}(Y) \le \dim \mu^n = n$. Theorem 4.2.15 shows that $\dim f_n^{-1}(Y) \ge n$.

Theorem 4.2.15 also guarantees that any map of an at most $n$-dimensional compactum into $f_n^{-1}(Y)$ can be arbitrarily closely approximated by embeddings. Therefore, by Theorem 4.1.19, $f_n^{-1}(Y)$ is a $\mu^n$-manifold (and is homeomorphic to $\mu^n$ if $Y \in C^{n-1}$). □

*Remark* 4.2.20. In fact a stronger result can be proved: for any $LC^{n-1}$-compactum $Y$ and for any map $\varphi\colon Y \to I^\omega$ (not only for embeddings, as in Theorem 4.2.19) the fibered product of $Y$ and $\mu^n$ with respect to $\varphi$ and $f_n$ is an $LC^{n-1}$-compactum.

Summarizing the results proved above, we have.

THEOREM 4.2.21. *There are maps* $f_n: \mu^n \to I^\omega$ *and* $g_n: \mu^n \to \mu^n$ *satisfying the following properties:*

(i) *The maps* $f_n$ *and* $g_n$ *are* $n$*-invertible,* $(n-1)$*-soft and polyhedrally* $n$*-soft.*
(ii) *All the fibers of the maps* $f_n$ *and* $g_n$ *are homeomorphic to* $\mu^n$*.*
(iii) *The inverse images of* $LC^{n-1}$*-compacta under the maps* $f_n$ *and* $g_n$ *are* $\mu^n$*-manifolds.*
(iv) *The maps* $f_n$ *and* $g_n$ *both satisfy the parametric version of* $DD^nP$*, that is, any two maps* $\alpha, \beta: I^n \to \mu^n$ *can be arbitrarily closely approximated by maps* $\alpha', \beta': I^n \to \mu^n$ *such that* $f_n\alpha' = f_n\alpha$*,* $f_n\beta' = f_n\beta$*,* $g_n\alpha' = g_n\alpha$*,* $g_n\beta' = g_n\beta$ *and* $im(\alpha') \cap im(\beta') = \emptyset$*.*

Using the map $f_n$ as a guide, additional considerations allow us to obtain the following result.

THEOREM 4.2.22. *Any metrizable* $A(N)E(n+1)$*-compactum is an* $UV^{n-1}$*-image of:*

(i) *An* $(n+1)$*-dimensional* $A(N)E$*-compactum.*
(ii) *The Hilbert cube* $I^\omega$ *(an* $I^\omega$ *-manifold, respectively).*
(iii) *The* $(2n+1)$*-dimensional cell ( a* $(2n+1)$*-dimensional topological manifold, respectively).*

A simple comparison of the major ingredients of the Hilbert cube manifold theory (see Chapter 2) with the corresponding results of Menger manifold theory presented so far, shows that from a certain point of view the $n$-dimensional analog of the Hilbert cube $Q$ should be considered to be, not the usual $n$-dimensional cube $I^n$, but the $n$-dimensional universal Menger compactum $\mu^n$ (moreover, the Hilbert cube itself may be viewed as the "infinite-dimensional Menger compactum"). In addition, one can observe a fairly deep analogy between the theories of $\mu^n$-manifolds and $Q$-manifolds themselves. On the other hand, at first glance it is not clear what is the analog of the operation of "taking the product by $Q$" in $\mu^n$-manifold theory – the operation which is involved in the formulations of triangulation (Theorem 2.3.28) and stability (Theorem 2.3.10) theorems for $Q$-manifolds. A decisive step in finding a "full" analog of this operation in $\mu^n$-manifold theory is based on Theorems 4.2.15 and 4.2.19. First observe that taking the product $X \times Q$ of a space $X$ and the Hilbert cube $Q$ may be interpreted as taking the inverse image $\pi_1^{-1}(X)$ of a space $X \subseteq Q$, where $\pi_1: Q \times Q \to Q$ denotes the natural projection onto the first coordinate. It turns out that the map $g_n = f_n/f_n^{-1}(\mu^n): \mu^n \to \mu^n$ in Theorem 4.2.15 may be thought of as the analog of the projection $\pi_1$ in the theory of $\mu^n$-manifolds. If this is agreed, everything then falls in place.

The following statement is a triangulation theorem in $\mu^n$-manifold theory.

THEOREM 4.2.23. *For any* $\mu^n$*-manifold* $M$*, there is an* $n$*-dimensional polyhedron* $P$ *such that for any embedding of* $P$ *into* $\mu^n$ *the inverse image* $g_n^{-1}(P)$ *is homeomorphic to* $M$*.*

PROOF. We consider the compact case. The locally compact case can be proved similarly. Take an $n$-dimensional finite polyhedron $P$ and a map $\varphi\colon P \to M$ which induces isomorphism of the $i$-dimensional homotopy groups for each $i \leq n-1$. Embed $P$ into $\mu^n$. It is easy to see that the composition

$$\varphi g_n/g_n^{-1}(P)\colon g_n^{-1}(P) \to P \to M$$

also induces isomorphism of the $i$-dimensional homotopy groups for each $i \leq n-1$. Observe also that the inverse image $g_n^{-1}(P)$ is a $\mu^n$-manifold (we use Theorem 4.2.19). Therefore, by Theorem 4.1.21, $M$ and $g_n^{-1}(P)$ are homeomorphic. □

Here is the promised stability theorem for $\mu^n$-manifolds.

THEOREM 4.2.24. *For any $\mu^n$-manifold $M$ in $\mu^n$, the inverse image $g_n^{-1}(M)$ is homeomorphic to $M$.*

PROOF. If $M$ is compact, then, by Theorem 4.2.19, the inverse image $g_n^{-1}(M)$ is a $\mu^n$-manifold. Since the restriction $g_n/g_n^{-1}(M)\colon g_n^{-1}(M) \to M$ is polyhedrally $n$-soft, we see that it induces isomorphism of the $i$-dimensional homotopy groups for each $i \leq n-1$. By Theorem 4.1.21, $g_n^{-1}(M)$ is homeomorphic to $M$. The proof of the non-compact case is similar. □

We conclude this section with the following statement which shows, in particular, that the map $f_n$ cannot be made $n$-soft.

THEOREM 4.2.25. *There is no $n$-soft map of an $n$-dimensional compactum onto a higher dimensional cube (the Hilbert cube in particular).*

PROOF. If $n = 0$, the validity of our statement is obvious: the open image of a zero-dimensional compactum is zero-dimensional. Thus we may assume that $n \geq 1$. If such a map $g\colon Z \to I^m$ existed, then $Z$ must be connected (notice that 1-soft maps are monotone). Recall that cell-like maps cannot raise cohomological dimension (see Remark 3.1.11) and that the cohomological dimension of any cube coincides with its Lebesgue dimension. Therefore, $g$ cannot be cell-like. It is clear now that it suffices to prove the following claim.

**Claim.** *Any non-constant $n$-soft map of a connected $n$-dimensional compactum is cell-like.*

*Proof of Claim.* Let $f\colon X \to Y$ be an $n$-soft map, where $X$ is connected, $dim X = n$ and $| Y | > 1$. Since the fiber $f^{-1}(y) \in LC^{n-1} \cap C^{n-1}$ for each $y \in Y$ we see that $H_k(f^{-1}(y)) = \check{H}_k(f_k^{-1}(y)) = 0$ for all $k < n$. Here $H_k$ denotes the singular homology with respect to the group of integers and $\check{H}_k$ the Čech-homology. Then it follows that $\check{H}^k(f^{-1}(y)) = 0$ for all $k < n$. Let us show that $\check{H}^n(f^{-1}(y)) = 0$. If so, using the standard criterion that a finite-dimensional Peano continuum with trivial cohomology has trivial shape, we obtain the desired conclusion.

Assume the contrary. Namely, that $\check{H}^n(f^{-1}(y)) \neq 0$ for some $y \in Y$. Take a map $\varphi\colon f^{-1}(y) \to K(\mathbb{Z}, n)$ (the Eilenberg-Maclane complex) which is not homotopic to a constant map. Consider a point $x \in Y$ different from $y$. Without loss of generality we may assume that $| f^{-1}(x) |= 1$ (otherwise shrink the fiber $f^{-1}(x)$ into a point). Since $dim X = n$, there is an extension $\bar{\varphi}\colon X \to K(\mathbb{Z}, n)$. Next denote by $Y_0$ the set of all those points $z \in Y$ for which the restriction $\bar{\varphi}/f^{-1}(z)$ is homotopic to a constant map. Obviously, $Y_0$ is not empty. Indeed, $x \in Y_0$. Let us show that $Y_0$ is an open set in $Y$. Take a point $z \in Y_0$, and let $h\colon f^{-1}(z) \times [0,1] \to K(\mathbb{Z}, n)$ be a homotopy connecting $\bar{\varphi}/f^{-1}(z)$ with a constant map to some point $c \in K(\mathbb{Z}, n)$, i.e.

$$h/(f^{-1}(z) \times \{0\}) = \bar{\varphi}/f^{-1}(z) \text{ and } h/(f^{-1}(z) \times \{1\}) = c.$$

Consider the closed subset $A = (X \times \{0,1\} \cup (f^{-1}(z) \times [0,1])$ of the product $X \times [0,1]$ and define the map $h_1\colon A \to K(\mathbb{Z}, n)$ by letting

$$h_1(X \times \{1\}) = c, h_1/(X \times \{0\}) = \bar{\varphi} \text{ and } h_1/(f^{-1}(z) \times [0,1]) = h.$$

Since the $CW$-complex $K(\mathbb{Z}, n)$ is an absolute extensor with respect to the class of compact spaces, we can extend $h_1$ to a map $\bar{h}\colon U \to K(Z,n)$, where $U$ is a neighborhood of $A$ in the product $X \times [0,1]$. Then $U$ contains an open set of the form $G \times [0,1]$, where $G$ is a neighborhood of the point $z$ in $Y$. Clearly $G$ is contained in $Y_0$ (see the definition of $Y_0$). Since $z$ was an arbitrarily chosen point of $Y_0$, we conclude that $Y_0$ is open in $Y$.

Next observe that the complement $Y - Y_0$ is also a non-empty set, since it contains $y$. Let us now show that this complement is also open in $Y$. Assume the contrary. Then there exists a sequence $\{z_k\}$ of points of $Y_0$ such that $z = \lim z_k \in Y - Y_0$. We construct a sequence $\{g_k\colon f^{-1}(z) \to f^{-1}(z_k)\}$ of maps, which converges to the identity map $id_{f^{-1}(z)}$ in the space $C(f^{-1}(z), X)$. Since the space $F = f^{-1}(z) \times (\{z_k\colon k \in N\} \cup \{z\})$ is at most $n$-dimensional, $n$-softness of the map $f$ guarantees the existence of a map $\psi\colon F \to X$ such that

$$\psi/(f^{-1}(z) \times \{z\}) = id_{f^{-1}(z)}$$

and

$$\psi(f^{-1}(z) \times \{z_k\}) \subseteq f^{-1}(z_k) \text{ for each } k \in N.$$

Let $g_k = \psi/(f^{-1}(z) \times \{z_k\})$. Thus we have a sequence $\{\bar{\varphi} g_k\}$of maps each of which is homotopic to a constant map and which converges to the map $\bar{\varphi}/f^{-1}(z)$. This contradicts the fact that sufficiently close maps into $ANR$-space are homotopic. Consequently, $Y - Y_0$ is open in $Y$. Connectedness of $Y$ shows that this is impossible □

*Historical and bibliographical notes* 4.2. Theorem 4.2.1 and Proposition 4.2.2 (a weaker version of it – without stating an $n$-invertibility) were proved in [**33**]. The construction of $n$-invertible, $(n-1)$-soft maps presented in this Section, as well as Theorems 4.2.23 and 4.2.24 in the compact case, are taken from [**127**](see [**90**] for the non-compact case). This result completes a circle of works of various authors concerning the existence of dimension raising maps. Historically, the first example of an open map (of the one-dimensional compactum onto the two-dimensional "Pontryagin surface") of this sort was constructed by Kolmogorov in **[196]** in 1937. Further examples, with some additional properties, have been constructed in **[11]**, **[12]**, **[15]**, **[319]**, **[320]**, **[189]**, **[190]**, **[191]**, **[200]**, **[211]**, **[245]**, **[248]**, **[304]** etc. Theorems 4.2.22 and 4.2.25 are taken from **[129]** and **[125]**. Theorem 4.2.22 extends an earlier result from **[71]**.

## 4.3. $n$-soft mappings of Polish spaces, raising dimension

It has already been remarked in Section 4.2 that the maps $f_n$ and $g_n$, constructed in that section, cannot be made $n$-soft. Also, both of them fail to satisfy the property of preservation of $Z$-sets in the inverse direction. In this section we construct a map with the last property which is "almost" $n$-soft.

We begin with the following technical statement.

LEMMA 4.3.1. *If $f: X \to Y$ is an $n$-soft map, then the inverse image of each $Z_n$-set in $Y$ is a $Z_n$-set in $X$.*

For each simplex $\sigma$, denote the first and second barycentric subdivisions of $\sigma$ by $\beta\sigma$ and $\beta^2\sigma$ respectively. $M_\sigma$ denotes the closed star of the barycenter $v_\sigma$ of $\sigma$ in the triangulation $\beta^2\sigma$: $M_\sigma = St(v_\sigma, \beta^2\sigma)$. We put $N_\sigma = Int(M_\sigma)$. Finally let $r_\sigma\colon (|\sigma| - \{v_\sigma\}) \to |\partial\sigma|$ denote the canonical deformation retraction.

LEMMA 4.3.2. *For each $n \geq 0$ and each countable locally-finite simplicial complex $K$, there exist a countable locally finite simplicial complex $B_K^n$ and proper simplicial maps $f_K^n\colon |B_K^n| \to |K|$ and $g_K^n\colon |B_K^n| \to |K^{(n+1)}|$ satisfying the following conditions:*

(i) *$f_K^n$ is an $(n+1)$-invertible $UV^n$-map.*
(ii) *If $\sigma$ is a simplex of $K$, then $g_K^n((f_K^n)^{-1}(|\sigma|)) \subseteq |\sigma^{(n+1)}|$.*
(iii) *$f_K^n/(f_K^n)^{-1}(|K^{(n+1)}|) = g_K^n/(f_K^n)^{-1}(|K^{(n+1)}|)$.*
(iv) *There exists a subspace $A_K^n$ of $|B_K^n|$ such that the restriction $f_K^n/A_K^n\colon A_K^n \to |K|$ is an $(n+1)$-soft map and the complement $|B_K^n| - A_K^n$ is a $\sigma Z$-set in $|B_K^n|$.*
(v) *If $Z$ is a $Z_{n+1}$-set in $|K|$, then $(f_K^n)^{-1}(Z)$ is a $Z_{n+1}$-set in $|B_K^n|$.*

PROOF. Let us consider an arbitrary simplex $\sigma$ and define two compact-valued retractions $\varphi_\sigma^n, \psi_\sigma^n\colon |\sigma| \to |\sigma^{(n+1)}|$. Definitions are given by induction on the $(n+1+i)$-dimensional skeleta of $\sigma$.

If $x \in |\sigma^{(n+1)}|$, then we put $\varphi_\sigma^n(x) = x = \psi_\sigma^n(x)$.

If $\tau$ is an $(n+2)$-dimensional face of $\sigma$, then

$$\varphi_\sigma^n(x) = \begin{cases} |\partial\tau|, & \text{if } x \in N_\tau, \\ r_\tau(x), & \text{if } x \in |\tau| - N_\tau \end{cases}$$

and

$$\psi_\sigma^n(x) = \begin{cases} |\partial\tau|, & \text{if } x \in M_\tau, \\ r_\tau(x), & \text{if } x \in |\tau| - M_\tau \end{cases}$$

If $\tau$ is an $(n+3)$-dimensional face of $\sigma$, then

$$\varphi_\sigma^n(x) = \begin{cases} |(\partial\tau)^{(n+1)}|, & \text{if } x \in N_\tau, \\ \varphi_\sigma^n(r_\tau(x)), & \text{if } x \in |\tau| - N_\tau \end{cases}$$

and

$$\psi_\sigma^n(x) = \begin{cases} |(\partial\tau)^{(n+1)}|, & \text{if } x \in M_\tau, \\ \psi_\sigma^n(r_\tau(x)), & \text{if } x \in |\tau| - M_\tau. \end{cases}$$

Continuing this process, we obtain the desired retractions of $|\sigma|$ onto $|\sigma^{(n+1)}|$. Finally, if $K$ is an arbitrary locally-finite simplicial complex, then the retractions $\varphi_K^n, \psi_K^n\colon |K| \to |K^{(n+1)}|$ are defined as the unions of $\varphi_\sigma^n$ and $\psi_\sigma^n$, $\sigma \in K$.

Now denote by $B_K^n$ the standard triangulation (induced by $K$) of the polyhedron $\{(x,y) \in |K| \times |K^{(n+1)}|\colon \psi_K^n(x) \ni y\}$. We put also $f_K^n = \pi_1/|B_K^n|$ and $g_K^n = \pi_2/|B_K^n|$, where $\pi_1\colon |K| \times |K^{(n+1)}| \to |K|$ and $\pi_2\colon |K| \times |K^{(n+1)}| \to |K^{(n+1)}|$ denote the natural projections.

Straightforward verification shows that:

(a) the compact-valued retraction $\varphi_K^n\colon |K| \to |K^{(n+1)}|$ is lower semi-continuous.

(b) the compact-valued retraction $\psi_K^n\colon |K| \to |K^{(n+1)}|$ is upper semi-continuous.

(c) $\varphi_K^n(x) \subseteq \psi_K^n(x)$ for each point $x \in |K|$.

(d) the collection $\{\varphi_K^n(x)\colon x \in |K|\}$ is connected and uniformly locally connected in all dimensions less than $n+1$ (we consider the standard metric on $|K|$).

Therefore, properties (i)-(iii) of the lemma are satisfied.

Let us verify condition *(iv)*. First of all, consider the subspace

$$A_K^n = \{(x,y) \in |K| \times |K^{(n+1)}|\colon y \in \varphi_K^n(x)\}$$

of $|B_K^n|$ and note that the complement $|B_K^n| - A_K^n$ is an $F_\sigma$-subset of $|B_K^n|$. By conditions (a), (c) and Theorem 2.1.15, the restriction $f_K^n/A_K^n$ is an $(n+1)$-soft map. Let us show that the complement $|B_K^n| - A_K^n$ is a $\sigma Z$-set in $|B_K^n|$. Clearly it is sufficient to show that the last fact is true for the $(n+2)$-dimensional simplex $\sigma$. It follows from the construction that $|B_K^n| - A_K^n = \partial M_\sigma \times |\sigma^{(n+1)}| - T$, where $T = \{(x,y) \in |\sigma| \times |\sigma^{(n+1)}|\colon \varphi_\sigma^n(x) = y\}$. Since $\partial M_\sigma$ is a $Z$-set in $M_\sigma$, we can conclude that $\partial M_\sigma \times |\sigma^{(n+1)}|$ is a $Z$-set in $M_\sigma \times |\sigma^{(n+1)}|$. Consequently, $|B_\sigma^n| - A_\sigma^n$ is a $\sigma Z$-set in $M_\sigma \times |\sigma^{(n+1)}|$. Consider now an open subspace $U = |B_\sigma^n| - T$ of the polyhedron $|B_\sigma^n|$. Evidently, $|B_\sigma^n| - A_\sigma^n \subseteq U \subseteq M_\sigma \times |\sigma^{(n+1)}|$. Then the complement $|B_\sigma^n| - A_\sigma^n$ is a $\sigma Z$-set in $U$. Finally, for the same reason, we can conclude that $|B_\sigma^n| - A_\sigma^n$ is a $\sigma Z$-set in $|B_\sigma^n|$.

The last condition is an easy consequence of condition (iv). Indeed, let $Z$ be a $Z_{n+1}$-set in $|K|$. Then, by (iv) and Lemma 4.3.1, the set

$$C_1 = \{h \in C(I^{n+1}, A^n_K) : im(h) \cap ((f^n_K)^{-1}(Z) \cap A^n_K) = \emptyset\}$$

is a dense $G_\delta$-subset of $C(I^{n+1}, A^n_K)$. At the same time, by (iv), the set

$$C_2 = \{h \in C(I^{n+1}, |B^n_K|) : im(h) \subseteq A^n_K\}$$

is also a dense $G_\delta$-subset of $C(I^{n+1}, |B^n_K|)$. A Baire category argument finishes the proof. □

THEOREM 4.3.3. *Let $n \geq 0$ and let $X$ be the limit space of an inverse sequence $\mathcal{S}_X = \{|X_i|, p_i^{i+1}\}$ all spaces of which are locally finite polyhedra and all bonding maps of which are proper, simplicial and $(n+1)$-soft. Then there exists an $(n+1)$-invertible proper $UV^n$-surjection $f^n_X : M_X \to X$ of some $\mu^{n+1}$-manifold $M_X$ onto $X$ satisfying the following conditions:*

(i) *For each $(n+1)$-dimensional locally compact space $Z$, closed subset $Z_0$ of $Z$, open cover $\mathcal{U}$ of $M_X$, and proper map $h : Z \to M_X$, such that $h/Z_0$ is a fibered $Z$-embedding (with respect to $f^n_X$), there exists a fibered $Z$-embedding (with respect to $f^n_X$) $g : Z \to M_X$ which is $\mathcal{U}$-close to $h$ and such that $f^n_X g = f^n_X h$ and $g/Z_0 = h/Z_0$.*

(ii) *There exists a subspace $A_X$ of $M_X$ such that the restriction $f^n_X/A_X : A_X \to X$ is an $(n+1)$-soft map and the complement $M_X - A_X$ is a $\sigma Z$-set in $M_X$.*

(iii) *If $Z$ is a $Z_{n+1}$-set in $X$, then $(f^n_X)^{-1}(Z)$ is a $Z$-set in $M_X$.*

PROOF. For simplicity we consider only the compact case. The general case can be handled similarly. A standard Baire category argument reduces the proof to the case when $Z_0 = \emptyset$. After making these assumptions, we proceed as follows. We construct another polyhedral inverse sequences $\mathcal{S}_M = \{|K_i|, q_i^{i+1}\}$, all bonding maps of which are simplicial $UV^n$-surjections. Further, we shall construct a family of $(n+1)$-invertible simplicial $UV^n$-surjections $f_i : |K_i| \to |X_i|$ which forms a strictly commutative $(n+1)$-invertible $UV^n$-morphism $\{f_i : |K_i| \to |X_i|\} : \mathcal{S}_M \to \mathcal{S}_X$. This means that if we fix an index $i$ and consider the naturally arising diagram

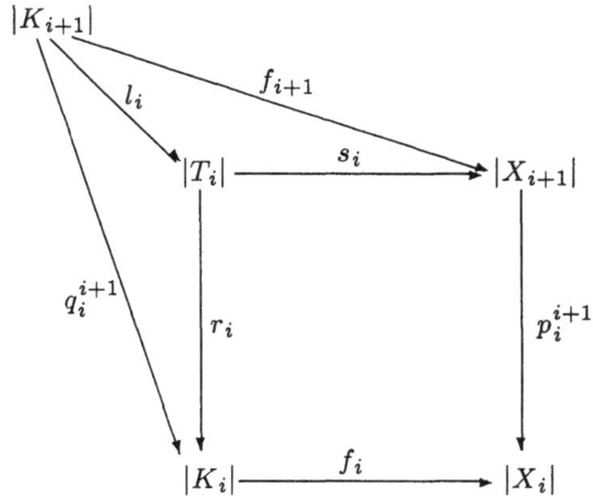


then the *characteristic map* $l_i = q_i^{i+1} \triangle f_{i+1}$ of the diagram is an $(n+1)$-invertible $UV^n$-surjection (here $|T_i|$ denotes the fibered product of $|K_i|$ and $|X_{i+1}|$ with respect to maps $f_i$ and $p_i^{i+1}$; $s_i$ and $r_i$ denote the corresponding projections of this fibered product). Then we shall obtain the space $M_X$ as the limit of $\mathcal{S}_M$ and the map $f_X^n$ as the limit map of the morphism $\{f_i\}$.

Let $K_1 = B_{X_1}^n$ and $f_1 = f_{X_1}^n \colon |K_1| \to |X_1|$ (see Lemma 4.3.2). We can suppose, of course, that $mesh(K_1) < \frac{1}{2}$ (if not, then we consider a sufficiently fine subdivision of $K_1$ and denote it again by $K_1$). Suppose now that we have already constructed finite simplicial complexes $K_m$ with $mesh(K_m) < \frac{1}{2^m}$ and simplicial $UV^n$-surjections $f_m \colon |K_m| \to |X_m|$ and $q_{m-1}^m \colon |K_m| \to |K_{m-1}|$, $m \leq i$, in such a way that the characteristic maps of all the naturally arising rectangular diagrams are $(n+1)$-invertible $UV^n$-surjections. In order to perform the inductive step, consider the fibered product $|T_i|$ of $|K_i|$ and $|X_{i+1}|$ with respect to the maps $f_i$ and $p_i^{i+1}$ and denote by $r_i \colon |T_i| \to |K_i|$ and $s_i \colon |T_i| \to |X_{i+1}|$ the corresponding projections. Now consider the map $f_{L_i}^n \colon |B_{L_i}^n| \to |L_i|$, where $L_i$ denotes the natural triangulation of the polyhedron $|T_i| \times I^{2n+3}$ (we again use Lemma 4.3.2). Let $K_{i+1} = B_{L_i}^n$, assuming at the same time that $mesh(K_{i+1}) < \frac{1}{2^{i+1}}$. Also define $f_{i+1} = s_i l_i$ and $q_i^{i+1} = r_i l + i$, where $l_i = \pi_1 f_{L_i}^n$ and $\pi_1 \colon |T_i| \times I^{2n+3} \to |T_i|$ denotes the projection onto the first coordinate. One can easily verify that all our requirements are satisfied and consequently the inductive step is complete.

As already remarked, we let $M_X = \lim \mathcal{S}_M$ and $f_X^n = \lim\{f_i\}$.

Since all mentioned rectangular diagrams strictly commute (because, by the construction, their characteristic maps – $l_i$'s – are surjective) and all $f_i$'s are $UV^n$-surjections, we can conclude that their limit map $f_X^n \colon M_X \to X$ is also a $UV^n$-surjection. The standard argument (see Chapter 6) shows that $(n+1)$-invertibility of all $f_i$'s and all $l_i$'s implies $(n+1)$-invertibility of $f_X^n$.

Let us now verify condition (i). Fix an index $m$ and an open cover $\mathcal{U}_m$ of $|K_m|$ such that $q_m^{-1}(\mathcal{U}_m)$ refines $\mathcal{U}$ (recall that $q_i \colon M_X \to |K_i|$ and $p_i \colon X \to |X_i|$ denote the limit projections of the spectra $\mathcal{S}_M$ and $\mathcal{S}_X$ respectively). We are going to construct maps $g_i\colon Z \to |K_i|$ in such a way that the following conditions are satisfied:

(a) If $i > m$, then $g_i$ is an embedding.

(b) $q_i^{i+1} g_{i+1} = g_i$.

(c) $f_i g_i = p_i f_X^n h$.

(d) If $i > m$ and $\alpha_i\colon I^{n+1} \to |K_i|$ is an arbitrary map, then there exists a map $\beta_i\colon I^{n+1} \to |K_i|$ such that $f_i\beta_i = f_i\alpha_i$, $q_{i-1}^i\beta_i = q_{i-1}^i\alpha_i$ and $im(\beta_i) \cap im(g_i) = \emptyset$.

Let $g_i = q_i h$ for each $i \leq m$. We now indicate how the map $g_{i+1}$ can be constructed. First of all fix an embedding $u\colon Z \to I^{2n+3}$ (we use the inequality $dim Z \leq n+1$). Now consider the diagonal product $g_m \triangle p_{m+1} f_X^n h$, which maps $Z$ into $|T_m|$. Then the diagonal product $(g_m \triangle p_{m+1} f_X^n h)\triangle u$ will be an embedding of $Z$ into the product $|L_m| = |T_m| \times I^{2n+3}$. By Lemma 4.3.2, the map $f_{L_m}^n : |K_{m+1}| \to |L_m|$ is $(n+1)$-invertible. Consequently, there exists an embedding $g_{m+1}\colon Z \to |K_{m+1}|$ such that $f_{L_m}^n g_{m+1} = (g_m \triangle p_{m+1} f_X^n h)\triangle u$. A simple verification shows that conditions (b) and (c) are satisfied. Now consider any map $\alpha_{m+1}\colon I^{n+1} \to |K_{m+1}|$. Since $u$ is an embedding and $\dim Z \leq n+1$, there is a point $a \in I^{2n+3}$ such that $a \notin im(u)$. Let the same letter $a$ denote the constant map which sends the whole cube $I^{n+1}$ to the point $a$. Consider the map $(l_m\alpha_{m+1}\triangle a)\colon I^{n+1} \to |L_m|$ and observe that the image of $I^{n+1}$ under this map does not intersect the image of $Z$ under the composition $f_{L_m}^m g_{m+1}$. Consider now any lifting $\beta_{m+1}\colon I^{n+1} \to |K_{m+1}|$ of the product $(l_m\alpha_{m+1}\triangle a)$. Again, straightforward verification shows that condition (d) is also satisfied. Therefore, continuing this process we obtain the maps $g_i$ for each $i$ By (b), the diagonal product $g$ of all $g_i$'s maps $Z$ into $M_X$. By (a), $g$ is an embedding. By (c), the desired equality $f_X^n g = f_X^n h$ also holds. The choice of an index $m$ and the equalities $q_m g = g_m = q_m h$ (which are true by our construction) show that $g$ and $h$ are $\mathcal{U}$-close. Finally, let us show that $im(g)$ is a fibered $Z$-set with respect to the map $f_X^n$. Fix an open cover $\mathcal{V}$ of $M_X$ and any map $\alpha\colon I^{n+1} \to M_X$. Clearly we can assume that there exist an index $j \geq m$ and an open cover $\mathcal{V}_j$ of $|K_j|$ such that $q_j^{-1}(\mathcal{V}_j)$ refines $\mathcal{V}$. As above we shall inductively construct maps $\beta_i\colon I^{n+1} \to |K_i|$ in such a way that the following conditions are satisfied:

(e) $q_i^{i+1}\beta_{i+1} = \beta_i$.

(f) $f_i\beta_i = p_i f_X^n \alpha$.

(g) $im(\beta_{j+1}) \cap im(g_{j+1}) = \emptyset$.

(h) $\beta_j = q_j\alpha$.

We let $\beta_i = q_i\alpha$ for each $i \leq j$ (consequently, the last condition is automatically satisfied). Let us construct the map $\beta_{j+1}$. By (d) (assuming that $i = j+1$), we obtain a map $\beta_{j+1}\colon I^{n+1} \to | K_{j+1} |$ such that $im(\beta_{j+1}) \cap im(g_{j+1}) = \emptyset$ (i.e. condition (g) is satisfied), $q_j^{j+1}\beta_{j+1} = \beta_j$ and $f_{j+1}\beta_{j+1} = p_{j+1} f_X^n \alpha$. For $i \geq j+2$, we can construct maps $\beta_i\colon I^{n+1} \to | K_i |$ in a similar way to the construction of $\beta_{j+1}$. It only remains to remark that if $\beta$ is the diagonal product of all $\beta_i$'s, then one can easily verify that $im(\beta) \cap im(g) = \emptyset$, $f_X^n\beta = f_X^n\alpha$ and that $\beta$ and $\alpha$ are

$\mathcal{V}$-close. This finishes the verification of condition (i).

Let us now show that $M_X$ is an $\mu^{n+1}$-manifold. Since $mesh(K_i) \to 0$ we can conclude (conditions (ii) and (iii) of Lemma 4.3.2) that $M_X$ admits small maps into $(n+1)$-dimensional polyhedra. Consequently, $dim M_X \leq n+1$. The inverse inequality is obvious, because, by the above verified condition (i), $M_X$ contains a copy of every $(n+1)$-dimensional compactum. Since $M_X$ is, by the construction, the limit space of the polyhedral inverse sequence $\mathcal{S}_M$ all bonding maps of which are $UV^n$-surjections we conclude that $M_X$ is an $LC^n$-compactum. Again, by (i), it follows immediately that $M_X$ satisfies the disjoint $(n+1)$-disk property and hence, by Theorem 4.1.19, $M_X$ is an $\mu^{n+1}$-manifold.

Next, we verify condition (ii). By Lemma 4.3.2, there exists a subspace $A_1$ of $|K_1|$ such that the restriction $f_1/A_1 \colon A_1 \to |X_1|$ is an $(n+1)$-soft map and the complement $|K_1| - A_1$ is a $\sigma Z_{n+1}$-set of $|K_1|$. It follows from our construction that the fibered product $R_1$ of $A_1$ and $| X_2 |$ with respect to the maps $f_1/A_1$ and $p_1^2$ is a subspace of $|T_1|$. Clearly the natural projections of $R_1$ onto $A_1$ and $|X_2|$ coincide with the restrictions $r_1/R_1$ and $s_1/R_1$ respectively. Since $f_1/A_1$ is $(n+1)$-soft, we conclude (see Lemma 6.2.5) that $s_1/R_1$ is also $(n+1)$-soft. Consider the set $R_1 \times I^{2n+3}$ and denote by $A_2$ its inverse image under the $(n+1)$-soft map $f_{L_1}^n$ (see Lemma 4.3.2). Then the restriction $f_2/A_2 \colon A_2 \to |X_2|$ is $(n+1)$-soft. Since the bonding map $p_1^2$ is $(n+1)$-soft, using again Lemma 6.2.5, we see that the map $r_1$ is also $(n+1)$-soft. Consequently, by Lemma 4.3.1, the set $|T_1| - R_1$ is a $\sigma Z_{n+1}$-set in $|T_1|$. In this situation one can easily observe that, again by Lemma 4.3.1, the complement $|K_2| - A_2$ is a $\sigma Z_{n+1}$-set in $|K_2|$. Continuing in such a manner we obtain an inverse sequence $\mathcal{S}_A = \{A_i, q_i^{i+1}/A_{i+1}\}$ and an $(n+1)$-soft morphism (in the sense of Chapter 6) morphism $\{f_i/A_i\}$ of $\mathcal{S}_A$ to $\mathcal{S}_X$. Clearly, the limit space $A_X$ of the spectrum $\mathcal{S}_A$ is a subspace of $M_X$ and the restriction $f_X^n/A_X \colon A_X \to X$, which coincides with the limit map of the morphism $\{f_i/A_i\}$, is an $(n+1)$-soft map (here we use Lemma 6.2.7). It only remains to remark that the complement $M_X - A_X$ is a $\sigma Z$-set in $M_X$.

Condition (iii) is a direct consequence of condition (ii) and Lemma 4.3.1 (compare with the proof pf Lemma 4.3.2). The proof is complete. □

We also need the following statement.

PROPOSITION 4.3.4. *Each $\mu^{n+1}$manifold $M$ can be represented as the limit space of an inverse sequence $\mathcal{S}_M = \{M_i, p_i^{i+1}\}$ consisting of locally compact polyhedra and proper simplicial $(n+1)$-invertible, $n$-soft and polyhedrally $(n+1)$-soft bonding maps.*

PROOF. By Proposition 4.1.10, there exists a proper $n$-homotopy equivalence $\alpha \colon | K | \to M$, where $K$ is an at most $(n+1)$-dimensional countable locally finite simplex. Let $K_0 = K$. Suppose that the countable locally finite simplicial simplexes $K_i$ and proper simplicial maps $p_{i-1}^i \colon |K_i| \to |K_{i-1}|$, satisfying conditions

(i) - (iii) from Lemma 4.3.2, have already been constructed for each $i \leq m$. We may also assume that the map $p^i_{i-1}\colon |K_i| \to |K_{i-1}|$ has the following property:

- For any map $\alpha\colon I^{n+1} \to |K_{i-1}|$ there exist two maps $\beta_1, \beta_2\colon I^{n+1} \to |K_i|$ such that $p^i_{i-1}\beta_j = \alpha$, $j = 1, 2$, and $im(\beta_1) \cap im(\beta_2) = \emptyset$.

In order to construct these objects for the next step, we apply Lemma 4.3.2 to a sufficiently fine triangulation of the polyhedron $|K_m| \times [0,1]$. In this way we get the next simplicial complex $K_{m+1} = B^n_{K_m \times [0,1]}$. We let $p^{m+1}_m = \pi_1 f^n_{K_m \times [0,1]}$, where $\pi_1\colon |K_m| \times [0,1] \to |K_m|$ is the projection. As in the proof of Theorem 4.3.3, we can verify that the limit space $M'$ of the inverse sequence $\mathcal{S}_M = \{M_i, p^{i+1}_i\}$ is a $\mu^{n+1}$-manifold. Clearly, the limit projection $p_0\colon M' \to |K|$, being a proper $UV^n$-map between at most $(n+1)$-dimensional locally finite $LC^n$-spaces, is a proper $n$-homotopy equivalence. Therefore $M$ and $M'$ are properly $n$-homotopy equivalent. Theorem 4.1.21 finishes the proof. □

Applying Theorem 4.3.3 and Proposition 4.3.4 we get the following statement.

THEOREM 4.3.5. *Let $n \geq 0$ and $X \in \{\mu^k\colon k \geq n+2\} \cup \{I^\omega\}$. Then there exists an $(n+1)$-invertible $UV^n$-surjection $f^n_X\colon \mu^{n+1} \to X$, satisfying conditions (i) - (iii) of Theorem* 4.3.3.

We conclude this section with the following statement.

THEOREM 4.3.6. *Let $n \geq 0$. For each locally compact polyhedron $K$, there exists a proper $(n+1)$-invertible $UV^n$-surjection $h^n_K\colon M^{n-1}_K \to K$ of some $\mu^{n+1}$-manifold $M^{n+1}_K$ onto $K$ satisfying the following conditions:*

(i) *If $L$ is a closed subpolyhedron of $K$, then its inverse image $(h^n_K)^{-1}(L)$ is a $\mu^{n+1}$-manifold.*

(ii) *If $L$ is a closed subpolyhedron of $K$ and $Z$ is a $Z$-set in $L$, then the inverse image $(h^n_K)^{-1}(Z)$ is a $Z$-set in $(h^n_K)^{-1}(L)$.*

PROOF. Repeat the proof of Theorem 4.3.3, first observing that if $L$ is a subcomplex of a countable locally finite simplicial complex $K$, then the simplicial complex $B^n_L$ from Lemma 4.3.2 is a subcomplex of the complex $B^n_K$ and the map $f^n_L$ coincides with the restriction $f^n_K / \mid B^n_L \mid$. □

*Historical and bibliographical notes* 4.3. Lemma 4.3.1 was proved in [**279**]. All other results of this Section were obtained by the author [**79**], [**84**], [**88**], [**97**], [**98**]. The non-separable case was considered in [**108**].

## 4.4. Further properties of Menger manifolds

Using the existence of the dimension raising maps constructed in Sections 4.2 and 4.3, we now discuss other major ingredients of Menger manifold theory.

### 4.4.1. $n$-homotopy kernel and Open Embedding Theorem.

The open embedding theorem for $I^\omega$-manifolds states (see Chapter 2) that for each $I^\omega$-manifold $X$, the product $X \times [0,1)$ can be embedded into $I^\omega$ as an open subspace. Observe that identifying $X$ with $X \times [0,1]$ (stability of $I^\omega$-manifolds), the product $X \times [0,1)$ may be viewed as the complement of the image of an appropriately chosen $Z$-embedding of $X$ into itself. Using this remark as a guide, we introduce the following notion.

Consider a $\mu^{n+1}$-manifold $M$ and two $Z$-embeddings $f, g\colon M \to M$ each of which is properly $n$-homotopic to the identity map $id_M$. Then the homeomorphism $gf^{-1}\colon f(M) \to g(M)$ is properly $n$-homotopic to $id_{f(M)}$ and, consequently, by Theorem 4.1.15, there exists a homeomorphism $h\colon M \to M$ extending $gf^{-1}$. Clearly the restriction $h/(M - f(M))$ is a homeomorphism between the complements $M - f(M)$ and $M - g(M)$. This shows that the following definition does not depend on the choice of a $Z$-embedding.

DEFINITION 4.4.1. *The* $n$-homotopy kernel $Ker_n(M)$ *of a* $\mu^{n+1}$*-manifold* $M$ *is defined to be the complement* $M - f(M)$*, where* $f\colon M \to M$ *is an arbitrary* $Z$*-embedding properly* $n$*-homotopic to* $id_M$.

PROPOSITION 4.4.2. *Let* $M$ *and* $N$ *be* $\mu^{n+1}$*-manifolds. Then the following conditions are equivalent:*

(i) $N$ *admits a proper* $UV^n$*-surjection onto the product* $M \times [0,1)$.
(ii) $N$ *is homeomorphic to* $Ker_n(M)$.

PROOF. It suffices to show that $Ker_n(M)$ also admits a proper $UV^n$-surjection onto the product $M \times [0,1)$. Take a proper $UV^n$-surjection $f\colon M_1\colon M \times [0,1]$, where $M_1$ is also a $\mu^{n+1}$-manifold (see Theorem 4.2.1). Consider the quotient space $M_2$ of $M_1$ with respect to the partition whose nontrivial elements are fibers $f^{-1}(m,1)$ over the $Z$-set $M \times \{1\}$ of the product $M \times [0,1]$. Clearly, by Theorem 4.1.19, $M_2$ is a $\mu^{n+1}$-manifold. Moreover, if we consider the naturally induced proper $UV^n$-surjection of $M_1$ onto $M_2$, then we conclude, by Theorem 4.1.21, that $M_1$ and $M_2$ are even homeomorphic. Next, it is easy to see that the set $g^{-1}(M \times \{1\})$ is a $Z$-set in $M_2$ and the restriction $g/g^{-1}(M \times \{1\})$ is a homeomorphism. By Theorem 4.1.20, the composition $\pi_M g$, where $\pi_M\colon M \times [0,1] \to M$ is the natural projection, can be arbitrarily closely approximated by homeomorphisms. In particular, there exists a homeomorphism $h\colon M_2 \to M$ that is properly $n$-homotopic to $\pi_M g$. Then the map $r = hg^{-1}i$, where

$i\colon M \to M \times \{1\}$ is a natural homeomorphism, is a $Z$-embedding properly $n$-homotopic to $id_M$. Indeed, $r = hg^{-1}i \simeq^n_p \pi_M g g^{-1} i = \pi_M i = id_M$. Thus, by Definition 4.4.1, we conclude that the complement $M - r(M)$ is homeomorphic to $Ker_n(M)$. Consequently, the space $M_2 - g^{-1}(M \times \{1\}) = h^{-1}(M - r(M))$ is also homeomorphic to $Ker_n(M)$. It only remains to note that the space $M_2 - g^{-1}(M \times \{1\})$ admits a proper $UV^n$-surjection onto the product $M \times [0,1)$. □

Now we are ready to prove the open embedding theorem for $\mu^{n+1}$-manifolds.

THEOREM 4.4.3. *The $n$-homotopy kernel of each $\mu^{n+1}$-manifold admits an open embedding into $\mu^{n+1}$.*

PROOF. First of all let us show that every $\mu^{n+1}$-manifold $M$ admits a proper $UV^n$-map onto a certain $I^\omega$-manifold $X$. For this we take a proper $UV^n$-map $\varphi\colon M \to P$, where $P$ is a locally compact polyhedron (see Theorem 4.2.23). Clearly the product $P \times I^\omega = X$ is a $I^\omega$-manifold (see Chapter 2). By Theorem 4.2.1, there is a proper $UV^n$-surjection $\psi\colon M' \to M \times I^\omega$ of some $\mu^{n+1}$-manifold onto the locally compact $LC^n$-space $M \times I^\omega$. Since the composition $\pi_M\psi\colon M' \to M$ is also a proper $UV^n$-map, we conclude, by Theorem 4.1.20, that $M'$ and $M$ are homeomorphic. It only remains to observe that the required proper $UV^n$-surjection $h\colon M \to X$ is given by the composition $(\varphi \times id_{I^\omega})\psi$.

Now we proceed with the direct proof of our theorem. Take a proper $UV^n$-surjection $h\colon M \to X$, where $X$ is a $I^\omega$-manifold. By the open embedding theorem for $I^\omega$-manifolds (see Theorem 2.3.25), one may suppose that the product $X \times [0,1)$ lies in $I^\omega$ as an open subspace. Consider a proper $(n+1)$-invertible $UV^n$-map $g\colon M_1 \to M \times [0,1]$, where $M_1$ is a $\mu^{n+1}$-manifold (we use Theorem 4.2.1). Since the map $g$ is $(n+1)$-invertible and $\dim M = n+1$, we may assume, without loss of generality, that the restriction $g/g^{-1}(M \times \{1\})$ is a homeomorphism and $A = g^{-1}(M \times \{1\})$ is a $Z$-set in $M_1$. As above, we conclude, using Theorem 4.1.20, that the $\mu^{n+1}$-manifolds $M$ and $M_1$ are homeomorphic. Therefore, it only remains to show that the complement $M_1 - A$ (which is obviously homeomorphic to $Ker_n(M)$) admits an open embedding into $\mu^{n+1}$.

Consider the map $f\colon \mu^{n+1} \to I^\omega$ of Theorem 4.3.5 and let $M_2$ denote the $\mu^{n+1}$-manifold $f^{-1}(X \times [0,1))$, which is open in $\mu^{n+1}$. All that remains to be shown is that the $\mu^{n+1}$-manifolds $M_2$ and $M_1 - A$ are homeomorphic. The last fact can be observed in the following way: both $\mu^{n+1}$-manifolds $M_1 - A$ and $M_2$ admit proper $UV^n$-maps onto $X \times [0,1)$ (consider the maps $(h \times id)g$ and $f$). Therefore, by Theorem 4.1.21, they are homeomorphic. The following diagram helps to reconstruct the complete argument.

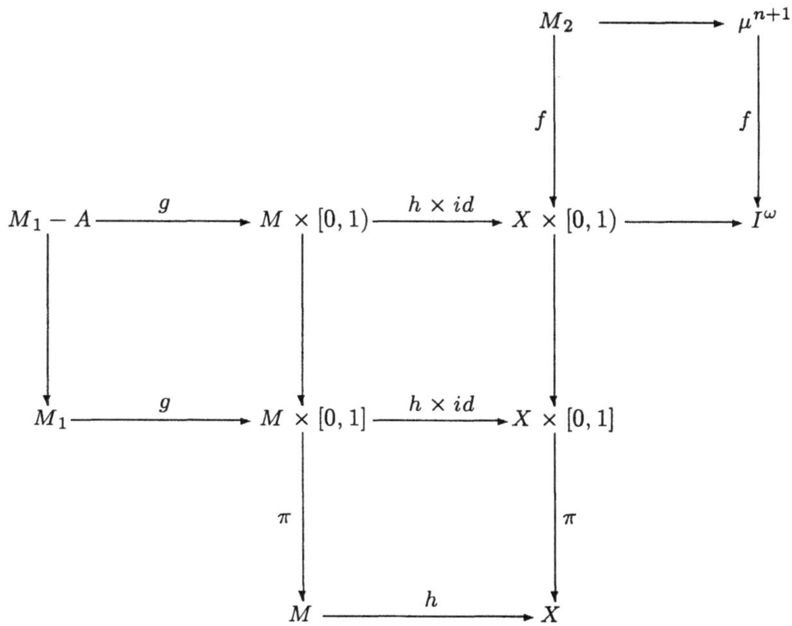


This finishes the proof. □

**4.4.2. $n$-homotopy Classification Theorem.** Theorem 4.1.21 completely describes the proper $n$-homotopy classification of $\mu^{n+1}$-manifolds. In particular, $n$-homotopy equivalent compact $\mu^{n+1}$-manifolds are homeomorphic. Obviously, the last fact is incorrect in the non-compact case: compare $\mu^{n+1}$ and $\mu^{n+1} - \{pt\}$. The main result of this subsection solves the $n$-homotopy classification problem of arbitrary $\mu^{n+1}$manifolds (compare with Theorem 2.3.26).

PROPOSITION 4.4.4. *For each $\mu^{n+1}$-manifold $M$, the spaces $Ker_n(M)$ and $Ker_n(Ker_n(M))$ are homeomorphic.*

PROOF. As above, take a proper $UV^n$-surjection $g\colon M \to X$ of $M$ onto a $I^\omega$ -manifold $X$. By Proposition 4.4.2, there exists a proper $UV^n$-surjection $f\colon Ker_n(M) \to M \times [0,1)$. For the same reason, there exists a proper $UV^n$-surjection $h\colon Ker_n(Ker_n(M)) \to Ker_n(M) \times [0,1)$. Consequently, we have two proper $UV^n$-surjections:

$$p = (g \times id)f\colon Ker_n(M) \to M \times [0,1)$$

and

$$q = (p \times id)h\colon Ker_n(Ker_n(M)) \to Ker_n(M) \times [0,1),$$

where $id$ denotes the identity map of $[0,1)$. Since $X$ is a Hilbert cube manifold, the product $X \times [0,1]$ is homeomorphic to $X$ (see Chapter 2). Remarking that the spaces $[0,1) \times [0,1)$ and $[0,1] \times [0,1)$ are homeomorphic, we have

$$X \times [0,1) \times [0,1) \approx X \times [0,1] \times [0,1) \approx X \times [0,1).$$

Consequently, the $\mu^{n+1}$-manifolds $Ker_n(M)$ and $Ker_n(Ker_n(M))$ admit proper $UV^n$-surjections onto the same $I^\omega$-manifold $X \times [0,1)$. Therefore they are properly $n$-homotopy equivalent. Theorem 4.1.21 finishes the proof. □

PROPOSITION 4.4.5. *Let $M$ be a $\mu^{n+1}$-manifold. If $A$ is a $Z$-set in $Ker_n(M)$, then the spaces $Ker_n(M)$ and $Ker_n(Ker_n(M) - A)$ are homeomorphic.*

PROOF. As in the proof of Proposition 4.4.4 consider three proper $UV^n$-surjections:

$$g\colon M \to X,\ f\colon Ker_n(M) \to M \times [0,1) \text{ and } p = (g \times id)f\colon Ker_n(M) \to X \times [0,1),$$

where $X$ is a $I^\omega$-manifold and $id$ denotes the identity map of $[0,1)$. Now we redefine the map $p$ in such a way that the set $p(A)$ will be a $Z$-set in $X \times [0,1)$. For this, consider any $Z$-embedding $r\colon A \to X \times [0,1)$ properly $n$-homotopic to the restriction $p/A$. By Proposition 4.2.2, there exists a proper $UV^n$-surjection $q\colon Ker_n(M) \to X \times [0,1)$ such that $q/A = r$. Moreover, as in the proof of Proposition 4.4.2, we can additionally suppose that $A = q^{-1}q(A)$. Consequently, the restriction

$$q/(Ker_n(M) - A)\colon Ker_n(M) - A \to X \times [0,1) - q(A)$$

is a proper $UV^n$-surjection. As above (compare with the proof of Proposition 4.4.4) this implies that $Ker_n(Ker_n(M) - A)$ admits a proper $UV^n$-surjection onto the product $(X \times [0,1) - q(A)) \times [0,1)$. By the same argument, there exists a proper $UV^n$-surjection of $Ker_n(Ker_n(M))$ onto the product $X \times [0,1) \times [0,1)$. By Proposition 4.4.4, $Ker_n(M)$ and $Ker_n(Ker_n(M))$ are homeomorphic. Therefore, by Theorem 4.1.21, it suffices to show that the spaces

$$(X \times [0,1) - q(A)) \times [0,1) \quad \text{and} \quad X \times [0,1) \times [0,1)$$

are homeomorphic. Indeed, since $q(A)$ is a $Z$-set in $X \times [0,1)$, we conclude that the $I^\omega$-manifolds $X \times [0,1)$ and $X \times [0,1) - q(A)$ are homotopy equivalent. Then, by the homotopy classification theorem for $I^\omega$-manifolds (Theorem 2.3.26), the products $X \times [0,1) \times [0,1)$ and $(X \times [0,1) - q(A)) \times [0,1)$ are homeomorphic. This finishes the proof. □

PROPOSITION 4.4.6. *Let a $\mu^{n+1}$-manifold $M$ be a $Z$-set of a $\mu^{n+1}$-manifold $N$, and suppose the inclusion $i\colon M \hookrightarrow N$ is an $n$-homotopy equivalence. Then there exists a $Z$-set $A$ in $N$ such that the complement $N - A$ is homeomorphic to $Ker_n(M)$.*

PROOF. By Theorem 4.2.23, there exist $(n+1)$-invertible proper $UV^n$-surjections $\varphi\colon M \to L$ and $\psi\colon N \to K_1$, where $L$ and $K_1$ are $(n+1)$-dimensional locally compact polyhedra. The $(n+1)$-invertibility of $\varphi$ implies the existence of a proper map $s\colon L \to M$ with $\varphi s = id_L$. Then the composition $\psi i s\colon L \to K_1$ is properly homotopic to some proper piecewise linear (PL) map $p\colon L \to K_1$. Form the mapping cylinder $M(p) = K$ of the map $p$. Recall that this is the space obtained from the disjoint union $(L \times [0,1]) \oplus K_1$, by identifying $(l,1)$ with $p(l)$ for each $l \in L$. At the same time we identify $L$ with $L \times \{0\}$. Clearly $L \times [0,1)$ can be considered as an open subspace of $K$. Since $p$ is a proper PL-map, $L \times \{0\}$ and $K_1$ are subpolyhedra of the polyhedron $K$. Let $c\colon K \to K_1$ be the collapse to the base, i.e. the natural retraction defined by sending $(l,t)$ to $p(l)$. Clearly, $c$ is a proper $CE$-surjection that is a homotopy equivalence.

Now we consider an $(n+1)$-invertible proper $UV^n$-surjection $f^n_K\colon M^{n+1}_K \to K$, satisfying the conditions of Theorem 4.3.6. Clearly, the composition $cf^n_K\colon M^{n+1}_K \to K_1$ is a proper $UV^n$-surjection, and hence, by Theorem 4.1.21, the $\mu^{n+1}$-manifolds $M^{n+1}_K$ and $N$ are homeomorphic. By Theorem 4.3.6, the inverse image $(f^n_K)^{-1}(L \times \{0\})$ is a $\mu^{n+1}$-manifold that, again by Theorem 4.1.21, is homeomorphic to $M$. One can easily verify, using the assumption and the specifics of the above construction, that the natural inclusion of $(f^n_K)^{-1}(L \times \{0\})$ into $M^{n+1}_K$ is an $n$-homotopy equivalence. Moreover, by Theorem 4.3.6, the above inverse image is a $Z$-set in $M^{n+1}_K$ (since $L \times \{0\}$ is a $Z$-set in $K$). Now redefining the above objects for simplicity, we have the following situation.

A proper $UV^n$-surjection $f\colon N \to K$, satisfying the conditions of Theorem 4.3.6, is given, $M = f^{-1}(L \times \{0\})$ is a $Z$-set in $N$ and the inclusion $M \hookrightarrow N$ is an $n$-homotopy equivalence. Clearly, $K - K_1 = L \times [0,1)$, and hence the inverse image $f^{-1}(L \times [0,1))$ admits a proper $UV^n$-surjection onto $L \times [0,1)$. On the other hand, $Ker_n(M)$ admits a proper $UV^n$-surjection onto $M \times [0,1)$, and hence onto the product $L \times [0,1)$ as well. Thus, by Theorem 4.1.21, the inverse image $f^{-1}(L \times [0,1))$ and $Ker_n(M)$ are homeomorphic. Consequently, to finish the proof it remains to construct an open embedding $h\colon f^{-1}(L \times [0,1)) \to N$ such that the complement $A = N - h(f^{-1}(L \times [0,1))$ is a $Z$-set in $N$.

Since $N$ is a $\mu^{n+1}$-manifold, by Theorem 4.1.19, there is a countable dense subset $\{\varphi_k\colon k = 1,2,\dots\}$ of $C(I^{n+1}, N)$ consisting of $Z$-embeddings. As observed above, $f^{-1}(L \times \{0\})$ is a $Z$-set in $N$ and the inclusion $f^{-1}(L \times \{0\}) \hookrightarrow N$ is an $n$-homotopy equivalence. It easily follows from Proposition 4.2.2 that in this case there exists a retraction $r_1\colon N \to f^{-1}(L \times \{0\})$ that is $n$-homotopic to $id_N$. Consider the restriction

$$s_1 = r_1/\dots\colon f^{-1}(L \times \{0\}) \cup \varphi_1(I^{n+1}) \to f^{-1}(L \times [0, \frac{1}{2})).$$

Clearly $s_1$ is a proper map. By Theorem 4.1.19, $s_1$ is properly $n$-homotopic to a $Z$-embedding

$$g_1\colon f^{-1}(L \times \{0\}) \cup \varphi_1(I^{n+1}) \to f^{-1}(L \times [0, \frac{1}{2}))$$

that coincides with the identity map on $f^{-1}(L \times \{0\})$. By Theorem 4.1.15, there exists a homeomorphism $G_1\colon N \to N$ extending $g_1$. Put $h_1 = G_1^{-1}$. Then $h_1$ is a homeomorphism such that

(a) $h_1/f^{-1}(L \times \{0\}) = id$.

(b) $\varphi_1(I^{n+1}) \subseteq h_1 f^{-1}(L \times [0, \frac{1}{2}))$.

Consider now the polyhedron $K - (L \times [0, \frac{1}{2}))$. Since the set $L \times \{\frac{1}{2}\}$ is a $Z$-set in $K - (L \times [0, \frac{1}{2}))$, we can conclude, by Theorem 4.3.6, that the set $h_1 f^{-1}(L \times \{\frac{1}{2}\})$ is a $Z$-set in a $\mu^{n+1}$-manifold $N - h_1 f^{-1}(L \times [0, \frac{1}{2}))$. Moreover, since the inclusion

$$L \times \{\frac{1}{2}\} \hookrightarrow K - L \times [0, \frac{1}{2}))$$

is a homotopy equivalence, we conclude that the inclusion

$$h_1 f^{-1}(L \times \{\frac{1}{2}\}) \hookrightarrow N - h_1 f^{-1}(L \times [0, \frac{1}{2}))$$

is an $n$-homotopy equivalence. Again, using the above construction, we see that there is a homeomorphism

$$h_2'\colon N - h_1 f^{-1}(L \times [0, \frac{1}{2})) \to N - h_1 f^{-1}(L \times [0, \frac{1}{2})),$$

which is the identity on $h_1 f^{-1}(L \times \{\frac{1}{2}\})$ and for which

$$\varphi_2(I^{n+1}) \cap (N - h_1 f^{-1}(L \times [0, \frac{1}{2}))) \subseteq h_2' h_1 f^{-1}(L \times [\frac{1}{2}, 2 \cdot \frac{1}{3})).$$

Extend $h_2'$ to a homeomorphism $h_2$ defined on $N$ by defining $h_2 = id$ on $h_1 f^{-1}(L \times [0, \frac{1}{2}])$. Then we have

$$\varphi_1(I^{n+1}) \cup \varphi_2(I^{n+1}) \subseteq h_2 h_1 f^{-1}(L \times [0, 2 \cdot \frac{1}{3})).$$

Inductively continuing this process, we construct homeomorphisms $h_k\colon N \to N$ in such a way that $h_{k+1} = id$ on $h_k f^{-1}(L \times [0, k \cdot \frac{1}{k+1}])$ and

$$\varphi_1(I^{n+1}) \cup \cdots \cup \varphi_k(I^{n+1}) \subseteq h_k h_{k-1} \cdots h_1 f^{-1}(L \times [0, k \cdot \frac{1}{k+1})).$$

Define an open embedding $h\colon f^{-1}(L \times [0,1)) \to N$ by $h(x) = \lim_{k\to\infty} h_k \cdots h_1(x)$ for each $x \in f^{-1}(L \times [0,1))$. Clearly

$$\cup\{\varphi_k(I^{n+1})\colon k = 1, 2, \dots\} \subseteq hf^{-1}(L \times [0,1))$$

and consequently, by the choice of the family $\{\varphi_k\colon k = 1, 2, \dots\}$, the complement $N - hf^{-1}(L \times [0,1))$ is a $Z$-set in $N$. □

Now we are ready to prove the $n$-homotopy classification theorem for $\mu^{n+1}$-manifolds.

THEOREM 4.4.7. *$\mu^{n+1}$-manifolds are $n$-homotopy equivalent if and only if their $n$-homotopy kernels are homeomorphic.*

PROOF. Let $M$ and $N$ be $n$-homotopy equivalent $\mu^{n+1}$-manifolds. Take maps $\alpha\colon M \to N$ and $\beta\colon N \to M$ such that $\beta\alpha \overset{n}{\simeq} id_M$ and $\alpha\beta \overset{n}{\simeq} id_N$. As above, we can find $(n+1)$-invertible proper $UV^n$-surjections $f\colon Ker_n(M) \to M \times [0,1)$ and $g\colon Ker_n(N) \to N \times [0,1)$. Let $\varphi\colon M \times [0,1) \to N \times [0,1)$ and $\psi\colon N \times [0,1) \to M \times [0,1)$ be a proper maps such that $\varphi$ is homotopic to $\alpha \times id$ and $\psi$ is homotopic to $\beta \times id$, where $id$ denotes the identity map of $[0,1)$ (compare with [**69**, Lemma 21.1]) . Since $g$ is $(n+1)$-invertible and $\dim Ker_n(M) = n+1$, there exists a proper map $r\colon Ker_n(M) \to Ker_n(N)$ such that $gr = \varphi f$. Similarly we have a proper map $s\colon Ker_n(N) \to Ker_n(M)$ such that $fs = \psi g$. In this situation one can verify directly that $sr \overset{n}{\simeq} id_{Ker_n(N)}$. Moreover, by Theorem 4.1.19, we can additionally suppose that $r$ and $s$ are $Z$-embeddings.

Now consider the $Z$-set $r(Ker_n(M))$ in a $\mu^{n+1}$-manifold $Ker_n(N)$. It follows immediately from the above construction that the inclusion map

$$r(Ker_n(M)) \hookrightarrow Ker_n(N)$$

is an $n$-homotopy equivalence. By Proposition 4.4.6, there is a $Z$-set $A$ in $Ker_n(N)$ such that $Ker_n(N) - A$ is homeomorphic to $Ker_n(r(Ker_n(M))) = Ker_n(Ker_n(M))$ (recall that $r$ is an embedding). Then, by Propositions 4.4.4 and 4.4.5,

$$Ker_n(M) \approx Ker_n(Ker_n(Ker_n(M))) \approx Ker_n(Ker_n(N) - A) \approx Ker_n(N)$$

as desired.

The second part of the theorem is trivial. □

**4.4.3. $n$-shape and the Complement Theorem.** The famous Complement Theorem for $I^\omega$-manifolds [69] states that if $X$ and $Y$ are $Z$-sets in $I^\omega$, then their complements $I^\omega - X$ and $I^\omega - Y$ are homeomorphic if and only if the shapes of $X$ and $Y$ coincide, i.e. $Sh(X) = Sh(Y)$. The obvious form of The Complement Theorem fails for $\mu^{n+1}$. The equality of shapes of two $Z$-sets $X$ and $Y$ in $\mu^{n+1}$ is sufficient, but far from necessary, for the complements $\mu^{n+1} - X$ and $\mu^{n+1} - Y$ to be homeomorphic. Indeed, it can be easily seen that if the $(n+1)$-dimensional sphere $S^{n+1}$ is embedded into $\mu^{n+1}$ as a $Z$-set, then $\mu^{n+1} - S^{n+1}$ is homeomorphic to $\mu^{n+1} - \{pt\}$. At the same time $Sh(S^{n+1}) \neq Sh(pt)$.

The problem was solved in [**87**] (see also [**90**], [**92**]) where the notion of $n$-shape was introduced. The relation between $n$-SHAPE and $n$-HOMOTOPY categories is of the same nature as that between the categories of SHAPE and HOMOTOPY. Roughly, $n$-SHAPE is a "spectral completion" of $n$-HOMOTOPY. The main result in this direction is the following.

THEOREM 4.4.8. *Let $X$ and $Y$ be $Z$-sets in $\mu^{n+1}$. The complements $\mu^{n+1} - X$ and $\mu^{n+1} - Y$ are homeomorphic if and only if $n - Sh(X) = n - Sh(Y)$.*

We would like to mention some corollaries of this theorem and the definition of $n$-shape itself.

COROLLARY 4.4.9. *If* $Sh(X) = Sh(Y)$, *then* $n - Sh(X) = n - Sh(Y)$.

COROLLARY 4.4.10. *If* $X$ *and* $Y$ *are at most* $n$*-dimensional, then* $Sh(X) = Sh(Y)$ *if and only if* $n - Sh(X) = n - Sh(Y)$.

COROLLARY 4.4.11. *If* $Z$*-sets* $X$ *and* $Y$ *in* $\mu^{n+1}$ *are* $\mu^{n+1}$*-manifolds, then the complements* $\mu^{n+1} - X$ *and* $\mu^{n+1} - Y$ *are homeomorphic if and only if the compacta* $X$ *and* $Y$ *are homeomorphic.*

Let us emphasize that the notion of $n$-equivalence, introduced by Ferry [**150**] as a generalization of Whitehead's notion of $n$-type, coincides in several important cases with the notion of $n$-shape. Relations between these two concepts have been studied in [**97**].We conclude this section by noting that Theorem 4.4.8 was extended [**280**] to a larger class of subspaces than $Z$-sets. These are the so-called weak $Z$-sets.

### 4.4.4. Menger manifolds with boundaries.

The problem of putting a boundary on various types of manifolds were considered in [**57**] (PL manifolds), [**282**] (smooth manifolds) and [**70**] ($I^\omega$ - manifolds). It was proved in [**70**] that if an $I^\omega$ -manifold $M$ satisfies certain minimal necessary homotopy-theoretical conditions (finite type and tameness at $\infty$), then there are two obstructions $\sigma_\infty(M)$ and $\tau_\infty(M)$ to $M$ having a boundary. The first one is an element of the group $\varprojlim\{\mathcal{K}_0\pi_1(M - A) \colon A \subseteq M\ A \text{ is compact}\}$, where $\mathcal{K}_0\pi_1$ is the projective class group functor. If $\sigma_\infty(M) = 0$, then the second obstruction can be defined as an element of the first derived limit of the inverse system $\varprojlim\{\mathcal{W}h\pi_1(M-A) \colon A \subseteq M\ A \text{ is compact}\}$, where $\mathcal{W}h\pi_1$ is the Whitehead group functor. It was shown in [**70**] that the different boundaries that can be put on $M$ constitute a whole shape class and that a classification of all possible ways of putting boundaries on $M$ can be done in terms of the group $\varprojlim\{\mathcal{W}h\pi_1(M - A) \colon A \subseteq M\ A \text{ is compact}\}$. It should be emphasized that the above mentioned obstructions essentially involve the Wall's finiteness obstruction [**303**].

The natural analog of Wall's obstruction vanishes in the $n$-homotopy category. This is exactly what was stated in 4.1.12. We will see that this observation significantly simplifies the situation for $\mu^{n+1}$-manifolds. First of all we need the following corollary of Proposition 4.1.12.

PROPOSITION 4.4.12. *If a* $\mu^{n+1}$*-manifold* $M$ *is* $n$*-homotopy dominated by an at most* $(n+1)$*-dimensional* $LC^n$*-compactum, then* $M$ *is* $n$*-homotopy equivalent to a compact* $\mu^{n+1}$*-manifold.*

We say that a $\mu^{n+1}$-manifold $M$ admits a *boundary* if there exists a compact $\mu^{n+1}$- manifold $N$ such that $M = N - Z$, where $Z$ is a $Z$-set in $N$. In this case we shall say that $N$ is a compactification of $M$ corresponding to the boundary $Z$, and conversely, $Z$ is a boundary of $M$ corresponding to the compactification $N$. We also need the following definition [**100**].

DEFINITION 4.4.13. *A space $X$ is said to be $n$-*tame at $\infty$ *if for each compactum $A \subseteq X$ there exists a larger compactum $B \subseteq X$ such that the inclusion $X - B \hookrightarrow X - A$ factors up to $n$-homotopy through an at most $(n+1)$-dimensional finite polyhedron.*

PROPOSITION 4.4.14. *If a $\mu^{n+1}$-manifold $M$ is $n$-tame at $\infty$, then $M$ is $n$-homotopy equivalent to a compact $\mu^{n+1}$-manifold.*

PROOF. Take a proper $UV^n$-retraction $r\colon M \to P$ of the given $\mu^{n+1}$-manifold $M$ onto some $(n+1)$-dimensional locally compact polyhedron $P$. It follows from elementary properties of proper $UV^n$-maps that $P$ is $n$-tame at $\infty$ as well. Using Proposition 4.1.8, one can easily see that $P$ is $n$-homotopy dominated by an at most $(n+1)$-dimensional compact polyhedron. Proposition 4.4.12 finishes the proof. □

Let us recall that an $I^\omega$-manifold $M$ lying in a larger $I^\omega$-manifold $N$ is said to be *clean* if $M$ is closed in $N$ and the topological frontier of $M$ in $N$ is collared both in $M$ and $N - IntM$. For obvious dimensional reasons we cannot directly define the corresponding notion for $\mu^{n+1}$-manifolds. Nevertheless, the following notion is sufficient for us.

DEFINITION 4.4.15. *A $\mu^{n+1}$-manifold $M$ lying in a $\mu^{n+1}$-manifold $N$ is said to be $n$-*clean *in $N$ provided that $M$ is closed in $N$ and there exists a closed subspace $\delta(M)$ of $M$ such that the following conditions are satisfied:*

(i) *$\delta(M)$ is a $\mu^{n+1}$-manifold.*
(ii) *$(N - M) \cup \delta(M)$ is a $\mu^{n+1}$-manifold.*
(iii) *$\delta(M)$ is a $Z$-set in $M$.*
(iv) *$\delta(M)$ is a $Z$-set in $(N - M) \cup \delta(M)$.*
(v) *$M - \delta(M)$ is open in $N$.*

Sometimes we say that $M$ is $n$-clean with respect to $\delta(M)$.

Let us indicate the standard situation in which $n$-clean submanifolds arise naturally. Suppose that $L$ is submanifold of a combinatorial PL-manifold $P$. Consider a proper $UV^n$-surjection $f\colon N \to P$ of a $\mu^{n+1}$-manifold $N$ from Theorem 4.3.6. Using the properties of $f$, it is easy to see that $M = f^{-1}(L)$ is an $n$-clean submanifold of $N$ with $\delta(M) = f^{-1}(\partial L)$. Generally speaking, $f$ is not an open map and consequently $\delta(M)$ does not necessarily coincide with the topological frontier of $M$ in $N$.

LEMMA 4.4.16. *Let $N$ be a $\mu^{n+1}$-manifold which is $n$-tame at $\infty$. Suppose that $M$ is a compact and $n$-clean submanifold of $N$. Then the $\mu^{n+1}$-manifold $(N - M) \cup \delta(M)$ is $n$-homotopy equivalent to a compact $\mu^{n+1}$-manifold.*

PROOF. By Proposition 4.4.14, it suffices to show that the $\mu^{n+1}$-manifold $(N-M)\cup\delta(M)$ is $n$-tame at $\infty$. Let $A$ be a compact subspace of $(N-M)\cup\delta(M)$. Clearly, $K_1 = A \cup M$ is compact. Since $N$ is $n$-tame at $\infty$, there exists a compactum $K_2$ such that $K_1 \subseteq K_2 \subseteq N$ and the inclusion $N - K_2 \hookrightarrow N - K_1$ factors up to $n$-homotopy through an at most $(n+1)$-dimensional compact polyhedron. Let $B = ((N - M) \cup \delta(M)) \cap K_2$. Clearly $B$ is a compactum and $A \subseteq B$. Note that

$$((N - M) \cup \delta(M)) - B = N - K_2 \text{ and } N - K_1 \subseteq ((N - M) \cup \delta(M)) - A.$$

Consequently, the inclusion

$$((N - M) \cup \delta(M)) - B \hookrightarrow ((N - M) \cup \delta(M)) - A$$

factors up to $n$-homotopy through an at most $(n+1)$-dimensional compact polyhedron. Hence $(N - M) \cup \delta(M)$ is $n$-tame at $\infty$. □

LEMMA 4.4.17. *Any $\mu^{n+1}$-manifold $M$ can be written as a union $M = \cup_{i=1}^{\infty} M_i$ such that all $M_i$'s are compact and $n$-clean and $M_i \subseteq M_{i+1} - \delta(M_{i+1})$ for each $i = 1, 2, \ldots$.*

PROOF. It suffices to show that for each compactum $K \subseteq M$, there exists a compact and $n$-clean $M_1 \subseteq M$ such that $K \subseteq M_1 - \delta(M_1)$. As before, take a proper $UV^n$-surjection $g\colon M \to X$, where $X$ is a $I^\omega$-manifold. There is a compact and clean $Y \subseteq X$ such that $g(K) \subseteq Int_X(Y)$ (see [70]). By the relative triangulation theorem for $I^\omega$-manifolds (see Theorem 2.3.31), there exists a polyhedron $P$ which can be written as a union of two subpolyhedra $P_1$ and $P_2$ such that $X = P \times I^\omega$, $Y = P_1 \times I^\omega$, $X - Int_X(Y) = P_2 \times I^\omega$ and $Bd_X(Y) = (P_1 \cap P_2) \times I^\omega$. Note also that the subpolyhedron $P_1 \cap P_2$ is a $Z$-set both in $P_1$ and $P_2$.

Consider now a proper $UV^n$-surjection $f\colon N \to P$ of a $\mu^{n+1}$-manifold $N$ onto the polyhedron $P$ satisfying the conditions of Theorem 4.3.6. Consequently, we have two proper $UV^n$-surjections $f\colon N \to P$ and $\pi_P g\colon M \to P$ (here $\pi_P\colon P \times I^\omega \to P$ denotes the natural projection) of two $\mu^{n+1}$-manifolds onto the polyhedron $P$. Consider an open cover $\mathcal{U} = \{P - \pi_P g(K), Int_P(P_1)\}$ of $P$. By Theorem 4.1.20, there exists a homeomorphism $h\colon M \to N$ such that the compositions $\pi_P g$ and $fh$ are $\mathcal{U}$-close. Let $M_1 = h^{-1}f^{-1}(P_1)$ and $\delta(M_1) = h^{-1}f^{-1}(P_1 \cap P_2)$. By the properties of the map $f$, $M_1$ is compact and $n$-clean. It only remains to note that $K \subseteq M_1 - \delta(M_1)$. This finishes the proof. □

We also need the following statement, which is a direct consequence of the characterization Theorem 4.1.19.

PROPOSITION 4.4.18. *Let a space $M$ be the union of two closed subspaces $M_1$ and $M_2$. If $M_1$, $M_2$ and $M_0 = M_1 \cap M_2$ are $\mu^{n+1}$-manifolds and $M_0$ is a $Z$-set both in $M_1$ and $M_2$, then $M$ is a $\mu^{n+1}$-manifold.*

PROOF. It suffices to show that for any map $f\colon X \to M$ of an at most $(n+1)$-dimensional compactum $X$ into $M$, and any open cover $\mathcal{U} \in cov(M)$, there exists an embedding $g\colon X \to M$, $\mathcal{U}$-close to $f$. Let us consider the case when $f(X) \cap M_i \neq \emptyset$ for each $i = 0, 1, 2$. All other cases are trivial. By Proposition 4.1.7, there exists an open cover $\mathcal{V} \in cov(M)$ refining $\mathcal{U}$ such that the following condition is satisfied:

$(*)_n$ for any at most $(n+1)$-dimensional compactum $B$, closed subspace $A$ of $B$, and any two $\mathcal{V}$-close maps $\alpha_1, \alpha_2\colon A \to M$ such that $\alpha_1$ has an extension $\beta_1\colon B \to M$, it follows that $\alpha_2$ also admits an extension $\beta_2\colon B \to M$ which is $\mathcal{U}$-close to $\beta_1$.

Let $X_i = f^{-1}(M_i)$, $i = 0, 1, 2$. Since $M_0$ is a $\mu^{n+1}$-manifold, there is a $Z$-embedding $g_0\colon X_0 \to M_0$ such that $g_0$ and $f/X_0$ are $\mathcal{V}$-close. By $(*)_n$, there is an extension $h\colon X \to M$ of $G-)$ such that $h$ and $f$ are $\mathcal{V}$-close. Since $M_0$ is a $Z$-set in both $M_1$ and $M_2$ we conclude that $g_0(X_0)$ is a $Z$-set both in $M_1$ and $M_2$. Consequently, by Theorem 4.1.19, for each $i = 1, 2$ there is a $Z$-embedding $g_i\colon X_i \to M_i$ such that $g_i/X_0 = g_0$ and $g_i$ is $\mathcal{U}$-close to $h/X_i$. At the same time, without loss of generality we can assume that one of these maps, say $g_1$, has the following property: $g_1(X_1 - X_0) \cap M_0 = \emptyset$ (we once again use the fact that $M_0$ is a $Z$-set in $M_1$). Then the map $g$, coinciding with $g_i$ on $X_i$ ($i = 1, 2$), is an embedding. It only remains to note that $g$ and $f$ are $\mathcal{U}$-close. □

LEMMA 4.4.19. *If a $\mu^{n+1}$-manifold $M$ is $n$-tame at $\infty$, then we can write $M = \cup_{i=1}^{\infty} M_i$ such that all $M_i$'s are compact and $n$-clean, $M_i \subseteq M_{i+1} - \delta(M_{i+1})$ and the inclusion $\delta(M_i) \hookrightarrow (M_{i+1} - M_i) \cup \delta(M_i)$ is $n$-homotopy equivalence for each $i = 1, 2, \ldots$.*

PROOF. Choose any compact and $n$-clean submanifold $A$ of $M$. By Lemma 4.4.17, it suffices to find a compact and $n$-clean submanifold $B$ of $M$ such that $A \subseteq B - \delta(B)$ and the inclusion $\delta(B) \hookrightarrow (M - B) \cup \delta(B)$ is an $n$-homotopy equivalence. By Lemma 4.4.16, the $\mu^{n+1}$-manifold $(M - A)\delta(A)$ is $n$-homotopy equivalent to some compact $\mu^{n+1}$-manifold $X$. Fix the corresponding $n$-homotopy equivalence $\psi_1\colon (M - a) \cup \delta(A) \to X$ and its $n$-homotopy inverse $\varphi_1\colon X \to (M - A) \cup \delta(A)$. Obviously there is a map $\psi_2\colon (M - A)\delta(A) \to X$ such that $\psi_2/\delta(A)\colon \delta(A) \to X$ is a $Z$-embedding and $\psi_2$ is as close to $\psi_1$ as we wish. Similarly, there is a $Z$-embedding $\varphi_2\colon X \to (M - A) \cup \delta(A)$ which is as close to $\varphi_1$ as we wish. In particular, we can assume that $\psi_2$ and $\varphi_2$ are $n$-homotopy equivalences. If $\psi_2$ and $\varphi_2$ were chosen sufficiently close to $\psi_1$ and $\varphi_1$ respectively, then, by Theorem 4.1.18, there exists a homeomorphism $h\colon (M - A) \cup \delta(A) \to (M - A) \cup \delta(A)$ which extends the homeomorphism $\varphi_2\psi_2/\delta(A)\colon \delta(A) \to \varphi_2\psi_2(\delta(A))$ and which is sufficiently close to the identity map of $(M - A) \cup \delta(A)$. In particular, we can assume that $h$ is $n$-homotopic to $id_{(M-A)\cup\delta(A)}$. Then the $n$-homotopy equivalence $\varphi = h^{-1}\varphi_2 \colon X \to (M - A) \cup \delta(A)$ is a $Z$-embedding and $\delta(A) \subseteq \varphi(X) \equiv Y$. Since $Y$ is a compact $\mu^{n+1}$-manifold, there exists a $UV^n$-retraction $s \colon Y \to K$ onto a finite $(n+1)$-dimensional polyhedron $K$ (see Theorem 4.2.23). Similarly, take a proper $UV^n$-retraction $r \colon (M - A) \cup \delta(A) \to T$, where T is a polyhedron. Let $i \colon Y \hookrightarrow (M - A) \cup \delta(A)$ denote the inclusion map and $j \colon K \to Y$ be a section of $s$ (i.e. $sj = id_K$). Note that $i$ is an $n$-homotopy equivalence. Let $p \colon K \to T$ be a PL-map homotopic to the composition $rij$. Form the mapping cylinder $M(p) \equiv P$ of the map $p$. For the reader's convenience, we again recall that $P$ is the space obtained from the disjoint union $(K \times [0,1]) \oplus T$, by identifying $(k,1)$ with $p(k)$, $k \in K$. At the same time we identify $K$ with $K \times \{0\}$. Since $p$ is a PL-map, $K \times \{0\}$ and $T$ are subpolyhedra of the polyhedron $P$. Let $c \colon P \to T$ be the collapse to the base, i.e. the natural retraction defined by sending $(k,t)$ to $p(k)$ for each $(k,t) \in K \times [0,1]$. Obviously, $c$ is a proper cell-like map that is a proper homotopy equivalence. Now consider a proper $UV^n$-surjection $f \colon N \to P$ of some $\mu^{n+1}$-manifold $N$ onto $P$, satisfying the conditions of Theorem 4.3.6. The compact $\mu^{n+1}$-manifolds $Y$ and $N_1 = f^{-1}(K \times \{0\})$ admit $UV^n$-surjections $s \colon Y \to K \times \{0\}$ and $f/N_1 \colon N_1 \to K \times \{0\}$ onto the same polyhedron. Consequently, by Theorem 4.1.21, there exists a homeomorphism $g_1 \colon Y \to N_1$ such that $fg_1 \overset{n}{\simeq} s$. Similarly, we have two proper $UV^n$-surjections $r \colon (M - A) \cup \delta(A) \to T$ and $cf \colon N \to T$. As above, there is a homeomorphism $g_2 \colon (M - A) \cup \delta(A) \to N$ such that $cfg_2 \overset{n}{\simeq}_p r$. By the construction and the corresponding properties of proper $UV^n$-surjections, we have

$$cfg_1 \overset{n}{\simeq} cs = ps \simeq rijs \overset{n}{\simeq} ri \overset{n}{\simeq} cfg_2 i.$$

Since $cf$ is a proper $n$-homotopy equivalence, we conclude that $g_1 \colon Y \to N$ and $g_2/Y \colon Y \to N$ are $n$-homotopic. Consider the homeomorphism

$$\alpha = g_1 g_2^{-1}/g_2(Y) \colon g_2(Y) \to N_1.$$

Clearly

$$\alpha \overset{n}{\simeq} g_2 g_2^{-1}/g_2(Y) = id_{g_2(Y)}.$$

By the properties of the map $f$, $N_1$ is a $Z$-set in $N$. Note also that, by our construction, $g_2(Y)$ is a $Z$-set in $N$ as well. By Theorem 4.1.15, we can find a homeomorphism $G \colon N \to N$ extending $\alpha$. Let $H = Gg_2$. Note that $H(Y) = Gg_2(Y) = \alpha g_2(Y) = N_1$. Finally, let

$$B = A \cup H^{-1}(f^{-1}(K \times [0, \tfrac{1}{2}])) \text{ and } \delta(B) = H^{-1}(f^{-1}(K \times \{\tfrac{1}{2}\})).$$

It follows from the properties of the map $f$ and Proposition 4.4.18 that $B$ is a compact and $n$-clean submanifold of $M$, $A \subseteq B - \delta(B)$, and the inclusion $\delta(B) \hookrightarrow (M - B) \cup \delta(B)$ is an $n$-homotopy equivalence. To see this, observe that the map $p$, and consequently the inclusion $K \times \{\frac{1}{2}\} \hookrightarrow P - (K \times [0, \frac{1}{2}])$, is an $n$-homotopy equivalence. □

LEMMA 4.4.20. *Let a $\mu^{n+1}$-manifold $M$ be a $Z$-set in a compact $\mu^{n+1}$-manifold $N$. If the inclusion $i\colon M \hookrightarrow N$ is an $n$-homotopy equivalence, then there exists a $UV^n$-retraction of $N$ onto $M$.*

PROOF. Let $j\colon N \to M$ be an $n$-homotopy inverse of $i$. By Theorem 4.1.21, there is a homeomorphism $h\colon N \to M$ such that $h \overset{n}{\simeq} j$. Then $hi \overset{n}{\simeq} ji \overset{n}{\simeq} id_M$. Consequently, by Proposition 4.2.2, there is a $UV^n$-surjection $r\colon N \to M$ such that $ri = id_M$. □

The following result gives us a characterization of $\mu^{n+1}$-manifolds with boundaries.

THEOREM 4.4.21. *A $\mu^{n+1}$-manifold admits a boundary if and only if it is $n$-tame at $\infty$.*

PROOF. Let $M$ be a $\mu^{n+1}$-manifold which is $n$-tame at $\infty$. By Lemma 4.4.19, we can represent $M$ as a union $M = \cup_{i=1}^{\infty} M_i$ such that all the $M_i$'s are compact and $n$-clean, $M_i \subseteq M_{i+1} - \delta(M_{i+1})$ and the inclusion $\delta(M_i) \hookrightarrow (M_{i+1} - M_i) \cup \delta(M_i)$ is an $n$-homotopy equivalence for each $i = 1, 2, \ldots$. By Lemma 4.4.20, for each $i$ there exists a $UV^n$-retraction $f_i\colon (M_{i+1} - M_i) \cup \delta(M_i) \to \delta(M_i)$. Let the $UV^n$-retraction $r_i\colon M_{i+1} \to M_i$ coincide with $f_i$ on $M_{i+1} - M_i$ and with the identity on $M_i$. Then we have an inverse sequence $\mathcal{S} = \{M_i, r_i\}$ consisting of compact $\mu^{n+1}$-manifolds and $UV^n$-retractions. By Theorem 4.1.20, $r_i$ is a near-homeomorphism for each $i$. By [**58**], each limit projection of the spectrum $\mathcal{S}$ is a near-homeomorphism as well. Consequently, $N = \lim \mathcal{S}$, being homeomorphic to $M_1$, is a compact $\mu^{n+1}$-manifold. Since $\delta(M_i)$ is a $Z$-set in $M_i$ for each $i$, we conclude that the subset $Z = \lim\{\delta(M_{i+1}), r_i/\delta(M_{i+1})\}$ is a $Z$-set in $N$. It only remains to note that $N - Z$ is naturally homeomorphic to $M$.

Conversely, suppose that the $\mu^{n+1}$-manifold $M$ admits a boundary. This means that there are a compact $\mu^{n+1}$-manifold $N$ and a $Z$-set $Z$ in $N$ such that $M = N - Z$. Let us show that $M$ is $n$-tame at $\infty$. Let $A$ be a compact subspace of $M$. As in the proof of Lemma 4.4.17, there exists a compact and $n$-clean submanifold $B$ of $M$ such that $A \subseteq B - \delta(B)$. It suffices to show that $(M - B) \cup \delta(B)$ is $n$-homotopy equivalent to an at most $(n+1)$-dimensional finite polyhedron. Indeed, it is easy to see that $(M - B) \cup \delta(B)$ is $n$-homotopy equivalent to a compact $\mu^{n+1}$-manifold $(N - B) \cup \delta(B)$. It only remains to apply Proposition 4.1.10. □

Not all $\mu^{n+1}$-manifolds admit boundaries in the above sense. To see this, consider the 3-dimensional (topological) manifold $W$ (constructed by Whitehead) which is defined as the complement in $S^3$ of a continuum $Wh$ which, in turn, is the intersection of a nested sequence of tori in $S^3$. The manifold $W$ has an infinitely generated fundamental group at $\infty$. Let $n \geq 1$, and consider a $\mu^{n+1}$-manifold $M$ and a proper $UV^n$-surjection $f\colon M \to W$. Since $n+1 \geq 2$, we see that $f$ induces an isomorphism of fundamental groups of ends. Then it is easy to see that $M$ is not 1-tame at $\infty$ and, therefore, cannot have a boundary.

On the other hand, it can shown that the Freudenthal compactification of any connected $\mu^1$-manifold contains its end as a $Z$-set. Consequently, the Freudenthal compactification of any connected $\mu^1$manifold is homeomorphic to $\mu^1$. In other words, any connected $\mu^1$-manifold has a boundary.

We also mention two related results.

PROPOSITION 4.4.22. *If the compactum $X$ is a boundary for a $\mu^{n+1}$-manifold $M$, then the compactum $Y$ is also a boundary for $M$ if and only if* $\dim Y \leq n+1$ *and* $n - Sh(Y) = n - Sh(X)$.

Two compactifications $N$ and $T$ of the same space $M$ are said to be *equivalent* if for every compactum $A \subseteq M$ there is a homeomorphism of $N$ onto $T$ fixing $A$ point wise.

Of course, if the $\mu^{n+1}$-manifolds $N$ and $T$ are compactifications of a $\mu^{n+1}$-manifold $M$, then the inclusions $M \hookrightarrow N$ and $M \hookrightarrow T$ are $n$-homotopy equivalences (because, $N - M$ and $T - M$ are $Z$-sets in $N$ and $T$ respectively). Consequently, $N$ and $T$ are homeomorphic as $n$-homotopy equivalent compact $\mu^{n+1}$-manifolds (Theorem 4.1.21). A stronger result can be obtained.

PROPOSITION 4.4.23. *Every two $\mu^{n+1}$-manifold compactifications of a given $\mu^{n+1}$-manifold are equivalent in the above sense.*

The problem of whether a $\mu^{n+1}$-manifold has a boundary which is itself a $\mu^{n+1}$-manifold was also considered in [**104**].

DEFINITION 4.4.24. *A proper map $f\colon Y \to X$ between at most $(n+1)$-dimensional locally compact spaces is an $n$*-domination *near $\infty$ provided that there exists a cofinite subspace $X_1$ of $X$ ( i.e. $X_1$ is closed and $X - X_1$ has compact closure) and a proper map $g\colon X_1 \to Y$ such that $fg$ is properly $n$-homotopic to the inclusion map $X_1 \hookrightarrow X$. If, in addition, for some cofinite subspace $Y_1$ of $Y$ the composition $gf/Y_1$ is properly $n$-homotopic to the inclusion map $Y_1 \hookrightarrow Y$, then we say that $f$ is an equivalence near $\infty$.*

DEFINITION 4.4.25. *A space $X$ is finitely $n$-dominated near $\infty$ if there exists a finite polyhedron $P$ and an $n$-domination near $\infty$, $f\colon (P \times [0,1))^{(n+1)} \to X$. If, in addition, $f$ is an equivalence near $\infty$, then we say that $X$ has finite $n$-type near $\infty$.*

The following statement corresponds to Proposition 4 1.12 in proper $n$-homotopy category.

PROPOSITION 4.4.26. *Each finitely $n$-dominated near $\infty$, at most $(n+1)$-dimensional, locally compact $LC^n$-space has finite $n$-type near $\infty$.*

Using Proposition 4.4.26, it is possible to prove [**104**] the following theorem.

THEOREM 4.4.27. *A $\mu^{n+1}$-manifold $M$ has a boundary which itself is a $\mu^{n+1}$-manifold if and only if $M$ is finitely $n$-dominated near $\infty$.*

It follows from Proposition 4.4.22 and Theorem 4.1.21 that if such a boundary exists, then it is uniquely determined.

*Historical and bibliographical notes* 4.4. Theorem 4.4.27 (as well as Proposition 4.4.26) was proved in [**104**]. All other results of this Section are due to the author. More precisely, the concept of the $n$-homotopy kernel of a $\mu^{n+1}$-manifold was introduced in [**98**]. Theorem 4.4.3 and 4.4.7 also appear in [**98**]. A weaker version of Theorem 4.4.3 was proved earlier in [**90**]. The complement theorem for $Z$-sets in Menger compacta (Theorem 4.4.8 and its corollaries) were proved in [**87**] (also, see the related papers [**90**] and [**92**]). The results of Subsection 4.4.4 appear in [**100**].

## 4.5. Homeomorphism Groups

Let $M$ be a Menger manifold and $\mathrm{Auth}(M)$ its group of autohomeomorphisms. The study of $\mathrm{Auth}(M)$ is well-developed. It is in their spaces of autohomeomorphisms that we see one of the major differences between $I^\omega$-manifolds on the one hand, and Menger manifolds on the other. Homeomorphism groups of $I^\omega$-manifolds are $l_2$-manifolds (see Chapter 2), while homeomorphism groups of $\mu^n$-manifolds $M$, as it will be shown below, are totally disconnected. If $n = 0$, then $\mathrm{Auth}(M)$ is 0-dimensional, and if $n \geq 1$, then $\mathrm{Auth}(M)$ is 1-dimensional. Hence, the only compact Lie groups to act effectively on a $\mu^n$-manifold are finite groups. On the other hand, all compact 0-dimensional metric groups act effectively on all Menger manifolds. The algebraic properties of homeomorphism groups of Menger manifolds are rather similar to the properties of homeomorphism groups of $n$- and $I^\omega$-manifolds. For an autohomeomorphism $H\colon X \to X$, $\operatorname{supp} H = \mathrm{cl}\{x \in X : H(x) \neq x\}$. If $\operatorname{supp} H$ is contained in a subset $A$, we say that $H$ is supported on $A$.

### 4.5.1. Dimension of $Auth(\mu^n)$. 

We begin with the following Definition.

DEFINITION 4.5.1. *A continuum $X$ is* locally setwise homogeneous *if there exists a basis $\mathcal{U}$ of connected open subsets of $X$ and a dense subset $B \subseteq X$ such that for each $E \in \mathcal{U}$ and $a, b \in B \cap E$ there exists $h \in \mathrm{Auth}(X)$ supported on $E$ such that $h(a) = b$.*

Clearly, a locally setwise homogeneous continuum is locally connected. The Sierpinski curve $M_1^2$ is locally setwise homogeneous but not homogeneous. The solenoids are homogeneous but not locally setwise homogeneous.

Suppose that $X$ is a locally setwise homogeneous continuum and $\epsilon > 0$ is a sufficiently small number. Let $U$ be a neighborhood of $\mathrm{id}_X$ in $\mathrm{Auth}(X)$ with $\mathrm{diam}\, U < \varepsilon$. Let $x, y \in X$ with $d(x,y) = \varepsilon$, and let $A$ be a small diameter arc from $x$ to $y$ in $X$. One can define, using local setwise homogeneity, a convergent sequence $\{h_i\}$ in $\mathrm{Auth}(X)$ "sliding $x$ towards $y$ along $A$" and such that $h = \lim h_i \in \mathrm{Auth}(X)$ and $h \in Bd(U)$. Thus, each $\varepsilon$-neighborhood of $\mathrm{id}_X$ has non-empty boundary, and so $\dim \mathrm{Auth}(X) \geq 1$. Therefore we have the following statement.

THEOREM 4.5.2. *Let $X$ be a locally setwise homogeneous continuum. Then $\mathrm{Auth}(X)$ is at least 1-dimensional.*

Strong local homogeneity of $\mu^n$-manifolds (Corollary 4.1.17), implies their local setwise homogeneity. Consequently, we have the following.

COROLLARY 4.5.3. *If $M$ is a compact $\mu^n$-manifold, then* $\dim \mathrm{Auth}(M) \geq 1$.

Recall that a space $X$ is *almost 0-dimensional* if it has a basis $\mathcal{B}$ of open sets such that for each $B \in \mathcal{B}$, $X - \mathrm{cl}\, B = \cup\{U_i : i \in N\}$ where each $U_i$ is both open and closed. Clearly, each 0-dimensional space is almost 0-dimensional. Also, it is easy to see that every almost 0-dimensional space is totally disconnected. The complete Erdös space $\mathcal{E} = \{x \in l_2 \colon x_i$ is irrational for each $i\}$ is a 1-dimensional space which is almost 0-dimensional.

PROPOSITION 4.5.4. *Each almost 0-dimensional space is at most 1-dimensional.*

SKETCH OF PROOF. Let $X$ be an almost 0-dimensional space and let $\mathcal{B}$ be a countable basis witnessing this fact. Let $\mathcal{F} = \{f_i : i \in N\}$ be a collection of continuous functions $f_i : X \to \{0,1\}$ such that if $B, B' \in \mathcal{B}$ with $\mathrm{cl}(B) \cap \mathrm{cl}(B') = \emptyset$, there is a $f_i \in \mathcal{F}$ with $f_i(B) = 0$ and $f_i(B') = 1$. Let $\rho$ be the metric on $X$ given by $\rho(x,y) = \sum_i 2^{-i} \mid f_i(x) - f_i(y) \mid$. Let $d$ be a totally bounded metric on $X$. It suffices to show that the metric dimension $\mu \dim(X,d) \leq 1$. Now, $d' = d + \rho$ is also a totally bounded metric on $X$. Let $Y$ be the completion of $X$ with respect to $d'$. It suffices to show that for each $t > 0$, there is an open set $U$ of $Y$ containing $X$ such that each continuum in $U$ has diameter less than $t$.

Let $\mathcal{U} = \{U$ open in $Y$ : $\mathrm{diam}\, U < t/3$ and $U \cap X \in \mathcal{B}\}$. Let $C$ be any continuum in $W = \cup\mathcal{U}$. If $B, B' \in \mathcal{B}$ with $B \cap C \neq \emptyset \neq B' \cap C$, then $\mathrm{cl}(B) \cap \mathrm{cl}(B') \neq \emptyset$, for otherwise there is $f_i \in \mathcal{F}$ with $f_i(B) = 0$ and $f_i(B') = 1$. But $f_i(X) = \{0,1\}$ and so, by the definition of $d', Y = \mathrm{cl}_Y(f_i^{-1}(0)) \cup \mathrm{cl}_Y(f_i^{-1}(1))$, where $\mathrm{cl}_Y(f_i^{-1}(0))$ and $\mathrm{cl}_Y(f_i^{-1}(1))$ are disjoint closed sets. Hence $C$, being a continuum, cannot meet both $\mathrm{cl}_Y(f_i^{-1}(0))$ and $\mathrm{cl}_Y(f_i^{-1}(1))$ which is a contradiction. It follows that $\mathrm{diam}\, C < t$. $\square$

PROPOSITION 4.5.5. *If $M$ is an $M_n^k$-manifold with $0 \leq n < k < \infty$, then $\mathrm{Auth}(M)$ is almost 0-dimensional.*

SKETCH OF PROOF. Let $g \in \text{Auth}(M)$ and $\varepsilon > 0$. Let $h \in \text{Auth}(M)$ with $d(g,h) > \varepsilon$. We shall show that there is an open and closed set $U$ containing $h$ such that $d(g,j) > \varepsilon$ for each $j \in U$. Now, $d(g,h) = \varepsilon + 4\delta$ for some $\delta > 0$, and there is $x \in M$ so that $d(g,h) = d(g(x), h(x))$. Choose an $n$-sphere $S$ in $M$ such that $g(S) \subseteq N(g(x), \delta)$ and $h(S) \subseteq N(h(x), \delta)$. Since $\dim M = n$, there is a retraction $r : M \to h(S)$ such that $r(M - N(h(x), 2\delta))$ is constant. Let $U = \{f \in \text{Auth}(M) \colon rf/S \not\simeq *\}$. Then $U$ is both open and closed in $\text{Auth}(M)$ because close maps into $S$ are homotopic. Also $h \in U$. Let $f \in N(g, \varepsilon)$. Then $f(S) \subseteq N(g(x), \varepsilon + \delta)$. Hence, $r \cdot f(S)$ is a point and $f \notin U$. □

COROLLARY 4.5.6. *Let $1 \leq n < \infty$. If $M$ is a compact $\mu^n$-manifold, then* $\dim \text{Auth}(M) = 1$.

PROOF. Apply Corollary 4.5.3 and Propositions 4.5.4 and 4.5.5. □

### 4.5.2. Simplicity.

Anderson [**16**], [**17**] originated a technique for identifying minimal, non-trivial normal subgroups of $\text{Auth}(X)$ for spaces with certain dilation and homogeneity properties.

DEFINITION 4.5.7. *Let $X$ be a space. A subset $A$ of $X$ is* deformable *if for every non-empty open set $U$ in $X$, there is $h \in \text{Auth}(X)$ with $h(A) \subseteq U$. Let $U$ be an open set. A collection $(\{B_i : i \in N\}, h)$ is called a* dilation system *in $U$ if $\{B_i\}$ is a sequence of disjoint non-empty open sets in $U$ with $\lim B_i = \{p\}$ for some $p \in U$ and $h \in \text{Auth}(X)$ supported on $U$ such that $h(B_{i+1}) = B_i$ for each $i$.*

PROPOSITION 4.5.8. *Let $X$ be a metrizable space in which each non-empty open set contains a dilation system. Let $G$ be a subgroup of $\text{Auth}(X)$ generated by all homeomorphisms which are supported by deformable subsets of $X$. If $G \neq \{e\}$, then $G$ is the smallest non-trivial normal subgroup of $\text{Auth}(X)$.*

If $X$ is a finite-dimensional manifold without boundary, then $\text{Auth}_0(X)$, the subgroup of homeomorphisms isotopic to the identity, is simple (see [**151**] and [**141**]). It is also known (see [**220**] and [**323**]) that $\text{Auth}(l_2)$ and $\text{Auth}(I^\omega)$ are simple.

DEFINITION 4.5.9. *Let $M$ be a $\mu^{n+1}$-manifold. A pair $(W, \delta(W))$ is an* $n$-clean pair *if $W$ is $n$-clean with respect to $\delta(W)$ in the sense of Definition 4.4.15 and if, in addition, both $W$ and $\delta(W)$ are homeomorphic to $\mu^{n+1}$.*

By Theorem 4.1.15 and the existence of $n$-clean pairs in $\mu^{n+1}$, (compare with [**184**]), it follows that every open set in $\mu^{n+1}$ has a dilation system. Also, every proper closed set in $\mu^{n+1}$ is deformable. Since every element of $\text{Auth}(\mu^{n+1})$ is stable (see Theorem 4.5.13 below) we have the following.

THEOREM 4.5.10. $\text{Auth}(\mu^{n+1})$ *is simple.*

**4.5.3. Stability of homeomorphisms.** An autohomeomorphism of a space $X$ is said to be *stable* [**32**] if it can be expressed as the composition of finitely many autohomeomorphisms each of which is the identity on some non-empty open subspace of $X$. It is well-known that all autohomeomorphisms of the Hilbert cube $I^\omega$ and the Hilbert space $l_2$ are stable (see, for example, [**32**]). Every orientation-preserving homeomorphism of $R^n$ is stable [**194**].

LEMMA 4.5.11. *For each $\mu^{n+1}$-manifold $M$, there is a $Z$-embedding $\alpha\colon M \to M$ which is properly $n$-homotopic to $id_M$ and which satisfies the following condition:*

(*) *If $F \in \mathrm{Auth}(M)$ and $F/\alpha(M) = id$, then $F$ can be expressed as the composition of two autohomeomorphisms of $M$ each of which is the identity on some open subspace of $M$.*

PROOF. Take a proper $UV^n$surjection $g\colon M \to K$, where $K$ is an at most $(n+1)$-dimensional locally compact polyhedron (Theorem 4.2.23). Consider also a proper $UV^n$-surjection $f\colon M_1\colon K \times [-1, 2]$, satisfying conditions of Theorem 4.3.6. It can easily be checked that the inverse image $M = f^{-1}(K \times [0, 2])$ is also a copy of the $\mu^{n+1}$-manifold $M$. Moreover, since the composition $\pi_1 f/M \to K \times \{0\}$, where $\pi_1\colon K \times [0, 2] \to K \times \{0\}$ denotes the projection, is a proper $UV^n$-surjection, we conclude that there exists a homeomorphism $\alpha\colon M \to f^{-1}(K \times \{0\})$ such that $f\alpha \simeq^n_p \pi_1 f/M$. Then $\alpha$ is properly $n$-homotopic to $id_M$. By the properties of $f$, we see that $\alpha\colon M \to M$ is a $Z$-embedding. Consequently, it only remains to show that if $F \in \mathrm{Auth}(M)$ and $F/f^{-1}(K \times \{0\}) = id$, then $F$ can be expressed as the composition of two autohomeomorphisms of $M$ each of which is the identity on some open subspace of $M$.

Let $U = F^{-1}(f^{-1}(K \times (1, 2]))$. Since $F/\alpha(M) = id$ we conclude that $U \cap f^{-1}(K \times \{0\}) = \emptyset$. Consequently, there exists a sufficiently small $t_1 > 0$ such that $f^{-1}(K \times [0, t_1]) \cap (U \cup F(U)) = \emptyset$. Let $V = f^{-1}(K \times [0, t_0))$, where $t_0 = \frac{1}{4}t_1\frac{1}{1+t_1}$.

Consider the homeomorphism $\varphi\colon [0, t_1] \to [-1, t_1]$ defined as $\varphi(x) = \frac{t_1}{1+t_1}x - 1$. Consider also an open cover

$$\mathcal{U} = \left\{K \times \left[-1, -\frac{1}{2}\right), K \times \left(-\frac{3}{4}, t_1\right]\right\}$$

of the product $K \times [-1, t_1]$. Again, by the properties of $f$, the inverse images $N_1 = f^{-1}(K \times [0, t_1])$ and $N_2 = f^{-1}(K \times [-1, t_1])$ are $\mu^{n+1}$-manifolds. Clearly these admit proper $UV^n$-surjections

$$f_1 = (id \times \varphi)f/N_1\colon N_1 \to K \times [-1, t_1] \text{ and } f_2 = f/N_2\colon K \times [-1, t_1]$$

onto the polyhedron $K \times [-1, t_1]$. Then we see that the inverse image $f^{-1}(K \times \{t_1\})$ is a $Z$-set both in $N_1$ and $N_2$. Moreover, since, $\varphi(t_1) = t_1$, we conclude that the identity homeomorphism $h$ of $f^{-1}(K \times \{t_1\})$ satisfies the equality $f_2 h = f_1/f^{-1}(K \times \{t_1\})$. Consequently, there exists a homeomorphism $T\colon N_1 \to N_2$ such that $T/f^{-1}(K \times \{t_1\}) = id$ and the composition $f_2 T$ is $\mathcal{U}$-close to $f_1$. Now consider the homeomorphism $H\colon M \to M_1$ such that $H/f^{-1}(K \times [0,t_1]) = T$ and $H/f^{-1}(K \times [t_1,2]) = id$. It is easy to verify that $H(V) \cap M = \emptyset$ and $H/(U \cup F(U)) = id$.

Let

$$G(x) = \begin{cases} F(x), & \text{if } x \in M \\ x, & \text{if } x \in M_1 - M. \end{cases}$$

Since $F/f^{-1}(K \times \{0\}) = id$, we conclude that $G \in \mathrm{Auth}(M_1)$. Now let $F_2 = H^{-1}GH \in \mathrm{Auth}(M)$. If $x \in U$, then $H(x) = x$. Hence, $GH(x) = F(x) \in F(U)$. Consequently, $H^{-1}GH(x) = F(x)$. In other words, $F_2/U = F/U$. Similarly, if $x \in V$, then $H(x) \in M_1 - M$. Hence $GH(x) = H(x)$ and $H^{-1}GH(x) = H^{-1}H(x) = x$. This means that $F_2/V = id$. Obviously, $F = F_2F_1$, where $F_1 = F_2^{-1}F$. It only remains to note that $F_1/U = id$ and $F_2/V = id$. □

LEMMA 4.5.12. *Let $M$ be an $\mu^{n+1}$-manifold and $\beta\colon M \to M$ be a $Z$-embedding properly $n$-homotopic to $id_M$. If $G \in \mathrm{Auth}(M)$ and $G/\beta(M) = id$, then $G$ can be expressed as the composition of two autohomeomorphisms each of which is the identity on some open subspace of $M$.*

PROOF. Take a $Z$-embedding $\alpha$ satisfying condition $(*)$ of Lemma 4.5.11. Then the homeomorphism $\beta\alpha^{-1}\colon \alpha(M) \to \beta(M)$ is properly $n$-homotopic to the inclusion $\alpha(M) \hookrightarrow M$. Consequently, by Theorem 4.1.15, there exists a homeomorphism $h\colon M \to M$ extending $\beta\alpha^{-1}$. Consider the homeomorphism $h^{-1}Gh \in \mathrm{Auth}(M)$. Evidently, $h^{-1}Gh/\alpha(M) = id$. By the choice of $\alpha$, there exist $F_1, F_2 \in \mathrm{Auth}(M)$ and open subspaces $V_1$ and $V_2$ of $M$ such that $F_i/V_i = id$, $i = 1,2$, and $h^{-1}Gh = F_2F_1$. Then $G = G_2G_1$, where $G_i = hF_ih^{-1}$, $i = 1,2$. It only remains to note that $G_i/h(V_i) = id$, $i = 1,2$. □

THEOREM 4.5.13. *Let $M$ be a $\mu^{n+1}$-manifold and $F \in \mathrm{Auth}(M)$. If $F$ is properly $n$-homotopic to $id_M$, then $F$ is stable.*

PROOF. Let $\beta\colon M \to M$ be a $Z$-embedding properly $n$-homotopic to $id_M$. Let $\alpha\colon M \to M$ be another $Z$-embedding properly $n$-homotopic to $id_M$ and such that $\alpha(M) \cap (\beta(M) \cup F(\beta(M))) = \emptyset$. Consider the homeomorphism

$$f\colon \alpha(M) \cup \beta(M) \to \alpha(M) \cup F(\beta(M))$$

which coincides with the identity on $\alpha(M)$ and with $F$ on $\beta(M)$. Clearly, $f$ is properly $n$-homotopic to the inclusion $\alpha(M) \cup \beta(M) \hookrightarrow M$. Therefore there exists an extension $F_2 \in \mathrm{Auth}(M)$ of $f$. Let $F_1 = F_2^{-1}F$. By Lemma 4.5.12, $F = F_2F_1$ can be expressed as the composition of four autohomeomorphisms each of which is the identity on some open subspace of $M$. □

Since each map of $\mu^{n+1}$ into itself is (properly) $n$-homotopic to the identity, we obtain the following statement.

COROLLARY 4.5.14. *Every autohomeomorphism of $\mu^{n+1}$ is stable.*

**4.5.4. Group actions on Menger manifolds.** The well-known Hilbert-Smith conjecture asks whether every compact group acting effectively on a manifold is a Lie group. This is equivalent to asking whether the group $A_p$ of $p$-adic integers acts effectively on a manifold. This long standing problem is still open.

The situation is rather different for Menger manifolds. For instance, it is known [**13**] that any compact metrizable zero-dimensional topological group $G$ acts freely on $\mu^1$ so that the orbit space $\mu^1/G$ is homeomorphic to $\mu^1$. There are several constructions of group actions on $\mu^n$-manifolds.

THEOREM 4.5.15. *Let $M$ be a $\mu^n$-manifold. Then:*

(i) *Every compact zero-dimensional metrizable group $G$ acts on $M$ so that the orbit space $M/G$ is homeomorphic to $M$*
(ii) *$A_p$ acts freely on $M$ so that* $\dim M/A_p = n+1$
(iii) *$A_p$ acts on $M$ so that* $\dim M/A_p = n+2$

There are unexpected ties between group actions on Menger compacta and the Hilbert-Smith problem mentioned above. Namely, a positive solution to the following conjecture would prove that there is no free $A_p$-action on a connected (topological) manifold $M$ with $\dim M/A_p < \infty$.

CONJECTURE 4.5.16. *Let $m$ and $n$ be positive integers and $G$ be a zero-dimensional compact metric group. If $\mu^{m+n}$ and $\mu^n$ are free $G$-spaces, then there is no equivariant map $\mu^{m+n} \to \mu^n$.*

*Historical and bibliographical notes* 4.5. Theorem 4.5.2 was proved in [**54**]. Corollary 4.5.6 appears in [**244**]. Corollary 4.5.14 is due to the author [**94**]. It was shown later that all autohomeomorphisms of all connected Menger manifolds are stable [**266**]. Item (i) of Theorem 4.5.15 was proved in [**131**] (see also [**265**]). Items (ii) and (iii) appear in [**219**] (the last statement is based on the work [**257**] ). Conjecture 4.5.16 appears in [**1**].

## 4.6. $\omega$-soft map of $\sigma$ onto $\Sigma$

In this Section, using Theorem 4.2.21, we show that there exists an "almost soft" map from $\sigma$ onto $\Sigma$.

DEFINITION 4.6.1. *A map $f: X \to Y$ is called $\omega$-soft if for each strongly countable-dimensional*[1] *space $B$, closed subspace $A$ of $B$ and any two maps $g: A \to X$ and $h: B \to Y$ with $fg = h/A$, there exists a map $k \to B \to X$ such that $k/A = g$ and $fk = h$, i.e. if the following diagram*

[1]Recall that a space is strongly countable-dimensional if it can be represented as the countable union of finite-dimensional closed subspaces.

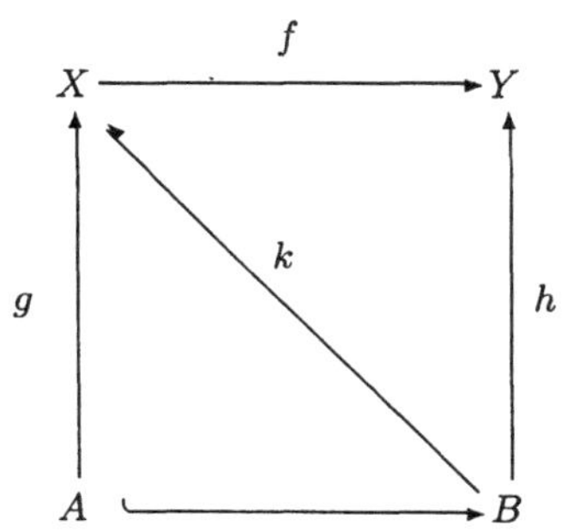


*commutes.*

Obviously, every $\omega$-soft map is $n$-soft for each $n \in \omega$. In particular, every $\omega$-soft map is surjective and open (compare with the proof of implication $(ii) \Longrightarrow (i)$ in Proposition 2.1.34).

LEMMA 4.6.2. *Let*

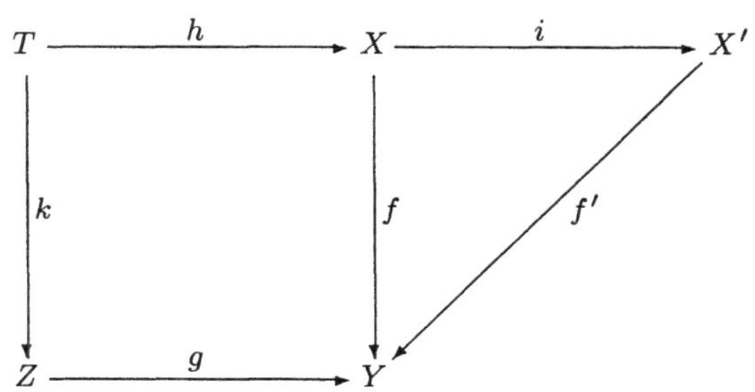


*be a commutative diagram, consisting of compact spaces, where $i$ is an embedding and* $\dim T \leq n$. *Then there exists a commutative $n$-soft diagram*[2]

[2]This means that the diagonal product $k'\triangle h'$, considered as the map of $T'$ onto its image, is $n$-soft (see page 159).

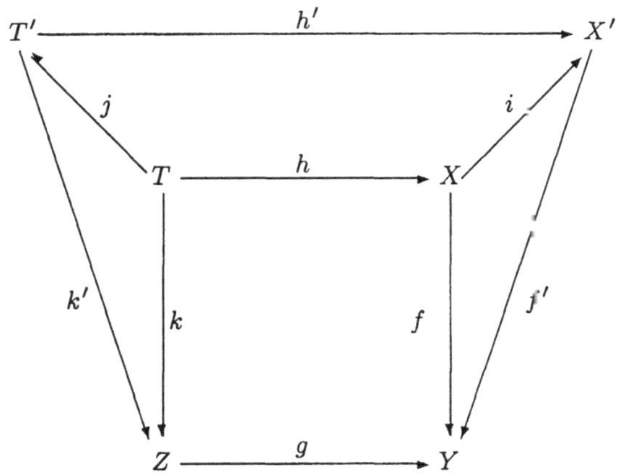


*where $j$ is an embedding and* $\dim T' \leq n+1$.

PROOF. Let $K$ be the fibered product of $Z$ and $X'$ with respect to the maps $g\colon Z \to Y$ and $f'\colon X' \to Z$. Let $\varphi\colon K \to Z$ and $\psi\colon K \to X'$ denote the restrictions of the natural projections $Z \times X' \to Z$ and $Z \times X' \to X'$ onto $K$ respectively. By Theorem 4.2.21, there exists an $n$-soft map $m\colon T' \to K$, where $\dim T' \leq n+1$. Since $\dim T \leq n$ we can find an embedding $j\colon T \to T'$ such that $mj = k\Delta ih$. It only remains to observe that the desired maps are $h' = \psi m$ and $k' = \varphi m$. □

PROPOSITION 4.6.3. *There exists an $\omega$-soft map $f\colon X \to I^\omega$, where $X$ is a $\sigma$-compact strongly countable-dimensional $AR$-space.*

PROOF. We consider the Hilbert cube $I^\omega$ to be the limit space of an inverse sequence $\mathcal{S}' = \{Y_i, q_i^{i+1}\}$, where $Y_i = I^i$ and $q_i^{i+1}\colon I^{i+1} \to I^i$ denotes the projection onto the corresponding subproduct. We now construct an inverse sequence $\mathcal{S} = \{X_i, p_i^{i+1}\}$, a collection of compact subspaces $Z_{i,1} \subseteq Z_{i,2} \subseteq \cdots \subseteq Z_{i,i} = X_i$ for each $i$ and a morphism

$$\{\alpha_i\colon X_i \to Y_i\}\colon \mathcal{S} \to \mathcal{S}'$$

such that the following conditions are satisfied:

(i) $\dim Z_{i,k} \leq k+1$, whenever $k < i$.

(ii) The diagram

$$\mathcal{D}_{i,k} = \begin{array}{ccc} Z_{i+1,k} & \xrightarrow{\alpha_{i+1}/Z_{i+1.k}} & Y_{i+1} \\ {\scriptstyle p_i^{i+1}/Z_{i+1,k}}\downarrow & & \downarrow{\scriptstyle q_i^{i+1}} \\ Z_{i,k} & \xrightarrow{\alpha_i/Z_{i,k}} & Y_i \end{array}$$

is $k$-soft.

(iii) The diagram

$$\mathcal{D}_i = \begin{array}{ccc} X_{i+1} & \xrightarrow{\alpha_{i+1}} & Y_{i+1} \\ {\scriptstyle p_i^{i+1}}\downarrow & & \downarrow{\scriptstyle q_i^{i+1}} \\ X_i & \xrightarrow{\alpha_i} & Y_i \end{array}$$

is soft.

Let $X_1 = I \times I$ and $\alpha_1 = \pi_1 \colon I \times I \to I$. Assume that for each $i \leq j$, the maps $\alpha_i$ and the spaces $Z_{i,k}$, $k = 1, 2, \ldots, i$, have already been defined. Let $Z'_{j+1,1}$ be the fibered product of the spaces $Z_{j,1}$ and $Y_{j+1}$ with respect to the maps $\alpha_j/Z_{j,1}$ and $q_j^{j+1}$, and $\varphi \colon Z'_{j+1,1} \to Z_{j,1}$ and $\psi \colon Z'_{j+1,1} \to Y_{j+1}$ be the corresponding projections of the fibered product. By Theorem 4.2.21, there exists a 1-soft map $m \colon Z_{j+1,1} \to Z'_{j+1,1}$, where $\dim Z_{j+1,1} = 2$. We define the maps $p_j^{j+1}$ and $\alpha_{j+1}$ by means of their restrictions to the subspaces $Z_{j+1,i}$. Let $p_j^{j+1}/Z_{j+1,1} = \varphi m$ and $\alpha_{j+1}/Z_{j+1,1} = \psi m$.

Using Lemma 4.6.2, we get the sequence of spaces

$$Z_{j+1,1} \subseteq Z_{j+1,2} \subseteq \cdots \subseteq Z_{j+1,j}$$

with $\dim Z_{j+1,k} \leq k + 1$, and maps

$$r_k = p_j^{j+1}/Z_{j+1,k} \colon Z_{j+1,k} \to Z_{j,k}, \quad \beta_k = \alpha_{j+1}/Z_{j+1,k} \colon Z_{j+1,k} \to Y_{j+1}$$

such that $r_k/Z_{k-1} = r_{k-1}$, $\beta_k/Z_{k-1} = \beta_{k-1}$ and the diagrams $\mathcal{D}_{j,k}$ are $k$-soft.

If $k = j$, let $l$ be an embedding of the $(j+1)$-dimensional space $Z_{j+1,j}$ into the cube $I^{2j+3}$. Let

$$X_{j+1} = X_j \times Y_{j+1} \times I^{2j+3}$$

and assume that the space $Z_{j+1,j}$ is embedded into $X_{j+1}$ by the map $l' = r_j \triangle \beta_j \triangle l$. Let $p_j^{j+1}\colon X_{j+1} \to X_j$ be the restriction of the projection onto the first coordinate and $\alpha_{j+1}\colon X_{j+1} \to Y_{j+1}$ be the restriction of the projection onto the second coordinate. It is easy to see that conditions (i)-(iii) are satisfied.

For each natural number $k \in N$, let

$$Z_k = \lim\{Z_{i,k}, pi + 1_i/Z_{i+1,k}, i\}, \quad X = \cup_{k=1}^{\infty} Z_k$$

and

$$f = \lim\{\alpha_i\}/X : X \to \lim \mathcal{S}' = I^\omega .$$

Obviously, the space $X$ is a strongly countable-dimensional $\sigma$-compact space.

Next we show that the map $f$ is $\omega$-soft. Consider the commutative diagram

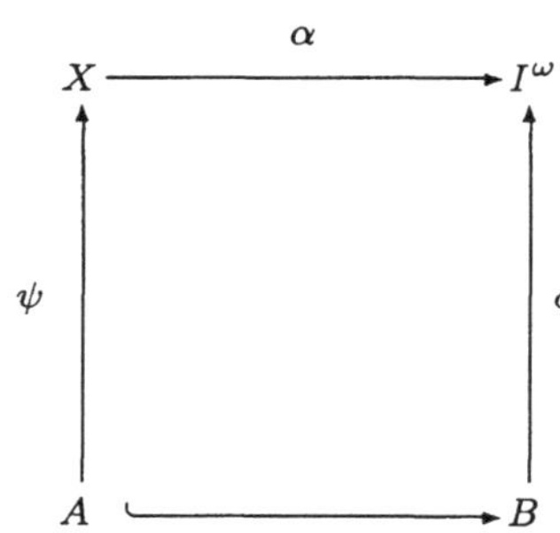


where $A$ is a closed subset of a strongly countable-dimensional $\sigma$-compact space $B$.

Let $\psi = \{\psi_i\}$ and $\phi = \{\phi_i\}$, where $\psi_i\colon A \to X_i$ and $\phi_i\colon B \to Y_i$ are the coordinate maps. We also let $B = \cup_{j=1}^{\infty} B_j$, where the $B_j$ are at most $j$-dimensional subsets of $B$ and $B_1 \subseteq B_2 \subseteq \cdots$.

Let $A_j = A \cap B_j$. Then $A = \cup_{j=1}^{\infty} A_j$. Note that, without loss of generality, we can assume that $\psi(A_j) \subseteq Z_j$. Now we are going to construct a map $\Phi\colon B \to X$ as $\Phi = \{\Phi_i\}$, where $\Phi_i\colon B \to X_i$ are coordinate maps.

We constructs the $\Phi_i$'s by induction. Let $\Phi_1\colon B \to X_1$ be a map such that $\alpha_1 \Phi_1 = \phi_1$ and $\Phi_1/A_1 = \psi_1$. Now assume that for every $i \le j$ map $\Phi_i$ has already been constructed so that $\Phi_i(B_k) \subseteq Z_{i,k}$, $k \le i$. Using the 1-softness of the square diagram $\mathcal{D}_{j,1}$ we choose a map $\Phi_{j+1}^{(1)}\colon B_1 \to Z_{j+1,1}$ such that the following conditions are satisfied:

- $p_j^{j+1} \Phi_{j+1}^{(1)} = \Phi_j/B_1$.
- $\alpha_{j+1} \Phi_{j+1}^{(1)} = \phi_{j+1}/B_1$.
- $\Phi_{j+1}^{(1)}/\psi_{j+1}^{-1}(A \cap Z_{j+1,1}) = \psi_{j+1}/\psi_{j+1}^{-1}(A \cap Z_{j+1,1})$.

Using 2-softness of the square diagram $\mathcal{D}_{j,2}$ we choose a map $\Phi^{(2)}_{j+1} \colon B_2 \to Z_{j+1,2}$ such that

- $p^{j+1}_j \Phi^{(2)}_{j+1} = \Phi_j / B_2$.
- $\alpha_{j+1} \Phi^{(2)}_{j+1} = \phi_{j+1} / B_2$.
- $\Phi^{(2)}_{j+1} / \psi^{-1}_{j+1}(A \cap Z_{j+1,2}) = \psi_{j+1} / \psi^{-1}_{j+1}(A \cap Z_{j+1,2})$.
- $\Phi^{(2)}_{j+1} / B_1 = \Phi^{(1)}_{j+1}$.

Continuing this process we obtain a map $\Phi^{(j)}_{j+1} \colon B_j \to Z_{j+1,j}$ satisfying the conditions:

- $p^{j+1}_j \Phi^{(j)}_{j+1} = \Phi_j / B_j$.
- $\alpha_{j+1} \Phi^{(j)}_{j+1} = \phi_{j+1} / B_j$.
- $\Phi^{(j)}_{j+1} / \psi^{-1}_{j+1}(A \cap Z_{j+1,j}) = \psi_{j+1} / \psi^{-1}_{j+1}(A \cap Z_{j+1,j})$.
- $\Phi^{(j)}_{j+1} / B_{j-1} = \Phi^{(j-1)}_{j+1}$.

Finally, using softness of the diagram $\mathcal{D}_j$ we can choose a map $\Phi_{j+1} \colon B \to X_{j+1}$ such that

- $p^{j+1}_j \Phi_{j+1} = \Phi_j$.
- $\alpha_{j+1} \Phi_{j+1} = \phi_{j+1}$.
- $\Phi_{j+1} / B_j = \Phi^{(j)}_{j+1}$.
- $\Phi_{j+1} / A = \psi_{j+1}$.

It follows easily from the construction that the map $\Phi = \{\Phi_j\}$ is well defined, $\Phi / A = \psi$ and $f\Phi = \phi$. It only remains to observe that the $\omega$-softness of $f$ guarantees that $X$ is an $AR$-space. □

THEOREM 4.6.4. *There exists an $\omega$-soft map $\phi \colon \sigma \to \Sigma$.*

PROOF. Let $f \colon X \to I^\omega$ be the map from Proposition 4.6.3 and let $X' = f^{-1}(\Sigma)$. Clearly, $X'$ is a strongly countable-dimensional $\sigma$-compact $AR$-space. By Proposition 2.5.5, the product $X' \times \sigma$ is homeomorphic to $\sigma$. The desired map is $\phi = f\pi \colon \sigma \to \Sigma$, where $\pi \colon X' \times \sigma \to X'$ denotes the projection onto the first coordinate. □

*Historical and bibliographical notes* 4.6. Theorem 4.6.4 was proved in [**325**]. A somewhat weaker version of it was obtained earlier in [**229**].

CHAPTER 5

# Nöbeling Manifolds

There is a deep relationship between Menger compacta and their non-compact counterparts – *universal Nöbeling spaces.* To some extent the Nöbeling space $N_n^{2n+1}$ can be viewed as an $n$-dimensional Hilbert space $l_2 \approx R^\omega$ (compare with the compact case: $\mu^n$ is an $n$-dimensional analog of the Hilbert cube $I^\omega$ ). In this Chapter we establish some basic facts in this direction. In particular, we show that $N_n^{2n+1}$ can be identified with the so-called pseudo-interior $\nu^n$ of the Menger compactum $\mu^n$.

Let us first recall the definition of the space $N_n^k$ $(0 \leq n < k \leq \infty)$. This is the subspace of $k$-dimensional Euclidean space $R^k$ consisting of points at most $n$ coordinates of which are rational. In what follows, we consider only the cases $0 < 2n+1 \leq k \leq \infty$. For example, $N_0^1$ is the space of irrational numbers and $N_\infty^\infty$ may be regarded as the separable Hilbert space $l_2 \approx R^\omega$ . The topological characterizations of Nöbelong spaces in these two cases have already been discussed (see Theorems 1.1.5 and 2.4.1). Moreover, these two are the only cases in which the corresponding universal Nöbeling spaces have the structure of an infinite product. The situation is quite different in all remaining cases $(1 \leq n < \infty)$, where until very recently nothing was known except the definition mentioned above. For instance, it can easily be shown (and we leave this to the reader as a manageable exercise) that the space $N_n^{2n+1}$ has all the properties listed in the following central problem.

CONJECTURE 5.0.5. *Characterize topologically the universal Nöbeling space $N_n^{2n+1}$. In particular, is a space $X$ homeomorphic to $N_n^{2n+1}$ provided that $X$ satisfies the following conditions:*

(a) *$X$ is a Polish space.*
(b) $\dim X = n$.
(c) $X \in AE(n) = LC^{n-1} \cap C^{n-1}$.
(d) *$X$ is strongly $\mathcal{A}_{\omega,n}$-universal, i.e. any map of any at most $n$-dimensional Polish space into $X$ can be arbitrarily closely approximated by closed embeddings.*

This problem is still open.

There are several constructions of strongly $\mathcal{A}_{\omega,n}$-universal spaces. We briefly recall some of these constructions. Each of these has its own advantages. For example, the first three constructions produce topologically homogeneous spaces, whereas the space obtained by the last construction admits a nice map onto the separable Hilbert space.

- *The universal n-dimensional pseudo-interior* $s_n^{2n+1}$ *of* $R^{2n+1}$ is the complement (in $R^{2n+1}$) of the so called $n$-dimensional universal pseudo-boundary $\sigma_n^{2n+1}$ of $R^{2n+1}$. The space $\sigma_n^{2n+1}$ is an absorber (even a skeletoid by [**120**]) for the class of all at most $n$-dimensional tame compacta in $R^{2n+1}$ (see [**158**]).
- *The polyhedral n-dimensional pseudo-interior of* $R^{2n+1}$ is the complement of the polyhedral $n$-dimensional pseudo-boundary of $R^{2n+1}$, which is an absorber with respect to the class of all compact $n$-dimensional, tame polyhedra of $R^{2n+1}$ (see [**158**]). The space $N_n^{2n+1}$ can easily be identified with the polyhedral $n$-dimensional pseudo-interior of $R^{2n+1}$ by noticing that $R^{2n+1} - N_n^{2n+1}$ forms an absorber with respect to the same class as the polyhedral pseudo-boundary. The last fact was first observed in [**295**] (see Theorem 5.5.2 below).
- The $n$-dimensional universal Menger compactum $\mu^n$ has an *n-dimensional pseudo-interior* $\nu^n$ as well as an *n-dimensional pseudo-boundary* $\Sigma^n$ which behave very much like the pseudo-interior and the pseudo-boundary of $I^\omega$ respectively. In particular, $\nu^n$ is strongly $\mathcal{A}_{\omega,n}$-universal (see [**90**]).
- A strongly $\mathcal{A}_{\omega,n}$-universal space is obtained from the geometric definition of the universal Menger compactum $\mu^n$. The space with required properties is defined as the complement of some $\sigma Z$-set in $\mu^n$ (see [**52**]).
- A strongly $\mathcal{A}_{\omega,n}$-universal space is constructed (see [**88**]) as the limit of an inverse sequence (see below). This construction allows us to represent the separable Hilbert space $l_2$ as an $n$-soft image of a strongly $\mathcal{A}_{\omega,n}$-universal space. Such spaces exist even in non-separable cases ([**108**]).

## 5.1. Strongly $\mathcal{A}_{\omega,n}$-universal spaces

In this Section we discuss properties of strongly $\mathcal{A}_{\omega,n}$-universal spaces. We also present the last construction of such spaces mentioned above.

PROPOSITION 5.1.1. *Let $X$ be a strongly $\mathcal{A}_{\omega,n}$-universal Polish $LC^n$-space. Then* $\dim X \geq n+1$.

PROOF. If $n = 0$, then the statement is trivial. Therefore we can assume that $n > 0$. Assume the contrary. Then there exists an $m$-dimensional strongly $\mathcal{A}_{\omega,n}$-universal Polish $LC^n$-space $X$ with $m \leq n$. In this situation, it easy to see that $X$ is an $ANR$-space. Let us show that $X$ is, in fact, an $R^\omega$-manifold and, consequently, is infinite dimensional. This will finish the proof. Since $X$

is an $ANR$-space, it is enough, by Theorem 2.4.1, to show that $X$ is strongly $\mathcal{A}_{\omega,\infty}$-universal, i.e. that $X$ has the discrete cells approximation property. Let $\mathcal{U}$ be an open cover of the space $X$, and let a map $f\colon N \times I^{\omega} \to X$ be given. Our goal is to find a map $g\colon N \times I^{\omega} \to X$, $\mathcal{U}$-close to $f$, such that the collection $\{g(\{k\} \times I^{\omega})\colon k \in N\}$ is discrete in $X$. Consider the projection $\pi\colon N \times X \to X$. Since the space $X$ is strongly $\mathcal{A}_{\omega,n}$-universal and $\dim(N \times X) \leq n$, there exists a closed embedding $\varphi\colon N \times X \to X$ which is $\mathcal{U}$-close to the projection $\pi$. Clearly the collection $\{\varphi(\{k\} \times X)\colon k \in N\}$ is discrete in $X$. It only remains to note that the required map $g$ can be defined by letting $g(k,a) = \varphi(k, f(k,a))$, $(k,a) \in N \times I^{\omega}$. □

The proofs of the following two statements are similar to the proofs of Propositions 2.2.3 and 2.2.6 respectively (hint: use $n$-homotopies instead of usual homotopies).

PROPOSITION 5.1.2. *Let $A$ be a closed subset of a Polish $ANE(n)$-space $X$. Then the following conditions are equivalent:*

(i) *$A$ is a $Z$-set in $X$.*
(ii) *For each at most $n$-dimensional locally compact polyhedron $P$, the set $\{f \in C(P,X)\colon f(P) \cap A = \emptyset\}$ is dense in the space $C(P,X)\}$.*
(iii) *For each at most $n$-dimensional Polish space $Y$, the set*

$$\{f \in C(Y,X)\colon f(Y) \cap A = \emptyset\}$$

*is dense in the space $C(Y,X)$.*

PROPOSITION 5.1.3. *Let $A$ be a closed subset of a Polish $ANE(n)$-space $X$. Then the following conditions are equivalent:*

(i) *$A$ is a strong $Z$-set in $X$.*
(ii) *The set $\{f \in C(N \times I^n, X)\colon cl_X f(N \times I^n) = \emptyset\}$ is dense in the space $C(N \times I^n, X)$.*
(iii) *For each at most $n$-dimensional locally compact polyhedron $P$, the set $\{f \in C(P,X)\colon cl_X f(P) \cap A = \emptyset\}$ is dense in the space $C(P,X)$.*
(iv) *For each at most $n$-dimensional Polish space $Y$, the set*

$$\{f \in C(Y,X)\colon cl_X f(Y) \cap A = \emptyset\}$$

*is dense in the space $C(Y,X)$.*

We also mention the following corollary of Proposition 4.1.7.

COROLLARY 5.1.4. *Let $X$ be an $ANE(n)$-space and $A$ be a closed subspace of an at most $n$-dimensional space $B$. Then the "restriction" map $C(B,X) \to C(A,X)$ is open.*

COROLLARY 5.1.5. *Every $Z$-set in an at most $n$-dimensional strongly $\mathcal{A}_{\omega,n}$-universal Polish $ANE(n)$-space is a strong $Z$-set.*

PROOF. Let $A$ be a $Z$-set in an at most $n$-dimensional strongly $\mathcal{A}_{\omega,n}$-universal Polish $ANE(n)$-space $X$. Then the space $C(N \times I^n, X)$ contains a dense (and obviously $G_\delta-$) subspace $F$, consisting of all closed embeddings. Since $A$ is a $Z$-set in $X$, by Corollary 5.1.4 and Proposition 5.1.2, the following sets are open and dense in $C(N \times I^n, X)$:

$$B_k = \{f : f(\{k\} \times I^n) \cap A = \emptyset\}, \quad k \in N.$$

By the Baire property in $C(N \times I^n, X)$ (see Proposition 2.1.7), the set

$$F \cap \bigcap_{k=1}^{\infty} B_k$$

is also dense in $C(N \times I^n, X)$. Proposition 5.1.3 finishes the proof. □

COROLLARY 5.1.6. *Each compact subset of an at most $n$-dimensional strongly $\mathcal{A}_{\omega,n}$-universal Polish $ANE(n)$-space is a strong $Z$-set.*

PROOF. Let $K$ be a compact subset of an at most $n$-dimensional strongly $\mathcal{A}_{\omega,n}$-universal Polish $ANE(n)$-space $X$. By Corollary 5.1.5, it suffices to show that $K$ is a $Z$-set in $X$. Let $\mathcal{U} \in cov(X)$ and $f : I^n \to X$ be a map. Denote by $\pi : N \times X \to X$ the natural projection. Strong $\mathcal{A}_{\omega,n}$-universality of $X$ guarantees the existence of a closed embedding $h : N \times X \to X$ which is $\mathcal{U}$-close to the composition $\pi f$. Obviously, the collection $\{h(\{k\} \times I^n) : k \in N\}$ is discrete in $X$. Therefore, by compactness of $K$, there exists an index $k_0 \in N$ such that $h(\{k_0\} \times I^n) \cap K = \emptyset$. Now note that the map $g : I^n \to X$ defined by letting $g(a) = h(k_0, a)$, $a \in I^n$, is $\mathcal{U}$-close to $f$ and $g(I^n) \cap K = \emptyset$. Proposition 5.1.2 finishes the proof. □

PROPOSITION 5.1.7. *Let $X$ be an at most $n$-dimensional strongly $\mathcal{A}_{\omega,n}$-universal Polish $ANE(n)$-space, and $A$ be a closed subspace of an at most $n$-dimensional Polish space $B$. Then each map $f : B \to X$, such that the restriction $f/A$ is a $Z$-embedding, can be arbitrarily closely approximated by $Z$-embeddings coinciding with $f$ on $A$. In particular, the set of all $Z$-embeddings of $B$ into $X$ is a dense $G_\delta$-subset of $C(B, X)$.*

PROOF. We consider only the second part of the statement, i.e. assume that $A = \emptyset$ (the general case follows from this one). Take a countable dense subset $\{\varphi\}$ of the space $C(I^n, X)$. By Corollary 5.1.6, the set

$$A_k = \{f : cl_X f(B) \cap \varphi_k(I^n) = \emptyset\}$$

is open and dense in the space $C(B, X)$ for each $k \in N$. By the strong $\mathcal{A}_{\omega,n}$-universality of $X$, the set $F$ of all closed embeddings of $B$ into $X$ is a dense $G_\delta$-subset of $C(B, X)$. Proposition 2.1.7 guarantees that the intersection $\cap_{k=1}^{\infty} A_k \cap F$, consisting obviously of $Z$-embeddings, is also dense and $G_\delta$-subset of $C(B, X)$. □

The following statement emphasizes the close tie between strongly $\mathcal{A}_{\omega,n}$-universal spaces and universal Menger compacta.

PROPOSITION 5.1.8. *Let $X$ be an $n$-dimensional strongly $\mathcal{A}_{\omega,n}$-universal Polish $AE(n)$-space, and $Y$ be a an $n$-dimensional $ANE(n)$-metrizable compactum containing $X$ in such a way that the complement $Y - X$ is a $Z_\sigma$-set in $Y$. Then $Y$ is homeomorphic to $\mu^n$.*

PROOF. If $n = 0$, the situation is trivial. Assume that $n \geq 1$. By Theorem 4.1.19, we need to show that $Y \in C^{n-1}$ and that $Y$ is strongly $\mathcal{B}_{n,\omega}$-universal.

Let $g\colon S^{k-1} \to Y$ be a map of the boundary of the $k$-dimensional disk $B^k$, $k \leq n$, into $Y$. Take an open cover $\mathcal{V} \in cov(Y)$ satisfying condition $(*)_n$ from Proposition 4.1.7 with respect to the one-element open cover $\{Y\} \in cov(Y)$. Since the complement $Y - X$ is a $Z_\sigma$-set in $Y$, there exists a $\mathcal{V}$-close to $g$ map $f\colon S^{k-1} \to X$. By our assumption, $X \in AE(n)$. Therefore, $f$ has an extension $F\colon B^k \to X$. By the choice of the cover $\mathcal{V}$, the map $g$ also has an extension $G\colon B^k \to Y$. This shows that $Y \in AE(n)$.

Consider now an arbitrary map $f\colon B \to Y$, where $B$ is an at most $n$-dimensional compactum. Consider also an open cover $\mathcal{U} \in cov(Y)$. Let $\mathcal{V}$ be a star-refinement of $\mathcal{U}$. Since $Y - X$ is a $Z_\sigma$-set in $Y$, there is a map $g\colon B \to X$ which is $\mathcal{V}$-close to $f$. The strong $\mathcal{A}_{\omega,n}$-universality of $X$ guarantees that there is an embedding $h\colon B \to X$, $\mathcal{V}$-close to $g$. It only remains to observe that the embedding $h$ is $\mathcal{U}$-close to $f$. □

**5.1.1. Universal maps.** In this Subsection we show that any Polish space is an $n$-soft image of a strongly $\mathcal{A}_{\omega,n}$-universal space. We begin with the following definition.

DEFINITION 5.1.9. *We say that a map $f\colon X \to Y$ between Polish spaces is* $n$-full, *$n = 0, 1, \ldots,$ if for each map $g\colon Z \to Y$ of any at most $n$-dimensional Polish space $Z$ into $Y$, there exists a closed embedding $h\colon Z \to X$ such that $fh = g$.*

For instance, the natural projection $\pi_1\colon R^\omega \times R^\omega \to R^\omega$ is $n$-full for each $n$. First we need the following statement.

THEOREM 5.1.10. *Each Polish space is an $n$-soft image of an at most $n$-dimensional Polish space.*

PROOF. Apply Theorem 4.3.5. □

PROPOSITION 5.1.11. *Suppose that all short projections of an inverse sequence $\mathcal{S} = \{X_k, p_k^{k+1}, \omega\}$, consisting of Polish spaces, are $n$-soft and $n$-full. Then the limit projection $p_0\colon X = \lim \mathcal{S} \to X_0$ is* strongly $\mathcal{A}_{\omega,n}$-universal *in the following sense:*

- *For each map $f: Y \to X$ of an at most $n$-dimensional Polish space $Y$, and for each open cover $\mathcal{U} \in cov(X)$, there is a closed embedding $g: Y \to X$ such that $p_0 g = p_0 f$ and $g$ is $\mathcal{U}$-close to $f$. In particular, $X$ is strongly $\mathcal{A}_{\omega,n}$-universal.*

PROOF. On the space $X_k$ consider a metric $d_k$ bounded by $2^{-k}$, and on the limit space $X$ of the spectrum $\mathcal{S}$ consider the metric

$$d(\{x_k\}, \{x'_k\}) = \max\{d_k(x_k, x'_k) : k \in \omega\}.$$

Let $Y$ be an at most $n$-dimensional Polish space. Suppose also that maps $f: Y \to X$ and $\alpha: X \to (0,1)$ are given. For our purposes, it suffices to construct a closed embedding $g: Y \to X$ such that $p_0 g = p_0 f$ and $d(f(y), g(y)) \le \alpha f(y)$ for each point $y \in Y$ (see Subsection 2.1.1).

Let $i_{k+1}: X_{k+1} \to R^\omega$ denote a closed embedding, $k \in \omega$. Since $R^\omega$ is a Polish absolute retract, there exists a continuous map

$$h: R^\omega \times R^\omega \times [0, \infty) \to R^\omega$$

such that

$$h(a, b, t) = a \quad \text{for each } t \le 1$$

and

$$h(a, b, t) = b \quad \text{for each } t \ge 2.$$

By Theorem 5.1.10, there is an $n$-soft map $q_{k+1}: Z_{k+1} \to X_k \times R^\omega$ of some at most $n$-dimensional Polish space $Z_{k+1}$. By the $n$-softness of the projection $p_k^{k+1}$ there exists a map $r_{k+1}: Z_{k+1} \to X_{k+1}$ such that

$$p_k^{k+1} r_{k+1} = \pi_k q_{k+1} \quad \text{and} \quad r_{k+1}/A_{k+1} = q_{k+1}/A_{k+1},$$

where $A_{k+1} = q_{k+1}^{-1}((p_k^{k+1} \triangle i_{k+1})(X_{k+1}))$ and $\pi_k: X_k \times R^\omega \to X_k$ is the projection onto the first coordinate.

We let $g_0 = p_o f$. By assumption, the projection $p_0^1$ of the spectrum $\mathcal{S}$ is $n$-full. Consequently, there exists a closed embedding $\varphi_1: Y \to X_1$ such that $p_0^1 \varphi_1 = g_0$. Consider the map

$$g_0 \triangle h(i_1 p_1 f \triangle i_1 \varphi_1 \triangle 2\alpha f): Y \to X_0 \times R^\omega .$$

The $n$-softness of the map $q_1$, coupled with the inequality $\dim Y \le n$, is enough to conclude that there exists a map $s_1: Y \to Z_1$ such that

$$q_1 s_1 = g_0 \triangle h(i_1 p_1 f \triangle i_1 \varphi_1 \triangle 2\alpha f).$$

Now we define the map $g_1: Y \to X_1$ as the composition $g_1 = r_1 s_1$. Straightforward verification shows that $p_0^1 g_1 = g_0$. Also, if $\alpha f(y) \le 2{-}1$, then, by the properties of the map $h$, we have $g_1(y) = p_1 f(y)$.

Let us also describe construction of the map $g_2\colon Y \to X_2$. Since the projection $p_1^2$ of the spectrum $\mathcal{S}$ is $n$-full, it follows that there exists a closed embedding $\varphi_2\colon Y \to X_2$ such that $p_1^2\varphi_2 = g_1$. Consider the map

$$g_1\triangle h(i_2p_2f\triangle i_2\varphi_2\triangle 2^2\alpha f)\colon Y \to X_1 \times R^\omega .$$

Since $\dim Y \leq n$, $n$-softness of the map $q_2$ implies the existence of a map $s_2\colon Y \to Z_2$ such that

$$q_2s_2 = g_1\triangle h(i_2p_2f\triangle i_2\varphi_2\triangle 2^2\alpha f).$$

We let $g_2 = r_2s_2$. As above, one can see, that $p_1^2g_2 = g_1$ and if $\alpha f(y) \leq 2^{-2}$, then $g_2(y) = p_2f(y)$. Additionally, if $\alpha f(y) \geq 2^{-1}$, then it follows from the construction that $g_2(y) = \varphi_2(y)$.

Let us assume now that the maps $g_i\colon Y \to X_i$ and closed embeddings $\varphi\colon Y \to X_i$ have already been constructed for each $i$, with $2 < i \leq k$, in such a way that the following conditions are satisfied:

$(a)_i$ $p_{i-1}^i g_i = g_{i-1}$.

$(b)_i$ $p_{i-1}^i \varphi_i = g_{i-1}$.

$(c)_i$ If $y \in Y$ and $\alpha f(y) \leq 2^{-i}$, then $g_i(y) = p_if(y)$.

$(d)_i$ If $y \in Y$ and $\alpha f(y) \geq 2^{-i+1}$, then $g_i(y) = \varphi_i(y)$.

We are now going to construct a map $g_{k+1}\colon Y \to Y_{k+1}$. Consider a closed embedding $\varphi_{k+1}\colon Y \to X_{k+1}$ such that $p_k^{k+1}\varphi_{k+1} = g_k$ (recall that the projection $p_k^{k+1}$ of the spectrum $\mathcal{S}$ is $n$-full). Since $\dim Y \leq n$, we have, by $n$-softness of the map $q_{k+1}$, a map $s_{k+1}\colon Y \to Z_{k+1}$ such that

$$q_{k+1}s_{k+1} = g_k\triangle h(i_{k+1}p_{k+1}f\triangle i_{k+1}\varphi_{k+1}\triangle 2^{k+1}\alpha f).$$

Then one can see that the maps $g_{k+1} = r_{k+1}s_{k+1}$ and $\varphi_{k+1}$ satisfy conditions $(a)_{k+1} - -(d)_{k+1}$.

It follows from conditions $(a)_k$ that the diagonal product

$$g = \triangle\{g_k\colon k \in \omega\}\colon Y \to X$$

satisfies the equality $p_kg = g_k$ for each $k \in \omega$. In particular, we have $p_0g = g_0 = p_0f$.

If $y \in Y$, then there exists $m \in \omega$ such that $2^{-(m+1)} \leq \alpha f(y) \leq 2^{-m}$. Therefore, by the conditions $(c)_k$ for $k \leq m$ we have $g_k(y) = p_kf(y)$. This implies that

$$d(f(y), g(y)) = \max\{d_k(p_kf(y), p_kg(y))\colon k = m+1, \dots\} \leq 2^{-(m+1)} \leq \alpha f(y).$$

Let us now show that $g$ is an injection. Let $y_1, y_2 \in Y$ and $y_1 \neq y_2$. If $m \in \omega$ is sufficiently large, then we have $\alpha f(y_i) \geq^{-m+1}$ for each $i = 1, 2$ and consequently, by the condition $(d)_m$,

$$p_mg(y_i) = g_m(y_i) = \varphi_m(y_i), \quad i = 1, 2.$$

Since $\varphi_m$ is an embedding, we conclude that $g(y_1) \neq g(y_2)$.

Next we show that $g$ is a closed embedding. Consider an arbitrary sequence $\{y_i\}$ of points in $Y$ such that the sequence $\{g(y_i)\}$ converges in $X$ to a point $x = \lim g(y_i)$. Obviously, it suffices to show that the sequence $\{y_i\}$ also converges in $Y$ to some point $y \in Y$ (indeed, this will imply, first of all, that the map $g\colon Y \to g(Y)$ is open and, secondly, closedness of the set $g(Y)$ in $X$).

First, assume that $0 < \epsilon = \inf\{\alpha f(y_i)\}$. Choose a number $m \in \omega$ so that $\epsilon \geq 2^{-m+1}$. Then, by condition $(d)_m$, we have $p_m g(y_i) = g_m(y_i) = \varphi_m(y_i)$ for each $i$. Continuity of the projection $p_m$ of the spectrum $\mathcal{S}$ guarantees that $p_m(x) = \lim \varphi_m(y_i)$. But $\varphi_m$ is a closed embedding. Therefore, the sequence $\{y_i\}$ converges in $Y$ to the point $y = \varphi_m^{-1}(p_m(x))$.

Thus it only remains to show that $\epsilon > 0$. Assume to the contrary that $\epsilon = 0$. Then there exists a subsequence $\{y_{i_k}\}$ of the sequence $\{y_i\}$ such that $\lim \alpha f(y_{i_k}) = x$. But this equality, coupled with the continuity of $\alpha$, leads us to the contradicting conclusion that $\alpha(x) = 0$. This completes the proof. □

THEOREM 5.1.12. *For each Polish space $X$ and each $n \in \omega$, there exists an $n$-dimensional Polish space $Y_X^n$ and a strongly $\mathcal{A}_{\omega,n}$-universal $n$-soft map $f_X^n\colon Y_X^n \to X$.*

PROOF. Let $X_0 = X$. Let $X_1$ be an $n$-dimensional Polish space which admits an $n$-soft map $h_1\colon X_1 \to X_0 \times R^\omega$ onto the product $X_0 \times R^\omega$ (the existence of $X_1$ and $h_1$ is guaranteed by Theorem 5.1.10). Let $p_0^1 = \pi_0 h_1$, where $\pi_0\colon X_0 \times R^\omega \to X_0$ denotes the natural projection onto the first coordinate. In general, if an $n$-dimensional Polish space $X_k$ has already been constructed as the space $X_{k+1}$ we take any $n$-dimensional Polish space which admits an $n$-soft map $h_{k+1}\colon X_{k+1} \to X_k \times R^\omega$. We also define the map $p_k^{k+1}$ as the composition $p_k^{k+1} = \pi_k h_{k+1}$, where $\pi_k\colon X_k \times R^\omega \to X_k$ denotes the projection onto the first coordinate. Therefore we have an inverse sequence $\mathcal{S} = \{X_k, p_k^{k+1}, \omega\}$ all of whose short projections are $n$-soft and $n$-full. Let $Y_X^n = \lim \mathcal{S}$ and $f_X^n = p_0$. Proposition 5.1.11 guarantees that the limit projection $f_X^n\colon Y_X^n \to X$ is strongly $\mathcal{A}_{\omega,n}$-universal. Since all spaces in our spectrum are $n$-dimensional and Polish, we easily conclude that $Y_X^n$ is an at most $n$-dimensional Polish space. Strong $\mathcal{A}_{\omega,n}$-universality of $f_X^n$ implies that $Y_X^n$ contains a topological copy of any $n$-dimensional Polish space. Therefore, $\dim Y_X^n = n$. This finishes the proof. □

*Historical and bibliographical notes* 5.1. The results of Subsection 5.1.1 are taken from [**88**]. The other results of this Section are well-known among experts.

## 5.2. Pseudo-interiors and pseudo-boundaries of Menger compacta

In this section, we prove the existence of $n$-dimensional pseudo-boundaries in the $k$-dimensional universal Menger compactum $\mu^k$ $(0 \leq n \leq k \leq \infty)$. Recall that $\mu^\infty$ denotes the Hilbert cube. Denote by $\mathcal{Z}(k,n)$ the collection of all $Z$-sets in $\mu^k$ of dimension at most $n$. It follows easily from the $Z$-set unknotting theorem (see 4.1.15 for $k < \infty$ and 2.2.30 for $k = \infty$) that the collection $\mathcal{Z}(k,n)$ is a perfect collection in the sense of Definition 2.2.21.

**5.2.1. Existence of $\mathcal{Z}(k,n)$-skeletoids.** For simplicity, the sup-metric generated by a metric $d$ is denoted by the same symbol $d$.

THEOREM 5.2.1. *Let $0 \leq k \leq \infty$. Then for each $n$ with $0 \leq n \leq k$, there exists a $\mathcal{Z}(k,n)$-skeletoid $\Sigma(k,n)$ in $\mu^k$.*

PROOF. Take a map $f: \mu^n \to \mu^k$, the existence of which is guaranteed by Theorem 4.3.5, and fix a metric $d$ on $\mu^k$. Inductively, we construct a sequence $\{g_j: \mu^n \to \mu^k\}$ of $Z$-embeddings, a sequence $\{A_j\}$ of $Z$-sets of $\mu^k$ and a sequence $\{i_j: A_j \to \mu^n\}$ of $Z$-embeddings as follows.

By Theorem 4.1.19, there exists a $Z$-embedding $g_1: \mu^n \to \mu^k$ such that $d(f,g_1) < 2^{-1}$. Let $A_1 = g_1(\mu^n)$. Since $f$ is $n$-invertible and $\dim A_1 = n$, there exists a map $i_1: A_1 \to \mu^n$ such that $fi_1 = id$. By the properties of the map $f$, we may assume that $i_1$ is a $Z$-embedding. Note that $f/i_1(A_1): i_1(A_1) \to \mu^k$ is a $Z$-embedding as well.

Suppose $m \geq 1$ and that for each $j \leq m$ we have already constructed $g_j: \mu^n \to \mu^k$, $A_j = g_j(\mu^n)$ and $i_j: A_j \to \mu^n$ so that the following conditions are satisfied:

($1_j$) $d(f,g_j) < 2^{-j}$.

($2_j$) $g_j/i_{j-1}(A_{j-1}) = f/i_{j-1}(A_{j-1})$.

($3_j$) $fi_j = id_{A_j}$ (and hence, $f/i_j(A_j)$ is a $Z$-embedding).

Let us construct $g_{m+1}$ and $i_{m+1}$. By ($3_m$) and by Theorems 4.1.19 and 2.3.18, there exists a $Z$-embedding $g_{m+1}: \mu^n \to \mu^k$ such that $d(f,g_{m+1}) < 2^{-(m+1)}$ and $g_{m+1}/i_m(A_m) = f/i_m(A_m)$. Let $A_{m+1} = g_{m+1}(\mu^n)$. Using the $n$-invertibility of $f$ and the properties of $f$ (see Theorem 4.3.5), we can take a $Z$-embedding $i_{m+1}: A_{m+1} \to \mu^n$ such that $fi_{m+1} = id_{A_{m+1}}$. This completes the inductive step.

Let us show that the collection $\{A_m\}$ is a $\mathcal{Z}(k,n)$-skeleton. Obviously, each $A_m$ is a $Z$-set and $A_m \subsetneq A_{m+1}$. By Theorem 2.2.22, it suffices to verify the following condition:

(∗) for each $\epsilon > 0$, for each $m \in N$, and for each at most $n$-dimensional $Z$-set $Z$ in $\mu^k$, there exists an integer $t > m$ and a $Z$-embedding $e: Z \to A_t$ such that $e/(Z \cap A_m) = id$ and $d(e, id_Z) < \epsilon$.

Let $t > m$ be so large that $3{\cdot}2^{-(t+1)} < \epsilon$. At the $t$-stage of the above construction, we have a $Z$-embedding $i_t: A_t \to \mu^n$ such that $fi_t = id_{A_t}$. Since $f$ is a $UV^{n-1}$-map between $LC^{n-1}$ compacta and $\dim Z \leq n$, there is a map $r: Z \to \mu^n$ such that $r/(Z \cap A_t) = i_t/(Z \cap A_t)$ and $d(fr, id_Z) < 2^{-(t+1)}$. Since $i_t$ is a $Z$-embedding, there exists a $Z$-embedding $s: Z \to \mu^n$ such that $s/(Z \cap A_t) = i_t/(Z \cap A_t)$ and $d(fs, fr) < 2^{-(t+1)}$. Finally, let $i = g_{t+1}s: Z \to A_{t+1}$. Simple verifications show that $d(i, id_Z) < \epsilon$ and $i/(Z \cap A_t) = id$. In particular, $i/(Z \cap A_m) = id$.

Thus $\{A_m : m \in N\}$ is a $\mathcal{Z}(k,n)$-skeleton in $\mu^k$. The corresponding $\mathcal{Z}(k,n)$-skeletoid is the union $\Sigma(k,n) = \cup_{m=1}^{\infty} A_m$. □

*Remark* 5.2.2. It follows easily from the above construction that if $A$ is a given $\sigma Z$-set in $\mu^k$, then there is a $\mathcal{Z}(k,n)$-skeletoid which does not intersect $A$.

In what follows, we consider only the case $k = n$. In order to simplify notations, we write $\mathcal{Z} = \mathcal{Z}(n,n)$.

The spaces $\Sigma^n = \Sigma(n,n)$ and $\nu^n = \mu^n - \Sigma^n$ are called the *categorical pseudo-boundary* and the *categorical pseudo-interior* of $\mu^n$ respectively. In the remainder of this paper, $\mathcal{Z}$-skeletoids and their complements in $\mu^n$ are referred as *pseudo-boundaries* and *pseudo-interiors* of $\mu^n$ respectively.

Applying the general theory of skeletoids (see Subsection 2.2.2) to our situation, we easily obtain the following statements (recall that $\mathcal{Z}$ is a perfect collection in $\mu^n$).

PROPOSITION 5.2.3. *If $A$ and $B$ are pseudo-boundaries in $\mu^n$, then the set $\{h \in Auth(\mu^n) : h(A) = B\}$ is dense in $Auth(\mu^n)$.*

PROPOSITION 5.2.4. *Let $A$ be a pseudo-boundary of $\mu^n$ and $B$ be a $\sigma Z$-set in $\mu^n$. Then $A \cup B$ is also a pseudo-boundary in $\mu^n$.*

COROLLARY 5.2.5. *If $A$ is a pseudo-boundary in $\mu^n$ which is contained in a $\sigma Z$-set $B$ of $\mu^n$, then the set $\{h \in Auth(\mu^n) : h(A) = B\}$ is dense in $Auth(\mu^n)$.*

**5.2.2. General properties of pseudo-interiors.** Let us establish some important properties of $\nu^n$ and $\Sigma^n$.

PROPOSITION 5.2.6. $\nu^n \in LC^{n-1} \cap C^{n-1}$.

PROOF. Since $\mu^n \in LC^{n-1} \cap C^{n-1}$ and the complement of $\nu^n$ $(= \Sigma^n)$ is a $\sigma Z$-set, the conclusion follows easily. □

PROPOSITION 5.2.7. *Let $P$ be an at most $n$-dimensional Polish space. Then the set of all $Z$-embeddings of $P$ into $\nu^n$ is a dense $G_\delta$-subset of $C(P, \nu^n)$.*

PROOF. Clearly, the set of all $Z$-embeddings of $P$ into $\nu^n$ is a $G_\delta$-set in $C(P,\nu^n)$. Therefore, it only remains to show that this set is dense in $C(P,\nu^n)$. Let $f : P \to \nu^n$ be a map and $\mathcal{U}$ be an open cover of $\nu^n$. Take a collection $\mathcal{V}$ of open sets of $\mu^n$ such that $\mathcal{V}/\nu^n = \mathcal{U}$ and let $M = \cup\mathcal{V}$. By Proposition 1.3.11, $f$ has an extension $F : \tilde{P} \to \mu^n$, where $\tilde{P}$ is a metrizable compactification of $P$ such that $\dim\ \tilde{P} = \dim\ P \leq n$. Let $Q = f^{-1}(M)$ and $g = F/Q$. Clearly, $g$ is a proper map defined on a locally compact metrizable space $Q$. Since $M$ is a $\mu^n$-manifold and $\dim\ Q \leq n$, we can find a $Z$-embedding $g_1 : Q \to M$ which is $\mathcal{V}$-close to $g$. Since $M \cap \Sigma^n$ is a $\sigma Z$-set in $M$, we may assume that $g_1(Q)$ is contained in $\nu^n$. Note also that $g_1(Q) - g_1(P) = g_1(Q - P)$ is $\sigma$-compact (recall that $P$, being a Polish space, is $G_\delta$ in $Q$). By Proposition 5.2.4, $\Sigma^n \cup g_1(Q - P)$

is a $Z$-skeletoid in $\mu^n$. By Theorem 2.2.23, $(M \cap \Sigma^n) \cup g_1(Q - P)$ is a $\mathcal{Z}(M)$-skeletoid in $M$. For the same reason, $M \cap \Sigma^n$ is also a $\mathcal{Z}(M)$-skeletoid. By Theorem 2.2.20, there is an $h \in Auth(M)$ which is $\mathcal{V}$-close to $id_M$ and is such that $h((M \cap \Sigma^n) \cup g_1(Q - P)) = M \cap \Sigma^n$. Then the embedding $g_2 = hg_1/P \colon P \to \nu^n$ is $st(\mathcal{V})$-close to $f$, and obviously $g_2(P)$ is a $Z$-set in $\nu^n$. □

COROLLARY 5.2.8. *The space $\nu^n$ has the Discrete n-cell Approximation Property, i.e. each map $f \colon \oplus_j I_j^n \to \nu^n$, where each $I_j^n$ is an n-cell, can be arbitrarily closely approximated by a map $g \colon \oplus_j I_j^n \to \nu^n$ such that the collection $\{g(I_j^n)\}_{j=1}^{\infty}$ is discrete.*

COROLLARY 5.2.9. $\dim \nu^n = n$.

PROPOSITION 5.2.10. *Let $A$ be a $\sigma Z$-set in $\nu^n$. Then the inclusion map $i \colon \nu^n - A \to \nu^n$ is a near homeomorphism.*

PROOF. Let $\mathcal{U}$ be an open cover of $\nu^n$ and $\mathcal{V}$ be a collection of open subsets of $\mu^n$ such that $\mathcal{V}/\nu^n = \mathcal{U}$. Let $M = \cup\mathcal{V}$. As in the proof of Proposition 5.2.7, we can conclude that $A \cup (\Sigma^n \cap M)$ is a $\mathcal{Z}(M)$-skeletoid. Obviously, $\Sigma^n \cap M$ is also a $\mathcal{Z}(M)$-skeletoid. Therefore, by Theorem 2.2.20, there exists a $h \in Auth(M)$ which is $\mathcal{V}$-close to $id_M$ and such that $h(A \cup (\Sigma^n \cap M)) = \Sigma^n \cap M$. Clearly,

$$h(\nu^n - A) = h(M - (A \cup (\Sigma^n \cap M)) = M - h(A \cup (\Sigma^n \cap M)) = M - (\Sigma^n \cap M) = \nu^n.$$

It only remains to note that $h/(\nu^n - A)$ is $\mathcal{U}$-close to $i$. □

LEMMA 5.2.11. *Each compact subset of $\nu^n$ is a strong $Z$-set.*

PROOF. Apply Corollaries 5.1.6, 5.2.9 and Propositions 5.2.6, 5.2.7. □

PROPOSITION 5.2.12. *Each homeomorphism between compact subsets of $\nu^n$ can be extended to an autohomeomorphism of $\nu^n$.*

PROOF. Each compact subset of $\nu^n$ is a $Z$-set in $\mu^n$. Consequently, the conclusion follows from Theorems 4.1.15, 2.2.20 and Proposition 2.2.24. □

COROLLARY 5.2.13. *The space $\nu^n$ is topologically homogeneous.*

The following statement will be used in the sequel.

PROPOSITION 5.2.14. $\Sigma^n \in LC^{n-1} \cap C^{n-1}$.

PROOF. We prove that for each $\epsilon > 0$, there is a $\delta > 0$ such that

(*) for each map $\alpha \colon S^p \to \Sigma^n$, $0 \le p \le n-1$, with $diam(im(\alpha)) < \delta$, there is an extension $\alpha^* \colon B^{p+1} \to \Sigma^n$ such that $diam(im(\alpha^*)) < \epsilon$.

Let $\{A_i\}$ be a skeleton such that $\cup_{i=1}^{\infty} A_i = \Sigma^n$. First we show that

(1) for each $\epsilon > 0$, there is a $\delta_1 = \delta_1(\epsilon) > 0$ such that (*) is true for each map $\alpha \colon S^p \to \Sigma^n$ such that $diam(im(\alpha)) < \delta_1$ and $im(\alpha)$ is a $Z$-set which is contained in some $A_i$.

*Proof of (1).* For a given $\epsilon > 0$, let $\delta_1 > 0$ be a number such that for each map $\alpha\colon S^p \to \mu^n$, $0 \le p \le n-1$, with $diam(im(\alpha)) < \delta_1$, there is an extension $\alpha^*\colon B^{p+1} \to \mu^n$ such that $diam(im(\alpha^*)) < \frac{\epsilon}{4}$. To prove that this $\delta_1$ is the desired number, take any map $\alpha\colon S^p \to A_i$ as above. There is an extension $\alpha'\colon B^{p+1} \to \mu^n$ such that $diam(im(\alpha')) < \frac{\epsilon}{4}$. By Theorem 4.1.19, we may assume that $im(\alpha')$ is a $Z$-set. Applying Theorem 2.2.22 to $im(\alpha')$, $i$ and $\frac{\epsilon}{4}$, we get an embedding $f\colon im(\alpha') \to A_j$ into some $A_j$ which is $\frac{\epsilon}{4}$-close to $id$ and such that $f/(im(\alpha') \cap A_i) = id$. Then $f\alpha'\colon B^{p+1} \to A_j$ is an extension of $\alpha$ such that $diam(im(f\alpha')) < \epsilon$. This completes the proof of (1).

Next we prove that

(2) for each $\epsilon > 0$, there is a $\delta_2 = \delta_2(\epsilon) > 0$ such that $(*)$ is true for each map $\alpha\colon S^p \to \Sigma^n$, $0 \le p \le n-1$, such that $diam(im(\alpha)) < \delta_2$ and $im(\alpha)$ is a $Z$-set.

*Proof of (2).* For a given $\epsilon > 0$, we show that $\delta_2 = \frac{\delta_1(\frac{\epsilon}{4})}{2}$ is the required number. Take a sequence $\{\alpha_m\colon S^p \to A_{i_m}\}$ such that $\lim_{m\to\infty} \alpha_m = \alpha$ and $im(\alpha_m$ is a $Z$-set for each $m$ (apply Theorem 2.2.22). Noticing that $\delta_1$ in (1) does not depend on $A_i$, we can take an $\epsilon_m$-homotopy $H_m\colon S^p \times [0,1] \to \Sigma^n$ such that $H_m(x,0) = \alpha_m(x)$ and $H_m(x,1) = \alpha_{m+1}(x)$, where $\lim_{m\to\infty}\epsilon_m = 0$. Let $H\colon S^p \times [0,1] \to \Sigma^n$ be a map defined by

$$H(x,t) = \begin{cases} H_m(x, m(m+1)\left(t - 1 + \frac{1}{m}\right)), & \text{if } t \in \left[1 - \frac{1}{m}, 1 - \frac{1}{m+1}\right] \\ \alpha(x), & \text{if } t = 1. \end{cases}$$

Then $H$ is continuous and $H/S^p \times [1 - \frac{1}{m}, 1]$ gives a homotopy between $\alpha_m$ and $\alpha$. Since, for sufficiently large $m$, $\alpha_m$ is null homotopic in $\Sigma^n$ by an $\frac{\epsilon}{4}$-homotopy (see (1)), we conclude that $\alpha$ is null homotopic in $\Sigma^n$ by an $\epsilon$-homotopy.

A similar argument allows us to obtain the final conclusion. □

*Remark* 5.2.15. For a similar situation see [**51**, Theorem 3.6]

PROPOSITION 5.2.16. *Let $\{A_i\}$ be an increasing sequence of compacta in $\mu^n$ satisfying the following properties:*

(i) *each $A_i$ is homeomorphic to $\mu^n$ and is a $Z$-set both in $A_{i+1}$ and in $\mu^n$.*

(ii) *$\{A_i\}$ is an equi $-$ $LC^{n-1}$-family.*

(iii) *for each $\epsilon > 0$, there is an integer $m > 0$ such that for each $i > m$, $A_i$ is $\epsilon$-dense.*

*Then $\{A_i\}$ is a pseudo-boundary.*

PROOF. By Theorem 2.2.22, it suffices to check the following absorption property:

- for each $Z$-set $Z$ in $\mu^n$, each $\epsilon > 0$ and each $k$, there exist an integer $l > k$ and an embedding $h\colon Z \to A_l$ such that $d(h, id) < \epsilon$ and $h/(A_k \cap Z) = id$.

Since the collection $\{A_i\}$ forms an $equi - LC^{n-1}$-family and $\dim Z \le n$, using the proof of [**125**, Lemma 2.1] and the nerve replacement trick, we can find a map

$f\colon Z \to A_l$ which is $\frac{\epsilon}{2}$-close to $id$ such that $f/A_k = id$. By the $DD^nP$, we can approximate $f$ by an embedding $h$ which is $\frac{\epsilon}{2}$-close to $f$ so that $h/A_k = id$. □

*Historical and bibliographical notes* 5.2. The categorical pseudo-boundary $\Sigma^n$ and the categorical pseudo-interior $\nu^n$ of the Menger compactum $\mu^n$ were constructed by the author in [**90**]. Theorem 5.2.1 appears in [**106**] and generalizes the original construction of the above spaces. $n$-dimensional skeletoids in the Hilbert cube were constructed earlier in [**119**].

## 5.3. Geometric pseudo-boundaries

In this section, using Menger's original construction, we define the geometric pseudo-boundary of the $n$-dimensional Menger compactum $M_n^p$ ($p \geq 2n+1$). Also, "digging out larger holes," we give another description of the categorical pseudo-boundary. This special construction will be important in the next section.

As a metric on $R^p$, we use the maximum metric, i.e.

$$d((x_i),(y_i)) = \max\{|\, x_i - y_i \,| : i = 1,\ldots,p\}.$$

The closed $\epsilon$-neighborhood of a subset $A$ of $R^p$ is denoted by $N(A,\epsilon)$. For a manifold $M$, $\partial M$ denotes its manifold boundary.

For the reader's convenience we recall Menger's construction of $M_n^p$ from Chapter 4.

For each integer $i \geq 0$, $\mathcal{L}_i(p)$ denotes the cell complex structure of $I^p$ whose $p$-cells are of the form

$$\prod_{t=1}^{p}\left[\frac{l_t}{3^i}, \frac{l_t+1}{3^i}\right], \text{where } l_t = 0,\ldots,3^i-1.$$

For a subcomplex $K$ of $\mathcal{L}_i(p)$, $st(K,\mathcal{L}_i(p))$ denotes the star of $K$ in $\mathcal{L}_i(p)$. The subdivision $sdK$ of $K$ by $\mathcal{L}_{i+1}(p)$ means the cell complex structure of $K$ consisting of all cells of $\mathcal{L}_{i+1}(p)$ contained in $K$. The $k$-skeleton of $\mathcal{L}_i(p)$, that is, the union of all $k$-dimensional cells of $\mathcal{L}_i(p)$, is denoted by $\mathcal{L}_i^{(k)}(p)$.

Under the above notation, we define the $n$-dimensional Menger compactum $M_n^p$ ($p \geq 2n+1$) as follows. Fix $p$ and $n$ so that $p \geq 2n+1$. Let $M_0(p) = I^p$ and for each integer $i \geq 0$, let

$$M_{i+1}(p) = st(sd\mathcal{L}_i^{(n)}(p)/M_i(p), \mathcal{L}_{i+1}(p)).$$

In this way we obtain a decreasing sequence of compacta, and

$$M_n^p = \bigcap_{i=0}^{\infty} M_i(p).$$

By Theorem 4.1.19, $M_n^p$ is homeomorphic to $\mu^n$. The indices $n$ and $p$ will be omitted if there is no confusion. We will also use another description of $M_n^p$. Let

$$V_i = \left\{ \frac{2t+1}{2 \cdot 3^{i-1}} : t = 0, 1, \ldots, 3^{i-1} \right\} \text{ and } V = \bigcup_{i=1}^{\infty} V_i.$$

Also

$$B_i = \overbrace{V_i \times V_i \times \cdots \times V_i}^{p-factors}$$

is the set of centers of $p$-cells of $\mathcal{L}_i$. Let $\mathcal{P}$ be the finite collection of homeomorphisms of $R^p$ defined by permutations of coordinates. For each $i$, we define

$$D_i = \bigcup \{ \alpha(\{c\} \times I^{p-n-1}) \colon \alpha \in \mathcal{P} \text{ and } c \in V_i^{n+1} \}$$

and

$$N_i = N \left( D_i, \frac{1}{2 \cdot 3^i} \right).$$

Then $D_i$ can be regarded as the "dual $(p-n-1)$-skeleton" of $\mathcal{L}_i$ and $N_i$ as the regular neighborhood of $D_i$. It is easy to see that

$M_n^p = I^p - \cup_{i=1}^{\infty} int(N_i)$, $I^p = M_{i+1}(p) \cup N_i$ and $\partial M_i = M_{i+1}(p) \cap (N_i \cup \partial I^p)$.

**5.3.1. Geometric pseudo-boundary and pseudo-interior.** Here we consider the Menger compactum $M_n^p$, where $p \geq 2n+1$, and define the geometric pseudo-boundary $B(M_n^p)$ and the geometric pseudo-interior $I(M_n^p)$. Recall that $M_n^p$ is homeomorphic to $\mu^n$. The *geometric pseudo-boundary* is defined by

$$B(M_n^p) = M_n^p \cap \bigcup_{i=1}^{\infty} \partial M_i$$

and the *geometric pseudo-interior* is defined by

$$I(M_n^p) = M_n^p - B(M_n^p).$$

It is convenient to consider the subset $F(M_n^p)$ of $B(M_n^p)$ defined by

$$F(M_n^p) = B(M_n^p) \cap int(I^p).$$

We define a sequence of $p$-cells $I_1 \subseteq I_2 \subseteq I_3 \subseteq \ldots$ of $I^p$ of the form $I_t = [a_t, b_t]^p$, where $0 < a_t < b_t < 1$, $\lim_{t\to\infty} a_t = 0$ and $\lim_{t\to\infty} b_t = 1$. In particular, $\lim I_t = I^p$. Let $F_i = I_i \cap \partial M_i \cap M_n^p$. Then $\{F_i\}$ is an increasing sequence of compacta such that $\cup_{i=1}^{\infty} F_i = F(M_n^p)$. Further, let $\partial M_n^p = B(M_n^p) \cap \partial I^p$. Then $\partial M_n^p$ is a $Z$-set and $F(M_n^p) \cup \partial M_n^p = B(M_n^p)$.

THEOREM 5.3.1. *$B(M_n^p)$ and $F(M_n^p)$ are $\sigma$-compact $LC^{n-1}$- and $C^{n-1}$-spaces.*

PROOF. First we prove that

(1) *each* $\partial M_i$ *is* $C^{n-1}$.

Let $A = cl(\partial M_i \cap int(M_{i-1}))$ and $B = \partial M_i \cap \partial M_{i-1}$. Then $\partial M_i = A \cup B$. Also notice the following:

$$\begin{aligned} A &= (\text{the regular nbhd of the } n\text{-skeleton of } \mathcal{L}_{i-1}/M_{i-1}) \cap \\ &\quad (\text{the regular nbhd of the dual } (p-n-1)\text{-skeleton of } \mathcal{L}_{i-1}), \end{aligned}$$

$$\begin{aligned} B = \partial M_{i-1} &- (\text{the regular nbhd of the dual } (p-n-2)\text{-skeleton of} \\ &\quad \mathcal{L}_{i-1}/\partial M_{i-1}) \end{aligned}$$

and

$$\begin{aligned} A \cap B &= (\text{the regular nbhd of the dual } (p-n-2)\text{-skeleton of } \mathcal{L}_{i-1}/\partial M_{i-1}) \\ &\quad \cap (\text{the regular nbhd of the } n\text{-skeleton of } \mathcal{L}_{i-1}/\partial M_{i-1}). \end{aligned}$$

By Van-Kampen's Theorem, the Mayer-Vietoris sequence for singular homologies, and the Hurewicz Theorem (see, for example, [**287**] or [**291**]) it suffices to prove that

(2) $A$ *and* $B$ *are* $(n-1)$*-connected and* $A \cap B$ *is* $(n-2)$*-connected.*

Take a map $\alpha\colon S^q \to A$, where $0 \le q \le n-1$, and extend it to a map $\alpha^*\colon B^{q+1} \to I^p$. Since $p \ge 2n+1$, by general position, we may assume that $im(\alpha^*)$ does not intersect either the $n$-skeleton or the dual $(p-n-1)$-skeleton of $\mathcal{L}_{i-1}$. Then $im(\alpha^*)$ can be deformed into $A$, the intersection of their regular neighborhoods, keeping $im(\alpha)$ fixed. Hence, $A \in C^{n-1}$. Similar arguments can be applied to see that $B \in C^{n-1}$ and $A \cap B \in C^{n-2}$. This proves the assertion (2) and, consequently, assertion (1).

Since $mesh\ \mathcal{L}_j \to 0$ as $j \to \infty$, we can easily see that $\{\partial M_i\}$ forms an $equi-LC^{n-1} \cap C^{n-1}$-family. By the construction, $\{\partial M_i \cap M_n^p\}$ is also an $equi-LC^{n-1}$-family. As in Proposition 5.2.14, one can show that $B(M_n^p) \in LC^{n-1} \cap C^{n-1}$.

Since $F(M_n^p) = B(M_n^p) - \partial M$ and $\partial M$ is a $Z$-set, we conclude that $F(M_n^p) \in LC^{n-1} \cap C^{n-1}$. □

*Remark* 5.3.2. If $p \ge 2n+2$, then it is easy to see that $\partial M_i \cap M$ is homeomorphic to $\mu^n$. If $p = 2n+1$, then $\partial M_i \cap M$ is essentially a finite union of Menger compacta of "type $M_n^{2n}$".

THEOREM 5.3.3. *If* $p \ge 2n+2$, *then* $B(M_n^p)$ *is a* $Z$*-skeletoid. Therefore,* $\nu^n$ *is homeomorphic to* $I(M_n^p)$.

PROOF. From the definition, $B(M_n^p) = \cup_i(\partial M_i \cap M_n^p)$ and each $\partial M_i \cap M_n^p$ is a $Z$-set in both $M_n^p$ and $\partial M_{i-1} \cap M_n^p$. Since $p \ge 2n+2$, $\partial M_i \cap M_n^p$ is homeomorphic to $\mu^n$ (see the above Remark). As in the proof of Theorem 5.3.1, the collection

$\{\partial M_i \cap M_n^p\}$ is an $equi-LC^{n-1}$-family. Then Proposition 5.2.16 gives the desired conclusion. □

**5.3.2. Geometric description of $\nu^n$.** In this subsection, we fix $p$ and $n$ so that $p = 2n + 1$. For an index $i \geq 0$, take an $\epsilon_i > 0$ satisfying the following conditions:

(a) For each $(u, v) \in V_i^{n+1} \times V_j^{n+1}$ and for each pair $(i, j)$ of indices, either $N(u, \epsilon_i) \cap N(v, \epsilon_j) = \emptyset$, or one of these neighborhoods contains the other in its interior.

(b) Each pair of members of

$$\mathcal{K}_i = \{N(u, \epsilon_i + \frac{1}{2 \cdot 3^{i+1}}) \colon u \in V_i^{n+1}\} \cup \{N(\partial I^{n+1}, \epsilon_0)\}$$

is mutually disjoint, or one of these members contains the other in its interior.

(c) $\epsilon_i < \frac{1}{3^{i+2}}$.

Let $\mathcal{E} = \{\epsilon_i\}$ and let $mesh\ \mathcal{E} = \sup_i\{\epsilon_i\}$.

Let $N_1^* = int\ N(N_1, \epsilon_1)$ and $\mathcal{N}_1 = \mathcal{K}_1$. Inductively, suppose that an open subset $N_i^*$ of $I^{2n+1}$ and a subcollection $\mathcal{N}_i$ of $\mathcal{K}_i$ have already been defined. Let $\mathcal{N}_{i+1}$ be the maximal subcollection of $\mathcal{K}_{i+1}$ which is disjoint from

$$cl(\bigcup_{j=1}^{i}(\cup\mathcal{N}_j)) \cup N(\partial I^{n+1}, \epsilon_0).$$

Then

$$N_{i+1}^* = \bigcup\{\alpha(int\ N(v, \epsilon_{i+1} + \frac{1}{2 \cdot 3^{i+2}}) \times I^n) \colon N(v, \epsilon_{i+1} + \frac{1}{2 \cdot 3^{i+2}}) \in \mathcal{N}_{i+1}, \alpha \in \mathcal{P}\}.$$

We define

$$M(\mathcal{E}) = I^{2n+1} - (int\ N(\partial I^{2n+1}, \epsilon_0) \cup \bigcup_{i=1}^{\infty} N_i^*).$$

Clearly $M(\mathcal{E})$ is homeomorphic to $M = M_n^{2n+1}$.

The next Lemma is crucial for the construction.

LEMMA 5.3.4. (i) *$M(\mathcal{E})$ is a $Z$-set in $M$.*

(ii) *For each $i \geq 1$ and for each $(2n+1)$-cell $e$ of $\mathcal{L}_i$, $e \cap M(\mathcal{E})$ is homeomorphic to $M$ if $e \cap M(\mathcal{E})$ is not empty and $Fr(e \cap M(\mathcal{E}))$ is a $Z$-set in $M(E) \cap e$.*

PROOF. (1) Since each $\partial M_j$ is an $(n-1)$-connected $ANR$ and dim $M = n$, we can find a map $r_j \colon M \to \partial M_j$ such that $d(r_j, id) \to 0$ as $j \to \infty$. Noticing that $\partial M_j \cap M(\mathcal{E}) = \emptyset$, it is easy to see that the conclusion of the lemma holds.

(2) This is clear from the construction. □

For each $i > 0$, take a sequence $\mathcal{E}_i$ satisfying condition (a)-(c) so that $mesh\ \mathcal{E}_i \to 0$ as $i \to \infty$. By Lemma 5.3.4, we have a tower $M(\mathcal{E}_1) \subseteq M(\mathcal{E}_2) \subseteq M(\mathcal{E}_3) \subseteq \dots$ of $Z$-sets.

THEOREM 5.3.5. *The sequence* $\{M(\mathcal{E}_i)\}_{i\geq 1}$ *is a* $Z$*-skeleton in* $M$.

PROOF. From the proof of Lemma 5.3.4, it is clear that $M(\mathcal{E}_{i+1})$ is a $Z$-set both in $M$ and $M(\mathcal{E}_i)$. It is easy to see that $\{M(\mathcal{E}_i)\}$ forms an $equi-LC^{n-1}$-family. Applying Proposition 5.2.16, we obtain the desired conclusion. □

COROLLARY 5.3.6. $(\mu^n, \Sigma^n)$ *is homeomorphic to* $(M, \cup_{i=1}^{\infty} M(\mathcal{E}_i))$. *In particular,* $\nu^n$ *is homeomorphic to* $M - \cup_{i=1}^{\infty} M(\mathcal{E}_i)$.

*Remark* 5.3.7. We used the Menger construction for ease of description, but we can apply essentially the same arguments as above to the Lefshetz construction. However, these arguments do not apply to the Bestvina construction, because the corresponding geometric pseudo-boundary in his construction yields a countable union of $(n-1)$-dimensional Menger compacta $\mu^{n-1}$.

*Historical and bibliographical notes* 5.3. The results of this Section originally appear in [**106**].

## 5.4. Equivalence of categorical and geometric pseudo-interiors

This section is devoted to the proof of the following theorem.

THEOREM 5.4.1. $I(M_n^p) = M - B(M_n^p)$, $p \geq 2n+1$, *is homeomorphic to* $\nu^n$.

Recall that $B(M_n^p) = F(M_n^p) \cup \partial M_n^p$ and $\partial M_n^p$ is a $Z$-set. By Proposition 5.2.4, we may assume that $\nu^n \supseteq \partial M_n^p$, and $\nu^n - \partial M_n^p$ is homeomorphic to $\nu^n$ by Proposition 5.2.10. Thus, Theorem 5.4.1 is a direct consequence of the next result. In what follows, we assume that $p = 2n+1$ and omit the indices for simplicity.

THEOREM 5.4.2. *Let* $M = M_n^{2n+1}$. *There is a continuous surjection* $f\colon M \to M$ *satisfying the following properties:*

(i) $f^{-1}(F(M)) = \Sigma^n$.
(ii) $f/\nu^n\colon M - \Sigma^n \to M - F(M)$ *is a homeomorphism.*
(iii) $f(\partial M) = \partial M$.

Recall the notation of Section 5.3. In particular, $M = \cap_{i=1}^{\infty} M_i$, where $M_{i+1} = st(sdL_i^{(n)}/M_i, \mathcal{L}_{i+1})$ for the standard cell complex structure $\mathcal{L}_i$ of $I^{2n+1}$. Also, $M$ can be represented as $M = I^{2n+1} - \cup_{i=1}^{\infty} int(N_i)$, where $N_i$ is the regular neighborhood of the "dual $n$-skeleton" of $\mathcal{L}_i$. Also, we have that $I^{2n+1} = M_i \cup N_i$ and $\partial M_i = M_i \cap (N_i \cup \partial I^{2n+1})$. The set of all $(2n+1)$-cells of $\mathcal{L}_i/M_i$ naturally determines a partition of $M_i$, denoted by $\mathcal{U}_i$. The restriction of $\mathcal{U}_i$

to $M$ determines the partition $\mathcal{U}_i^M$ of $M$ consisting of homeomorphic copies of $M$. $F(M)$ has a tower structure, $F(M) = \cup_{i=1}^{\infty} F_i$. If $\partial M = M \cap \partial I^{2n+1}$, then $F(M) = B(M) - \partial M$.

It is convenient to use the $Z$-skeletoid constructed in Subsection 5.3.2. Fix a sequence $(\mathcal{E}_i)$ in the construction and let $X_i = M(\mathcal{E}_i)$. The following property of $X_i$, stated in Lemma 5.3.4, will be important in the sequel.

(∗) For each $U \in \mathcal{U}_j$ with $int(U) \cap X_i \neq \emptyset$, we have that $U \cap X_i$ is homeomorphic to $M$ and $Fr(U) \cap X_i$ is a $Z$-set in $U \cap X_i$.

The required map $f: M \to M$ is constructed as the limit of a sequence of surjections. More precisely, we construct a sequence of continuous surjections $\{f_i: M \to M\}$ and a sequence $\{Z_i\}$ of compacta satisfying the following properties:

(a) $f_i$ is a $UV^{n-1}$-map such that $f_i^{-1}(Z_i) = X_i$, $f_i(\partial M) = \partial M$ and $f_i/\dots : M - X_i \to M - Z_i$ is a homeomorphism.
(b) $d(f_i, f_{i+1}) < 2^{-(i+1)}$.
(c) $f_{i+1}/X_i = f_i/X_i$
(d) $F_i \subseteq Z_i \subseteq \cup_{j=1}^{\infty} F_j$ and, hence, $\cup_{i=1}^{\infty} Z_i = \cup_{i=1}^{\infty} F_i = F(M)$.

Condition (b) guarantees that $f = \lim_{i\to\infty} f_i$ exists. To ensure that $f$ satisfies the desired conditions, the following control will be involved. There is a continuous function $\delta_i: M \to [0, 2^{-(i+1)})$ such that:

(e) $\delta_i^{-1}(0) = Z_i$ and $\delta_i(x) < d(x, Z_i)/3$ for each $x \in M$.
(f) For each subset $A$ of $M$ with $diam\ A < sup\{\delta_i(x): x \in A\}$, we have $diam f_i^{-1}(A) < 2^{-i}$.
(g) $d(f_i(x), f_{i+j}(x)) < \delta_i(f_i(x))$ for each $x$ and each $j \geq 0$.

It follows from (a), (c) and (d) that $f(\cup_{i=1}^{\infty} X_i) = \cup_{i=1}^{\infty} Z_i = \cup_{i=1}^{\infty} F_i$. Conditions (b) and (g) ensure that no points of $M - \cup_{i=1}^{\infty} X_i$ are mapped into $\cup_{i=1}^{\infty} F_i$ by $f$. Thus $f^{-1}(\cup_{i=1}^{\infty} F_i) = \cup_{i=1}^{\infty} X_i$. Take any point $y$ of $M - \cup_{i=1}^{\infty} F_i$. Then, by (g), we have $f^{-1}(y) \subseteq \cap_{i=1}^{\infty} f_i^{-1}(N(y, \delta_i))$, where $\delta_i = 2\delta_i(y)$. Applying condition (f), we see that $f^{-1}(y)$ is a point. Thus, the map $f$ has the desired properties.

For the construction of these sequences, we need two lemmas.

LEMMA 5.4.3. *Let $A$ be a $Z$-set in a Menger manifold $N$ and $U$ be an open set containing $A$. For each $\epsilon > 0$, there is a $\delta > 0$ such that*

- *for each $UV^{n-1}$-map $f: A \to B$ between $Z$-sets in $U$ with $d(f, id) < \delta$, there is an extension $F: N \to N$ such that $d(F, id) < \epsilon$, $F/(N - U) = id$ and $F/: N - A \to N - B$ is a homeomorphism.*

PROOF. Suppose that $f: A \to B$ is a $UV^{n-1}$-map between $Z$-sets in an open set $U$. We first prove that $f$ extends to a $UV^{n-1}$-map $F: N \to N$ so that $F/\dots : N - A \to N - B$ is a homeomorphism and $F/(N - U = id$.

Consider the adjunction space $N \cup_f B$, and let $\pi: N \to N \cup_f B$ be the projection. By Theorems 4.1.19 and 4.1.20, $N \cup_f B$ is a Menger manifold and $\pi$ is a strong near homeomorphism. It is easy to see that $\pi(A)$ is a $Z$-set in $N \cup_f B$

which is contained in the open set $\pi(U)$. Take a homeomorphism $h\colon N \to N \cup_f B$ such that $h/(N - U) = \pi/(N - U)$ and define a homeomorphism $k\colon \pi(A) \to h(B)$ by $k = hf\pi^{-1}/\pi(A)$. Apply the $Z$-set unknotting theorem to obtain an extension $K\colon N \cup_f B \to N \cup_f B$ of $k$ so that $K/(N \cup_f B - \pi(U)) = id$. Then it is easy to see that $F = h^{-1}K\pi$ is the desired extension.

To obtain the controlled version as required in the lemma, suppose that $d(f, id) < \delta$ for sufficiently small $\delta$ so that $A \cup B$ is contained in the interior of a Menger manifold $V$ in $U$. Take an extension $F\colon N \to N$ of the function $f$ constructed above, and choose a connected Menger manifold $W$ which is a finite union of elements of $\mathcal{U}_i$ for a large $i$ such that

(i) $A \subseteq intW \subseteq W \subseteq V$.
(ii) $FrW$ is a $Z$-set in $W$ and in $(V - W) \cup FrW$.
(iii) $d(F/W, id) < \delta$.

By taking sufficiently small $\delta$, we can proceed as in [**33**, 3.1.1-3] to construct an $\epsilon$-homeomorphism $j\colon V - W \to V - F(W)$ so that $j/FrV = id$ and $j/FrW = F$. Then $j$ naturally extends to a homeomorphism $J\colon N - W \to N - F(W)$ by defining $J/(N - W) = id$. Then $F^*\colon N \to N$, defined by $F^*/(N - W) = J$ and $F^*/W = F$ is the desired extension. □

*Remark* 5.4.4. The argument in [**33**] allows us to obtain a slightly stronger conclusion. That is, for each continuous function $\epsilon\colon M \to [0, \infty)$ such that $\epsilon^{-1}(0) = A$, we can find another continuous function $\delta\colon M \to [0, \infty)$ satisfying the same conclusion as in Lemma 5.4.3 above.

Recall that $\mathcal{U}_i^M$ is the partition of $M$ consisting of the restrictions of $(2n+1)$-cells of $M_i$ and $F_i = I_i \cap M \cap \partial M_i$. A compactum $K$ of $M$ is called a *nice* compactum of $M$ (of $F(M)$ respectively) if it is the union of some (possibly infinitely many) elements of $\cup\mathcal{U}_i^M$ $(\cup(\mathcal{U}_i^M \cap F(M))$ respectively). For a subset $A$ of $M$, $St(A, \mathcal{U}_i^M)$ denotes the collection of members of $\mathcal{U}_i^M$ which intersect $A$ and $st(A, \mathcal{U}_i^M) = \cup St(A, \mathcal{U}_i^M)$.

LEMMA 5.4.5. *Let $f\colon M \to M$ be a $UV^{n-1}$-surjection and $A$ be a nice compactum of $F(M)$ such that*

*(i) $A$ is a $Z$-set in $M$, and $f^{-1}(A) = X_s$ for some $s$, and*
*(ii) $f/\ldots\colon M - X_s \to M - A$ is a homeomorphism.*

*For each $t \geq 1$ and for each continuous function $\epsilon\colon M \to [0, 1)$, with $A = \epsilon^{-1}(0)$, there exist a nice compactum $X$ in $M$, a nice compactum $Z$ in $F(M)$ and a $UV^{n-1}$-surjection $g\colon f(X) \to Z$ such that*

*(3) $X_s \subseteq X \subseteq \cup_{l=1}^{\infty} X_l$ and $X_s$ is a $Z$-set in $X$; $A \cup F_t \subseteq Z \subseteq F(M)$ and $A \cup F_t$ is a $Z$-set in $Z$, and*
*(4) $d(g, id) < \epsilon$ and $g/A = id$.*

PROOF. Define a sequence $\{U_i\}$ of subcomplexes of $\{M_i\}$ by $U_0 = I^{2n+1}$, $U_i = st(A, \mathcal{U}_i)$, and define a sequence $\{U_i^M\}$ of nice compacta of $M$ by $U_0^M = M$, $U_i^M = st(A, \mathcal{U}_i^M) = U_i \cap M$ for each $i$. For simplicity, let $E_i = cl(U_i - U_{i+1})$

and $E_i^M = E_i \cap M$. Further, let $\epsilon_i = \min\{\epsilon(x) : x \in E_i^M\} > 0$ and take a subsequence $\{\mathcal{U}_{k(i)}\}$ of $\{\mathcal{U}_i\}_{i \geq j}$ and an increasing subsequence $\{j(i)\}$ so that

(a) $meshf(St(f^{-1}(E_i^M), \mathcal{U}_{j(i)})) < \frac{\epsilon_i}{3}$.

(b) for each $U \in St(E_i^M, \mathcal{U}_{k(i)})$, there is an element $V_U \in \mathcal{U}_{j(i)}$ such that $f^{-1}(U) \subseteq st(V_U, \mathcal{U}_{j(i)})$.

Define a partition $\mathcal{P}$ of $M - A$ by

(c) $\mathcal{P}/E_i^M = \mathcal{U}_{k(i)+1}/E_i^M$ for $i \geq 1$.

The nice compactum $Z$ is defined by

(d) $Z = A \cup \cup_{i=1}^{\infty} F_{k(i)+t+1} \cap E_i^M$.

Let us show that

(e) $Z$ is an $LC^{n-1}$- and $C^{n-1}$-compactum and $A$ is a $Z$-set in $Z$.

*Proof of (e).* First we prove the following assertion.

(f) $Z - A$ is uniformly $LC^{n-1}$.

Take a map $\alpha: S^q \to Z - A$ $(0 \leq q \leq n-1)$ with a small image, and extend it to a map $\alpha^*: B^{q+1} \to M$ with a small image by the $LC^{n-1}$ property of $M$. Since $A \cup \partial M$ (recall that $\partial M = M \cap \partial I^{2n+1}$) is a $Z$-set of $M$, we may assume that $im(\alpha^*)$ does not intersect $A \cup \partial M$, hence it is contained in $\cup_{i=0}^{m} E_i^M$ for some $m$. Regarding $\alpha^*$ as a map into $\cup_{i=0}^{m} M_{k(i)+t+1} \cap E_i$, take a $PL$ approximation $\tilde{\alpha}$ of $\alpha^*$ such that

$$\tilde{\alpha}(S^q) \subseteq \cup_{i=0}^{m} \partial M_{k(i)+t+1} \cap E_i \cap I_{k(0)+t+1}.$$

By general position and the property of regular neighborhoods, we can absorb $im(\tilde{\alpha})$ into $(E_0 \cap \partial M_{k(0)+t+1} \cap I_{k(0)+t+1}) \cup \cup_{i=0}^{m}(M_{k(i)+t+1} \cap E_i)$ keeping $\tilde{\alpha}(S^q)$ fixed. Let $A_0$ be the image of $im(\tilde{\alpha})$ under this absorption. Next we absorb $A_0$ into $I_{k(0)+k+1} \cap (\cup_{j=0}^{1}(\partial M_{k(0)+t+j} \cap E_j) \cup \cup_{i=2}^{m}(M_{k(i)+t+1} \cap E_i))$ . Continuing this process $m$ times, we obtain a map $\alpha': B^{q+1} \to I_{k(0)+k+1} \cap (\cup_{i=0}^{m} \partial M_{k(i)+t+1} \cap E_i)$ which is homotopic to $\tilde{\alpha} rel. S^q$. Since $\tilde{\alpha}$ is a $PL$ map and the above absorption can be performed by $PL$ homotopies, we may assume that $im(\alpha')$ is tame in $\cup_{i=0}^{m} M_{k(i)+t+1}$ in the sense of Štanko [**288**]. By the proof of [**289**, Theorem 2], we may further absorb $im(\alpha')$ into $\cup_{i=0}^{m} F_{k(i)+t+1} \cap E_i^M$ keeping $\tilde{\alpha}(S^q)$ fixed. The resulting map is an extension of $\tilde{\alpha}/S^q$ into $Z$ with a small image. An examination of the"size of the above absorption moves" shows that we have proved the following statement:

- For each $\epsilon > 0$ there exists a $\delta > 0$ such that, for each map $\alpha: S^q \to Z - A$ with $diam(im(\alpha)) < \delta$ and for each $\beta > 0$, there exists a map $\bar{\alpha}: B^{q+1} \to Z - A$ such that $d(\bar{\alpha}/S^q, \alpha) < \beta$ and $diam(im(\bar{\alpha})) < \epsilon$.

Using the above, a method similar to that of Proposition 5.2.14 gives us the desired assertion (f).

To prove the $LC^{n-1}$ property of $Z$, take a map $\alpha: S^q \to Z$ $(0 \leq q \leq n-1)$ with a small image. If $im(\alpha)$ does not intersect $A$, (f) immediately gives an extension to $B^{q+1}$ with small image. Suppose that $im(\alpha) \cap A \neq \emptyset$. Then, since $A$ is a $Z$-set, there is a sequence $\{\alpha_i : S^q \to M - A\}$ of maps such that $\lim_{i\to\infty} \alpha_i = \alpha$. The"absorption technique" described in the proof of (f) enables us to assume that $im(\alpha_i) \subseteq \cup_{s=0}^{m_i} E_s$ for some $m_i$. Again, the method of Proposition 5.2.14 gives us an extension of $\alpha$ to $B^{q+1}$ with a small image.

The same argument can be applied to prove the $(n-1)$-connectedness of $Z$. Any map $\alpha: I^n \to M$ can be approximated by another map $\alpha': I^n \to M - A$, because $A$ is a $Z$-set in $M$. Then the"absorption technique" can be applied again to approximate $\alpha'$ by a map $\tilde{\alpha}': I^n \to Z - A$. This shows that $A$ is a $Z$-set in $Z$ and finishes the proof of (e).

Observe that $Z$ is a $Z$-set in $M$ and $Z - A$ has a locally finite partition of the form $\cup_{i=0}^{\infty}\{\mathcal{U}_{k(i)+1}/Z_i \cap E_i^M\}$ and $F_t \subseteq Z$.

Next we define a nice compactum $X$. Take an increasing sequence $(m_j)$ with $m_1 > k$ such that

(g) $meshf(\mathcal{U}_{m_i}) < \frac{\epsilon_i+1}{3}$ and for each $U \in \mathcal{U}_{k(i)+1}$, $f(X_{m_i}) \cap int(U) \neq \emptyset$.

Inductively, we define $(W_i)$ as follows. $W_0 = st(f^{-1}(cl(M - U_1)), \mathcal{U}_{m_0})$, and if $W_i$ is defined, then $W_{i+1} = st(cl(f^{-1}(E_{i+1}) - W_i), \mathcal{U}_{m_{i+1}})$. Then $X = X_s \cup \cup_{i=0}^{\infty} W_i \cap X_{m_i}$. By an argument similar to the proof of (e), we can prove that

(h) $X$ is homeomorphic to $M$ and $X_s$ is a $Z$-set in $X$.

We claim that

(i) each map $\alpha: S^q \to U \cap f(X)$, where $U \in \mathcal{U}_{k(i)} - St(A, \mathcal{U}_{k(i)})$ is null homotopic by an $\epsilon_i$-homotopy $(0 \leq q \leq n-1)$.

*Proof of (i).* By (a) and (b), there is an element $V_U \in U_{j(i)}$ such that $f^{-1}(U) \subseteq st(V_U, U_{j(i)})$ and $diamf(st(V_U, U_{j(i)})) < \frac{\epsilon_i}{3}$. Since $st(V_U, U_{j(i)})$ is $(n-1)$-connected, $f^{-1}\alpha \simeq 0$ in $st(V_U, U_{j(i)})$. Thus, $\alpha$ is $\epsilon_i$-homotopic to a constant map.

From the construction, $f(X) \cap int(U) \neq \emptyset$ for each $U \in \mathcal{U}_{k(i)+1}/(M - U_{k+j+1})$. Since $Z$ is an $LC^{n-1}$- and $C^{n-1}$-compactum and $dim f(X) = n$, there is a map $r: f(X) \to Z$ such that $r(f(X)) \cap int(U) \neq \emptyset$ for each $U \in \mathcal{U}_{k(i)+1}/(M - U_{i+1})$. Using $r$ as a "guide map" and using the above mentioned partition of $Z$, we proceed as in [**33**, 2.4-2.7 and Appendix]. Using (g) we construct a $UV^{n-1}$-map $g: f(X) \to Z$ such that $g(f(X) \cap U) \subseteq st^2(U, \mathcal{U}_{k(i)+1})$ for each $U \in \mathcal{U}_{k(i)+1}/(M - U_{i+1})$. By the choice of $\mathcal{U}_{k(i)+1}$, we have $d(g, id) < \epsilon$. This finishes the verification of (i) and completes the proof of Lemma 5.4.5. □

Now we are ready to prove Theorem 5.4.2. Inductively, we define the required sequence as follows.

$i = 1$. Let $Z_1 = F_1$ and take a $UV^{n-1}$-surjection $g_1: X_1 \to Z_1$. The existence of such a map is guaranteed by Theorem 4.2.1. By Lemma 5.4.3, we have an extension $f_1: M \to M$ which maps $M - X_1$ homeomorphically onto $M - Z_1$ so that $f_1(\partial M) = \partial M$. Let $X_{i_1} = X_1$, take an index $j_1$ such that $mesh\mathcal{U}_{j_1} < 2^{-3}$, and take a continuous function $\delta_1: M \to [0, 2^{-1})$ for $Z_1$ as in (e)-(g). For $\frac{\delta_1}{2}$, take $\gamma_1: M \to [0, 2^{-1})$ as in Remark 5.4.4.

$i = 2$. This step describes the general procedure. Applying Lemma 5.4.5 to $X_s = X_{i_1}$, $A = Z_1$, $t = 2$ and $f = f_1$, we obtain a nice $C^{n-1}$-compactum $X_{i_2}$ in $M$, $Z_2$ in $F(M)$ and a $UV^{n-1}$-map $g_1: f_1(X_{i_2}) \to Z_2$ such that $X_{i_1} \subseteq X_{i_2} \subseteq$

$\cup_{i=1}^{\infty} X_i$, $Z_1 \cup F_2 \subseteq Z_2$ and $d(g_1, id) < \gamma_1$, $\gamma_1/Z_1 = id$.

By Lemma 5.4.3, we can define an extension $g_1' : M \to M$ which maps $M - f_1(X_{i_2})$ homeomorphically onto $M - Z_2$ so that $d(g_1', id) < \delta_1$ and $g_1'/\partial M = id$. Let $f_2 = g_1' f_1 : M \to M$. Then $f_2$ satisfies (a) - (e).

Continuing this process, we obtain the desired sequence $\{f_i\}$. This completes the proof.

*Historical and bibliographical notes* 5.4. The results of this Section originally appeared in [**106**].

## 5.5. Equivalence of the Nöbeling space and the pseudo-interior of $\mu^n$

Recall that the Nöbeling space $N_n^p$ is defined as follows:

$$N_n^p = \{x \in R^p : \text{ at most } n \text{ coordinates of } x \text{ are rational}\}.$$

It is easy to see that $N_n^p \cap Int(I^p)$ is homeomorphic to $N_n^p$ (we have a homeomorphism between $Int(I^p)$ and $R^p$ preserving rational points). The purpose of this section is to prove the following:

THEOREM 5.5.1. *If $p \geq 2n+1$, the geometric pseudo-interior $I(M_n^p)$ is homeomorphic to $N_n^p$.*

By the above remark, it suffices to establish a homeomorphism between $I(M_n^p)$ and $N_n^p \cap Int(I^p)$. From the definition of $N_n^p$, $Int(I^p) - N_n^p = \cup_{i=1}^{\infty} L_i$, where each $L_i$ is an open $(p-n-1)$-dimensional cell. The following theorem due to Torunczyk [**295**] is an essential ingredient of the proof. A polyhedron $K$ in $Int(I^p)$ is said to be *tame* if there is a homeomorphism $h: Int(I^p) \to Int(I^p)$ such that $h(K)$ is a subcomplex of the standard $PL$ structure of $I^p$.

THEOREM 5.5.2. *Suppose that $p \geq 2n+1$. Then $\{L_i\}$ forms an absorber with respect to all $n$-dimensional, locally compact, tame polyhedra in $Int(I^p)$.*

For the proof of Theorem 5.5.1, we first construct a surjection $f: M_n^p \to I^p$ which maps $I(M_n^p)$ homeomorphically onto $I^p - f(B(M_n^p))$. From the construction of the map and from the proof of Theorem 5.4.2, we see that $f(B(M_n^p)) \cap Int(I^p) = f(F(M_n^p))$ forms an absorber with respect to the same class as does $\{L_i\}$. Then the uniqueness of absorbers (Theorem 2.2.15) gives us a homeomorphism between $Int(I^p) - f(B(M_n^p))$ and $N_n^p \cap Int(I^p)$, which finishes the proof.

Therefore, in order to prove Theorem 5.5.1 it suffices to prove the following statement. In what follows, we fix $n$ and $p$ so that $p \geq 2n+1$ and omit these indices.

THEOREM 5.5.3. *There is a continuous surjection $f: M \to I^p$ such that:*

(i) $f(F(M))$ *is an absorber with respect to all locally compact, n-dimensional tame polyhedra in* $Int(I^p)$.

(ii) $f^{-1}(f(F(M))) = F(M)$, $f^{-1}(\partial I^p) = \partial I^p$ *and* $f/(M - B(M)) \colon M - B(M) \to Int(I^p) - f(F(M))$ *is a homeomorphism.*

PROOF. The desired map is defined as the restriction (to $M$) of the limit of a sequence $\{f_i \colon I^p \to I^p\}$ of surjections. Recall the notations of Section 5.3. In particular, $D_i = D_i(p)$ is the "dual $(p-n-1)$-skeleton" of $L_i$ and $N_i = N_i(p)$ is its regular neighborhood. $\{D_i\}$ and $\{N_i\}$ are increasing sequences. For a sequence $\{\epsilon_i\}$, with $0 < \epsilon_i < 3^{-(i+1)}$, we define $f_i$ as follows.

*Construction of* $f_1$. Since $N_1$ is a regular neighborhood of $D_1$ with respect to $\mathcal{L}_1$, there is a map $f_1 \colon I^p \to I^p$ such that

(1.1) $d(f_1, id) < 3^{-1}$ and $supp f_1 \subseteq N(N_1, \epsilon_1)$.

(1.2) $f_1(N_1) = D_1$.

Note that the map $f_1$ is induced by the standard retraction of $N_1$ onto $D_1$.

*Construction of* $f_2$. Recall that $N_2$ is a regular neighborhood of $D_2$. From the construction of $f_1$, it is easy to see that $f_1(N_2)$ retracts onto $D_2$. Let $r_2 \colon I^p \to I^p$ be a retraction of $f_1(N_2)$ onto $D_2$ and let $f_2 = r_2 f_1$. By a careful choice of $r_2$, we may assume that $f_2$ satisfies the following conditions:

(2.1) $d(f_2, f_1) < 3^{-2}$ and $supp f_2 \subseteq N(N_2, \epsilon_2)$.

(2.2) $f_2^{-1}(f_2(N_2)) = N_2$ and $f_2(N_1) = f_1(N_1) = D_1$.

(2.3) if $x, y \in I^p - N_2$ and $d(x,y) > 3^{-1} + \epsilon_2$, then $d(f_2(x), f_2(y)) > \frac{5}{2 \cdot 3^2} > \frac{2}{3^2}$.

Continuing this process, we obtain a sequence $\{f_i \colon I^p \to I^p\}$ such that

(i.1) $d(f_{i+1}, f_i) < 3^{-i}$ and $supp f_i \subseteq N(N_i, \epsilon_i)$.

(i.2) $f_i^{-1}(f_i(N_i)) = N_i$ and $f_i(N_j) = f_j(N_j) = D_j$ for each $j < i$.

(i.3) if $x, y \in I^p - N_i$ and $d(x,y) > 3^{-(i-1)} + 2 \cdot \epsilon_i$, then $d(f_i(x), f_i(y)) > \frac{5}{2 \cdot 3^i} > \frac{2}{3^i}$.

Then $f = \lim_{i \to \infty} f_i$ exists (by (i.1)) and is a surjection. Condition (i.2) guarantees that $f(\cup_{i=1}^{\infty} N_i) = \cup_{i=1}^{\infty} D_i$. Taking sufficiently small $\epsilon_i$'s, we may assume that, for any point $x \in I^p - \cup_{i=1}^{\infty} N_i$, $f(x) \in I^p - \cup_{i=1}^{\infty} D_i$, that is, $f^{-1}(f(\cup_{i=1}^{\infty} N_i)) = \cup_{i=1}^{\infty} N_i$. For each $x \in I^p$, by (i.1),

$$d(f(x), f_i(x)) < \Sigma_{j \geq i+1} \frac{1}{3^j} < \frac{1}{2 \cdot 3^i}.$$

It follows from (i.3) that $f/\ldots \colon I^p - \cup_{i=1}^{\infty} N_i \to I^p - f(\cup_{i=1}^{\infty} N_i)$ is a homeomorphism. Since $F(M) = Int(I^p) \cap M \cap \cup_{i=1}^{\infty} N_i$, we have the desired condition (2).

From the construction of $f$ and from the proof of Theorem 5.4.2, we can see that $f(F(M))$ is an absorber with respect to all $n$-dimensional tame polyhedra. This completes the proof of Theorem 5.5.3. □

*Remark* 5.5.4. It follows from the construction that $f/M \colon M \to I^p$ is a $UV^{n-1}$-map.

Combining Theorems 5.4.1 and 5.5.1, we have the following theorem.

THEOREM 5.5.5. *If $p \geq 2n+1$, $\nu^n$ is homeomorphic to $N_n^p$.*

*Remark* 5.5.6. As in Section 5.3, we can perform essentially the same argument for the Lefschetz construction, but not for the Bestvina construction.

*Historical and bibliographical notes* 5.5. The results of this Section originally appear in [**106**]. Theorem 5.5.5 is very important for the whole theory of Nöbeling manifolds. It allows us to use the results and tools of Menger manifold theory in this situation.

## 5.6. Further properties of Nöbeling spaces

In this section, we give some applications of our results.

THEOREM 5.6.1. *Let $0 \leq n < k \leq \infty$ and let $f\colon \mu^n \to \mu^k$ be an $n$-invertible $UV^{n-1}$-map satisfying the properties of Theorem 4.3.5. Then there exists a pseudo-boundary $A$ of $\mu^k$ such that $f^{-1}(A)$ is a pseudo-boundary.*

PROOF. Consider a pseudo-boundary $B$ in $\mu^k$ and its inverse image $f^{-1}(B)$. By property (3) of Theorem 4.3.5, $f^{-1}(B)$ is a $\sigma Z$-set in $\mu^n$. Let $C$ be a pseudo-boundary of $\mu^n$ such that $f^{-1}(B) \cap C = \emptyset$ (see Remark 5.2.2). By Proposition 5.2.4, $C \cup f^{-1}(B)$ is a pseudo-boundary of $\mu^n$. Consider the set $f(C)$. Since $f(C) \subseteq \mu^k - B$, we see that $f(C)$ is a $\sigma Z$-set in $\mu^k$. By Proposition 5.2.4, $A = f(C \cup f^{-1}(B)) = f(C) \cup B$ is a pseudo-boundary in $\mu^k$. Again by Proposition 5.2.4, we see that $f^{-1}(A)$ is a pseudo-boundary in $\mu^n$. □

COROLLARY 5.6.2. *Let $0 \leq n < k \leq \infty$ and let $f\colon \mu^n \to \mu^k$ be the map of Theorem 4.3.5. For any pair $\Sigma^n$, $\Sigma^k$ of pseudo-boundaries of $\mu^n$ and $\mu^k$, there exist autohomeomorphisms $F\colon \mu^n \to \mu^n$ and $H\colon \mu^k \to \mu^k$ such that $(HfF)^{-1}(\Sigma^k) = \Sigma^n$.*

PROOF. This follows from Proposition 5.2.3 (the uniqueness of pseudo-boundaries) and the above theorem. □

COROLLARY 5.6.3. *For each $n \geq 0$ there exists a proper $n$-invertible $UV^{n-1}$-surjection $f\colon N_n^{2n+1} \to R^\omega$.*

PROOF. Apply Corollary 5.6.2 and Theorem 5.5.5. □

Let $f_n\colon \mu^n \to \mu^\infty = I^\omega$ be the $n$-invertible, $(n-1)$-soft, $UV^{n-1}$-map constructed in Theorem 4.2.21. The Hilbert space $R^\omega$ is naturally identified with the pseudo-interior $s$ of $I^\omega$. It is natural to ask [**123**, Problem 612] whether the inverse images $f_n^{-1}(R^\omega)$ and $f_n^{-1}(I^\omega - R^\omega)$ are homeomorphic to $N_n^{2n+1}$ and $\sigma_n^{2n+1}$ (=the universal $n$-dimensional pseudo-boundary of $R^{2n+1}$) respectively. If we could show that $f_n^{-1}(I^\omega - R^\omega)$ is a pseudo-boundary, then Theorem 5.5.5 would provide a positive answer to the first question. However it is known that the inverse image of a $Z$-set of $I^{\omega}$ by $f_n$ is not necessarily a $Z$-set in $\mu^n$. Therefore, it is not clear whether $f_n^{-1}(I^{\omega} - R^{\omega})$ is a pseudo-boundary of $\mu^k$. Of course, even if the set $f_n^{-1}(I^{\omega} - R^{\omega})$ is not a pseudo-boundary as a subset of $\mu^n$, a positive answer is still possible. On the other hand, Theorem 5.6.1 and its corollaries assert that we can construct a proper $n$-invertible $UV^{n-1}$-surjection $g\colon \mu^k \to I^{\omega}$ such that $g^{-1}(I^{\omega} - R^{\omega})$ is a pseudo-boundary and $g^{-1}(R^{\omega})$ is homeomorphic to $N_n^{2n+1}$

Notice also that in Corollary 5.6.3, we cannot make the map $n$-soft (this essentially follows from Theorem 4.2.25). However, if we do not require the map to be proper, we obtain the following generalization of the classical Hausdorff theorem (see Theorem 1.1.7) stating that $R^{\omega}$ is an open ( = 0-soft) image of the space of irrational numbers ($= N_0^1$).

THEOREM 5.6.4. *Let $n \geq 0$. Each Polish absolute extensor (in particular, the separable Hilbert space $l_2 \approx R^{\omega}$ ) is an $n$-soft image of $N_n^{2n+1}$.*

PROOF. The Hilbert space $R^{\omega}$ is identified with the pseudo-interior of the Hilbert cube $\mu^{\infty}$. By Corollary 5.6.2, there is a map $f\colon \mu^n \to \mu^{\infty}$ which has the properties of Theorem 4.3.5 and satisfies $f^{-1}(R^{\omega}) = \nu^n$. By property (2) of Theorem 4.3.5, there is a subspace $A_n$ of $\mu^n$ such that $f/A_n\colon A_n \to \mu^{\infty}$ is $n$-soft and the complement $\mu^n - A_n$ is a $\sigma Z$-set in $\mu^n$. Clearly, $f^{-1}(\mu^{\infty} - R^{\omega}) \cup (\mu^n - A_n)$ is an $n$-dimensional pseudo-boundary of $\mu^n$ and, hence, its complement $f^{-1}(R^{\omega}) \cap A_n$ is a pseudo-interior. Notice that

$$g = f/f^{-1}(R^{\omega}) \cap A^n\colon f^{-1}(R^{\omega}) \cap A^n \to R^{\omega}$$

is $n$-soft as the restriction of the $n$-soft map $f/A^n$ to an inverse image. Then the uniqueness of pseudo-interiors and Theorem 5.5.5 finish the proof for the Hilbert space $R^{\omega}$. The general case is obtained by noticing that if $X$ is a Polish absolute extensor, then $X \times R^{\omega}$ is homeomorphic to $R^{\omega}$. □

PROPOSITION 5.6.5. *For each $0 \leq n \leq k$, the space $N_k^{2k+1}$ is an $n$-soft image of $N_n^{2n+1}$.*

PROOF. Fix an $n$-soft map $g\colon N_n^{2n+1} \to R^{2k+1}$ as guaranteed by Theorem 5.6.4. Clearly, $R^{2k+1} - N_k^{2k+1}$ is a $\sigma Z_n$-set in $R^{2k+1}$. Since $n \leq k$ and $g$ is $n$-soft, $g^{-1}(R^{2k+1} - N_k^{2k+1})$ is a $\sigma Z$-set in $N_n^{2n+1}$ (see Lemma 4.3.1). Consequently, by Proposition 5.2.10, $g^{-1}(N_k^{2k+1}) = N_n^{2n+1} - (\sigma Z\text{-set}) \approx N_n^{2n+1}$. It only remains to note that the restriction of an $n$-soft map to the inverse image of a subset is also $n$-soft. □

We conjecture the following.

CONJECTURE 5.6.6. *Every Polish $LC^{n-1}$- and $C^{n-1}$-space is an $n$-soft image of the universal Nöbeling space $N_n^{2n+1}$.*

As an application of the map constructed in Section 5.5, we prove that $\nu^n$ is homeomorphic to the universal $n$-dimensional pseudo-interior $s_n^p$, $p \geq 2n+1$ (see the beginning of this Chapter).

THEOREM 5.6.7. *$\nu^n$ is homeomorphic to $s_n^p$ ($p \geq 2n+1$).*

PROOF. By [**121**], it suffices to establish the existence of a homeomorphism between $\nu^n$ and $s_n^{2n+1}$. Let $M = M_n^{2n+1}$ be the Menger compactum obtained by the Menger construction in $I^{2n+1}$ and let $f\colon M \to I^{2n+1}$ be the $UV^{n-1}$-map constructed in Section 5.5. From that construction it is easy to see that $f$ has the following properties.

(i) $f^{-1}(\partial I^{2n+1}) = \partial I^{2n+1} \cap M$ is a $Z$-set in $M$.

(ii) $S(f) = \{x \in I^{2n+1} \colon \mid f^{-1}(x) \mid \geq 2\}$ is a $\sigma Z_n$-set in $I^{2n+1}$ and $f^{-1}(S(f))$ is a $\sigma Z$-set in $M$.

Let $\{A_i \colon i \in N\}$ be the $Z$-skeleton of $M$ constructed in Subsection 5.3.2. Then, again from the construction, we have

(3) $A_i$ is homeomorphic to $M$ and $f^{-1}(f(A_i)) = A_i$, for each $i$, and

(4) $\cup_{i=1}^{\infty} A_i \cap (f^{-1}(\partial I^{2n+1}) \cup f^{-1}(S(f))) = \emptyset$.

Observe that

(5) $\nu^n$ is homeomorphic to $M - \cup_{i=1}^{\infty} A_i \cup f^{-1}(\partial I^{2n+1} \cup S(f))$.

Consider the restriction $f/\ldots \colon M \cap IntI^{2n+1} \to IntI^{2n+1}$. We prove

(6) $\{f(A_i)\}$ is a $Z_n$-skeleton of $IntI^{2n+1}$.

*Proof of (6).* Since, by (4), $A_i$ does not intersect $f^{-1}(S(f))$, $A_i$ is homeomorphic to $M$. From the construction of $f$ it follows that $A_i$ is a tame compactum for each $i$. To prove the absorption property, take a $Z_n$-set $Z$ in $IntI^{2n+1}$, $\epsilon > 0$ and an integer $i > 0$. Let $i_Z \colon Z \to IntI^{2n+1}$ be the inclusion. Since $f/A_i \colon A_i \to f(A_i)$ is a homeomorphism (by (4)), we have a partial lift $(f/A_i)^{-1}/(Z \cap f(A_i)) \colon Z \cap f(A_i) \to M$ of $i_Z$. Notice that if $\lim_{k\to\infty} y_k = y$, $y_k \in IntI^{2n+1}$ and $y \in Z \cap f(A_i)$, then $\lim_{k\to\infty} diam f^{-1}(y_k) = 0$. By the $UV^{n-1}$-property of $f$, the nerve replacement trick and the above observation, we can define a map $t\colon Z \to M$ such that $ft$ is $\epsilon$-close to $id$ and $t/(Z \cap f(A_i)) = (f/A_i)^{-1}/(Z \cap f(A_i))$. Since $t(Z \cap f(A_i))$ is a $Z$-set, we may assume that $t$ is a $Z$-embedding. By the absorption property of $\{A_i\}$, we have an integer $j > i$ and a small embedding $h\colon t(Z) \to A_j$ such that $h(t(Z)) \subseteq A_j$ and $h/t(Z \cap f(A_i)) = id$. Since $f/A_j$ is injective, $e = (f/A_j)ht \colon Z \to f(A_j)$ is an embedding such that $e/(Z \cap f(A_i) = id$. By taking $h$ close enough to $id$, we can additionally assume that $e$ is $\epsilon$-close to $id$. Thus, $\{f(A_i)\}$ is a $Z_n$-skeletoid in $IntI^{2n+1}$. This finishes the verification of (6).

By Theorem 2.2.20, $IntI^{2n+1} - \cup_{i=1}^{\infty} f(A_i)$ is homeomorphic to $s_n^{2n+1}$. Since $S(f)$ is a $\sigma Z_n$-set by (2), Theorem 2.2.20 and Proposition 2.2.25 imply that

(7) $IntI^{2n+1} - \{S(f) \cup \cup_{i=1}^{\infty} f(A_i)\}$ is homeomorphic to $s_n^{2n+1}$.

Then we have the desired conclusion as follows.

$$\nu^n \approx M - \{f^{-1}(S(f) \cup \partial I^{2n+1}) \cup \cup_{i=1}^{\infty} A_i\} \overset{\text{by (5)}}{\approx} IntI^{2n+1} - \{S(f) \cup \cup_{i=1}^{\infty} f(A_i)\}.$$

Observe that

$$f/\ldots : M - \{f^{-1}(S(f) \cup \partial I^{2n+1}) \cup \cup_{i=1}^{\infty} A_i\} \to Int I^{2n+1} - \{S(f) \cup \cup_{i=1}^{\infty} f(A_i)\}$$

is a homeomorphism. Therefore, by (7), $\nu^n \approx s_n^{2n+1}$. □

Next, we prove that the pseudo-boundary $\Sigma^n$ of $\mu^n$ is homeomorphic to the universal $n$-dimensional pseudo-boundary $\sigma_n^{2n+1}$ of $R^{2n+1}$. We start with the following result. Let $\tilde{M} = \mu^n - \{\text{point}\}$.

PROPOSITION 5.6.8. *There is a proper $n$-invertible $UV^{n-1}$-surjection $g\colon \tilde{M} \to R^{2n+1}$ such that $g^{-1}(\sigma_n^{2n+1}) = \Sigma^n$.*

PROOF. Fix an $n$-invertible $UV^{n-1}$-surjection $f\colon \mu^n \to I^{2n+1}$ as constructed in Theorem 4.3.6 such that

(i) for each $Z_n$-set $Z$ in $I^{2n+1}$, $f^{-1}(Z)$ is a $Z$-set in $\mu^n$, and

(ii) for each subpolyhedron $L$ of $I^{2n+1}$, $f^{-1}(L)$ is a $\mu^n$-manifold.

Below we identify $R^{2n+1}$ with $Int(I^{2n+1})$. Since $\partial I^{2n+1}$ is a $Z$-set in $I^{2n+1}$, we see,by (1), that $f^{-1}(\partial I^{2n+1})$ is a $Z$-set. By (2) and the $UV^{n-1}$-property of $f$, $f^{-1}(\partial I^{2n+1})$ is an $(n-1)$-connected $\mu^n$-manifold, and hence is homeomorphic to $\mu^n$. The complement theorem 4.4.8 implies that $\mu^n - f^{-1}(\partial I^{2n+1})$ is homeomorphic to $\tilde{M}$. Fix a homeomorphism $\phi\colon \tilde{M} \to \mu^n - f^{-1}(\partial I^{2n+1})$. We prove that $g = f\phi$ is the desired map. For simplicity, we identify $\tilde{M}$ with $\mu^n - f^{-1}(\partial I^{2n+1})$ by $\phi$ and regard it as a subset of $\mu^n$.

Since $\sigma_n^{2n+1}$ is a $\sigma Z_n$-set in $R^{2n+1}$, $g^{-1}(\sigma_n^{2n+1})$ is a $\sigma Z$-set in $\tilde{M}$ by(1). By Remark 5.2.2, there is a pseudo-boundary $\Sigma^n$ of $\mu^n$ which is contained in $\tilde{M} - g^{-1}(\sigma_n^{2n+1})$. Clearly, $g(\Sigma^n)$ is contained in $R^{2n+1} - \sigma_n^{2n+1} = s_n^{2n+1}$, the universal $n$-dimensional pseudo-interior of $R^{2n+1}$. Thus, each compact subset $K$ of $g(\Sigma^n)$ is an at most $n$-dimensional, tame compactum (see [**121**, Lemma 2.6]). Hence $g(\Sigma^n)$ is a countable union of at most $n$-dimensional compacta. By the absorption property of $\sigma_n^{2n+1}$ and general position, we see that $g(\Sigma^n)$ is a $\sigma Z_n$-set in $R^{2n+1}$ (hence, the union $g(\Sigma^n) \cup \sigma_n^{2n+1}$ is again an $n$-dimensional pseudo-boundary in $R^{2n+1}$). Condition (1) implies that the set $g^{-1}(g(\Sigma^n) \cup \sigma_n^{2n+1})$ is a $\sigma Z$-set. This set contains $\Sigma^n$, thus we conclude that it is a pseudo-boundary of $\mu^n$. The identifications

$$\Sigma^n = g^{-1}(g(\Sigma^n) \cup \sigma_n^{2n+1})$$

and

$$\sigma_n^{2n+1} = g(\Sigma^n) \cup \sigma_n^{2n+1}$$

complete the proof. □

The second problem we are interested in (see [**105**], [**121**], [**123**], [**315**]) is closely related to Conjecture 5.0.5.

CONJECTURE 5.6.9. *Characterize topologically the universal n-dimensional pseudo-boundary $\sigma_n^k$, $k \geq 2n+1$, of the Euclidean space $R^k$. In particular, is a space $X$ homeomorphic to $\sigma_n^{2n+1}$ provided that $X$ satisfies the following conditions:*

(a) *$X$ is a metrizable $\sigma$-compact space.*
(b) $\dim X = n$.
(c) $X \in AE(n) = LC^{n-1} \cap C^{n-1}$.
(d) *$X$ is a countable union of strong $Z$-sets.*
(e) *For every at most n-dimensional compactum $C$ and for every closed subset $D \subseteq C$, each map $f: C \to X$ such that $f|D: D \to X$ is a $Z$- embedding can be approximated arbitrarily closely by $Z$-embeddings coinciding with $f$ on $D$.*

This problem is also open.

THEOREM 5.6.10. *The spaces $\Sigma^n$ and $\sigma_n^{2n+1}$ are homeomorphic.*

PROOF. Take a map $g: \tilde{M} \to R^{2n+1}$ as constructed in Proposition 5.6.8. We show that $h = g/\ldots: \Sigma^n \to \sigma_n^{2n+1}$ is a near homeomorphism. We use the following shrinking criterion, which is an analog of the one for $\sigma$-compact $ANR$'s (see Proposition 2.5.3). For any compactum $A$ in $\sigma_n^{2n+1}$, consider the decomposition of $\Sigma^n$ whose non-trivial elements are of the form $h^{-1}(a)$, $a \in A$. Let $(\Sigma^n, h)_A$ be its decomposition space and let $h_A: \Sigma^n \to (\Sigma^n, h)_A$ be the projection. Our shrinking criterion can be stated as follows:

*If the following condition is satisfied, then h is a near-homeomorphism:*

($*$) *For every pair $(A, B)$ of compacta in $\sigma_n^{2n+1}$ with $B \subseteq A$, $h_A/\ldots: (\Sigma^n - h^{-1}(B)) \to (\Sigma^n, h)_A - B$ is a near-homeomorphism.*

*Remark* 5.6.11. The proof of the above statement is same as that of Proposition 2.5.3 except for minor changes, and hence, is omitted. The $ANR$ condition can be replaced by $LC^{n-1}$ (and $C^{n-1}$) properties of $\Sigma^n$ and $\sigma_n^{2n+1}$, coupled with the $n$-dimensionality of these spaces. This allows us to replace ordinary homotopies by $(n-1)$-homotopies. Further, $\Sigma^n$ has the estimated extension property for compacta and $\sigma_n^{2n+1}$ is strongly universal for $n$-dimensional compacta (compare with the corresponding definitions in Section 2.5). In addition, the fact that $h$ is a proper map simplifies our situation.

To verify ($*$), we need the following claim. For a $\mu^n$-manifold $L$, $\mathcal{Z}_L$ denotes the collection of all compact $Z$-sets of $L$.

*Claim. Let $N$ be a $\mu^n$-manifold and $Z$ be a compact $Z$-set in $N$. If $A$ is a $\mathcal{Z}_{(N-Z)}$-skeletoid, then $A$ is a $\mathcal{Z}_N$-skeletoid.*

*Proof of Claim.* Let $\{A_i: i \in N\}$ be a $\mathcal{Z}_{(N-Z)}$-skeleton such that $A = \cup_{i=1}^{\infty} A_i$. Let us show that $\{A_i: i \in N\}$ is a $\mathcal{Z}_N$-skeleton. Since the collection of compact $Z$-sets of any $\mu^n$-manifold is perfect, it suffices to verify the condition of Theorem 2.2.22. Let $d$ be a metric and let $K$ be a compact $Z$-set of $N$, $i \in N$ and $\epsilon > 0$. Since $Z$ is a $Z$-set, there is a $Z$-embedding $\alpha: K \to M - Z$ such that

$\alpha/(K \cap A_i = id$ and $d(\alpha, id) < \frac{\epsilon}{2}$. Since $\{A_i\}$ is a $\mathcal{Z}_{(N-Z)}$-skeleton, there is a $j > i$ and a $Z$-embedding $\beta\colon \alpha(K) \to A_j$ such that $\beta/(\alpha(K) \cap A_j) = id$ and $d(\beta, id) < \frac{\epsilon}{2}$. Clearly, the composition $\beta\alpha\colon K \to A_j$ is a $Z$-embedding such that $d(\beta\alpha, id) < \epsilon$ and $\beta\alpha/(K \cap A_i) = id$. The Claim is proven.

Let $(A, B)$ be a pair of compacta in $\sigma_n^{2n+1}$. To prove condition $(*)$, it suffices to prove that the map $h_A/\ldots : \Sigma^n - h^{-1}(B) \to (\Sigma^n, h)_A - B$ is a near-homeomorphism. Consider the decomposition of $\tilde{M}$ whose non-trivial elements are of the form $g^{-1}(a)$, $a \in A$ and let $\tilde{M}'$ be the decomposition space and $g_A\colon \tilde{M} \to \tilde{M}'$ be the projection. It follows from Theorem 4.1.20 that $g_A$ is a near-homeomorphism and $\tilde{M}'$ is also a $\mu^n$-manifold. Since $g$ is a proper $UV^{n-1}$-surjection, so is $g_A$. By the construction, we have that

$$\Sigma^n = g_A^{-1}(g_A(\Sigma^n)), g^{-1}(A) = g_A^{-1}(g_A(g^{-1}(A))), g^{-1}(B) = g_A^{-1}(g_A(g^{-1}(B)))$$

and $h_A = g_A/\Sigma^n$. Observe also that $g_A(g^{-1}(A))$ is a compact $Z$-set in $\tilde{M}'$.

Take any open cover $\mathcal{U}$ of $g_A(\Sigma^n) - (g_A(g^{-1}(B)) = (\Sigma^n, h)_A - B$ and let $\mathcal{V}$ be a collection of open sets in $\tilde{M}' - g_A g^{-1}(B)$ such that $\mathcal{V}/(g_A(\Sigma^n) - (g_A(g^{-1}(B))) = \mathcal{U}$. Let $V = \cup\mathcal{V}$ and $\mathcal{W}$ be an open cover of $V$ which is a star-refinement of $\mathcal{V}$. Clearly, $V$ and $g_A^{-1}(V)$ are $\mu^n$-manifolds and $g_A/\ldots : g_A^{-1}(V) \to V$ is a proper $UV^{n-1}$-surjection. By Theorem 4.1.20, $g_A$ is a near-homeomorphism. Choose a homeomorphism $H : g_A^{-1}(V) \to V$ which is $\mathcal{W}$-close to $g_A/g_A^{-1}(V)$. By Theorem 5.1.12, $\Sigma^n - g^{-1}(B)$ is a $\mathcal{Z}_{g_A^{-1}(V)}$-skeletoid in $g_A^{-1}(V)$. Since $H$ is a homeomorphism, $H(\Sigma^n - g^{-1}(B))$ is a $\mathcal{Z}_V$-skeletoid in $V$.

Next, we prove that $g_A(\Sigma^n) - g_A(g^{-1}(B))$ is also a $\mathcal{Z}_V$-skeletoid in $V$. To see this, first notice that

- $g_A(\Sigma^n)$ is a $\mathcal{Z}_{\tilde{M}'}$-skeletoid.

Indeed, by Theorem 5.1.12, $\Sigma^n - g^{-1}(A)$ is a $\mathcal{Z}_{\tilde{M}-g^{-1}(A)}$-skeletoid. Since $g_A$ maps $\tilde{M} - g^{-1}(A)$ homeomorphically onto $\tilde{M}' - g_A(g^{-1}(A))$, we see that $g_A(\Sigma^n) - g_A(g^{-1}(A))$ is a $\mathcal{Z}_{\tilde{M}'-g^{-1}(A)}$-skeletoid in $\tilde{M}' - g_A(g^{-1}(A))$. By the construction, $g_A(g^{-1}(A))$ is a compact $Z$-set in $\tilde{M}'$. By the above Claim, $g_A(\Sigma^n) - g_A(g^{-1}(A))$ is a $\mathcal{Z}_{\tilde{M}'}$-skeletoid. Then, by Proposition 2.2.25,

$$g_A(\Sigma^n) = (g_A(\Sigma^n) - g_A(g^{-1}(A))) \cup g_A(g^{-1}(A))$$

is also a $\mathcal{Z}_{\tilde{M}'}$-skeletoid. This proves ($\bullet$).

Again, by Theorem 5.1.12, $g_A(\Sigma^n) - g_A(g^{-1}(B))$ is a $\mathcal{Z}_V$-skeletoid as desired.

Thus, both $g_A(\Sigma^n) - g_A(g^{-1}(B))$ and $H(\Sigma^n - g^{-1}(B))$ are $\mathcal{Z}_V$-skeletoids. Theorem 2.2.20 ensures that there is an autohomeomorphism $G\colon V \to V$ which is $\mathcal{W}$-close to $id_V$ such that $G(H(\Sigma^n - g^{-1}(B)) = g_A(\Sigma^n) - g_A(g^{-1}(B))$. It only remains to note that $GH/\ldots : \Sigma^n - g^{-1}(B) \to g_A(\Sigma^n) - g_A(g^{-1}(B)) = (\Sigma^n, h)_A - B$ is a homeomorphism which is $\mathcal{U}$-close to $h_A/(\Sigma^n - g^{-1}(A))$. This completes the proof of Theorem 5.6.10. □

*Historical and bibliographical notes* 5.6. The results of this Section originally appear in [**106**]. *Problem 611* in [**123**] asks whether $\mu^n$ can be represented as the union $X \cup Y$, where $X \approx \sigma_n^{2n+1}$, $Y \approx N_n^{2n+1}$ and both of $X$ and $Y$ are locally $n$-homotopy negligible in $\mu^n$ (for the definition, see [**123**, p.413]). Expressing $\mu^n$ as the union $\Sigma^n \cup \nu^n$, Theorems 5.5.5 and 5.6.10 give a positive solution of this problem. Theorem 5.6.10 also gives a positive solution of *Problem 610* in [**123**].

It is known [**121**] that $\sigma_n^{2n+1} \approx \sigma_n^k$ for each $k \geq 2n+1$. If $n = 0, \infty$, then Conjecture 5.6.9 has a positive solution. This follows from [**114**] and Theorem 2.5.1 (or Theorem 2.5.2).

## 5.7. Open subspaces of Nöbeling spaces

The results presented in this Chapter so far show that all of the major geometrical constructions of strongly $\mathcal{A}_{\omega,n}$-universal spaces discussed above produce the same (topologically) space $N_n^{2n+1}$. On the other hand, there is another natural way of producing strongly $\mathcal{A}_{\omega,n}$-universal spaces. Namely, it can easily be seen that $(n-1)$-connected open subspaces of $N_n^{2n+1}$ have all of the mentioned properties. Are they also homeomorphic to $N_n^{2n+1}$ ?

Below we show (Corollary 5.7.11) that this is indeed the case.

We also show (Corollary 5.7.13), as in the case of $N_n^{2n+1}$, that $(n-1)$-connected open subspaces of $\sigma_n^{2n+1}$ are homeomorphic to the ambient space $\sigma_n^{2n+1}$.

**5.7.1. Characterization of pseudo-interiors and pseudo-boundaries of Menger manifolds.** The collection $Z(M)$ of compact Z-sets of a Menger manifold $M$ is a perfect collection in the sense of Definition 2.2.21. Therefore, pseudo-boundaries of Menger manifolds are $Z(M)$-skeletoids, and they can be studied using the general theory of skeletoids. In particular, we have the following statements (compare with the corresponding results from Subsection 2.2.2).

PROPOSITION 5.7.1. *Any two pseudo-boundaries of a $\mu^n$-manifold are homeomorphic. Moreover, if $A$ and $B$ are pseudo-boundaries of $M$, then the identity map $id_M$ of $M$ can be approximated by autohomeomnorphisms of $M$ which send $A$ onto $B$.*

PROPOSITION 5.7.2. *If $A$ is a pseudo-boundary of a $\mu^n$-manifold $M$ and $Z$ is a $\sigma Z$-set in $M$, then $A \cup Z$ is a pseudo-boundary of $M$.*

PROPOSITION 5.7.3. *If $A$ is a pseudo-boundary of a $\mu^n$-manifold $M$ and $U$ is open in $M$, then $U \cap A$ is a pseudo-boundary of $U$.*

Let $M$ be a $\mu^n$-manifold (not necessarily compact). We say that a subset $A$ is a *pseudo-boundary* of $M$ if $M = \cup_{k=1}^{\infty} A_k$, where $A_k$ is a compact Z-set in $M$ such that $A_k \subseteq A_{k+1}$ and the following condition is satisfied[1]:

(*) for each $\epsilon > 0$, each $k = 1, 2, \ldots$, and each compact subset $K$ of $M$, there is an integer $m \geq k$ and an embedding $f : K \to A_m$ such that $f/(K \cap A_k) = id$ and $f$ and $id_M$ are $\epsilon$-close.

In the case of $I^\omega$-manifolds, pseudo-boundaries in the above sense have been considered in [**64**] under the name of *cap-sets* (=compact absorption property).

We begin with the following statement.

PROPOSITION 5.7.4. *Each $\mu^n$-manifold $M$ has a pseudo-boundary $\Sigma^n(M)$.*

PROOF. By Theorem 4.4.3, $Ker_{n-1}(M)$ can be identified with an open subspace of $\mu^n$. By Proposition 5.7.3, $A = \Sigma^n \cap Ker_{n-1}(M)$ is a pseudo-boundary of $Ker_{n-1}(M)$. Therefore, $A = \cup_{k=1}^{\infty} A_k$, where $\{A_k\}_{k=1}^{\infty}$ is an increasing sequence of compact Z-sets of $Ker_{n-1}(M)$ satisfying condition (*). Let us show that $A$ is a pseudo-boundary of $M$ as well. Fix $\epsilon > 0$, an integer $k$ and a compact subset $K$ of $M$. Let $\rho$ be a metric on $M$. By definition, $Ker_{n-1}(M) = M - Z$, where $Z$ is a Z-set in $M$. Consequently, there is a Z-embedding $h \colon K \to M$ such that $h(K) \cap Z = \emptyset$, $h/(K \cap A_k) = id$ (recall that $A \cap Z = \emptyset$) and $\rho(id, h) < \epsilon/2$. Since $A$ is a pseudo-boundary of $Ker_{n-1}(M)$ and $h(K)$ is compact, there is an integer $m \geq k$ and an embedding $g : h(K) \to A_m$ such that $g/((h(K) \cap A_k) = id$ and $\rho(id, g) < \epsilon/2$. Let $f = g \cdot h$. It only remains to note that $f/(K \cap A_k) = id$, $f(K) \subseteq A_m$ and $\rho(id, f) < \epsilon$. □

For any $\mu^n$-manifold $M$, the complement $M - \Sigma^n(M)$ of its pseudo-boundary will be denoted by $\nu^n(M)$ and called the *pseudo-interior* of $M$. Recall that if $M = \mu^n$, we use the simpler notation: $\Sigma^n = \Sigma^n(M)$ and $\nu^n = \nu^n(M)$.

PROPOSITION 5.7.5. *For any space $X$, the following conditions are equivalent:*

(i) *$X$ is homeomorphic to an open subspace of the universal n-dimensional Nöbeling space $N_n^{2n+1}$.*

(ii) *$X$ is homeomorphic to an open subspace of $\nu^n$.*

(iii) *$X$ is homeomorphic to the pseudo-interior $\nu^n(M)$ of some $\mu^n$- manifold $M$.*

PROOF. (i) ⇒ (ii). By Theorem 5.5.5, $N_n^{2n+1}$ and $\nu^n$ are homeomorphic.

(ii) ⇒ (iii). Choose an open subspace M of $\mu^n$ such that $M \cap \nu^n = X$. By Proposition 5.7.3, $M \cap \Sigma^n$ is a pseudo-boundary of $M$. Consequently, $X$ is a pseudo-interior of $M$.

(iii) ⇒ (i). Suppose that $X = M - \Sigma^n(M)$, where $\Sigma^n(M)$ is a pseudo-boundary of a $\mu^n$-manifold $M$. Consider an (n-1)-homotopy kernel $Ker_{n-1}(M)$ of $M$. Choose a Z-embedding $f : M \to M$ such that $f$ is (n-1)- homotopic to the identity map $id_M$ and $f(M) \cap \Sigma^n(M) = \emptyset$ (this is possible because $\Sigma^n(M)$ is

[1] Compare with Theorem 2.2.22.

a $\sigma Z$-set in $M$). Then $\Sigma^n(M) \subseteq M - f(M) = Ker_{n-1}(M)$. By Theorem 4.4.3, $Ker_{n-1}(M)$ can be identified with an open subspace of $\mu^n$. By Proposition 5.7.3, $Ker_{n-1}(M) \cap \Sigma^n$ is a pseudo-boundary of $Ker_{n-1}(M)$. For the same reason, $\Sigma^n$ is a pseudo-boundary of $Ker_{n-1}(M)$. Uniqueness of pseudo-boundaries (Proposition 5.7.1) implies that the complements $Ker_{n-1}(M) - \Sigma^n$ and $Ker_{n-1}(M) - \Sigma^n(M)$ are homeomorphic. But $Ker_{n-1}(M) - \Sigma^n = Ker_{n-1}(M) \cap \nu^n$ is an open subspace of $\nu^n$. Therefore it only remains to show that the second complement $Y = Ker_{n-1}(M) - \Sigma^n(M)$ is homeomorphic to $X$. Consider the complement $M - Y = \Sigma^n(M) \cup f(M)$ and $M - X = \Sigma^n(M)$ (recall that $X = \nu^n(M) = M - \Sigma^n(M)$). Obviously, $f(M)$ is a $\sigma$-compact $\sigma Z$-set in $M$. Consequently, by Proposition 5.7.2, the union $\Sigma^n(M) \cup f(M)$ is a pseudo- boundary of $M$ as well. Proposition 5.7.1 finishes the proof. □

The next statement can be proved by using the same argument.

PROPOSITION 5.7.6. *The following conditions are equivalent for any space $X$:*
(i) *$X$ is homeomorphic to an open subspace of $\sigma_n^{2n+1}$.*
(ii) *$X$ is homeomorphic to an open subspace of $\Sigma^n$.*
(iii) *$X$ is homeomorphic to the pseudo-boundary $\Sigma^n(M)$ of some $\mu^n$- manifold $M$.*

The following statement is a direct consequence of Propositions 5.7.1 and 5.7.2.

PROPOSITION 5.7.7. *Let $Z$ be a $\sigma Z$-set in the pseudo-interior $\nu^n(M)$ of a $\mu^n$-manifold $M$. Then the inclusion map $i : \nu^n(M) - Z \to \nu^n(M)$ is a near-homeomorphism.*

THEOREM 5.7.8. *Let $\nu^n(M_1)$ and $\nu^n(M_2)$ be pseudo-interiors of the $\mu^n$-manifolds $M_1$ and $M_2$ respectively. If there is a map $f\colon \nu^n(M_1) \to \nu^n(M_2)$ which induces an isomorphism of the $i$-dimensional homotopy groups for each $i \leq n-1$, then $\nu^n(M_1)$ and $\nu^n(M_2)$ are homeomorphic.*

PROOF. Denote by $i_k\colon \nu^n(M_k) \hookrightarrow M_k$, $k = 1, 2$, the inclusion map. Clearly, $i_k$ is an $(n-1)$-homotopy equivalence. Let $j_k\colon M_k \to \nu^n(M_k)$ be an $(n-1)$-homotopy inverse of $i_k$. Then the composition $g = i_2 f j_1\colon M_1 \to M_2$ induces an isomorphism of the $i$-dimensional homotopy groups for each $i \leq n-1$ and, consequently, is an $(n-1)$-homotopy equivalence. Therefore, by Theorem 4.4.7, the $(n-1)$-homotopy kernels of $M_1$ and $M_2$ are homeomorphic. Denote by

$$h\colon Ker_{n-1}(M_1) \to Ker_{n-1}(M_2)$$

the corresponding homeomorphism. Since $M_k - \nu^n(M_k)$ is a $\sigma Z$-set in $M_k$, we can write $Ker_{n-1}(M_k) = M_k - Z_k$, where $Z_k$ is a $Z$-set in $M_k$ such that $Z_k \cap (M_k - \nu^n(M_k)) = \emptyset$, $k = 1, 2$. Further, by Proposition 5.7.3, $M_k - \nu^n(M_k)$ is a pseudo-boundary in $Ker_{n-1}(M_k)$. Therefore, by Proposition 5.7.1, $\nu^n(M_1) - Z_1 = Ker_{n-1}(M_1) - (M_1 - \nu^n(M_1))$ is homeomorphic to $Ker_{n-1}(M_2) - (M_2 -$

$\nu^n(M_2)) = \nu^n(M_2) - Z_2$. By Proposition 5.7.7, $\nu^n(M_k) - Z_k \approx \nu^n(M_k)$, $k = 1, 2$. Consequently, $\nu^n(M_1) \approx \nu^n(M_2)$. □

The following statement can be proved in a similar way.

THEOREM 5.7.9. *Let $\Sigma^n(M_1)$ and $\Sigma^n(M_2)$ be pseudo-boundaries of the $\mu^n$-manifolds $M_1$ and $M_2$ respectively. If the map $f\colon \Sigma^n(M_1) \to \Sigma^n(M_2)$ induces an isomorphism of the i-dimensional homotopy groups for each $i \leq n-1$, then $\Sigma^n(M_1)$ and $\Sigma^n(M_2)$ are homeomorphic.*

*Remark* 5.7.10. In fact, stronger versions of Theorems 5.7.8 and 5.7.9 are correct. Namely, the homeomorphism between $\nu^n(M_1)$ and $\nu^n(M_2)$ (as well as between $\Sigma^n(M_1)$ and $\Sigma^n(M_2)$) can be chosen to be (n-1)-homotopic to the originally given map $f$. Technically the proof of the last fact is more complicated, but the idea itself is the same. Loosely speaking all we need is the existence of a homeomorphism $h : Ker_{n-1}(M_1) \to Ker_{n-1}(M_2)$ which is (n-1)-homotopic to $g/Ker_{n-1}(M_1)$ (we maintain the notations given in the above proof). This appears implicitly in [**98**], or one can observe that $Ker_{n-1}(Ker_{n-1}(M))$ is homeomorphic to $Ker_{n-1}(M)$ (Proposition 4.4.4) and then apply [**182**, Theorem III].

COROLLARY 5.7.11. *An $(n-1)$-connected open subspace of the n-dimensional universal Nöbeling space $N_n^{2n+1}$ is homeomorphic to $N_n^{2n+1}$.*

PROOF. Apply Proposition 5.7.5 and Theorem 5.7.8 □

COROLLARY 5.7.12. *A connected open subspace of the one-dimensional universal Nöbeling space $N_1^3$ is homeomorphic to $N_1^3$.*

Similarly, we have the following statements.

COROLLARY 5.7.13. *An $(n-1)$-connected open subspace of $\Sigma^n$ $(\sigma_n^{2n+1})$ is homeomorphic to $\Sigma^n$ $(\sigma_n^{2n+1})$.*

PROOF. Apply Proposition 5.7.6 and Theorem 5.7.9 □

COROLLARY 5.7.14. *A connected open subspace of $\Sigma^1$ $(\sigma_1^3)$ is homeomorphic to $\Sigma^1$ $(\sigma_1^3)$.*

The following statement can be considered to be the converse of Theorem 5.6.4.

PROPOSITION 5.7.15. *Each open subspace of $N_n^{2n+1}$ admits an n-soft map onto an $l_2$-manifold as well as onto an n-dimensional locally finite polyhedron.*

PROOF. Let $U$ be an open subspace of $N_n^{2n+1}$. By Proposition 5.7.5, $U$ can be identified with the pseudo-interior $\nu^n(M)$ of some $\mu^n$-manifold $M$. By the triangulation theorem for $\mu^n$-manifolds, there is a proper $UV^{n-1}$-map $f\colon M \to K$ of $M$ onto an $n$-dimensional locally finite polyhedron $K$. Embed the product $K \times l_2$ into $l_2$ as an open subspace. By Theorem 5.6.4, there is an $n$-soft map $g\colon N_n^{2n+1} \to l_2$. Let $V = g^{-1}(K \times l_2)$. Clearly, $g/V\colon V \to K \times l_2$ (as well as the composition $\pi_K g/V : V \to K \times l_2 \to K$) is an $n$-soft map. Therefore it only remains to show that $U$ and $V$ are homeomorphic.

Since $\Sigma^n(M)$ is a $\sigma Z$-set in $M$, the inclusion map $i \colon \nu^n(M) \hookrightarrow M$ is an $(n-1)$-homotopy equivalence. Note that the map $f$, as a proper $UV^{n-1}$-map between locally compact $\mathcal{L}C^{n-1}$-spaces, is an $(n-1)$-homotopy equivalence. By the $n$-softness of $g$ and the equality $\dim K = n$, there is a map $j \colon K \to V$ such that $\pi_K g j = id_K$. In this situation it can be easily seen that the composition $jfi \colon U \to V$ induces an isomorphism of $i$-dimensional homotopy groups for each $i \leq n-1$. By Theorem 5.7.8, $U$ and $V$ are homeomorphic. □

**5.7.2. Finite-dimensional absorbing sets.** In this Subsection we discuss some problems concerning absorbing sets in various classes of spaces.

Let $\mathcal{K}$ be a class of spaces that is topological, finitely additive and hereditary with respect to closed subspaces class of spaces. A space $X$ is *strongly $\mathcal{K}$-universal* if, for every map $f \colon C \to X$ from a space $C \in \mathcal{K}$ into $X$, for every closed subspace $D \subseteq C$ such that $f/D \colon D \to X$ is a $Z$-embedding and for every open cover $\mathcal{U} \in cov(X)$, there exists a $Z$-embedding $g \colon C \to X$ such that $g/D = f/D$ and $g$ is $\mathcal{U}$-close to $f$.

The class consisting of countable unions of members of $\mathcal{K}$ is denoted by $\mathcal{K}_\sigma$.

Let $n \in \omega \cup \{\infty\}$. An $n$-dimensional separable metrizable space $X$ is a *$\mathcal{K}$-absorbing set* if:

- $X \in AE(n)$.
- $X$ is a countable union of strong $Z$-sets.
- $X \in \mathcal{K}_\sigma$.
- $X$ is strongly $\mathcal{K}$-universal.

Several examples of $\mathcal{K}$-absorbing sets (for various classes $\mathcal{K}$) can be found in [**123**].

Let us recall that for each space $X$ and for each ordinal $\alpha < \omega_1$, we can define two classes of subspaces of $X$ – the *additive Borelian class* $\alpha$, $\mathcal{A}_\alpha(X)$, and the *multiplicative Borelian class* $\alpha$, $\mathcal{M}_\alpha(X)$, – as follows: $\mathcal{A}_0(X)$ is the collection of all open subsets of $X$ and $\mathcal{M}_0(X)$ is the collection of all closed subsets of $X$. Assuming that for each ordinal $\beta < \alpha$, where $\alpha < \omega_1$, the classes $\mathcal{A}_\beta$ and $\mathcal{M}_\beta$ have already been constructed, we proceed as follows: the class $\mathcal{A}_\alpha$ consists of countable unions of elements of $\cup\{\mathcal{M}_\beta \colon \beta < \alpha\}$ and the class $\mathcal{M}_\alpha$ consists of countable intersections of elements of $\cup\{\mathcal{A}_\beta \colon \beta < \alpha\}$.

Further, let $\alpha < \omega_1$ and $X$ be a separable metrizable space. We say that $X$ belongs to the *absolute additive Borelian class* $\mathcal{A}_\alpha$, if for any embedding $i \colon X \to Y$ into any separable metrizable space $Y$, we have $i(X) \in \mathcal{A}_\alpha(Y)$. Similarly, $X$ belongs to the *absolute multiplicative Borelian class* $\mathcal{M}_\alpha$ if for any embedding $i \colon X \to Y$ into any separable metrizable space $Y$, we have $i(X) \in \mathcal{M}_\alpha(Y)$. It is well-known that: (a) $X \in \mathcal{A}_\alpha$, $\alpha \geq 2$, if and only if $X \in \mathcal{A}_\alpha(l_2)$ and (b) $X \in \mathcal{M}_\alpha$, $\alpha \geq 1$, if and only if $X \in \mathcal{M}_\alpha(l_2)$.

Obviously, $\mathcal{A}_0 = \emptyset$ and $\mathcal{M}_0$ coincides with the class of all metrizable compacta. Further, $\mathcal{A}_1 = \{\sigma\text{-compact spaces}\}$, $\mathcal{M}_1 = \{\text{Polish spaces}\}$, etc.

Let $\mathcal{A}_\alpha(n) = \{X \in \mathcal{A}_\alpha \colon \dim X \leq n\}$ and $\mathcal{M}_\alpha(n) = \{X \in \mathcal{M}_\alpha \colon \dim X \leq n\}$ for each $n \in \omega$ and $\alpha < \omega_1$.

The main problem to be considered in this section is the following [**123**, Problem 551]:

- Do there exist $\mathcal{A}_\alpha(n)$- and/or $\mathcal{M}_\alpha(n)$-absorbing sets?

Since, by the construction, the pseudo-boundary $\Sigma^n$ of the $n$-dimensional Menger compactum $\mu^n$ is a $Z$-skeletoid in $\mu^n$, we see that $\Sigma^n$ serves as an example of $\mathcal{A}_1(n)$-absorbing set. Let us show that $\mathcal{M}_1(n)$-absorbing sets also exist. The corresponding space is a countable increasing union of copies of the $n$-dimensional Nöbeling space $N_n^{2n+1}$.

Fix $n \in N$ and consider the $\mathcal{Z}(\infty, n)$-skeletoid $\Sigma(\infty, n)$ (see Theorem 5.2.1) of the Hilbert cube $I^\omega$. By the construction, $\Sigma(\infty, n)$ is the union of an increasing family $\{A_m \colon m \in N\}$, where each $A_m$ is a copy of $\mu^n$. Note also that $A_m$ is a $Z$-set in $A_{m+1}$ for each $m$. Let $D_m$ be a pseudo-boundary of $A_m$ such that $A_{m-1} \cap D_m = \emptyset$ (see Remark 5.2.2) and let $B_m = \cup_{i=1}^m D_i$. By Proposition 5.2.4, $B_m$ is a pseudo-boundary of $A_m$ and, therefore, $L_m = A_m - B_m$ is a pseudo-interior of $A_m$ for each $m \in N$. Finally, let $L^n = \cup_{m=1}^\infty L_m$. We prove the following statement.

Proposition 5.7.16. *$L^n$ is a $\mathcal{M}_1(n)$-absorbing set.*

Proof. The proof of Proposition 5.2.14 can be adapted to prove both the local and global $(n-1)$-connectedness of $L^n$. Therefore $L^n \in LC^{n-1} \cap C^{n-1}$. To verify the strong $\mathcal{M}_1(n)$-universality, we first prove that

- any map $f \colon C \to L^n$, where $C \in \mathcal{M}_1(n)$, can be approximated by closed $Z$-embeddings.

Since the proof is similar to that of Proposition 5.2.7, we give only a brief sketch. Take a metrizable compactification $\tilde{C}$ of $C$, with dim $\tilde{C} \leq n$, which admits an extension $F \colon \tilde{C} \to \mu^\infty$ of $f$ (Proposition 1.3.11). Take an open cover $\mathcal{U}$ of $L^n$, extend it to a collection $\mathcal{V}$ of open sets of $I^\omega$, and let $M = \cup\mathcal{V}$ and $N = F^{-1}(M)$. Since $M$ is an $I^\omega$-manifold, we can $\mathcal{V}$-approximate $F/N$ by a $Z$-embedding $g \colon N \to M$. By the properties of skeletoids and the $\sigma$-compactness of $N$, we can change $g$ slightly into a map $h \colon N \to M$ so that $h(N)$ is contained in $\Sigma(\infty, n)$. Since $C$ is a $G_\delta$-set in $N$, $h(N - C)$ is a $\sigma Z$-set in $\Sigma(\infty, n)$. Using Theorem 2.2.20, we can again change $h$ slightly into a map $k \colon N \to \Sigma(\infty, n)$ so that $k(N - C) \subseteq \cup_{m=1}^\infty D_m$. Then $f' = k/C$ is the required embedding.

To prove the general case, let $f \colon C \to L^n$ be a map of a Polish space $C$ with dim $C \leq n$ and suppose that $f/D \colon D \to L^n$ is a $Z$-embedding, where $D$ is a closed set in $C$. As above, take an extension $F \colon \tilde{C} \to I^\omega$ of $f$, and take any open cover $\mathcal{U}$ of $L^n$ and an extension $\mathcal{V}$ of $\mathcal{U}$ to $I^\omega$. Consider $M = \cup\mathcal{V}$ and let $N = F^{-1}(M)$, $D^* = cl_N(D)$. Since $\{M \cap A_m\}$ is an absorber in $M$ (Theorem

5.1.12) and $D^* - D$ is $\sigma$-compact, $F(D^* - D)$ can be absorbed into $M \cap \bigcup_{m=1}^{\infty} A_m$ by a small homeomorphism of $M$. Further, using the $\sigma$-compactness of $N - D^*$ and the fact that $\bigcup_{m=1}^{\infty} B_m$ is a $\sigma Z$-set, we can change $F/N$ slightly into $F'\colon N \to \bigcup_{m=1}^{\infty} A_m$ so that $F'(D^*) \cap L^n = f(D)$. Then it is easy to see that $F'(D')$ is a $Z$-set in $\mu^n$. The proof of the $Z$-set approximation theorem and the absorption property of $\{M \cap A_m\}$ allow us to approximate $F'$ by $G\colon N \to M \cap \bigcup_{m=1}^{\infty} A_m$ so that

- $G$ is $\mathcal{V}$-close to $F'$ and $G/D^* = F/D^*$,
- $G(N - D^*) \cap F(D^*) = \emptyset$ and $G/(N - D^*)$ is an embedding, and
- $im(G)$ is a $Z$-set.

Since $G(N - D^*)$ is a $\sigma Z$-set, we can "push $G(N - D)$ into $\bigcup_{m=1}^{\infty} D_m$" by a small homeomorphism of $M$ keeping $D^*$ fixed (Theorem 2.2.20). Then the desired embedding is $G/C$. This completes the proof. □

*Remark* 5.7.17. It is not hard to see that the same construction can be performed to obtain an $\mathcal{M}_1(n)$-absorbing set in $\mu^n$.

The space $L^n$ of Proposition 5.7.16 is an increasing union of closed subspaces each of which is a copy of the $n$-dimensional Nöbeling space $N_n^{2n+1}$. A similar construction produces the space $L^\omega$, which is an increasing union of increasing-dimensional Nöbeling spaces and serves as an $\mathcal{M}_1(\omega)$-absorbing set. Indeed, represent the skeletoid $\sigma$ with respect to the collection of all finite dimensional $Z$-sets in $I^\omega$ as a union $\sigma = \bigcup_{n=1}^{\infty} A_n$, where $A_n$ is a topological copy of $\mu^n$ and is a $Z$-set in $A_{n+1}$. As above, in each $A_n$, we fix $D_n$, a pseudo-boundary, so that $A_{n-1} \cap D_n = \emptyset$. Let $B_n = \bigcup_{i=1}^{n} D_i$ and let $L_m = A_m - B_m$ and $L^\omega = \bigcup_{m=1}^{\infty} L_m$. We claim that $L^\omega$ is an $\mathcal{M}_1(\omega)$-absorbing set. The straightforward verification of this fact is left to the reader.

Now we show that the above problem has a positive solution for each $n \in \omega$ and $\alpha < \omega_1$ with $\alpha \geq 2$ (the case $\alpha = 1$ was discussed above).

Both (additive and multiplicative) cases can be proved simultaneously by the same argument for all $n \in \omega$ and $\alpha < \omega_1$ with $\alpha \geq 2$. For this reason, below we consider only the case of additive Borel classes.

LEMMA 5.7.18. *Let $n \in \omega$ and $\alpha < \omega_1$ with $\alpha \geq 2$. For each space $Y \in \mathcal{A}_\alpha$ there exists an $n$-soft map $f\colon X \to Y$ such that $X \in \mathcal{A}_\alpha(n)$. If, additionally, $Y \in \mathcal{A}_\alpha \cap AE(n)$, then $X \in \mathcal{A}_\alpha(n) \cap AE(n)$.*

PROOF. Let $g\colon N_n^{2n+1} \to l_2$ be an $n$-soft map. We can assume that $Y \subseteq l_2$. Take $X = g^{-1}(Y)$ and $f = g/X : X \to Y$. Clearly, $X \in \mathcal{A}_\alpha(n)$. If $Y \in AE(n)$, then the $n$-softness of $g$ guarantees that $X \in AE(n)$. □

*Remark* 5.7.19. This statement is not true for $\alpha = 1$. Indeed, since the maps under consideration are not proper, we cannot conclude that inverse images of $\sigma$-compact sets are $\sigma$-compact.

LEMMA 5.7.20. *Let $n \in \omega$ and $\alpha < \omega_1$ with $\alpha \geq 2$. Suppose that $X_1 \in \mathcal{A}_\alpha(n) \cap AE(n)$. Then there exist an $n$-soft map $p\colon X_2 \to X_1$ and a $Z$- embedding $i\colon X_1 \to X_2$ such that:*

(i) $X_2 \in \mathcal{A}_\alpha(n) \cap AE(n)$.

(ii) $p \cdot i = id_{X_1}$.

(iii) *For each space $C \in \mathcal{A}_\alpha(n)$, its closed subspace $D \subseteq C$ and each map $\varphi\colon C \to X_2$ such that $\varphi/D$ is a closed embedding with $\varphi(D) \subseteq i(X_1)$, there is a $Z$-embedding $\bar{\varphi}\colon C \to X_2$ such that $\bar{\varphi}/D = \varphi/D$ and $p \cdot \bar{\varphi} = p \cdot \varphi$.*

PROOF. Clearly, $X_1 \times \Lambda_\alpha \in \mathcal{A}_\alpha \cap AE(n)$ (recall that $\Lambda_\alpha \in AR$). By Lemma 5.7.18, there is an $n$-soft map $h\colon X_2 \to X_1 \times \Lambda_\alpha$ such that $X_2 \in \mathcal{A}_\alpha(n) \cap AE(n)$. Let $p = \pi_1 \cdot h\colon X_2 \to X_1$, where $\pi_1\colon X_1 \times \Lambda_\alpha \to X_1$ denotes the projection.

Let us now define the section $i\colon X_1 \to X_2$. Since $\Lambda_\alpha$ is strongly $\mathcal{A}_\alpha$-universal and $X_1 \in \mathcal{A}_\alpha(n)$, there is a $Z$-embedding $j\colon X_1 \to \Lambda_\alpha$. Then the diagonal product $id_{X_1} \triangle j\colon X_1 \to X_1 \times \Lambda_\alpha$ is also a $Z$-embedding. Since $\dim X_1 \leq n$ and $h$ is $n$-soft, there is a map $i\colon X_1 \to X_2$ such that $h \cdot i = id_{X_1} \triangle j$. In particular, $p \cdot i = \pi_1 \cdot h \cdot i = \pi_1 \cdot (id_{X_1} \triangle j) = id_{X_1}$. Observe that, since $id_{X_1} \triangle j$ is a $Z$-embedding, the map $i$ is also a closed embedding. Further, inverse images of $Z$-sets under $n$-soft maps with at most $n$-dimensional domains are $Z$- sets. Therefore, $i$ is a $Z$-embedding.

Let us verify condition (3). Let $C \in \mathcal{A}_\alpha(n)$, $D$ be a closed subspace of $C$ and $\varphi\colon C \to X_2$ be a map such that $\varphi/D$ is a closed embedding with $\varphi(D) \subseteq i(X_1)$. Consider the composition $\psi = \pi_2 \cdot h \cdot \varphi\colon C \to \Lambda_\alpha$, where $\pi_2\colon X_1 \times \Lambda_\alpha \to \Lambda_\alpha$ denotes the projection onto the second coordinate. One can easily verify that in our situation $\psi/D\colon D \to \Lambda_\alpha$ is a $Z$-embedding. Since $C \in \mathcal{A}_\alpha(n)$ we conclude, using the strong $\mathcal{A}_\alpha$-universality of $\Lambda_\alpha$, that there is a $Z$-embedding $\bar{\psi}\colon C \to \Lambda_\alpha$ such that $\bar{\psi}/D = \psi/D$. Then $\phi = (\pi_1 \cdot h \cdot \varphi) \triangle \bar{\psi}\colon C \to X_1 \times \Lambda_\alpha$ is also a $Z$-embedding extending $h \cdot \varphi/D$. Finally, the $n$- softness of $h$ (recall that $\dim C \leq n$) guarantees the existence of a map $\bar{\varphi}\colon C \to X_2$ such that $\bar{\varphi}/D = \varphi/D$ and $h \cdot \bar{\varphi} = \phi$. It only remains to note that $\bar{\varphi}$ is a $Z$-embedding and $p \cdot \bar{\varphi} = p \cdot \varphi$. □

THEOREM 5.7.21. *Let $n \in \omega$ and $\alpha < \omega_1$. Then there exist an $\mathcal{A}_\alpha(n)$-absorbing set $\Lambda_\alpha(n)$ and an $\mathcal{M}_\alpha(n)$-absorbing set $\Omega_\alpha(n)$.*

SKETCH OF PROOF. The statement has already been proved for $\alpha = 1$. Therefore, below we assume that $\alpha \geq 2$. Further, the proofs of both (additive and multiplicative) cases are identical, and for this reason we consider only the first one.

First we construct the space $\Lambda_\alpha(n)$, and then we outline the proof of all the desired properties of this space.

Let us construct an inverse sequence $\mathcal{S} = \{X_k, p_k^{k+1}, \omega\}$, all spaces $X_k$ of which belong to the class $\mathcal{A}_\alpha(n) \cap AE(n)$ and all bonding maps $p_k^{k+1}\colon X_{k+1} \to X_k$ of which are $n$-soft. As the first space $X_0$ take any element of the class

$\mathcal{A}_\alpha(n) \cap AE(n)$ (we can do this by Lemma 5.7.18 for example). For each $k$, by Lemma 5.7.20, we can define an $n$-soft map $p_k^{k+1}\colon X_{k+1} \to X_k$ satisfying conditions (i), (ii) and (iii) of that Lemma. In particular, at each stage we have $Z$-embeddings $i_k^{k+1}\colon X_k \to X_{k+1}$ such that:

$(i)_k$ $p_k^{k+1} \cdot i_k^{k+1} = id_{X_k}$ for each $k$.

$(ii)_k$ For each space $C \in \mathcal{A}_\alpha(n)$, closed subspace $D \subseteq C$ and each map $\varphi\colon C \to X_2$ such that $\varphi/D$ is a closed embedding with $\varphi(D) \subseteq i_k^{k+1}(X_k)$, there is a $Z$-embedding $\bar{\varphi}\colon C \to X_2$ such that $\bar{\varphi}/D = \varphi/D$ and $p_k^{k+1} \cdot \bar{\varphi} = p_k^{k+1} \cdot \varphi$.

Let $X = \lim \mathcal{S}$. For each $k$ denote by $p_k\colon X \to X_k$ the $k$-th limit projection of the spectrum $\mathcal{S}$. Since all the $p_k^{k+1}$'s are $n$-soft, we see that all the $p_k$'s are also $n$-soft (this shows, by the way, that $X \in AE(n)$). For a given $k$ and arbitrary $m \geq k$, consider the composition $i_k^m = i_{m-1}^m \cdots i_k^{k+1}\colon X_k \to X_m$. Clearly, for each pair $(k, m)$ of indexes with $k \leq m$, we have $p_m^{m+1} \cdot i_k^{m+1} = i_k^m$. Therefore, the diagonal product $i_k = \triangle\{i_k^m\colon m \geq k\}$ maps $X_k$ into $X$ and $p_k \cdot i_k = id_{X_k}$ for each $k$. Therefore, $Y_k = i_k(X_k)$ is closed in $X$. Note also that $Y_k \subseteq Y_{k+1}$ for each $k$.

Let $\Lambda_\alpha(n) = \cup_{k=1}^\infty Y_k$.

First of all, observe that $\Lambda_\alpha(n)$ is dense in $X$ (because $p_k(\Lambda_\alpha(n)) = X_k$ for each $k$). Since $\dim X = n$, we have $\dim \Lambda_\alpha(n) \leq n$. The inequality $\dim \Lambda_\alpha(n) \geq n$ follows from the fact that $Y_2 \subseteq \Lambda_\alpha(n)$ contains a copy of the $n$-dimensional disk (condition $(ii)_k$). Thus $\dim \Lambda_\alpha(n) = n$. Since each $Y_k$, as a copy of $X_k$, belongs to the class $\mathcal{A}_\alpha(n)$, we see that $\Lambda_\alpha(n) \in \mathcal{A}_\alpha(n)_\sigma$.

Applying essentially the same argument which was used to show that $\Sigma^n \in AE(n)$ (Proposition 5.2.14), one can prove that $X - \Lambda_\alpha(n)$ is homotopy $n$- negligible in $X$ (the later means that for every open set $U \subseteq X$ the inclusion $U \cap \Lambda_\alpha(n) \hookrightarrow U$ induces an isomorphism of $m$-dimensional homotopy groups for each $m$ with $m < n$). As was noted above, $X \in AE(n)$. Therefore, by [**297**], $\Lambda_\alpha(n) \in AE(n)$.

Finally, combining the arguments used in the proofs of [**35**, Proposition 2.5] and Proposition 5.1.11, one can show that the space $\Lambda_\alpha(n)$ is strongly $\mathcal{A}_\alpha(n)$-universal. □

*Historical and bibliographical notes* 5.7. The results of this Section were presented by the author at the 10-th Summer Conference on General Topology and Applications, Amsterdam, August, 1994. Theorem 5.7.21 was independently obtained by M. Zarichnyi (private communication).

CHAPTER 6

# General Theory of Absolute Extensors in Dimension n and n-soft Mappings

## 6.1. $ANE(n)$-spaces and $n$-soft mappings

**6.1.1. Discussion of the concept of Absolute Retract.** It is well known (and easy to see) that a retract of a Hausdorff space is *closed* in the space under consideration. It is for this reason that an absolute retract in some class $\mathcal{P}$ of Hausdorff spaces is defined to be a space $X \in \mathcal{P}$ such that for any *closed* embedding of $X$ into an arbitrary space $Y \in P$, there exists a retraction of $Y$ onto $X$. This definition of the concept of an absolute retract has an essential deficiency of a logical nature. The fact of the matter is that retracts are not only closed, but are also $C$-embedded in the ambient spaces. But, by the Brouwer-Tietze-Urysohn extension theorem, the property of closed subspaces being $C$-embedded characterizes normal spaces, and therefore the classical definition of the concept of an absolute retract is satisfactory only for various subclasses of the class of normal spaces. However, the indicated deficiency already manifests itself fully in the class of Tychonov spaces. This is confirmed by a result of Hanner [**162**] to the effect that each absolute retract in the class of all Tychonov spaces is compact. Consequently, under this definition neither the Euclidean space $R^n$, nor the Hilbert space $l_2$, is an absolute retract in the class of Tychonov spaces (although both are such in the class of Polish spaces) and this makes it impossible to use the theory of retracts in studying non-metrizable topological linear spaces and the associated problems of infinite-dimensional topology and functional analysis.

Because of this, it becomes necessary to add the requirement of being $C$-embedded to the classical definition of the concept of an absolute retract. Moreover, it turns out that in a number of cases we can even omit the requirement of being closed and confine ourselves to the requirement of being $C$-embedded. This is possible, for instance, in the case of spaces for which any $C$-embedding is automatically closed. What are the spaces characterized by the last property? These are, according to Proposition 1.1.16, precisely the realcompact spaces, i.e. the spaces homeomorphic to closed subspaces of arbitrary powers of the real line $R$ (the class of realcompact spaces is sufficiently broad: it contains all the compact Hausdorff spaces and all the Polish spaces; at the same time, it is sufficiently geometric, and in the study of it such classical objects of topology as infinite powers $R^\tau$ and $N^\tau$ of a Hilbert space and of a Baire space are brought to the forefront). But what happens outside the class of realcompact spaces? It may seem that a logical inconsistency analogous to that considered above emerges when the requirement of closeness is waived. But this is not so, since each absolute retract (under the altered definition) in the class of Tychonov spaces turns out to be realcompact. A more precise interpretation can be given to the above discussion. Consider the category **TYCH** of Tychonov spaces and their continuous maps on the one hand, and the category **VECT** of vector spaces their linear maps on the other. The contravariant functor $\mathcal{C}$: **TYCH** $\to$ **VECT** assigning to each space $X$ the vector space $C(X)$ of all continuous real-valued functions defined on $X$, and to each continuous map $f\colon X \to Y$ the induced linear map $C(f)\colon C(Y) \to C(X)$, enables us to single out the class of morphisms of **TYCH** corresponding to the epimorphisms of **VECT**. It can be shown that $C(f)$ is epimorphic if and only if $f$ is a $C$-embedding (i.e., an embedding with $C$-embedded image). In other words, $C$-embeddings turn out to be "monomorphisms" in **TYCH** (we use the contravariance of the functor $\mathcal{C}$). But since the concept of an absolute retract is categorical, we obtain the following definition.

DEFINITION 6.1.1. *A Tychonov space $X$ is said to be an* absolute retract *(notation: $X \in AR$) if for any $C$-embedding of $X$ into an arbitrary Tychonov space $Y$, there exists a retraction of $Y$ onto $X$.*

DEFINITION 6.1.2. *A Tychonov space $X$ is said to be an* absolute neighborhood retract *(notation: $X \in ANR$) if for any $C$-embedding of $X$ into an arbitrary Tychonov space $Y$, there exist a functionally open neighborhood $U$ of $X$ in $Y$ and a retraction of $U$ onto $X$.*

**6.1.2. Definition of $ANE(n)$-space.** The notion of $A(N)R$-space admits another approach as well. Before we give the corresponding definitions, let us call a functionally open neighborhood $U$ of a subset $A$ in a space $X$ *stable* if there is a functionally closed subset $Z$ of $X$ such that $A \subseteq Z \subseteq U$. Observe that if a subset $A$ is $C$-embedded in $X$, then every functionally open neighborhood of $A$ is stable (compare with [**159**]).

DEFINITION 6.1.3. *A space $X$ is called an* absolute (neighborhood) extensor in dimension $n$ *(shortly, $A(N)E(n)$-space), $n = 0, 1, \dots, \infty$, if for each at most $n$-dimensional space $Z$ and each subspace $Z_0$ of $Z$, any map $f\colon Z_0 \to X$, such that $C(f)(C(X)) \subseteq C(Z)/Z_0$, can be extended to (some stable functionally open neighborhood of $Z_0$) $Z$. An $A(N)E(\infty)$-space is called an* absolute (neighborhood) extensor, *or shortly, an $A(N)E$-space.*

PROPOSITION 6.1.4. *The class of $A(N)R$-spaces coincides with the class of $A(N)E$-spaces.*

PROOF. We prove only the parenthetical case. The proof of the absolute version is completely analogous. First of all, we note that each $ANE$-space is an $ANR$-space. This immediately follows from the corresponding definitions. Suppose now that $X$ is an $ANR$-space. We can assume that $X$ is $C$-embedded in $R^\tau$ for some cardinal $\tau$ (see Subsection 1.1.2). Take a functionally open neighborhood $V$ of $X$ in $R^\tau$ and a retraction $r\colon V \to X$. Consider also a space $Z$, its subspace $Z_0$, and a map $f\colon Z_0 \to X$ such that $C(f)(C(X)) \subseteq C(Z)/Z_0$. Since, as is easily seen, the space $R^\tau$ is an $AE$-space, we see that there exists a map $g\colon Z \to R^\tau$ such that $g/Z_0 = f$. Since $X$ is $C$-embedded in $R^\tau$, the functionally open neighborhood $U = g^{-1}(V)$ of $Z_0$ in $Z$ is stable. It only remains to note that the map $rg/U : U \to X$ is the desired extension of $f$. □

PROPOSITION 6.1.5. *Each $ANE(0)$-space is an $AE(0)$-space.*

PROOF. Let $X$ be an $ANE(0)$-space and $f\colon Z_0 \to X$ be a map, defined on a subspace $Z_0$ of a zero-dimensional space $Z$, such that $C(f)(C(X)) \subseteq C(Z)/Z_0$. Then $f$ has an extension $g\colon U \to X$, where $U$ is a stable functionally open neighborhood of $Z_0$ in $Z$. The stability of $U$ guarantees that there is a functionally closed subset $F$ in $Z$ such that $Z_0 \subseteq F \subseteq U$. Since $Z$ is zero-dimensional, there exists an open and closed subset $G$ in $Z$ such that $F \subseteq G \subseteq U$.Obviously the map $g/G\colon G \to X$ has an extension $h\colon Z \to X$ onto $Z$. It only remains to note that $h$ extends the map $f$. □

COROLLARY 6.1.6. *Each $ANE(n)$-space is an $AE(0)$-space.*

PROPOSITION 6.1.7. *Each $AE(0)$-space is realcompact.*

PROOF. Embed $X$ into $R^\tau$ as a $C$-embedded subspace for some cardinal $\tau$. Take a proper map $g\colon N^\tau \to R^\tau$, and consider its restriction $f = g/g^{-1}(X)$ onto the inverse image of $X$. Since $X$ is $C$-embedded in $R^\tau$, we can conclude that $C(f)(C(X)) \subseteq C(N^\tau)/g^{-1}(X)$. But $\dim N^\tau = 0$. Therefore there exists an extension $h\colon N^\tau \to X$ of the map $f$. By the properness of $g$, the many-valued map

$$r = hg^{-1}\colon R^\tau \to X$$

is compact-valued and upper semi-continuous. Since $r$ is the identity on $X$, we see that $X$ is closed in $R^\tau$. Proposition 1.1.13 finishes the proof. □

PROPOSITION 6.1.8. *Each $AE(0)$-space is perfectly $\kappa$-normal and has a countable Suslin number.*

PROOF. As in the proof of the preceding statement, embed an $AE(0)$-space $X$ into $R^\tau$ as a $C$-embedded subspace. Take a proper map $g\colon N^\tau \to R^\tau$ and consider a map $h\colon N^\tau \to X$ which extends the map $g/g^{-1}(X)$. The surjectivness of $g$ implies the surjectivness of $h$. Therefore, since the Suslin number of $N^\tau$ is countable, we conclude that $X$ also has a countable Suslin number.

Let us show now that $X$ is perfectly $\kappa$-normal. Take an arbitrary open subset $V$ of $X$ and let $U = X - \text{cl}_X V$. The properness of the map $g$ guarantees that the sets

$$V' = R^\tau - gh^{-1}(X - V) \text{ and } U' = R^\tau - gh^{-1}(X - U)$$

are open in $R^\tau$. Observe also that

$$V' \cap X = V, \quad U' \cap X = U \quad \text{and} \quad V' \cap U' = \emptyset.$$

As was remarked just before Proposition 1.1.21, the space $R^\tau$ is perfectly $\kappa$-normal. Consequently, the set $Z = \text{cl}_{R^\tau} V'$ is functionally closed in $R^\tau$. Also, $Z \cap U' = \emptyset$. It only remains to note that $\text{cl}_X V = Z \cap X$. □

PROPOSITION 6.1.9. *A space $X$ is an $A(N)E(n)$-space if and only if $X$ is realcompact and the following condition is satisfied:*

- *For each at most $n$-dimensional realcompact space $Z$, each closed subspace $Z_0$ of $Z$, any map $f\colon Z_0 \to X$, such that $C(f)(C(X)) \subseteq C(Z)/Z_0$, admits an extension onto (a stable functionally open neighborhood of $Z_0$ in $Z$) $Z$.*

PROOF. By Proposition 6.1.7, each $ANE(n)$-space is realcompact and obviously satisfies the above condition.

Let us prove the converse. Let $X$ be a realcompact space satisfying the above formulated condition. Consider an arbitrary space $Z$ of dimension at most $n$, an arbitrary subspace $Z_0$ of $Z$ and a map $f\colon Z_0 \to X$ such that $C(f)(C(X)) \subseteq C(Z)/Z_0$. Denote by $A_0$ the closure of the set $Z_0$ in the Hewitt realcompactification $\nu Z$ of $Z$. Since $Z$ is $C$-embedded in $\nu Z$, we have $C(f)(C(X)) \subseteq C(\nu Z)/Z_0$. By Proposition 1.3.12, the map $f$ can be extended to a map $g\colon A_0 \to X$ so that $C(g)(C(X)) \subseteq C(\nu Z)/A_0$. Since $\dim \nu Z = \dim Z \leq n$, our assumption guarantees the existence of a stable functionally open neighborhood $V$ of the set $A_0$ in $\nu Z$ and a map $h\colon V \to X$ with $h/A_0 = g$. It only remains to observe that $U = V \cap Z$ is a stable functionally open neighborhood of $Z_0$ in $Z$ and the map $h/U$ extends the originally given map $f$. □

PROPOSITION 6.1.10. *Each Polish space is an $AE(0)$-space. If $n \geq 1$, then the class of Polish $ANE(n)$-spaces (respectively, $AE(n)$-spaces) coincides with the class of Polish $LC^{n-1}$-spaces (respectively, $LC^{n-1} \cap C^{n-1}$-spaces).*

PROOF. Let $X$ be a Polish space and $Z_0$ be a closed subset of a zero-dimensional realcompact space $Z$. Consider a map $f\colon Z_0 \to X$ satisfying the inclusion $C(f)(C(X)) \subseteq C(Z)/Z_0$.By Theorem 1.3.10, the space $Z$ can be represented as the limit space of a factorizing $\omega$-spectrum $\mathcal{S} = \{Z_\alpha\ p_\alpha^\beta, A\}$ consisting of zero-dimensional Polish spaces. Since $Z_0$ is closed in $Z$, the limit space of the induced spectrum

$$\mathcal{S}_0 = \{\mathrm{cl}_{Z_\alpha} p_\alpha(Z_0), p_\alpha^\beta, A\}$$

coincides with $Z_0$ (Corollary 1.2.6). By Proposition 1.3.13, there exist an index $\alpha \in A$ and a map

$$f_\alpha \colon \mathrm{cl}_{Z_\alpha} p_\alpha(Z_0) \to X$$

such that $f = f_\alpha p_\alpha / Z_0$. Since $\dim Z_\alpha = 0$, the map $f_\alpha$ has an extension $g_\alpha$ onto $Z_\alpha$ (see Proposition 2.1.13). Then the composition $g = g_\alpha p_\alpha$ is the desired extension of $f$ onto $Z$. Consequently, by Proposition 6.1.9, $X$ is an $AE(0)$-space. The remaining part of our statement can be proved in a completely similar way (compare with Theorem 2.1.12). □

PROPOSITION 6.1.11. *Let $X$ be a compact space. Then $X$ is an $A(N)E(n)$-space if and only if the following condition is satisfied:*

- *For each at most $n$-dimensional compactum $Z$ and for each closed subspace $Z_0$ of $Z$, any map $f \colon Z_0 \to X$ has an extension to (a neighborhood of $Z_0$ in $Z$) $Z$.*

PROOF. Obviously, if $X$ is an $A(N)E(n)$-compactum, then it satisfies the above formulated condition.

Let us show that the converse is also true. Suppose that a compactum $X$ satisfies the above condition. We need to show that $X$ is an $A(N)E(n)$-space in the sense of Definition 6.1.3. Consider an arbitrary space $Z$ of dimension at most $n$, its arbitrary subspace $Z_0$, and a map $f \colon Z_0 \to X$ such that $C(f)(C(X)) \subseteq C(Z)/Z_0$. Denote by $A_0$ the closure of $Z_0$ in the Stone-Čech compactification $\beta Z$ of the space $Z$. Following the proof of Proposition 1.3.12, we can conclude that the map $f$ can be extended to a map $g \colon A_0 \to X$. By our assumption and the equality $\dim \beta Z = \dim Z$, the map $g$ can be extended to a map $h \colon V \to X$, where $V$ is a functionally open neighborhood of the set $A_0$ in $\beta Z$. Let $U = V \cap Z$. Clearly, $U$ is a stable functionally open neighborhood of $Z_0$ in $Z$ and the map $h/U \colon U \to X$ extends the originally given map $f$. □

As follows from the last two statements, the above defined notion of $A(N)E(n)$-space is completely compatible with the known definition of this notion in the classes of Polish and compact spaces. Of course, within Polish spaces (and, in general, in the class of metrizable spaces) these concepts have been exploited for decades, and an extensive literature is devoted to their study. On the other hand, until the middle of the nineteen seventies almost nothing, except the definition itself, was known about non-metrizable $A(N)E(n)$-compacta. The first serious investigation in this direction was the work of Haydon [**164**] in which the spectral characterization of non-metrizable $AE(0)$-compact spaces was obtained. Moreover, it was shown in [**164**] that the class of $AE(0)$-compacta coincides with the class of *Dugundji compacta* (the latter class, in different terms, was defined in [**249**]). In Section 6.3 we consider this problem more formally.

Concluding this Subsection, we remark that the compactness of $X$ is absolutely essential in Proposition 6.1.11. More formally:

- There exists a Polish space, satisfying the condition of Proposition 6.1.11 (for $n = \infty$), which is not an absolute retract.

This fact follows from Corollary 3.2.16. As already mentioned in Subsection 2.1.2, this was observed by van Mill (see [**22**]).

**6.1.3. Definition of $n$-soft mappings.** The following definition shows that the notion of an $n$-soft map is a categorical counterpart of the notion of $AE(n)$-space.

DEFINITION 6.1.12. *A map $f\colon X \to Y$ is said to be* $n$-soft, $n = 0, 1, \ldots, \infty$, *if for each at most $n$-dimensional space $Z$, its two subspaces $Z_0$ and $Z_1$ with $Z_0 \subseteq Z_1$ and any two maps $g\colon Z_0 \to X$ and $h\colon Z_1 \to Y$ such that $C(g)(C(X)) \subseteq C(Z)/Z_0$, $C(h)(C(Y)) \subseteq C(Z)/Z_1$ and $fg = h/Z_0$, there exists a map $k\colon Z_1 \colon X$ such that $fk = h$, $g = k/Z_0$ and $C(k)(C(X)) \subseteq C(Z)/Z_1$. Maps that are $\infty$-soft are called* soft

*The following diagram helps to understand the situation described in the above Definition.*

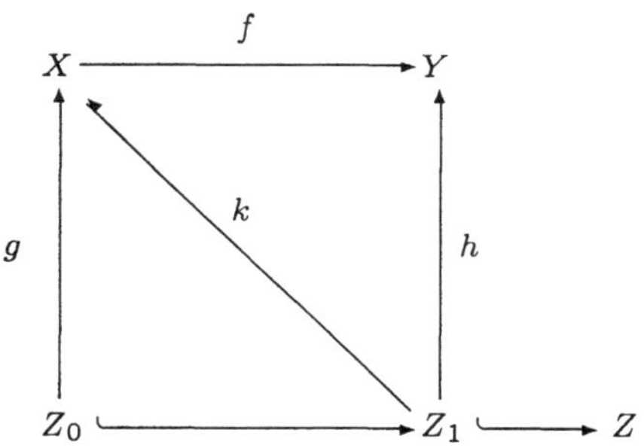


We are now going to establish some elementary properties of $n$-soft maps. We begin with the following statement, the proof of which is trivial and hence omitted.

LEMMA 6.1.13. *Each $n$-soft map is surjective.*

LEMMA 6.1.14. *A finite composition of $n$-soft maps is $n$-soft.*

PROOF. It suffices to consider the case of two maps. Let $f_1\colon X_1 \to X_2$ and $f_2\colon X_2 \to X_3$ be $n$-soft maps. Consider an at most $n$-dimensional space $Z$, its subspaces $Z_0$ and $Z_1$ with $Z_0 \subseteq Z_1$, and two maps $g\colon Z_0 \to X_1$ and $h\colon Z_1 \to X_3$ satisfying the conditions

$$C(g)(C(X_1)) \subseteq C(Z)/Z_0,\ C(h)(C(X_3)) \subseteq C(Z)/Z_1 \text{ and } f_2 f_1 g = h/Z_0.$$

Straightforward verification shows that $C(f_1g)(C(X_2)) \subseteq C(Z)/Z_0$. Consequently, by the $n$-softness of the map $f_2$, there exists a map $k'\colon Z_1 \to X_2$ such that

$$f_2k' = h, \quad f_1g = k'/Z_0 \text{ and } C(k')(C(X_2)) \subseteq C(Z)/Z_1.$$

Further, the $n$-softness of the map $f_1$ guarantees the existence of a map $k\colon Z_1 \to X_1$ satisfying the following conditions:

$$f_1k = k' \ k/Z_0 = g \text{ and } C(k)(C(X_1)) \subseteq C(Z)/Z_1.$$

It only remains to observe that $f_2f_1k = f_2k' = h$. □

LEMMA 6.1.15. *Let $f = f_2f_1$ and $f_1$ and $f$ be $n$-soft maps. Then $f_2$ is also an $n$-soft map.*

PROOF. Consider two maps $f_1\colon X_1 \to X_2$ and $f_2\colon X_2 \to X_3$ and assume that $f_1$ and $f = f_2f_1$ are $n$-soft. Consider also an at most $n$-dimensional space $Z$, its subspaces $Z_0$ and $Z_1$ with $Z_0 \subseteq Z_1$, and two maps $g\colon Z_0 \to X_2$ and $h\colon Z_1 \to X_3$ satisfying the following conditions:

$$C(g)(C(X_2)) \subseteq C(Z)/Z_0, \quad C(h)(C(X_3)) \subseteq C(Z)/Z_1 \text{ and } f_2g = h/Z_0.$$

Consider an arbitrary point $z_0 \in Z_0$. By Lemma 6.1.13, there is a point $x_1 \in X_1$ such that $f_1(x_1) = g(z_0)$. By the $n$-softness of the map $f_1$, there exists a map $g'\colon Z_0 \to X_1$ satisfying, in particular, the following conditions:

$$C(g')(C(X_1)) \subseteq C(Z)/Z_0, \text{ and } f_1g' = g.$$

The $n$-softness of the map $f$ guarantees the existence of a map $k'\colon Z_1\colon X_1$ such that

$$C(k')(C(X_1)) \subseteq C(Z)/Z_1, \quad k'/Z_0 = g' \text{ and } fk' = h.$$

Finally observe that the map $k = f_1k'\colon Z_1 \to X_2$ has all the required properties from Definition 6.1.12 with respect to the map $f_2$. □

PROPOSITION 6.1.16. *A map $f\colon X \to Y$ of a realcompact space $X$ onto an $ANE(n)$-space $Y$ is $n$-soft if and only if the following condition is satisfied:*

- *For each at most $n$-dimensional realcompact space $Z$, for its closed subspace $Z_0$, and for any two maps $g\colon Z_0 \to X$ and $h\colon Z \to Y$ such that $fg = h/Z_0$ and $C(g)(C(X)) \subseteq C(Z)/Z_0$, there exists a map $k\colon Z \to X$ such that $k/Z_0 = g$ and $fk = h$.*

PROOF. Obviously every $n$-soft map satisfies the above stated condition. Suppose now that $f$ is the map given in our statement. Let us show its $n$-softness.

Let $Z$, $Z_0$, $Z_1$, $g$ and $h$ satisfy the conditions of Definition 6.1.12 with respect to the map $f$. Consider the Hewitt realcompactification $\nu Z$ of the space $Z$, and the closures $A_0$ and $A_1$ of $Z_0$ and $Z_1$, respectively, in $\nu Z$. Since every space is $C$-embedded in its Hewitt realcompactification, we have the following two inclusions:

$$C(g)(C(X)) \subseteq C(\nu Z)/Z_0 \text{ and } C(h)(C(Y)) \subseteq C(\nu Z)/Z_1.$$

By Propositions 6.1.7 and 1.3.12, the maps $g$ and $h$ have extensions $\bar{g}\colon A_0 \to X$ and $\bar{h}\colon A_1 \to Y$ such that

$$C(\bar{g})(C(X)) \subseteq C(\nu Z)/A_0 \text{ and } C(\bar{h})(C(Y)) \subseteq C(\nu Z)/A_1.$$

Since $f\bar{g}/Z_0 = \bar{h}/Z_0$, and since $Z_0$ is dense in $A_0$, we conclude that $f\bar{g} = \bar{h}/A_0$. Since $Y$ is an $ANE(n)$-space, and since $\dim \nu Z = \dim Z \leq n$, there exists a map $h'\colon U \to Y$, defined on a stable functionally open neighborhood $U$ of $A_1$ in $\nu Z$, extending $\bar{h}$. Corollary 1.1.15, Lemma 1.1.18 and Proposition 1.3.18 imply that $U$ is an at most $n$-dimensional realcompact space. Observe also that $A_0$ is closed in $U$. Consequently, according to our assumption, there exists a map $k'\colon U \to X$ such that $fk' = h'$ and $k'/A_0 = \bar{g}$. The restriction $k'/Z_1$ will be denoted by $k$. It follows from the construction that $fk = h$ and $k/Z_0 = g$.

All that remains to be verified is the inclusion $C(k)(C(X)) \subseteq C(Z)/Z_1$. Let $\varphi \in C(X)$. By the construction, $U$ is a stable functionally open neighborhood of the set $A_1$, and hence of the set $Z_1$, in $\nu Z$. Take a functionally closed subset $F$ of $\nu Z$ such that $Z_1 \subseteq F \subseteq U$. The function $\varphi k \in C(Z_1)$ can be extended to a function $\varphi k' \in C(U)$. In this situation, Proposition 1.1.23 guarantees that the function $\varphi k'/F \in C(F)$ has an extension onto $\nu Z$. Denote this extension by $\psi$. Obviously, the restriction $\psi/Z$ is an extension of $\varphi k$ onto $Z$. This completes the proof. □

COROLLARY 6.1.17. *Let $f\colon X \to Y$ be an $n$-soft map between realcompact spaces and $Y_0$ be an $ANE(n)$-subspace of $Y$. Then the restriction*

$$f/f^{-1}(Y_0)\colon f^{-1}(Y_0) \to Y_0$$

*is also $n$-soft.*

PROOF. It is easy to see that the inverse image $f^{-1}(Y_0)$ is realcompact. Therefore we can use Proposition 6.1.16. Let $Z_0$ be a closed subspace of an at most $n$-dimensional realcompact space $Z$. Consider two maps $g\colon Z_0 \to f^{-1}(Y_0)$ and $h\colon Z \to Y_0$, satisfying the conditions

$$fg = h \text{ and } C(g)(C(f^{-1}(Y_0))) \subseteq C(Z)/Z_0.$$

One can easily verify that

$$C(g)(C(X)) \subseteq C(Z)/Z_0.$$

Consequently, by the $n$-softness of the map $f$, there exists a map $k\colon Z \to X$ such that $k/Z_0 = g$ and $fk = h$. The last equality guarantees that $k(Z) \subseteq f^{-1}(Y_0)$. □

Combining the proofs of Proposition 6.1.11 and 6.1.16, we obtain the following statement.

PROPOSITION 6.1.18. *A map $f: X \to Y$ between compact spaces is $n$-soft if and only if the following condition is satisfied:*

- *For each at most $n$-dimensional compactum $Z$, its closed subspace $Z_0$, and any two maps $g: Z_0 \to X$ and $h: Z \to Y$ with $fg = h/Z_0$, there exists a map $k \to Z \to X$ such that $k/Z_0 = g$ and $fk = h$.*

Combining the proofs of Proposition 6.1.10 and 6.1.16 we get

PROPOSITION 6.1.19. *A map between Polish spaces in $n$-soft if and only if the following condition is satisfied:*

- *For each at most $n$-dimensional Polish space $Z$, its closed subspace $Z_0$, and any two maps $g: Z_0 \to X$ and $h: Z \to Y$ with $fg = h/Z_0$, there exists a map $k \to Z \to X$ such that $k/Z_0 = g$ and $fk = h$.*

*Remark* 6.1.20. If the compact spaces $X$ and $Y$ in Proposition 6.1.18 are metrizable, then we can assume that the compacta $Z$ and $Z_0$ are also metrizable. Therefore the last two Propositions, coupled with the fact just mentioned, show that in the class of Polish spaces the definition $n$-soft map (Definition 6.1.12) coincides with the definition of the same notion given in Chapter 2 (see Definition 2.1.33). It is also useful to observe the following simple fact:

- in Corollary 6.1.17 the assumption that $Y_0$ is an $ANE(n)$-space can be removed if
  - (a) $X$, $Y$ and $Y_0$ are compact spaces, or
  - (b) $X$, $Y$ and $Y_0$ are Polish spaces.

PROPOSITION 6.1.21. *Let $f: X \to Y$ be an $n$-soft map. Then $X$ is an $A(N)E(n)$-space if and only if $Y$ is an $A(N)E(n)$-space.*

PROOF. First assume that $X$ is an $A(N)E(n)$-space. Consider an arbitrary at most $n$-dimensional space $Z$, its subspace $Z_0$, and a map $g: Z_0 \to Y$ such that $C(g)(C(Y)) \subseteq C(Z)/Z_0$. Take a point $z_0 \in Z_0$ and, using Lemma 6.1.13, consider a point $x_0 \in X$ such that $f(x_0) = g(z_0)$. Let $g_0(z_0) = x_0$. The $n$-softness of the map $f$ guarantees the existence of a map $g'\colon Z_0 \to X$ such that $fg' = g$ and $C(g)(C(X)) \subseteq C(Z)/Z_0$. By our assumption, there exist a stable functionally open neighborhood $U$ of the set $Z_0$ in $Z$ and a map $h'\colon U \to X$, extending $g'$. Finally, observe that the composition $h = fh'\colon U \to Y$ is an extension of the originally given map $g$. Therefore $Y$ is an $A(N)E(n)$-space.

Conversely, assume that $Y$ is an $A(N)E(n)$-space. Consider an arbitrary at most $n$-dimensional space $Z$, its subspace $Z_0$, and a map $g: Z_0 \to X$ such that $C(g)(C(X)) \subseteq C(Z)/Z_0$. Then, by our assumption, there exist a stable functionally open neighborhood $U$ of the set $Z_0$ in $Z$ and a map $h'\colon U \to Y$ such that $h'/Z_0 = fg$. By Lemma 1.1.18 and Proposition 1.3.18, we have $\dim U \leq n$.

Consequently, the $n$-softness of the map $f$ guarantees the existence of a map $h: U \to X$ with $h/Z_0 = g$. This shows that $X$ is an $A(N)E(n)$-space. □

Next we introduce some additional concepts needed in the sequel.

DEFINITION 6.1.22. *A map $f: X \to Y$ is called* $n$-invertible *if for any map $h: Z \to Y$ of any at most $n$-dimensional space $Z$ into $Y$, there is a map $g: Z \to X$ such that $fg = h$.*

Obviously, every $n$-soft map is $n$-invertible. In turn, every $n$-invertible map is surjective. It is known [**171**] that each 0-invertible map between compact spaces has a so-called *regular averaging operator* (see Definition 6.1.24) and that for maps between *metrizable* compacta the converse is also true. Every metrizable compactum is a 0-invertible image of the Cantor cube $D^\omega$ (see [**233**], [**171**]). In Chapter 4 we have already seen that a more general statement is true: each metrizable compactum is an $n$-invertible image of an $n$-dimensional compactum (the universal Menger compactum $\mu^n$ for example; see Theorem 4.2.21). In the meantime, the map constructed in Theorem 4.2.21 has several additional properties (for instance, that map is open for each $n \geq 1$). If we do not require the presence of these additional properties, then the above fact can be obtained as an immediate consequence of Theorem 1.3.10. Indeed, take an arbitrary metrizable compactum $X$ and consider all possible maps $g_\alpha: Z_\alpha \to X$ of at most $n$-dimensional metrizable compacta $Z_\alpha$ into $X$. Let $Z$ denote the discrete sum of these compacta $Z_\alpha$. The maps $g_\alpha$ naturally induce the map $g: Z \to X$. Consider the Stone-Čech extension $g': \beta Z \to X$ of $g$. Since $\dim \beta Z = n$ we can, applying Theorem 1.3.10, find an $n$-dimensional metrizable compactum $Y$ and two maps $p: \beta Z \to Y$ and $f: Y \to X$ such that $g' = fp$. It only remains to observe that the map $f$ is $n$-invertible.

DEFINITION 6.1.23. *A map $f: X \to Y$ is said to be* functionally open *(respectively,* functionally closed*) if the image of any functionally open (respectively, functionally closed) subset of $X$ is functionally open (respectively, functionally closed) in $Y$*

DEFINITION 6.1.24. *We say that a surjection $f: X \to Y$ admits a* regular averaging operator *if there exists a linear continuous*[1] *map $v: C(X) \to C(Y)$ satisfying the following conditions:*

(a) $v(1_X) = 1_Y$.
(b) *If $\varphi \geq 0$, then $v(\varphi \geq 0$.*
(c) $v \cdot C(f) = id_{C(Y)}$.

The following statement will be exploited below.

[1] Here $C(X)$ and $C(Y)$ are endowed with the compact-open topology.

PROPOSITION 6.1.25. *For each infinite cardinal number $\tau$, there exists a functionally closed, proper and 0-invertible map $g\colon N^\tau \to R^\tau$ admitting a regular averaging operator.*

PROOF. For each $\alpha \in \tau$, take a 0-invertible map $h_\alpha\colon D^\omega_\alpha \to [0,1]_\alpha$ of the Cantor cube $D^\omega_\alpha$ onto the closed interval $[0,1]_\alpha$ (see the above discussion). Let $L_\alpha = h_\alpha^{-1}((0,1)_\alpha)$ and $g_\alpha = h_\alpha/L_\alpha$, $\alpha \in \tau$. Denote by $g$ the product of $g_\alpha$'s, i.e.

$$g = \times\{g_\alpha\colon \alpha \in \tau\}.$$

Since $L_\alpha$ is a zero-dimensional, non-compact, locally compact space with a countable basis, we conclude (recall that $\tau$ is infinite cardinal) that the product $\prod\{L_\alpha\colon \alpha \in \tau\}$ is homeomorphic to $N^\tau$ (use Theorem 1.1.5). In turn, the product $\prod\{(0,1)_\alpha\colon \alpha \in \tau\}$ is obviously homeomorphic to $R^\tau$. The 0-invertibility of the map $g$ follows from the 0-invertibility of the $g_\alpha$'s, $\alpha \in \tau$. It is easy to see that the map $g$ is proper and functionally closed. Thus, it only remains to show that $g$ admits a regular averaging operator.

It easily follows from the above construction that we are in the following situation. A 0-invertible map (namely the product of $h_\alpha$'s, $\alpha \in \tau$) $h\colon D^\tau \to I^\tau$ is given, $R^\tau \subseteq I^\tau$, $h^{-1}(R^\tau)$ is homeomorphic to $N^\tau$ and the map $g\colon N^\tau \to R^\tau$ coincides with the restriction $h/h^{-1}(R^\tau)$. Take a regular averaging operator $u$ for the map $h$. It follows from elementary properties of regular averaging operators [**249**] that for each compactum $K$ in $R^\tau$, the map $h_K = h/h^{-1}(K)$ also admits a regular averaging operator $u_K$. These operators are compatible in the following sense:

- If $K_1$ and $K_2$ are compacta in $R^\tau$ and $K_1 \subseteq K_2$, then for each function $\varphi \in C(h^{-1}(K_2)$ we have

$$u_{K_2}(\varphi)/K_1 = u_{K_1}(\varphi/h^{-1}(K_1)).$$

Using these properties of operators $u_K$, and also the fact that both spaces $N^\tau$ and $R^\tau$ are $f_K$-*spaces*[2], it is not hard to see that by letting

$$v(\varphi)(y) = u_{\{y\}}(\varphi/g^{-1}(y)),\quad y \in R^\tau \quad \text{and} \quad \varphi \in C(N^\tau)$$

we define a regular averaging operator for the map $g$. $\square$

PROPOSITION 6.1.26. *Every 0-soft map between $AE(0)$-spaces is functionally open.*

PROOF. Let $f\colon X \to Y$ be a 0-soft map and $Y$ be an $AE(0)$-space. Without loss of generality we can assume that for some infinite cardinal number $\tau$ the space $Y$ is $C$-embedded into $R^\tau$, the space $X$ is $C$-embedded into the product $R^\tau \times R^\tau$, and the map $f$ coincides with the restriction of the projection $\pi_1\colon R^\tau \times R^\tau \to R^\tau$ onto $X$. Take a map $g_1\colon N^\tau \to R^\tau$ with the properties of Proposition

[2] $X$ is an $f_K$-space if any real-valued function, defined on $X$ and having continuous restrictions on every compact subspace of $X$, is continuous on $X$; see [**193**]

6.1.25. It follows from the constructions presented in the proof of Proposition 6.1.25 that the product

$$g_2 = g_1 \times g_1 \colon N^\tau \times N^\tau \to R^\tau \times R^\tau$$

is also proper and functionally closed (moreover, $g_2$ is 0-invertible and admits a regular averaging operator). Observe also that $\pi_1 g_2 = g_1\lambda_1$, where $\lambda_1 \colon N^\tau \times N^\tau \to N^\tau$ denotes the projection onto the first coordinate. Since $Y$ is a $C$-embedded $AE(0)$-subspace of $R^\tau$, there exists a map $h_1 \colon N^\tau \to Y$ coinciding with $g_1$ on the set $g_1^{-1}(Y)$. Similarly, since $X$ is a $C$-embedded $AE(0)$-subspace of the product $R^\tau \times R^\tau$, and since the map $f$ is 0-soft, we conclude that there is a map $h_2 \colon N^\tau \times N^\tau \to X$ coinciding with $g_2$ on the set $\mathrm{E}g_2^{-1}(X)$ and such that $fh_2 = h_1\lambda_1$. Now consider compact-valued and upper semi-continuous retractions

$$r_X = h_2 g_2^{-1} \colon R^\tau \times R^\tau \to X \text{ and } r_Y = h_1 g_1^{-1} \colon R^\tau \to Y.$$

It follows from the constructions that $fr_X = r_Y\pi_1$. In particular, since $r_Y$ is the identity on $Y$, we see that for each point $b \in \pi_1^{-1}(Y)$ we have the equality $f(r_X(b) = \pi_1(b)$. The functional closedness of the map $g_2$ guarantees that the set

$$r_X^{-1}(U) = \{b \in R^\tau \times R^\tau : r_X(b) \subseteq U\}$$

is functionally open in the product $R^\tau \times R^\tau$ for each functionally open subset $U$ of $X$. It is also easy to see that the projection $\pi_1$ is functionally open. Consequently, the set

$$V = \pi_1(r_X^{-1}(U)) \cap Y$$

is functionally open in $Y$. Thus, in order to complete the proof, it only remains to show that $V = f(U)$.

Since $U \subseteq r_X^{-1}(U) \cap \pi_1^{-1}(Y)$, we see that

$$f(U) = \pi_1(U) \subseteq \pi_1(r_X^{-1}(U) \cap \pi_1^{-1}(Y)) = \pi_1(r_X^{-1}(U)) \cap Y = V.$$

If $y \in V$, then there is a point $b \in r_X^{-1}(U) \cap \pi_1^{-1}(Y)$ such that $\pi_1(b) = y$. As remarked above, for the point $b$ we have $fr_X(b) = \pi_1(b)$. But $r_X(b) \subseteq U$. Consequently,

$$y = \pi_1(b) = fr_X(b) \subseteq f(U).$$

This implies the desired equality $V = f(U)$. □

It is useful to compare the following immediate consequence of Proposition 6.1.26 with Proposition 2.1.34.

COROLLARY 6.1.27. *A surjection between Polish spaces is 0-soft if and only if it is open.*

*Historical and bibliographical notes* 6.1. Definitions 6.1.22 and 6.1.24 appeared in [**171**] and [**39**] respectively. The other notions and results of this Section are due to the author (see, for example, [**79**], [**84**], [**80**]).

## 6.2. Morphisms of spectra and square diagrams

Every morphism between spectra induces the limit map between a limit spaces of given spectra (see Subsection 1.2.2). On the other hand, for "good" spectra the converse is also true: every map between the limit spaces of these spectra is induced by some morphism (see Section 1.3). Consequently, every topological property of these limit maps can be described in terms of morphisms of these spectra. From this point of view, the study of the structure of morphisms of spectra themselves becomes very important. There are several possible approaches to this problem. One of them deals with the description of properties of morphisms using the language of square diagrams, consisting of elements of morphisms and short projections of the corresponding spectra.

We begin with the following definition.

DEFINITION 6.2.1. *A* characteristic map *of a commutative square diagram*

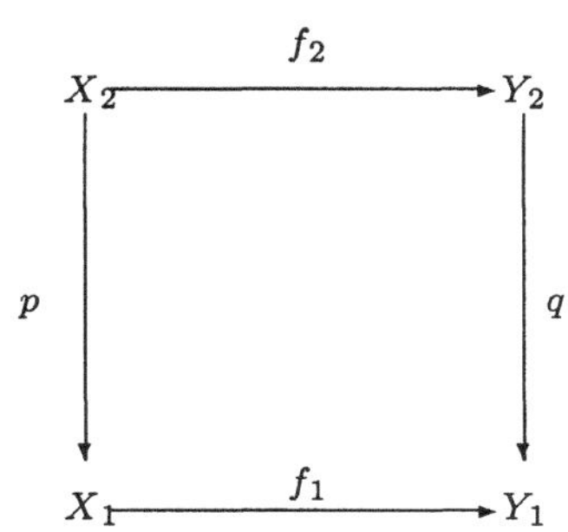


*is the diagonal product* $p\triangle f_2$*, considered as a map from the space* $X_2$ *into the fibered product (i.e. pullback) of the spaces* $X_1$ *and* $Y_2$ *with respect to the maps* $f_1\colon X_1 \to Y_1$ *and* $q\colon Y_2 \to Y_1$.

Square diagrams with surjective characteristic maps are called *bicommutative* (see, for example, [**202**], [**245**], [**278**]). A much finer way of distinguishing square diagrams is contained in the following definition.

DEFINITION 6.2.2. *A commutative square diagram is called* $n$-soft, $n = 0, 1, ..$ $, \infty$*, if its characteristic map is* $n$*-soft. The* $\infty$*-soft diagrams are called* soft.

DEFINITION 6.2.3. *A commutative square diagram is said to be a* Cartesian square *if its characteristic map is a homeomorphism.*

LEMMA 6.2.4. *Suppose that the spaces $Y_1$ and $Y_2$ in the above square diagram, consisting of realcompact spaces, are $ANE(n)$-spaces. If the map $f_1$ is $n$-soft and the diagram is a Cartesian square, then $f_2$ is also $n$-soft.*

PROOF. By Proposition 6.1.16, it suffices to consider an arbitrary at most $n$-dimensional realcompact space $Z$, its closed subspace $Z_0$, and two maps $g_2 \colon Z_0 \to X_2$ and $h_2 \colon Y_2$ satisfying the following conditions

$$f_2 g_2 = h_2/Z_0 \text{ and } C(g_2)(C(X_2)) \subseteq C(Z)/Z_0,$$

and to prove the existence of a map $k_2 \colon Z \to X_2$ such that $f_2 k_2 = h_2$ and $g_2 = k_2/Z_0$.

Let $g_1 = pg_2$ and $h_1 = qh_2$. Clearly

$$f_1 g_1 = h_1/Z_0 \text{ and } C(g_1)(C(X_1)) \subseteq C(Z)/Z_0.$$

The $n$-softness of the map $f_1$ guarantees the existence of a map $k_1 \colon Z \to X_1$ satisfying the following two conditions: $f_1 k_1 = h_1$ and $g_1 = k_1/Z_0$. Since our diagram is a Cartesian square, the diagonal product $k = k_1 \triangle h_2$ of the maps $k_1$ and $h_2$ maps the space $Z$ into the space $X_2$. It only remains to observe that $f_2 k_2 = h_2$ and $g_2 = k_2/Z_0$. □

LEMMA 6.2.5. *Let an $n$-soft square diagram, consisting of realcompact spaces, be given. If the spaces $Y_1$ and $Y_2$ are $ANE(n)$-spaces and the map $f_1$ is $n$-soft, then $f_2$ is also $n$-soft. Moreover, for any at most $n$-dimensional realcompact space $Z$, its closed subspace $Z_0$, and any three maps $g_2 \colon Z_0 \to X_2$, $h_2 \colon Z \to Y_2$ and $k_1 \colon Z \to X_1$ satisfying the conditions*

$$f_1 k_1 = qh_2,\ f_2 g_2 = h_2/Z_0,\ pg_2 = k_1/Z_0 \text{ and } C(g_2)(C(X_2)) \subseteq C(Z)/Z_0,$$

*there exists a map $k_2 \colon Z \to X_2$ such that $g_2 = k_2/Z_0$, $h_2 = f_2 k_2$ and $pk_2 = k_1$.*

PROOF. Let $X$ denote the fibered product of the spaces $X_1$ and $Y_2$ with respect to the maps $f_1 \colon X_1 \to Y_1$ and $q \colon Y_2 \to Y_1$. Let $\varphi \colon X \to X_1$ and $\psi \colon X \to Y_2$ denote the natural projections of the fibered product. Since the diagram

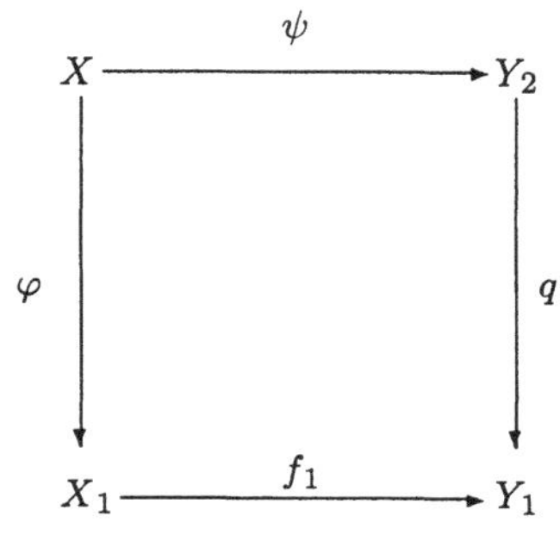

is a Cartesian square, we conclude, by Lemma 6.2.4, that the map $\psi$ is $n$-soft. The characteristic map $h = p\triangle f_2$ of the given diagram is $n$-soft by our assumption. Consequently, by Lemma 6.1.14, the map $f_2$ (being the composition of the maps $h$ and $\psi$) is also $n$-soft.

The validity of the second part of our statement can be shown as follows. First, observe that the diagonal product $k_1\triangle h_2\colon Z \to X$ satisfies the following equalities:

$$\varphi(k_1\triangle h_2) = k_1, \quad \psi(k_1\triangle h_2) = h_2 \text{ and } hg_2 = (k_1\triangle h_2)/Z_0.$$

The $n$-softness of the map $h$ guarantees the existence of a map $k_2\colon Z \to X_2$ such that $g_2 = k_2/Z_0$ and $hk_2 = (k_1\triangle h_2)$. It only remains to note that

$$f_2k_2 = \psi hk_2 = \psi(k_1\triangle h_2) = h_2$$

and

$$pk_2 = \psi hk_2 = \varphi(k_1\triangle h_2) = k_1.$$

This completes the proof. □

LEMMA 6.2.6. *Let $\mathcal{S}_X = \{X_\alpha, p_\alpha^\beta, \tau\}$ be a well ordered continuous spectrum, consisting of realcompact spaces $X_\alpha$ and $n$-soft short projections $p_\alpha^{\alpha+1}$. If $X_0$ is an $ANE(n)$-space, then the limit projection $p_0\colon \lim \mathcal{S}_X \to X_0$ is $n$-soft.*

PROOF. The limit space of the spectrum $X = \mathcal{S}_X$ is realcompact (see Subsection 1.1.2). Therefore we can use Proposition 6.1.16. Take an arbitrary at most $n$-dimensional realcompact space $Z$, a closed subspace $Z_0$ of $Z$, and any two maps $g\colon Z_0 \to X$ and $h\colon Z \to X_0$ such that

$$p_0g = h/Z_0 \text{ and } C(g)(C(X)) \subseteq C(Z)/Z_0.$$

Let $k_0 = h$. Suppose that for each $\alpha < \beta$ we have already constructed maps $k_\alpha\colon Z \to X_\alpha$ such that

$$p_\alpha g = k_\alpha/Z_0 \text{ whenever } \alpha < \beta$$

and

$$k_\alpha = p_\alpha^\gamma k_\gamma \text{ whenever } \alpha < \gamma < \beta.$$

We now construct a map $k_\beta\colon Z \to X_\beta$. If $\beta$ is a limit ordinal, then, by the continuity of the spectrum $\mathcal{S}_X$, the diagonal product

$$k_\beta = \triangle\{k_\alpha\colon \alpha < \beta\}$$

obviously satisfies the required properties. If $\beta = \alpha + 1$, then the existence of the map $k_\beta$ follows immediately from the $n$-softness of the short projection $p_\alpha^\beta$. Thus, the maps $k_\alpha\colon Z \to X_\alpha$ are constructed for each $\alpha < \tau$ and satisfy the conditions

$$p_\alpha g = k_\alpha/Z_0 \text{ and } k_\alpha = p_\alpha^\beta k_\beta$$

whenever $\alpha < \beta < \tau$. It only remains to remark that the diagonal product $k = \triangle\{k_\alpha\colon \alpha < \tau\}$ satisfies the required equalities $g = k/Z_0$ and $p_0k = h$. □

LEMMA 6.2.7. *Let $\mathcal{S}_X = \{X_\alpha, p_\alpha^\beta, \tau\}$ and $\mathcal{S}_Y = \{Y_\alpha, q_\alpha^\beta, \tau\}$ be two well ordered continuous spectra consisting of realcompact spaces. Let*

$$\{f_\alpha \colon X_\alpha \to Y_\alpha; \alpha \in \tau\}$$

*be a morphism between these spectra such that all arising adjacent square diagrams are $n$-soft. If $Y_0$ is an $ANE(n)$-space and the maps $f_0$ and $q_\alpha^{\alpha+1}$, $\alpha < \tau$ are $n$-soft, then the limit map of the morphism $\{f_\alpha\}$ is also $n$-soft.*

PROOF. Let $X = \lim \mathcal{S}_X$, $Y = \lim \mathcal{S}_Y$ and $f = \lim\{f_\alpha\}$. By Proposition 6.1.21, $X$ is a realcompact space and, by Lemma 6.2.6, $Y$ is an $ANE(n)$-space. Consider an arbitrary at most $n$-dimensional realcompact space $Z$, a closed subspace $Z_0$ of $Z$, and any two maps $g \colon Z_0 \to X$ and $h \colon Z \to Y$ satisfying the conditions

$$fg = h/Z_0 \text{ and } C(g)(C(X)) \subseteq C(Z)/Z_0.$$

By Proposition 6.1.16, in order to establish the $n$-softness of the map $f$, it suffices to construct a map $k \colon Z \to X$ such that $g = k/Z_0$ and $fk = h$. Let $g_\alpha = p_\alpha g$ and $h_\alpha = q_\alpha h$ for each $\alpha < \tau$. Observe that

$$f_\alpha g_\alpha = h_\alpha/Z_0 \text{ and } C(g_\alpha)(C(X_\alpha)) \subseteq C(Z)/Z_0 \text{ whenever } \alpha < \tau.$$

By our assumption, the map $f_0$ is $n$-soft and, consequently, there exists a map $k_0 \colon Z \to X_0$ such that $g_0 = k_0/Z_0$ and $f_0 k_0 = h_0$. Suppose that for each $\alpha < \beta$, where $\beta < \tau$, we have already constructed maps $k_\alpha \colon Z \to X_\alpha$ so that the following conditions are satisfied:

(a) $f_\alpha k_\alpha = h_\alpha$.
(b) $g_\alpha = k_\alpha/Z_0$.
(c) $k_\alpha = p_\alpha^\gamma k_\gamma$ whenever $\alpha < \gamma < \beta$.

Let us construct the map $k_\beta \colon Z \to X_\beta$ with the required properties.

If $\beta$ is a limit ordinal, then let

$$k_\beta = \triangle\{k_\alpha \colon \alpha < \beta\}.$$

Obviously, by continuity of the spectra $\mathcal{S}_X$ and $\mathcal{S}_Y$, we have

$$f_\beta k_\beta = h_\beta, \;\; g_\beta = k_\beta/Z_0 \text{ and } k_\alpha = p_\alpha^\beta k_\beta \text{ whenever } \alpha < \beta.$$

If $\beta = \alpha + 1$, then, by the assumption, the diagram

$$\begin{array}{ccc} X_\beta & \xrightarrow{f_\beta} & Y_\beta \\ {\scriptstyle p_\alpha^\beta}\downarrow & & \downarrow{\scriptstyle q_\alpha^\beta} \\ X_\alpha & \xrightarrow{f_\alpha} & Y_\alpha \end{array}$$

is $n$-soft. Therefore, by Lemma 6.2.5, there exists a map $k_\beta \colon Z \to X_\beta$ such that

$$g_\beta = k_\beta / Z_0, \quad f_\beta k_\beta = h_\beta \text{ and } k_\alpha = p_\alpha^\beta k_\beta.$$

Thus, the maps $k_\alpha$ are now constructed for each $\alpha < \tau$. We need only to observe that the map

$$k = \triangle\{k_\alpha \colon \alpha < \tau\}$$

satisfies all the desired properties. □

*Remark* 6.2.8. The proofs of Lemmas 6.2.4 — 6.2.7, coupled with Proposition 6.1.18, allow us to conclude that if all participating (in these Lemmas) spaces are compact, then we can drop the requirement that the spaces $X_0$, $Y_0$, $Y_1$ and $Y_2$ are $ANE(n)$-spaces. We have the same situation when all spaces are Polish and the length of the spectra, i.e. the cardinal number $\tau$, does not exceed $\omega_1$.

We call a morphism between spectra *bicommutative* if all naturally arising square diagrams are bicommutative in the above sense.

PROPOSITION 6.2.9. *Let $\mathcal{S}_X = \{X_\alpha, p_\alpha^\beta, A\}$ and $\mathcal{S}_Y = \{Y_\alpha, q_\alpha^\beta, A\}$ be two factorizing $\omega$-spectra with surjective limit projections and $f \colon X \to Y$ be a proper map between their limit spaces. Then the following conditions are equivalent:*

(i) *$f$ is functionally closed.*

(ii) *$f$ is induced by a bicommutative cofinal and $\omega$-closed morphism, consisting of proper maps.*

PROOF. First assume that $f$ is the limit map of some bicommutative morphism

$$\{f_\alpha \colon X_\alpha \to Y_\alpha, A'\} \colon \mathcal{S}_X/A' \to \mathcal{S}_Y/A',$$

where $A'$ is a cofinal and $\omega$-closed subset of the indexing set $A$ and all maps $f_\alpha$, $\alpha \in A'$, are proper. We wish to show that $f$ is a functionally closed map. Consider a functionally closed subset $Z$ of $X$. Since $\mathcal{S}_X$ is a factorizing spectrum, there exist an index $\alpha \in A'$ and a closed subset $Z_\alpha$ of the space $X_\alpha$ such that $Z = p_\alpha^{-1}(Z_\alpha)$. The bicommutativity of the morphism $\{f_\alpha\}$ implies the equality

$$f(Z) = f(p_\alpha^{-1}(Z)) = q_\alpha^{-1}(f_\alpha(Z_\alpha)).$$

Using the continuity of the projection $q_\alpha$ and the properness of the map $f_\alpha$, we see from the above equality that the set $f(Z)$ is indeed a functionally closed set.

Suppose now that $f$ is functionally closed. We are going to find cofinal and $\omega$-closed subspectra of the spectra $\mathcal{S}_X$ and $\mathcal{S}_Y$, and we shall construct a bicommutative morphism between these subspectra. Without loss of generality, applying Theorem 1.3.4, we can assume from the very beginning that a morphism

$$\{f_\alpha\colon X_\alpha \to Y_\alpha, A\}\colon \mathcal{S}_X \to \mathcal{S}_Y$$

with $f = \lim\{f_\alpha\}$ is already given. We are now going to perform a spectral search (see Subsection 1.3.2) with respect to the following relation $L \subseteq A^2$:

$L = \{(\alpha,\beta) \in L\colon \alpha \leq \beta$ and there exists a closed basis $\{Z_t^\alpha\colon t \in T_\alpha\}$ of the space $X_\alpha$, containing finite intersections of its elements, such that for each $t \in T_\alpha$ there is a closed subset $F_t^\beta$ of the space $X_\beta$ with $fp_\alpha^{-1}(Z_t^\alpha) = q_\beta^{-1}(F_t^\beta)\}$

The verification of the majorantness and $\omega$-closeness conditions of Proposition 1.1.29 is straightforward. Let us check the existence condition. Let $\alpha \in A$. Take an arbitrary countable closed basis $\{Z_n^\alpha\colon n \in \omega\}$ in the space $X_\alpha$, containing finite intersections of its elements. Functional closedness of the map $f$ guarantees that each of the sets $f(p_\alpha^{-1}(Z_n^\alpha)$, $n \in \omega$, is functionally closed in $Y$. Since the spectrum $\mathcal{S}_Y$ is factorizing we conclude, by $\omega$-completeness of the indexing set $A$, that there exist an index $\beta \in A$, $\beta \geq \alpha$, and closed subsets $F_n^\beta$, $n \in \omega$, of the space $Y_\beta$ such that

$$fp^{-1}(Z_n^\alpha) = q_\beta^{-1}(F_n^\beta), \quad \text{for each } n \in \omega.$$

By Proposition 1.1.29, the set $A'$ of all $L$-reflexive elements is cofinal and $\omega$-closed in $A$.

Let us show that the morphism $\{f_\alpha\colon \alpha \in A'\}$ is bicommutative. Obviously it suffices to show that for each $\alpha \in A'$ and for any two points $x_\alpha \in X_\alpha$ and $y \in Y$ such that $f_\alpha(x_\alpha) = q_\alpha(y)$, we necessarily have the equality

$$p_\alpha^{-1}(x_\alpha) \cap f^{-1}(y) \neq \emptyset.$$

Assume the contrary. Then for some index $\alpha \in A'$, and for some points $x_\alpha \in X_\alpha$ and $y \in Y$ with $f_\alpha(x_\alpha) = q_\alpha(y)$, we have

$$p_\alpha^{-1}(x_\alpha) \cap f^{-1}(y) = \emptyset.$$

Represent the point $x_\alpha$ as the intersection $\cap\{Z_{n_i}^\alpha\colon i \in \omega\}$, where $\{Z_{n_i}^\alpha\colon i \in \omega\}$ is a decreasing sequence of elements of the basis $\{Z_n^\alpha\colon n \in \omega\}$. Then, obviously,

$$p_\alpha^{-1}(x_\alpha) = \cap\{p_\alpha^{-1}(Z_{n_i}^\alpha)\colon i \in \omega\}.$$

The properness of the map $f$ guarantees that

$$f(\cap\{p_\alpha^{-1}(Z_{n_i}^\alpha)\colon i \in \omega\}) = \cap\{f(p_\alpha^{-1}(Z_{n_i}^\alpha))\colon i \in \omega\}.$$

Therefore $y \notin \cap\{f(p_\alpha^{-1}(Z_{n_i}^\alpha)) : i \in \omega\}$. By the $L$-reflexivity of the index $\alpha$, for each $i \in \omega$ we have the equality

$$f(p_\alpha^{-1}(Z_{n_i}^\alpha) = q_\alpha^{-1}(F_{n_i}^\alpha).$$

Consequently, $q_\alpha(y) \notin \cap\{F_{n_i}^\alpha : i \in \omega\}$. On the other hand,

$$q_\alpha(y) = f_\alpha(x_\alpha) \in \cap\{f_\alpha(Z_{n_i}^\alpha) : i \in \omega\} = \cap F_{n_i}^\alpha : i \in \omega\}.$$

This contradiction shows the bicommutativity of the morphism $\{f_\alpha\}$.

The properness of the map $f$ and the bicommutativity of the morphism $\{f_\alpha\}$ imply the compactness of the fibers of all maps $f_\alpha$ with $\alpha \in A'$. Finally, let us show that each such $f_\alpha$ is a closed map. Take a closed subset $Z$ in the space $X_\alpha$, and represent it as the intersection $\cap\{Z_{n_i}^\alpha\}$ of a decreasing sequence of elements of the basis $\{Z_n^\alpha : n \in \omega\}$. The compactness of the fibers of $f_\alpha$ guarantees that

$$f_\alpha(Z) = f_\alpha(\cap\{Z_{n_i}^\alpha : i \in \omega\}) = \cap\{f_\alpha(Z_{n_i}^\alpha) : i \in \omega\}.$$

Consequently, $f_\alpha(Z)$ is closed in $Y_\alpha$. □

A similar argument proves the following statement.

PROPOSITION 6.2.10. *Under assumptions of Proposition* 6.2.9, *the map $f$ is open if and only if it is induced by a bicommutative, cofinal and $\omega$-closed morphism, consisting of open (and proper) maps.*

The next result extends Theorem 4.2.21 to non-metrizable compact spaces.

THEOREM 6.2.11. *Let $n \geq 1$. Every compactum of weight $\tau \geq \omega$ is an $n$-invertible, $(n-1)$-soft image of an $n$-dimensional compactum of weight $\tau$.*

PROOF. We proceed by induction on $\tau$. For metrizable compacta (i.e. in the case $\tau = \omega$) our statement is contained in Theorem 4.2.21. Suppose that the theorem has already been proved for all compacta with weight $< \tau$, and consider an arbitrary compactum $X$ with $w(X) = \tau$. Represent $X$ as the limit space of a continuous well ordered spectrum $\mathcal{S}_X = \{X_\alpha, p_\alpha^\beta, \lambda\}$, consisting of compacta of weight $< \tau$ (and surjective projections).

Next we construct, by induction, a continuous well ordered inverse spectrum $\mathcal{S}_Y = \{Y_\alpha, q_\alpha^\beta, \lambda\}$, "parallel" to the above indicated spectrum $\mathcal{S}_X$ and consisting of $n$-dimensional compacta, and a morphism $\{f_\alpha\} : \mathcal{S}_X \to \mathcal{S}_Y$ consisting of $n$-invertible and $(n-1)$-soft maps.

By the inductive hypothesis, there exist an $n$-dimensional compactum $Y_0$ of weight $w(Y_0) = w(X_0)$ and an $n$-invertible and $(n-1)$-soft map $f_0 : Y_0 \to X_0$.

Suppose now that for each $\alpha < \gamma$ (where $\gamma < \lambda$) we have already constructed

(a) an $n$-dimensional compactum $Y_\alpha$ of weight $w(Y_\alpha) = w(X_\alpha)$,

(b) an $n$-invertible and $(n-1)$-soft map $f_\alpha : Y_\alpha \to X_\alpha$, and

(c) maps $q_\alpha^\beta : Y_\beta \to Y_\alpha$ (here $\alpha < \beta < \gamma$), satisfying the following conditions:

  (i) $q_\alpha^\beta = q_\alpha^\delta q_\delta^\beta$ whenever $\alpha \leq \delta \leq \beta < \gamma$.

(ii) all square diagrams of the form

$$\begin{array}{ccc} Y_{\alpha+1} & \xrightarrow{f_{\alpha+1}} & X_{\alpha+1} \\ \Big\downarrow q_{\alpha}^{\alpha+1} & & \Big\downarrow p_{\alpha}^{\alpha+1} \\ Y_{\alpha} & \xrightarrow{f_{\alpha}} & X_{\alpha} \end{array}$$

are $n$-invertible and $(n-1)$-soft (whenever $\alpha + 1 < \gamma$).

Let us now construct the compactum $Y_\gamma$ and maps

$$f_\gamma\colon Y_\gamma \to X_\gamma \text{ and } q_\alpha^\gamma\colon Y_\gamma \to Y_\alpha, \text{ for } (\alpha < \gamma).$$

If $\gamma$ is a limit ordinal, then

$$Y_\gamma = \lim\{Y_\alpha, q_\alpha^\beta, \gamma\} \text{ and } f_\gamma = \lim\{f_\alpha\colon \alpha < \gamma\}.$$

Obviously, $\dim Y_\gamma = n$ and $w(Y_\gamma) = w(X_\gamma)$. By Lemma 6.2.7 and Remark 6.2.8, the map $f_\gamma$ is $(n-1)$-soft. Essentially the same argument (in fact, an even simpler version of it) shows that $f_\gamma$ is $n$-invertible as well. The maps $q_\alpha^\gamma$ are naturally defined as the limit projections of the above spectrum, the limit space of which was $Y_\gamma$.

Now suppose that $\gamma = \alpha + 1$.

Consider the fibered product $Z$ of the compacta $Y_\alpha$ and $X_\gamma$ with respect to the maps $f_\alpha$ and $p_\alpha^\gamma$. Denote by $\varphi\colon Z \to Y_\alpha$ and $\psi\colon Z \to X_\gamma$ the canonical projections of this fibered product. By Lemma 6.2.4, the map $\psi$ is $(n-1)$-soft. It is not hard to see the $n$-invertibility of $\psi$ as well. Since the compactum $Z$ is contained in the product of the spaces $Y_\alpha$ and $X_\gamma$, we see that $w(Z) < \tau$. Consequently, by the inductive hypothesis, there exist an $n$-dimensional compactum $Y_\gamma$ of weight $w(Y_\gamma) = w(Z)$ and an $n$-invertible and $(n-1)$-soft map $h\colon Y_\gamma \to Z$. Let $f_\gamma = \psi h$ and $q_\alpha^\gamma = \varphi h$. Obviously, the map $f_\gamma$ is $n$-invertible and $(n-1)$-soft (Lemma 6.1.14). The newly formed square diagram, the characteristic map of which coincides with $h$, is, by definition, $n$-invertible and $(n-1)$-soft. This completes the construction of the spectrum $\mathcal{S}_Y$ and of the morphism $\{f_\alpha\}$, satisfying all the required properties.

Finally, let $Y = \lim \mathcal{S}_Y$ and $f = \lim\{f_\alpha\}$. Clearly, $Y$ is an $n$-dimensional compactum of weight $\tau$. It can be shown as above that $f\colon Y \to X$ is $n$-invertible and $(n-1)$-soft. □

Denote by $D_n^\tau$ an $n$-dimensional compactum of weight $\tau$, admitting an $n$-invertible and $(n-1)$-soft map $f\colon D_n^\tau \to I^\tau$ onto the Tychonov cube $I^\tau$ (apply

Theorem 6.2.11). The $(n-1)$-softness of this map guarantees (Proposition 6.1.21) that $D_n^\tau$ is an $AE(n-1)$-compactum. If $X$ is an at most $n$-dimensional compactum of weight $w(X) \leq n$, then there obviously exists an embedding $i\colon X \to I^\tau$. Since $\dim X \leq n$ and $f$ is $n$-invertible, we conclude that there is a map $j\colon X \to D_n^\tau$ such that $fj = i$. Clearly, in this situation $j$ is an embedding. Consequently, the compactum $D_n^\tau$ contains a topological copy of every at most $n$-dimensional compactum of weight $\leq \tau$.

Below, factorizing $\omega$-spectra consisting of Polish spaces and surjective projections will be called *Polish*. A space is *spectrally complete* if it can be represented as the limit space of at least one Polish spectrum.

An embedding of $R^A$ into $I^A$, where $A$ is an arbitrary set, is said to be *standard*, if $t_B(R^A) = R^B$ and $t_B/R^A = \pi_B$ for each subset $B \subseteq A$ (here $t_B\colon I^A \to I^B$ and $\pi_B\colon R^A \to R^B$ denote the projections onto the corresponding subproducts).

LEMMA 6.2.12. *Let $f\colon T \to I^A$ be a 0-soft map of a compactum $T$ of weight $|A| > \omega$. Suppose that $R^A$ is standardly embedded in $I^A$. Then the space $K = f^{-1}(R^A)$ is spectrally complete. Moreover, $K$ can be represented as the limit space of some factorizing Polish spectrum, the indexing set of which coincides with $exp_\omega A$.*

PROOF. By Proposition 6.1.21, $T$ is an $AE(0)$-compactum. Consequently, by Proposition 6.1.8, $T$ is perfectly $\kappa$-normal. Represent $T$ as the limit space of an $\omega$-spectrum

$$\mathcal{S}_T = \{T_B, p_C^B, exp_\omega A\},$$

consisting of metrizable compacta and surjective projections. In turn, let

$$\mathcal{S}_I = \{I^B, t_C^B, exp_\omega A\} \text{ and } \mathcal{S}_R = \{R^B, \pi_C^B, exp_\omega A\}$$

be the standard factorizing $\omega$-spectra associated with $I^A$ and $R^A$ respectively. By Proposition 6.1.26, the map $f$ is open. Consequently, by Proposition 6.2.10, we may assume without loss of generality that the map $f$ is the limit map of a bicommutative morphism

$$\{f_B\colon T_B \to I^B; exp_\omega A\}\colon \mathcal{S}_T \to \mathcal{S}_I,$$

consisting of open maps $f_B$. Let $K_B = p_B(K)$ for each $B \in exp_\omega A$ and $q_C^B = p_C^B/K_B$ whenever $C \subseteq B$ and $C, B \in exp_\omega A$. Let us show that the spectrum

$$\mathcal{S}_K = \{K_B, q_C^B, exp_\omega A\}$$

is a factorizing Polish spectrum, the limit space of which coincides with $K$.

First we show that $\lim \mathcal{S}_K = K$. In order to establish this fact it suffices to show that $K = \cap\{p_B^{-1}(K_B)\colon B \in exp_\omega A\}$ (see Proposition 1.2.4). Since $R^A$ is standardly embedded into $I^A$, and since the morphism $\{f_B\}$ is bicommutative, we conclude that

$$K_B = p_B(K) = p_B f^{-1}(R^A) = f_B^{-1} t_B(R^A) = f_B^{-1}(R^B)$$

for each $B \in exp_\omega A$. Therefore,

$$p_B^{-1}(K_B) = p_B^{-1} f^{-1} B(R^B) = f^{-1} t_B^{-1}(R^B), \quad B \in exp_\omega A.$$

But in this situation we have

$$\bigcap\{p_B^{-1}(K_B) \colon B \in exp_\omega A\} = f^{-1}(\bigcap\{t_B^{-1}(R^B) \colon B \in exp_\omega A\} = f^{-1}(R^A) = K.$$

Thus, $K = \lim \mathcal{S}_K$. A completely similar argument shows the continuity of the spectrum $\mathcal{S}_K$. Obviously, all the spaces $K_B$, by the equalities, $K_B = f_B^{-1}(R^B)$, $B \in exp_\omega A$, are Polish. The construction immediately implies that the projections of the spectrum $\mathcal{S}_K$ are surjective. Therefore, $\mathcal{S}_K$ is a Polish spectrum.

Let us show now that $\mathcal{S}_K$ is a factorizing spectrum. Since $f$ is open, $K$ is dense in $T$. Therefore, by Proposition 1.1.21, $K$ is $z$-embedded in $T$. By Proposition 1.1.22, for each function $\varphi \in C(K)$ there exist a set $L$, which can be represented as the countable intersection of functionally open subsets of $T$, and a function $\psi \in C(L)$ such that $K \subseteq L$ and $\varphi = \psi/K$. Clearly, $L$ is a Lindelöf space. Observe also that since the spectrum $\mathcal{S}_T$ is factorizing, there exists a $G_\delta$-subset $L_{B_0}$ of the space $T_{B_0}$ (for some $B_0 \in exp_\omega A$) such that $L = p_{B_0}^{-1}(L_{B_0})$. Then $L = \lim \mathcal{S}_L$, where

$$\mathcal{S}_L = \{L_B = p_B(L), p_C^B, B \supseteq B_0, B \in exp_\omega A\}.$$

Obviously, the spectrum $\mathcal{S}_L$ is Polish (with proper projections). Its limit space (i.e. the space $L$), as was already mentioned, is Lindelöf. Consequently, by Corollary 1.3.2, $\mathcal{S}_L$ is a factorizing spectrum. This means that there exist an element $B \in exp_\omega A$, $B \supseteq B_0$, and a function $\psi_B \in C(L_B)$, such that $\psi = \psi_B p_B/L$. Since $L_B \supseteq K_B$, we have $\varphi_B = \psi_B/K_B \in C(K_B)$. It only remains to observe that $\varphi = \varphi_B q_B$. □

PROPOSITION 6.2.13. *Let $\tau \geq \omega$ and $n = 0, 1, \dots, \infty$. Then there exist spectrally complete realcompact spaces $K_1$ and $K_2$ and maps $\lambda \colon K_2 \to K_1$, $g_1 \colon K_1 \to R^\tau$ and $g_2 \colon K_2 \to R^\tau \times R^\tau$, satisfying the following conditions:*

(a) *The maps $g_i$, $i = 1, 2$, are proper, functionally closed and $n$-invertible.*
(b) *For each function $\varphi \in C(K_1)$ there is a function $\psi \in C(R^\tau)$ such that if $\varphi$ is constant on the fiber $g_1^{-1}(y)$, $y \in R^\tau$, then $\psi(y) = \varphi(g_1^{-1}(y))$.*
(c) *For each function $\varphi \in C(K_2)$, there is a function $\psi \in C(R^\tau \times R^\tau)$ such that if $\varphi$ is constant on the fiber $g_2^{-1}(y)$, $y \in R^\tau \times R^\tau$, then $\psi(y) = \varphi(g_2^{-1}(y))$.*
(d) $\dim K_i = n$ *for each* $i = 1, 2$.
(e) $R - w(K_i) = w(K_i) = \tau$ *for each* $i = 1, 2$.

(f) *The square diagram*

$$\begin{array}{ccc} K_2 & \xrightarrow{\ g_2\ } & R^\tau \times R^\tau \\ {\scriptstyle\lambda}\downarrow & & \downarrow{\scriptstyle\pi_1} \\ K_1 & \xrightarrow{\ g_1\ } & R^\tau \end{array}$$

*is commutative.*

PROOF. If $n = \infty$, then the validity of our statement is obvious. Indeed, it suffices to let $K_1 = R^\tau$, $K_2 = R^\tau \times R^\tau$, $g_1 = id_{R^\tau}$, $g_2 = id_{R^\tau \times R^\tau}$ and $\lambda = \pi_1$. If $n = 0$, the statement is a consequence of Proposition 6.1.25. Thus we only have to consider the case when $n$ is a natural number.

Take an $n$-dimensional compactum $T_1$ of weight $\tau$ and an $n$-invertible and $(n-1)$-soft map $f_1 \colon T_1 \to I^\tau$ from Theorem 6.2.11. We may assume that $R^\tau$ is standardly embedded in $I^\tau$. Denote by $K_1$ the inverse image $f_1^{-1}(R^\tau)$, and by $g_1$ the restriction of $f_1$ on $K_1$. By Lemma 6.2.12, the space $K_1$ is a spectrally complete realcompact space. The map $g_1$ is open, proper and, consequently, functionally closed. For each function $\varphi \in C(K_1)$, the function $\psi \in C(R^\tau)$, defined by letting

$$\psi(y) = \max\{\varphi(x) \colon x \in g_1^{-1}(y)\}, \quad y \in R^\tau,$$

is continuous (here we use the fact that the map $g_1$ is open and closed) and obviously satisfies condition (b). As remarked in the proof of Lemma 6.2.12, the space $K_1$ is $z$-embedded in $T_1$. Consequently, by Proposition 1.3.18, $\dim K_1 \leq n$. The reverse inequality follows from the $n$-invertibility of the map $g_1$ and the universality of the space $R^\tau$. Observe also that since the space $K_1$ is contained in a compactum of weight $\tau$ and also admits an open map onto a space of weight $\tau$ (namely, onto $R^\tau$), we conclude that the weight of the space $K_1$ equals $\tau$. Since the weight of any space cannot exceed its $R$-weight we see, applying Lemmas 1.3.15 and 6.2.12, that $R - w(K_1) = \tau$.

Consider the following diagram

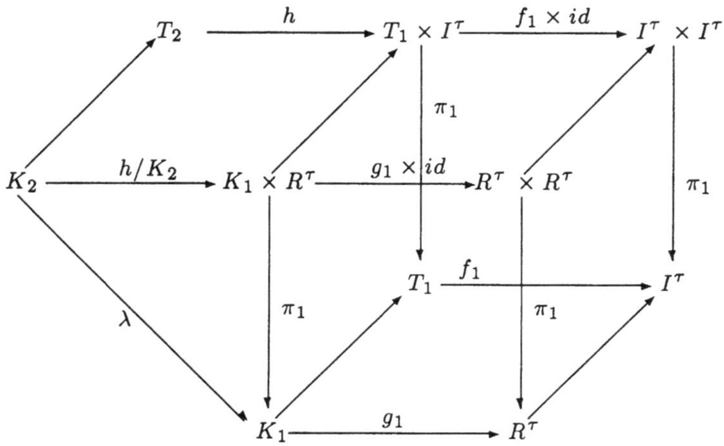


Here $h\colon T_2 \to T_1 \times I^\tau$ denotes an $n$-invertible and $(n-1)$-soft map of an $n$-dimensional compactum $T_2$ of weight $\tau$ onto the product $T_1 \times I^\tau$ (Theorem 6.2.11). Denote by $K_2$ the inverse image of the product $K_1 \times R^\tau$ under the map $h$. Finally, $\lambda$ denotes the composition of the map $h/K_2$ and the projection $\pi_1\colon K_1 \times R^\tau \to K_1$ and $g_2$ denotes the composition of the restriction $h/K_2$ and the product $g_1 \times id$. Since $T_1$ is an $AE(0)$-compactum, the product $T_1 \times I^\tau$ is also an $AE(0)$-space. Further, by Proposition 6.1.21, the space $T_2$ is an $AE(0)$-space and, hence, by Proposition 6.1.8, is perfectly $\kappa$-normal. An argument, completely similar to that presented above, shows that all the constructed objects satisfy the desired conditions. Finally, we remark that the characteristic map of the diagram in condition (f) coincides with the map $h/K_2$ and, therefore, is surjective by the construction. □

*Historical and bibliographical notes* 6.2. The concept of an $n$-soft diagram was introduced by the author [**78**], [**79**]. The main result of this Section — Theorem 6.2.11 — was proved in [**125**]. In the case $n = 1$, weaker results have been obtained in [**148**] and [**78**] (see also [**245**]). Proposition 6.2.13 was proved in [**84**]. Propositions 6.2.9 and 6.2.10, for the class of compact spaces, was obtained in [**278**]. It has already been mentioned in the main text that the compacta $D_n^\tau$ are universal with respect to the class of all at most $n$-dimensional compact spaces of weight $\le \tau$. The first results in this direction were simultaneously and independently obtained in [**246**], [**247**] and [**324**].

## 6.3. Spectral characterizations of $n$-soft mappings

We say that a map $f\colon X \to Y$ has a *Polish kernel* if there exists a Polish space $P$ such that $X$ is $C$-embedded in the product $Y \times P$ so that $f$ coincides with the restriction $\pi_Y/X$ of the projection $\pi_Y\colon Y \times P \to Y$ onto $X$. Observe that any map between Polish spaces has a Polish kernel.

The following theorem is one of the main results of this Chapter.

THEOREM 6.3.1. *Let $n = 0, 1, \ldots, \infty$. A map $f\colon X \to Y$ between $A(N)E(n)$-spaces is $n$-soft if and only if there exist factorizing well ordered continuous spectra $\mathcal{S}_X = \{X_\alpha, p_\alpha^\beta, \tau\}$ , $\mathcal{S}_Y = \{Y_\alpha, q_\alpha^\beta, \tau\}$ and a morphism $\{f_\alpha\}\colon \mathcal{S}_X \to \mathcal{S}_Y$, satisfying the following conditions:*

(i) $X = \lim \mathcal{S}_X$, $Y = \lim \mathcal{S}_Y$ *and* $f = \lim\{f_\alpha\}$.

(ii) *The spaces $X_0$ and $Y_0$ are Polish $A(N)E(n)$-spaces and the map $f_0\colon X_0 \to Y_0$ is an $n$-soft map between them.*

(iii) *The spaces $X_\alpha$ and $Y_\alpha$ are $A(N)E(n)$-spaces and the map $f_\alpha\colon X_\alpha \to Y_\alpha$ is an $n$-soft map between them, $\alpha < \tau$.*

(iv) *All short projections in the spectra $\mathcal{S}_X$ and $\mathcal{S}_Y$ are $n$-soft and have Polish kernels.*

(v) *All adjacent square diagrams (generated by the short projections of the spectra $\mathcal{S}_X$ and $\mathcal{S}_Y$ and the corresponding elements of the given morphism) are $n$-soft and their characteristic maps have Polish kernels.*

(vi) *If the map $f$ itself has a Polish kernel, then all square diagrams indicated in (v) are Cartesian squares.*

PROOF. One part of our theorem follows from Lemmas 6.2.6 and 6.2.7. Let us prove the converse, i.e. show that $n$-softness of the map $f$ implies the existence of the indicated spectra and morphism with the corresponding properties. We consider only the absolute case. The remaining part can be proved in a completely similar way.

Let $\tau = \max\{R - w(X), R - w(Y)\}$. The non-trivial case is the case of uncountable $\tau$. Take any set $A$ of cardinality $|A| = \tau$. Without loss of generality, we may assume that the space $Y$ is $C$-embedded in $R^A$, the space $X$ is $C$-embedded in the product $R^A \times R^A$ and the map $f\colon X \to Y$ coincides with the restriction $\pi_1/X$ of the projection $\pi_1\colon R^A \times R^A \to R^A$ onto $X$. Next consider spectrally complete realcompact spaces $K_1$, $K_2$ and maps $g_1$, $g_2$ and $\lambda$ from Proposition 6.2.13. Since $Y$ is $C$-embedded in $R^A$ and $\dim K_1 = n$, there exists a map $h_1\colon K_1 \to Y$ such that $h_1/g_1^{-1}(Y) = g_1/g_1^{-1}(Y)$. Similarly, the fact that $X$ is $C$-embedded in $R^A \times R^A$, the equality $\dim K_2 = n$, the $n$-softness of the map $f$, and the obvious equality $fg_2/g_2^{-1}(X) = h_1\lambda/g_2^{-1}(X)$ guarantee the existence of a map $h_2\colon K_2 \to X$ such that $h_2/g_2^{-1}(X) = g_2/g_2^{-1}(X)$ and $fh_2 = h_1\lambda$. Consequently, the diagram

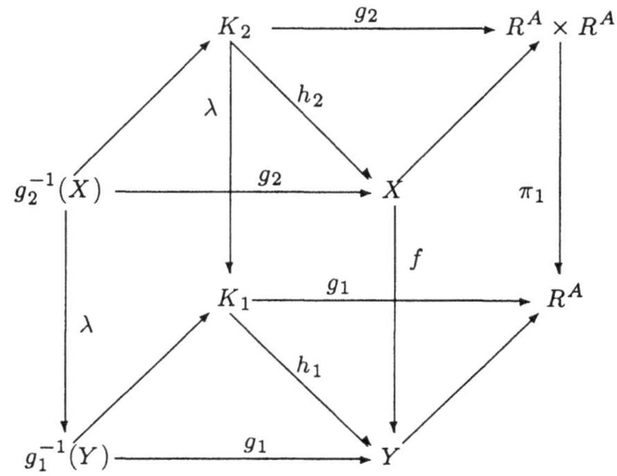


is commutative.

It follows from these constructions that the compact-valued and upper semi-continuous retractions $r_i = h_i g_i^{-1}$, $i = 1, 2$, satisfy the equality $fr_2 = r_1\pi_1$.

Let us now introduce some notations. If $C \subseteq B \subseteq A$, then

$$\pi_C^B \colon R^B \to R^C \text{ and } \pi_B \colon R^A \to R^B$$

denote the projections onto the corresponding subproducts. By $Y_B$ (respectively, by $X_B$) we denote the space $\pi_B(Y)$ (respectively, the space $(\pi_B \times \pi_B)(X)$), and by $Y(B)$ (respectively, by $X(B)$) we denote the inverse image $(\pi_B^{-1}(Y_B)$ (respectively, the inverse image $(\pi_B \times \pi_B)^{-1}(X_B))$. The restriction of the projection $\pi_C^B$ onto $Y_B$ (respectively, the restriction of $\pi_C^B \times \pi_C^B$ onto $X_B$) is denoted by $q_C^B$ (respectively, by $p_C^B$). A similar meaning is assigned to the symbols $q_B$ and $p_B$. Finally, by $f_B$ we denote the restriction of the projection $\pi_1^B \colon R^B \times R^B \to R^B$ onto $X_B$.

A subset $B \subseteq A$ is called *admissible* if for each point $x \in X(B)$ we have the equality

$$p_B r_2(x) = (\pi_B \times \pi_B)(x).$$

Straightforward verification shows that the union of an arbitrary collection of admissible subsets is admissible. Let us investigate other properties of admissible sets.

First of all, observe that for each admissible set $B$ and for each point $y \in Y(B)$ we have the following equality: $q_B r_1(y) = \pi_B(y)$. Indeed, let $y \in Y(B)$. Take a point $x \in X(B)$ such that $\pi_1(x) = y$. Then

$$q_B r_1(y) = q_B h_1 g_1^{-1}(y) = q_B h_1 g_1^{-1}(\pi_1(x)).$$

By condition (f) of Proposition 6.2.13, we have

$$g_1^{-1}(\pi_1(x)) = \lambda g_2^{-1}(x).$$

Consequently,

$$q_B r_1(y) = q_B h_1 \lambda g_2^{-1}(x) = q_B f h_2 g_2^{-1}(x) = q_B f r_2(x) =$$

$$f_B p_B r_2(x) = f_B(\pi_B \times \pi_B)(x) = \pi_B \pi_1(x) = \pi_B(y).$$

Let us now show that for each admissible subset $B$, the subspace $Y_B$ is closed and $C$-embedded in $R^B$. Take a section $i_B \colon R^B \to R^A$ of the projection $\pi_B$ and let $r_1^B = q_B r_1 i_B$. Since $r_1^B$ is a compact-valued upper semi-continuous retraction of $R^B$ onto $Y_B$, we conclude that $Y_B$ is closed in $R^B$ and, consequently, is realcompact. Let $\varphi \in C(Y_B)$. Then $\varphi q_B h_1 \in C(K_1)$. Take a function $\psi \in C(R^A)$, satisfying condition (b) of Proposition 6.2.13 with respect to the composition $\varphi q_B h_1$. The function $\bar{\varphi}$, defined by letting

$$\bar{\varphi}(y) = \psi(i_B(y)) \quad y \in R^B,$$

extends $\varphi$ to $R^B$. This shows that $Y_B$ is $C$-embedded in $R^B$. A similar argument proves that $X_B$ is $C$-embedded into $R^B \times R^B$ for the admissible set $B$.

Next we show that for each admissible set $B$ the space $Y_B$ is an $AE(n)$-space. The realcompactness of $Y_B$ allows us to use Proposition 6.1.9. Consider an arbitrary at most $n$-dimensional realcompact space $Z$, its closed subspace $Z_0$, and a map $s_0 \colon Z_0 \to Y_B$ satisfying the inclusion $C(s_0)(C(Y_B)) \subseteq C(Z)/Z_0$. Since $R^B$ is an $AE$-space, there is an extension $s_2 \colon Z \to R^B$ of the map $s_0$, considered as a map of $Z_0$ into $R^B$. By condition (a) of Proposition 6.2.13, $g_1$ is an $n$-soft map. Therefore there exists a map $s_1 \colon Z \to K_1$ such that $g_1 s_1 = i_B s_2$. Let $s = q_B h_1 s_1$. Using the above indicated properties of admissible sets, we see that the map $s \colon Z \to Y_B$ is an extension of $s_0$. A completely similar argument shows that $X_B$ is also an $AE(n)$-space for each admissible set $B$.

Now we are going to prove that for each admissible set $B$ the map $q_B$ is $n$-soft. Since the space $Y$, by our assumption, and the space $Y_B$, by the above proof, are both $AE(n)$-spaces, it suffices, in order to show the $n$-softness of $q_B$, to apply Proposition 6.1.16. Thus, let $Z$ be an arbitrary at most $n$-dimensional realcompact space, $Z_0$ be its closed subspace, and $s_0 \colon Z_0 \to Y$ and $s \colon Z \to Y_B$ be arbitrary maps satisfying the conditions

$$q_B s_0 = s/Z_0 \text{ and } C(s_0)(C(Y)) \subseteq C(Z)/Z_0.$$

Since the projection $\pi_B$ is a soft map, there is a map $k_2 \colon Z \to R^A$ such that

$$\pi_B k_2 = s \text{ and } k_2/Z_0 = s_0.$$

The $n$-invertibility of the map $g_1$, coupled with the inequality $\dim Z \leq n$, guarantees the existence of a map $k_1 \colon Z \to K_1$ such that $g_1 k_1 = k_2$. Let $k = h_1 k_1$. Straightforward verification shows that $k$ is the required lifting of the map $s$, extending the map $s_0$. This proves the $n$-softness of the map $q_B$. A similar argument shows the $n$-softness of the map $p_B$ for each admissible set $B$.

The $n$-softness of the maps $f$, $p_B$ and $q_B$ imply, by Lemmas 6.1.14 and 6.1.15, the $n$-softness of the maps $f_B$, $p_C^B$ and $q_C^B$ for admissible sets $C$ and $B$ with $C \subseteq B$.

Suppose now that $C$ and $B$ both are admissible subsets of $A$ and $C \subseteq B$. Denote by $T(C, B)$ the following commutative diagram:

$$\begin{array}{ccc} X_B & \xrightarrow{\ f_B\ } & Y_B \\ {\scriptstyle p_C^B}\downarrow & & \downarrow{\scriptstyle q_C^B} \\ X_C & \xrightarrow{\ f_C\ } & Y_C \end{array}$$

Obviously this diagram is the subdiagram of the following commutative diagram, denoted by $R(C, B)$:

$$\begin{array}{ccc} R^B \times R^B & \xrightarrow{\ \pi_1^B\ } & R^B \\ {\scriptstyle \pi_C^B \times \pi_C^B}\downarrow & & \downarrow{\scriptstyle \pi_C^B} \\ R^C \times R^C & \xrightarrow{\ \pi_1^C\ } & R^C \end{array}$$

Our first goal is to establish the $n$-softness of the diagram $T(C, B)$. First, let us show that the diagram $T(C, B)$ is bicommutative. Let $x_C \in X_C$, $y_B \in Y_B$ and $f_C(x_C) = q_C^B(y_B)$. Since the diagram $R(C, B)$ is obviously bicommutative (it is not hard to see that this diagram is even soft), there is a point $z_0 \in R^B \times R^B$ such that

$$(\pi_C^B \times \pi_C^B)(z_0) = x_C \quad \text{and} \quad \pi_1^B(z_0).$$

Clearly there is a point $z \in X(C)$ such that $z_0 = (\pi_B \times \pi_B)(z)$. Consider a point $x$ of the compactum $r_2(z)$ and let $x_B = p_B(x)$. It is easy to see that

$$p_C^B(x_B) = x_C \quad \text{and} \quad f_B(x_B) = y_B.$$

This shows that $T(C, B)$ is indeed a bicommutative diagram.

Denote by $\chi_0\colon X_B \to T_0$ the characteristic map of the diagram $T(C,B)$ (here $T_0$ stands for the corresponding fibered product) and by $\chi\colon R^B \times R^B \to T$ — the characteristic map of the diagram $R(C,B)$ (recall that, as remarked above, $\chi$ is a soft map; here $T$ is the corresponding fibered product, which as can easily be seen, itself is a power of the real line and, hence, is an absolute extensor). Since $T(C,B)$ is a subdiagram of the diagram $R(C,B)$, one can see that $T_0$ is a subspace of $T$ and that the map $\chi_0$ coincides with the restriction of $\chi$ to $X_B$. Summarizing the above data, we arrive at the following diagram:

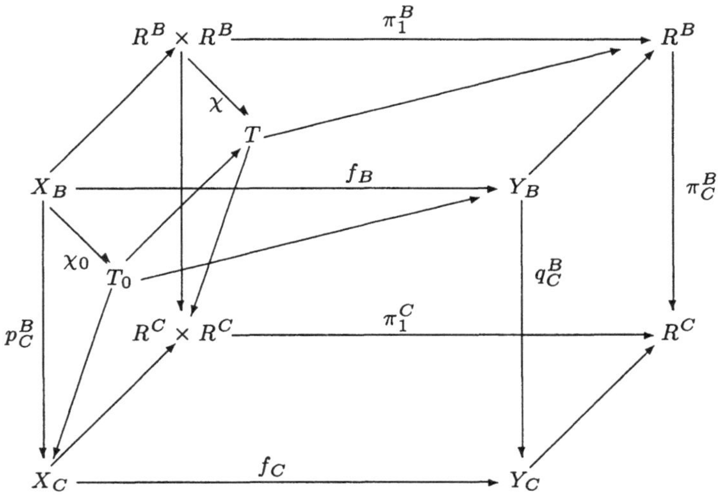


By Proposition 6.1.21 and Lemma 6.2.4, $T_0$ is an $AE(n)$-space and, consequently, in order to show the $n$-softness of the map $\chi_0$ we can use Proposition 6.1.16. Let $Z$ be an arbitrary at most $n$-dimensional realcompact space, $Z_0$ be a closed subspace of $Z$, and $s_0\colon Z_0 \to X_B$ and $s\colon Z \to T_0$ be maps satisfying the conditions

$$\chi_0 s_0 = s/Z_0 \quad\text{and}\quad C(s_0)(C(X_B)) \subseteq C(Z)/Z_0.$$

The softness of the map $\chi$ implies the existence of a map $k_2\colon Z \to R^B \times R^B$ such that $s_0 = k_2/Z_0$ and $s = \chi k_2$. Consider a section $i_B\colon R^B \to R^A$ of the projection $\pi_B$. Since the map $g_2$ is $n$-invertible, there is a map $k_1\colon Z \to K_2$ such that

$$g_2 k_1 = (i_B \times i_B)k_2.$$

Let $k = p_B h_2 k_1$. Straightforward verification shows that $k$ is the desired lifting of the map $s$, which extends $s_0$. Therefore $T(C,B)$ is $n$-soft diagram.

If the map $f$ has a Polish kernel, then a similar argument shows that the characteristic map of the diagram $T(C,B)$ is a homeomorphism. Indeed, in this case instead of diagram $R(C,B)$ we deal with a diagram of the type

$$\begin{array}{ccc} R^B \times P & \xrightarrow{\pi_1} & R^B \\ \Big\downarrow{\scriptstyle \pi_C^B \times id_P} & & \Big\downarrow{\scriptstyle \pi_C^B} \\ R^C \times P & \xrightarrow{\pi_1} & R^C \end{array}$$

where $P$ is a Polish space witnessing the fact that $f$ has a Polish kernel. It is easy to see that the above diagram is the Cartesian square, i.e. the corresponding characteristic map is a homeomorphism. This fact, coupled with the bicommutativity of the diagram $T(C, B)$, obviously implies that $T(C, B)$ is also the Cartesian square.

We return to the general situation. Represent the product $R^A \times R^A$ as the limit space of the standard factorizing Polish spectrum

$$\mathcal{S} = \{R^B \times R^B, \pi_C^B \times \pi_C^B, exp_\omega A\}.$$

Then the space $X$ is the limit space of the induced factorizing $\omega$-spectrum

$$\mathcal{S}_X = \{\mathrm{cl}_{R^B \times R^B} X_B, (\pi_C^B \times \pi_C^B)/ \mathrm{cl}_{R^B \times R^B} X_B, exp_\omega A\}.$$

Since the $r$-weight of a spectrally complete realcompact space $K_2$ equals $|A|$ (see condition (d) from Proposition 6.2.13), we conclude that $K_2 = \lim \mathcal{S}_{K_2}$, where

$$\mathcal{S}_{K_2} = \{k_2^B, t_C^B, exp_\omega A\}$$

is a factorizing Polish spectrum. Since $g_2$ is proper and functionally closed (condition (a) from Proposition 6.2.13), by Proposition 5.2.9, we may assume that $g_2 = \lim g_2^B$, where

$$\{g_2^B\} \colon \mathcal{S}_{K_2} \to \mathcal{S}$$

is a bicommutative morphism, consisting of proper maps. By Spectral Theorem 1.3.6, the map $h_2 \colon K_2 \to X$ can also represented as the limit map of a morphism

$$\{h_2^B\} \colon \mathcal{S}_{K_2} \to \mathcal{S}_X.$$

Therefore, the collection of those element $B \in exp_\omega A$ for which the diagram

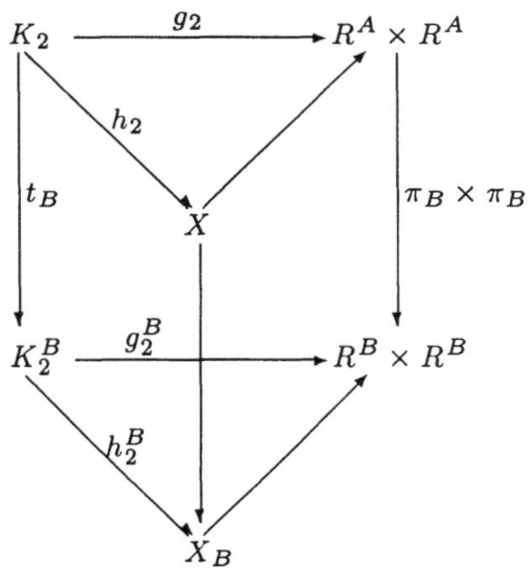


is commutative is $\omega$-closed and cofinal in $exp_\omega A$ (apply Proposition 1.1.27). The bicommutativity of the morphism $\{g_2^B\}$ implies the equality

$$h_2^B/(g_2^B)^{-1}(X_B) = g_2^B/(g_2^B)^{-1}(X_B).$$

We claim that each of the indicated countable subsets $B \subseteq A$ is admissible. Indeed, let $x \in X(B)$. Then

$$p_B r_2(x) = p_B h_2 g_2^{-1}(x) = h_2^B t_B g_2^{-1}(x).$$

Consequently,

$$p_B r_2(x) = h_2^B(g_2^B)^{-1}(\pi_B \times \pi_B)(x) = g_2^B(g_2^B)^{-1}(\pi_B \times \pi_B)(x) = (\pi_B \times \pi_B)(x).$$

This shows that $exp_\omega A$ contains an $\omega$-closed and cofinal subcollection, consisting of admissible subsets of $A$.

Since $|A| = \tau$, we can write $A = \{a_\alpha \colon \alpha < \tau\}$. Since the collection of countable admissible subsets of $A$ is cofinal in $exp_\omega A$, each point $a_\alpha \in A$ is contained in a countable admissible subset $B_\alpha \subseteq A$. Let $A_\alpha = \cup\{B_\beta \colon \beta \leq \alpha\}$ and

$$X_\alpha = X_{A_\alpha}, p_\alpha^{\alpha+1} = p_{A_\alpha}^{A_{\alpha+1}}, Y_\alpha = Y_{A_\alpha}, q_\alpha^{\alpha+1} = q_{A_\alpha}^{A_{\alpha+1}} \text{and} f_\alpha = f_{A_\alpha}, \alpha < \tau.$$

It follows, from the properties of admissible subsets proved above, that the spectra

$$\mathcal{S}_X = \{X_\alpha, p_\alpha^\beta, \tau\} \quad \mathcal{S}_Y = \{Y_\alpha, q_\alpha^\beta, \tau\} \quad \text{and the morphism } \{f_\alpha\} \colon \mathcal{S}_X \to \mathcal{S}_Y$$

satisfy all of the required properties. This completes the proof of Theorem 6.3.1. □

While proving the above theorem, we in fact have obtained the another spectral characterization of $n$-soft maps. Sometimes this version is more convenient.

THEOREM 6.3.2. *Let $n = 0, 1, \dots, \infty$. A map $f\colon X \to Y$ between $A(N)E(n)$-spaces is $n$-soft if and only if there exist factorizing Polish spectra $\mathcal{S}_X = \{X_\alpha, p_\alpha^\beta, A\}$, $\mathcal{S}_Y = \{Y_\alpha, q_\alpha^\beta, A\}$, consisting of Polish $A(N)E(n)$-spaces, and a morphism $\{f_\alpha\}\colon \mathcal{S}_X \to \mathcal{S}_Y$, consisting of $n$-soft maps, such that the following conditions are satisfied:*

(i) $X = \lim \mathcal{S}_X$, $Y = \lim \mathcal{S}_Y$ *and* $f = \lim\{f_\alpha\}$.

(ii) *All limit projections of the spectra $\mathcal{S}_X$ and $\mathcal{S}_Y$ are $n$-soft.*

(iii) *All limit square diagrams, generated by limit projections of spectra $\mathcal{S}_X$ and $\mathcal{S}_Y$, by elements of the morphism $\{f_\alpha\}$ and by the map $f$, are $n$-soft.*

(iv) *If the map $f$ has a Polish kernel, then all square diagrams indicated in (iii) are the Cartesian squares.*

As an immediate corollary of Theorem 6.3.2 we have the following statement.

PROPOSITION 6.3.3. *Let $n = 0, 1, \dots, \infty$ and $\tau$ be an infinite cardinal number. If $f\colon X \to Y$ is an $n$-soft map between $A(N)E(n)$-spaces, then there exist factorizing $\tau$-spectra $\mathcal{S}_X = \{X_\alpha, p_\alpha^\beta, A\}$, $\mathcal{S}_Y = \{Y_\alpha, q_\alpha^\beta, A\}$, consisting of $A(N)E(n)$-spaces, and a morphism $\{f_\alpha\}\colon \mathcal{S}_X \to \mathcal{S}_Y$, consisting of $n$-soft maps, such that the following conditions are satisfied:*

(i) $X = \lim \mathcal{S}_X$, $Y = \lim \mathcal{S}_Y$ *and* $f = \lim\{f_\alpha\}$.

(ii) *All limit projections of the spectra $\mathcal{S}_X$ and $\mathcal{S}_Y$ are $n$-soft.*

(iii) *All limit square diagrams, generated by limit projections of spectra $\mathcal{S}_X$ and $\mathcal{S}_Y$, by elements of the morphism $\{f_\alpha\}$ and by the map $f$, are $n$-soft.*

PROOF. The desired spectra can be obtained by taking $\tau$-completions of the $\omega$-spectra from Theorem 6.3.2. The desired morphism is then generated by the morphism from that Theorem. □

Let us now consider a particular case of the above statement. Namely, assume that $X = Y$ and that $f$ is the identity map. Then we obtain the following two results.

PROPOSITION 6.3.4. *A space $X$ is an $A(N)E(n)$-space if and only if it is the limit space of a factorizing well ordered continuous spectrum $\mathcal{S}_X = \{X_\alpha, p_\alpha^\beta, \tau\}$, all short projections of which are $n$-soft and have Polish kernels, and the first element $X_0$ of which is a Polish $A(N)E(n)$-space.*

PROPOSITION 6.3.5. *A space $X$ is an $A(N)E(n)$-space if and only if it is the limit space of a factorizing $\tau$-spectrum consisting of $A(N)E(n)$-spaces and $n$-soft limit projections.*

*Remark* 6.3.6. As a useful (and manageable) exercise we suggest that the reader verifies that if the spaces $X$ or $Y$ in the above statements are Čech complete and Lindelöf, then all projections in the corresponding spectra are proper.

*Remark* 6.3.7. In the case when all spaces are compact and $n = 0$, Proposition 6.3.4 was obtained by Haydon [**164**]. He first investigated spectra with open projections having *metrizable kernel* (observe that in the case of compact spaces, the latter class of maps coincides with those with Polish kernels in our sense). Such spectra are called *Haydon spectra.* Thus, $AE(0)$-compacta are precisely the limit spaces of Haydon spectra having a point (or any other *metrizable* compactum) as their first element. In general, the investigation of $AE(0)$-compacta by means of analysis of the corresponding Haydon spectra is similar to Pontrjagin's method of investigation of (locally) compact topological groups by means of their spectral representations in terms of Lie series (we will see in Chapter 8 that the mentioned analogy between Haydon spectra and Lie series leads us in fact to the proof of the fact that each compact topological group is an $AE(0)$-compactum).

Consider some of corollaries of the above results.

COROLLARY 6.3.8. *The weight and the $R$-weight of a non-discrete $AE(0)$-space coincide.*

PROOF. Let $X$ be a non-discrete $AE(0)$-space of weight $\tau$. Since $X$ is non-discrete, $\tau \geq \omega$. By Proposition 6.3.5, the space $X$ can be represented as the limit space of a factorizing $\tau$-spectrum $\mathcal{S}_X = \{X_\alpha, p_\alpha^\beta, A\}$ , consisting of $AE(0)$-spaces of $R$-weight $\tau$ and 0-soft limit projections. By Theorem 1.3.6, there exist an index $\alpha \in A$ and a map $i_\alpha \colon X_\alpha \to X$ such that $i_\alpha p_\alpha = id_X$. Being a 0-soft map, $f$ is surjective (Lemma 6.1.13) and open (Proposition 6.1.26). Consequently, $p_\alpha$ is a homeomorphism. But then $R - w(X) = R - w(X_\alpha) = \tau$. □

COROLLARY 6.3.9. *If a space $X$ admits an $n$-soft map with a Polish kernel onto an uncountable product $\prod\{Y_\alpha \colon \alpha \in A\}$ of non-trivial*[3] *Polish $AE(n)$-spaces, then $X$ is homeomorphic to the product*

$$X_0 \times \prod\{Y_\alpha \colon \alpha \in A - A_0\},$$

*where $A_0$ is a countable subset of the indexing set $A$ and $X_0$ is a Polish space, admitting an $n$-soft map onto the product $\prod\{Y_\alpha \colon \alpha \in A_0\}$.*

PROOF. The product $\prod\{Y_\alpha \colon \alpha \in A\}$ is an $AE(n)$-space. Consequently, by Proposition 6.1.21, $X$ is also an $AE(n)$-space. Represent $Y$ as the limit space of the standard factorizing Polish spectrum $\mathcal{S} = \{Y_B, \pi_C^B, exp_\omega A\}$, where $Y_B = \prod\{Y_\alpha \colon \alpha \in B\}$ and $\pi_C^B \colon Y_B \to Y_C$ are the natural projections ($C \subseteq B$). Consider also factorizing Polish spectra $\mathcal{S}_X$ and $\mathcal{S}_Y$ from Theorem 6.3.2. Since the $R$-weight of the space $Y$ coincides with $|A|$ and since $f$ has a Polish kernel, one can easily conclude that the $R$-weight of the space $X$ is also $\tau$. Therefore, without loss of generality, we may assume that the indexing sets of the spectra $\mathcal{S}_X$ and $\mathcal{S}_Y$ coincide with $exp_\omega A$. Further, by Theorem 1.3.6, again without lost of generality, we may assume that the spectra $\mathcal{S}$ and $\mathcal{S}_Y$ are identical. Consider now the morphism $\{f_B\}$ from Theorem 6.3.2. Then the square diagram

[3] i.e. containing at least two points.

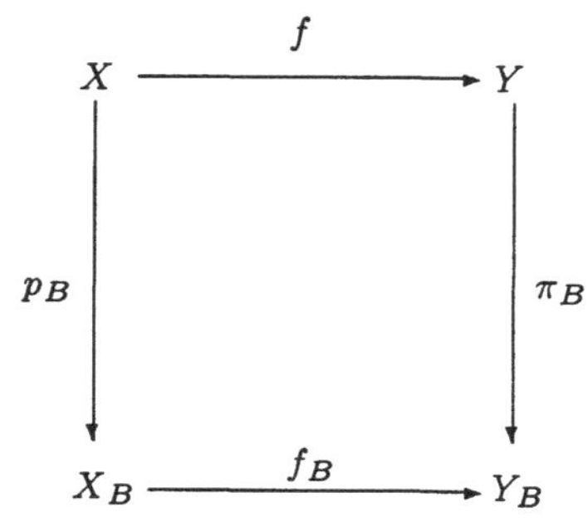


is a Cartesian square. Denote $X_B$ by $X_0$ and let $A_0 = B$. Since the projection $\pi_B\colon Y \to Y_B$ is a trivial bundle with fiber $\prod\{Y_\alpha\colon \alpha \in A - A_0\}$, we see that (since the above diagram is the Cartesian square) the projection $p_{A_0}$ of the spectrum $\mathcal{S}_X$ is also a trivial bundle with the same fiber. □

COROLLARY 6.3.10. *If a compactum admits an $n$-soft with a Polish kernel map onto the Tychonov cube $I^\tau$, $\tau > \omega$, then it is homeomorphic to the product of the cube $I^\tau$ and a metrizable compactum which admits an $n$-soft map onto the Hilbert cube $I^\omega$.*

It should be observed that the analog of Corollary 6.3.10 (as well as of Corollary 6.3.9) is not true in the metrizable case. Indeed, we saw in Chapter 4 that the Hilbert cube is an $n$-soft image of the universal $(n+1)$-dimensional Menger compactum $\mu^{n+1}$.

*Historical and bibliographical notes* 6.3. The results of this Section form the foundation of the general theory of non-metrizable $ANE(n)$-spaces. In particular, they solve the so-called adequacy problem (of $ANE(n)$-spaces and $n$-soft maps) posed in [**278**]. Without fear of overstatement one might say that the entire circle of problems has its roots in the work of Haydon [**164**], which in fact contains the proof of the indicated adequacy within the class of compacta for $n = 0$ (i.e. Proposition 6.3.4 for $n = 0$). The case $n = \infty$ in Proposition 6.3.4 was considered in [**272**] and the case $n = 1$ in [**148**] and [**242**]. The remaining cases of Proposition 6.3.4 (as well as Proposition 6.3.5 in the class of compacta) have been obtained in [**125**]. The remaining results in this direction, covering the non compact case and even the case of maps (Theorems 6.3.1 and 6.3.2, Propositions 6.3.3 — 6.3.5), belong to the author [**79**], [**84**]. Corollaries 6.3.8 — 6.3.10 were also obtained by the author [**79**], [**80**].

## 6.4. Further properties of $AE(0)$-spaces

In this Section, based on the results obtained in Section 6.3, we establish several important properties of $AE(0)$-spaces. These properties will be later used in the development of $I^\tau$- and $R^\tau$-manifold theories.

**6.4.1. General properties of $AE(0)$-spaces.** We begin with the following statement.

LEMMA 6.4.1. *Let $f\colon X \to Y$ and $g_1\colon N^\tau \to Y$ be 0-soft maps, and suppose that $f$ has a Polish kernel. Then there exists a 0-soft map $g_2\colon N^\tau \times N^\omega \to X$ such that $fg_2 = g_1\pi_1$, where $\pi_1\colon N^\tau \times N^\omega \to N^\omega$ is the projection. Moreover, the resulting square diagram*

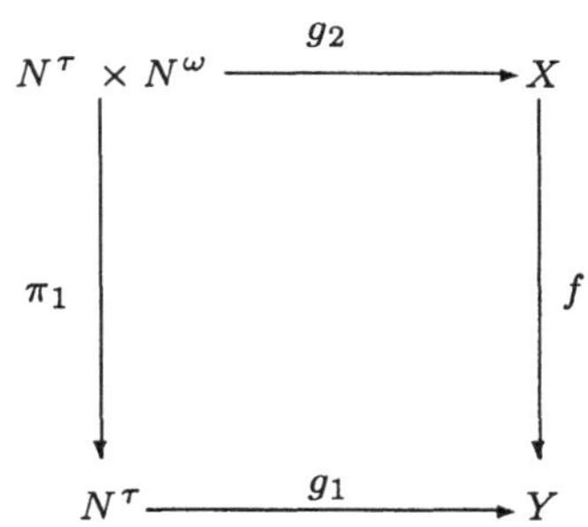


*is 0-soft.*

PROOF. First we consider the case $\tau = \omega$. In this case, the spaces $X$ and $Y$ are both Polish, and the maps $f$ and $g_1$ are open and surjective. Let $Z$ be the fibered product of spaces $N^\omega$ and $X$ with respect to maps $g_1\colon N^\omega \to Y$ and $f\colon X \to Y$. Denote by $\psi\colon Z \to N^\omega$ and $\varphi\colon Z \to X$ the canonical projections. Our goal will be attained if we construct an open surjection $h\colon N^\omega \times N^\omega \to Z$ satisfying the equality $\pi_1 = \psi h$. It is clear that the maps $\psi$ and $\varphi$ are open and $Z$ is a Polish space. Take any complete metric $d$ on $Z$ and construct a sequence $\{\mathcal{U}_k\}$ of countable open covers of $Z$ with the following properties:

(i) $\mathcal{U}_k = \{U_{i_1,\dots,i_k}\colon i_j \in N\}$, $j \in N$, $k \in N$.
(ii) $\mathrm{cl}_Z\, U_{i_1,\dots,i_k,i_{k+1}} \subseteq U_{i_1,\dots,i_k}$, $i_{k+1} \in N$.
(iii) $U_{i_1,\dots,i_k} = \cup\{U_{i_1,\dots,i_k,i_{k+1}}\colon i_{k+1} \in N\}$.
(iv) $\mathrm{diam}_d(U_{i_1,\dots,i_k} \cap \psi^{-1}(y)) < \frac{1}{2^k}$, $y \in N^\omega$.
(v) $\psi(U_{i_1,\dots,i_k}) = N^\omega$, $i_j \in N$, $k \in N$.

To prove that such a sequence of open covers exists it suffices to show that if $U$ is an open subset of $Z$ such that $\psi(U) = N^\omega$, then for any $\epsilon > 0$ and any point $x_0 \in U$ there is an open subset $V$ of $Z$ such that:

(a) $x_0 \in V$.
(b) $\mathrm{cl}_Z\, V \subseteq U$.

(c) $\psi(V) = N^{\omega}$.

(d) $\mathrm{diam}_d(V \cap \psi^{-1}(y)) < \epsilon$ for any point $y \in N^{\omega}$.

Fix a map $s\colon N^{\omega} \to U$ such that $\psi s = id_{N^{\omega}}$ and $x_0 \in A \equiv s(N^{\omega})$. The set $A$ is closed in $Z$ and, consequently, there exists an open neighborhood $OA$ of $A$ in $Z$ whose closure (in $Z$) is contained in $U$. For any point $a \in A$ we take an open neighborhood $Oa$ of $a$ of diameter less than $\epsilon$ which is contained in $OA$. Choose a countable subfamily $\{Oa_i\colon i \in \omega\}$ of the collection $\{Oa\colon a \in A\}$ that covers $A$. In a combinatorial manner, we refine the open cover $\{\psi(Oa_i)\colon i \in \omega\}$ of $N^{\omega}$ by a countable disjoint open cover $\{G_i\colon i \in \omega\}$ with the condition that $\psi(x_0) \in G_0$. Then we let

$$V = \cup\{Oa_i \cap \psi^{-1}(G_i)\colon i \in \omega\}.$$

The reader can easily verify that $V$ is the desired set.

Let us now define the map $h\colon N^{\omega} \times N^{\omega} \to Z$ as follows. For any point $(a, b) = (a, (i_1, \dots, i_k, \dots)) \in N^{\omega} \times N^{\omega}$ let

$$h(a, b) = \psi^{-1}(a) \cap \bigcap\{U_{i_1,\dots,i_k}\colon k \in N\}.$$

It follows from the completeness of $Z$ and from properties (ii) and (iv) that $h(a, b)$ is non-empty and consists of precisely one point. The easy verifications of the continuity, surjectivity and openness of $h$ are left to the reader.

Next we consider the case of uncountable $\tau$. Clearly $X$ and $Y$ are $AE(0)$-spaces (Proposition 6.1.21). By Theorem 6.3.2, applied to both of the maps $f$ and to $g_1$, there exist Polish spaces $X_0$ and $Y_0$, open surjections $f_0\colon X_0 \to Y_0$ and $g_0\colon N^{\omega} \to Y_0$, and 0-soft maps $p\colon X \to X_0$ and $q\colon Y \to Y_0$ such that the following conditions hold:

(e) The square diagram

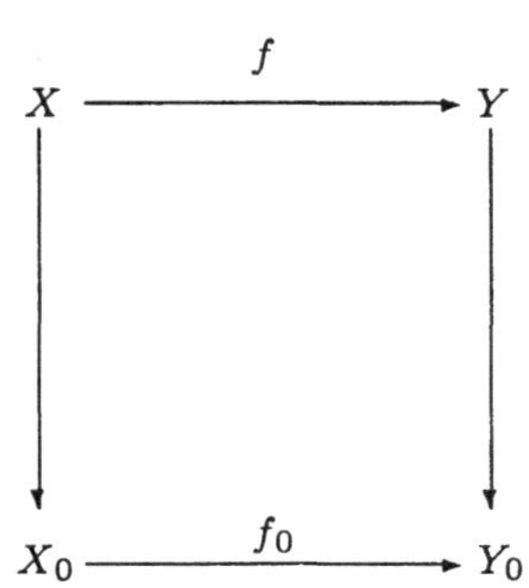


is a Cartesian square.

(f) The square diagram

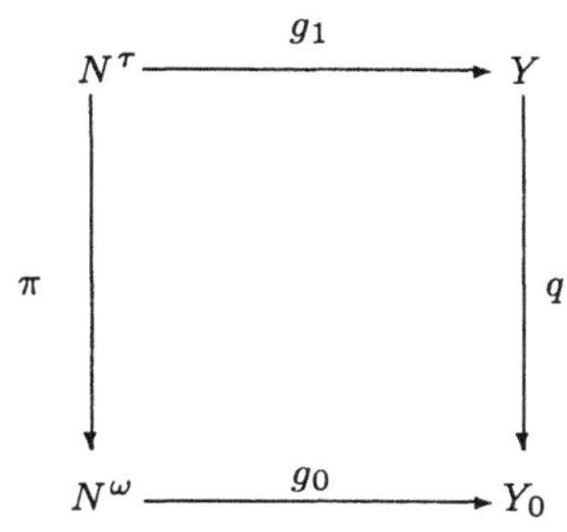


where $\pi\colon N^\tau \to N^\omega$, is 0-soft.

Let $Z$ denote the fibered product of $X$ and $N^\tau$ with respect ro $f$ and $g_1$, and let $\varphi\colon Z \to X$ and $\psi\colon Z \to N^\tau$ be the canonical projections. The symbols $Z_0$, $\varphi_0$ and $\psi_0$ have analogous meaning. The following diagram helps us to summarize the situation:

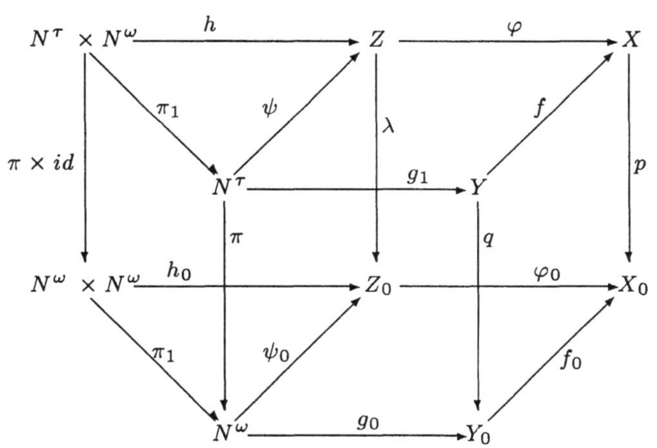


Here $\lambda$ denotes the restriction of the product $p \times \pi$ onto $Z$. Note that $\lambda$ is surjective.

Observe also that the construction and property (a) imply that:

(g) The square diagram

$$\begin{array}{ccc} Z & \xrightarrow{\psi} & N^{\tau} \\ \lambda \downarrow & & \downarrow \pi \\ Z_0 & \xrightarrow{\psi_0} & N^{\omega} \end{array}$$

is a Cartesian square (only general categorical arguments are needed while making this conclusion [**212**]).

By the case already considered above, there exists an open surjection

$$h_0 \colon N^{\omega} \times N^{\omega} \to Z_0$$

such that $\pi_1 = \psi_0 h_0$. Obviously,

(h) The square diagram

$$\begin{array}{ccc} N^{\tau} \times N^{\omega} & \xrightarrow{\pi_1} & N^{\tau} \\ \pi \times id \downarrow & & \downarrow \pi \\ N^{\omega} \times N^{\omega} & \xrightarrow{\pi_1} & N^{\omega} \end{array}$$

is a Cartesian square.

Since all the diagrams constructed up to now are commutative, one can easily see that the map

$$h = \pi_1 \triangle h_0 (\pi \times id_{N^{\omega}}) \colon N^{\tau} \times N^{\omega} \to Z$$

makes the whole diagram commutative. By properties (g) and (h), the square diagram

$$\begin{array}{ccc} N^{\tau} \times N^{\omega} & \xrightarrow{\ h\ } & Z \\ \Big\downarrow{\scriptstyle \pi \times id} & & \Big\downarrow{\scriptstyle \lambda} \\ N^{\omega} \times N^{\omega} & \xrightarrow{\ h_0\ } & Z_0 \end{array}$$

is a Cartesian square. Consequently, by Lemma 6.2.4, $h$ is a 0-soft map. It only remains to note that the composition $g_2 = \varphi h$ satisfies all the required properties. This completes the proof of the lemma. □

THEOREM 6.4.2. *The class of* $AE(0)$*-spaces of weight* $\leq \tau$ *coincides with the class of* 0*-soft images of* $N^{\tau}$ .

PROOF. One part of this statement is a direct consequence of Proposition 6.1.21 (coupled with the fact that $N^{\tau}$ is an $AE(0)$-space).

Let us prove the remaining part, i.e. that each $AE(0)$-space of weight $\leq \tau$ is a 0-soft image of $N^{\tau}$ . If $\tau = \omega$, as already observed in Subsection 1.1.1 (see Theorem 1.1.7), the theorem is true. Suppose that $\tau > \omega$ and consider an $AE(0)$-space $X$ of weight $\tau$. By Corollary 6.3.8, $R - w(X) = \tau$. Therefore, by Proposition 6.3.4, the space $X$ is homeomorphic to the limit space of a factorizing well ordered continuous spectrum $\mathcal{S}_X = \{X_{\alpha}, p_{\alpha}^{\beta}, \tau\}$ , all short projections $p_{\alpha}^{\alpha+1}$, $\alpha < \tau$, of which are 0-soft and have Polish kernels, and such that $X_0$ is a Polish space. Let $f_0\colon N^{\omega} \to X_0$ be an open surjection, the existence of which is guaranteed by Theorem 1.1.7. Further, suppose that for each $\alpha$, with $\alpha < \beta < \tau$, we have already constructed 0-soft maps

$$f_{\alpha}\colon (N^{\omega})^{\alpha} \to X_{\alpha}$$

such that the square diagram

$$\begin{array}{ccc} (N^{\omega})^{\gamma} & \xrightarrow{\ f_{\gamma}\ } & X_{\gamma} \\ \Big\downarrow{\scriptstyle \pi_{\alpha}^{\gamma}} & & \Big\downarrow{\scriptstyle p_{\alpha}^{\gamma}} \\ (N^{\omega})^{\alpha} & \xrightarrow{\ f_{\alpha}\ } & X_{\alpha} \end{array}$$

is 0-soft whenever $\alpha, \gamma < \beta$ (here $\pi^\gamma_\alpha$ denotes the corresponding natural projection).

We are now going to construct a map $f_\beta\colon (N^\omega)^\beta \to X_\beta$.

If $\beta$ is a limit ordinal, then the desired map can be obtained by using Lemma 6.2.7. Observe, in the meantime, that all newly arising square diagrams are also 0-soft.

If $\beta = \alpha+1$, then the existence of a map $f_\beta$, satisfying the required properties, follows from Lemma 6.4.1. Thus, the 0-soft map $f_\alpha\colon (N^\omega)^\alpha \to X_\alpha$ is constructed for each $\alpha < \tau$. Consider the corresponding 0-soft (by construction) morphism

$$\{f_\alpha\}\colon \mathcal{S} \to \mathcal{S}_X,$$

where $\lim \mathcal{S} = N^\tau$ and let

$$f = \lim\{f_\alpha\}\colon N^\tau \to X.$$

It only remains to note that, by Lemma 6.2.7, $f$ is a 0-soft map. □

COROLLARY 6.4.3. *Let $X$ be an $AE(0)$-space of weight $w(X) > \tau \geq \omega$ and $\mathcal{F} = \{F_t\colon t \in T\}$ be an arbitrary family of subsets of $X$ each of which can be written as the intersection of not more than $\tau$ functionally open subsets of $X$. Then there exists a subfamily $\mathcal{F}' \subseteq \mathcal{F}$ of cardinality $\leq \tau$ such that $\mathrm{cl}_X \cup \mathcal{F} = \mathrm{cl}_X \cup \mathcal{F}'$. Moreover, the set $\mathrm{cl}_X \cup \mathcal{F}$ can also be represented as the intersection of not more than $\tau$ functionally open subsets of $X$.*

PROOF. By Theorem 6.4.2, there is a 0-soft map $f\colon N^{w(X)} \to X$. The family $f^{-1}(\mathcal{F}) = \{f^{-1}(F_t)\colon t \in T\}$ consists of sets that can be represented as the intersection of not more than $\tau$ functionally open subsets of $N^{w(X)}$. By the result of [**251**], there is a subfamily $\mathcal{F}' \subseteq \mathcal{F}$ of cardinality $\leq \tau$ such that

$$\mathrm{cl}_{N^{w(X)}} \cup f^{-1}(\mathcal{F}) = \mathrm{cl}_{N^{w(X)}} \cup f^{-1}(\mathcal{F}').$$

Since $f$ is an open map (Proposition 6.1.26), we have

$$\mathrm{cl}_X \cup \mathcal{F} = \mathrm{cl}_X \cup \mathcal{F}'.$$

Let us now prove the second part of our statement. Represent $X$ as the limit space of a factorizing $\tau$-spectrum $\mathcal{S}_X = \{X_\alpha, p^\beta_\alpha, A\}$, consisting of $AE(0)$-spaces and 0-soft limit projections (Proposition 6.3.5). It is not hard to see that there exists an index $\alpha \in A$ such that $F_t = p_\alpha^{-1}(p_\alpha(F_t))$ for each $F_t \in \mathcal{F}'$. Since $w(X_\alpha) \leq \tau$, we easily see that the set

$$F = \mathrm{cl}_{X_\alpha} \cup \{p_\alpha(F_t)\colon F_t \in \mathcal{F}'\}$$

can be represented as the intersection of not more than $\tau$ functionally open subsets of the space $X_\alpha$. But then its inverse image $p_\alpha^{-1}(F)$ also can be represented as the intersection of not more than $\tau$ functionally open subsets of $X$. The openness of the limit projection $p_\alpha$ guarantees that $\mathrm{cl}_X \cup \mathcal{F} = p_\alpha^{-1}(F)$. □

COROLLARY 6.4.4. *Let $X$ be an $AE(0)$-space of uncountable weight. Then an arbitrary family of functionally closed subsets of $X$ contains a countable subfamily with the same closure. Moreover, this closure itself is functionally closed.*

COROLLARY 6.4.5. *An $AE(0)$-space with a dense subset, consisting of points of countable pseudocharacter, is Polish.*

COROLLARY 6.4.6. *The closure of any $G_\delta$-subset of an arbitrary $AE(0)$-space is functionally closed.*

COROLLARY 6.4.7. *Each $G_\delta$-subset of an arbitrary $AE(0)$-space is $z$-embedded.*

COROLLARY 6.4.8. *Each functionally closed subset of an arbitrary $AE(0)$-space is $C$-embedded.*

PROPOSITION 6.4.9. *The property of being an $AE(0)$-space is hereditary with respect to subspaces that can be represented as countable intersections of functionally open subsets.*

PROOF. Let $X$ be an $AE(0)$-space and $Y = \cap\{G_n : n \in \omega\}$, where each of the sets $G_n$ is functionally open in $X$. Represent $X$ as the limit space of a factorizing Polish spectrum $\mathcal{S}_X = \{X_\alpha, p_\alpha^\beta, A\}$ with 0-soft limit projection. Then there is an index $\alpha \in A$ such that the space $X_\alpha$ contains a $G_\delta$-subset $Y_\alpha$ with the property $Y = p_\alpha^{-1}(Y_\alpha)$. It only remains to note that the restriction $p_\alpha/Y : Y \to Y_\alpha$ is 0-soft (Corollary 6.1.17), and then to apply Propositions 6.1.10 and 6.1.21. □

Since the space $N^\tau$ has the Baire property (i.e. the intersection of countable many open dense subsets is dense), Theorem 6.4.2 implies the following result.

COROLLARY 6.4.10. *Each $AE(0)$-space has the Baire property.*

COROLLARY 6.4.11. *An open realcompact subspace of an arbitrary $AE(0)$-space is functionally open.*

PROOF. Let $U$ be an open realcompact subspace of an $AE(0)$-space $X$. It follows from the results of Subsection 1.1.2 that the complement $X - U$ can be represented as the union of functionally closed subsets of $X$. By Corollary 6.4.4, $X - U$ is functionally closed in $X$. □

A bit finer argument proves the following.

COROLLARY 6.4.12. *A realcompact $G_\delta$-subspace of an arbitrary $AE(0)$-space can be represented as the intersection of a countable collection of functionally open subsets.*

**6.4.2. Dimensional properties of $AE(0)$-spaces.** It was shown in Section 6.2 that for each $n \geq 0$ and for each $\tau \geq \omega$, there exists an $(n+1)$-dimensional compactum $D^{\tau}_{n+1}$ of weight $\tau$ admitting an $n$-soft map onto the Tychonov cube $I^{\tau}$. Obviously, this compactum serves as an example of an $(n+1)$-dimensional $AE(n)$-compactum of weight $\tau$. It can be shown [**125**] that we cannot lower the dimension of such compacta. More precisely:

- If $n \geq 1$, then each $n$-dimensional $AE(n)$-compactum is metrizable (if $n = 0$ this statement is not true — take the Cantor cube $D^{\tau}$ of any weight $\tau$).

The reason for this lies in the metrizable case. Namely in Theorem 4.2.25, which states that there is no $n$-soft map of an $n$-dimensional metrizable compactum onto the Hilbert cube $I^{\omega}$. On the other hand, as shown in Theorem 5.1.10, every Polish space is an $n$-soft image of an $n$-dimensional Polish space. Analysis of this situation leads us to the hypothesis that perhaps, in contrast to the compact case, for each $n \geq 0$ and for each $\tau \geq \omega$ there exist $n$-dimensional $AE(n)$-spaces of weight $\tau$. In order to establish this principal fact, we need two preliminary results.

LEMMA 6.4.13. *Let $\mathcal{S} = \{X_k, p^{k+1}_k, \omega\}$ be an inverse sequence, consisting of $n$-dimensional $ANE(n)$-spaces and $n$-soft projections. Then* $\dim(\lim \mathcal{S}) \leq n$.

PROOF. In our situation each of the spaces $X_k$, $k \in \omega$, can be $C$-embedded in $X = \lim \mathcal{S}$. Consequently, $R - w(X_k) \leq R - w(X)$ for each $k \in \omega$. Having in mind this observation, we can, based on Proposition 6.3.5 represent the spaces $X_k$, $k \in \omega$, as well as the space $X$, as the limit spaces of factorizing $\omega$-spectra

$$\mathcal{S}_k = \{X^{\alpha}_k, q^{\beta}_{\alpha,k}, A\} \text{ and } \mathcal{S}_X = \{X_{\alpha}, p^{\beta}_{\alpha}, A\},$$

consisting of $ANE(n)$-spaces and $n$-soft limit projections, and with the same indexing set $A$. Further, by Theorems 1.3.6 and 1.3.10, we can without loss of generality assume that all spaces in the spectra $\mathcal{S}_k$ are at most $n$-dimensional. Applying Theorem 6.3.2 to each of the short projections of the sequence $\mathcal{S}$, one can see that for each $k \in \omega$ there is a bicommutative (even $n$-soft) morphism

$$M^{k+1}_k = \{p^{k+1,\alpha}_k\} \colon \mathcal{S}_{k+1} \to \mathcal{S}_k$$

the limit map of which coincides with the projection $p^{k+1}_k$. In this situation, for each index $\alpha \in A$ we get an inverse sequence

$$\mathcal{S}_{\alpha} = \{X^{\alpha}_k, p^{k+1,\alpha}_k, \omega\}$$

consisting of $n$-dimensional Polish spaces (and $n$-soft projections). The bicommutativity of all naturally occuring square diagrams guarantees that $X_{\alpha} = \lim \mathcal{S}_{\alpha}$ for each $\alpha \in A$. Consequently, each of the spaces $X_{\alpha}$ $\alpha \in A$, as the limit of the sequence $\mathcal{S}_{\alpha}$, consisting of $n$-dimensional Polish spaces, is $n$-dimensional. Applying Lemma 1.3.7, we conclude that $\dim X \leq n$. The reverse inequality follows from the remark made at the beginning of the proof (that $X$ contains a $C$-embedded copy of each $X_k$). □

LEMMA 6.4.14. *Let $\mathcal{S}_X = \{X_\alpha, p_\alpha^\beta, \tau\}$ be a continuous well ordered spectrum, consisting of $n$-dimensional $ANE(n)$-spaces and $n$-soft projections. Then* $\dim(\lim \mathcal{S}_X) = n$.

PROOF. We consider two cases. First, if the spectrum $\mathcal{S}_X$ contains a countable cofinal subspectrum, then our statement follows from Lemma 6.4.13.

Suppose now than there is no countable cofinal subspectrum in $\mathcal{S}_X$. Then this spectrum is $\omega$-complete. By Lemma 6.2.6 and Proposition 6.1.21, the limit space $\lim \mathcal{S}_X$ is an $ANE(n)$-space. Consequently, by Proposition 6.1.8, its Suslin number is countable. Therefore, based on Proposition 1.3.3, we can conclude that $\mathcal{S}_X$ is a factorizing spectrum. Now the inequality $\dim(\lim \mathcal{S}_X) \leq n$ follows from Lemma 1.3.7. The reverse inequality is obvious. □

THEOREM 6.4.15. *Let $n = 0, 1, \ldots$ and $\tau \geq \omega$. Each $ANE(n)$-space of weight $\tau$ is an $n$-soft image of $n$-dimensional space of weight $n$.*

PROOF. We proceed by induction on $\tau$. For $ANE(n)$-spaces of countable weight, i.e. for Polish $ANE(n)$-spaces (see Proposition 6.1.10), our statement is contained in Theorem 5.1.10. Suppose that the theorem has already been proved for $ANE(n)$-spaces of weight $< \tau$ and consider an arbitrary $ANE(n)$-space $X$ of weight $\tau$. By Proposition 6.3.4, $X$ can be represented as the limit space of a continuous well ordered spectrum

$$\mathcal{S}_X = \{X_\alpha, p_\alpha^{\alpha+1}, \lambda\}$$

consisting of $ANE(n)$-spaces of weight $< \tau$ and $n$-soft projections.

By induction, let us now construct a continuous well ordered spectrum

$$\mathcal{S}_Y = \{Y_\alpha, q_\alpha^{\alpha+1}, \lambda\}$$

"parallel" to the given one and consisting of $n$-dimensional $ANE(n)$-spaces. We also construct a morphism

$$\{f_\alpha\} : \mathcal{S}_Y \to \mathcal{S}_X,$$

consisting of $n$-soft maps.

By the inductive hypothesis, there exist an $n$-dimensional $ANE(n)$-space $Y_0$ of weight $w(Y_0) = w(X_0)$ and an $n$-soft map $f_0 : Y_0 \to X_0$. Suppose that for each $\alpha < \gamma$, where $\gamma < \lambda$, we have already constructed:

(a) an $n$-dimensional $ANE(n)$-space $Y_\alpha$ of weight $w(Y_\alpha) = w(X_\alpha)$.
(b) an $n$-soft map $f_\alpha : Y_\alpha \to X_\alpha$.
(c) $n$-soft maps $q_\alpha^\beta : Y_\beta \to Y_\alpha$ ($\alpha \leq \beta < \gamma$, satisfying the following two conditions:
   - $q_\alpha^\beta = q_\alpha^\delta q_\delta^\beta$ for each $\alpha, \delta, \beta$ with $\alpha \leq \delta \leq \beta < \gamma$.

– All square diagrams of the type

$$\begin{array}{ccc} Y_{\alpha+1} & \xrightarrow{f_{\alpha+1}} & X_{\alpha+1} \\ \downarrow {\scriptstyle q_\alpha^{\alpha+1}} & & \downarrow {\scriptstyle p_\alpha^{\alpha+1}} \\ Y_\alpha & \xrightarrow{f_\alpha} & X_\alpha \end{array}$$

are $n$-soft ($\alpha + 1 < \gamma$).

We now construct the space $X_\gamma$ and maps $f_\gamma$ and $q_\alpha^\gamma$ with $\alpha < \gamma$.

If $\gamma$ is a limit ordinal, then

$$X_\gamma = \lim\{Y_\alpha, q_\alpha^\beta, \gamma\} \quad \text{and} \quad f_\gamma = \lim\{f_\alpha \colon \alpha < \gamma\}.$$

By Lemmas 6.2.6 and 6.4.14, $X_\gamma$ is an $n$-dimensional $ANE(n)$-space of weight $w(Y_\gamma) = w(X_\gamma)$. Lemma 6.2.7 guarantees the $n$-softness of the map $f_\gamma \colon Y_\gamma \to X_\gamma$. We define the maps $q_\alpha^\gamma$ as the limit projections of the spectrum the limit space of which coincides with $Y_\gamma$. These maps are $n$-soft by Lemma 6.2.6. It is easy to see that all naturally occuring square diagrams are $n$-soft.

Next we consider the case $\gamma = \alpha + 1$.

Consider the fibered product $X$ of spaces $Y_\alpha$ and $X_\gamma$ with respect to the maps $f_\alpha$ and $p_\alpha^\gamma$. Denote by $\varphi \colon Z \to Y_\alpha$ and $\psi \colon Z \to X_\gamma$ the canonical projections of this fibered product. By Lemma 6.2.4, $\varphi$ and $\psi$ are $n$-soft maps. Consequently, the space $Z$ is an $ANE(n)$-space (Proposition 6.1.21). Since the space $X$ is contained in the product $Y_\alpha \times X_\gamma$, its weight is less than $\tau$. Consequently, by inductive hypothesis, there exist an $n$-dimensional $ANE(n)$-space $Y_\gamma$ of weight $w(Z)$ and an $n$-soft map $h \colon Y_\gamma \to Z$. Let $f_\gamma = \psi h$ and $q_\alpha^\gamma = \varphi h$. Obviously, these maps are $n$-soft (Lemma 6.1.14) and the newly arising square diagram is also $n$-soft (because its characteristic map coincides, by construction, with $n$-soft map $h$). Thus the spectrum $\mathcal{S}_Y$ and the morphism $\{f_\alpha\}$, satisfying all the required properties, is constructed.

Finally, let

$$Y = \lim \mathcal{S}_Y \quad \text{and} \quad f = \lim\{f_\alpha\}.$$

Applying Lemmas 6.2.6, 6.2.7 and 6.4.14, we conclude that $Y$ is an $n$-dimensional $ANE(n)$-space of weight $w(X)$ and $f \colon Y \to X$ is an $n$-soft map. □

Consider now an $n$-dimensional space $R_n^\tau$ of weight $\tau$ which admits an $n$-soft map onto $R^\tau$ (we have just proved the existence of such spaces in Theorem 6.4.15). If $n = 0$ we can, by Theorem 6.4.2, assume that $R_n^\tau = N^\tau$. These spaces have specific universality properties. We will investigate these in the next Section.

Thus, in contrast with the compact case, for each $n \geq 0$ there are $n$-dimensional $AE(n)$-spaces of any weight. Is it possible to lower the dimension of such spaces? In order to answer this question we need the following lemma.

LEMMA 6.4.16. *Let $F: X \to Y$ be an $n$-soft map between Polish spaces and* $\dim Y = k < n$. *If $|f^{-1}(y)| \geq 2$ for each point $y \in Y$, then* $\dim X > k$.

PROOF. Since $\dim Y = k$, there is a point $y_0 \in Y$ such that every neighborhood of $y_0$ in $Y$ is exactly $k$-dimensional. By our assumption, the fiber $f^{-1}(y_0)$ contains at least two points, say $x_0$ and $x_1$. By the $n$-softness of the map $f$, we can find two maps $i_0, i_1: Y \to X$ satisfying the following conditions:

(a) $fi_m = id_Y$ for each $m = 0, 1$.

(b) $i_m(y_0) = x_m$ for each $m = 0, 1$.

Let $U = \{y \in Y: i_0(y) \neq i_1(y)\}$. Obviously, $U$ is an open neighborhood of the point $y_0$ in $Y$. By the choice of $y_0$, we have $\dim U = k$. Let $V = f^{-1}(U)$ and $g = f/V$. Then $g: V \to U$ is an $n$-soft map with two sections (namely: $j_m = i_m/U$, $m = 0, 1$) with disjoint images. Since $\dim U = k$, a theorem on partitions (see, for example, [**146**, page 67]) shows that there exists a collection $T = \{(A_p, B_p): 1 \leq p \leq k\}$ of closed subsets of $U$ such that $A_p \cap B_p = \emptyset$ for each $p = 1, 2, \ldots, k$ and for any choice of partitions $C_p$ (in $U$) between $A_p$ and $B_p$ their intersection $\cap\{C_p: 1 \leq p \leq k\}$ is non-empty. It is not hard to verify (we leave the corresponding details to the reader) that the collection

$$T' = \{(A_p \times [0,1], B_p \times [0,1]): 1 \leq p \leq k\} \cup \{U \times \{0\}, U \times \{1\})$$

of closed subsets of $U \times [0,1]$ also has the mentioned property. Namely, for each choice of partitions (in $U \times [0,1]$) between elements of pairs of the collection $T'$, their intersection is non-empty.

Next consider in the space $V$ the following collection of pairs of disjoint closed subsets:

$$\{g^{-1}(A_p), g^{-1}(B_p)): 1 \leq p \leq k\} \cup (j_0(U), j_1(U)).$$

Suppose that $C_p$ is a partition in $V$ between the sets $g^{-1}(A_p)$ and $g^{-1}(B_p)$, $p = 1, 2, \ldots, k$, and $C$ is a partition in $V$ between the sets $j_0(U)$ and $j_1(U)$. Consider the map $\varphi: U \times \{0,1\} \to V$, defined by letting

$$\varphi(y,t) = j_t(y) \text{ for each } (y,t) \in U \times \{0,1\}.$$

Since $k < n$, we have $\dim(U \times [0,1]) \leq n$ and, consequently, by the $n$-softness of the map $g$, there is a map $h: U \times [0,1] \to V$ such that

$$h/(U \times \{0,1\}) = \varphi \text{ and } gh = \pi_U,$$

where $\pi_U: U \times [0,1] \to U$ is the natural projection. The sets $h^{-1}(C_p)$, $p = 1, 2, \ldots, k$ and $h^{-1}(C)$ are partitions in $U \times [0,1]$ between the elements of the corresponding pair of the collection $T'$. Consequently, as noted above,

$$h^{-1}(C) \cap \bigcap \{h^{-1}(C_p) \colon 1 \leq p \leq k\} \neq \emptyset.$$

Then the intersection $C \cap \bigcap \{C_p \colon 1 \leq p \leq k\}$ is also non-empty. By the above mentioned theorem on partitions, we have $\dim X \geq \dim V > k$. □

PROPOSITION 6.4.17. *If $X$ is an non-metrizable $ANE(n)$-space, then* $\dim X \geq n$.

PROOF. Obviously, our statement is true for $n = 0$. It is also clear that it suffices to consider only the case of spaces of weight $\omega_1$. Thus, let $X$ be an $ANE(n)$-space of weight $\omega_1$, $n \geq 1$ and suppose the contrary, i.e. assume that $\dim X = k < n$. Without loss of generality, according to Corollaries 6.4.4 and 6.4.5, we may assume that there is no point of countable pseudo--character in $X$. Represent $X$ as the limit space of a factorizing Polish spectrum $\mathcal{S}_X = \{X_\alpha, p_\alpha^\beta, A\}$ , consisting of Polish $ANE(n)$-spaces and $n$-soft limit projections (Proposition 6.3.5). By Theorems 1.3.6 and T:3.2.25, we may additionally assume that $\dim X_\alpha = k$ for each $\alpha \in A$. Take an arbitrary index $\alpha \in A$. If $x$ is a point in $X_\alpha$, then the fiber $p_\alpha^{-1}(x)$ is functionally closed in $X$. Consequently, the absence of points of countable pseudo-character in $X$ guarantees that this fiber contains at least two points. But then there is an index $\beta_x \in A$ with $\beta_x > \alpha$ such that the fiber $(p_\alpha^{\beta_x})^{-1}(x)$ is also non-trivial. The openness of the projection $p_\alpha^{\beta_x}$ guarantees the existence of an open neighborhood $V_x$ of the point $x$ in $X_\alpha$ such that $|(p^{\beta_x}{}_\alpha)^{-1}(y)| \geq 2$ for each point $y \in V_x$. Take a countable subcover $\{V_{x_m} \colon m \in \omega\}$ of the open cover $\{V_x \colon x \in X_\alpha\}$ of the Polish space $X_\alpha$ and let $\beta = \sup \beta_{x_m}$. Next observe that for each point $x \in X_\alpha$ we have $|(p_\alpha^\beta)^{-1}(x)| \geq 2$. By Lemma 6.4.16, $\dim X_\beta > k$. This contradiction completes the proof. □

COROLLARY 6.4.18. *Every finite-dimensional $ANR$-space is metrizable.*

*Historical and bibliographical notes* 6.4. The results of this Section are due to the author [**80**], [**84**], [**85**]. A particular case of Corollary 6.4.3 can be found in [**251**]. In the compact case Corollaries 6.4.3, 6.4.5, and 6.4.18, as well as Lemma 6.4.16 were obtained in [**275**], [**276**]. For the same class of spaces, Proposition 6.4.9 appears in [**171**]. The metrizability of 1-dimensional $ANR$-compacta was established earlier by Isbell [**179**].Characterizations of $AE(0)$-spaces in terms of many-valued maps are given in [**109**].

## 6.5. Strongly universal spaces

**6.5.1. Spaces of maps (continuation).** Let $X$ and $Y$ be arbitrary Tychonov spaces and $\tau$ be an arbitrary infinite cardinal number. We are going to introduce a topology, depending on $\tau$, on the set $C(Y, X)$ of all continuous maps from $Y$ into $X$. The space thus obtained will be denoted by $C_\tau(Y, X)$. We recall that $cov(X)$ denotes the collection of all countable functionally open covers of the space $X$. For each map $f: Y \to X$ the sets of the form

$$B(f, \{\mathcal{U}_t : t \in T\}) = \{g \in C_\tau(Y, X) : g \text{ is } \mathcal{U}_t - \text{close to } f \text{ for each } t \in T\},$$

where $|T| < \tau$ and $\mathcal{U}_t \in cov(X)$ for each $t \in T$, are declared to be open basic neighborhoods of the point $f$ in $C_\tau(Y, X)$.

The maps, contained in the neighborhood $B(f, \{\mathcal{U}_t : t \in T\})$ are called $\{\mathcal{U}_t : t \in T\}$-close to $f$.

If a space $X$ has a countable basis, then, obviously, the space $C_\omega(Y, X)$ (for any space $Y$) coincides with the space $C(Y, X)$, endowed with the topology introduced in Subsection 2.1.1.

We will make use of the following simple lemma.

LEMMA 6.5.1. *Let $X$ be an $AE(o)$-space of infinite weight $\tau$. Then there exists a subcollection $\{\mathcal{U}_t : T\}$ of the collection $cov(X)$ such that for any space $Y$ and for any two maps $f, g: Y \to X$ the following conditions are equivalent:*

(i) $f = g$.

(ii) *The maps $f$ and $g$ are $\{\mathcal{U}_t : t \in T\}$-close.*

PROOF. If $X$ is a Polish space, then take a metric $d$ generating the topology of $X$, and denote by $\mathcal{U}_n$ an arbitrary countable open cover of $X$, the diameters of whose elements are less than $\frac{1}{n}$. Obviously, $\{\mathcal{U}_n\}$ is the desired collection.

Next consider the case when $X$ is a non-metrizable $AE(0)$-space. Represent $X$ as the limit space of a factorizing Polish spectrum $\mathcal{S}_X = \{X_t, p_t^{t'}, T\}$. Here the indexing set $T$ of the spectrum $\mathcal{S}_X$ satisfies the equality $|T| = w(X)$. In each of the Polish spaces $X_t$, take a countable collection $\{\mathcal{U}_t^n : n \in N\}$ as considered above. Then the collection

$$\{p_t^{-1}(\mathcal{U}_n^t) : n \in N, t \in T\}$$

has all the required properties. □

The above topology on the set of continuous maps allows us to give a precise meaning to the notion of strong universality in the general case (compare with Definition 2.3.12). Consider an arbitrary subclass $\mathcal{P}$ of the class of all Tychonov spaces. If $\tau \geq \omega$ and $n = 0, 1, \ldots, \infty$, we let

$$\mathcal{P}_{\tau,n} = \{X \in \mathcal{P} : R - w(X) \leq \tau \text{ and } \dim X \leq n\}.$$

A space $X \in \mathcal{P}_{\tau,n}$ is said to be *strongly $\mathcal{P}_{\tau,n}$-universal* if for each space $Y \in \mathcal{P}_{\tau,n}$ the set of all $C$-embeddings is dense in the space $C_\tau(Y,X)$. We consider only two classes: the class $\mathcal{A}$ of all Tychonov spaces and the class $\mathcal{B}$ of all compact spaces.

It should be observed that if $\omega \leq \tau < \tau'$, then the $\tau'$-topology on the set $C(Y,X)$ is stronger than the $\tau$-topology for any spaces $X$ and $Y$. In other words, the identity map

$$id\colon C_{\tau'}(Y,X) \to C_\tau(Y,X)$$

is continuous. Moreover, the fundamental meaning of the hierarchy of the concept of strong universality generated above is, in fact, an increased refinement of the concept of "closeness". For instance, as shown in Subsection 2.3.2 (Proposition 2.3.13), the Hilbert cube $I^\omega$ is strongly $\mathcal{B}_{\omega,\infty}$-universal, i.e. every map of any metrizable compactum into $I^\omega$ can be arbitrarily closely (in the space $C_\omega(\cdot, I^\omega))$ approximated by embeddings (recall that every compactum is $C$ embedded in any ambient space). One can easily verify that this fact remains valid for the Tychonov cube $I^\tau$ as well. This means that the set of all embedding of any compactum $Y$ of weight $\leq \tau$ into $I^\tau$ is dense in the space $C_\omega(Y, I^\tau)$. It turns out that if $\tau > \omega$, the Tychonov cube $I^\tau$ satisfies a much stronger version of the strong universality property. Namely: the cube $I^\tau$ is strongly $\mathcal{B}_{\tau,\infty}$-universal. Let us see why. Take a map $f\colon Y \to I^\tau$, where $Y$ is a compactum of weight $\leq \tau$. Let $Of$ be an arbitrary neighborhood of the point $f$ in the space $C_\tau(Y, I^\tau)$. Our goal is to show that the neighborhood $Of$ contains at least one embedding. Without loss of generality we may assume that $Of = B(f, \{\mathcal{U}_t \colon t \in T\})$, where $|T| < \tau$ and $\mathcal{U}_t \in cov(I^\tau)$ for each $t \in T$. Let $A$ be an arbitrary set of cardinality $\tau$. We identify the cube $I^\tau$ with $I^A$. Since $|T| < \tau$, there exist a subset $B \subseteq A$ of cardinality $\lambda = \max\{\omega, |T|\} < \tau$ and functionally open covers $\mathcal{U}_t^B \in cov(I^B)$ such that $\mathcal{U}_t = \pi_B^{-1}(\mathcal{U}_t^B)$ for each $t \in T$ (here $\pi_B \colon I^A \to I^B$ denotes the natural projection onto the corresponding subproduct). Since $\lambda < \tau$, obviously, $I^{A-B} \approx I^\tau$ and, consequently, there is an embedding $j\colon Y \to I^{A-B}$ (recall that $w(Y) \leq \tau$). Next, consider the embedding

$$i = \pi_B f \triangle j \colon Y \to I^B \times I^{A-B} = I^A.$$

Since $\pi_B i = \pi_B f$ we see, by the choice of $B$, that $i \in Of$. Therefore we have proved the following statement.

LEMMA 6.5.2. *Let $\tau \geq \omega$. Then the Tychonov cube $I^\tau$ is strongly $\mathcal{B}_{\tau,\infty}$-universal.*

In a similar way, one can prove the 0-dimensional counterpart of the above result.

LEMMA 6.5.3. *Let $\tau \geq \omega$. Then the Cantor cube $D^\tau$ is strongly $\mathcal{B}_{\tau,0}$-universal.*

Next we consider the non-compact case.

LEMMA 6.5.4. *Let $\tau \geq \omega$. Then the space $N^\tau$ is strongly $\mathcal{A}_{\tau,0}$-universal.*

PROOF. First consider the case $\tau = \omega$. Let $f: Y \to N^\omega$ be a map of an arbitrary zero-dimensional Polish space $Y$. Let $\mathcal{U} = \{U_n : n \in \omega\}$ be a countable open cover of $N^\omega$. Our goal is to construct a closed embedding $g: Y \to N^\omega$, $\mathcal{U}$-close to $f$. Without loss of generality, we may assume that $\mathcal{U}$ consists of pairwise disjoint sets. Let $Y_n = f^{-1}(U_n)$, $n \in \omega$. The space $Y_n$ is a zero-dimensional Polish space, $n \in \omega$. The space $U_n$ (as an open subspace of $N^\omega$) does not contain an open compact subspace and, consequently, by Theorem 1.1.5, is homeomorphic to $N^\omega$, $n \in \omega$. By Proposition 1.1.6, there exists a closed embedding $g_n : Y_n \to U_n$ for each $n \in \omega$. It only remains to note that the map $g: Y \to N^\omega$, coinciding with $g_n$ on $Y_n$, $n \in \omega$, is a closed embedding $\mathcal{U}$-close to $f$.

Let us now consider the case $\tau > \omega$. Take an arbitrary set $A$ of cardinality $\tau$ and let $Y$ be a zero-dimensional space of $R$-weight $< \tau$. Consider also a map $f: Y \to N^A$ and its neighborhood $Of$ in the space $C_\tau(Y, N^A)$. Without loss of generality we may assume that

$$Of = \{g \in C_\tau(Y, N^A) : \pi_B g = \pi_B f\}$$

where $B$ is a subset of $A$ of cardinality $< \tau$ and $\pi_B : N^A \to N^B$ stands for the natural projection onto the corresponding subproduct. By Theorem 6.4.2, the space $Y$ admits a $C$-embedding into $N^\tau$[4]. Since $|A - B| = \tau$, it follows that there is a $C$-embedding $j: Y \to N^{A-B}$. Then the desired $C$-embedding of $Y$ into $N^A$ can be defined by letting

$$i = \pi_B f \triangle j.$$

This completes the proof. □

LEMMA 6.5.5. *Let $\tau \geq \omega$. Then the space $R^\tau$ is strongly $\mathcal{A}_{\tau,\infty}$-universal.*

PROOF. If $\tau = \omega$, then the conclusion follows from Proposition 2.3.14. An argument similar to that used in Lemma 6.5.4 verifies the remaining cases (i.e. $\tau > \omega$). □

Thus, if $n = 0$ or $n = \infty$, strongly $\mathcal{A}_{\tau,n}$-universal spaces exist for each $\tau \geq \omega$. In the meantime, as shown in Theorem 5.1.12, strongly $\mathcal{A}_{\omega,n}$-universal spaces exist for each $n \in \omega$. Do we have a similar situation in the non-metrizable case? The following statement answers this question.

PROPOSITION 6.5.6. *Let $\tau > \omega$ and $n = 0, 1, \ldots, \infty$. Each $n$-dimensional space of weight $\tau$, admitting an $n$-soft map onto $R^\tau$, is strongly $\mathcal{A}_{\tau,n}$-universal.*

[4] Indeed, consider a 0-soft map $\varphi: N^\tau \to R^\tau$. Since $R-w(Y) \leq \tau$, we may assume that $Y$ is $C$-embedded in $R^\tau$. Since $\dim Y = 0$, we conclude, by the 0-softness of $\varphi$, that there is a map $\psi: Y \to N^\tau$ such that $\varphi\psi = id_Y$. It only remains to note that $\psi$ is a $C$-embedding.

PROOF. Let $f: X \to R^\tau$ be an $n$-soft map, where $X$ is an $n$-dimensional space of weight $\tau$. Consider an arbitrary at most $n$-dimensional space $Y$ of $R$-weight $\leq \tau$. Consider also a map $g: Y \to X$ and its neighborhood

$$B(g, \{\mathcal{U}_t : t \in T\}),\ |T| < \tau,\ \mathcal{U}_t \in cov(X),\ t \in T$$

in the space $C_\tau(Y, X)$. Let $\lambda = \max\{\omega, |T|\}$. Since $\tau$ is uncountable, we have $\lambda < \tau$. Represent the space $R^\tau$ as the limit space of the standard (factorizing) $\lambda$-spectrum

$$\mathcal{S} = \{Z_\alpha, \pi_\alpha^\beta, A\}$$

where each of the spaces $Z_\alpha$ is homeomorphic to $R^\lambda$ and each of the limit projections $\pi_\alpha: R^\tau \to Z_\alpha$ is the trivial bundle with fiber $R^\tau$. Since $X$, as an $n$-soft preimage of $R^\tau$, is an $AE(n)$-space, we conclude, applying Proposition 6.3.3, that there exist a factorizing $\lambda$-spectrum $\mathcal{S}_X = \{X_\alpha, p_\alpha^\beta, A\}$, consisting of $AE(n)$-spaces of weight $\lambda$, and a morphism $\{f_\alpha\}: \mathcal{S}_X \to \mathcal{S}$, consisting of $n$-soft maps, such that

(i) $X = \lim \mathcal{S}_X$ and $f = \lim\{f_\alpha\}$.

(ii) All the naturally occuring[5] limit square diagrams are $n$-soft.

Since $\mathcal{S}_X$ is a factorizing $\lambda$-spectrum and $|T| \leq \lambda$, there exist an index $\alpha \in A$ and functionally open countable covers $\mathcal{U}_t^\alpha \in cov(X_\alpha)$, $t \in T$, such that $\mathcal{U}_t = p_\alpha^{-1}(\mathcal{U}_t^\alpha)$ for each $t \in T$. Consider the corresponding limit square diagram:

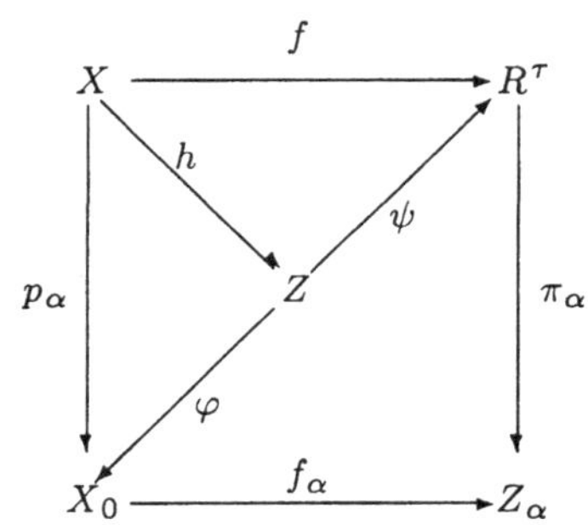


Here $Z$ denotes the fibered product of the spaces $X_\alpha$ and $R^\tau$ with respect to the maps $f_\alpha$ and $\pi_\alpha$. As usual $\psi: Z \to R^\tau$ and $\varphi: Z \to X_\alpha$ denote the canonical projections of this product. The map $h: X \to Z$ is the characteristic map of the indicated diagram, which is $n$-soft by condition (ii). The map $\pi_\alpha$ is the trivial bundle with fiber $R^\tau$. Consequently, by the inequality $R - w(Y) \leq \tau$, there is a $C$-embedding $i: Y \to R^\tau$ such that $\pi_\alpha i = \pi_\alpha f g$. Then

$$f_\alpha p_\alpha g = \pi_\alpha f g = \pi_\alpha i.$$

[5]Generated by the limit projections of the spectra $\mathcal{S}_X$, $\mathcal{S}$, elements of the morphism $\{f_\alpha\}$ and by the map $f$.

But in this situation the diagonal product $j = p_\alpha g \triangle i$ maps the space $Y$ into $Z$. Since $i$ is a $C$-embedding, we see that $j$ is also a $C$-embedding. Observe also that

$$\psi j = i \text{ and } \varphi j = p_\alpha g.$$

Since $\dim Y \leq n$, the $n$-softness of the characteristic map $h$ guarantees the existence of a map $g'\colon Y \to X$ such that $hg' = j$. Clearly, $g'$ is a $C$-embedding. It only remains to show that

$$g' \in B(g, \{\mathcal{U}_t \colon t \in T\}).$$

This fact follows immediately from the following equalities

$$p_\alpha g' = \varphi h g' = \varphi j = p_\alpha g$$

and from the choice of the index $\alpha \in A$. The proof is completed. □

A similar argument proves the following statement.

PROPOSITION 6.5.7. *Let $\tau > \omega$. Each compactum of weight $\tau$, admitting a soft map onto $I^\tau$, is strongly $\mathcal{B}_{\tau,\infty}$-universal.*

THEOREM 6.5.8. *Let $\tau \geq \omega$ and $n = 0, 1, \dots, \infty$. Then there exists a strongly $\mathcal{A}_{\tau,n}$-universal $AE(n)$-space $R_n^\tau$.*

PROOF. If $n = 0$ or $n = \infty$, then the result follows from Lemmas 6.5.4 and 6.5.5. If $\tau = \omega$ and $n$ is a positive integer, it suffices to apply Theorem 5.1.12. Finally, if $\tau > \omega$, apply Theorem 6.4.15 and Proposition 6.5.6. □

*Historical and bibliographical notes* 6.5. All the results of this Section are due to the author [**84**], [**83**].

CHAPTER 7

# Topology of Non-Metrizable Manifolds

## 7.1. Non-metrizable manifolds

**7.1.1. Definition of manifolds.** We are now ready to define the fundamental concepts of $R^\tau$-manifolds and $I^\tau$-manifolds for $\tau > \omega$. First let us ask: is it possible to extend the usual definition of $R^\omega$- (or $I^\omega$-) manifold to the non-metrizable case? In other words, is the following definition, stating that $R^\tau$- ($I^\tau$-) manifolds are precisely those spaces locally homeomorphic to $R^\tau$ (respectively, $I^\tau$), satisfactory? Let us discuss this question more formally. Consider the simplest case. Let $\tau > \omega$ and $x$ be an arbitrary point in $R^\tau$. Let $U = R^\tau - x$. Obviously, $U$ is locally homeomorphic to $R^\tau$. In the meantime, $U$ is not an $ANR$-space and, moreover, is not realcompact[1]. But under a "good" definition, each $R^\tau$-manifold must be realcompact (since the model space $R^\tau$ is an absolute retract; compare with Corollary 6.1.6 and Proposition 6.1.7). Therefore the standard definition has to be adapted to the non-metrizable case. In order to understand exactly what modifications have to be performed, we continue our analysis and ask the next question: which open subspaces of $R^\tau$ are realcompact? The complete answer to this question is provided by Corollary 6.4.11 — these are precisely the functionally open subspaces of $R^\tau$. In turn, each functionally open subspace $U$ of $R^\tau$ can be represented as the product $V \times R^\tau$, where $V$ is an open subspace of $R^\omega$, i.e. an $R^\omega$-manifold. Obviously, $V$ has a countable functionally open cover each element of which is homeomorphic to $R^\omega$. Therefore, the space $U$ also admits a countable functionally open cover each element of which is homeomorphic to $R^\omega \times R^\tau \approx R^\tau$. With these observations in mind, we arrive at the following definition.

DEFINITION 7.1.1. *A space is called an $R^\tau$-manifold ($\tau \geq \omega$), if it has a countable functionally open cover each element of which is homeomorphic to $R^\tau$.*

[1] The last fact immediately follows from Proposition 1.1.24.

Of course, in the case $\tau = \omega$ the above definition coincides with the usual definition of $R^\omega$-manifolds. In addition, it is obvious that each functionally open subspace of $R^\tau$ is an $R^\tau$-manifold.

A similar discussion leads to the following definition.

DEFINITION 7.1.2. *A space is called an $I^\tau$-manifold ($\tau \geq \omega$), if it has a countable functionally open cover each element of which is homeomorphic to a functionally open subspace of $I^\tau$.*

Obviously, each $I^\tau$-manifold is locally compact and Lindelöf.

Our next goal is to show that $R^\tau$- and $I^\tau$-manifolds are $ANR$-spaces. This allows us to apply the spectral technique developed in Chapter 6.

The proof of the following statement is trivial and is left to the reader.

LEMMA 7.1.3. *A functionally open subspace of an $ANR$-space is an $ANR$-space.*

PROPOSITION 7.1.4. *If a space $X$ is represented as a countable union of its functionally open subspaces, each of which is an $ANR$-space, then $X$ is also an $ANR$-space.*

PROOF. If $X$ is metrizable, then the statement follows from the well-known result of [**161**]. Consider the case of non-metrizable $X$. It follows from Lemma 1.1.18 and Proposition 1.1.26 that the space $X$ is realcompact. Therefore, we can assume that $X$ is a closed and $C$-embedded subspace of the space $R^\tau$ for some uncountable $\tau$. Let $X = \cup\{X_i\colon i \in N\}$, where each $X_i$ is a functionally open $ANR$-subspace of $X$. The proof of the statement is divided into three parts.

**Case 1.** First consider the case when there are only finitely many members in the above representation of $X$. In this case, without loss of generality, we can assume that $X = X_1 \cup X_2$. Let $\varphi\colon X \to [1,2]$ be a continuous function such that $\varphi^{-1}(i) = X - X_i$, $i = 1,2$. Since $X$ is $C$-embedded in $R^\tau$, there is a function $\bar{\varphi}\colon R^\tau \to [1,2]$ such that $\bar{\varphi}/X = \varphi$. The sets $\bar{\varphi}^{-1}(i)$, $i = 1,2$, are disjoint functionally closed subsets of $R^\tau$. Consequently, there exist functionally open subsets $O_1$ and $O_2$ in $R^\tau$ such that $\bar{\varphi}^{-1}(i) \subseteq O_i$ and $O_1 \cap O_2 = \emptyset$, $i = 1,2$. Let $Z_i = R^\tau - O_i$, $i = 1,2$. Evidently the sets $Z_1$ and $Z_2$ are functionally closed in $R^\tau$ and satisfy the following conditions: $Z_1 \cup Z_2 = R^\tau$ and $Z_0 \cap X \subseteq X_0$, where $Z_0 = Z_1 \cap Z_2$ and $X_0 = X_1 \cap X_2$. Let us show that $X_0$ is $C$-embedded in $X_0 \cup Z_0$. In order to do this we have to verify the following two properties:

(i) $X_0$ is $z$-embedded in $X_0 \cup Z_0$.

(ii) If $Z$ is functionally closed in $X_0 \cup Z_0$ and $Z \cap X_0 = \emptyset$, then $Z$ and $X_0$ can be functionally separated in $X_0 \cup Z_0$.

Let $Z$ be a functionally closed subset of $X_0$. Since $X_0$ is $z$-embedded in $X$, there is a functionally closed subset $Z'$ in $X$ such that $Z = Z' \cap X_0$. Since $X$ is even $C$-embedded in $R^\tau$, there is a functionally closed subset $Z''$ in $R^\tau$ such that $Z'' \cap X = Z'$. Let $F = Z'' \cap (X_0 \cup Z_0)$. It can easily be seen that $F$ is functionally closed in $X_0 \cup Z_0$ and $F \cap X_0 = Z$. This finishes the verification of property $(i)$.

Let us now verify property $(ii)$. If $Z$ is a functionally closed subset of $X_0 \cup Z_0$ and $Z \cap X_0 = \emptyset$, then $Z \subseteq Z_0 - X$. Since, by Corollary 6.4.8, every functionally closed subset of $R^\tau$ is $C$-embedded in $R^\tau$, we conclude that $Z$ is functionally closed in $R^\tau$. But $X$ is $C$-embedded in $R^\tau$. Consequently, $X$ and $Z$ are functionally separated in $R^\tau$. This shows that $X_0$ and $Z$ are functionally separated in $X_0 \cup Z_0$ and finishes the verification of property $(ii)$.

Thus, $X_0$ is $C$-embedded in $X_0 \cup Z_0$. By Lemma 7.1.3, $X_0$ is an $ANR$-space. Therefore there exists a retraction $r_0\colon V \to X_0$, where $V$ - is a functionally open neighborhood of $X_0$ in $X_0 \cup Z_0$. Obviously the set $(X_0 \cup Z_0) - V$ is functionally closed in $X_0 \cup Z_0$ and $(X_0 \cup Z_0) - V \subseteq Z_0 - X$. Therefore, $(X_0 \cup Z_0) - V$ is functionally closed in $R^\tau$. Once again using $C$-embeddedness of $X$ in $R^\tau$, we can conclude that there exists a functionally open subset $U$ of $R^\tau$ such that $X \subseteq U$ and $U \cap (X_0 \cup Z_0) \subseteq V$. It can now be easily seen that by setting

$$r_i(a) = \begin{cases} r_0(a), & \text{if } a \in (X_0 \cup Z_0) \cap Z_i \cap U \\ a, & \text{if } a \in X_i - X_0 \end{cases}$$

we obtain a map $r_i\colon (X \cap Z_0) \cup (Z_0 \cap U) \to X_i$ such that $C(r_i)(C(X_i)) \subseteq C(U \cap Z_i)/((X \cap Z_i) \cup (Z_0 \cap U))$, $i = 1, 2$. Since $X_i$ is an $ANR$-space, there exists an extension $\tilde{r}_i\colon G_i \to X_i$ of the map $r_i$, where $G_i$ is a functionally open neighborhood of the set $((X \cap Z_i) \cup (Z_0 \cap U))$ in $U \cap Z_i$, $i = 1, 2$. A straightforward argument shows that the set $G = G_1 \cup G_2$ is a functionally open neighborhood of $X$ in $R^\tau$. It only remains to observe that the map $r\colon G \to X$, defined by setting $r(a) = \tilde{r}_i(a)$ for $a \in G_i$ $(i = 1, 2)$, is a retraction.

**Case 2.** Now we consider the case when $X = \cup\{X_i\colon i \in N\}$ and $X_i \cap X_j = \emptyset$ for $i \neq j$, $i, j \in N$. Consider a function $\varphi\colon X \to R$ defined by setting $\varphi(X_i) = i$, $i \in N$. Since $X$ is $C$-embedded in $R^\tau$, there is a function $\psi\colon R^\tau \to R$ extending $\varphi$. The pairwise disjoint sets $U_i = \psi^{-1}((i - \frac{1}{3}, i + \frac{1}{3}))$ are functionally open in $R^\tau$, and $U_i \cap X = X_i$ for each $i \in N$. Obviously, $X_i$ is $C$-embedded in $U_i$, $i \in N$. Consequently, since each $X_i$ is an $ANR$-space, there is a retraction $r_i\colon V_i \to X_i$, where $V_i$ is a functionally open in $U_i$ (and, hence, in $R^\tau$) neighborhood of $X_i$. Let $V = \cup\{V_i\colon i \in N\}$ and $r(a) = r_i(a)$ for each $a \in V_i$ and $i \in N$. Clearly, $r\colon V \to X$ is a retraction and $V$ is a functionally open neighborhood of $X$ in $R^\tau$.

**Case 3.** We are now in position to consider the general case. Let $G_i = X_1 \cup \cdots \cup X_i$, $i \in N$. The sets $G_i$ are functionally open in $X$ and $\cup\{G_i\colon i \in N\} = X$. By $(a)$, $G_i$ is an $ANR$-space, $i \in N$. The $C$-embeddedness of $X$ in $R^\tau$ guarantees the existence of a countable subset $T \subseteq \tau$ and open subsets $O_i$ in $\pi_T(X)$ such that $\pi_T^{-1}(O_i) \cap X = G_i$ ($\pi_T\colon R^\tau \to R^T$ denotes the natural projection onto the corresponding subproduct), $i \in N$. Since $T$ is countable, $\pi_T(X)$ has a countable base and, consequently, we can fix a metric $d$ which induces the topology of

$\pi_T(X)$. Let

$$F_i = \{x \in \pi_T(X) : d(x, \pi_T(X) - O_i) > \frac{1}{i}\}.$$

It is easy to see that $cl(F_i) \subseteq F_{i+1}$ and $\pi_T(X) = \cup\{F_i : i \in N\}$. Let

$$K_i = \begin{cases} F_i, & \text{if } i = 1, 2 \\ F_i - cl(F_{i-2}), & \text{if } i = 3, 4, \cdots \end{cases}$$

Let $L_i = \pi_T^{-1}(K_i) \cap X_i$, $i \in N$. Then $X = \cup\{L_i : i \in N\}$, and $L_j \cap L_{i+2} = \emptyset$ for each $i \in N$ and $j \leq i$. Obviously, $L_i$ is functionally open in $G_i$ and consequently, by Lemma 7.1.3, is an $ANR$-space, $i \in N$. By Case 2, the spaces $A = \cup\{L_{2i-1} : i \in N\}$ and $B = \cup\{L_{2i} : i \in N\}$ are $ANR$-spaces. Observe now that $A$ and $B$ are functionally open subspaces of $X$ and $X = A \cup B$. Therefore, by Case 1, X is an $ANR$-space. □

COROLLARY 7.1.5. *Every $R^\tau$ -manifold is an $ANR$-space.*

COROLLARY 7.1.6. *Every $I^\tau$ -manifold is an $ANR$-space.*

LEMMA 7.1.7. *Let $Y \in \{R^\tau, [-1,1]^\tau\}$, $\tau \geq \omega$. Then there exists a homeomorphism $f: Y^N \times I \to Y^N \times I$ such that the following conditions are satisfied:*

(i) *$\pi_I f(y,t) = t$ for each $(y,t) \in Y^N \times I$, where $\pi_I : Y^N \times I \to I$ denotes the projection onto the second factor.*
(ii) *$f((0,\cdots,0,\cdots),t) = ((0,\cdots,0,\cdots),t)$ for each $t \in I$.*
(ii) *$f(y,0) = (y,0)$ for each $y \in Y^N$.*
(iv) *$f((y_1,y_2,y_3,\cdots,y_n,\cdots),1) = ((y_2,y_3,\cdots,y_n,\cdots),1)$ for each $\{y_n\} \in Y^N$.*

PROOF. Let $g: Y \times Y \times I \to Y \times Y$ be a reflection isotopy from Lemma 2.3.4. The desired homeomorphism $f: Y^N \times I \to Y^N \times I$ can be defined as follows:

$$f((y_1,y_2,y_3,\cdots,y_n,\cdots),t) = ((g(y_1,y_2,t),y_3,\cdots,y_n,\cdots),t).$$

□

The following Lemma is an immediate consequence of the above statement.

LEMMA 7.1.8. *Let $Y \in \{R^\tau, [-1,1]^\tau\}$, $\tau \geq \omega$. Then there exists a map $\varphi: Y^N \times Y^N \times I \to Y^N \times I$ such that the following conditions are satisfied:*

(i) *$\pi_I \varphi(x,y,t) = t$ for each $(y,x,t) \in Y^N \times Y^N \times I$.*
(ii) *The restriction $\varphi / Y^N \times Y^N \times (0,1) : Y^N \times Y^N \times (0,1) \to Y^N \times (0,1)$ is a homeomorphism.*
(ii) *$\varphi(y,x,0) = (y,0)$ for each $(y,x) \in Y^N \times Y^N$.*
(iv) *$\varphi(y,x,1) = (x,1)$ for each $(y,x) \in Y^N \times Y^N$.*
(v) *$\varphi((0,\cdots,0,\cdots),(0,\cdots,0,\cdots),t) = ((0,\cdots,0,\cdots),t)$ for each $t \in I$.*

(vi) *For each $n \geq 1$, there exists $\epsilon \in (0,1)$ such that if $t \in [0,\epsilon]$ and $1 \leq i \leq n$, then*

$$\pi_i\varphi((y_1,\cdots,y_i,\cdots,y_n,\cdots),x,t) = (y_1,\cdots,y_i)$$

*for each $(\{y_n\},x,t) \in Y^N \times Y^N \times I$ ($\pi_i\colon Y^N \times I \to Y^i$ denotes the projection onto the product of the first $i$ factors).*

(vii) *For each $n \geq 1$, there exists $\epsilon \in (0,1)$ such that if $t \in [1-\epsilon,1]$ and $1 \leq i \leq n$, then*

$$\pi_i\varphi(y,(x_1,\cdots,x_i,\cdots,x_n,\cdots),t) = (x_1,\cdots,x_i)$$

*for each $(y,\{x_n\},t) \in Y^N \times Y^N \times I$.*

COROLLARY 7.1.9. *Let $Y \in \{R^\tau,[-1,1]^\tau\}$, $\tau \geq \omega$. Suppose that for a Polish space $X$ and a map $\alpha\colon X \to I$, there is a homeomorphism $f\colon X \times Y^N \to X \times Y^N$ such that $\pi_X f(x,y) = x$ for each $(x,y) \in X \times Y^N$. Then there exists a homeomorphism $f_\alpha\colon X \times Y^N \to X \times Y^N$ satisfying the following conditions:*

(i) *$\pi_X f_\alpha(x,y) = x$ for each $(x,y) \in X \times Y^N$.*
(ii) *If $x \in \alpha^{-1}(0)$, then $f_\alpha(x,y) = f(x,y)$ for each $y \in Y^N$.*
(ii) *If $x \in \alpha^{-1}(1)$, then $f_\alpha(x,y) = (x,y)$ for each $y \in Y^N$.*

PROOF. The formula $i(x,y) = (x,y,\alpha(x))$, $(x,y) \in X \times Y^N$ defines the map $i\colon X \times Y^N \to X \times Y^N \times I$. The desired homeomorphism $f_\alpha$ can be now defined as the composition of the five maps indicated in the diagram

$$X \times Y^N \xrightarrow{i} X \times Y^N \times I \xrightarrow{id_X \times \varphi^{-1}} X \times Y^N \times Y^N \times I \xrightarrow{f \times id_{Y^N} \times id_I}$$

$$\to X \times Y^N \times Y^N \times I \xrightarrow{id_X \times \varphi} X \times Y^N \times I \xrightarrow{\pi_{12}} X \times Y^N,$$

where $\varphi$ denotes the homeomorphism from Lemma 7.1.8. □

THEOREM 7.1.10. *Let $p\colon E \to B$ be a locally trivial bundle with fiber $Y \in \{R^\tau, I^\tau\}$, $\tau \geq \omega$. If the base space $B$ is Polish, then $p$ is trivial.*

PROOF. Take a countable, locally finite open cover $\{U_n\colon n \in \omega\}$ of the base space $B$ such that for each $n \in \omega$ there is a homeomorphism $h_n\colon U_n \times Y^N \to p^{-1}(U_n)$ satisfying the equation $ph_n = \pi_{U_n}$. We wish to construct a homeomorphism $h\colon B \times Y^N \to E$ such that $ph = \pi_B$.

Choose a closed cover $\{C_n\colon n \in \omega\}$ of $B$ which refines $\{U_n\colon n \in \omega\}$ combinatorially. Consider the set $\mathcal{A}$ consisting of all possible pairs $(A, h(A))$, where $A \subseteq \omega$ and

$$h(A)\colon \cup\{C_n\colon n \in A\} \times Y^N \to p^{-1}(\cup\{C_n\colon n \in A\})$$

is a homeomorphism satisfying the equation $ph(A) = \pi_{\cup\{C_n:\, n\in A\}}$. Let us introduce a partial order on the set $\mathcal{A}$. We say that $(A_1, h(A_1)) \preceq (A_2, h(A_2))$ if and only if $A_1 \subseteq A_2$ and the homeomorphisms $h(A_1)$ and $h(A_2)$ coincide on the set

$$(\cup\{C_n : n \in A_1\} - \cup\{U_n : n \in A_2 - A_1\}) \times Y^N.$$

Let us now consider an arbitrary chain $\mathcal{B}$ in $\mathcal{A}$ and show that it has an upper bound in $\mathcal{A}$.

Denote by $T$ the collection $\{A : (A, h(A)) \in \mathcal{B}\}$, and let $A^* = \cup\{A : A \in T\}$ and $C(A) = \cup\{C_n : n \in A\}$, where $A \in T \cup \{A^*\}$. For each point $b \in C(A^*)$ choose an element $A_b \in T$ such that $b \in U_b \subseteq C(A_b)$, where $U_b$ is open in $C(A^*)$ and $U_b \cap U_n = \emptyset$ for each $n \in A^* - A_b$. Obviously the relations

$$h(A^*)/U_b \times Y^N = h(A_b)/U_b \times Y^N, b \in C(A^*)$$

define a homeomorphism $h(A^*) : C(A^*) \times Y^N \to p^{-1}(C(A^*))$ such that $ph(A^*) = \pi_{C(A^*)}$. It is not hard to see that $(A^*, h(A^*))$ is an upper bound of the chain $\mathcal{B}$ in $\mathcal{A}$.

Since the chain $\mathcal{B}$ was chosen arbitrarily, we conclude, by Zorn's Lemma, that the partially ordered set $\mathcal{A}$ has a maximal element. Denote it by $(A, h(A))$. Obviously it only remains to show that $A = \omega$. Assume, to the contrary, that this is not the case. Therefore there is an element $m \in \omega - A$. Let $C = \cup\{C_n : n \in A\}$ and $h = (h(A))^{-1} h_m/(C \cap U_m) \times Y^N$. Consider a function $\beta : C \to I$ such that $C \cap C_m \subseteq \beta^{-1}(0)$ and $K = cl(\beta^{-1}([0, 1]) \subseteq C \cap U_m$. Let $\alpha = \beta/C \cap U_m$. By Corollary 7.1.9 (replacing in it $X$ by $C \cap U_M$ and $f$ by $h$), we have a homeomorphism $h_\alpha : (C \cap U_m) \times Y^N \to (C \cap U_m) \times Y^N$ satisfying the following two conditions

$$h_\alpha/(C \cap C_m) \times Y^N = h/(C \cap C_m) \times Y^N$$

and

$$h_\alpha/((C \cap U_m) - K) \times Y^N = id.$$

Then $h(A)h_\alpha : (C \cap U_m) \times Y^N \to p^{-1}(C \cap U_m)$ is a homeomorphism such that

$$h(A)h_\alpha/(C \cap C_m) \times Y^N = h_m/(C \cap C_m) \times Y^N$$

and

$$h(A)h_\alpha/(C \cap U_m) - K) \times Y^N = h(A)/(C \cap U_m) - K) \times Y^N.$$

Consequently, the formula

$$g(b, y) = \begin{cases} h(A)(b, y), & \text{if } (b, y) \in (C - K) \times Y^N \\ h(A)h_\alpha(b, y), & \text{if } (b, y) \in (C \cap U_m) \times Y^N \\ h_m(b, y), & \text{if } (b, y) \in C_m \times Y^N \end{cases}$$

defines a homeomorphism $g\colon (C \cup C_m) \times Y^N \to p^{-1}(C \cup C_m)$. Straightforward verification shows that $(A \cup \{m\}, g) \in \mathcal{A}$ and $(A, h(A)) \prec (A \cup \{m\}, g)$. Since the last relation contradicts the maximality of the element $(A, h(A))$, the proof is complete. □

### 7.1.2. $R^\tau$-manifolds.

THEOREM 7.1.11. *The following conditions are equivalent for any space $X$:*

(i) *$X$ is an $R^\tau$-manifold.*

(ii) *There is an $R^\omega$-manifold (unique up to a homeomorphism) $P$ such that $X$ is homeomorphic to the product $P \times R^\tau$.*

PROOF. For $\tau = \omega$, the statement follows from the results of Chapter 1. Consequently we assume below that $\tau > \omega$. First note that it immediately follows from Definition 7.1.1 that the product $P \times R^\tau$ is an $R^\tau$-manifold for each $R^\omega$-manifold $P$.

Let us prove the implication $(i) \Rightarrow (ii)$. By Corollary 7.1.5 and Proposition 6.3.5, $X$ is homeomorphic to the limit space of some factorizing $\omega$-spectrum $\mathcal{S}_X = \{X_\alpha, p_\alpha^\beta, A\}$ consisting of Polish $ANR$-spaces and soft limit projections. Let $X = \cup\{X_i \colon i \in N\}$, where each $X_i$ is functionally open in $X$ and is homeomorphic to $R^\tau$. The factorizability and $\omega$-completeness of the spectrum $\mathcal{S}_X$ implies the existence of an index $\alpha_0 \in A$ such that $X_i = p_{\alpha_0}^{-1}(X_i^{\alpha_0})$, where $X_i^{\alpha_0} = p_{\alpha_0}(X_i)$, $i \in N$. For each $i \in N$ consider the spectrum

$$\mathcal{S}_i = \{p_\alpha(X_i), p_\alpha^\beta, \alpha \in A, \alpha \geq \alpha_0\}.$$

Obviously the spectra $\mathcal{S}_i$ are $\omega$-complete. All projections of these spectra are open, since the projections of the spectrum $\mathcal{S}_X$ are open (see Proposition 6.1.26). Proposition 1.3.3 allows us to conclude that the spectra $\mathcal{S}_i$, $i \in N$, are factorizing. Note also that $X_i = \lim \mathcal{S}_i$, $i \in N$. By our assumption, $X_i \approx R^\tau$ for each $i \in N$. Consequently, representing $R^\tau$ as the limit space of the standard factorizing $\omega$-spectrum, consisting of countable subproducts of $R^\tau$ and the corresponding subproducts, and applying Theorem 1.3.6 (countably many times), we conclude that for some index $\alpha \in A$, $\alpha \geq \alpha_0$, the spaces $p_\alpha(X_i)$, $i \in N$, are all homeomorphic to $R^\omega$, and the restrictions $p_\alpha/X_i \colon X_i \to p_\alpha(X_i)$, $i \in N$ are all trivial bundles with fiber $R^\tau$. Therefore, the space $X_\alpha$ is an $R^\omega$-manifold (because it is represented as the union of open subspaces $p_\alpha(X_i)$, $i \in N$, each of which is homeomorphic to $R^\omega$) and the limit projection $p_\alpha \colon X \to X_\alpha$ is a locally trivial bundle with fiber $R^\tau$ (because, for each $i \in N$, the restriction of the projection $p_\alpha$ onto the subspace $X_i$ is a trivial bundle with fiber $R^\tau$). By Theorem 7.1.10, the projection $p_\alpha \colon X \to X_\alpha$ is a trivial bundle with fiber $R^\tau$. Therefore, the space $X$ is homeomorphic to the product $X_\alpha \times R^\tau$.

Suppose now that $P \times R^\tau \approx L \times R^\tau$, where $P$ and $L$ are $R^\omega$-manifolds. Represent $R^\tau$ as the limit space of the standard factorizing $\omega$-spectrum $\mathcal{S} =$

$\{(R^\omega)^\alpha, \pi_\alpha^\beta, exp_\omega\tau\}$. Then the products $P \times R^\tau$ and $L \times R^\tau$ are the limit spaces of the factorizing $\omega$-spectra

$$\mathcal{S}_P = \{P \times (R^\omega)^\alpha, id_P \times \pi_\alpha^\beta, exp_\omega\tau\} \text{ and } \mathcal{S}_L = \{L \times (R^\omega)^\alpha, id_L \times \pi_\alpha^\beta, exp_\omega\tau\}$$

respectively. Applying Theorem 1.3.6 to these spectra, we conclude that the products $P \times R^\omega$ and $L \times R^\omega$ are also homeomorphic. The stability of $R^\omega$-manifolds guarantees that, in this case, $P \approx L$. This shows the uniqueness (up to a homeomorphism) of our representation. □

COROLLARY 7.1.12. *If $X$ is an $R^\tau$-manifold, then $X \approx X \times R^\tau$.*

PROOF. By Theorem 7.1.11, $X \approx P \times R^\tau$, where $P$ is an $R^\omega$-manifold. Consequently, $X \times R^\tau \approx P \times R^\tau \times R^\tau \approx P \times R^\tau \approx X$. □

COROLLARY 7.1.13. *Every $R^\tau$-manifold is triangulable, i.e. can be represented as the product $K \times R^\tau$, where $K$ is a locally compact polyhedron.*

PROOF. The case $\tau = \omega$ was considered in Chapter 1. If $\tau > \omega$, then Theorem 7.1.11 allows us to represent a given $R^\tau$-manifold $X$ as the product $P \times R^\tau$, where $P$ is an $R^\omega$-manifold. By the triangulation theorem for $R^\omega$-manifolds, there exists a locally compact polyhedron $K$ such that $P \approx K \times R^\omega$. Obviously, $X \approx P \times R^\tau \approx K \times R^\omega \times R^\tau \approx K \times R^\tau$. □

COROLLARY 7.1.14. *The following conditions are equivalent for any space $X$:*
(i) *$X$ is an $R^\tau$-manifold.*
(ii) *$X$ is homeomorphic to a functionally open subspace of $R^\tau$.*

PROOF. The case $\tau = \omega$ was considered in Chapter 1. Consider the case $\tau > \omega$. As indicated above, each functionally open subspace of $R^\tau$ is an $R^\tau$-manifold.

Conversely, suppose that $X$ is an $R^\tau$-manifold. By Theorem 7.1.11, $X \approx P \times R^\tau$, where $P$ is an $R^\omega$-manifold. Obviously, $P$ is homeomorphic to an open subspace of $R^\omega$. Consequently, $X$ is homeomorphic to a functionally open subspace of the product $R^\omega \times R^\tau \approx R^\tau$. □

COROLLARY 7.1.15. *Homotopy equivalent $R^\tau$-manifolds are homeomorphic.*

PROOF. The case $\tau = \omega$ was considered in Chapter 1. Consider the case $\tau > \omega$. Let $X$ and $Y$ be homotopy equivalent $R^\tau$-manifolds. By Theorem 7.1.11, $X \approx P \times R^\tau$ and $Y \approx L \times R^\tau$, where $P$ and $L$ are $R^\omega$-manifolds. Since homotopy equivalent $R^\omega$-manifolds are homeomorphic, it suffices to show that $P$ and $L$ are also homotopy equivalent. Represent $X$ as the limit space of factorizing $\omega$-spectrum $\mathcal{S}_X = \{X_\alpha, p_\alpha^\beta, A\}$, all spaces $X_\alpha$ of which are homeomorphic to the product $P \times R^\omega$ and all limit projections $p_\alpha$ of which are soft (even trivial bundles with fiber $R^\tau$; compare with the proof of Theorem 7.1.11). Similarly, let $Y = \mathcal{S}_Y$, where $\mathcal{S}_Y = \{Y_\alpha, q_\alpha^\beta, A\}$ is a factorizing $\omega$-spectrum consisting of spaces $Y_\alpha$, homeomorphic to the product $L \times R^\omega$, and soft limit projections. Consider two additional factorizing $\omega$-spectra: $\mathcal{S}_1 = \{X_\alpha \times I, p_\alpha^\beta, A\}$ and $\mathcal{S}_2 = \{Y_\alpha \times I, q_\alpha^\beta, A\}$. It can easily be seen that $\lim \mathcal{S}_1 = X \times I$ and $\lim \mathcal{S}_2 = Y \times I$. By our assumption, there are maps $f\colon X \to Y$, $g\colon Y \to X$, $H\colon X \times I \to X$ and $F\colon Y \times I \to Y$ satisfying the following conditions:

(i) $H(x,0) = x$ for each $x \in X$.
(ii) $H(x,1) = gf(x)$ for each $x \in X$.
(ii) $F(y,0) = y$ for each $y \in Y$.
(iv) $F(y,1) = fg(y)$ for each $y \in Y$.

Applying Theorem 1.3.6 to the maps $f, g, H$ and $F$, we can assume without loss of generality that for some index $\alpha \in A$ we have maps $f_\alpha\colon X_\alpha \to Y_\alpha$, $g_\alpha\colon Y_\alpha \to X_\alpha$, $H_\alpha\colon X_\alpha \times I \to X_\alpha$ and $F_\alpha\colon Y_\alpha \times I \to Y_\alpha$ satisfying the following conditions:

(v) $H_\alpha(x_\alpha, 0) = x_\alpha$ for each $x_\alpha \in X_\alpha$.
(vi) $H_\alpha(x_\alpha, 1) = g_\alpha f_\alpha(x_\alpha)$ for each $x_\alpha \in X_\alpha$.
(vii) $F_\alpha(y_\alpha, 0) = y_\alpha$ for each $y_\alpha \in Y_\alpha$.
(viii) $F_\alpha(y_\alpha, 1) = f_\alpha g_\alpha(y_\alpha)$ for each $y_\alpha \in Y_\alpha$.

Therefore $g_\alpha f_\alpha \simeq id_{X_\alpha}$ and $f_\alpha g_\alpha \simeq id_{Y_\alpha}$, i.e. the products $P \times R^\omega$ and $L \times R^\omega$ are homotopy equivalent. But $P \approx P \times R^\omega$ and $L \approx L \times R^\omega$. Consequently, $P$ and $L$ are homotopy equivalent as desired. □

COROLLARY 7.1.16. *A contractible $R^\tau$-manifold is homeomorphic to $R^\tau$.*

COROLLARY 7.1.17. *Let $\tau > \omega$ and $Y$ be a $G_\delta$-subset of an $R^\tau$-manifold. Then the following conditions are equivalent:*

(i) *$Y$ is an $R^\tau$-manifold.*
(ii) *$Y$ is an $ANR$-space.*

PROOF. The implication $(i) \Rightarrow (ii)$ follows from Corollary 7.1.5.

Let us show the validity of the implication $(ii) \Rightarrow (i)$. Suppose that $Y$ is a $G_\delta$-subset of an $R^\tau$-manifold $X$. By (ii) and Proposition 6.1.7, $Y$ is realcompact. Consequently, by Corollary 6.4.12, $Y$ can be represented as an intersection of countably many functionally open subsets of $X$. We can also assume that $X = P \times R^\tau$, where $P$ is an $R^\omega$-manifold (Theorem 7.1.11). In this situation, $Y = \pi^{-1}\pi(Y)$, where $\pi\colon (P \times R^\omega) \times R^\tau \to P \times R^\omega$ denotes the projection. The softness of the projection $\pi$, coupled with Proposition 6.1.21, guarantees that $\pi(Y)$ is a Polish $ANR$-space. Therefore, $Y \approx \pi(Y) \times R^\tau \approx \pi(Y) \times R^\omega \times R^\tau$. By Theorem 2.3.9, the product $\pi(Y) \times R^\omega$ is an $R^\omega$-manifold. Consequently, by Theorem 7.1.11, $Y$ is an $R^\tau$-manifold. □

COROLLARY 7.1.18. *Each $R^\tau$-manifold is strongly $\mathcal{A}_{\tau,\infty}$-universal.*

PROOF. Apply Theorem 7.1.11 and Lemma 6.5.5. □

**7.1.3. $I^\tau$ -manifolds.** The following statement also has several important corollaries.

THEOREM 7.1.19. *The following conditions are equivalent for any space $X$:*
(i) *$X$ is an $I^\tau$ -manifold.*
(ii) *There is an $I^\omega$ -manifold (unique up to a homeomorphism) $P$ such that $X$ is homeomorphic to the product $P \times I^\tau$.*

The proof of this theorem is similar to the Proof of Theorem 7.1.11, and is therefore omitted.

COROLLARY 7.1.20. *If $X$ is an $I^\tau$ -manifold, then $X \approx X \times I^\tau$.*

COROLLARY 7.1.21. *Every $I^\tau$-manifold is triangulable, i.e. can be represented as the product $K \times I^\tau$, where $K$ is a locally compact polyhedron.*

COROLLARY 7.1.22. *If $X$ is an $I^\tau$ -manifold, then the product $X \times [0,1)$ is homeomorphic to a functionally open subspace of $I^\tau$ .*

PROOF. Represent $X$ as the product $P \times T^\tau$, where $P$ is an $I^\omega$ -manifold (Theorem 7.1.19). By Theorem 2.3.25, the product $P \times [0,1)$ is homeomorphic to an open subspace of the Hilbert cube $I^\omega$ . It only remains to note that the product $X \times [0,1) \approx P \times [0,1) \times I^\tau$ is functionally open in $I^\omega \times I^\tau \approx I^\tau$. □

The converse to Corollary 7.1.22 will be proved in the next section.

COROLLARY 7.1.23. *The $I^\tau$ -manifolds $X$ and $Y$ are homotopy equivalent if and only if the products $X \times [0,1)$ and $Y \times [0,1)$ are homeomorphic.*

PROOF. Obviously the projections $\pi_X\colon X \times [0,1) \to X$ and $\pi_Y\colon Y \times [0,1) \to Y$ are homotopy equivalences, and consequently, if the products $X \times [0,1)$ and $Y \times [0,1)$ are homeomorphic we can conclude that the spaces $X$ and $Y$ are homotopy equivalent.

Suppose now that the $I^\tau$-manifolds $X$ and $Y$ are homotopy equivalent. Represent, by Theorem 7.1.19, the spaces $X$ and $Y$ as the products $P \times I^\tau$ and $L \times I^\tau$ respectively, where $P$ and $L$ are $I^\omega$ -manifolds. As in the proof of Corollary 7.1.15 it can be seen that the $I^\omega$ -manifolds $P$ and $L$ are homotopy equivalent. Consequently, by Theorem 2.3.26, the products $P \times [0,1)$ and $L \times [0,1)$ are homeomorphic. If so, it only remains to note that $X \times [0,1) \approx P \times [0,1) \times I^\tau$ and $Y \times [0,1) \approx L \times [0,1) \times I^\tau$. □

COROLLARY 7.1.24. *Every contractible compact $I^\tau$ -manifold is homeomorphic to $I^\tau$ .*

PROOF. Represent a given contractible compact $I^\tau$ -manifold $X$ as the product $P \times I^\tau$, where $P$ is a compact $I^\omega$ -manifold. Obviously, $P$ is contractible and consequently, $P \approx I^\omega$. Then $X \approx P \times I^\tau \approx I^\omega \times I^\tau \approx I^\tau$. □

We conclude this section with the following two $I^\tau$-counterparts of Corollaries 7.1.17 and 7.1.18.

COROLLARY 7.1.25. *Let $\tau > \omega$ and $Y$ be a $G_\delta$-subset of an $I^\tau$-manifold. Then the following conditions are equivalent:*

(i) *$Y$ is an $I^\tau$ -manifold.*

(ii) *$Y$ is an $ANR$-space.*

COROLLARY 7.1.26. *Each $I^\tau$ -manifold is strongly $\mathcal{B}_{\tau,\infty}$-universal.*

*Historical and bibliographical notes* 7.1. The definitions of $R^\tau$ - and $I^\tau$ -manifolds presented here were given in [**84**], although the initial definition of $I^\tau$ -manifolds was given by Ščepin in [**277**]. Theorem 7.1.10 is due to Chapman [**67**]. The results of subsection 7.1.2 were obtained in [**84**]. The results of subsection 7.1.3, except Corollaries 7.1.23 and 7.1.26, are due to Ščepin [**277**]. Corollaries 7.1.23 and 7.1.26 were proved in [**83**].

## 7.2. Topological characterization of $I^\tau$ -manifolds

We begin this Section with some auxiliary considerations.

DEFINITION 7.2.1. *A space $X$ is said to be* homogeneous with respect to pseudo-character *if any two points $x_1$ and $x_2$ in $X$ have the same pseudo character, i.e. $\psi(x_1, X) = \psi(x_2, X)$.*

Since the weight of any non-discrete $AE(0)$-space coincides with its pseudo-character (this easily follows from the results of Section 6.4), we get the following simple observation.

LEMMA 7.2.2. *If $X$ is a non-discrete, homogeneous with respect to pseudo-character, $AE(0)$-space of weight $\tau > \omega$, then $\psi(x, X) = \tau$ for each point $x \in X$.*

LEMMA 7.2.3. *Let $\tau \geq \omega$ and $\mathcal{S}_X = \{X_\alpha, p_\alpha^\beta, A\}$ be a factorizing $\tau$-spectrum, consisting of $ANR$-spaces and soft projections. Suppose that the limit space $X$ of the spectrum $\mathcal{S}_X$ is homogeneous with respect to pseudo-character and $w(X) > \tau$. Then for each index $\alpha \in A$ there exist an index $\alpha' \in A$, $\alpha' > \alpha$, and two sections $i_\alpha^0, i_\alpha^1 \colon X_\alpha \to X_{\alpha'}$ of the projection $p_\alpha^{\alpha'}$, such that $i_\alpha^0(X_\alpha) \cap i_\alpha^1(X_\alpha) = \emptyset$.*

PROOF. Let $\alpha_0 = \alpha$. Take an arbitrary point $z \in X_{\alpha_0}$ and consider its inverse image $p_{\alpha_0}^{-1}(z)$. Since the weight of the space $X_{\alpha_0}$ does not exceed $\tau$, the pseudo-character of any point in $X_{\alpha_0}$ is $\leq \tau$. Consequently, the pseudo-character of the set $p_{\alpha_0}^{-1}(z)$ in $X$ is also $\leq \tau$. By our assumption, $X$ is homogeneous with respect to pseudo-character. Therefore, by Lemma 7.2.2, we may conclude that the fiber $p_{\alpha_0}^{-1}(z)$ contains at least two points. Then, obviously, there exists an index $\alpha_1 \in A$ such that $\alpha_1 > \alpha_0$ and $|(p_{\alpha_0}^{\alpha_1})^{-1}(z)| > 1$. The softness of the projection $p^{\alpha_1}_{\alpha_0}$ guarantees that it has two sections $i^0_1, i^1_1\colon X_{\alpha_0} \to X_{\alpha_1}$ such that $i^0_1(z) \neq i^1_1(z)$. Let

$$V_1 = \{x \in X_{\alpha_0}\colon i^0_1(x) \neq i^1_1(x)\}.$$

Clearly, $V_1$ is a non-empty[2] open subset of $X_{\alpha_0}$.

Let $\gamma < \tau^+$ (by $\tau^+$ we denote the minimal cardinal number greater than $\tau$). Suppose that for each $\beta$, $1 \leq \beta < \gamma$, we have already constructed an index $\alpha_\beta \in A$, an open subset $V_\beta$ in $X_{\alpha_0}$ and two sections $i^0_\beta, i^1_\beta\colon X_{\alpha_0} \to X_\beta$ of the projection $p^\beta_\alpha\colon X_\beta \to X_{\alpha_0}$, satisfying the following conditions:

(i) $\alpha_\delta < \alpha_\beta$, whenever $1 \leq \delta < \beta < \gamma$.
(ii) $\alpha_\beta = \sup\{\alpha_\delta\colon \delta < \beta\}$, whenever $\beta$ is a limit ordinal.
(iii) $V_\delta \subseteq V_\beta$, whenever $1 \leq \delta < \beta < \gamma$.
(iv) $V_\beta = \cup\{V_\delta\colon \delta < \beta\}$, whenever $\beta$ is a limit ordinal.
(v) $i^k_\delta = p^{\alpha_\beta}_{\alpha_\delta} i^k_\beta$, whenever $1 \leq \delta < \beta < \gamma$ and $k = 0, 1$.
(vi) $i^k_\beta = \Delta\{i^k_\delta\colon \delta < \beta\}$, whenever $\beta$ is a limit ordinal and $k = 0, 1$.
(vii) $V_\beta = \{x \in X_{\alpha_0}\colon i^0_\beta(x) \neq i^1_\beta(x)\}$.

We shall now construct an index $\alpha_\gamma \in A$, an open subset $V_\gamma$ of $X_{\alpha_0}$ and two sections $i^k_\gamma\colon X_{\alpha_0} \to X_\gamma$, $k = 0, 1$, of the projection $p^{\alpha_\gamma}_{\alpha_0}$, satisfying all the required properties.

First assume that $\gamma$ is a limit ordinal. It follows from conditions (i) and (ii) that the collection $\{\alpha_\beta\colon \beta < \gamma\}$ forms a chain in the indexing set $A$. Since $\gamma < \tau^+$, the cardinality of this chain does not exceed $\tau$. By the $\tau$-completeness of $A$ there exists an index

$$\alpha_\gamma = \sup\{\alpha_\beta\colon \beta < \gamma\} \in A.$$

By the continuity of the spectrum $\mathcal{S}_X$, the space $X_\gamma$ is naturally homeomorphic to the limit space of the spectrum $\{X_{\alpha_\beta}, p^{\alpha_\beta}_{\alpha_\delta}, \beta, \delta < \gamma\}$. Therefore, by conditions (v) and (vi), the maps $i^k_\gamma\colon X_{\alpha_0} \to X_{\alpha_\gamma}$, $k = 0, 1$, satisfying the equalities $p^{\alpha_\gamma}_{\alpha_\beta} i^k_\gamma = i^k_\beta$ for each $k = 0, 1$ and each $\beta$, with $1 \leq \beta < \gamma$, are well defined. In particular, the maps $i^k_\gamma$, $k = 0, 1$ are sections of the projection $p^{\alpha_\gamma}_{\alpha_0}$.

Let $V_\gamma = \{x \in X_{\alpha_0}\colon i^0_\gamma(x) \neq i^1_\gamma(x)\}$. We shall verify conditions (iii) and (iv). Let $\beta < \gamma$. Consider an arbitrary point $x \in V_\beta$. By condition (vii), $i^0_\beta(x) \neq i^1_\beta(x)$. But, by construction, $p^{\alpha_\gamma}_{\alpha_\beta} i^k_\gamma = i^k_\beta$, $k = 0, 1$. Consequently,

$$p^{\alpha_\gamma}_{\alpha_\beta} i^0_\gamma(x) = i^0_\beta(x) \neq i^1_\beta(x) = p^{\alpha_\gamma}_{\alpha_\beta} i^1_\gamma(x)$$

and obviously $i^0_\gamma(x) \neq i^1_\gamma(x)$. This proves the inclusion $V_\beta \subseteq V_\gamma$ (condition (iii)). Consequently, $V_\gamma \supseteq \cup\{V_\beta\colon \beta < \gamma\}$. Let us prove the converse inclusion (and finish the verification of condition (iv)). Let $x \in V_\gamma$. By definition, $i^0_\gamma(x) \neq i^1_\gamma(x)$. It is not hard to see that there exists an index $\beta < \gamma$ such that $i^0_\beta(x) \neq i^1_\beta(x)$. Consequently, by condition (vii), $x \in V_\beta \subseteq \cup\{V_\beta\colon \beta < \gamma\}$. Thus the above constructed objects satisfy all the required properties.

[2]Since $z \in V_1$.

Consider now the case $\gamma = \beta + 1$. Suppose that $V_\beta \neq X_{\alpha_0}$. Take a point $z \in X_{\alpha_0} - V_\beta$, and let

$$x_\beta = i^0_\beta(z) \in i^0_\beta(X_{\alpha_0}) \subseteq X_{\alpha_\beta}.$$

Consider the fiber $p^{-1}_{\alpha_\beta}(x_\beta)$. Since the weight of the space $X_{\alpha_0}$ does not exceed $\tau$, and the space $X$ is homogeneous with respect to pseudo-character and has weight strictly more than $\tau$, we conclude, by Lemma 7.2.2, that the fiber $p^{-1}_{\alpha_\beta}(x_\beta)$ contains at least two different points. Then there is an index $\alpha_\gamma \in A$ such that $\alpha_\gamma > \alpha_\beta$ and

$$|(p^{\alpha_\gamma}_{\alpha_\beta})^{-1}(x_\beta)| > 1.$$

Softness of the projection $p^{\alpha_\gamma}_{\alpha_\beta}$ guarantees the existence of sections

$$j^0_\beta, j^1_\beta \colon X_{\alpha_\beta} \to X_{\alpha_\gamma}$$

such that $j^0_\beta(x_\beta) \neq j^1_\beta(x_\beta)$. The sections $i^0_\gamma, i^1_\gamma \colon X_{\alpha_0} \to X_{\alpha_\gamma}$ of the projection $p^{\alpha_\gamma}_{\alpha_0}$ can now be defined by letting

$$i^k_\gamma = j^k_\beta i^k_\beta, \quad k = 0, 1.$$

As above, if we let

$$V_\gamma = \{x \in X_{\alpha_0} \colon i^0_\gamma(x) \neq i^1_\gamma(x)\},$$

then $V_\beta \subseteq V_\gamma$. Observe also that, by construction, $z \in V_\gamma - V_\beta$.

Thus the construction of the above objects, satisfying conditions (i) — (vii), can be continued and carried out for each $\beta < \tau^+$. Then we obtain an increasing (by conditions (iii) and (iv)) collection $\{V_\beta \colon \beta < \tau^+\}$ of open subsets of $X_{\alpha_0}$. Since $w(X_{\alpha_0}) \leq \tau$, this collection must stabilize, i.e. there must exist an index $\beta_0 < \tau^+$ such that $V_{\beta_0} = V_\beta$ whenever $\beta_0 \leq \beta\ \tau^+$. It follows from the above construction that this is possible only if $V_{\beta_0} = X_{\alpha_0}$. Let $\alpha' = \beta_0$. It only remains to note that the sections $i^k_{\alpha'}$, $k = 0, 1$, of the projection $p^{\alpha'}_{\alpha_0}$ have disjoint images in $X_{\alpha'}$. □

LEMMA 7.2.4. *Let $f \colon X \to Y$ be a soft map between $ANR$-spaces. If the space $X$ is homogeneous with respect to pseudo-character and $w(X) > w(Y) \geq \omega$, then $f$ has two sections with disjoint images.*

PROOF. Let $\tau = w(Y)$ and represent $X$ as the limit space of a factorizing $\tau$-spectrum $\mathcal{S}_X = \{X_\alpha, p^\beta_\alpha, A\}$ , consisting of $ANR$-spaces and soft limit projections (Proposition 6.3.5). By Theorem 1.3.4, there exist an index $\alpha \in A$ and a map $f_\alpha \to Y$ such that $f = f_\alpha p_\alpha$. It follows from Lemma 7.2.3 that the projection $p_\alpha \colon X \to X_\alpha$ of the spectrum $\mathcal{S}_X$ has two sections $i_0$ and $i_1$ with disjoint images. Softness of $f$ implies softness of the map $f_\alpha$ (Lemma 6.1.15). Therefore $f_\alpha$ has a section $i \colon Y \to X_\alpha$. Then the desired sections of the map $f$ can be defined as the compositions $j_0 = i_0 i$ and $j_1 = i_1 i$. □

LEMMA 7.2.5. *Let a square diagram*

$$\begin{array}{ccc} X & \xrightarrow{p} & Y \\ {\scriptstyle q_0}\downarrow & & \downarrow{\scriptstyle q} \\ X_0 & \xrightarrow{p_0} & Y_0 \end{array}$$

*be a Cartesian square. If $q$ is a retraction and the map $p$ has two sections with disjoint images, then the map $p_0$ also has two sections with disjoint images.*

PROOF. Let $r: Y_0 \to Y$ be a section of the retraction $q$ and $i_1, i_2: Y \to X$ be sections of the map $p$ with disjoint images. Let $j_k = q_0 i_k r$, $k = 1, 2$. Obviously the maps $j_k$, $k = 1, 2$ are sections of the map $p_0$. Suppose that

$$j_1(Y_0) \cap j_2(Y_0) \neq \emptyset,$$

and take a point $x_0$ from this intersection. Then there is a point $y_0 \in Y_0$ such that $j_1(y_0) = x_0 = j_2(y_0)$. Since the diagram is a Cartesian square, there is a unique point $x \in X$ with the properties

$$q_0(x) = x_0 \text{ and } p(x) = r(y_0).$$

At the same time, it is easy to check that the two different points $i_1(r(y_0))$ and $i_2(r(y_0))$ of $X$ also have the properties

$$q_0(i_k(r(y_0))) = j_k(y_0) = x_0 \text{ and } p(i_k(r(y_0))) = r(y_0), \; k = 1, 2.$$

This contradictions shows that the sections $j_1$ and $j_2$ have disjoint images. □

LEMMA 7.2.6. *Let $\mathcal{S} = \{X_n, p_n^{n+1}, \omega\}$ be an inverse sequence, consisting of locally compact and Lindelöf $ANR$-spaces and soft proper short projections with Polish kernels. If each of these projections has two sections with disjoint images, then the limit projection*

$$p_0: \lim \mathcal{S} \to X_0$$

*is the trivial bundle with fiber the Hilbert cube $I^\omega$.*

PROOF. If $w(X_0) = \omega$, the statement has already been proved in Subsection 2.4.4 (Proposition 2.4.38). Therefore, we may assume that $w(X_0) = \tau > \omega$. Each of the spaces $X_n$, $n \in \omega$ can be represented as the limit space of a factorizing $\omega$-spectrum $\mathcal{S}_n = \{X_\alpha^n, q_\alpha^{\beta,n}, A\}$ consisting of locally compact $ANR$-spaces and soft limit projections (Proposition 6.3.5 and Remark 6.3.6). In the meantime, observe that these spectra have the same indexing set $A$. This follows from the equalities $w(X_n) = \tau$, $n \in \omega$. Since all short projections $p_n^{n+1}$ of the spectrum $\mathcal{S}$ are proper, soft and have Polish kernels, we may assume without loss of generality (by Theorem 6.3.2) that for each $n \in \omega$ the projection $p_n^{n+1}$ is the limit map of a Cartesian morphism

$$M_n^{n+1} = \{p_n^{n+1,\alpha} \colon X_\alpha^{n+1} \to X_\alpha^n, A\} \colon \mathcal{S}_{n+1} \to \mathcal{S}_n$$

consisting of proper and soft maps between locally compact $ANR$-spaces with countable bases.

Take an arbitrary index $\alpha \in A$ and consider the countable inverse spectrum $\mathcal{S}_\alpha = \{X_\alpha^n, p_n^{n+1,\alpha}, \omega\}$. Let $X_\alpha = \lim \mathcal{S}_\alpha$ and $X = \lim \mathcal{S}$. The limit map of the morphism $\{q_\alpha^n\} \colon \mathcal{S} \to \mathcal{S}_\alpha$ is denoted by $q_\alpha$. Thus we have the following infinite commutative diagram

$$\begin{array}{ccccccccc}
X & \longrightarrow & \cdots & \longrightarrow & X_{n+1} & \xrightarrow{p_n^{n+1}} & X_n & \longrightarrow \cdots \longrightarrow & X_0 \\
\downarrow{\scriptstyle q_\alpha} & & & & \downarrow{\scriptstyle q_\alpha^{n+1}} & & \downarrow{\scriptstyle q_\alpha^n} & & \downarrow{\scriptstyle q_\alpha^n} \\
X_\alpha & \longrightarrow & \cdots & \longrightarrow & X_\alpha^{n+1} & \xrightarrow{p_n^{n+1,\alpha}} & X_\alpha^n & \longrightarrow \cdots \longrightarrow & X_\alpha^0
\end{array}$$

Straightforward verification shows that all the square subdiagrams of the above diagram are Cartesian squares. In particular, the diagram

$$\begin{array}{ccc}
X & \xrightarrow{p_0} & X_0 \\
\downarrow{\scriptstyle q_\alpha} & & \downarrow{\scriptstyle q_\alpha^0} \\
X_\alpha & \xrightarrow{p_0^\alpha} & X_\alpha^0
\end{array}$$

is also a Cartesian square. By Lemma 7.2.5, each of the maps $p_n^{n+1,\alpha}$, $n \in \omega$, has two sections with disjoint images. In this situation, the limit projection $p_0^\alpha \colon X_\alpha \to X_\alpha^0$ of the spectrum $\mathcal{S}_\alpha$ is the trivial bundle with fiber the Hilbert cube $I^\omega$ (Proposition 2.4.38). Since the last diagram is a Cartesian square, we easily conclude that the limit projection $p_0 \colon X \to X_0$ of the spectrum $\mathcal{S}$ is also the trivial bundle with fiber $I^\omega$. $\square$

We need one more auxiliary lemma.

LEMMA 7.2.7. *Let $X$ be a locally compact and Lindelöf space of weight $\tau > \omega$. Then $X$ admits a closed embedding into a product $P \times I^\tau$, where $P$ is an $I^\omega$-manifold.*

PROOF. Consider the one-point compactification $\alpha X$ of the space $X$. Identify $\alpha X$ with a (closed) subspace of the cube $I^\tau$. Obviously $X$ is functionally open in $X$. Therefore there is a functionally open subset $V$ of $I^\tau$ such that $\alpha X \cap V = X$. Then $X$ is closed in $V$. It only remains to note that every functionally open subspace of $I^\tau$ admits a representation as a product $P \times I^\tau$, where $P$ is a $I^\omega$-manifold. □

We are now ready to prove the following characterization theorem.

THEOREM 7.2.8. *Let $\tau \geq \omega$. Then the following conditions are equivalent for every locally compact and Lindelöf $ANR$-space $X$ of weight $\tau$:*

(i) *$X$ is an $I^\tau$-manifold.*
(ii) *$X$ is strongly $\mathcal{B}_{\tau,\infty}$-universal.*
(iii) *For each compactum $Y$ of weight $< \tau$, the set of embeddings is dense in the space $C_\tau(Y, X)$.*
(iv) *The set of embeddings is dense in the space $C_\tau(D, X)$.*[3]
(v) *$X$ is homogeneous with respect to the pseudo-character.*

PROOF. The validity of the implications (i) $\Longrightarrow$ (ii) has been established in Corollary 7.1.26. Implications (ii) $\Longrightarrow$ (iii), (iii) $\Longrightarrow$ (iv) and (iv) $\Longrightarrow$ (v) are obvious.

Let us prove that (v) $\Longrightarrow$ (i). Recalling Lemma 7.2.7, we may assume that $X$ is a closed subspace of the product $P \times I^A$, where $P$ is an $I^\omega$-manifold and $A$ is an arbitrary set of cardinality $\tau$. Since every closed subset of any normal space in $C$-embedded, we see that there is a retraction $r \colon V \to X$, where $V$ is a functionally open neighborhood of $X$ in the product $P \times I^A$. The space $V$ can be represented in the same manner — as the product $L \times I^A$, where $L$ is a $I^\omega$-manifolds.

Thus we have the following situation: The space $X$ is closed in the product $L \times I^A$ (where $L$ is a $I^\omega$-manifold) and there is a retraction $r \colon L \times I^A \to X$.

We need some additional notations. If $C \subseteq B \subseteq A$, then

$$\lambda^B_C \colon L \times I^B \to L \times I^C \text{ and } \lambda_B \colon L \times I^A \to L \times I^B$$

denote the maps $id_L \times \pi^B_C$ and $id_L \times \pi_B$ respectively. Here $\pi^B_C \colon I^B \to I^C$ and $\pi_B \colon I^A \to I^B$ denote the natural projections onto the corresponding subproducts. The subspace $\lambda_B(X)$ of the product $L \times I^B$ shall be denoted by $X_B$. The properness of the map $\lambda_B$ guarantees that $X_B$ is closed in $L \times I^B$. Let

[3]Recall that $D$ stands for the two-point discrete space.

$X(B) = \lambda_B^{-1}(X_B)$. The restriction of the map $\lambda_C^B$ onto $X_B$ is denoted by $p_C^B$. Similarly, $p_B = \lambda_B/X$.

A subset $B \subseteq A$ is said to be *admissible* if for each point $x \in X(B)$ we have

$$p_B r(x) = \lambda_B(x).$$

As in the proof of Theorem 6.3.1 one can see that admissible subsets of $A$ have the following two properties:

(a) the union of an arbitrary collection of admissible subsets is admissible.

(b) each countable subset of $A$ is contained in a countable admissible subset of $A$.

(c) if $C \subseteq B \subseteq A$, and $C$ and $B$ are admissible subsets of $A$, then the maps $p_B \colon X \to X_B$ and $p_C^B \colon X_B \to X_C$ are proper and soft.

Next consider an arbitrary admissible subset $B$ of $A$ of cardinality $|B| < \tau$. We are now going to show that there is a countable admissible subset $B'$ of $A$ such that the map

$$p_B^{B\cup B'} \colon X_{B\cup B'} \to X_B$$

is the trivial bundle with fiber the Hilbert cube $I^\omega$. Since the map $p_B \colon X \to X_B$ is soft, and since the space $X$ is homogeneous with respect to pseudo-character, we conclude by Lemma 7.2.4, that there are two sections $i_0, i_1 \colon X_B \to X$ of the map $p_B$ such that

$$i_0(X_B) \cap i_1(X_B) = \emptyset.$$

Obviously, the sets $i_0(X_B)$ and $i_1(X_B)$ are closed in the product $L \times I^A$. Consequently, there is a countable subset $B_1$ of $A$ such that

$$\lambda_{B_1}(i_0(X_B)) \cap \lambda_{B_1}(i_1(X_B)) = \emptyset.$$

By property (b) of admissible sets, we may assume that $B_1$ is an admissible subset of $A$. In this situation, the map

$$p_{B\cup B_0}^{B\cup B_1} \colon X_{B\cup B_1} \to X_{B\cup B_0}$$

has two sections (namely, $p_{B\cup B_1} i_0$ and $p_{B\cup B_1} i_1$) with disjoint images (here we assume that $B_0 = \emptyset$). Continuing this process by induction, we construct countable admissible subsets $B_n$ of $A$ such that the map

$$p_{B\cup B_0\cup\cdots\cup B_n}^{B\cup B_0\cup\cdots\cup B_n\cup B_{n+1}} \colon X_{B\cup B_0\cup\cdots\cup B_{n+1}} \to X_{B\cup B_0\cup\cdots\cup B_n}$$

has two sections with disjoint images, $n \in \omega$. The countability of each of the sets $B_n$ ensures that each of the indicated maps $p_{B\cup B_0\cup\cdots\cup B_n}^{B\cup B_0\cup\cdots\cup B_{n+1}}$ has Polish kernel and, by properties (a) and (c), are proper and soft. Therefore we obtain the inverse sequence

$$\{X_{B\cup B_0\cup\cdots\cup B_n}, p_{B\cup B_0\cup\cdots\cup B_n}^{B\cup B_0\cup\cdots\cup B_{n+1}}, n \in \omega\}$$

consisting of locally compact and Lindelöf $ANR$-spaces and proper and soft short projections with Polish kernels admitting two sections with disjoint images.

Let $B' = \cup\{B_n : n \in \omega\}$. The limit space of the above inverse sequence is obviously homeomorphic to the space $X_{B \cup B'}$, and the limit projection[4] of that sequence coincides with the map

$$p_B^{B \cup B'} : X_{B \cup B'} \to X_B.$$

Lemma 7.2.6 allows us to conclude that the map $p_B^{B \cup B'}$ is the trivial bundle with fiber $I^\omega$. Observe also that the set $B \cup B'$ is, by property (a), also an admissible subset of $A$.

Further, by transfinite induction (repeating the above described process of passing from $B$ to $B \cup B'$) we construct a continuous well ordered spectrum $\mathcal{S} = \{X_\alpha, p_\alpha^{\alpha+1}, \tau\}$ of length $\tau$, satisfying the following conditions:

(i) $X = \lim \mathcal{S}$.

(ii) all spaces $X_\alpha$ are locally compact and Lindelöf $ANR$-spaces.

(iii) all short projections $p_\alpha^{\alpha+1}$ are trivial bundles with fiber $I^\omega$.

(iv) the space $X_0$ is a locally compact $ANR$-space of countable weight.

In this situation, the space $X$ is homeomorphic to the product $X_0 \times I^\tau$. By condition (iv), the product $X_0 \times I^\omega$ is an $I^\omega$-manifold (Theorem 2.3.21). It only remains to note that, by Theorem 7.1.19, the space $X$ is an $I^\tau$-manifold. □

The following statement, providing a topological characterization of the Tychonov cube $I^\tau$, follows directly from Theorem 7.2.8.

THEOREM 7.2.9. *Let $\tau > \omega$. The following conditions are equivalent for each compact $AR$-space of weight $\tau$:*

(i) *$X$ is homeomorphic to $I^\tau$.*

(ii) *$X$ is strongly $\mathcal{B}_{\tau,\infty}$-universal.*

(iii) *for each compactum $Y$ of weight $< \tau$, the set of embeddings is dense in the space $C_\tau(Y, X)$.*

(iv) *the set of embeddings is dense in the space $C_\tau(D, X)$.*

(v) *$X$ is homogeneous with respect to pseudo-character.*

It follows from Theorem 2.4.18 that the equivalences (i) $\Longleftrightarrow$ (ii) in Theorems 7.2.8 and 7.2.9 are valid in the case $\tau = \omega$ as well. The other conditions from Theorems 7.2.8 and 7.2.9 are not equivalent to condition (i). For instance, the closed unit interval is homogeneous with respect to pseudo-character and is an $AR$-compactum.

The theorems proved above have several corollaries.

COROLLARY 7.2.10. *Let $\tau \geq \omega$. The product $X \times I^\tau$ is an $I^\tau$-manifold if and only if $X$ is a locally compact and Lindelöf $ANR$-space of weight $\leq \tau$.*

PROOF. If $\tau = \omega$, then the statement is true (Theorem 2.3.21). Assume that $\tau > \omega$. If the product $X \times I^\tau$ is an $I^\tau$-manifold, then the space $X$, as a retract of an $I^\tau$-manifold, is a locally compact and Lindelöf $ANR$-space of weight $\leq \tau$.

[4]which maps $X_{B \cup B'}$ onto the space $X_B$

Conversely, since the product $X \times I^\tau$ is a locally compact and Lindelöf $ANR$-space of weight $\tau$, it suffices, by Theorem 7.2.8, to verify the homogeneity with respect to pseudo-character of the product $X \times I^\tau$. The last fact is an immediate consequence of an easy observation that the Tychonov cube $I^\tau$ is homogeneous with respect to pseudo-character. □

Similarly we have the following.

COROLLARY 7.2.11. *Let $\tau \geq \omega$. The product $X \times I^\tau$ is homeomorphic to the cube $I^\tau$ if and only if $X$ is a compact $AR$-space of weight $\leq \tau$.*

The following two statements do not have metrizable counterparts.

COROLLARY 7.2.12. *Let $\tau > \omega$. If the product of two spaces is an $I^\tau$-manifold (respectively, is homeomorphic to the cube $I^\tau$), then at least one of these spaces is an $I^\tau$-manifold (respectively, is homeomorphic to $I^\tau$).*

PROOF. Suppose that $X_1 \times X_2$ is an $I^\tau$-manifold. Then each of the spaces $X_1$ and $X_2$ is a locally compact and Lindelöf $ANR$-space of weight $\leq \tau$. Assume, contradicting our hypothesis, that neither $X_1$ nor $X_2$ is an $I^\tau$-manifold. Then, by Theorem 7.2.8, there are points $x_1 \in X_1$ and $x_2 \in X_2$ such that $\psi(x_1, X_1) < \tau$ and $\psi(x_2, X_2) < \tau$. It only remains to note that in this situation the pseudo-character of the point $(x_1, x_2)$ in $X_1 \times X_2$ is strictly less than $\tau$. This contradiction completes the proof. □

COROLLARY 7.2.13. *If the square of a space $X$ is homeomorphic to the Tychonov cube $I^\tau$, $\tau > \omega$, then $X$ is homeomorphic to $I^\tau$.*

We are now in position to prove the converse to Corollary 7.1.22.

COROLLARY 7.2.14. *If the product $X \times [0,1)$ can be embedded in the Tychonov cube $I^\tau$, $\tau > \omega$, as a functionally open subspace, then $X$ is an $I^\tau$-manifold.*

PROOF. Obviously the product $X \times [0,1)$, as a functionally open subspace of $I^\tau$, is an $I^\tau$-manifold. By Corollary 7.2.12, $X$ must be an $I^\tau$-manifold. □

We conclude this Section by presenting the following statement.

COROLLARY 7.2.15. *Let $\tau > \omega$. Each compactum of weight $\tau$, admitting a soft map onto the Tychonov cube $I^\tau$, is homeomorphic to $I^\tau$.*

PROOF. It suffices to apply Proposition 6.5.7 and Theorem 7.2.9. □

*Historical and bibliographical notes* 7.2. The results of this Section are basically obtained in [**277**]. The equivalence of conditions (ii) and (iii) with conditions (i) and (iv) in Theorems 7.2.8 and 7.2.9 was observed in [**83**]. Corollaries 7.2.14 and 7.2.15 also are due to the author.

## 7.3. Topological characterization of $R^{\tau}$-manifolds

We begin with the following statement.

LEMMA 7.3.1. *Let $\mathcal{S} = \{X_n, p_n^{n+1}, \omega\}$ be an inverse sequence, consisting of $ANR$-spaces and soft short projections with Polish kernels. Suppose that for each $n \in \omega$, the space $X_{n+1}$ contains a $C$-embedded copy of the product $X_n \times R^{\omega}$ such that $p_n^{n+1}/(X_n \times R^{\omega}) = \pi_{X_n}$, where $\pi_{X_n} \colon X_n \times R^{\omega} \to X_n$ is the projection onto the first coordinate. Then the limit projection $p_0 \colon X = \lim \mathcal{S} \to X_0$ is the trivial bundle with fiber $R^{\omega}$.*

PROOF. The case $w(X_0) = \omega$ was considered in Chapter 2 (see Proposition 2.4.40). Consider the case $w(X_0) = \tau > \omega$. Since each sort projection of the spectrum $\mathcal{S}$ has Polish kernel, we conclude that $w(X_n) = \tau$ for each $n \in \omega$. Represent the space $X_n$, $n \in \omega$, as the limit space of a factorizing $\omega$-spectrum $\mathcal{S}_n = \{X_{\alpha}^n, q_{\alpha}^{\beta,n}, A\}$, consisting of Polish $ANR$-spaces and soft limit projections. Observe, in the meantime, that the indexing sets of all these spectra coincide with $A$ (which has cardinality $\tau$). Since all short projections $p_n^{n+1}$ of the spectrum $\mathcal{S}$ are soft and have Polish kernels, we see, by Theorem 6.3.2, that $p_n^{n+1}$, $n \in \omega$, is the limit of some Cartesian morphism

$$M_n^{n+1} = \{p_n^{n+1,\alpha} \colon X_{\alpha}^{n+1} \to X_{\alpha}^n, A\} \colon \mathcal{S}_{n+1} \to \mathcal{S}_n,$$

consisting of soft maps between Polish $ANR$-spaces.

For each $\alpha \in A$, consider the inverse sequence $\mathcal{S}_{\alpha} = \{X_{\alpha}^n, p_n^{n+1,\alpha}, \omega\}$, and let $X_{\alpha} = \lim \mathcal{S}_{\alpha}$. If $\beta \geq \alpha$, then there is a Cartesian morphism

$$M_{\alpha}^{\beta} = \{q_{\alpha}^{\beta,n} \colon X_{\beta}^n \to X_{\alpha}^n, \omega\} \colon \mathcal{S}_{\beta} \to \mathcal{S}_{\alpha}$$

consisting of soft maps $q_{\alpha}^{\beta,n}$, $n \in \omega$. Denote by $q_{\alpha}^{\beta}$ the limit map of the morphism $M_{\alpha}^{\beta}$. Thus, the following infinite commutative diagram[5] arises:

$$\begin{array}{ccccccccc}
X & \longrightarrow & \cdots & \longrightarrow & X_{n+1} & \xrightarrow{p_n^{n+1}} & X_n & \longrightarrow \cdots \longrightarrow & X_0 \\
\downarrow{\scriptstyle q_{\alpha}} & & & & \downarrow{\scriptstyle q_{\alpha}^{n+1}} & & \downarrow{\scriptstyle q_{\alpha}^{n}} & & \downarrow{\scriptstyle q_{\alpha}^{n}} \\
X_{\alpha} & \longrightarrow & \cdots & \longrightarrow & X_{\alpha}^{n+1} & \xrightarrow{p_n^{n+1,\alpha}} & X_{\alpha}^{n} & \longrightarrow \cdots \longrightarrow & X_{\alpha}^{0}
\end{array}$$

[5]Compare with the proof of Lemma 7.2.6.

Straightforward verification shows that the limit space of the spectrum $\mathcal{S}' = \{X_\alpha, q_\alpha^\beta, A\}$ coincides with the space $X$, and that all newly arising square diagrams are also Cartesian squares. By assumption, the space $X_{n+1}$ contains a $C$-embedded copy of the product $X_n \times R^\omega$ so that $p_n^{n+1}/(X_n \times R^\omega) = \pi_{X_n}$. Consider the factorizing $\omega$-spectrum

$$\mathcal{S}'_n = \{X_\alpha^n \times R^\omega, q_\alpha^{\beta,n} \times id_{R^\omega}, A\},$$

the limit space of which is the product $X_n \times R^\omega$. Since $X_n \times R^\omega$ is $C$-embedded in $X_{n+1}$, we see that the spectrum

$$\mathcal{S}_{n+1}/(X_n \times R^\omega) = \{\mathrm{cl}_{X_\alpha^{n+1}} q_\alpha^{n+1}(X_n \times R^\omega), q_\alpha^{\beta,n+1}, A\},$$

also having the product $X_n \times R^\omega$ as its limit, is factorizing. Consequently, by Theorem 1.3.6, we may assume that the spectra $\mathcal{S}'_n$ and $\mathcal{S}_{n+1}/(X_n \times R^\omega)$ are isomorphic. Analyzing the maps $p_n^{n+1}$ and $\pi_{X_n}$ (using results of Section 6.3), we conclude that the indexing set $A$ contains a cofinal and $\omega$-closed subset $A_n$, $n \in \omega$, such that for each $\alpha \in A_n$ the space $X_\alpha^{n+1}$ contains a closed copy of the product $X_\alpha^n \times R^\omega$ so that

$$p_n^{n+1,\alpha}/(X_\alpha^n \times R^\omega) = \pi_{X_\alpha^n},$$

where $\pi_{X_\alpha^n} : X_\alpha^n \times R^\omega \to X_\alpha^n$ denotes the projection onto the first coordinate, $n \in \omega$. Let $A' = \cap\{A_n : n \in \omega\}$. By Proposition 1.1.27, the set $A'$ is cofinal and $\omega$-closed in $A$. In particular, $A' \neq \emptyset$. Take $\alpha \in A'$. Then the limit map $p_0^\alpha : X_\alpha \to X_\alpha^0$ of the spectrum $\mathcal{S}_\alpha$ is, by the case considered above (recall that $w(X_\alpha^0 = \omega)$, the trivial bundle with fiber $R^\omega$. Using the fact that the diagram

$$\begin{array}{ccc} X & \xrightarrow{p_0} & X_0 \\ \downarrow{\scriptstyle q_\alpha} & & \downarrow{\scriptstyle q_\alpha^0} \\ X_\alpha & \xrightarrow{p_0^\alpha} & X_\alpha^0 \end{array}$$

is a Cartesian square, we see that the limit projection $p_0$ of the spectrum $\mathcal{S}$ is also the trivial bundle with fiber $R^\omega$. □

LEMMA 7.3.2. *Let $X$ be an $ANR$-space of weight $\tau > \omega$, and suppose that for each space $Y$ of $R$-weight $< \tau$ the set of $C$-embeddings is dense in the space $C_\tau(Y, X)$. Suppose, in addition, that a soft map $p: X \to Z$, where $w(Z) < \tau$, is also given. Then $X$ contains a $C$-embedded copy of the product $Z \times R^\omega$ so that $p/(Z \times R^\omega) = \pi_Z$, where $\pi_Z : Z \times R^\omega \to Z$ denotes the projection onto the first coordinate.*

PROOF. Since the map $p$ is soft, $Z$ is an $ANR$-space (Proposition 6.1.21). Then the product $Z \times R^\omega$ is also an $ANR$-space and, consequently, by Corollary 6.3.8, the $R$-weight of the product $Z \times R^\omega$ is strictly less than $\tau$. By Lemma 6.5.1, there is a collection $\{\mathcal{U}'_t : t \in T\}$, $|T| = w(Z)$, of countable functionally open covers of the space $Z$, satisfying the conditions of that Lemma. Let $\mathcal{U}_t = p^{-1}(\mathcal{U}'_t)$, $t \in T$. By the softness of the map $p$, there is a section $i: Z \to X$ of $p$. Then the neighborhood

$$B(i\pi_Z, \{\mathcal{U}_t : t \in T\})$$

of the point $i\pi_Z : Z \times R^\omega \to X$, in the space $C_\tau(Z \times R^\omega, X)$, contains at least one $C$-embedding. Denote it by $g$. By the choice of the above neighborhood, we have $pg = \pi_Z$. Therefore, $g$ is the desired $C$-embedding of the product $Z \times R^\omega$ into the space $X$. □

Now we are ready to prove the characterization theorem for $R^\tau$-manifolds (and, in particular, of the space $R^\tau$ itself).

THEOREM 7.3.3. *Let $\tau > \omega$. The following conditions are equivalent for any $A(N)R$-space $X$ of weight $\tau$:*

(i) *$X$ is homeomorphic to $R^\tau$ (respectively, $X$ is an $R^\tau$-manifold).*
(ii) *$X$ is strongly $\mathcal{A}_{\tau,\infty}$-universal.*
(iii) *For each space $Y$ of $R$-weight $< \tau$, the set of $C$-embeddings is dense in the space $C_\tau(Y, X)$.*

PROOF. Implication (i) $\Longrightarrow$ (ii) was proved in Corollary 7.1.18. Implication (ii) $\Longrightarrow$ (iii) is trivial.

Let us prove implication (iii) $\Longrightarrow$ (i). We consider only the absolute case (the remaining case can be proved in a similar way).

Since $X$ is an $AR$-space of weight $\tau$, we may assume that $X$ is a retract of the space $R^A$, where $A$ is an arbitrary set of cardinality $\tau$. Let $B$ be an arbitrary admissible subset of $A$ (see the proof of Theorem 6.3.1) of cardinality $< \tau$. Then the map $\pi_B/X : X \to \pi_B(X)$, where $\pi_B : R^A \to R^B$ is the projection onto the corresponding subproduct, is soft. Consequently, by Lemma 7.3.2, the space $X$ contains a $C$-embedded copy of the product $\pi_B(X) \times R^\omega$ so that $\pi_B/(\pi_B(X) \times R^\omega) = \pi_{\pi_B(X)}$, where $\pi_{\pi_B(X)} : \pi_B(X) \times R^\omega \to \pi_B(X)$ is the projection onto the first coordinate. Since the $R$-weight of the product $\pi_B(X) \times R^\omega$ is less than $\tau$, we see (using results of Section 6.3) that there is a countable admissible subset $B'$ of $A$, satisfying the following conditions:

(a) The restriction of the projection $\pi_{B\cup B'}$ onto $\pi_B(X) \times R^\omega$ (considered as a $C$-embedded subspace of the space $R^A$) is a homeomorphism.
(b) The subspace $\pi_{B\cup B'}(\pi_B(X) \times R^\omega)$ is $C$-embedded in $\pi_{B\cup B'}(X)$.

In this situation the space $\pi_{B\cup B'}(X)$ contains a $C$-embedded copy of the product $\pi_B(X) \times R^\omega$ so that

$$\pi_B^{B\cup B'}/(\pi_B(X) \times R^\omega) = \pi_{\pi_B(X)},$$

where

$$\pi_B^{B\cup B'}: R^{B\cup B'} \to R^B$$

is the projection onto the corresponding subproduct. Keeping in mind that the union of an arbitrary collection of admissible subsets is admissible, by transfinite induction (repeating the corresponding part of the proof Theorem 7.2.8) we construct a continuous well ordered spectrum $\mathcal{S}_X = \{X_\alpha, p_\alpha^\beta, \tau\}$ of length $\tau$ so that the following conditions are satisfied:

(1) $X = \lim \mathcal{S}_X$.
(2) All spaces $X_\alpha$ are $AR$-spaces.
(3) All short projections $p_\alpha^{\alpha+1}$ are soft and have Polish kernels (since the above constructed sets $B'$ are countable).
(4) For each $\alpha < \tau$ the space $X_{\alpha+1}$ contains a $C$-embedded copy of the product $X_\alpha \times R^\omega$ so that $p_\alpha^{\alpha+1}/(X_\alpha \times R^\omega) = \pi_{X_\alpha}$, where $\pi_{X_\alpha}: X_\alpha \times R^\omega \to X_\alpha$ is the projection onto the first coordinate.
(5) $X_0$ is a Polish $AR$-space.

Applying transfinite induction one more time and using Lemma 7.3.1, we see that $X$ is homeomorphic to the product $X_0 \times R^\tau \approx R^\tau$. The proof is complete. □

This result has several consequences.

COROLLARY 7.3.4. *The product $X \times R^\tau$, $\tau \geq \omega$, is an $R^\tau$-manifold if and only if $X$ is an $ANR$-space of weight $\leq \tau$.*

PROOF. If $\tau = \omega$, then the statement has already been proved in Chapter 2 (see Theorem 2.3.22). Suppose that $\tau > \omega$. If the product $X \times R^\tau$ is an $R^\tau$-manifold, then the space $X$, as a retract of $R^\tau$-manifold, is an $ANR$-space of weight $\leq \tau$. Let us prove the sufficiency of our condition. Thus, let $X$ be an $ANR$-space of weight $\leq \tau$. Obviously, the product $X \times R^\tau$ is an $ANR$-space of weight $\leq \tau$. Consequently, by Theorem 7.3.3, it suffices to verify that for each space $Y$ of $R$-weight $\leq \tau$, the set of $C$-embeddings is dense in the space $C_\tau(Y, X \times R^\tau)$. Take an arbitrary set $A$ of cardinality $\tau$, and consider a map $f: Y \to X \times R^A$ and an arbitrary basic neighborhood

$$O = B(f, \{\mathcal{U}_t : t \in T\}), |T| < \tau,$$

of the point $f$ in the space $C_\tau(Y, X \times R^A)$. Let $\lambda = \max\{\omega, |T|\}$. The uncountability of $\tau$ implies that $\lambda < \tau$. Represent $X$ as the limit space of a factorizing $\lambda$-spectrum $\mathcal{S}_X = \{X_B, p_C^B, exp_\lambda A\}$, consisting of $ANR$-spaces of weight $\lambda$ and soft limit projections. The space $R^A$ can also be represented as the limit space of the standard factorizing $\lambda$-spectrum $\mathcal{S} = \{R^B, \pi_C^B, exp_\lambda A\}$. Then the product $X \times R^A$ is the limit space of the factorizing $\lambda$-spectrum

$$\mathcal{S}_X \times \mathcal{S} = \{X_B \times R^B, p_C^B \times \pi_C^B, exp_\lambda A\}.$$

In this situation, by Lemma 6.5.1, there is an element $B \in exp_\lambda A$ such that

$$O' = \{g \in C_\tau(Y, X \times R^A): (p_B \times \pi_B)g = (p_B \times \pi_B)f\}$$

is a neighborhood of $f$ in the space $C_\tau(Y, X \times R^A)$, contained in $O$. Since $R-w(Y) \leq \tau$, there is a $C$-embedding $j\colon Y \to R^{A-B}$. Let

$$j_B\colon X_B \times R^B \to X \times R^B$$

be a section of the map $p_B \times id_{R^B}$. Then the desired $C$-embedding of $Y$ into $X \times R^A$, contained in $O'$ (and, consequently, in $O$), can be defined by letting

$$i = j_B(p_B \times \pi_B)f \triangle j\colon Y \to X \times R^B \times R^{A-B}.$$

This completes the proof. □

A similar argument proves the following statement.

COROLLARY 7.3.5. *The product $X \times R^\tau$ is homeomorphic to $R^\tau$, $\tau \geq \omega$, if and only $X$ is an $AR$-space of weight $\leq \tau$.*

If $\tau > \omega$, then Corollary 7.3.5 can be proved using a different approach. Namely, we can apply the following statement to the projection $\pi_X\colon X \times R^\tau \to R^\tau$.

COROLLARY 7.3.6. *Let $\tau > \omega$. Every space of weight $\tau$, admitting a soft map onto $R^\tau$, is homeomorphic to $R^\tau$.*

PROOF. Apply Theorem 7.3.3 and Proposition 6.5.6. □

*Historical and bibliographical notes* 7.3. All results of this Section are due to the author [**84**]. As already shown (Corollary 7.1.14; compare with Corollary 7.1.22), the class of functionally open subspaces of $R^\tau$ coincides with the class of $R^\tau$-manifolds. The problem of the satisfactory topological characterization of open (not necessarily functionally open) subspaces of $R^\tau$ is still open. Corollary 6.4.10 and Proposition 6.1.7 show that if an open subspace of $R^\tau$ is not functionally open, then it is not even an $AE(0)$-space and, consequently, we cannot apply the spectral theory developed here. A similar problem with respect to open subspaces of Tychonov cubes also remains open.

## 7.4. Trivial bundles

In this Section we extend Theorems 2.4.32 and 2.4.33 to the non-metrizable case.

**7.4.1. Trivial bundles with fiber $R^\tau$.** Let a cardinal number $\tau \geq \omega$ and a map $f\colon X \to Y$ be given. Then for each map $g\colon Z \to X$, we denote by $C^g_\tau(Z,X)$ the subspace of the space $C_\tau(Z,X)$, consisting of maps $h\colon Z \to X$ such that $fh = fg$.

THEOREM 7.4.1. *Let $\tau > \omega$ and $f\colon X \to Y$ be a soft map between ANR-spaces with $w(X) = \tau$. Then the following conditions are equivalent:*

(i) *$f$ is the trivial bundle with fiber $R^\tau$.*

(ii) *For each space $Z$ of $R$-weight $\leq \tau$ and for any map $g\colon Z \to X$, the set of $C$-embeddings is dense in the space $C^g_\tau(Z,X)$.*

PROOF. We begin with implication (i) $\Longrightarrow$ (ii). Consider the projection $\pi_Y\colon Y \times R^A \to Y$, where $A$ is an arbitrarily chosen set of cardinality $\tau$. Take a space $Z$ of $R$-weight $\leq \tau$ and any map $g\colon Z \to Y \times R^A$. Let $\{\mathcal{U}_t\colon t \in T\} \subseteq cov(Y \times R^A)$, $|T| < \tau$. Let $\lambda = \max\{\omega|T|\}$ and observe that $\lambda < \tau$. From the results of Section 6.3 and Lemma 6.5.1 it follows that there exists a subset $B$ of $A$ of cardinality $\lambda$ such that the set

$$O = \{h \in C^g_\tau(Z, Y \times R^A)\colon (id_Y \times \pi_B)h = (id_Y \times \pi_B)g\}$$

is a neighborhood of the point $g$ in the space $C^g_\tau(Z, Y \times R^A)$, contained in the neighborhood of the same point generated by the covers $\mathcal{U}_t$, $t \in T$. Also, consider a $C$-embedding $\varphi\colon Z \to R^{A-B}$. Then the desired $C$-embedding,contained in the neighborhood $O$, can be defined by letting

$$h = (id_Y \times \pi_B)g \triangle \varphi.$$

Let us now prove the implication (ii) $\Longrightarrow$ (i). Consider a factorizing well ordered continuous inverse spectra

$$\mathcal{S}_X = \{X_\alpha, p^{\alpha+1}_\alpha, 1 \leq \alpha < \tau\},\ \mathcal{S}_Y = \{Y_\alpha, q^{\alpha+1}_\alpha, 1 \leq \alpha < \tau\}$$

and a morphism

$$\{f_\alpha\colon X_\alpha \to Y_\alpha\}\colon \mathcal{S}_X \to \mathcal{S}_Y$$

between them, satisfying conditions of Theorem 6.3.1. We now need to introduce some additional notation:

(a) $Z^{\alpha+1}_\alpha$ denotes the fibered product of the spaces $X_\alpha$ and $Y_{\alpha+1}$ with respect to the maps $f_\alpha$ and $q^{\alpha+1}_\alpha$; the canonical projections of this product are denoted by $\varphi^{\alpha+1}_\alpha\colon Z^{\alpha+1}_\alpha \to X_\alpha$ and $\psi^{\alpha+1}_\alpha\colon Z^{\alpha+1}_\alpha \to Y_{\alpha+1}$, $1 \leq \alpha < \tau$.

(b) $Z_\alpha$ denotes the fibered product of the spaces $X_\alpha$ and $Y$ with respect to the maps $f_\alpha$ and $q_\alpha$; the corresponding canonical projections are denoted by $\varphi_\alpha\colon Z_\alpha \to X_\alpha$ and $\psi_\alpha\colon Z_\alpha \to Y$, $1 \leq \alpha < \tau$.

(c) $h^{\alpha+1}_\alpha\colon X_{\alpha+1} \to Z^{\alpha+1}_\alpha$ denotes the diagonal product of the maps $p^{\alpha+1}_\alpha$ and $f_{\alpha+1}$, $1 \leq \alpha < \tau$.

(d) $h_\alpha\colon X \to Z_\alpha$ denotes the diagonal product of the maps $p_\alpha$ and $f$, $1 \leq \alpha < \tau$.

(e) $Z_0 = Y$ and $t^1_0 = \psi_1\colon Z_1 \to Z_0$.

(f) $t_\alpha^{\alpha+1}\colon Z_{\alpha+1} \to Z_\alpha$ denotes the diagonal product of the maps $p_\alpha^{\alpha+1}\varphi_{\alpha+1}$ and $\psi_{\alpha+1}$, $1 \leq \alpha < \tau$.

(g) $r_\alpha\colon Z_\alpha \to Z_\alpha^{\alpha+1}$ denotes the diagonal product of the maps $\varphi_\alpha$ and $q_{\alpha+1}\psi_\alpha$, $1 \leq < \tau$.

Straightforward verification shows that the generated inverse spectrum

$$\mathcal{S} = \{Z_\alpha, t_\alpha^{\alpha+1}, \tau\}$$

has the space $X$ as its limit, and the limit projection $t_0\colon \lim \mathcal{S} \to Z_0$ coincides with the originally given map $f$. It follows from the above discussion that the square diagram

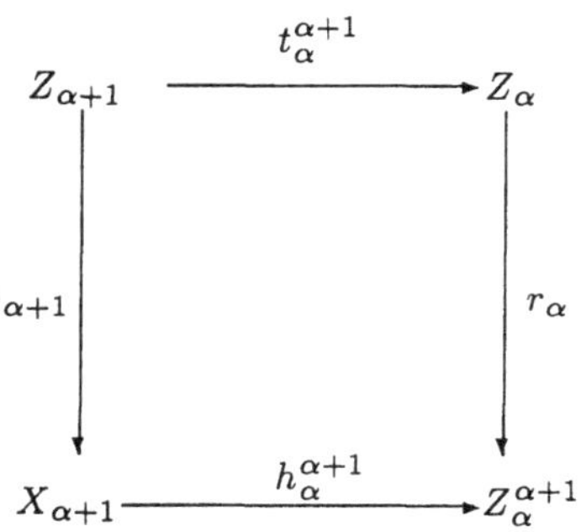


is a Cartesian square, $1 \leq \alpha < \tau$. consequently, the maps $t_\alpha^{\alpha+1}$, $\alpha \in \tau$, are soft and have Polish kernels (since, by Theorem 6.3.1, the maps $h_\alpha^\alpha$ and $f_1$ have these properties).

Let us show that for each $\alpha$, $1 \leq \alpha < \tau$, there is an index $\beta(\alpha) > \alpha$ such that the projection $t_\alpha^{\beta(\alpha)}\colon Z_{\beta(\alpha)} \to Z_\alpha$ of the spectrum $\mathcal{S}$ has a Polish kernel and the space $Z_{\beta(\alpha)}$ contains a $C$-embedded copy of the product $Z_\alpha \times R^\omega$ so that $t_\alpha^{\beta(\alpha)}/(Z_\alpha \times R^\omega) = \pi_{Z_\alpha}$, where, as usual, $\pi_{Z_\alpha}\colon Z_\alpha \times R^\omega \to Z_\alpha$ is the projection onto the first coordinate. As follows from the proof of Theorem 6.3.1 in order to show the last fact it suffices to prove the existence in $X$ of a $C$-embedded copy of the product $Z_\alpha \times R^\omega$ so that $h_\alpha/(Z_\alpha \times R^\omega) = \pi_{Z_\alpha}$. Since $h_\alpha$ is a soft map, there is a map $i\colon Z_\alpha \times R^\omega \to X$ such that $h_\alpha i = \pi_{Z_\alpha}$. By Lemma 6.5.1, there is a subcollection $\{\mathcal{U}_t' \in T \in T\}$ of cardinality $|T| = w(X_\alpha) < \tau$ of the collection $cov(X_\alpha)$ such that the conditions of Lemma 6.5.1 are satisfied. Let $\mathcal{U}_t = p_\alpha^{-1}(\mathcal{U}_t')$, for each $t \in T$. Then, by our assumption, there is a $C$-embedding $g\colon Z_\alpha \times R^\omega \to X$ such that the following conditions hold:

(h) The maps $g$ and $i$ are $\{\mathcal{U}_t\colon t \in T\}$-close.

(i) $fg = fi$.

We claim that $g$ is the desired $C$-embedding. Indeed, it suffices to observe that

$$h_\alpha g = (p_\alpha \triangle f)g = p_\alpha g \triangle fg = p_\alpha i \triangle fi = (p_\alpha \triangle f)i = h_\alpha i = \pi_{Z_\alpha}.$$

Thus indexes $\beta(\alpha)$, satisfying the above conditions, exist for each $\alpha$. In this situation, applying Lemma 7.3.1, by straightforward transfinite induction (compare with the proof of Theorem 7.3.3) we reach the final conclusion — $f$ is the trivial bundle with fiber $R^\tau$. □

The following statement is an immediate consequence of the above Theorem.

COROLLARY 7.4.2. *Let $\tau > \omega$, $f: X \to Y$ be a soft map between ANR-spaces, and $w(X) \leq \tau$. Then the composition $f\pi_X$, where $\pi_X: X \times R^\tau \to X$ is the projection onto the first coordinate, is the trivial bundle with fiber $R^\tau$.*

COROLLARY 7.4.3. *Let $\tau > \omega$ and $f: X \to Y$ be a soft map of an $R^\tau$-manifold $X$ onto a space $Y$ of weight $< \tau$. Then $f$ is the trivial bundle with fiber $R^\tau$.*

PROOF. By Theorem 7.1.11, the $R^\tau$-manifold $X$ is homeomorphic to the product $P \times R^\tau$, where $P$ is an $R^\omega$-manifold. Represent the product $P \times R^\tau$ as the limit space of the standard factorizing $\lambda$-spectrum

$$\mathcal{S} = \{P \times (R^\lambda)^\alpha, id_P \times \pi^\beta_\alpha, exp_\lambda \tau\}$$

where $\pi^\beta_\alpha: (R^\lambda)^\beta \to (R^\lambda)^\alpha$ denotes the projection onto the corresponding subproduct and $\lambda = \max\{\omega, w(Y)\}$. Then there are an index $\alpha \in exp_\lambda \tau$ and a map

$$f_\alpha: P \times (R^\lambda)^\alpha \to Y$$

such that $f = f_\alpha(id_P \times \pi_\alpha)$, where $\pi_\alpha: R^\tau \to (R^\lambda)^\alpha$ is the projection. The softness of the maps $f$ and $id_P \times \pi_\alpha$ guarantees the softness of the map $f_\alpha$ (Lemma 6.1.15). Since the map $id_P \times \pi_\alpha$ is the trivial bundle with fiber $R^\tau$, Corollary 7.4.2 completes the proof. □

**7.4.2. Trivial bundles with fiber $I^\tau$.** Replacing Lemma 7.3.1 by Lemma 7.2.6 in the proof of Theorem 7.4.1, we obtain the following result.

THEOREM 7.4.4. *Let $\tau > \omega$ and $f: X \to Y$ be a soft map between ANR-spaces with $w(X) = \tau$. Then the following conditions are equivalent:*

(i) *$f$ is the trivial bundle with fiber $I^\tau$.*
(ii) *For each compact space $Z$ of weight $\leq \tau$ and for any map $g: Z \to X$, the set of embeddings is dense in the space $C^g_\tau(Z, X)$.*
(iii) *The set of embeddings is dense in the space $C^\pi_\tau(X \times D, X)$, where $\pi: X \times D \to X$ is the projection.*

COROLLARY 7.4.5. *Let $\tau > \omega$, $f: X \to Y$ be a proper and soft map between locally compact and Lindelöf ANR-spaces, and $w(X) \leq \tau$. Then the composition $f\pi_X$, where $\pi_X: X \times I^\tau \to X$ is the projection onto the first coordinate, is the trivial bundle with fiber $I^\tau$.*

COROLLARY 7.4.6. *Let $\tau > \omega$ and $f\colon X \to Y$ be a proper and soft map of an $I^\tau$-manifold $X$ onto a space $Y$ of weight $< \tau$. Then $f$ is the trivial bundle with fiber $I^\tau$.*

In connection with Theorem 7.4.4, the following question arises naturally [**277**]:

- Is the soft map between compact $ANR$-spaces, whose fibers are all homeomorphic to $I^\tau$ and which satisfies condition (iii) for $\tau = \omega$, the trivial bundle with fiber $I^\tau$.

Theorem 7.4.4 allows us to answer the above question negatively. Indeed, consider an arbitrary soft map $f\colon I^{\omega_1} \to I^{\omega_1}$, all fibers of which are homeomorphic to $I^{\omega_1}$ and which does not admit two sections with disjoint images (such maps exist: take an $\omega_1$ power of the soft map $\varphi\colon I^\omega \to I^\omega$ of the Hilbert cube onto itself, all fibers of which are the copies of the Hilbert cube and which, nevertheless, is not the trivial bundle with fiber $I^\omega$ [**300**]). Let $\pi\colon I^{\omega_1} \times I^\omega \to I^{\omega_1}$ be the projection onto the first coordinate. We claim the desired map is the composition $f\pi$. Obviously, all fibers of the map $f\pi$ are copies of the Tychonov cube $I^{\omega_1}$. Since the projection $\pi$ satisfies condition (iii) of Theorem 7.4.4, the composition $f\pi$ also does. We are now going to outline why $f\pi$ is not the trivial bundle. Assume the contrary, i.e. suppose that $f\pi$ is the trivial bundle with fiber $I^\tau$. Take a section $i\colon I^{\omega_1} \to I^{\omega_1}$ of the map $f$ and a section $j\colon I^{\omega_1} \to I^{\omega_1} \times I^\omega$ of the projection $\pi$. Also take a countable collection $\{\mathcal{U}_i\colon i \in \omega\}$ of open covers of $I^\omega$ satisfying the conditions of Lemma 6.5.1. Let $\mathcal{V}_i = \{I^{\omega_1} \times \mathcal{U}_1,\ i \in \omega\}$. Then by condition (iii) of Theorem 7.4.4, there exist maps $h_k\colon I^{\omega_1} \to I^{\omega_1} \times I^\omega$, $k = 1, 2$, $\{\mathcal{V}_i\colon i \in \omega\}$-close to the composition $ji$ such that $h_1(I^\omega) \cap I^{\omega_1} = \emptyset$ and $f\pi h_k = f\pi ji = id_{I^{\omega_1}}$, $k = 1, 2$. Let $g_k = \pi h_k$, $k = 1, 2$. Obviously, the maps $g_1$ and $g_2$ are sections of the map $f$. It is also not hard to see that $g_1(I^{\omega_1}) \cap g_2(I^{\omega_1}) = \emptyset$. This contradicts the choice of $f$. Therefore, the composition $f\pi$ is not the trivial bundle.

**7.4.3. Approximation by homeomorphisms.** A map $f\colon X \to Y$ is said to be a $\tau$-*near-homeomorphism*, $\tau \geq \omega$, if $f$ is in the closure of the set of homeomorphisms in the space $C_\tau(X, Y)$. In other words, $f$ is a $\tau$-near-homeomorphism if for each collection $\{\mathcal{U}_t\colon t \in T\} \subseteq cov(Y)$ with $|T| < \tau$ there is a homeomorphism $h\colon X \to Y$ $\{\mathcal{U}_t\colon t \in T\}$-close to $f$.

In Subsection 2.1.1 (see Theorem 2.1.8) we have already discussed $\omega$-near-homeomorphisms (i.e. near-homeomorphisms in the usual sense). Near-homeomorphisms between $R^\omega$- and $I^\omega$-manifolds also have been characterized (Propositions 2.4.20 and 2.4.21 respectively). In order to describe the general situation we need the following definition.

DEFINITION 7.4.7. *Let $\tau \geq \omega$. A map $f\colon X \to Y$ between realcompact spaces is called* $\tau$-approximatively $n$-soft *($n = 0, 1, \ldots, \infty$) if for each at most $n$-dimensional realcompact space $Z$, arbitrary closed subset $Z_0$ of $Z$, collection $\{\mathcal{U}_t\colon t \in T\} \subseteq cov(Y)$, where $|T| < \tau$, and maps $g\colon Z_0 \to X$ and $h\colon Z \to Y$*

*such that* $fg = h/Z_0$ *and* $C(g)(C(X)) \subseteq C(Z)/Z_0$, *there is a map* $k\colon Z \to X$ *such that* $g = k/Z_0$ *and the composition* $fk$ *is* $\{\mathcal{U}_t\colon t \in T\}$*-close to* $h$. *The* $\tau$*-approxitively* $\infty$*-soft maps are called* $\tau$-approxitively soft.

Obviously, every $n$-soft map is $\tau$-approximatively $n$-soft for each $\tau \geq \omega$. Observe also that if, as in the case of $n$-soft maps (see Proposition 6.1.18), the spaces $X$ and $Y$ in Definition 7.4.7 are also compact, then the spaces $Z$ and $Z_0$ also can be assumed to be compact. In this case the inclusion $C(g)(C(X)) \subseteq C(Z)/Z_0$ is automatically satisfied.

LEMMA 7.4.8. *A map* $f\colon X \to Y$ *between Polish spaces is* $\omega$*-approximatively* $n$*-soft* ($n = 0, 1, \ldots, \infty$) *if and only if the following condition is satisfied:*

- *For each at most* $n$*-dimensional Polish space* $Z$, *for each closed subset* $Z_0$ *of* $Z$, *for each open cover* $\mathcal{U} \in cov(Y)$ *and for any two maps* $g\colon Z_0 \to X$ *and* $h\colon Z \to Y$ *with* $fg = h/Z_0$, *there is a map* $k\colon Z \to X$ *such that* $g = k/Z_0$ *and the composition* $fk$ *is* $\mathcal{U}$*-close to* $h$.

PROOF. Obviously every $\omega$-approximatively $n$-soft map between Polish spaces satisfies the above condition. Let us prove the converse. Thus, suppose that a map $f\colon X \to Y$ has the above property. Take an arbitrary at most $n$-dimensional realcompact space $Z$, closed subset $Z_0$ of $Z$, and a collection $\{\mathcal{U}_t\colon t \in T\} \subseteq cov(Y)$ with $|T| < \omega$. Also take two maps $g\colon Z_0 \to X$ and $h\colon Z \to Y$ such that $fg = h/Z_0$ and $C(g)(C(X)) \subseteq C(Z)/Z_0$. Represent the space $Z$ as the limit space of factorizing $\omega$-spectrum

$$\mathcal{S} = \{Z_\alpha, p_\alpha^\beta, A\}$$

consisting of at most $n$-dimensional Polish spaces (Theorem 1.3.10). Since $Z_0$ is closed in $Z$, the limit space of the induced spectrum

$$\mathcal{S}_0 = \{\mathrm{cl}_{Z_\alpha}, p_\alpha^\beta, A\}$$

coincides with $Z_0$. By Theorem 1.3.6, there exist an index $\alpha_1 \in A$ and a map $h_{\alpha_1}\colon Z_{\alpha_1} \to Y$ such that $h = h_{\alpha_1} p_{\alpha_1}$. Further, by Proposition 1.3.13, there exist an index $\alpha_2 \in A$ and a map $g_{\alpha_2}\colon \mathrm{cl}_{Z_{\alpha_2}} \to X$ such that $g = g_{\alpha_2} p_{\alpha_2}/Z_0$. Without loss of generality we may assume that $\alpha_1 = \alpha_2 = \alpha$. It is not hard to see that the maps $h_\alpha = h_{\alpha_1}$ and $g_\alpha = g_{\alpha_2}$ have the following properties:

$$h = h_\alpha p_\alpha, \quad g = g_\alpha p_\alpha/Z_0 \quad \text{and} \quad fg_\alpha = h_\alpha/\operatorname{cl}_{Z_\alpha} p_\alpha(Z_0).$$

By our assumption, there is a map $k_\alpha\colon Z_\alpha \to X$ such that $g_\alpha = k_\alpha/\operatorname{cl}_{Z_\alpha} p_\alpha(Z_0)$ and the composition $fk_\alpha$ is $\mathcal{U}$-close to $h_\alpha$, where $\mathcal{U}$ refines each of the covers $\mathcal{U}_t$, $t \in T$ (recall that $T$ is finite). Then the desired map $k$ can be defined as $k = k_\alpha p_\alpha$. □

THEOREM 7.4.9. *Let* $\tau \geq \omega$. *A map between* $R^\tau$*-manifolds is a* $\tau$*-near-homeomorphism if and only if it is* $\tau$*-approximatively soft.*

PROOF. If $\tau = \omega$, then the statement follows from Lemma 7.4.8 and Proposition 2.4.20. Consider the case $\tau > \omega$. Let $f\colon X \to Y$ be a $\tau$-approximatively soft map between $R^\tau$-manifolds. Consider an arbitrary collection $\{\mathcal{U}_t\colon t \in T\} \subseteq cov(Y)$ with $|T| < \tau$. We shall construct a homeomorphism $h\colon X \to Y$ which is $\{\mathcal{U}_t \in T\}$-close to $f$. Let $\lambda = \max\{\omega, |T|\}$. Clearly, $\lambda < \tau$. By Theorem 7.1.11, the $R^\tau$-manifolds $X$ and $Y$ are homeomorphic to the limit spaces of factorizing $\lambda$-spectra $\mathcal{S}_X = \{X_\alpha, p_\alpha^\beta, A\}$ and $\mathcal{S}_Y = \{Y_\alpha, q_\alpha^\beta, A\}$ respectively. Moreover, these spectra consist of $R^\lambda$-manifolds and soft limit projections (even trivial bundles with fiber $R^\tau$). By Theorem 1.3.6, the map $f$ is the limit map of the morphism

$$\{f_\alpha\colon X_\alpha \to Y_\alpha, A\}\colon \mathcal{S}_X \to \mathcal{S}_Y.$$

A standard argument shows that there exist an index $\alpha \in A$ and open covers $\mathcal{U}_t^\alpha \in cov(Y_\alpha)$ such that $\mathcal{U}_t = q_\alpha^{-1}(\mathcal{U}_t^\alpha)$ for each $t \in T$. Take a subcollection $\{\mathcal{V}_t^\alpha\colon t \in T'\}$ of the collection $cov(Y_\alpha)$ of cardinality $|T'| = w(Y_\alpha) = \lambda < \tau$, satisfying the conditions of Lemma 6.5.1. Let $\mathcal{V}_t = q_\alpha^{-1}(\mathcal{V}_t^\alpha)$, $t \in T'$.

Let us show that the map $f_\alpha\colon X_\alpha \to Y_\alpha$ is soft. Take an arbitrary realcompact space $Z$, a closed subspace $Z_0$ of $Z$, and two maps $g_\alpha\colon Z_0 \to X_\alpha$ and $h_\alpha\colon Z \to Y_\alpha$ satisfying the conditions

$$f_\alpha g_\alpha = h_\alpha/Z_0 \text{ and } C(g_\alpha)(C(X_\alpha)) \subseteq C(Z)/Z_0.$$

Take a section $i\colon X_\alpha \to X$ of the limit projection $p_\alpha$ of the spectrum $\mathcal{S}_X$ and let $g = ig_\alpha$. Obviously,

$$q_\alpha f g = h_\alpha/Z_0 \text{ and } C(fg)(C(Y)) \subseteq C(Z)/Z_0.$$

Consequently, by softness of the limit projection $q_\alpha$ of the spectrum $\mathcal{S}_Y$, there is a map $h\colon Z \to Y$ such that $fg = h/Z_0$ and $g_\alpha h = h_\alpha$. By our assumption, $f$ is $\tau$-approximatively soft. Therefore there is a map $k\colon Z \to X$ such that $g = k/Z_0$ and the composition $fk$ is $\{\mathcal{V}_t\colon t \in T'\}$-close to $h$. Let $k_\alpha = p_\alpha k$. Keeping in mind the choice of the collection $\{\mathcal{V}_t\colon t \in T'\}$, one can easily verify that $g_\alpha = k_\alpha/Z_0$ and $f_\alpha k_\alpha = h_\alpha$. This shows the softness of the map $f_\alpha$. But then, by Lemma 6.1.14, the composition $f_\alpha p_\alpha\colon X \to Y_\alpha$ is also soft. By Corollary 7.4.3, this composition is the trivial bundle with fiber $R^\tau$. Similarly, the limit projection $q_\alpha$ of the spectrum $\mathcal{S}_Y$ is also the trivial bundle with fiber $R^\tau$. The bases of these trivial bundles coincide with the space $Y_\alpha$. Therefore there is a homeomorphism $s\colon X \to Y$ such that $q_\alpha s = f_\alpha p_\alpha$. The last equality shows, by the choice of the index $\alpha$, that the homeomorphism $s$ is $\{\mathcal{U}_t\colon t \in T\}$-close to the map $f$. Thus, $f$ is a $\tau$-near-homeomorphism.

Let us prove the converse. Let $f\colon X \to Y$ be a $\tau$-near-homeomorphism between $R^\tau$-manifolds. We are going to show that $f$ is $\tau$-approximatively soft. Consider an arbitrary realcompact space $Z$, a closed subspace $Z_0$ of $Z$, a collection $\{\mathcal{U}_t\colon t \in T\} \subseteq cov(Y)$ with $|T| < \tau$, and two maps $g\colon Z_0 \to X$ and $h\colon Z \to Y$ such that $fg = h/Z_0$ and $C(g)(C(X)) \subseteq C(Z)/Z_0$. As above, let $\lambda = \max\{\omega, |T|\}$. Obviously $\lambda < \tau$. Represent the space $Y$ as the limit space of a factorizing $\lambda$-spectrum $\mathcal{S}_Y = \{Y_\alpha, q_\alpha^\beta, A\}$ consisting of $R^\lambda$-manifolds and soft limit projections. As above, we can find an index $\alpha \in A$ and an open covers $\mathcal{U}_t^\alpha \in cov(Y_\alpha)$ such that $\mathcal{U}_t = q_\alpha^{-1}(\mathcal{U}_t^\alpha)$ for each $t \in T$. Applying Lemma 6.5.1, we get a collection $\{\mathcal{V}_t^\alpha : t \in T'\} \subseteq cov(Y_\alpha)$, $|T'| = \lambda < \tau$ with the properties of that Lemma. Let $\mathcal{V}_t = q_\alpha^{-1}(\mathcal{V}_t^\alpha)$, $t \in T'$. By our assumption, there is a homeomorphism $s \colon X \to Y$ $\{\mathcal{V}_t : t \in T'\}$-close to $f$. By the choice of the collection $\{\mathcal{V}_t : t \in T'\}$, we have $q_\alpha f = q_\alpha s$. Therefore, the map $q_\alpha f$, as the composition of soft maps $s$ and $q_\alpha$, is soft (Lemma 6.1.14). Then there is a map $k \colon Z \to X$ such that $g = k/Z_0$ and $q_\alpha f k = q_\alpha h$. The last equality, and the choice of the index $\alpha$, guarantee that the composition $fk$ is $\{\mathcal{U}_t : t \in T\}$-close to $h$. This completes the proof. □

A similar argument proves the following statement.

THEOREM 7.4.10. *Let $\tau \geq \omega$. A proper map between $I^\tau$-manifolds is a $\tau$-near-homeomorphism if and only if it is $\tau$-approximatively soft.*

COROLLARY 7.4.11. *Let $\tau \geq \omega$ and $X$ be an $R^\tau$-manifold. Then the projection $\pi_X \colon X \times R^\tau \to X$ is a $\tau$-near-homeomorphism.*

COROLLARY 7.4.12. *Let $\tau \geq \omega$ and $X$ be an $I^\tau$-manifold. Then the projection $\pi_X \colon X \times I^\tau \to X$ is a $\tau$-near-homemorphism.*

*Historical and bibliographical notes* 7.4. The results of this Section were obtained by the author [**83**], [**91**]. Corollaries 7.4.5 and 7.4.6 can be found in [**277**].

CHAPTER 8

# Applications

## 8.1. Uncountable powers of countable discrete spaces

In this Section we give topological characterizations of the spaces $N^\tau$ and $D^\tau$ for $\tau > \omega$. Recall that characterizations of the spaces $N^\omega$ and $D^\omega$ are given in Theorems 1.1.5 and 1.1.10, respectively. We also suggest that the reader compare the proofs presented below to those of Theorems 7.3.3 and 7.2.8.

**8.1.1. Characterization of $N^\tau$.** We begin with some auxiliary statements.

LEMMA 8.1.1. *Let $X$ be a zero-dimensional Polish space and let $G$ be open and closed in the product $X \times N^\omega$. Then the map $\pi/G\colon G \to \pi(G)$ is the trivial bundle with fiber $N^\omega$, where $\pi\colon X \times N^\omega \to X$ denotes the projection onto the first factor.*

PROOF. Without loss of generality it can be assumed that $\pi(G) = X$. Fix complete bounded metrics $\rho_1$ and $\rho_2$ on $X$ and $N^\omega$, respectively. On $X \times N^\omega$ we consider the metric $\rho$ defined by the formula

$$\rho((x_1, a_1), (x_2, a_2)) = 2^{-1}\rho_1(x_1, x_2) + 2^{-2}\rho_2(a_1, a_2)$$

where $(x_k, a_k) \in X \times N^\omega$, $k = 1, 2$. Using a standard argument we construct a collection $\{G_n\colon n \in N\}$ of open and closed subsets in $G$ such that

(a) $G = \cup\{G_n\colon n \in N\}$,
(b) $G_n \cap G_m = \emptyset$ whenever $n \neq m$,
(c) $\pi(G_n) = X$ for each $n \in N$, and
(d) $diam_\rho(\pi^{-1}(x) \cap G_n) < 2^{-1}$ for any $n \in N$ and $x \in X$.

Let $\mathcal{U}_1 = \{G_n\colon n \in N\}$. Repeating the procedure described above for each of the sets $G_n$ separately, and using induction, we construct a sequence $\{\mathcal{U}_k\colon k \in N\}$ of countably infinite open and closed disjoint covers of $G$ with the following properties (compare with the proof of Lemma 6.4.1):

(i) $\mathcal{U}_k = \{G_{n_1,\cdots,n_k}\colon n_i \in N, i \in N\}$, $k \in N$.
(ii) $G_{n_1,\cdots,n_k} = \cup\{G_{n_1,\cdots,n_k,n_{k+1}}\colon n_{k+1} \in N\}$.
(iii) $\pi(G_{n_1,\cdots,n_k}) = X$ for any $k$-tuple $(n_1, \cdots, n_k)$.

(iv) $diam_\rho(\pi^{-1}(x) \cap G_{n_1,\cdots,n_k}) < 2^{-k}$ for any $x \in X$ and any $k$-tuple $(n_1, \cdots, n_k)$.

We define a map $h\colon X \times N^\omega \to G$ as follows:

$$h(x,a) = \pi^{-1}(x) \cap \bigcap\{G_{n_1,\cdots,n_k} : k \in N\}$$

for any point $(x,a) = (x,(n_1,\cdots,n_k,\cdots)) \in X \times N^\omega$. It follows from the completeness of $\rho$ and properties (ii) and (iv) that $h(x,a)$ is non-empty and consists of precisely one point. Direct verification shows that $h$ is a homeomorphism. It remains to observe that, by the construction, $\pi h = \pi$. □

LEMMA 8.1.2. *An open surjection $p\colon X \to Y$ between zero-dimensional Polish spaces is a trivial bundle with fiber $N^\omega$ if and only if any open and closed subset $G$ of $X$ admits a countably infinite disjoint open cover $\{G_n : n \in N\}$ such that $p(G_n) = p(G)$ for each $n \in N$.*

PROOF. The necessity follows from Lemma 8.1.1. In order to prove the sufficiency we follow the argument used in the proof of Lemma 6.4.1. As in that proof (using, in addition, the zero-dimensionality of our spaces and the above condition) we can take a complete metric $d$ on $X$ and construct a sequence $\{\mathcal{U}_k\}$ of countably infinite disjoint open covers of $X$ with the following properties:

(i) $\mathcal{U}_k = \{U_{i_1,\ldots,i_k} : i_j \in N\}$, $j \in N$, $k \in N$.
(ii) $U_{i_1,\ldots,i_k,i_{k+1}} \subseteq U_{i_1,\ldots,i_k}$, $i_{k+1} \in N$.
(iii) $U_{i_1,\ldots,i_k} = \cup\{U_{i_1,\ldots,i_k,i_{k+1}} : i_{k+1} \in N\}$.
(iv) $\text{diam}_d(U_{i_1,\ldots,i_k} \cap p^{-1}(y)) < \frac{1}{2^k}$, $y \in Y$.
(v) $p(U_{i_1,\ldots,i_k}) = Y$, $i_j \in N$, $k \in N$.

The desired fiber preserving homeomorphism $h\colon Y \times N^\omega \to X$ can be defined by letting

$$h(y,a) = p^{-1} \cap \bigcap\{U_{i_1,\ldots,i_k} : k \in N\}$$

for each point $(y,a) = (y,(i_1,\ldots,i_k,\ldots)) \in Y \times N^\omega$. □

LEMMA 8.1.3. *Let $\mathcal{S} = \{X_n, p_n^{n+1}, \omega\}$ be an inverse sequence consisting of zero-dimensional $AE(0)$-spaces and 0-soft short projections with Polish kernels. Suppose that for each $n \in \omega$, the space $X_{n+1}$ contains a $C$-embedded copy of the product $X_n \times N^\omega$ such that $p_n^{n+1}/(X_n \times N^\omega) = \pi_{X_n}$, where $\pi_{X_n}\colon X_n \times N^\omega \to X_n$ is the projection onto the first coordinate. Then the limit projection $p_0\colon X = \lim \mathcal{S} \to X_0$ is the trivial bundle with fiber $N^\omega$.*

PROOF. Apply the argument presented in the proof of Lemma 7.3.1. □

Combining Lemmas 6.5.4, 8.1.3 and the proof of Theorem 7.3.3 we obtain the following characterization of the space $N^\tau$, $\tau > \omega$.

THEOREM 8.1.4. *Let $\tau > \omega$. The following conditions are equivalent for any zero-dimensional $AE(0)$-space $X$ of weight $\tau$:*

(i) *$X$ is homeomorphic to $N^\tau$.*
(ii) *$X$ is strongly $\mathcal{A}_{\tau,0}$-universal.*

(iii) *For each zero-dimensional space* $Y$ *of* $R$*-weight* $< \tau$ *the set of* $C$*-embeddings is dense in the space* $C_\tau(Y, X)$.

Applying Theorem 1.1.5 and the above characterization we get (compare with Corollary 7.3.4) the following statement.

COROLLARY 8.1.5. *The product* $X \times N^\tau$, $\tau \geq \omega$, *is homeomorphic to* $N^\tau$ *if and only if* $X$ *is a zero-dimensional* $AE(0)$*-space of weight* $\leq \tau$.

**8.1.2. Characterization of** $D^\tau$**.** Recall that the Cantor cube $D^\tau$ is strongly $\mathcal{B}_{\tau,0}$-universal (Lemma 6.5.3). In this Subsection we are going to show that this property in fact characterizes the space $D^\tau$.

THEOREM 8.1.6. *Let* $\tau > \omega$. *Then the following conditions are equivalent for every zero-dimensional compact (locally compact and Lindelöf)* $AE(0)$*-space* $X$ *of weight* $\tau$:

(i) $X$ *is homeomorphic to* $D^\tau$ *(to* $D^\tau \times N$*).*
(ii) $X$ *is strongly* $\mathcal{B}_{\tau,0}$*-universal.*
(iii) *For each zero-dimensional compactum* $Y$ *of weight* $< \tau$ *the set of embeddings is dense in the space* $C_\tau(Y, X)$.
(iv) *The set of embeddings is dense in the space* $C_\tau(D, X)$.
(v) $X$ *is homogeneous with respect to pseudo-character.*

The proof of this Theorem involves Lemma 6.5.3, and follows the proof of Theorem 7.2.8 with obvious changes of a technical nature (identical to those between the proofs of Theorems 7.3.3 and 8.1.4). In particular, the proof is based on the following statement (compare with Lemma 8.1.2).

LEMMA 8.1.7. *Let* $f \colon X \to Y$ *be an open surjection between zero-dimensional metrizable compacta. Then the following conditions are equivalent:*

(i) $f$ *is a trivial bundle with fiber* $D^\omega$.
(ii) *All fibers of* $f$ *are homeomorphic to* $D^\omega$.

Theorem 8.1.6 has several consequences.

COROLLARY 8.1.8. *The product* $X \times D^\tau$, $\tau \geq \omega$, *is homeomorphic to* $D^\tau$ *if and only if* $X$ *is a zero-dimensional* $AE(0)$*-compactum of weight* $\leq \tau$.

COROLLARY 8.1.9. *The Cantor cube* $D^{\omega_1}$ *is homeomorphic to its hyperspace* $expD^{\omega_1}$.

PROOF. Observe that if $f \colon X \to Y$ is an open surjection between zero-dimensional metrizable compacta, then the corresponding map

$$expf \colon expX \to expY$$

is also an open surjection. Next verify that $exp\pi$ is equivalent to $\pi$, where $\pi\colon D^{\omega} \times D^{\omega} \to D^{\omega}$ denotes the projection onto the first coordinate. Finally apply Theorem 8.1.6. □

COROLLARY 8.1.10. *The Cantor cube $D^{\omega_2}$ is not homeomorphic to its hyperspace $expD^{\omega_2}$.*

PROOF. Assuming the contrary and applying Theorems 8.1.6 and 1.3.4 we can easily conclude that the map $exp\pi$, where $\pi\colon D^{\omega_1} \times D^{\omega_1} \to D^{\omega_1}$ must be equivalent to $\pi$. Further observe that not all fibers of the map $exp\pi$ are homeomorphic to the Cantor cube $D^{\omega_1}$. This contradiction proves our statement. □

*Historical and bibliographical notes* 8.1. The results of Subsection 8.1.1 are due to the author. The equivalence of conditions (i) and (v) of Theorem 8.1.6 has been proved in [**272**]. The related result, Lemma 8.1.7, providing a characterization of trivial bundles with fiber $D^{\omega}$ was obtained in [**286**]. One can prove this statement by simply "parameterizing" the well known proof of Theorem 1.1.10. Another possible way involves Michael's Selection Theorem 2.1.14. Corollary 8.1.9 also appears in [**286**]. Corollary 8.1.10 was proved in [**271**].

## 8.2. Spectral representations of topological groups

A classical result of Pontrjagin [**253**] asserts that each compact topological group can be expressed as the inverse limit of some transfinite inverse spectrum in which all short projections are continuous homomorphisms whose kernels are compact Lie groups (so-called Lie series). Here kernel (i.e. the inverse image of the unit element) has algebraic meaning. Recall also Proposition 6.3.5 (and Remark 6.3.7) which states, in particular, that the $AE(0)$-compacta are precisely those that admit special spectral representations (the so-called Haydon series). A simple comparison of these two results immediately shows that Haydon's theorem is, in some sense, the analog of Pontrjagin's theorem in the category of compacta and continuous maps. This comparison can be extended to yield a proof of the fact that each compact topological group is an $AE(0)$-compactum. Moreover, noticing that each locally compact topological group is Lindelöf and applying Remark 6.3.6, one can easily see that every locally compact topological group is an $AE(0)$-space. This allows us to study locally compact topological groups from the point of view of the general theory of $AE(0)$-spaces and 0-soft maps. For instance, it immediately follows from Theorem 8.1.6 that each uncountable, zero-dimensional compact (locally compact and Lindelöf) group is topologically equivalent to $D^{\tau}$ (to the product $N \times D^{\tau}$) [**181**], [**204**]. In this Section we extend these characterizations to the non-locally compact case.

We begin with the following useful statement.

LEMMA 8.2.1. *Let $\lambda \geq \omega$. Each $AE(0)$ topological group $X$ of weight $\tau > \lambda$ is isomorphic to the limit space of a factorizing $\lambda$-spectrum $\mathcal{S}_X = \{X_\alpha, p_\alpha^\beta, A\}$ all spaces of which are $AE(0)$-groups of weight $\lambda$ and all projections of which are 0-soft homomorphisms.*

PROOF. By Proposition 6.3.5, the space $X$ can be represented as the limit space of a factorizing $\lambda$-spectrum $\mathcal{S}_X = \{X_\alpha, p_\alpha^\beta, A\}$ consisting of $AE(0)$-spaces of weight $\lambda$ and 0-soft limit projections. Let us show that this spectrum contains a $\lambda$-closed and cofinal subspectrum consisting of topological groups and limit projections that are (continuous) homomorphisms.

Since $X$ is a topological group, there is a continuous multiplication $\mu\colon X \times X \to X$. Clearly, $X \times X$ is the limit space of the spectrum

$$\mathcal{S}_X \times \mathcal{S}_X = \{X_\alpha \times X_\alpha, p_\alpha^\beta \times p_\alpha^\beta, A\}.$$

All projections of the spectrum $\mathcal{S}_X \times \mathcal{S}_X$ are 0-soft, and hence open (Proposition 6.1.26). The Suslin number of the product $X \times X$ is obviously countable. Consequently, by Proposition 1.3.3, the spectrum $\mathcal{S}_X \times \mathcal{S}_X$ is factorizing. Next we apply Theorem 1.3.4 to the spectra $\mathcal{S}_X \times \mathcal{S}_X$, $\mathcal{S}_X$ and to the map $\mu$ between their limit spaces. Then we get a $\lambda$-closed and cofinal subset $B$ of $A$ such that for each $\alpha \in B$ we have a continuous map $\mu_\alpha\colon X_\alpha \times X_\alpha \to X_\alpha$ such that

$$p_\alpha \mu = \mu_\alpha(p_\alpha \times p_\alpha).$$

Now we can define a continuous multiplication operation in $X_\alpha$ by letting

$$x_\alpha \cdot y_\alpha = \mu_\alpha(x_\alpha, y_\alpha) \text{ whenever } (x_\alpha, y_\alpha) \in X_\alpha \times X_\alpha.$$

It is easy to see that $X_\alpha$, $\alpha \in B$, becomes a topological group with respect to this operation. Moreover, for all $\alpha, \beta \in B$ with $\alpha \leq \beta$, the projection $p_\alpha^\beta\colon X_\beta \to X_\alpha$ becomes a homomorphism with respect to the above defined operations. □

COROLLARY 8.2.2. *If (in the notations of Lemma 8.2.1) $X$ is $n$-dimensional, then all elements of the spectrum $\mathcal{S}_X = \{X_\alpha, p_\alpha^\beta, A\}$ can also be assumed to be $n$-dimensional.*

PROOF. Apply Lemma 8.2.1 and Theorem 1.3.10. □

We also need the following two simple statements.

LEMMA 8.2.3. *Let $p\colon X \to Y$ be a 0-soft homomorphism between zero-dimensional $AE(0)$ topological groups. Then there exists a homeomorphism $h\colon Y \times \ker p \to X$ such that $ph = \pi_1$, where $\pi_1\colon Y \times \ker p \to Y$ denotes the natural projection onto the first factor.*

PROOF. Since $p$ is 0-soft and $\dim Y = 0$, there exists a continuous map $i\colon Y \to X$ such that $pi = id_Y$ and $i(e_Y) = e_X$ ($e_X$ and $e_Y$ denote neutral elements of the groups $X$ and $Y$ respectively). Define a homeomorphism $h\colon Y \times \ker p \to X$ by letting

$$h(y,a) = i(y) \cdot a, \quad \text{whenever} \quad (y,a) \in Y \times \ker p$$

(the dot $\cdot$ denotes the multiplication operation of the group $X$). Clearly $ph = \pi_1$. □

LEMMA 8.2.4. *Any zero-dimensional space of $R$-weight $\tau$ admits a $C$-embedding into $N^\tau$.*

PROOF. [1] Since any zero-dimensional completely metrizable space can be considered as a closed subspace of the space $N^\omega$ (Proposition 1.1.6), the Lemma is true for $\tau = \omega$. Consider now a zero-dimensional space $X$ of $R$-weight $\tau > \omega$. Obviously the Hewitt realcompactification $\nu X$ of $X$ is zero-dimensional and has the same $R$-weight as $X$. By Theorem 1.3.10, $\nu X$ can be represented as the limit space of some factorizing $\omega$-spectrum $\mathcal{S} = \{Y_\alpha, q_\alpha^\beta, A\}$ consisting of zero-dimensional Polish spaces. Since $\mathcal{S}$ is factorizing, it is easy to see that $\nu X$ is $C$-embedded (and closed) in the product $\prod\{Y_\alpha \colon \alpha \in A\}$. This product, in turn, can be $C$-embedded into $N^\tau$ (since each $Y_\alpha$ can be embedded into $N^\omega$ as a closed subspace and $|A| = \tau$). Consequently, $X$ can be $C$-embedded into $N^\tau$. □

In the following Theorem, $k(X)$ denotes the minimal cardinal $\tau$ such that $X$ contains a compact subspace $K$ whose pseudo-character in $X$, $\psi(K, X)$, is $\tau$. Note that if, for a topological group $X$, we have $k(X) < \omega$, then $X$ is locally

[1]One might formally apply Lemma 6.5.3.

compact.

THEOREM 8.2.5. *Let $X$ be a non-locally compact zero-dimensional $AE(0)$ topological group. Then $X$ is homeomorphic to the product $N^{k(X)} \times D^{w(X)}$.*

PROOF. We consider two cases separately.

*Case 1:* $k(X) = w(X)$. We use transfinite induction. The starting point of the induction (i.e. the case $k(X) = w(X) = \omega$) is trivial, because topologically $X$ is a zero-dimensional Polish space without open compact subspaces. Consequently, by Theorem 1.1.5, $X$ is homeomorphic to the space of irrationals $N^{\omega} \approx N^{\omega} \times D^{\omega}$. Thus we can suppose that in Case 1 our statement is true for all spaces of weight $\lambda$, where $\omega \leq \lambda < \tau$, and consider a zero-dimensional $AE(0)$ topological group $X$ with $k(X) = w(X) = \tau$. To prove that $X$ is homeomorphic to the product $N^{\tau} \times D^{\tau} \approx N^{\tau}$, we can use a topological characterization of the latter space (Theorem 8.1.4). Translating condition (iii) of Theorem 8.1.4 into spectral language, we see that it suffices to show the following:

Consider a zero-dimensional space $Y$ of $R$-weight $\lambda < \tau$ (without loss of generality, we may assume that $\lambda \geq \omega$), fix an arbitrary map $f\colon Y \to X$, fix a factorizing $\lambda$-spectrum $\mathcal{S}_X = \{X_\alpha, p_\alpha^\beta, A\}$, satisfying the conditions of Lemma 8.2.1, and any index $\alpha \in A$. Our goal is to prove the existence of a $C$-embedding $g\colon Y \to X$ such that $p_\alpha f = p_\alpha g$.

First, let us try to find an index $\beta > \alpha$ such that the projection $p_\alpha^\beta\colon X_\beta \to X_\alpha$ of the spectrum $\mathcal{S}_X$ is (topologically) a trivial bundle with fiber $N^\lambda$. Let $\alpha_0 = \alpha$. Since the weight of the space $X_{\alpha_0}$ is strictly less than the weight of the space $X$, we can conclude that the fiber $\ker p_{\alpha_0}$ (i.e. the inverse image $p_{\alpha_0}^{-1}(e_{\alpha_0})$, where $e_{\alpha_0}$ denotes the neutral element of the group $X_{\alpha_0}$) is non-compact. Indeed, otherwise we would have

$$k(X) \leq \psi(\ker p_{\alpha_0}, X) \leq \psi(e_{\alpha_0}, X_{\alpha_0}) \leq \lambda < \tau = w(X)$$

which contradicts our assumption $k(X) = w(X)$. Thus we can choose an index $\alpha_1 \in A$, with $\alpha_1 > \alpha_0$, such that $\ker p_{\alpha_0}^{\alpha_1}$ is non-compact. Suppose now that for every ordinal $\gamma$ with $\gamma < \delta$, where $\delta < \lambda$, we have already chosen indexes $\alpha_\gamma \in A$ in such a way that the following conditions are satisfied:

(a) If $\mu < \nu < \delta$, then $\alpha_\mu < \alpha_\nu$.

(b) If $\nu$ is a limit ordinal, then $\alpha_\nu = \sup\{\alpha_\mu \colon \mu < \nu\}$.

(c) The fiber $\ker p_{\alpha_\mu}^{\alpha_\nu}$ is non-compact whenever $\mu < \nu < \delta$.

We now construct an index $\alpha_\delta$ with the desired properties. If $\delta$ is a limit ordinal, then we put

$$\alpha_\delta = \sup\{\alpha_\gamma \colon \gamma < \delta\}.$$

By the inequality $\delta < \lambda$ and $\lambda$-completeness of the indexing set $A$ of the spectrum $\mathcal{S}_X$ we have $\alpha_\delta \in A$ (Corollary 1.1.28). If $\delta = \gamma + 1$ the index $\alpha_\delta$ can be constructed in precisely the same way as the index $\alpha_1$. Now consider

$$\beta = \sup\{\alpha_\gamma \colon \gamma < \lambda\}$$

and note that, as above, $\beta \in A$. Consider the well-ordered spectrum

$$\mathcal{S}' = \{X_\gamma, p_{\alpha_\gamma}^{\alpha_{\gamma+1}}, \lambda\}.$$

The $\lambda$-continuity of the spectrum $\mathcal{S}_X$ guarantees that the limit space of the spectrum $\mathcal{S}'$ is naturally isomorphic to the topological group $X_\beta$ and the limit projection

$$\lim \mathcal{S}' \to X_{\alpha_0} = X_\alpha$$

is isomorphic to the projection $p_\alpha^\beta \colon X_\beta \to X_\alpha$ of the spectrum $\mathcal{S}_X$. By lemma 8.2.3, $p_\alpha^\beta$ is topologically a trivial bundle with fiber $\ker p_\alpha^\beta$, i.e. there exists a homeomorphism

$$h_\beta \colon X_\alpha \times \ker p_\alpha^\beta \to X_\beta$$

such that $p_\alpha^\beta h_\beta = \pi_1$, where $\pi_1 \colon X_\alpha \times \ker p_\alpha^\beta \to X_\alpha$ denotes the natural projection onto the first factor. It follows from the properties of the spectrum $\mathcal{S}_X$ that $\ker p_\alpha^\beta$ is a zero-dimensional $AE(0)$ topological group. By the construction of the index $\beta$, we conclude immediately that $k(\ker p_\alpha^\beta) = w(\ker p_\alpha^\beta) = \lambda$. Consequently, by the inductive hypothesis, $\ker p_\alpha^\beta$ is homeomorphic ( as a topological space) to $N^\lambda$. Thus, by Lemma 8.2.4, there exists a $C$-embedding $j \colon Y \to \ker p_\alpha^\beta$. Let $g_\beta = h_\beta(p_\alpha f \triangle j)$. Obviously, $g_\beta \colon Y \to X_\beta$ is a $C$-embedding such that $p_\alpha^\beta g_\beta = p_\alpha f$. Choose an arbitrary section $i \colon X_\beta \to X$ of the 0-soft limit projection $p_\beta \colon X \to X_\beta$ of the spectrum $\mathcal{S}_X$ (this means that $p_\beta i = id_{X_\beta}$). Then the composition

$$g = i g_\beta \colon Y \to X$$

is the desired $C$-embedding of $Y$ into $X$. This completes the verification of the condition mentioned above, and hence $X$ is homeomorphic to $N^\tau$. The proof of Case 1 is complete.

*Case 2:* $k(X) < w(X)$. Since, by our assumption, $X$ is non-locally compact we may suppose that $\omega \leq k(X) < w(X) = \tau$. Take a factorizing $\lambda$-spectrum (where $\lambda = k(X)$) $\mathcal{S}_X = \{X_\alpha, p_\alpha^\beta, A\}$ satisfying the conditions of Lemma 8.2.1. Choose a compact subspace $K \subseteq X$ such that $\psi(K, X) = \lambda$. Without loss of generality, we may assume that $e \in K$, where $e$ denotes the neutral element of the group $X$. Since $\mathcal{S}_X$ is a factorizing $\lambda$-spectrum and $\psi(K, X) = \lambda$, there exist an index $\alpha \in A$ and a compact subspace $K_\alpha \subseteq X_\alpha$ such that $p_\alpha^{-1}(K_\alpha) = K$. Consequently, $\ker p_\alpha$ is homeomorphic to the Cantor cube $D^\tau$ (as a compact topological group of weight $\tau$). By Lemma 8.2.3, the limit projection $p_\alpha \colon X \to X_\alpha$ is topologically the natural projection $\pi_1 \colon X_\alpha \times D^\tau \to X_\alpha$.

Consequently, $k(X_\alpha) = w(X_\alpha) = \lambda$ (otherwise we would have $k(X) < \lambda$, a contradiction). But $X_\alpha$ is a zero-dimensional $AE(0)$ topological group. As was established in Case 1, $X_\alpha$ is homeomorphic to $N^\lambda$. Therefore $X$ is homeomorphic to the product $N^\lambda \times D^\tau$. $\square$

*Historical and bibliographical notes* 8.2. The results of this section are taken from [**29**]. Lemma 8.2.1 was first exploited in [**93**] (see Section 8.3).

## 8.3. Locally convex linear topological spaces

One of the fundamental results of infinite-dimensional topology and functional analysis – the well known theorem of Anderson-Kadets (see Proposition 2.4.30) – states that each separable infinite-dimensional Fréchet space is homeomorphic to the countable (infinite) power $R^\omega$ of the real line $R$. It is also well known [**32**] that each locally compact locally convex linear topological space is homeomorphic (even isomorphic) to some finite power $R^n$ of $R$ (here we consider only Hausdorff linear topological spaces over the field of real numbers). The following reformulation of these two facts immediately follows from the corresponding definitions.

THEOREM 8.3.1. *The following conditions are equivalent for each locally convex linear topological space $E$:*

(i) *$E$ is homeomorphic to the countable (finite or infinite) power of the real line.*
(ii) *$E$ is a Polish space.*
(iii) *$E$ is an $AE(0)$-space of weight $\omega$.*

Our main goal in this Section is to extend the above result to spaces of arbitrary weight.

We start with the following statement.

PROPOSITION 8.3.2. *Let $\lambda \geq \omega$. Each locally convex $AE(0)$-space is isomorphic to the limit space of some factorizing $\lambda$-spectrum consisting of locally convex $AE(0)$-spaces (of weight $\lambda$) and 0-soft linear limit projections. In particular, each locally convex $AE(0)$-space is isomorphic to the limit space of some factorizing $\omega$-spectrum consisting of separable Fréchet space spaces and 0-soft linear limit projections.*

PROOF. Let $E$ be a locally convex $AE(0)$-space. Obviously $E$ can be represented as the limit space of some factorizing $\lambda$-spectrum $\mathcal{S} = \{E_\alpha, p_\alpha^\beta, A\}$ consisting of $AE(0)$-spaces (of weight $\lambda$) and 0-soft limit projections. We are going to show that this spectrum contains a $\lambda$-closed and cofinal subspectrum consisting of locally convex spaces and linear limit projections. Obviously this will be sufficient for us.

Let $\mu\colon E \times E \to E$ denote the continuous addition operation given in $E$ as a linear topological space. Applying the argument of Lemma 8.2.1, we get a $\lambda$-closed and cofinal subset $A_1$ of $A$ and a morphism

$$\{\mu_\alpha\colon E_\alpha \times E_\alpha \to E_\alpha, A_1\}\colon (\mathcal{S} \times \mathcal{S})/A_1 \to \mathcal{S}/A_1$$

whose limit map coincides with $\mu$. We then define a continuous addition operation $+_\alpha$ in each $E_\alpha$, $\alpha \in A_1$, simply by letting

$$x_\alpha +_\alpha y_\alpha = \mu_\alpha(x_\alpha, y_\alpha) \text{ whenever } (x_\alpha, y_\alpha) \in E_\alpha \times E_\alpha.$$

Similarly, consider a factorizing $\lambda$-spectrum

$$id_R \times \mathcal{S} = \{R \times E_\alpha, id_R \times p^\beta_\alpha, A\}$$

and the continuous operation $\nu\colon R \times E \to E$ of multiplication by scalars. Applying Theorem 1.3.4 to the spectra $id_R \times \mathcal{S}$ and $\mathcal{S}$ (and to the map $\nu$ between their limit spaces), we get a $\lambda$-closed and cofinal subset $A_2$ of $A$ and a morphism

$$\{\nu_\alpha\colon R \times E_\alpha \to E_\alpha, A_2\}\colon (id_R \times \mathcal{S})/A_2 \to \mathcal{S}/A_2$$

with $\nu = \lim\{\nu_\alpha\colon \alpha \in A_2\}$. As above, we now define the continuous operation of multiplication by scalars in $E_\alpha$, $\alpha \in A_2$, by letting

$$t \cdot_\alpha x_\alpha = \nu_\alpha(t, x_\alpha) \text{ whenever } (t, x_\alpha) \in R \times E_\alpha.$$

The set $A' = A_1 \cap A_2$ is $\lambda$-closed and cofinal in $A$ (Proposition 1.1.27). It only remains to note that for the spaces $E_\alpha$, $\alpha \in A'$ we have defined both operations from the definition of linear topological spaces. Both of these are continuous, and consequently all spaces of the spectrum $\mathcal{S}/A'$ are linear topological spaces. Moreover, all limit projections in the latter spectrum are linear with respect to the above defined operations. □

COROLLARY 8.3.3. *Each locally convex $AE(0)$-space is isomorphic to a $C$-embedded (and closed) linear subspace of the product of separable Fréchet spaces.*

PROOF. Let $E$ be a locally convex $AE(0)$-space. By Proposition 8.3.2, $E$ is isomorphic to the limit space of some factorizing $\omega$-spectrum $\mathcal{S} = \{E_\alpha, p^\beta_\alpha, A\}$ consisting of separable Fréchet spaces and 0-soft linear limit projections. Consider the product $F = \prod\{E_\alpha\colon \alpha \in A\}$. Obviously, $E$ is isomorphic to a linear (and closed) subspace of $F$ (this subspace is the image of the space $E$ under the diagonal product of all limit projections of the spectrum $\mathcal{S}$). Since $\mathcal{S}$ is factorizing, the mentioned subspace is $C$-embedded in $F$. □

LEMMA 8.3.4. *Let $f\colon E \to F$ be a 0-soft linear map between locally convex $AE(0)$-spaces that has Polish kernel (in the sense of Section 6.3). Suppose that $f(K) = T$, where $K$ and $T$ are convex and $C$-embedded $AE(0)$subspaces of $E$ and $F$ respectively. Then the following conditions are equivalent:*

(i) *The restriction $f/K\colon K \to T$ is a 0-soft map.*
(ii) *The restriction $f/K\colon K \to T$ is a soft map.*

PROOF. The non trivial part is the implication (i) $\Longrightarrow$ (ii). By Proposition 8.3.2, there exist factorizing $\omega$-spectra

$$\mathcal{S}_E = \{E_\alpha, p^\beta_\alpha, A\} \text{ and } \mathcal{S}_F = \{F_\alpha, q^\beta_\alpha, A\}$$

consisting of separable Fréchet spaces and 0-soft linear projections, the limit spaces of which are isomorphic to the spaces $E$ and $F$ respectively. By Theorem 6.3.2 we can suppose, without loss of generality, that $f$ is the limit map of some morphism

$$\{f_\alpha \colon E_\alpha \to F_\alpha, A\} \colon \mathcal{S}_E \to \mathcal{S}_F$$

consisting of 0-soft maps and such that all the corresponding limit square diagrams are Cartesian squares.

Since $K$ is a $C$-embedded and closed subspace of $E$ we can conclude that the spectrum

$$\mathcal{S}_K = \{K_\alpha = \mathrm{cl}_{E_\alpha}(p_\alpha(K)), p_\alpha^\beta/K_\beta, A\},$$

the limit space of which coincides with $K$, is also a factorizing $\omega$-spectrum. Since $K$ is an $AE(0)$-space, we can apply Proposition 6.3.5 and Theorem 1.3.6 and suppose, without loss of generality, that for each $\alpha \in A$ the map $p_\alpha/K$ is 0-soft and $p_\alpha(K) = K_\alpha$. Similarly, we may assume that $q_\alpha(T) = \mathrm{cl}_{F_\alpha}(q_\alpha(T))$ and the map $q_\alpha/T$ is 0-soft whenever $\alpha \in A$. Finally, by Theorem 6.3.2, we can suppose (again without loss of generality) that each commutative diagram of the type

$$\begin{array}{ccc} K & \xrightarrow{\quad f/K \quad} & T \\ {\scriptstyle p_\alpha/K}\big\downarrow & & \big\downarrow{\scriptstyle q_\alpha/T} \\ p_\alpha(K) & \xrightarrow{\quad f_\alpha/p_\alpha(K) \quad} & q_\alpha(T) \end{array}$$

is a Cartesian square and each map $f_\alpha/p_\alpha(K)$ is 0-soft. In this situation, to complete the proof of the lemma it only remains to be proved that the last map is soft (then we can use the fact that the above diagram is a Cartesian square to obtain, by Lemma 6.2.4, the softness of the map $f/K$).

Linearity of the limit projections of spectra $\mathcal{S}_E$ and $\mathcal{S}_F$ implies that the spaces $p_\alpha(K)$ and $q_\alpha(T)$ are convex subspaces of the spaces $E_\alpha$ and $F_\alpha$, respectively. Moreover, it is easy to see that each map $f_\alpha$ is also linear. By Theorem 2.1.16, the map $f_\alpha/p_\alpha(K)$, as an open map between closed and convex subspaces of separable Fréchet spaces with closed convex fibers, is soft. The proof is complete. □

PROPOSITION 8.3.5. *Let $K$ be a $C$-embedded and convex subspace of a locally convex $AE(0)$-space. Then the following conditions are equivalent:*

(i) *$K$ is an $AE(0)$-space.*

(ii) *$K$ is an absolute retract.*

PROOF. It is sufficient to prove the implication (i) $\Longrightarrow$ (ii). By Corollary 8.3.3, we may suppose that $K$ is a convex and $C$-embedded subspace of some product $E = \prod\{E_a : a \in A\}$ of separable Fréchet spaces. If $|A| \leq \omega$, then $K$ is a convex and closed subspace of a separable Fréchet space and hence, by Theorem 2.1.17, is an absolute retract.

Now consider the case $|A| = \tau > \omega$. Without loss of generality we may suppose that each space $E_a$, $a \in A$, contains at least two points, and consequently the space $E$ is homeomorphic to $R^\tau$ (apply Proposition 2.4.23). Consider a functionally closed, proper and 0-invertible map $f : N^A \to E$ having a regular averaging operator (see Proposition 6.1.25). Since $\dim N^A = 0$ and $K$ is a $C$-embedded $AE(0)$-subspace of $E$, there exists a map $h : N^A \to K$ such that $h/f^{-1}(K) = f/f^{-1}(K)$.

We now introduce some notation. If $C \subseteq B \subseteq A$, then

$$\pi_C^B : \prod\{E_a : a \in B\} \to \prod\{E_a : a \in C\}$$

and

$$\pi_B : E \to \prod\{E_a : a \in B\}$$

denote the natural projections onto the corresponding subproducts. Also

$$K_B = \pi_B(K),\ K(B) = \pi_B^{-1}(K_B),\ p_B = \pi_B/K \text{ and } p_C^B = \pi_C^B/K_B.$$

We say that a subset $B$ of $A$ is *admissible* if

$$\pi_B(x) = p_B h f^{-1}(x) \text{ for each point } x \in K(B).$$

As in the proof of Theorem 6.3.1, we can verify the following properties of admissible subsets:

(a) The union of an arbitrary collection of admissible subsets is again an admissible subset.
(b) For any admissible subset $B$, the set $K_B$ is closed and $C$-embedded in the product $\prod\{E_a : a \in B\}$.
(c) For any admissible subset $B$ the space $K_B$ is an $AE(0)$-space.
(d) For any two admissible subsets $C$ and $B$ with $C \subseteq B$, the maps $p_B$ and $p_C^B$ are 0-soft.
(e) For any admissible subset $B$, the set $K_B$ is convex in the space $\prod\{E_a : a \in B\}$.
(f) The family of all countable admissible subsets of $A$ is $\omega$-closed and cofinal in the $\omega$-complete set $exp_\omega A$.

Since $|A| = \tau$, we may write $A = \{a_\alpha : \alpha < \tau\}$. By (f), each point $a_\alpha$ of $A$ is contained in some countable admissible subset $B_\alpha$. Put

$$A_\alpha = \cup\{B_\beta : \beta \leq \alpha\},\ K_\alpha = K_{A_\alpha} \text{ and } p_\alpha^{\alpha+1} = p_{A_\alpha}^{A_{\alpha+1}}\ (\alpha < \tau).$$

It follows, from the properties of admissible subsets listed above, that the limit space of the naturally induced well ordered continuous spectrum

$$\mathcal{S} = \{K_\alpha, p_\alpha^{\alpha+1}, \tau\}$$

coincides with the given space $K$. The countability of the sets $B_\alpha$ guarantees, by Lemma 8.3.4, the softness of the short projections of this spectrum. Obviously, the space $B_0$ is an absolute retract (Theorem 2.1.17). In this situation, Proposition 6.1.21 and Lemma 6.2.6 conclude the proof. □

We are now ready to prove the following statement, which extends Proposition 2.4.30 to the non-metrizable case.

THEOREM 8.3.6. *Let $\tau$ be an uncountable cardinal. Then the following conditions are equivalent for each locally convex linear topological space $E$ of weight $\tau$:*

(i) *$E$ is homeomorphic to $R^\tau$.*
(ii) *$E$ is an absolute retract.*
(iii) *$E$ is an $AE(0)$-space.*

PROOF. The implications (i) $\Longrightarrow$ (ii) and (ii) $\Longrightarrow$ (iii) are obvious. Let us prove the implication (iii) $\Longrightarrow$ (i). We use transfinite induction. If $\tau = \omega$, then $E$ is an infinite-dimensional separable Fréchet space and the conclusion follows from Proposition 2.4.30. Therefore we may suppose that our statement is true for all infinite-dimensional locally convex $AE(0)$-spaces of weight $\lambda$, $\omega \leq \lambda < \tau$, and consider a locally convex $AE(0)$-space $E$ of weight $\tau$. By Proposition 8.3.5, $E$ is an absolute retract. Consequently, by Theorem 7.3.3 it suffices to show that for any space $Z$ of $R$-weight $< \tau$, the set of $C$-embeddings is dense in the space $C_\tau(Z, E)$.

Consider an arbitrary space $Z$ of $R$-weight $\lambda$ (where $\omega \leq \lambda < \tau$), a map $g\colon Z \to E$, a factorizing $\lambda$-spectrum $\mathcal{S}_E = \{E_\alpha, p_\alpha^\beta, A\}$ the limit space of which coincides with $E$, and an index $\alpha \in A$. We wish to construct a $C$-embedding $h\colon Z \to E$ such that $p_\alpha h = p_\alpha g$. By Proposition 8.3.2, we may suppose without loss of generality that all spaces $E_\alpha$ are locally convex $AE(0)$-spaces of weight $\lambda$ and all limit projections $p_\alpha$ are linear 0-soft maps. Moreover, by Proposition 8.3.5, we see that all spaces $E_\alpha$ are absolute retracts. Finally, by Proposition 6.3.5, we may assume without loss of generality that all limit projections are (linear and) soft. Let us find an index $\beta \in A$ such that $\beta \geq \alpha$ and the projection $p_\alpha^\beta\colon E_\beta \to E_\alpha$ is a trivial bundle with fiber $R^\lambda$. Put $\alpha_0 = \alpha$. Since the weight of $E_{\alpha_0}$ is strictly less than the weight of $E$ and the projection $p_{\alpha_0}\colon E \to E_{\alpha_0}$ is linear and open (Proposition 6.1.26), we conclude that the fiber $\ker p_{\alpha_0}$ contains at least two points. Then we can choose an index $\alpha_1 > \alpha_0$ such that $\ker p_{\alpha_0}^{\alpha_1}$ also contains at least two points. Suppose now that for every ordinal $\gamma$ with $\gamma < \delta$, where $\delta < \lambda$, we have already chosen indexes $\alpha_\gamma \in A$ in such a way that the following conditions are satisfied:

(a) If $\mu < \nu < \delta$, then $\alpha_\mu < \alpha_\nu$.

(b) If $\nu$ is a limit ordinal, then $\alpha_\nu = \sup\{\alpha_\mu : \mu < \nu\}$.

(c) The fiber $\ker p^{\alpha_\nu}_{\alpha_\mu}$ contains at least two points ($\mu < \nu < \delta$).

Now we construct an index $\alpha_\delta$ with the desired properties. If $\delta$ is a limit ordinal, then we put $\alpha_\delta = \sup\{\alpha_\gamma : \gamma < \delta\}$. By the inequality $\delta < \lambda$ and the $\lambda$-completeness of the spectrum $\mathcal{S}$, we have that $\alpha_\delta \in A$. If $\delta = \gamma + 1$, then the index $\alpha_\delta$ can be constructed in precisely the same way as the index $\alpha - 1$. Now consider $\beta = \sup\{\alpha_\gamma : \gamma < \lambda\}$ and note that $\beta \in A$. Consider the well ordered spectrum

$$\mathcal{S}' = \{E_\gamma, p^{\alpha_{\gamma+1}}_{\alpha_\gamma}, \lambda\}.$$

The $\lambda$-continuity of the spectrum $\mathcal{S}$ guarantees that the limit space of the spectrum $\mathcal{S}'$ is naturally isomorphic to the space $E_\beta$, and the limit projection $\lim \mathcal{S}' \to E_{\alpha_0} = E_\alpha$ is isomorphic to the short projection $p^\beta_\alpha : E_\beta \to E_\alpha$ of the spectrum $\mathcal{S}$. As in Lemma 8.2.3, we can easily see that the map $p^\beta_\alpha$ is topologically a trivial bundle with fiber $\ker p^\beta_\alpha$. It is also easy to verify that the fiber $\ker p^\beta_\alpha$ is homeomorphic to the product

$$\prod\{\ker p^{\alpha_{\gamma+1}}_{\alpha_\gamma} : \gamma < \lambda\}.$$

Consequently, by conditions (a) – (c), the weight of $\ker p^\beta_\alpha$ is equal to $\lambda$. By the same conditions we can conclude that even if $\lambda = \omega$, the space is infinite-dimensional (otherwise we get a contradiction to condition (c)). Consequently, by our hypothesis, the space $\ker p^\beta_\alpha$ is homeomorphic to $R^\lambda$. Moreover, in this case the projection $p^\beta_\alpha$ can be topologically identified with the natural projection

$$\pi_1 : E_\alpha \times R^\lambda \to E_\alpha.$$

Since the $R$-weight of the space $X$ is equal to $\lambda$, there exists a $C$-embedding $i : Z \to R^\lambda$. Then the diagonal product

$$j = p_\alpha f \triangle i : Z \to E_\alpha \times R^\lambda = E_\beta$$

is a $C$-embedding satisfying the equality $p^\beta_\alpha j = p_\alpha f$. Take an arbitrary section $k : E_\beta \to E$ of the limit projection $p_\beta$ of the spectrum $\mathcal{S}$ (recall that $p_\beta$ is a soft map). Then the composition $h = kj$ is the desired $C$-embedding of $Z$ into $E$. □

COROLLARY 8.3.7. *Every closed convex body in a locally convex $AE(0)$-space of weight $\tau > \omega$ is homeomorphic to $R^\tau$.*

PROOF. Let $X$ be a closed and convex body in a locally convex $AE(0)$-space $E$ of weight $\tau > \omega$. Since $\operatorname{int} X =\neq \emptyset$ we may conclude that $X = \operatorname{cl}(\operatorname{int} X)$. In particular, $w(X) = \tau$. By Theorem 8.3.6, $E$ is homeomorphic to $R^\tau$ and consequently, by Proposition 6.1.8, $X$ is functionally closed in $E$. By Corollary 6.4.8, $X$ is $C$-embedded in $E$. By Propositions 6.4.9 and 8.3.5, $X$ is an absolute retract. Corollary 7.1.17 finishes the proof. □

COROLLARY 8.3.8. *Every linear $G_\delta$-subspace of a locally convex $AE(0)$-space of weight $\tau > \omega$ is homeomorphic to $R^\tau$.*

PROOF. Let $X$ be a linear $G_\delta$-subspace of a locally convex $AE(0)$-space $E$ of weight $\tau > \omega$. Put $Y = \mathrm{cl}\, X$. As in the proof of Corollary 8.3.7, we can conclude that $Y$ is homeomorphic to $R^\tau$. Consequently it is sufficient to show that $X = Y$. Indeed, if there exists a point $y \in Y - X$, then the translate $X + y$ would be a dense $G_\delta$-subspace of $Y$ and $(X + y) \cap Y = \emptyset$, which contradicts the Baire property in $Y \approx R^\tau$. □

COROLLARY 8.3.9. *The closed convex hall of a compact subset of a locally convex $AE(0)$-space is compact.*

PROOF. We only have to observe that each locally convex $AE(0)$-space, by Proposition 8.3.2, is complete. The rest is well known (see [**258**]). □

The concept of *projector* is important in the theory of linear topological spaces. Recall that a (continuous) linear map $f\colon E \to E$ is a projector if $f^2 = f$. Obviously, the image $f(E)$ is a linear continuous retract of $E$. Applying the argument used in the proof of Theorem 8.3.6 we obtain the following statement.

PROPOSITION 8.3.10. *The image of a continuous projector, defined on a product of separable Fréchet spaces, is isomorphic to a product of separable Fréchet spaces.*

*Historical and bibliographical notes* 8.3. All results of this Section are due to the author [**93**]. Corollaries 8.3.7 and 8.3.8 extend to the non-metrizable case the corresponding results of Klee-Corson and Mazur-Strenbach (see [**32**]).

## 8.4. Shape properties of non-metrizable compacta

**8.4.1. Spectral theorem in Shape category.** We start with the following definition.

DEFINITION 8.4.1. *An inverse spectrum $\mathcal{S}_X = \{X_\alpha, p_\alpha^\beta, A\}$ is said to be* homotopically stable with respect to the space $Y$ *if for each index $\alpha \in A$, each closed subspace $F$ of $X_\alpha$, and maps $f, g\colon F \to Y$, the relation $fp_\alpha/p_\alpha^{-1}(F) \simeq gp_\alpha/p_\alpha^{-1}(F)$ implies the relation $f \simeq g$. We say that $\mathcal{S}_X$ is (absolutely)* homotopically stable *if $\mathcal{S}_X$ is homotopically stable with respect to any metrizable $ANR$-compactum.*

LEMMA 8.4.2. *There exists a countable collection $\{G_n\colon n \in \omega\}$ of metrizable locally compact $ANR$-spaces such that any inverse spectrum which is homotopically stable with respect to each space $G_n$, $n \in \omega$, is absolutely homotopically stable.*

PROOF. As the desired collection, we may take an arbitrary countable open basis of the topology of the Hilbert cube containing unions of its finite subcollections. The verification of this fact is trivial and is left to the reader. □

LEMMA 8.4.3. *Let $p\colon X \to Y$ be a surjection and $\mathcal{F} = \{F_\alpha\}$ be a closed basis of the topology of $Y$ containing intersections of its finite subfamilies. Let $Z$ be an $ANR$-space and suppose that for each $F_\alpha \in \mathcal{F}$ and any maps $f, g\colon F_\alpha \to Z$, the condition $fp/p^{-1}(F_\alpha) \simeq gp/p^{-1}(F_\alpha)$ implies the condition $f \simeq g$. Then for each closed subspace $F$ of $Y$ and any maps $f, g\colon F \to Z$, the condition $fp/p^{-1}(F) \simeq gp/p^{-1}(F)$ implies the condition $f \simeq g$.*

PROOF. Since $Z$ is an $ANR$-space there exists a neighborhood $V$ of $F$ in $Y$ and extensions $\bar{f}, \bar{g}\colon V \to Z$ of $f$ and $g$ respectively. By our assumption, the restrictions $\bar{f}p/p^{-1}(V)$ and $\bar{g}p/p^{-1}(V)$ to $p^{-1}(F)$ are homotopic. Since $Z$ is an $ANR$-space, we can find a neighborhood $U$ of $p^{-1}(F)$ in $X$ contained in $p^{-1}(V)$ such that the restrictions $\bar{f}p/U$ and $\bar{g}p/U$ are homotopic. Clearly, we may suppose without loss of generality that the set $U$ has the form $U = p^{-1}(W)$, where $W$ is a neighborhood of $F$ in $Y$. Then the compactness of $Y$ and the corresponding property of $\mathcal{F}$ imply that there exists an element $F_\alpha \in \mathcal{F}$ such that $F \subseteq F_\alpha \subseteq W$. By our assumption $\bar{f}/p^{-1}(F_\alpha) \simeq \bar{g}/p^{-1}(F_\alpha)$. Consequently, $f \simeq g$. □

THEOREM 8.4.4. *Each $\omega$-spectrum, consisting of metrizable compacta, contains a homotopically stable $\omega$-closed and cofinal subspectrum.*

PROOF. Let $\mathcal{S}_X = \{X_\alpha, p_\alpha^\beta, A\}$ be an arbitrary $\omega$-spectrum consisting of metrizable compacta. Fix an integer $n \in \omega$. We shall show that the indexing set $A$ of our spectrum contains an $\omega$-closed and cofinal subset $A_n$ such that the corresponding subspectrum

$$\mathcal{S}_X/A_n \equiv \mathcal{S}_n = \{X_\alpha, p_\alpha^\beta, A_n\}$$

is homotopically stable with respect to the space $G_n$ from Lemma 8.4.2. Let

$$\begin{aligned}\mathcal{L}_n \;=\; & \{(\alpha,\beta) \in A^2\colon \alpha \leq \beta, \text{ and for each closed subspace } F \text{ of } X_\alpha \text{ and any} \\ & \text{two maps } f, g\colon F \to G_n \text{ the condition} fp_\alpha/p_\alpha^{-1}(F) \simeq gp_\alpha/p_\alpha^{-1}(F) \\ & \text{implies the condition} fp_\alpha^\beta/(p_\alpha^\beta)^{-1}(F) \simeq gp_\alpha/(p_\alpha^\beta)^{-1}(F)\}\end{aligned}$$

Let us verify that the existence condition of Proposition 1 1.29 is satisfied with respect to the relation $\mathcal{L}_n$. Fix $\alpha \in A$. Our goal is to show that there exists an index $\beta \in A$ such that $(\alpha, \beta) \in \mathcal{L}_n$. First of all, we shall fix a closed subspace $F$ of the compactum $X_\alpha$ and construct an index $\alpha_F \geq \alpha$ such that the following condition is satisfied:

$(*)_F$ for maps $f, g\colon F \to G_n$ the condition $fp_\alpha/p_\alpha^{-1}(F) \simeq gp_\alpha/p_\alpha^{-1}(F)$ implies the condition $fp_\alpha^{\alpha_F}/(p_\alpha^{\alpha_F})^{-1}(F) \simeq gp_\alpha^{\alpha_F}/(p_\alpha^{\alpha_F})^{-1}(F)$.

Fix a countable dense subset $\{h_i\colon i \in N\}$ of the space $C(F, G_n)$ (the compact-open topology is considered). Let $M$ denote the set of those pairs $(i,j) \in N^2$ for which the conditions $h_i p_\alpha / p_\alpha^{-1}(F) \simeq h_j p_\alpha / p_\alpha^{-1}(F)$ are satisfied. For each $(i,j) \in M$ we can choose an index $\alpha(i,j) \geq \alpha$ such that (see, for example, [**214**]):

$$h_i p_\alpha^{\alpha(i,j)} / (p_\alpha^{\alpha(i,j)})^{-1}(F) \simeq h_j p_\alpha^{\alpha(i,j)} / (p_\alpha^{\alpha(i,j)})^{-1}(F).$$

It follows easily from the construction that for each $(i,j) \in M$ and each $\gamma \in A$ we may suppose that $\alpha(i,j) \geq \gamma$. Consequently, well-ordering $M$ in an arbitrary way, we may assume that $\{\alpha(i,j)\colon (i,j) \in M\}$ forms a chain in $A$. Let us verify that the index $\alpha_F = \max\{\alpha(i,j)\colon (i,j) \in M\}$ is the desired one. Consider two maps $f, g\colon F \to G_n$ such that $f p_\alpha / p_\alpha^{-1}(F) \simeq g p_\alpha / p_\alpha^{-1}(F)$. By our assumption, $G_n$ is an $ANR$-space and, by the construction, $\{h_i\colon i \in N\}$ is a dense subset of the space $C(F, G_n)$. Consequently, there exist integers $i, j \in N$ such that $f \simeq h_i$ and $g \simeq h_j$. Then

$$h_i p_\alpha / p_\alpha^{-1}(F) \simeq f p_\alpha / p_\alpha^{-1}(F) \simeq g p_\alpha / p_\alpha^{-1}(F) \simeq h_j p_\alpha / p_\alpha^{-1}(F).$$

In the other words, $(i,j) \in M$. Hence, by the definition of the index $\alpha(i,j)$, we have

$$h_i p_\alpha^{\alpha_F} / (p_\alpha^{\alpha_F})^{-1}(F) = h_i p_\alpha^{\alpha(i,j)} p_{\alpha(i,j)}^{\alpha_F} / (p_\alpha^{\alpha_F})^{-1}(F) \simeq h_j p_\alpha^{\alpha(i,j)} p_{\alpha(i,j)}^{\alpha_F} / (p_\alpha^{\alpha_F})^{-1}(F)$$

$$= h_j p_\alpha^{\alpha_F} / (p_\alpha^{\alpha_F})^{-1}(F).$$

Then

$$f p_\alpha^{\alpha_F} / (p_\alpha^{\alpha_F})^{-1}(F) \simeq h_i p_\alpha^{\alpha_F} / (p_\alpha^{\alpha_F})^{-1}(F) \simeq h_j p_\alpha^{\alpha_F} / (p_\alpha^{\alpha_F})^{-1}(F) \simeq$$

$$g p_\alpha^{\alpha_F} / (p_\alpha^{\alpha_F})^{-1}(F).$$

Thus, for each closed subspace $F$ of the compactum $X_\alpha$ we have an index $\alpha_F$ satisfying the above condition $(*)_F$. Consider now an arbitrary countable basis $\{F_k\colon k \in \omega\}$ of the topology of $X_\alpha$ containing the intersections of its finite subfamilies. For each $k \in \omega$, fix an index $\alpha_k \equiv \alpha_{F_k}$. Obviously, by Corollary 1.1.28, there exists an index $\beta \in A$ such that $\beta \geq \alpha_k$ for each $k \in \omega$. It follows from the above construction, and from Lemma 8.4.3, that $(\alpha, \beta) \in \mathcal{L}_n$. The verification of the existence condition is completed.

The verification of the majorantness condition from Proposition 1.1.29 is trivial.

Finally we must verify the $\omega$-closeness condition from Proposition 1.1.29. This means that if we have a countable chain $\{\alpha_m\colon m \in \omega\}$ in $A$ and condition $(\alpha_m, \beta) \in \mathcal{L}_n$ is satisfied for each $m \in \omega$ and some $\beta \in A$, then the condition $(\alpha, \beta) \in \mathcal{L}_n$, where $\alpha = \sup\{\alpha_m\colon m \in \omega\}$, is also satisfied.

Indeed, since our spectrum is $\omega$-continuous, the compactum $X_\alpha$ is naturally homeomorphic to the limit space of the inverse sequence $\{X_{\alpha_m}, p_{\alpha_m}^{\alpha_{m+1}}, \omega\}$. Consequently, the sets of the form $(p_{\alpha_m}^{\alpha})^{-1}(F_m)$, where $F_m$ is closed in $X_{\alpha_m}$ and

$m \in \omega$, form a closed basis of the topology of $X_\alpha$ which contains the intersections of its finite subfamilies. Consequently, if we wish to prove the inclusion $(\alpha, \beta) \in \mathcal{L}_n$ it suffices, by Lemma 8.4.3, to show that if $F_m$ is closed in $X_{\alpha_m}$, $m \in \omega$, and the maps $f, g\colon F = (p^\alpha_{\alpha_m})^{-1}(F_m) \to G_n$ satisfy the condition $fp_\alpha/p_\alpha^{-1}(F) \simeq gp_\alpha/p_\alpha^{-1}(F)$, then the condition $fp^\beta_\alpha/(p^\beta_\alpha)^{-1}(F) \simeq gp^\beta_\alpha/(p^\beta_\alpha)^{-1}(F)$ will also be satisfied. Since $G_n$ is an $ANR$-space, there exist an integer $k \geq m$ and maps $f_1, g_1\colon p^\alpha_{\alpha_k}(F) \to G_n$ such that $f \simeq f_1p^\alpha_{\alpha)_k}$ and $g \simeq g_1 p^\alpha_{\alpha_k}$. Then we have

$$f_1 p_{\alpha_k}/p_{\alpha_k}^{-1}(F_k) \simeq g_1 p_{\alpha_k}/p_{\alpha_k}^{-1}(F_k)$$

where $F_k = p^\alpha_{\alpha_k}(F)$. By the assumption, $(\alpha_k, \beta) \in \mathcal{L}_n$. Consequently, we have

$$f_1 p^\beta_{\alpha_k}/(p^\beta_{\alpha_k})^{-1}(F_k) \simeq g_1 p^\beta_{\alpha_k}/(p^\beta_{\alpha_k})^{-1}(F_k).$$

It only remains to note that

$$(p^\beta_\alpha)^{-1}(F) = (p^\beta_\alpha)^{-1}((p^\alpha_{\alpha_m})^{-1}(F_m)) = (p^\beta_{\alpha_k})^{-1}(F_k)$$

and that

$$fp^\beta_\alpha/(p^\beta_\alpha)^{-1}(F) \simeq f_1 p^\alpha_{\alpha_k} p^\beta_\alpha/(p^\beta_\alpha)^{-1}(F) = f_1 p^\beta_{\alpha_k}/(p^\beta_\alpha)^{-1}(F) \simeq g_1 p^\beta_{\alpha_k}/(p^\beta_\alpha)^{-1}(F)$$

$$\simeq g_1 p^\alpha_{\alpha_k} p^\beta_\alpha/(p^\beta_\alpha)^{-1}(F) \simeq gp^\beta_\alpha/(p^\beta_\alpha)^{-1}(F).$$

Thus the $\omega$-closeness condition is also satisfied.

Now, by Proposition 1.1.29, we can conclude that the set $A_n$ of all $\mathcal{L}_n$-reflexive indexes from $A$ is $\omega$-closed and cofinal in $A$. At the same time we note that, by the corresponding definitions, this means that the subspectrum $\mathcal{S}_n$ of $\mathcal{S}_X$ is homotopically stable with respect to the space $G_n$.

Finally, let

$$A' = \cap\{A_n\colon n \in \omega\}.$$

By Proposition 1.1.27, this set is $\omega$-closed and cofinal in $A$ as well. Consequently, the corresponding subspectrum $\mathcal{S}' = \{X_\alpha, p^\beta_\alpha, A'\}$ is homotopically stable with respect to each space $G_n$, $n \in \omega$. Lemma 8.4.2 finishes the proof. □

*Remark* 8.4.5. It is easy to see that each homotopically stable spectrum is homotopically stable with respect to every $\mathcal{CW}$-complex. Let $\mathcal{S}_X = \{X_\alpha, p^\beta_\alpha, A\}$ be any homotopically stable spectrum and $K$ be a $\mathcal{CW}$-complex. Suppose that for some index $\alpha \in A$ and for some maps $f, g\colon F \to K$, where $F$ is a closed subspace of $X_\alpha$, the condition $fp_\alpha/p_\alpha^{-1}(F) \simeq gp_\alpha/p_\alpha^{-1}(F)$ is satisfied. Then, by compactness of $\lim \mathcal{S}_X$ and by the properties of the topology of $\mathcal{CW}$-complexes, there exists a metrizable $ANR$-compactum $K'$ lying in $K$, such that the maps $fp_\alpha/p_\alpha^{-1}(F)$ and $gp_\alpha/p_\alpha^{-1}(F)$ are homotopic in $K'$. By our assumption, we can conclude that $f$ and $g$ are homotopic in $K'$ and, consequently, in $K$ as well.

Let us now recall the definition of the *SHAPE category* over compacta. The class of objects of the *SHAPE* category is the class of all compacta. The morphisms, or more precisely, the elements of the set $Mor_{SHAPE}(X,Y)$ are defined to be the maps

$$\alpha \colon \bigcup\{[Y,P] \colon P \text{ is an } ANR\} \to \bigcup\{[X,P] \colon P \text{ is an } ANR\}$$

satisfying the following two conditions:

- If $P$ is an $ANR$-compactum, then $\alpha([Y,P]) \subseteq [X,P]$.
- If $P$ and $P'$ are $ANR$-compacta, $\psi \in [Y,P]$, $\varphi \in [Y,P']$ and $\gamma \in [P,P']$ are such that $\gamma\psi = \varphi$, then $\gamma\alpha(\psi) = \alpha(\varphi)$.

The natural (*fundamental*) functor

$$\mathcal{F} \colon HOMOT \to SHAPE$$

acts in the following fashion:

- $\mathcal{F}(X) = X$ for each compactum $X$.
- If $\varphi \in [X,Y]$, then $\mathcal{F}(\varphi) \colon X \to Y$ is given by $\mathcal{F}(\varphi)(\psi) = \psi\varphi$ for any $\psi \in [Y,P]$ where $P$ is an $ANR$-compactum.

The composition of the functors $\mathcal{H}$ and $\mathcal{F}$ is denoted by

$$Sh \colon COMP \to SHAPE$$

and called the *shape functor*. It is known [**175**], [**214**] that (unlike the homotopy functor) the shape functor is continuous.

LEMMA 8.4.6. *Let $\mathcal{S}_X = \{X_\alpha, p_\alpha^\beta, A\}$ be an $\omega$-spectrum consisting of metrizable compacta and $\varphi \colon \lim \mathcal{S}_X \to Y$ be a shape morphism, where $Y$ is a metrizable compactum. Then there exist an index $\alpha \in A$ and a shape morphism $\varphi_\alpha \colon X_\alpha \to Y$ such that $\varphi = \varphi_\alpha \operatorname{Sh}(p_\alpha)$.*

PROOF. Without loss of generality, by Theorem 8.4.4, we may assume that the spectrum $\mathcal{S}_X$ is homotopically stable. Represent $Y$ as the limit space of some inverse sequence $\mathcal{S}_Y = \{Y_n, q_n^{n+1}, \omega\}$ consisting of compact polyhedra.

The shape morphism $\varphi \colon X \to Y$, where $X = \lim \mathcal{S}_X$, naturally induces a homotopy class $\varphi([q_1]) \colon X \to Y_1$. Denote by $g_1 \colon X \to Y_1$ any representative of this class. Fix an index $\alpha_1 \in A$ and a map $g_{\alpha_1} \colon X_{\alpha_1} \to Y_1$ such that $g_1 = g_{\alpha_1} p_{\alpha_1}$. Now consider the map $q_2 \colon Y \to Y_2$. As above, let $g_2 \colon X \to Y_2$ be any representative of the homotopy class $\varphi([q_2]) \colon X \to Y_2$. Since $q_1 = q_1^2 q_2$ and $Y_1$ and $Y_2$ are compact polyhedra, we can conclude, by the definition of shape morphism [**214**], that the condition $\varphi([q_1]) = [q_1^2]\varphi([q_2])$ is satisfied. Consequently, $g_1 \simeq q_1^2 g_2$. As above, we choose an index $\alpha_2 \geq \alpha_1$ and a map $g_{\alpha_2} \colon X_{\alpha_2} \to Y_2$ such that $g_2 = g_{\alpha_2} p_{\alpha_2}$. Thus we have

$$g_{\alpha_1} p_{\alpha_1}^{\alpha_2} p_{\alpha_2} = g_{\alpha_1} p_{\alpha_1} = g_1 \simeq q_1^2 g_2 = q_1^2 g_{\alpha_2} p_{\alpha_2}.$$

Since $Y_1$ is an $ANR$-compactum and our spectrum is homotopically stable, we conclude that $g_{\alpha_1} p_{\alpha_1}^{\alpha_2} \simeq q_1^2 g_{\alpha_2}$.

Continuing this process, we get, for each $n \in \omega$, an index $\alpha_n \in A$ and a map $g_{\alpha_n}: X_{\alpha_n} \to Y_n$ satisfying the following conditions:

(i) $\alpha_n \leq \alpha_{n+1}$ whenever $n \in \omega$.

(ii) $g_{\alpha_n} p_{\alpha_n} \in \varphi([q_n])$ whenever $n \in \omega$.

(iii) $g_{\alpha_n} p_{\alpha_n}^{\alpha_{n+1}} \simeq q_n^{n+1} g_{\alpha_{n+1}}$ whenever $n \in \omega$.

We now let $\alpha = \sup\{\alpha_n : n \in \omega\}$ and define a shape morphism $\varphi_\alpha: X_\alpha \to Y$. First, we define the homotopy classes $\varphi_\alpha([q_n])$ by letting

$$\varphi_\alpha([q_n]) = [g_{\alpha_n} p_{\alpha_n}^{\alpha}], \quad n \in \omega.$$

Suppose now that $[f]: Y \to Z$ is an arbitrary homotopy class, where $Z$ is any metrizable $ANR$-compactum. We fix $n \in \omega$ and a map $f_n: Y_n \to Z$ such that $f \simeq f_n q_n$. Then we put

$$\varphi_\alpha([f]) = [f_n g_{\alpha_n} p_{\alpha_n}^{\alpha}], \quad n \in \omega.$$

It follows from conditions (i)—(iii) that the definition of $\varphi_\alpha([f])$ does not depend on the choice of $n \in \omega$ and $f_n$. Therefore, the shape morphism $\varphi_\alpha: X_\alpha \to Y$ is well defined.

Finally let us verify the equality $\varphi = \varphi_\alpha \operatorname{Sh}(p_\alpha)$. For this, it is sufficient to verify the equalities $\varphi([q_n]) = \varphi_\alpha([q_n])[p_\alpha]$, $n \in \omega$. But these equalities are trivially true, because by (ii) and by the definition of $\varphi_\alpha$,

$$\varphi([q_n]) = [g_{\alpha_n} p_{\alpha_n}] = [g_{\alpha_n} p_{\alpha_n}^{\alpha} p_\alpha] = \varphi_\alpha([q_n])[p_\alpha].$$

This completes the proof. □

Now we are ready to prove the main result of this Section, which is a shape analog of the spectral theorem for usual maps.

THEOREM 8.4.7. *Let $\mathcal{S}_X = \{X_\alpha, p_\alpha^\beta, A\}$ and $\mathcal{S}_Y = \{Y_\alpha, q_\alpha^\beta, A\}$ be $\omega$-spectra consisting of metrizable compacta and having the same indexing set. Let a shape morphism $\varphi: \lim \mathcal{S}_X \to \lim \mathcal{S}_Y$ be given. Then the set of those indexes $\alpha \in A$ for which there exist shape morphisms $\varphi_\alpha: X_\alpha \to Y_\alpha$ satisfying the conditions $\operatorname{Sh}(q_\alpha)\varphi = \varphi_\alpha \operatorname{Sh}(p_\alpha)$ is $\omega$-closed and cofinal in $A$.*

PROOF. Without loss of generality, by Theorem 8.4.4 we may assume that both given spectra are homotopically stable. Let

$$\mathcal{L} = \{(\alpha, \beta) \in A^2 : \alpha \leq \beta \text{ and there exists a shape morphism } \varphi_\alpha^\beta: X_\beta \to Y_\alpha \text{ such that } \operatorname{Sh}(q_\alpha)\varphi = \varphi_\alpha^\beta \operatorname{Sh}(p_\alpha)\}.$$

The validity of the existence condition of Proposition 1.1.29 follows directly from Lemma 8.4.6. The verification of the majorantness condition is trivial.

Let us verify the $\omega$-closeness condition. Consider an arbitrary countable chain $\{\alpha_n : n \in \omega\}$ in $A$, and suppose that $(\alpha_n, \beta) \in \mathcal{L}$ for each $n \in \omega$ and for some $\beta \in A$. We wish to show that $(\alpha, \beta) \in \mathcal{L}$, where $\alpha = \sup\{\alpha_n : n \in \omega\}$. By the definition of $\mathcal{L}$, there exist shape morphisms $\varphi^{\beta}_{\alpha_n}: X_\beta \to X_{\alpha_n}$ such that $\mathrm{Sh}(q_{\alpha_n})\varphi = \varphi^{\beta}_{\alpha_n}\,\mathrm{Sh}(p_\beta)$, $n \in \omega$.

Let us first show that for each $n \in \omega$, the equality

$$\varphi^{\beta}_{\alpha_n} = \mathrm{Sh}(q^{\alpha_{n+1}}_{\alpha_n})\varphi^{\beta}_{\alpha_{n+1}}$$

is true. Consider an arbitrary map $f\colon Y_{\alpha_n} \to Z$ into some metrizable $ANR$-compactum $Z$. Note that the equality

$$\varphi^{\beta}_{\alpha_n}\,\mathrm{Sh}(p_\beta) = \mathrm{Sh}(q_{\alpha_n})\varphi$$

implies the equality

$$\varphi^{\beta}_{\alpha_n}([f])[p_\beta] = \varphi([f q_{\alpha_n}]).$$

Similarly the equality

$$\varphi^{\beta}_{\alpha_{n+1}}\,\mathrm{Sh}(p_\beta) = \mathrm{Sh}(q_{\alpha_{n+1}})\varphi$$

implies the equality

$$\varphi^{\beta}_{\alpha_{n+1}}([f q^{\alpha_{n+1}}_{\alpha_n}])[p_\beta] = \varphi([f q^{\alpha_{n+1}}_{\alpha_n} q_{\alpha_{n+1}}]) = \varphi([f q_{\alpha_n}]).$$

Consequently,

$$\varphi^{\beta}_{\alpha_n}([f])[p_\beta] = \varphi^{\beta}_{\alpha_{n+1}}([f q^{\alpha_{n+1}}_{\alpha_n}])[p_\beta].$$

Since the spectrum $\mathcal{S}_X$ is homotopically stable, we conclude that

$$\varphi^{\beta}_{\alpha_n}([f]) = \varphi^{\beta}_{\alpha_{n+1}}([f q^{\alpha_{n+1}}_{\alpha_n}]).$$

This means that $\varphi^{\beta}_{\alpha_n} = \mathrm{Sh}(q^{\alpha_{n+1}}_{\alpha_n})\varphi^{\beta}_{\alpha_{n+1}}$ as desired. By the continuity of the SHAPE functor and by the $\omega$-continuity of the spectrum $\mathcal{S}_Y$, we conclude that in this situation there exists a shape morphism $\varphi^{\beta}_{\alpha}\colon X_\beta \to Y_\alpha$ such that $\varphi^{\beta}_{\alpha_n} = \mathrm{Sh}(q^{\alpha}_{\alpha_n})\varphi^{\beta}_{\alpha}$ for each $n \in \omega$. It only remains to note that the equality $\varphi^{\beta}_{\alpha}\,\mathrm{Sh}(p_\beta) = \mathrm{Sh}(q_\alpha)\varphi$ is also satisfied. The verification of all the conditions of Proposition 1.1.29 is finished. By this Proposition, the set of all $\mathcal{L}$-reflexive indexes is $\omega$-closed and cofinal in $A$. This completes the proof of our statement. □

COROLLARY 8.4.8. *If (in the notations of Theorem* 8.4.7*) $\varphi$ is a shape isomorphism, then the set of those $\alpha \in A$ for which $\varphi_\alpha\colon X_\alpha \to Y_\alpha$ is a shape isomorphism is $\omega$-closed and cofinal in $A$.*

PROOF. We use the notation of the proof of Theorem 8.4.7. Consider the shape inverse $\varphi^{-1}\colon Y \to X$, which exists by our assumption. By Theorem 8.4.7, the set $A_2$ of those indexes $\alpha \in A$ for which there exists a shape morphism $\psi_\alpha\colon Y_\alpha \to X_\alpha$ satisfying the condition $\psi_\alpha\,\mathrm{Sh}(q_\alpha) = \mathrm{Sh}(p_\alpha)\varphi^{-1}$ is $\omega$-closed and cofinal in $A$. Denote by $A_1$ the set of those indexes $\alpha \in A$ for which there exists a shape morphism $\varphi_\alpha\colon X_\alpha \to Y_\alpha$ satisfying the condition $\mathrm{Sh}(q_\alpha)\varphi = \varphi_\alpha\,\mathrm{Sh}(p_\alpha)$. By Theorem 8.4.7, this set is $\omega$-closed and cofinal in $A$. By Proposition 1.1.27, the intersection $B = A_1 \cap A_2$ is also $\omega$-closed and cofinal in $A$. Clearly, for each $\alpha \in B$ we have two shape morphisms $\varphi_\alpha\colon X_\alpha \to Y_\alpha$ and $\psi_\alpha\colon Y_\alpha \to X_\alpha$ satisfying the above mentioned conditions. The proof will be finished if we show that $\psi_\alpha\varphi_\alpha = 1_{X_\alpha}$ and $\varphi_\alpha\psi_\alpha = 1_{Y_\alpha}$, whenever $\alpha \in B$ ($1_X$ denotes the identity morphism of an object $X$ in the SHAPE category). Let $F\colon X_\alpha \to Z$ be an arbitrary map into a metrizable $ANR$-compactum. It follows from the above constructions that

$$\psi_\alpha\varphi_\alpha \operatorname{Sh}(p_\alpha) = \psi_\alpha \operatorname{Sh}(q_\alpha)\varphi = \operatorname{Sh}(p_\alpha)\varphi^{-1}\varphi = \operatorname{Sh}(p_\alpha).$$

Consequently, $\varphi_\alpha(\psi_\alpha([f]))[p_\alpha] = [f][p_\alpha]$. The homotopical stability of the spectrum $\mathcal{S}_X$ implies that $\varphi_\alpha(\psi_\alpha([f])) = [f]$, i.e. $\psi_\alpha\varphi_\alpha = 1_{X_\alpha}$. The verification of the second equality is similar. □

**8.4.2. Shape properties of non-metrizable compacta.** The following statement is a direct consequence of Lemma 8.4.6.

PROPOSITION 8.4.9. *Let $X$ be a compactum. Then the following conditions are equivalent:*

(i) *For each metrizable $ANR$-compactum $Y$, the set $[X,Y]$ is countable.*
(ii) *$X$ has a shape of a metrizable compactum.*
(iii) *$X$ can be represented as the limit space of an $\omega$-spectrum, all limit projections of which are shape equivalences.*
(iv) *$X$ admits a surjection onto a metrizable compactum and this surjection is a shape equivalence.*

PROOF. The equivalence of conditions (i) and (ii) was proved in [**310**].

*(ii)* $\Longrightarrow$ *(iii)*. By Theorem 8.4.4, we can represent $X$ as the limit space of some homotopically stable $\omega$-spectrum $\mathcal{S}_X = \{X_\alpha, p_\alpha^\beta, A\}$ . By (ii), there exists a shape isomorphism $\varphi\colon X \to Y$, where $Y$ is a metrizable compactum. By Lemma 8.4.6, there exist an index $\alpha_0 \in A$ and a shape morphism $\varphi_{\alpha_0}\colon X_{\alpha_0} \to Y$ such that $\varphi = \varphi_{\alpha_0} \operatorname{Sh}(p_{\alpha_0})$. For each $\alpha \geq \alpha_0$, define a shape morphism $\varphi_\alpha\colon X_\alpha \to Y$ by letting $\varphi_\alpha = \varphi_{\alpha_0} \operatorname{Sh}(p_{\alpha_0}^\alpha)$. Clearly, $\varphi = \varphi_\alpha \operatorname{Sh}(p_\alpha)$ for each $\alpha \geq \alpha_0$. Since the compactum $X$ is the limit space of the $\omega$-spectrum $\mathcal{S}' = \{X_\alpha, p_\alpha^\beta, A_0\}$, where $A_0 = \{\alpha \in A\colon \alpha \geq \alpha_0\}$, it suffices to show that each limit projection $p_\alpha\colon X \to X_\alpha$, $\alpha \geq \alpha_0$, is a shape equivalence. This means that for each metrizable $ANR$-compactum $Z$, the natural correspondence $[X_\alpha, Z] \to [X, Z]$, induced by the limit projection $p_\alpha$ by the rule $[f] \to [fp_\alpha]$, is a bijection. Since the spectrum $\mathcal{S}$ is homotopically stable, we immediately conclude that this correspondence is injective. Let us show that it is surjective as well. Consider an arbitrary homotopy class $[g] \in [X, Z]$. Then

$$\varphi^{-1}([g]) \in [Y, Z] \text{ and } \varphi_\alpha(\varphi^{-1}([g])) \in [X_\alpha, Z].$$

It only remains to note that

$$\varphi_\alpha(\varphi^{-1}([g]))[p_\alpha] = \varphi(\varphi^{-1}([g])) = [g].$$

$(iii) \Longrightarrow (iv)$. As the desired surjection, we may take any limit projection of the spectrum satisfying the properties of condition (iii).

The implication $(iv) \Longrightarrow (i)$ is trivial. □

PROPOSITION 8.4.10. *Let $X$ be a compactum. Then the following conditions are equivalent:*

(i) *For each closed $G_\delta$-subset $Z$ of $X$ and each metrizable $ANR$-compactum $P$, the set $[Z, P]$ is countable.*

(ii) *$X$ can be represented as the limit space of some $\omega$-spectrum, all limit projections of which are hereditary shape equivalences.*

PROOF. $(i) \Longrightarrow (ii)$. By Theorem 8.4.4, we can represent $X$ as the limit space of a homotopically stable $\omega$-spectrum $\mathcal{S}_X = \{X_\alpha, p_\alpha^\beta, A\}$ . Let

$$\mathcal{L} = \{(\alpha, \beta) \in A^2 \colon \alpha \leq \beta, \text{ and for each closed subspace } F \text{ of } X_\alpha \text{ the map } \varphi_\beta/p_\alpha^{-1}(F) \colon p_\alpha^{-1}(F) \to (p_\alpha^\beta)-1(F) \text{ is a shape equivalence}\}.$$

First, let us verify the existence condition of Proposition 1.1.29. Fix an index $\alpha \in A$ and a countable closed basis $\{F_n \colon n \in \omega\}$ of the topology of metrizable compactum $X_\alpha$ containing the intersections of its finite subcollections. By (i), for each $n \in \omega$ and for each metrizable $ANR$-compactum $P$, the set $[p_\alpha^{-1}(F_n)], P]$ is countable. Consequently, by Proposition 8.4.9, for each $n \in \omega$ the compactum $p_\alpha^{-1}(F_n)$ can be represented as the limit space of some $\omega$-spectrum all limit projections of which are shape equivalences. By Theorem 1.3.4, there exists an $\omega$-closed and cofinal subspectrum with the above property in every $\omega$-spectrum representing the compactum $p_\alpha^{-1}(F_n)$. In particular, this is true for the naturally induced (by $\mathcal{S}_X$) spectrum

$$\{(p_\alpha^\gamma)^{-1}(F_n), p_\gamma^\delta/(p_\gamma^\delta)^{-1}(F_n), \delta \geq \gamma \geq \alpha\}.$$

Hence there exists at least one index $\alpha_n \geq \alpha$ such that the map

$$p_{\alpha_n}/p_\alpha^{-1}(F_n) \colon p_\alpha^{-1}(F_n) \to (p_\alpha^{\alpha_n})^{-1}(F_n)$$

is a shape equivalence. Without loss of generality, we may assume that $\{\alpha_n \colon n \in \omega\}$ is a chain in $A$. Let $\beta = \sup\{\alpha_n \colon n \in \omega\}$. Note that for each $n \in \omega$, the map

$$p_\beta/p_\alpha^{-1}(F_n) \colon p_\alpha^{-1}(F_n) \to (p_\alpha^\beta)^{-1}(F_n)$$

is a shape equivalence. Let us show that $(\alpha, \beta) \in \mathcal{L}$. Let $F$ be an arbitrary closed subset of $X_\alpha$ and $f \colon p_\alpha^{-1}(F) \to P$ be a map into a metrizable $ANR$-compactum. As in the proof of Lemma 8.4.3, we can find an integer $n \in \omega$ and a map $f' \colon p_\alpha^{-1}(F_n) \to P$ where $F \subseteq F_n$ and $f = f'/p_\alpha^{-1}(F)$. By the definition of $\beta$, there exists a map $g' \colon (p_\alpha^\beta)^{-1}(F_n) \to P$ such that $f' \simeq g' p_\beta/p_\alpha^{-1}(F_n)$. Put $g = g'/(p_\alpha^\beta)^{-1}(F)$. Obviously, $f \simeq g p_\beta/p_\alpha^{-1}(F)$. Thus we have shown that the natural correspondence $[(p_\alpha^\beta)^{-1}(F), P] \to [p_\alpha^{-1}(F), P]$, induced by $p_\beta/p_\alpha^{-1}(F)$, is surjective. Since the spectrum $\mathcal{S}_X$ is homotopically stable we conclude that the above correspondence is bijective.

The verification of the majorantness condition from Proposition 1.1.29 is trivial.

Now let $\{\alpha_n : n \in \omega\}$ be a countable chain in $A$, and let $(\alpha_n, \beta) \in \mathcal{L}$ for each $n \in \omega$ and for some $\beta \in A$. Let us show that $(\alpha, \beta) \in \mathcal{L}$, where $\alpha = \sup\{\alpha_n : n \in \omega\}$. By the $\omega$-continuity of the spectrum $\mathcal{S}_X$, the compactum $X_\alpha$ is naturally homeomorphic to the limit space of the inverse sequence $\{X_{\alpha_n}, p_{\alpha_n}^{\alpha_{n+1}}, \omega\}$. Let $F$ be a closed subset of $X_\alpha$ and $f\colon p_\alpha^{-1}(F) \to P$ be an arbitrary map into a metrizable $ANR$-compactum. Fix a neighborhood $V$ of $p_\alpha^{-1}(F)$ in $X$ and an extension $f'\colon V \to P$ of $f$. Without loss of generality, we may assume that $V = p_\alpha^{-1}(U)$, where $U$ is a neighborhood of $F$ in $X_\alpha$. There exists an integer $n \in \omega$ such that

$$F \subseteq (p_{\alpha_n}^{\alpha})^{-1}(F_n) \subseteq U,$$

where $F_n = (p_{\alpha_n}^{\alpha})(F)$. Hence

$$p_\alpha^{-1}(F) \subseteq p_{\alpha_n}^{-1}(F_n) \subseteq V.$$

Since $(\alpha_n, \beta) \in \mathcal{L}$ there exists a map $g'\colon (p_{\alpha_n}^{\beta})(F_n) \to P$ such that

$$f'/p_{\alpha_n}^{-1}(F_n) \simeq g' p_\beta / p_{\alpha_n}^{-1}(F_n).$$

Clearly the map $g = g'/(p_\alpha^\beta)^{-1}(F)$ satisfies the condition $f \simeq g p_\beta / p_\alpha^{-1}(F)$. This shows that the natural correspondence $[(p_\alpha^\beta)^{-1}(F), P] \to [p_\alpha^{-1}(F), P]$ is surjective. The homotopy stability of $\mathcal{S}_X$ implies that this correspondence is bijective. Thus $(\alpha, \beta) \in \mathcal{L}$.

Now, by Proposition 1.1.29, the set $A'$ of $\mathcal{L}$-reflexive indexes of $A$ is $\omega$-closed and cofinal in $A$. It only remains to note that the $\mathcal{L}$-reflexibility of an index $\alpha \in A$ means precisely that the limit projection $p_\alpha\colon X \to X_\alpha$ is a hereditary shape equivalence. The proof of the implication $(i) \Longrightarrow (ii)$ is finished.

The verification of the converse implication is trivial. □

PROPOSITION 8.4.11. *Let $X$ be a finite-dimensional $1-UV$-compactum. Then the following conditions are equivalent:*

(i) *$X$ is shape equivalent to a finite polyhedron.*

(ii) *The Čech cohomology groups of $X$ are finitely generated.*

PROOF. For metrizable $X$ this statement was proved in [**157**]. Therefore, we may assume below that $X$ is a non-metrizable compactum.

$(ii) \Longrightarrow (i)$. Let $\mathcal{S}_X = \{X_\alpha, p_\alpha^\beta, A\}$ be an $\omega$-spectrum the limit space of which coincides with $X$. By (ii), Proposition 8.4.9 and Theorem 1.3.4, there is an $\omega$-closed and cofinal subset $A_1 \subseteq A$ such that the limit projection $p_\alpha\colon X \to X_\alpha$ is a shape equivalence whenever $\alpha \in A_1$. In particular, the Čech cohomology groups of $X_\alpha$, $\alpha \in A_1$, are finitely generated [**47**]. For the same reason, $X_\alpha$ is $1-UV$ for each $\alpha \in A_1$. By Theorem 1.3.10, there is an $\omega$-closed and cofinal subset $A_2 \subseteq A$ such that $\dim X_\alpha = \dim X < \infty$ for each $\alpha \in A_2$. Now, by Proposition 1.1.27, the set $B = A_1 \cap A_2$ is also $\omega$-closed and cofinal in $A$. In particular, $B \neq \emptyset$. Let $\alpha \in B$. Since $X_\alpha$ is metrizable, we conclude (see the beginning of the proof) that $X_\alpha$ is shape equivalent to a finite polyhedron. It only remains to note that $X$ is shape equivalent to $X_\alpha$.

The implication $(i) \Longrightarrow (ii)$ is trivial. □

*Historical and bibliographical notes* 8.4. The results of this Section are taken from [**99**]. The equivalence of conditions (i) and (ii) in Proposition 8.4.9 was proved in [**310**]. Proposition 8.4.11 for metrizable compacta was obtained in [**157**]. The axiomatic description of the shape functor appears in [**174**].

## 8.5. Fixed point sets of Tychonov cubes

The problem we are interested in is the following: Which closed subsets of the Tychonov cube $I^\tau$ coincide with the set of fixed points of continuous self-maps of $I^\tau$ ? The problem is motivated by the following two facts. First of all, each nonempty closed subset of the Hilbert cube $I^\omega$ coincides with the set of fixed points of some self-mapping (actually, autohomeomorphism) of $I^\omega$ [**216**]. On the other hand, for each uncountable $\tau$ there is a closed zero-dimensional subspace $K_\tau$ of the Tychonov cube $I^\tau$ which does not coincide with the set of fixed points of any self-map of $I^\tau$ [**197**]. The first fact says that the Hilbert cube $I^\omega$ has the *CIP* (*complete invariance property* [**308**]), whereas the second tells us that the Tychonov cube $I^\tau$ $(\tau > \omega)$ does not. Both results express a property of the ambient spaces $I^\omega$ and $I^\tau$ , $\tau > \omega$, respectively. But obviously, there is another dual point of view of this situation. Before we start a more formal discussion, let us (for simplicity) introduce appropriate terminology. Let us say that a compactum $X$ *can be fixed* in a compactum $Y$ if there is an embedding $i\colon X \to Y$ such that $i(X) = fix(f)$ for some map $f\colon Y \to Y$ ($fix(f)$ denotes the set of all fixed points of $f$). Respectively, we say that $X$ *cannot be fixed* in $Y$ if there is no such embedding of $X$ into $Y$. Of course, if we fix an ambient compactum $Y$, then the above properties become topological properties of $X$. Therefore, it does not follow from the above cited result of [**197**] that the compactum $K_\tau$ cannot be fixed in $I^\tau$ . So there still is a possibility that any compactum of weight $\leq \tau$ can be fixed in $I^\tau$ .

In this Section we give a complete solution of the above problem by providing a spectral characterization of those compacta which can (not) be fixed in $I^\tau$. The main result says that a compactum $X$ can be fixed in $I^\tau$ if and only if $X$ can be represented as the limit space of some transfinite spectrum $\mathcal{S}_X = \{X_\alpha, p_\alpha^\beta, \tau\}$ of length $\tau$ whose short projections $p_\alpha^{\alpha+1}\colon X_{\alpha+1} \to X_\alpha$ are *stable* and whose first element $X_0$ is metrizable. The notion of stability of a map $p\colon X \to Y$ involves two ingredients. The first says that there is an embedding $i\colon X \to Y \times I^\omega$ such that $\pi_Y i = p$, where $\pi_Y\colon Y \times I^\omega \to\!\!\to Y$ denotes the natural projection. Maps with this property are exactly maps with a metrizable kernel. The second ingredient says that there is a map $f\colon Y \times I^\omega \to Y \times I^\omega$ such that $fix(f) = i(X)$

and $\pi_Y f = \pi_Y$. In other words, $i(X)$ can be fixed in $Y \times I^\omega$ by a map $f$ which acts fiberwise (with respect to $\pi_Y$). Open retractions with metrizable kernels are stable. This allows us to conclude that a retract of any product of metrizable compacta can be fixed in Tychonov cubes of the corresponding weight. Going back to the compacta $K_\tau$, $\tau > \omega$, we strengthen the above mentioned result of [**197**] by showing that the compactum $K_{\omega_1}$ cannot be fixed in any non-metrizable ANR-compactum. Representing $K_{\omega_1}$ as the limit space of a transfinite inverse spectrum consisting of zero-dimensional metrizable compacta, and applying our characterization theorem together with a spectral theorem, we conclude that among the projections of this spectrum, only countably many can be stable. All others serve as examples of surjections between zero-dimensional metrizable compacta which are not stable. Obviously, these maps are not open (open surjections between zero-dimensional metrizable compacta are retractions, and therefore are stable). We also present an example of a stable retraction between metrizable AR-compacta (which is not open).

**8.5.1. Extension of homeomorphisms.** The following definition is motivated by the definition of (usual) Z-sets, and is adapted to the case of uncountable products.

DEFINITION 8.5.1. *Let $\tau \geq \omega$. We say that a closed subset $Z$ of a space $X$ is a $Z_\tau$-set in $X$ if for each collection $\{\mathcal{U}_t\colon t \in T\}$, where $\mathcal{U}_t \in cov(X)$ and $|T| < \tau$, there is a map $f\colon X \to X$ such that $f(X) \cap Z = \emptyset$ and $f$ is $\{\mathcal{U}_t\colon t \in T\}$-close to $id_X$.*

In other words, $Z$ is a $Z_\tau$-set in $X$ if

$$id_X \in \mathrm{cl}_{C_\tau(X,X)}\{f \in C_\tau(X,X)\colon f(X) \cap Z = \emptyset\}.$$

Note that the notion of $Z_\omega$-set coincides with the usual notion of $Z$-set.

LEMMA 8.5.2. *Let $X = \prod\{X_a\colon a \in A\}$ be a product of compact $I^\omega$-manifolds, $|A| = \tau > \omega$, and suppose there is an index $b \in A$ such that $X_a$ is a copy of the Hilbert cube $I^\omega$ for each $a \neq b$. Suppose that $Z$ and $F$ are closed subsets of $X$ such that $\pi_a(Z)$ and $\pi_a(F)$ are $Z$-sets in $X_a$ for each $a \in A$. Then for each homeomorphism $g\colon Z \to F$ which is homotopic (in $X$) to the inclusion map of $Z$, there exists an autohomeomorphism $G\colon X \to X$ such that $G/Z = g$ and $G \simeq id_X$.*

PROOF. Represent $X$ as the limit space of the standard $\omega$-spectrum $\mathcal{S}_X = \{X_B, \pi^B_C, exp_\omega A\}$ consisting of countable subproducts $X_B = \prod\{X_a\colon a \in B\}$ of $X$ and natural projections $\pi^B_C\colon X_B \to X_C$. Consider also the induced $\omega$-spectra

$$\mathcal{S}_Z = \{Z_B, p^B_C, exp_\omega A\} \text{ and } \mathcal{S}_F = \{F_B, q^B_C, exp_\omega A\}$$

where $Z_B = \pi_B(Z)$ , $p_C^B = \pi_C^B/Z_B$, $F_B = \pi_B(F)$ and $q_C^B = \pi_C^B/F_B$ $(C, B \in exp_\omega A$, $C \subseteq B)$. Obviously, $Z = \lim \mathcal{S}_Z$ and $F = \lim \mathcal{S}_F$. By Theorem 1.3.4, the homeomorphism $g$ is the limit map of some morphism

$$g = \{g_B \colon F_B \to Z_B, B \in \mathcal{K}_1\} \colon \mathcal{S}_Z/\mathcal{K}_1 \to \mathcal{S}_F/\mathcal{K}_1$$

consisting of homeomorphisms $g_B$'s, where $\mathcal{K}_1$ denotes a cofinal and $\omega$-closed subset of $exp_\omega A$. Consider an $\omega$-spectrum

$$\mathcal{S}_Z \times \mathcal{S}_I = \{Z_B \times I, p_C^B \times id_I, exp_\omega A\}$$

where $I = [0,1]$. Obviously $\lim(\mathcal{S}_Z \times \mathcal{S}_I) = Z \times I$. Fix a homotopy $H : Z \times I \to X$ connecting $g$ and the inclusion map $i \colon Z \to X$. Applying Theorem 1.3.4 to the spectra $\mathcal{S}_Z \times \mathcal{S}_I$, $\mathcal{S}_X$ and the homotopy $H$, we can represent $H$ as the limit map of some morphism

$$\mathcal{H} = \{H_B \colon Z_B \times I \to X_B, B \in \mathcal{K}_2\} \colon (\mathcal{S}_Z \times \mathcal{S}_I)/\mathcal{K}_2 \to \mathcal{S}_X/\mathcal{K}_2$$

where $\mathcal{K}_2$ denotes a cofinal and $\omega$-closed subset of $exp_\omega A$. By Proposition 1.1.27, $\mathcal{K} = \mathcal{K}_1 \cap \mathcal{K}_2$ is again a cofinal and $\omega$-closed subset of $exp_\omega A$. It is easy to see that for each $B \in \mathcal{K}$, the homeomorphism $g_B$ and the inclusion map $i_B \colon Z_B \to X_B$ are connected by the homotopy $H_B$.

Fix a well-ordering $A = \{a_\alpha \colon \alpha < \tau\}$ of $A$ such that $a_0 = b$. Let $B_0 \in \mathcal{K}$ and $a_0 \in B_0$ . By the properties of $Z$ and $F$, their projections $Z_{B_0}$ and $F_{B_0}$ are $Z$-sets in the Hilbert cube $X_{B_0}$. The homeomorphism $g_0 \equiv g_{B_0} \colon Z_{B_0} \to F_{B_0}$ is homotopic to the inclusion map $i_0 \equiv i_{B_0}$ (via the homotopy $H_0 \equiv H_{B_0}$). Consequently, by Theorem 2.3.19, $g_0$ can be extended to an autohomeomorphism $G_0 \colon X_{B_0} \to X_{B_0}$ which is homotopic (via some homotopy $R_0 \colon X_{B_0} \times I \to X_{B_0}$) to the identity map of $X_{B_0}$.

Suppose now that for each $\beta < \alpha$, where $\alpha < \tau$ and $\alpha > 0$, we have already constructed subsets $B_\beta$ of the index set $A$, homeomorphisms $g_\beta \colon Z_{B_\beta} \to F_{B_\beta}$, autohomeomorphisms $G_\beta$ of $X_{B_\beta}$ and homotopies $R_\beta \colon X_{B_\beta} \times I \to X_{B_\beta}$ in such a way that the following conditions are satisfied:

(a) $\{a_\delta \colon \delta < \beta\} \subseteq B_\beta$.
(b) $B_\delta \subseteq B_\beta$ whenever $\delta < \beta$.
(c) $B_\beta = \cup\{B_\delta \colon \delta < \beta\}$ if $\beta$ is a limit ordinal.
(d) $g_\delta \pi_{B_\delta}^{B_\beta} = \pi_{B_\delta}^{B_\beta} g_\beta$ whenever $\delta < \beta$.
(e) $g_\beta = \lim\{g_\delta \colon \delta < \beta\}$ if $\beta$ is a limit ordinal.
(f) $G_\delta \pi_{B_\delta}^{B_\beta} = \pi_{B_\delta}^{B_\beta} G_\beta$ whenever $\delta < \beta$.
(g) $G_\beta / Z_{B_\beta} = g_\beta$.
(h) $R_\delta(\pi_{B_\delta}^{B_\beta} \times id_I) = \pi_{B_\delta}^{B_\beta} R_\beta$ whenever $\delta < \beta$.
(i) $R_\beta = \lim\{R_\delta \colon \delta < \beta\}$ if $\beta$ is a limit ordinal.

Let us construct corresponding objects for the ordinal $\alpha$.

If $\alpha$ is a limit ordinal, let $B_\alpha = \cup\{B_\beta \colon \beta < \alpha\}$, $g_\alpha = \lim\{g_\beta \colon \beta < \alpha\}$ and $R_\alpha = \lim\{R_\beta \colon \beta < \alpha\}$.

Suppose now that $\alpha = \beta + 1$. Consider the smallest ordinal $\gamma < \tau$ such that $a_\gamma \notin B_\beta$. By (a), $\gamma \geq \beta$. By Corollary 1.3.17, there exist an element $C \in exp_\omega(A - B_\beta)$, containing $a_\gamma$, and a homeomorphism

$$g_\alpha \equiv g_{B_\beta \cup C} \colon Z_{B_\beta \cup C} \to F_{B_\beta \cup C}.$$

Define the set $B_\alpha$ as the union $B_\beta \cup C$. Obviously, $\{a_\beta. \beta < \alpha\} \subseteq B_\alpha$. By Corollary 1.3.17,

$$q_{B_\alpha} g = g_\alpha p_{B_\alpha} \text{ and, consequently, } q^{B_\alpha}_{B_\beta} g_\alpha = g_\beta p^{B_\alpha}_{B_\beta}.$$

Using Theorem 1.3.16, we may assume without loss of generality that a homotopy $H_\alpha \colon Z_{B_\alpha} \times I \to X_{B_\alpha}$ (connecting $g_\alpha$ and the inclusion map $i_\alpha \colon Z_{B_\alpha} \to X_{B_\alpha}$) such that

$$H_\beta(\pi^{B_\alpha}_{B_\beta} \times id_I) = \pi^{B_\alpha}_{B_\beta} H_\alpha$$

is also defined. By our assumptions, the projections $\pi_a(Z)$ and $\pi_a(F)$ are $Z$-sets in $X_a$ for each $a \in A$. Consequently, we may conclude that $Z_{B_\alpha}$ and $F_{B_\alpha}$ are fibered $Z$-sets with respect to the projection

$$\pi^{B_\alpha}_{B_\beta} \colon X_{B_\alpha} = X_{B_\beta} \times X_C \to X_{B_\beta}.$$

Now, $\pi^{B_\alpha}_{B_\beta}$ is a trivial bundle with the Hilbert cube $X_C$ as a fiber (because $b = a_0 \notin C$). Let $t$ denote the identity map of the fiber $X_C$ and consider the set $T = (G_\beta \times t)(Z_{B_\alpha})$. Clearly $T$ is a fibered $Z$-set with respect to $\pi^{B_\alpha}_{B_\beta}$. The homeomorphism

$$\psi = g_\alpha(G_\beta \times t)^{-1}/T \colon T \to F_{B_\alpha}$$

is fiber preserving, i.e. $\pi^{B_\alpha}_{B_\beta}\psi = \pi^{B_\alpha}_{B_\beta}$. A fiber preserving homotopy

$$\varphi = H_\alpha((G_\beta \times t) \times id_I)^{-1} \colon T \times I \to X_{B_\alpha}$$

connects $\psi$ and the inclusion map of $T$ into $X_{B_\alpha}$. By **[300]**, there is a homeomorphism $\Psi \colon X_{B_\alpha} \to X_{B_\alpha}$ such that $\Psi/T = \psi$ and $\pi^{B_\alpha}_{B_\beta}\Psi = \pi^{B_\alpha}_{B_\beta}$. Moreover, there is a homotopy

$$\Theta \colon X_{B_\alpha} \times I \to X_{B_\alpha}$$

which connects $\Psi$ and the identity map and satisfies the equality

$$\pi^{B_\alpha}_{B_\beta}\Theta(x,r) = \pi^{B_\alpha}_{B_\beta}(x) \text{ for each } (x,r) \in X_{B_\alpha} \times I.$$

Now we are in position to define the desired objects. Let $G_\alpha = \Psi(G_\beta \times t)$ and $R_\alpha(x,y,r) = \Theta((R_\beta(x,r),y,r)$ for each $(x,y,r) \in X_{B_\beta} \times X_C \times I = X_{B_\alpha} \times I$. It follows from the construction that $G_\alpha$ and $R_\alpha$ are fiber preserving and satisfy all the conditions (a) - (i) formulated for the ordinal $\alpha$. This completes the inductive step.

Let $G = \lim\{G_\alpha \colon \alpha < \tau\}$ and $R = \lim\{R_\alpha \colon \alpha < \tau\}$. Then $G$ is an autohomeomorphism of $X$ and extends $g$. The homotopy $R$ connects $G$ and $id_X$. $\square$

LEMMA 8.5.3. *Let $\{Z_n : n \in \omega\}$ be a countable collection of $Z_\tau$-sets in a product $X = \prod\{X_a : a \in A\}$ of metrizable compacta $X_a$, where $|A| = \tau > \omega$. Then there is a collection $\{A_\alpha : \alpha < \tau\}$ of countable subsets of $A$ such that:*

(i) $\cup\{A_\alpha : \alpha < \tau\} = A$.

(ii) $A_\alpha \cap A_\beta = \emptyset$ *whenever* $\alpha \neq \beta$.

(iii) $\pi_{A_\alpha}(Z_n)$ *is a $Z$-set in* $X_{A_\alpha}$ *for each* $\alpha < \tau$ *and each* $n \in \omega$.

PROOF. First we need the following:

*Claim. Let $B \subseteq A$, $|B| < \tau$ and $C \in exp_\omega(A - B)$. For $\mathcal{U} \in cov(X_C)$ there is a countable subset $D$ of $A - B$, containing $C$, such that for each $n \in \omega$ there is a map $\psi_n : X_D \to X_D$ satisfying the following conditions:*

(a) $\psi_n(X_D) \cap \pi_D(Z_n) = \emptyset$ *for each* $n \in \omega$

*and*

(b) $\pi^D_C \psi_n$ *and* $\pi^D_C$ *are* $\mathcal{U}$*-close.*

*Proof of Claim.* Consider the collection $\{\mathcal{U}_t : t \in T\}$ of open covers of $X_B$ from Lemma 6.5.1. Consider also the collection $\{\pi^{-1}(\mathcal{U}_t) : t \in T\} \cup \pi^{-1}(\mathcal{U})$ of open covers of $X$. Since $Z_n$ is a $Z_\tau$-set in $X$ and the cardinality of $B$ is strictly less than $\tau$, we can conclude that for each $n \in \omega$ there is a map $g_n : X \to X$ such that $g_n(X) \cap Z_n = \emptyset$ and $g_n$ is $(\{\pi^{-1}(\mathcal{U}_t) : t \in T\} \cup \{\pi^{-1}(\mathcal{U})\})$-close to the identity map of $X$. The latter condition implies the equality $\pi_B g_n = \pi_B$, $n \in \omega$ (by the properties of the collection $\{\mathcal{U}_t : t \in T\}$). This is enough to conclude that there is a countable subset $D$ of $A - B$ such that $C \subseteq D$ and $\pi_D(Z) \cap \pi_D g_n(X) = \emptyset$ for each $n \in \omega$. Consider a section $j : X_D \to X$ of the projection $\pi_D$ and define $\psi_n : X_D \to X_D$ as the composition $\psi_n = \pi_D g_n j$, $n \in \omega$. Clearly $\psi_n(X_D) \cap \pi_D(Z) = \emptyset$ and $\pi^D_C \psi_n$ is $\mathcal{U}$-close to $\pi^D_C$ (because $\pi_C g_n$ is $\mathcal{U}$-close to $\pi_C$) for each $n \in \omega$. The proof of Claim is complete.

Let us consider the following relation $\mathcal{L}_B$ on $exp_\omega(A - B)$:

$$\mathcal{L}_B = \{(C, D) \in (exp_\omega(A - B))^2 : C \subseteq D, \text{ and for each } \mathcal{U} \in cov(X_C)\ n \in \omega \text{ there is a map } \psi_n : X_D \to X_D \text{ such that } \psi_n(X_D) \cap \pi_D(Z_n) = \emptyset \text{ and } \pi^D_C \psi_n \text{ and } \pi^D_C \text{ are } \mathcal{U}\text{-close}\}.$$

We need the following three properties of this relation.

*Property 1. For each $C \in exp_\omega(A - B)$, there is a $D \in exp_\omega(A - B)$ such that $(C, D) \in \mathcal{L}_B$.*

*Proof of Property 1.* Fix a countable family $\{\mathcal{U}_i : i \in \omega\}$ of open covers of $X_C$ such that for each open cover $\mathcal{U}$ of $X_C$ there is an index $i$ with the property that $\mathcal{U}_i$ refines $\mathcal{U}$. By the Claim, for each $i \in \omega$ there is a countable subset $D_i$ of $B - A$ containing $C$, and maps

$$\psi_{i,n} : X_{D_i} \to X_{D_i},\ n \in \omega,$$

such that $\psi_{i,n}(X_{D_i}) \cap \pi_{D_i}(Z_n) = \emptyset$ and $\pi^{D_i}_C \psi_{i,n}$ and $\pi^{D_i}_C$ are $\mathcal{U}_i$-close. Let us show that $(C, D) \in \mathcal{L}_B$, where $D = \cup\{D_i : i \in \omega\}$. Indeed, for any given open cover $\mathcal{U}$

of $X_C$, fix an index $i$ such that $\mathcal{U}_i$ refines $\mathcal{U}$. Also fix any section $s_i\colon X_{D_i} \to X_D$ of the projection $\pi^D_{D_i}$. Then the maps

$$\psi_n = s_i\psi_{i,n}\pi^D_{D_i}\colon I^D \to I^D,\ n \in \omega$$

obviously verify the fact that $(C, D) \in \mathcal{L}_B$.

*Property 2. If $(C, D) \in \mathcal{L}_B$, $E \in exp_\omega(A - B)$ and $D \subseteq E$, then $(C, E) \in \mathcal{L}_B$.*

*Proof of Property 2.* Fix a section $s\colon X_D \to X_E$ of the projection $\pi^E_D\colon X_E \to X_D$. If an open cover $\mathcal{U} \in cov(X_C)$ is given we can, using the fact that $(C, D) \in \mathcal{L}_B$, find maps $\varphi_n\colon X_D \to X_D$, $n \in \omega$, such that $\varphi_n(X_D) \cap \pi_D(Z) = \emptyset$ and $\varphi_n\pi^D_C$ and $\pi^D_C$ are $\mathcal{U}$-close. Let

$$\psi_n = s\varphi_n\pi^E_D\colon X_E \to X_E, n \in \omega.$$

Clearly, $\psi_n(X_E) \cap \pi_E(Z_n) = \emptyset$ and $\pi^E_C\psi_n$ is $\mathcal{U}$-close to $\pi^E_C$, $n \in \omega$.

*Property 3. Let $(C_i, D) \in \mathcal{L}_B$ and $C_i \subseteq C_{i+1}$ for each $i \in \omega$. Then $(C, D) \in \mathcal{L}_B$, where $C = \cup\{C_i\colon i \in \omega\}$.*

*Proof of Property 3.* For any open cover $\mathcal{U} \in cov(X_C)$, there is an index $i \in \omega$ and an open cover $\mathcal{U}_i \in cov(X_{C_i})$ such that $(\pi^C_{C_i})^{-1}(\mathcal{U}_i)$ refines $\mathcal{U}$. Since $(C_i, D) \in \mathcal{L}_B$, we conclude that there are maps $\psi_n\colon X_D \to X_D$, $n \in \omega$, such that $\psi_n(X_D) \cap \pi_D(Z_n) = \emptyset$ and $\pi^D_{C_i}\psi_n$ is $\mathcal{U}_i$-close to $\pi^D_{C_i}$. It only remains to note that $\pi^D_C\psi_n$ and $\pi^D_C$ are $\mathcal{U}$-close.

We now continue with the proof of our Lemma. By the Properties 1,2 ,3 and Proposition 1.1.29, the collection $\mathcal{K}_B$ of all $C \in exp_\omega(A - B)$ with the property $(C, C) \in \mathcal{L}_B$ is cofinal and $\omega$-closed in $exp_\omega(A - B)$. Note that if $C \in \mathcal{K}_B$, then $\pi_C(Z_n)$ is a $Z$-set in $X_C$ for each $n \in \omega$.

Since $|A| = \tau$ we can write $A = \{a_\alpha\colon \alpha < \tau\}$. For a point $a_0$, fix a set $A_0 \in exp_\omega A$ such that $\pi_{A_0}(Z_n)$ is a $Z$-set in $X_{A_0}$ for each $n \in \omega$ (we use the cofinality of the collection $\mathcal{K}_B$ in $exp_\omega(A - B)$ with $B = \emptyset$). Suppose that for each $\beta \leq \alpha$, where $\alpha < \tau$, we have already constructed countable subsets $A_\beta$ of $A$ so that

$$\{a_\beta\colon \beta \leq \alpha\} \subseteq \cup\{A_\beta\colon \beta \leq \alpha\}$$

and $\pi_{A_\beta}(Z_n)$ is a $Z$-set in $X_{A_\beta}$ whenever $\beta \leq \alpha$ and $n \in \omega$. Let us construct $A_{\alpha+1}$. Let $\delta$ denote the smallest ordinal such that $a_\delta \notin B = \cup\{A_\beta\colon \beta \leq \alpha\}$. Clearly, $\delta > \alpha$. Note that $|B| < \tau$. Since the collection $\mathcal{K}_B$ is cofinal in $exp_\omega(A - B)$, there is an element $A_{\alpha+1} \in \mathcal{K}_B$ such that $a_\delta \in A_{\alpha+1}$. Clearly, $A_{\alpha+1} \cap A_\beta = \emptyset$ for each $\beta \leq \alpha$ (because $\mathcal{K}_B \subseteq exp_\omega(A - B)$) and $\{a_\beta\colon \beta \leq \alpha + 1\} \subseteq \{a_\beta\colon \beta \leq \delta\} \subseteq \cup\{A_\beta\colon \beta \leq \alpha + 1\}$. This completes the proof of Lemma 8.5.3. □

THEOREM 8.5.4. *Suppose that the homeomorphism $g\colon Z \to F$ between $Z_\tau$-sets of the compact $I^\tau$-manifold $X$ is homotopic to the inclusion map $Z \hookrightarrow X$ ($\tau > \omega$). Then there is an autohomeomorphism $G\colon X \to X$ which extends $g$ and is homotopic to $id_X$.*

PROOF. By Theorem 7.1.19, $X$ is homeomorphic to the product of a compact $I^\omega$-manifold and the Tychonov cube $I^\tau$. Consequently, we can represent $X$ as the product $\prod\{X_a: a \in A\}$, $|A| = \tau$, where $X_b$ is a compact $I^\omega$-manifold for some $b \in A$ and all other $X_a$'s are copies of the Hilbert cube $I^\omega$. Fix a collection $\{A_\alpha: \alpha < \tau\}$ of countable subsets of $A$ satisfying conditions (i)-(iii) of Lemma 8.5.3 with respect to the $Z_\tau$-sets $Z$ and $F$.

If $\alpha < \tau$ and $b \notin A_\alpha$, choose an arbitrary point $a_\alpha \in A_\alpha$. If $b \in A_\alpha$, then let $a_\alpha = b$. By the stability of $I^\omega$-manifolds (Theorem 2.3.10), there is a homeomorphism

$$h_\alpha\colon \prod\{X_a: a \in A_\alpha\} \to X_{a_\alpha}$$

for each $\alpha < \tau$. Clearly,

$$h = \times\{h_\alpha: \alpha < \tau\}\colon X \to Y$$

is a homeomorphism, where $Y = \prod\{X_{a_\alpha}: \alpha < \tau\}$. By Lemma 8.5.3 (condition (iii)), $h(Z)$ and $h(F)$ are closed subsets of $Y$ such that their projections on each coordinate space $X_{a_\alpha}$ are $Z$-sets. Define the homeomorphism $f\colon h(Z) \to h(F)$ to be the composition $f = hgh^{-1}/h(Z)$. Clearly, $f$ is homotopic to the inclusion map $h(Z) \hookrightarrow Y$. Consequently, by Lemma 8.5.2, there is a homeomorphism $\bar{f}\colon Y \to Y$ which extends $f$ and is homotopic to $id_Y$. It only remains to note that the composition $G = h^{-1}\bar{f}h$ is the desired autohomeomorphism of $X$. □

A similar argument shows the validity of the following statement (compare with Theorem 2.3.18).

THEOREM 8.5.5. *Let $X$ be a compact $I^\tau$-manifold, $\tau > \omega$. For each collection $\{\mathcal{U}_t: t \in T\} \subseteq cov(X)$, $|T| < \tau$, there exists another collection $\{\mathcal{V}_t: t \in T'\}$, $|T'| < \tau$, such that the following condition is satisfied:*

(∗) *For any homeomorphism $h\colon Z \to F$ between arbitrary $Z_\tau$-sets of $X$ which is $\{\mathcal{V}_t: t \in T'\}$-close to the inclusion map $Z \hookrightarrow X$, there is a homeomorphism $H\colon X \to X$ such that $H/Z = h$ and which is $\{\mathcal{U}_t: t \in T\}$-close to $id_X$.*

We conclude this Subsection with the following statement which allows us to recognize $Z_\tau$-sets for $\tau > \omega$.

PROPOSITION 8.5.6. *Let $X = \prod\{X_a: a \in A\}$, $|A| = \tau > \omega$, be a product of metrizable compacta such that all $X_a$'s with $a \neq b$ ($b \in A$) are homogeneous. Then the following conditions are equivalent for each closed subset $Z$ of $X$:*

(i) *$\psi(F, X) = \tau$ for each closed subset $F$ of $Z$.*

(ii) *$Z$ is a $Z_\tau$-set in $X$.*

PROOF. (i) $\Longrightarrow$ (ii). Suppose that the collection $\{\mathcal{U}_t: t \in T\}$, $|T| < \tau$, of open covers of $X$ is given. Our goal is the construction of a map $f\colon X \to X$ which is $\{\mathcal{U}_t: t \in T\}$-close to $id_X$ and which satisfies the equality $f(X) \cap Z = \emptyset$.

Since $|T| < \tau$, there exist a subset $B \subseteq A$ of cardinality $\kappa = \max\{\omega, |T|\}$ and a collection $\{\mathcal{V}_t : t \in T\}$ of open covers of $X_B$ such that $pi_B^{-1}(\mathcal{V}_t)$ refines $\mathcal{U}_t$ for each $t \in T$ . Without loss of generality we may assume that $b \in B$. Note also that $\kappa < \tau$. In order to proceed, we need the following statement.

*Claim. There exist a subset $D \subseteq A$ such that $B \subseteq D$, $|D| = \kappa$, and a section $i: X_B \to X_D$ such that $i(X_B) \cap \pi_D(Z) = \emptyset$.*

*Proof of Claim.* Let $B_0 = B$ and let $i_0$ denote the identity map of $X_{B_0}$. Suppose that for each $\beta < \alpha$, where $\alpha < \kappa^{+2}$, we have already constructed subsets $B_\beta$ of $A$ and sections $i_\beta : X_{B_0} \to X_{B_\beta}$ of the projection $\pi_{B_0}^{B_\beta} : X_{B_\beta} \to X_{B_0}$ satisfying the following conditions:

(a) $|B| = \kappa$.
(b) $B_\delta \subseteq B_\beta$ whenever $\delta \leq \beta$.
(c) $B_\beta = \cup\{B_\delta : \delta < \beta\}$ whenever $\beta$ is a limit ordinal.
(d) $\pi_{B_\delta}^{B_\beta} i_\beta = i_\delta$ whenever $\delta \leq \beta$.
(e) $i_\beta = \lim\{i_\delta : \delta < \beta\}$ whenever $\beta$ is a limit ordinal.

For each $\beta < \alpha$, let

$$V_\beta = \{x \in X_{B_0} : i_\beta(x) \notin \pi_{B_\beta}(Z)\}.$$

Clearly, $V_\beta$ is an open subset of $X_{B_0}$. It follows from the above conditions that:

(f) $V_\delta \subseteq V_\beta$ whenever $\delta \leq \beta$.
(g) $V_\beta = \cup\{V_\delta : \delta < \beta\}$ whenever $\beta$ is a limit ordinal.

Let us construct the set $B_\alpha$ and the section $i_\alpha$. If $\alpha$ is a limit ordinal let

$$B_\alpha = \cup\{B_\beta : \beta < \alpha\} \text{ and } i_\alpha = \lim\{i_\beta : \beta < \alpha\}.$$

Consider now the case $\alpha = \beta + 1$. Fix any section $j : X_{B_\beta} \to X$ of the projection $\pi_{B_\beta} : X \to X_{B_\beta}$. There are two possible cases: $j(X_{B_\beta}) \cap Z = \emptyset$ and $j(X_{B_\beta}) \cap Z \neq \emptyset$. We consider each case separately.

*Case 1 ($j(X_{B_\beta}) \cap Z = \emptyset$).* In this case there is a finite subset $C$ of $A$ such that

$$\pi_C(j(X_{B_\beta})) \cap \pi_C(Z) = \emptyset.$$

Let

$$B_\alpha = B_\beta \cup C \text{ and } i_\alpha = \pi_{B_\alpha} j i_\beta.$$

Clearly, $i\emptyset_\alpha(X_{B_0}) \cap \pi_{B_\alpha}(Z) = \emptyset$. Consequently, $D = B_\alpha$ and $i = i_\alpha$ verify the claim in this case.

*Case 2 ($j(X_{B_\beta}) \cap Z \neq \emptyset$).* Choose a point $z \in j(X_{B_\beta}) \cap Z$ and let $y = \pi_{B_\beta}(z)$. Note that $y \in \pi_{B_\beta}(Z)$, and hence $y \notin V_\beta$. Consider the set $\pi_{B_\beta}(y)$. Since $|B_\beta| = \kappa < \tau$, the character of $y$ in $X_{B_\beta}$ is $\leq \kappa$. Consequently, $\psi(\pi_{B_\beta}(y), X) \leq \kappa < \tau$. By (i), $\pi_{B_\beta}(y)$ is not contained in $Z$. Choose a point $z' \in \pi_{B_\beta}(y) - Z$. There exists a finite subset $C$ of $A$ such that $\pi_C(z') \notin \pi_C(Z)$. Let $B_\alpha = B_\beta \cup C$. Clearly

---

[2] $\kappa^+$ denotes the smallest cardinal greater than $\kappa$.

$$y_1 \equiv \pi_{B_\alpha}(j(y)) \neq \pi_{B_\alpha}(z') \equiv y_2 \text{ and } \pi_{B_\beta}(y_1) = y = \pi_{B_\beta}(y_2).$$

Consequently

$$y_1, y_2 \in (\pi_{B_\beta})^{-1}(y) \subseteq \{y\} \times X_{C-B_\beta}.$$

Since $X_{C-B_\beta}$ is homogeneous (as a product of homogeneous compacta; $b \notin C - B_\beta$) there is a homeomorphism

$$h\colon X_{C-B_\beta} \to X_{C-B_\beta}$$

such that $h(\pi_{C-B_\beta}(y_1)) = \pi_{C-B_\beta}(y_2)$. Let

$$H = id_{X_\beta} \times h.$$

Then $H$ is an autohomeomorphism of $X_{B_\alpha}$ such that $H(y_1) = y_2$ and $\pi^{B_\alpha}_{B_\beta} H = \pi^{B_\alpha}_{B_\beta}$. Let

$$i_\alpha = H\pi_{B_\alpha} j i_\beta\colon X_{B_0} \to X_{B_\alpha}.$$

It is easy to see that $V_\beta$ is a proper subset of $V_\alpha$ ($y \in V_\alpha - V_\beta$).

This completes the inductive step. Thus, subsets $B_\alpha$ and sections $i_\alpha$ satisfying the above conditions exist for each $\alpha < \kappa^+$. The family $\{V_\alpha\colon \alpha < \kappa^+\}$ of the open subsets of $X_{B_0}$ described above is increasing and has length $\kappa$. Since the weight of $X_{B_0}$ is $\kappa$, it follows that the collection $\{V_\alpha\colon \alpha < \kappa\}$ must stabilize, i.e. there must be an index $\beta < \kappa^+$ such that $V_\alpha = V_\beta$ for each $\alpha \geq \beta$. Because of our construction, this is possible only when $V_\beta = X_{B_0}$. Let $D = B_\beta$ and $i = i_\beta$. The proof of the claim is complete.

We now complete the proof of our proposition. Fix a subset $D$ and a section $i$ as in the Claim. Let $j\colon X_D \to X$ be any section of the projection $\pi_D\colon X \to X_D$. Then $f = ji\pi_B\colon X \to X$ is $\{\mathcal{U}_t\colon t \in T\}$-close to $id_X$ and $f(X) \cap Z = \emptyset$.

(ii) $\Longrightarrow$ (i). This is straightforward and so is omitted. □

COROLLARY 8.5.7. *Let $\tau > \omega$, and let $X$ be a closed subset of a compact $I^\tau$-manifold $Y$. If $\dim X < \infty$ or $w(X) < \tau$, then $X$ is a $Z_\tau$-set in $Y$.*

PROOF. Suppose that $X$ is not a $Z_\tau$-set in $Y$. By Proposition 8.5.6, there must exist a closed subset $F$ of $X$ such that $\psi(F, Y) < \tau$. In this case, $F$ depends on fewer than $\tau$ coordinates and, consequently, contains a copy of $I^\tau$. In either case this is impossible. □

**8.5.2. Fixed point sets of Tychonov cubes.** The main result of this Subsection (Theorem 8.5.9) gives a complete characterization of compacta which can be fixed in Tychonov cubes as limit spaces of transfinite inverse spectra whose first elements are metrizable compacta and all short projections of which are stable in the sense of Definition 8.5.8. Here is a simple scheme illustrating how this result might be used. Suppose that $X$ is a compactum of weight $\omega_1$. Consider any $\omega$-spectrum $\mathcal{S}_X$ whose limit space coincides with $X$. Converting $\mathcal{S}_X$ into a transfinite spectrum, we may assume from the outset that $\mathcal{S}_X = \{X_\alpha, p^{\alpha+1}_\alpha, \omega_1\}$

is a transfinite spectrum of length $\omega_1$ consisting of metrizable compacta. We now investigate the short projections of $\mathcal{S}_X$. Theorem 8.5.9 says that if for "many" $\alpha$'s, the projection $p_\alpha^{\alpha+1}$ is stable, then $X$ can be fixed in the Tychonov cube of weight $\omega_1$. This is in some sense the obvious part. The more important fact is that if for "many" $\alpha$'s, the projection $_\alpha^{a+1}$ is not stable, then $X$ cannot be fixed in that cube. Let us emphasize this point again. The spectrum $\mathcal{S}_X$ is chosen arbitrarily (there are many others representing $X$). Nevertheless, the behavior of the short projections (from the point of view of stability) of a given spectrum allows us to conclude the existence or non-existence of an embedding of $X$ into $I^{\omega_1}$ whose image is the fixed point set of some self-mapping of $I^{\omega_1}$. For example, any (uncountable) product of metrizable compacta can be represented as the limit space of the standard spectrum consisting of the corresponding subproducts and natural projections between them. Any closed subspace $X$ of such a product is then the limit space of an induced spectrum $\mathcal{S}_X$ . If, additionally, $X$ is a retract of that product, then we can conclude that "many" projections of $\mathcal{S}_X$ are open retractions (because the projections of the spectrum representing the whole product are obviously open retractions). Therefore they are stable (Proposition 8.5.13) and $X$ can be fixed in the Tychonov cube of the corresponding weight. This proves Proposition 8.5.14.

DEFINITION 8.5.8. *We say that a map $p\colon X \to Y$ is* stable *if there is an embedding $i\colon X \to Y \times I^\omega$ and a map $f\colon Y \times I^\omega \to Y \times I^\omega$ such that $\pi_Y i = p$, $\pi_Y f = \pi_Y$ and $fix(f) = i(X)$, where $\pi_Y\colon Y \times I^\omega \to Y$ is the projection.*

THEOREM 8.5.9. *Let $\tau > \omega$. The following conditions are equivalent for any compactum $X$:*

(a) *$X$ can be fixed in the Tychonov cube $I^\tau$ .*
(b) *$X$ can be fixed in any compact $I^\tau$ -manifold.*
(c) *$X$ can be fixed in some compact $I^\tau$ -manifold.*
(d) *$X$ can be represented as the limit space of some transfinite inverse spectrum $\mathcal{S}_X = \{X_\alpha, p_\alpha^{\alpha+1}, \tau\}$ whose short projections $p_\alpha^{\alpha+1}\colon X_{\alpha+1} \to X_\alpha$ are stable and whose first element $X_0$ is metrizable.*

PROOF. (a) $\Longrightarrow$ (b). Any compact $I^\tau$-manifold $Y$ contains a copy of $I^\tau$. Fix a retraction $r\colon Y \to I^\tau$ and a map $f\colon I^\tau \to I^\tau$ such that $fix(f) = X$. Then $fix(g) = X$, where $g = fr$.

(b) $\Longrightarrow$ (c). Trivial.

(c) $\Longrightarrow$ (d). Suppose that $Y$ is a compact $I^\tau$ -manifold and $fix(f) = X$ for some map $f\colon Y \to Y$. Represent $Y$ as a product $Y = \prod\{Y_a\colon a \in A\}$ of compact $I^\omega$ -manifolds, where all $Y_a$'s with $a \neq b$ are copies of the Hilbert cube $I^\omega$ ($b \in A$) and $A$ is a set of cardinality $\tau$.

Fix a well ordering $\{a_\alpha\colon \alpha < \tau\}$ of $A$ such that $a_0 = b$. By Theorem 1.3.16, there is a countable subset $A_0$ of $A$, containing $a_0$, and a map $f_0\colon Y_{A_0} \to Y_{A_0}$ such that $\pi_{A_0} f = f_0 \pi_{A_0}$. Let us show that $\pi_{A_0}(X) = fix(f_0)$. Indeed, if $x \in \pi_{A_0}(X)$, then there is a point $y \in X$ such that $\pi_{A_0}(y) = x$. Therefore

$$f_0(x) = f_0(\pi_{A_0}(y)) = \pi_{A_0}(f(y)) = \pi_{A_0}(y) = x$$

and $X_{A_0} \subseteq fix(f_0)$. Conversely, if $x \in fix(f_0)$, then the same equality $\pi_{A_0} f = f_0 \pi_{A_0}$ shows that $f(\pi_{A_0}^{-1}(x)) \subseteq \pi_{A_0}^{-1}(x)$. Being a topological copy of $I^\omega$, the fiber $\pi_{A_0}^{-1}(x)$ has the fixed point property (note that $b \notin A - A_0$). Therefore there is a point $y \in \pi_{A_0}^{-1}(x)$ such that $f(y) = y$. Then $y \in fix(f) = X$. It only remains to note that $x = \pi_{A_0}(y) \in \pi_{A_0}(X)$. Therefore $\pi_{A_0}(X) = fix(f_0)$.

Suppose now that for each $\beta < \alpha$, where $\alpha < \tau$, we have already constructed a subset $A_\beta$ of $A$ and a map $f_\beta \colon Y_{A_\beta} \to Y_{A_\beta}$ in such a way that the following conditions are satisfied:

(i) $A_\beta \subseteq A_{\beta+1}$ and $|A_{\beta+1} - A_\beta| \leq \omega$.
(ii) $A_\beta = \cup\{A_\delta \colon \delta < \beta\}$ whenever $\beta$ is a limit ordinal.
(iii) $\{a_\delta \colon \delta < \beta\} \subseteq A_\beta$ whenever $\beta > 0$.
(iv) $\pi_{A_\beta} f = f_\beta \pi_{A_\beta}$.
(v) $\pi_{A_\delta}^{A_\beta} f_\beta = f_\delta \pi_{A_\delta}^{A_\beta}$ whenever $\delta \leq \beta$.
(vi) $f_\beta = \lim\{f_\delta \colon \delta < \beta\}$ whenever $\beta$ is a limit ordinal.
(vii) $fix(f_\beta) = \pi_{A_\beta}(X)$.

Let us construct the set $A_\alpha$ and a map $f_\alpha \colon Y_{A_\alpha} \to Y_{A_\alpha}$ with the corresponding properties. If $\alpha$ is a limit ordinal, let

$$A_\alpha = \cup\{A_\beta \colon \beta < \alpha\} \text{ and } f_\alpha = \lim\{f_\beta \colon \beta < \alpha\}.$$

Suppose now that $\alpha = \beta+1$. Let $\gamma$ denote the smallest ordinal such that $a_\gamma \notin A_\beta$. By Theorem 1.3.16, there is a countable subset $C_\beta$ of $A - A_\beta$, containing $a_\gamma$, and a map

$$f_\alpha \colon Y_{A_\beta \cup C_\beta} \to Y_{A_\beta \cup C_\beta}$$

such that $\pi_{A_\beta \cup C_\beta} f = f_\alpha \pi_{A_\beta \cup C_\beta}$. Let $A_\alpha = A_\beta \cup C_\beta$. This obviously completes the inductive step of our construction. Consequently, the corresponding objects are defined for each $\alpha < \tau$.

Let $X_\alpha = \pi_{A_\alpha}(X)$ and $p_\alpha^{\alpha+1} = \pi_{A_\alpha}^{A_{\alpha+1}}/X_{\alpha+1}$ for each $\alpha < \tau$. Obviously, the limit space of the transfinite inverse spectrum $\mathcal{S}_X = \{X_\alpha, p_\alpha^{\alpha+1}, \tau\}$ coincides with $X$.

It only remains to note that $\pi_1 g = \pi_1$ and $fix(g) = X_{\alpha+1}$, where

$$\pi_1 \colon X_\alpha \times I^{C_\alpha - A_\alpha} \to X_\alpha$$

denotes the projection onto the first coordinate and

$$g = f_{\alpha+1}/(X_\alpha \times I^{C_\alpha - A_\alpha}).$$

This shows that $p_\alpha^{\alpha+1}$ is stable in the above sense.

(d) $\Longrightarrow$ (a). Since for each $\alpha < \tau$ the short projection $p_\alpha^{\alpha+1} \colon X_{\alpha+1} \to X_\alpha$ is stable, the compactum $X_{\alpha+1}$ can be identified with the subspace of the product

$X_\alpha \times I^\omega$ so that $p_\alpha^{\alpha+1}$ coincides with the restriction of the projection $\pi_1^\alpha\colon X_\alpha \times I^\omega \to X_\alpha$ and $fix(f_{\alpha+1}) = X_{\alpha+1}$ for some map

$$f_{\alpha+1}\colon X_\alpha \times I^\omega \to X_\alpha \times I^\omega$$

with $\pi_1^\alpha f_{\alpha+1} = \pi_1^\alpha$.

Embed $X_0$ into the Hilbert cube $Q^0$ and choose a map $g_0\colon Q^0 \to Q^0$ such that $fix(g_0) = X_0$ (see **[216]** for the possibility of choosing a homeomorphism such as $g_0$; compare with the next Section). We proceed by transfinite induction. Suppose that for each $\beta < \alpha$, where $0 < \alpha < \tau$, we have already made an identification of $X_\beta$ with a subspace of the Tychonov cube $Q^\beta$ and have constructed a map $g_\beta\colon Q^\beta \to Q^\beta$ such that:

$(i)_\beta$ $fix(g_\beta) = X_\beta$.

$(ii)_\beta$ $p_\delta^\beta = \pi_\delta^\beta / X_\beta$ whenever $\delta < \beta$ (here $\pi_\beta^\alpha\colon Q^\alpha \to Q^\beta$ denotes the standard projection onto the corresponding subproduct).

$(iii)_\beta$ $\pi_\delta^\beta g_\beta = g_\delta \pi_\delta^\beta$ whenever $\delta < \beta$.

$(iv)_\beta$ $g_\beta = \lim\{g_\delta\colon \delta < \beta\}$.

Let us carry out the inductive step. If $\alpha$ is a limit ordinal, let $g_\alpha = \lim\{g_\beta\colon \beta < \alpha\}$.

Conditions $(ii)_\alpha - -(iv)_\alpha$ are trivially satisfied. Let us verify condition $(i)_\alpha$.

The identifications made above allow us to represent $X_\alpha$ as the limit space of an inverse spectrum $\mathcal{S}_\alpha = \{X_\beta, p_\beta^{\beta+1}, \beta < \alpha\}$. Consequently, if $x \in X_\alpha$, then

$$g_\alpha(x) = \triangle\{g_\beta \pi_\beta^\alpha(x)\colon \beta < \alpha\}.$$

But $g_\beta \pi_\beta^\alpha(x) \in X_\beta$ and, therefore, by $(i)_\beta$, $g_\beta \pi_\beta^\alpha(x) = \pi_\beta^\alpha(x)$. This shows that $g_\alpha(x) = x$ and $X_\alpha \subseteq fix(g_\alpha)$. Conversely, if $x \notin X_\alpha$, then for some $\beta < \alpha$ we have $\pi_\beta^\alpha(x) \notin X_\beta$. Again, by $(i)_\beta$, this means that $g_\beta \pi_\beta^\alpha(x) \neq \pi_\beta^\alpha(x)$. Therefore $x \neq g_\alpha(x)$ and $x \notin fix(g_\alpha)$. This finishes the verification of condition $(i)_\alpha$.

Suppose now that $\alpha = \beta+1$. Since $\pi_1^\beta$ is a trivial bundle (with the Hilbert cube as a fiber) there exists a map $g_\alpha\colon Q^\alpha \to Q^\alpha$ which extends $f_\alpha\colon X_\beta \times Q \to X_\beta \times Q$ (i.e. $g_\alpha/(X_\beta \times Q) = f_\alpha$) and which commutes with $g_\beta$ (i.e. $\pi_\beta^\alpha g_\alpha = g_\beta \pi_\beta^\alpha$). Then $X_\alpha = fix(g_\alpha)$. Indeed, since $g_\alpha$ is an extension of $f_\alpha$ and $X_\alpha = fix(f_\alpha)$, it follows immediately that $X_\alpha \subseteq fix(g_\alpha)$. Conversely, suppose that $x \in fix(g_\alpha)$. Then $x = g_\alpha(x)$ and, consequently

$$\pi_\beta^\alpha(x) = \pi_\beta^\alpha(g_\alpha(x)) = g_\beta(\pi_\beta^\alpha(x)).$$

This means that $\pi_\beta^\alpha(x) \in fix(g_\beta) = X_\beta$. Therefore $x \in X_\beta \times Q$. But $g_\alpha$ and $f_\alpha$ coincide on this set and we have $x \in fix(f_\alpha) = X_\alpha$. This completes the inductive step.

In this situation, it can be easily seen that $X$, as a subset of $I^\tau$, can be fixed there by the map $g = \lim\{g_\alpha\colon \alpha < \tau\}$. □

LEMMA 8.5.10. *Suppose that the compactum $X$ can be fixed in $I^\tau$, where $\tau \geq \omega$ and $\tau > w(X)$. Then $X$ can be fixed in $I^\kappa$, where $\kappa = \max\{w(X), \omega\}$.*

PROOF. The statement is non-trivial when $\tau > \kappa$. It follows easily from Theorem 1.3.4, for $\kappa$-spectra, that there exists a subset $A$ of $\tau$ such that $|A| = \kappa$ and the restriction $\pi_B/X: X \to \pi_B(X)$ of the projection $\pi_B: I^\tau \to I^B$ is a homeomorphism for each $B$ with $A \subseteq B \subseteq \tau$. Let $f: I^\tau \to I^\tau$ be a map such that $fix(f) = X$. Using Theorem 1.3.4 for $\kappa$-spectra (with respect to the standard spectrum consisting of $\kappa$-subproducts of $I^\tau$ and natural projections) we can represent $f$ as the limit map of some morphism

$$\mathcal{F} = \{f_C: I^C \to I^C; C \in \mathcal{K}\}$$

where $\mathcal{K}$ is a cofinal and $\kappa$-closed subset of $exp_\kappa \tau$. Consequently, there exists $C \in \mathcal{K}$ such that $A \subseteq C$. Let us show that $\pi_C(X)$ (which is copy of $X$) is fixed in $I^C$ by the map $f_C$, i.e. $fix(f_C) = \pi_C(X)$. What we need is the equation $\pi_C f = f_C \pi_C$, which follows from the fact that $\mathcal{F}$ is a self-morphism of the above mentioned spectrum. Let $y \in \pi_C(X)$, and fix a point $x \in X$ such that $\pi_C(x) = y$. Then

$$f_C(y) = f_C\pi_C(x) = \pi_C f(x) = \pi_C(x) = y.$$

Consequently, $\pi_C(X) \subseteq fix(f_C)$. Conversely, if $y \in fix(f_C) - \pi_C(X)$, then, using the equality $\pi_C f = f_C \pi_C$, we may conclude that $f(\pi_C^{-1}(y)) \subseteq \pi_C^{-1}(y)$. But $\pi_C^{-1}(y)$ is, topologically, the Tychonov cube $I^\tau$ (because $\pi_C: I^\tau \to I^C$ is the natural projection onto the corresponding subproduct). Therefore there is a point $x \in \pi_C^{-1}(y)$ such that $f(x) = x$. The last equality implies that $x \in X$. On the other hand $x \notin X$ (because $y \notin \pi_C(X)$). This contradiction shows that $fix(f_C) - \pi_C(X) = \emptyset$ and consequently $\pi_C(X) = fix(f_C)$. We now need only to recall that $\pi_C(X)$ is homeomorphic to $X$. □

PROPOSITION 8.5.11. *There is a zero-dimensional compactum $K$ of weight $\omega_1$ which admits an embedding into every non-metrizable $ANR$-compactum but cannot be fixed in any of them.*

PROOF. Let $K$ denote the zero-dimensional closed subspace of $I^{\omega_1}$ constructed in [**197**]. Recall that $w(K) = \omega_1$ and that $K$ cannot be represented as the set of fixed points of any continuous self-map of $I^{\omega_1}$. Since every non-metrizable $ANR$-compactum contains a copy of the Tychonov cube $I^{\omega_1}$, we see that $K$ admits an embedding into any non-metrizable $ANR$-compactum. Suppose that $K$ can be fixed in some $ANR$-compactum. Since each $ANR$-compactum is a retract of some compact $I^\tau$-manifold, we conclude that $K$ can be fixed in a compact $I^\tau$-manifold (for some $\tau \geq \omega_1$) as well. By Theorem 8.5.9, $K$ can be fixed in $I^\tau$. By Lemma 8.5.10, $K$ can be fixed in $I^{\omega_1}$. Since $K$ is zero-dimensional, by Corollary 8.5.7, all copies of $K$ in $I^{\omega_1}$ are $Z_{\omega_1}$-sets. Therefore, by Theorem 8.5.4, all embeddings of $K$ in $I^{\omega_1}$ are equivalent. This contradiction finishes the proof □

COROLLARY 8.5.12. *There is no non-metrizable compact absolute retract with the complete invariance property.*

Applying Theorem 8.5.9 to the compactum $K$ from Proposition 8.5.11, we may conclude (see the beginning of this section) that there is a surjection $p\colon X \to Y$ between zero-dimensional metrizable compacta which is not stable in the above sense. On the other hand, the following statement provides a wide class of stable maps.

PROPOSITION 8.5.13. *Let $f\colon X \to Y$ be an open retraction between compacta. Suppose also that $X$ is embedded in the product $Y \times I^\omega$ so that the restriction $\pi_Y/X$ of the projection $\pi_Y\colon Y \times I^\omega \to Y$ coincides with $f$. Then there exists a map $g\colon Y \times I^\omega \to Y \times I^\omega$ such that the following conditions are satisfied:*

(i) $fix(g) = X$.
(ii) $\pi_Y g = \pi_Y$.

PROOF. By Theorem 6.3.1, the general case reduces to the case when $Y$ (and, consequently, $X$) is metrizable.

For any point $r \in [0,1]$, define a homotopy $H_r\colon [0,1] \times [0,1] \to [0,1]$ by the formula $H_r(x,t) = (1-t)x + rt$. Note that

(i) $H_r(x,0) = x$ for any $x \in [0,1]$.
(ii) $H_r(x,1) = r$ for any $x \in [0,1]$.
(iii) $H_r(x,t) \neq x$ whenever $t > 0$ and $x \neq r$.
(iv) The homotopies $H_r$ and $H_s$ are close (as maps of $[0,1]^2$ into $[0,1]$) whenever the points $r$ and $s$ are sufficiently close.

Suppose now that $p = \{p_i\colon i \in \omega\}$ is a point in the Hilbert cube $I^\omega$. Define the homotopy $H_p\colon I^\omega \times [0,1] \to I^\omega$ as follows:

$$H_p(\{x_i\colon i \in \omega\}, t) = \{H_p(x_i, t)\colon i \in \omega\}.$$

As above, note that:

(v) $H_p(x,0) = x$ for any $x \in I^\omega$.
(vi) $H_p(x,1) = p$ for any $x \in I^\omega$.
(vii) $H_p(x,t) \neq x$ whenever $t > 0$ and $x \neq p$.
(viii) The homotopies $H_p$ and $H_q$ are close (as maps of $I^\omega \times [0,1]$ into $I^\omega$) whenever the points $p$ and $q$ are sufficiently close (in $I^\omega$).

Fix any section $j\colon Y \to X$ of the retraction $f$ ( i.e. $fj = id_Y$). Also fix metrics $r_1$ and $r_2$ on $Y$ and $I^\omega$, respectively. On the product $Y \times I^\omega$, consider a metric $d$ bounded by 1 and equivalent to the metric $\sqrt{r_1^2 + r_2^2}$. It follows directly from the above listed properties that the homotopy

$$H\colon Y \times I^\omega \times [0,1] \to Y \times I^\omega$$

defined by the formula $H(y,x,t) = (y, H_{i(y)}(x,t))$ has the following properties:

(ix) $H(y,x,0) = (y,x)$ for any $(y,x) \in Y \times I^\omega$.
(x) $H(y,x,1) = (y,i(y))$ for any $(y,x) \in Y \times I^\omega$.
(xi) $H(y,x,t) \neq (y,x)$ whenever $t > 0$ and $x \neq i(y)$.

Define the map $g$ as follows:

$$g(y,x) = H(y,x,d(x,f^{-1}(y))) \text{ for any point } (y,x) \in Y \times I^\omega .$$

The openness of the map $f$ implies that $g$ is continuous. It only remains to note that $\pi_Y g = \pi_Y$ and $fix(g) = X$. □

PROPOSITION 8.5.14. *A retract of any product of an arbitrary family of metrizable compacta can be fixed in the Tychonov cube of the corresponding weight.*

PROOF. Apply Theorem 8.5.9 and Proposition 8.5.13. □

Finally, we present a simple example of a stable non-open retraction $p\colon X \to Y$. Let $Y$ be the closed segment $[-1,1]$. Consider the subspace

$$X = [-1,0] \times [0,1] \cup [0,1] \times \{0\}$$

of the product $Y \times [0,1]$ and let

$$A = [0,1] \times \{0\} \cup \{0\} \times [0,1].$$

The map $p$ coincides with the restriction of the projection $\pi_1\colon Y \times [0,1] \to Y$ onto $X$. Obviously, $p$ is a non-open retraction.

Define a deformation $H\colon [0,1] \times [0,1] \to [0,1]$ by $H(t,s) = t(1-s)$, $(t,s) \in [0,1]^2$. Let $g\colon Y \times [0,1] \to Y \times [0,1]$ be defined as follows:

$$g(y,t) = \begin{cases} (y, H(t, d((y,t),A))), & \text{if } (y,t) \in [0,1] \times [0,1] \\ (y,t), & \text{if } (y,t) \in [-1,0] \times [0,1]. \end{cases}$$

Here $d$ denotes the standard metric on the product $[0,1] \times [0,1]$. Clearly $\pi_1 g = \pi_1$ and $fix(g) = X$. Embed the segment $[0,1]$ into the Hilbert cube $I^\omega$ and fix a retraction $r\colon I^\omega \to [0,1]$. Finally, let

$$f = g(id_Y \times r)\colon Y \times I^\omega \to Y \times I^\omega .$$

Obviously, $\pi_Y f = \pi_Y$ (where $\pi_Y\colon Y \times I^\omega \to Y$ is the projection), $\pi_Y/X = p$ and $fix(f) = X$. This shows the stability of $f$.

*Historical and bibliographical notes* 8.5. The results of this Section originally appeared in [**107**].

## 8.6. Compact groups and fixed point sets

In this Section we investigate spaces with the *complete invariance property with respect to homeomorphisms*, shortly, CIPH. A space $X$ has the CIPH if for each non-empty closed subset $F$ of $X$ there is a homeomorphism $h_F\colon X \to X$ such that $F = \{x \in X : x = h_F(x)\}$. Recall that if, in the above definition, $h_F$ is only required to be a continuous map, then we say that $X$ has the CIP (*complete invariance property*; see Section 8.5). Closed surfaces, even-dimensional disks, positive-dimensional spheres and the Hilbert cube are all known to have CIPH (see [**268**], [**269**] and [**216**] respectively). In certain situations the existence of a free action of a "nice" group $G$ on a given space $X$ allows one to draw an even stronger conclusion: some closed subset $F \subseteq X$ is the fixed point set of each non-trivial homeomorphism $g \in G$ (see [**311**], [**182**]).

### 8.6.1. Structure theorems for compact abelian groups.

In this Subsection a general structure theorem for compact abelian groups is proven and this is used as a basis for a self-contained development yielding an explicit structure theorem for finite-dimensional compact abelian groups. These results are used to show that a nondegenerate compact metrizable group has the CIPH iff it is infinite. Furthermore, it is shown that the product of two metrizable spaces has the CIPH if one of the factors is a positive-dimensional compact group.

We assume as known the duality between the category of compact abelian groups and that of discrete abelian groups according to Pontryagin and van Kampen. For two topological abelian groups $A$ and $B$ we shall always denote by $\mathrm{Hom}(A, B)$ the topological abelian group of all continuous group homomorphisms from $A$ to $B$ equipped with the compact open topology. We use the notation $\mathbb{R}$ for the locally compact additive group of reals and $\mathbb{T} = \mathbb{R}/\mathbb{Z}$ for the circle group. We write $\widehat{G} = \mathrm{Hom}(G, \mathbb{T})$ for the group of characters $\chi\colon G \to \mathbb{T}$ (continuous in the category of compact abelian groups). The duality theorem states that the evaluation morphism $\eta_G\colon G \to \widehat{\widehat{G}}$, $\eta_G(g)(\chi) = \chi(g)$ is an isomorphism in all cases. One identifies $\widehat{\mathbb{R}}$ with $\mathbb{R}$ in a natural way.

DEFINITION 8.6.1. *For a compact topological group $G$ we write*

$$L(G) \stackrel{def}{=} \mathrm{Hom}(\mathbb{R}, G) \cong \mathrm{Hom}(\widehat{G}, \mathbb{R})$$

*and note that $L(G)$ is a real topological vector space with the topology of uniform convergence on compact sets. We set* $\exp\colon L(G) \to G$, $\exp X = X(1)$.

All one-parameter subgroups of $G$ are of the form $X = (r \mapsto \exp r \cdot X)\colon \mathbb{R} \to G$. We note that $\mathrm{Hom}(\widehat{G}, \mathbb{R}) \cong \mathrm{Hom}(\mathbb{R} \otimes \widehat{G}, \mathbb{R})$ is the vector space of *all* linear functionals on the real vector space $\mathbb{R} \otimes \widehat{G}$, i.e., the algebraic dual $(\mathbb{R} \otimes \widehat{G})^*$ given the weak-* topology. The only isomorphy invariant is the cardinal

$$d \stackrel{def}{=} \dim_{\mathbb{R}}(\mathbb{R} \otimes \widehat{G}) = \dim_{\mathbb{Q}}(\mathbb{Q} \otimes \widehat{G}) = \operatorname{rank} \widehat{G}.$$

Therefore, $L(G)$ is algebraically and topologically isomorphic to $\mathbb{R}^d$.

PROPOSITION 8.6.2. *For a compact abelian group $G$ there is a compact zero-dimensional subgroup $\Delta$ such that the homomorphism*

$$\phi\colon \Delta \times L(G) \to G, \quad \phi(d, X) = d \exp X$$

*satisfies the following conditions:*

(i) *$\phi$ is continuous, surjective and open, i.e., is a quotient morphism.*

(ii) *$\ker \phi$ is algebraically and topologically isomorphic to $D \overset{def}{=} \exp^{-1}(\Delta)$, and $D$ is a closed totally disconnected subgroup of $L(G)$. In particular, it does not contain any nonzero vector spaces.*

(iii) *$\phi(\{1\} \times L(G)) = \exp L(G)$ is dense in $G_0$, the identity component of $G$.*

PROOF. By Zorn's Lemma, the abelian group $\widehat{G}$ contains a free subgroup $F$ of maximal rank. Then $E \overset{def}{=} \widehat{G}/F$ is a torsion group. We set $\Delta = F^{\perp}$ (the annihilator of $F$ in $G$). Then, by duality, $\widehat{\Delta} \cong E$. Since $E$ is a torsion group, $\Delta$ is a totally disconnected subgroup. Clearly, $\phi\colon \Delta \times L(G) \to G$ defined by $\phi(d, X) = d \exp X = dX(1)$ is a well defined continuous homomorphism. Further, $T \overset{def}{=} \widehat{F}$, as the character group of a direct sum of copies of $\mathbb{Z}$, is a product of copies of $\mathbb{T} \cong \widehat{\mathbb{Z}}$, i.e., a torus. In particular, $\exp\colon L(T) \to T$ is surjective. Let $p\colon G \to T$ denote the quotient morphism which identifies $G/\Delta$ with $T$. The morphism

$$L(p) = \operatorname{Hom}(\mathbb{R}, p)\colon L(G) = \operatorname{Hom}(\mathbb{R}, G) \to \operatorname{Hom}(\mathbb{R}, T) = L(T)$$

is an isomorphism since $\operatorname{Hom}(\mathrm{incl}, \mathbb{R})\colon \operatorname{Hom}(\widehat{G}, \mathbb{R}) \to \operatorname{Hom}(F, \mathbb{R})$ is an isomorphism (as $\operatorname{Hom}(E, \mathbb{R}) = \{0\}$).

Proof of (i). *The surjectivity of $\phi$*: The exact sequence

$$0 \to F \xrightarrow{\mathrm{incl}} \widehat{G} \xrightarrow{\mathrm{quot}} E \to 0$$

gives an exact sequence of compact groups

$$0 \to \Delta \xrightarrow{\mathrm{incl}} G \xrightarrow{p} T \to 0.$$

Since

$$\begin{array}{ccc} L(G) & \xrightarrow{L(p)} & L(T) \\ {\scriptstyle \exp_G}\downarrow & & \downarrow{\scriptstyle \exp_T} \\ G & \xrightarrow{p} & T \end{array}$$

is commutative, $p(\exp_G L(G)) = T$. Hence $\Delta \exp L(G) = (\ker p) \exp L(G) = G$. Thus $\phi$ is surjective.

*The openness of* $\phi$: For every subgroup $F'$ of $F$ with finite index we get exact sequences

$$0 \to F' \xrightarrow{\text{incl}} \widehat{G} \xrightarrow{\text{quot}} E' \to 0$$

and

$$0 \to \Delta' \xrightarrow{\text{incl}} G \xrightarrow{p'} T' \to 0,$$

and $\Delta'$ has finite index in $\Delta$. We note that

- the family of all $\Delta'$ intersects in $\{1\}$,
- the argument showing $\Delta \exp L(G) = G$ above shows in exactly the same fashion that $\Delta' \exp L(G) = G$.

Since $F/F'$ is finite, we can write $F = F_1 \oplus F_2$ such that $F' = F_1 \oplus F_2'$ with $F_2/F_2' \cong F/F'$.

Accordingly, we have $T = T_1 \times T_2$, $T' = T_1 \times T_2'$ (with a natural identification), $\text{Hom}(F, \mathbb{R}) = \text{Hom}(F_1, \mathbb{R}) \oplus \text{Hom}(F_2, \mathbb{R})$, and thus $L(T) = L(T)_1 \oplus L(T)_2$. Correspondingly, $L(G) = L(G)_1 \oplus L(G)_2$. We write

$$0 \to \Delta' \xrightarrow{\text{incl}} G \xrightarrow{p'} T_1 \oplus T_2' \to 0.$$

Suppose that $p'(g) \in T_1$. Then there is an $X_1 \in L(G)_1$ such that

$$\exp_{T_1} L(p')(X_1) = p'(\exp_G X_1) = p'(g).$$

Thus $g \exp_G(-X_1) \in \ker p' = \Delta'$. Therefore,

$$\Delta' \exp_G L(G)_1 = (p')^{-1}(T_1).$$

Hence $\Delta' \exp_G L(G)_1$ is closed in $G$.

Let $U \times V$ be a product zero-neighborhood of $\Delta \times L(G)$. Then $U$ contains a subgroup $\Delta'$ of $\Delta$ which is open in $\Delta$ and has finite index in $\Delta$ and $V$ contains a vector subspace $L(G)_1$ of $L(G)$ with $\dim L(G)/L(G)_1 < \infty$.

It is no loss of generality to assume that $U = U\Delta'$ and $V = VL(G)_1$, i.e., that $U$ and $V$ are "saturated". We observe $(\Delta \times L(G))/(\Delta_1 \times L(G)_1) \cong \Delta/\Delta_1 \times L(G)/L(G)_1$ is a Lie group with finitely many components and $\phi$ induces a surjective homomorphism of Lie groups $\phi'\colon (\Delta \times L(G))/(\Delta_1 \times L(G)_1) \to G/\Delta' \exp L(G)_1$. By the Open Mapping Theorem for locally compact groups, $\phi'$ is open. The following diagram is commutative:

$$\begin{array}{ccc}
\Delta \times L(G) & \xrightarrow{\text{quot}} & \frac{\Delta \times L(G)}{\Delta' \times L(G)_1} \\
\phi \Big\downarrow & & \Big\downarrow \phi' \\
G & \xrightarrow{\text{quot}} & \frac{G}{\Delta \exp_G L(G)_1}
\end{array}$$

Since $U \times V$ is saturated we may conclude that $\phi(U \times V)$ is open in $G$.

Proof of (ii). We have $\ker \phi = \{(d, X) \in \Delta \times L(G) : d \exp X = 1\}$. The map $X \mapsto (\exp -X, X) : D \to \ker \phi$ is therefore bijective. This morphism has the inverse $(d, X) \mapsto X : \ker \phi \to D$. Hence it is an algebraic and topological isomorphism. The projection $\mathrm{pr}_\Delta : \Delta \times L(G) \to \Delta$ induces an injective morphism $j : \ker \phi \to \Delta$. Let $C \stackrel{def}{=} (\ker \phi)_0$ denote the identity component of $\ker \phi$. Then $j(C)$ is a connected subgroup of the totally disconnected group $\Delta$. It is therefore singleton and thus $C$ is singleton, i.e., $\ker \phi$ and thus $D$ are totally disconnected.

Proof of (iii). This statement is known. Here is a short argument: Set $H = \overline{\exp L(G)}$. The inclusion $j : H \to G$ induces an isomorphism $L(j) : L(H) \to L(G)$ in view of the definitions. Since $L(\cdot) = \mathrm{Hom}(\mathbb{R}, \cdot)$, the following sequence is exact:

$$0 \to L(H) \stackrel{L(j)}{\to} L(G) \to L(G/H) \to \mathrm{Ext}(\mathbb{R}, H).$$

But $H$ is divisible as the underlying group of a connected compact abelian group. Hence $H$ is injective and thus $\mathrm{Ext}(\mathbb{R}, H) = \{0\}$. Hence $L(G/H) = \{0\}$. Since every nondegenerate compact connected abelian group contains a nondegenerate one-parameter subgroup, $(G/H)_0 = \{0\}$. Therefore $H = G_0$. □

To the best of our knowledge Proposition 8.6.2 is new in the generality stated. It would be more valuable if we knew the totally disconnected closed subgroups of a topological vector space $\mathbb{R}^d$ for any cardinal $d$. In finite dimensions, this matter is no problem at all as we shall record in the following.

PROPOSITION 8.6.3. (Dixmier, see [**122**]) *For any compact abelian group $G$ the subgroup $\exp L(G)$ is exactly the arc component of* 0 *in $G$.*

We can write $\pi_0(G) = G/\exp L(G)$, and Dixmier has shown that $\pi_0(G) \cong \mathrm{Ext}(\widehat{G}, \mathbb{Z})$.

COROLLARY 8.6.4. *The morphism $\phi : \Delta \times L(G) \to G$ maps arc components onto arc components and $\pi_0(G) \cong \frac{\Delta}{\Delta \cap \exp L(G)}$ (algebraically).*

THEOREM 8.6.5. *The following conditions are equivalent for a compact abelian group $G$ and a natural number $n$:*

(i) $\operatorname{rank}\widehat{G} = \dim_{\mathbb{Q}}(\mathbb{Q}\otimes\widehat{G}) = n$.

(ii) *There is an exact sequence*

$$0 \to \operatorname{tor}(\widehat{G}) \to \widehat{G} \to \mathbb{Q}^n \to E \to 0$$

*with the torsion subgroup* $\operatorname{tor}(\widehat{G})$ *and some torsion group* $E$.

(iii) *There is an exact sequence*

$$0 \to \mathbb{Z}^n \to \widehat{G} \to E \to 0$$

*with some torsion group* $E$.

(iv) *There is a compact zero-dimensional subgroup* $Z$ *of* $(\widehat{\mathbb{Q}})^n$ *and an exact sequence*

$$0 \to Z \to (\widehat{\mathbb{Q}})^n \to G \to G/G_0 \to 0,$$

*where* $G_0$ *is the identity component of* $G$.

(v) *There is a compact zero-dimensional subgroup* $\Delta$ *of* $G$ *and an exact sequence*

$$0 \to \Delta \to G \to \mathbb{T}^n \to 0.$$

(vi) $\dim L(G) = n$.

(vii) *There is a compact zero-dimensional subgroup* $\Delta$ *of* $G$ *and quotient homomorphism* $\phi\colon \Delta\times\mathbb{R}^n \to G$ *which has a discrete kernel. In particular,* $\phi$ *yields a local isomorphism of* $\Delta\times\mathbb{R}^n$ *and* $G$.

(viii) *The identity of* $G$ *has a neighborhood basis each member of which is homeomorphic to* $D\times C^n$ *with some totally disconnected compact space* $D$ *and an* $n$*-cell* $C^n$.

PROOF. In the theory of abelian groups, conditions (1), (2) and (3) are all known to be equivalent to the statement that the (torsion free) rank of $\widehat{G}$ is $n$. By duality, (4) is equivalent to (2), and (5) is equivalent to (3).

Condition (6), saying $\dim \operatorname{Hom}(\mathbb{R}, G) = n$ in view of

$$\operatorname{Hom}(\mathbb{R}, G) \cong \operatorname{Hom}(\widehat{G}, \mathbb{R}) \cong \operatorname{Hom}(\mathbb{Q}\otimes\widehat{G}, \mathbb{R}) \cong \mathbb{R}^{\dim_{\mathbb{Q}}(\mathbb{Q}\otimes\widehat{G})},$$

is equivalent to (1).

(1) $\Longrightarrow$ (7). By (1) we have $L(G) \cong \mathbb{R}^n$. By Proposition 8.6.2, we obtain $\phi\colon \Delta\times\mathbb{R}^n \to G$ as asserted, because $\ker\phi \cong \exp^{-1}(\Delta)$ and this closed subgroup of $L(G)\cong\mathbb{R}^n$ does not contain vector subgroups, hence is discrete.

(7) $\Longrightarrow$ (8). Immediate consequence of (7).

(8) $\Longrightarrow$ (6). Let $U$ be an identity neighborhood of $G$ and $h\colon D\times C^n \to U$ a homeomorphism with a closed $n$-cell $C^n$.

We let $p\colon G \to T$ be as in the proof of Proposition 8.6.2 so that $L(p)\colon L(G) \to L(T)$ is an isomorphism. Let $K = \ker(L(p)\exp_T) = \ker(p\exp_G)$ as before. We may assume that $\Delta = \ker p$ is such that there is a compact identity neighborhood $V \subseteq U$ with $V\Delta = V$.

We recall the free subgroup $F$ of $\widehat{G}$ and let $(\epsilon_j)_{j\in J}$ be a basis of $F$. Set $\rho_0\colon \mathbb{R}^J \to \operatorname{Hom}(F,\mathbb{R})$, $\rho_0((r_j)_{j\in J})(\sum_{j\in J} r_j\cdot\epsilon_j) = \sum_{j\in J}\epsilon_j(r_j)$. Then $\rho_0$ is an isomorphism of topological groups, and since $\mathrm{Hom}(F,\mathbb{R}) \cong \mathrm{Hom}(\mathbb{R},T) \cong L(T)$ we have an isomorphism of topological groups $\rho\colon \mathbb{R}^J \to L(T)$. Moreover, if we write $F = \bigoplus_{j\in J}\mathbb{Z}\cdot\epsilon_j$ and, accordingly $T = \prod_{j\in J}\widehat{\mathbb{Z}\cdot\epsilon_j} \xrightarrow{\sigma} \mathbb{T}^J$, then $\exp' = \sigma\exp_T\rho\colon \mathbb{R}^J \to \mathbb{T}^J$ has the kernel $\mathbb{Z}^J$. It follows that $K = \ker\exp_T L(p)$ maps isomorphically onto $\mathbb{Z}^J$ under $\rho L(p)$. Now $p(V)$ is an identity neighborhood of $T \cong \mathbb{T}^J$. Then we find an $\frac{1}{2} > r > 0$ so that $S_r = L(p)^{-1}\rho^{-1}([-r,r]^J)$ satisfies $S_{2r}\cap\mathbb{Z}^J = \{0\}$ and $p(\exp_G S_r) = \exp_T L(p)(S_r) \subseteq p(V)$. Then $p\exp_G$ maps $S_r$ homeomorphically into $T$ and then a fortiori $\exp_G$ maps $S_r$ homeomorphically into $G$. But $V = V\Delta = p^{-1}p(V) \subseteq U$. Hence $\exp_G S_r \subseteq U$, and $e = \exp_G|S_r\colon S_r \to U$ is a homeomorphism onto the image. Then $h^{-1}e\colon S_r \to D\times C^n$ is a homeomorphism onto the image. If $(d,c) = h^{-1}(1)$, then $h(\{d\}\times C^n)$ is the connected component of 1 in $U$, and thus $h^{-1}e(S_r)$ is a homeomorphic copy of $S_r$ contained in the $n$-cell $\{d\}\times C^n$. But $S_r$ is homeomorphic to $[-r,r]^J$ hence to $[-1,1]^J$. Since $[-1,1]^J$ contains $[-1,1]^m$ for $m = 0,1,2,\ldots,|J|$, this entails $|J| \le n$ by the invariance of domain.

Thus $m \stackrel{def}{=} \dim L(G) = |J| \le n$. Then by "(6) $\Rightarrow$ (8)", there are arbitrarily small identity neighborhoods homeomorphic to $D'\times C^m$ with a totally disconnected compact space $D'$ and an $m$-cell $C^m$. Thus, by hypothesis (8), an $n$-cell must be contained in an $m$-cell, and this implies $n \le m$ by invariance of domain. □

DEFINITION 8.6.6. *Let $G$ be a compact abelian group. Then we set* $\dim G = \dim_{\mathbb{Q}}(\mathbb{Q}\otimes\widehat{G})$ *and call this cardinal the* dimension *of $G$. If* $\dim G$ *is finite, then $G$ is called* finite-dimensional *and otherwise* infinite-dimensional.

COROLLARY 8.6.7. *If for a compact abelian group $G$ there is a natural number $n$ such that the equivalent conditions of Theorem* 8.6.5 *are satisfied, then* $n = \dim G$.

Suppose that DIM is a dimension function defined for compact topological spaces such that

- it assigns to a product $D\times C^n$ of a compact totally disconnected space and a compact $n$-cell $C^n$ the dimension $n$,
- it assigns to a homogeneous space containing such a subspace with nonempty interior the dimension $n$, and
- it assigns to a compact space the dimension $\infty$ if it contains a homeomorphic copy of $[-1,1]^\omega$.

Then we will have

$$\mathrm{DIM}(G) = \begin{cases}\dim G & \text{if } G \text{ is finite-dimensional,}\\ \infty & \text{if } G \text{ is infinite-dimensional.}\end{cases}$$

PROPOSITION 8.6.8. (Pontryagin) *For a compact abelian group $G$ one has* $w(G) = |\widehat{G}|$, *where $w(G)$ denotes the weight of the space $G$.*

From Theorem 8.6.5 for an infinite compact group we note that

$$|\widehat{G}/\operatorname{tor}(\widehat{G})| = |\mathbb{Q} \otimes \widehat{G}| = \max\{\omega, \dim G\}.$$

We derive

PROPOSITION 8.6.9. *For an infinite compact abelian group $G$,*

$$w(G) = w(G/G_0)\max\{\omega, \dim G\}.$$

*In particular, a finite-dimensional connected compact abelian group is metric.*

More generally, a finite-dimensional compact abelian group is metrizable iff $w(G/G_0) \leq \omega$.

For the proof of the next proposition we need some facts on abelian groups.

LEMMA 8.6.10. (Pontryagin) *Let $A$ be a countable torsion free group and assume that every finite rank pure subgroup is free. Then $A$ is free.*

PROOF. The claim is clear if $\operatorname{rank} A < \infty$. We now assume $\operatorname{rank} A = \omega$. Let $\{e_1, e_2, \dots\}$ denote a maximal free set. Define $P_j$ to be the free pure subgroup generated by $\{e_1, \dots, e_j\}$. Inductively define a free set $\{f_1\ f_2, \dots\}$ such that $f_1, \dots, f_{j(n)}$ is a basis of $P_n$ for a suitable sequence of natural numbers $j(n)$. Indeed, if $f_1, \dots, f_j(n)$ is a basis of $P_n$, then $P_{n+1}$ is free and the pure subgroup $P_n$ is a direct summand (in view of the elementary divisor theorem). Hence we can complement the basis of $P_n$ to a basis $f_1, \dots, f_{j(n)}, f_{j(n)+1}\ \dots, f_{j(n+1)}$. This completes the recursion. The span of the free set $f_1, \dots$ contains all $P_n$ and thus all $e_n$, and thus is $A$. □

Now let $A$ be an abelian group. Let $\mathcal{K}$ denote the set of all subgroups $K$of an $A$ such that $A/K$ is free. Then $\mathcal{K}$ is a filter basis; for if $K_1$, $K_2 \in \mathcal{K}$ then $K = K_1 \cap K_2$ is the kernel of the map $a \mapsto (a + K_1, a + K_2) \to A/K_1 \times A/K_2$. The image of this homomorphism is a subgroup of a free group and thus is free by the Schreier Subgroup Theorem. We can form

$$K_\infty = K_\infty(A) = \bigcap \mathcal{K}.$$

Then all morphisms into free groups factor through $A \to A/K_\infty$ and the homomorphisms $A/K_\infty \to \mathbb{Z}$ separate the points. In particular, $A/K_\infty$ is torsion free and $\operatorname{tor}(A) \subseteq K_\infty$. Notably, $K_\infty$ is a pure subgroup.

For a subgroup $H$ of a torsion free abelian group $A$, the group

$$[H] \stackrel{\text{def}}{=} \{a \in A \colon \exists n \in \mathbb{N} \text{ such that } n \cdot a \in H\}$$

is the smallest pure subgroup containing $H$.

LEMMA 8.6.11. *If $K_\infty(A) = \{0\}$, then every finite rank pure subgroup of $A$ is free.*

PROOF. Let $P$ be a finite rank pure subgroup of $A$ and $F$ a maximal rank free subgroup contained in $P$. Then $P = [F]$. Since $K_\infty(A) = \{0\}$, there is a subgroup $K \in \mathcal{K}$ such that $F \cap K = \{0\}$. It follows that $P \cap K = \{0\}$, for if $p \in P \cap K$ then there is an $m \in \mathbb{N}$ such that $m \cdot p \in F \cap K = \{0\}$, whence $p = 0$ since $A$ is torsion free. The map $x \mapsto x + K: P \to A/K$ is therefore injective. But $A/K$ is free by the definition of $\mathcal{K}$, and thus $P$ is free by the Schreier Subgroup Theorem, as we wanted to show. □

LEMMA 8.6.12. (Main Lemma) *Let $A$ be an abelian group such that $A/K_\infty(A)$ is countable. Then $A$ contains a free subgroup $F$ such that $A = F \oplus K_\infty(A)$. Moreover, $K_\infty(A)$ does not have any nondegenerate free quotients.*

PROOF. The group $A/K_\infty(A)$ is torsion free, countable, and the morphisms into free groups separate the points. Hence $K_\infty\big(A/K_\infty(A)\big) = \{0\}$. Thus from Lemma 8.6.11 we know that every finite rank pure subgroup is free. Then, by Lemma 8.6.10, the quotient $A/K_\infty(A)$ is free. Since free groups are projective, this implies the existence of $F$ as asserted. Again any free quotient of $K_\infty(A)$ splits, so $K_\infty(A) = F' \oplus K$ with a free $F'$ isomorphic to the free quotient. But then $F \oplus F'$ is free and thus $K \in \mathcal{K}$. It follows that $K_\infty(A) \subseteq K$ and that, as a consequence, $F'$ is degenerate. □

THEOREM 8.6.13. (i) *If $T$ is a torus subgroup (a product of circle groups) of a compact abelian group $G$, then there is a (not necessarily unique) subgroup $C$ such that*

$$(t, c) \mapsto tc: T \times C \to G$$

*is an isomorphism of compact groups.*

(ii) *Every compact abelian group $G$ contains a (fully characteristic) unique smallest closed subgroup $M = M(G)$ containing all circle subgroups, and $M^\perp = K_\infty(\widehat{G})$. The subgroup $M(G)$ is always locally connected.*

(iii) *If $M(G)$ satisfies the first (and hence the second) axiom of countability, i.e., is metric, then $M(G)$ is a torus and $G$ contains a closed subgroup $C$ not containing circle groups such that*

$$(m, c) \mapsto mc: M(G) \times C \to G$$

*is an isomorphism of compact groups, i.e., $G$ is a direct product of a metric torus and a torus free compact subgroup.*

(iv) *Every metric compact group $G$ is the direct product of a fully characteristic maximal torus subgroup $M(G)$ and some torus free closed subgroup.*

PROOF. Part (i) is restating the fact, by duality, that in the category of abelian groups the free groups are the projectives and that a homomorphism onto a projective splits.

Proof of part (ii): We let $M$ be the closure of the group generated by the union of all circle groups. This is the smallest closed subgroup of $G$ containing all circle groups and is, therefore, fully characteristic. The product of two tori is a torus (It is clearly a compact subgroup, and the first factor splits by (i); the complementary factor is a homomorphic image of the second factor and is, therefore, a torus). Hence the set of all tori in $G$ is upwards directed, and $M$ is the closure of its union. Dually, the annihilator of $M$ is the intersection of all $T^\perp$ as $T$ ranges through all tori. By duality, the $T^\perp$ are exactly the members of $\mathcal{K}(\widehat{G})$. Thus $M(G)^\perp = K_\infty(\widehat{G})$. In particular, $\widehat{M(G)} \cong \widehat{G}/K_\infty(\widehat{G})$. Thus by Lemma 8.6.11, in the character group of $M(G)$ every finite rank pure subgroup is free. By a theorem of Pontryagin [**253**, II, Satz 48, Seite 33], this property characterizes locally connected compact abelian groups.

(iii) Now suppose that $M(G)$ is metric. Then $\widehat{G}/M(G)^\perp \cong \widehat{G}$ is countable. By (ii) we have $M(G)^\perp = K_\infty(\widehat{G})$. Now Lemma 8.6.12 implies $\widehat{G} = F \oplus M(G)^\perp$ with a (countable) free group $F$. We set $C = F^\perp \subseteq G$ and obtain $G \cong M(G) \times C$ as stated in (iii). Since $M(G)$ contains all circle groups, $C$ does not contain any circle groups. (Equivalently: $K_\infty(\widehat{G})$ does not have any free quotients by Lemma 8.6.12.)

(iv) If $G$ is metric, then, in particular, (iii) applies and proves the assertion. □

It is known, that for metric compact connected groups, arc connectivity, local connectivity, and being a torus group are equivalent properties. The character group $G$ of $\mathbb{Z}^\omega$ is connected, locally connected, but not arcwise connected, let alone a torus group. But for this $G$ we have $G = M(G)$. This shows that metrizability in (iii) is essential.

Dixmier [**122**] has observed (using the Axiom of Choice) that there is an abelian group $A$ containing a subgroup $Z$ isomorphic to $\mathbb{Z}$ such that $A/Z \cong \mathbb{Z}^\omega$ and that $Z$ does not split. If $g \in Z$ is a generator and if there existed a homomorphism $f\colon A \to F$ into a free group $F$ with $f(g) \neq 0$, then $[f(Z)] \cong \mathbb{Z}$ and $F = F_1 \oplus [f(Z)]$; hence there would exist a morphism $\phi\colon A \to \mathbb{Z}$ with $\phi(Z) = \mathbb{Z}$, yielding $A = H \oplus Z$ and thus contradicting the fact that $Z$ does not split. Hence $Z \subseteq K_\infty(A)$. Since $A/Z \cong \mathbb{Z}^\omega$, the free quotients separate the points of $A/Z$ whence $Z = K_\infty(A)$. One notes that every finitely generated pure subgroup of $A$ is free. As a consequence $G \overset{\text{def}}{=} \widehat{A}$ is *a compact connected locally connected (not arcwise connected) group for which $G/M(G)$ is a circle group.* I.e., $M(G/M(G))$ need *not* be zero in general! One notes that, in the absence of metrizability, this is the starting point of a transfinite ascending recursion process which we shall not pursue in this paper.

However, in view of Theorem 8.6.13(iv), in the metric case, the hypothesis that $G$ be torus free is frequently no restriction of generality.

PROPOSITION 8.6.14. *Let $\Gamma$ be a compact finite-dimensional abelian group. We have the following conclusions:*

(i) *$\Gamma \cong \mathbb{T}^m \times G$ with a unique maximal torus subgroup $\mathbb{T}^m \times \{1\}$, and some torus free compact $n$-dimensional group $G$, $m + n = \dim \Gamma$.*

(ii) *There is a compact zero-dimensional subgroup* $\Delta$ *of* $G$ *and a quotient homomorphism* $\phi\colon \mathbb{R}^n \times \Delta \to G$ *with a discrete kernel isomorphic to a lattice* $\mathbb{Z}^p$ *with* $p \leq n = \dim G$.

(iii) *The arc component of* 1 *in* $G$ *is* $G_a = \exp L(G) = \phi(\mathbb{R}^n \times \{1\})$ *The arc components of* $G$ *are the sets* $d = G_a = G_a d = \phi(\mathbb{R}^n \times \{d\}) = d \exp L(G)$ *and* $\phi$ *maps the arc component* $\mathbb{R}^n \times \{d\}$ *continuously and bijectively onto* $dG_a$.

(iv) *If* $\Gamma/\Gamma_0$ *is metric, then we may endow* $\mathbb{R}^{\dim \Gamma} \times \Delta = \mathbb{R}^m \times \mathbb{R}^n \times \Delta$ *with an invariant product metric* $d = d_0 \times d_1$ *with a Hilbert space metric* $d_0$ *on* $\mathbb{R}^{\dim \Gamma}$ *and an ultrametric* $d_1$ *on* $\Delta$, *and give* $\Gamma$ *the quotient metric* $d_\Gamma$ *for the homomorphism*

$$\Phi\colon \mathbb{R}^{\dim \Gamma} \times \Delta \to \Gamma,$$

$$(X, Y, d) \mapsto (X + \mathbb{Z}^n, \phi(d, Y))\colon \mathbb{R}^m \times \mathbb{R}^n \times \Delta \to \mathbb{T}^m \times G \cong \Gamma.$$

*Then* $(\Gamma, d_\Gamma)$ *is locally isometric to* $\mathbb{R}^{\dim \Gamma} \times \Delta$ *under* $\Phi$.

PROOF. (i) By Proposition 8.6.9, the component $\Gamma_0$ is metric, hence $M(\Gamma) = M(\Gamma_0)$ is metric. Thus Theorem 8.6.13(iii) proves $\Gamma \cong M(\Gamma) \times G$ with a torus $M(\Gamma)$ and a torus free compact group $G$. We have $\dim \Gamma = \dim M(\Gamma) + \dim G$. This proves (i).

(ii) From Theorem 8.6.5 we may identify $L(G)$ with $\mathbb{R}^n$ and we get $\phi\colon L(G) \times \Delta \to G$, $\phi(X, d) = d \exp X$, a local isomorphism. (Recall from the proof of Theorem 8.6.5 that totally disconnected closed subgroups of $\mathbb{R}^n$ are discrete lattices.)

(iii) From Corollary 8.6.4 we know that $\phi$ maps arc components as stated. If $\exp X = 1$, then $\exp \mathbb{R} \cdot X$ is a circle group or is trivial. But $G$ does not contain any circle groups. Hence $X = 0$ and $\phi$ maps $\mathbb{R}^n \times \{1\}$ injectively. Thus (iii) is proved.

(iv) is a straightforward consequence of (i) and (ii). □

DEFINITION 8.6.15. *A* solenoid *is a* 1*-dimensional compact connected group. A* proper solenoid *is a solenoid which is not a circle group.*

PROPOSITION 8.6.16. *A compact abelian group* $G$ *is a solenoid iff its character group is isomorphic to a nonzero subgroup of* $\mathbb{Q}$. *It is proper iff this subgroup is not free.*

PROPOSITION 8.6.17. *There is a* 2*-dimensional compact connected group* $G$ *which is not a product of two solenoids.*

PROOF. Pontryagin (see [**253**, I, Beispiel 15, Seite 48 and II, Beispiel 68, Seite 39]) has exhibited subgroups $A$ of $\mathbb{Q}^2$ with $\mathbb{Z}^2 \subseteq A \subseteq \mathbb{Q}^2$ which have no nondegenerate direct summands. For such a group, $G = \widehat{A}$ is a 2-dimensional compact connected group which is not a product of two solenoids. □

A *continuum* is a compact connected space. It is called *decomposable* if it is the union of two proper subcontinua; otherwise, it is *indecomposable*. The proper solenoids are known to be indecomposable metric continua. The following proposition shows that this situation cannot occur for compact groups having dimension greater than one.

THEOREM 8.6.18. *For a compact group $G$ the following statements are equivalent:*

(i) *$G$ is a proper solenoid.*

(ii) *The underlying space of $G$ is an indecomposable continuum.*

PROOF. We just noted $(i) \implies (ii)$ and observe that the circle is decomposable. Therefore we must prove that $\dim G > 1$ implies the decomposability of $G$. This is certainly the case if $G$ is disconnected, so we assume $G$ to be connected.

(a) We preface the proof by the following remark. Assume that $N$ is a compact and *connected* normal subgroup of $G$ and that $p\colon G \to G/N$ is the quotient morphism. If $X \subseteq G/N$ is connected, then $p^{-1}(X)$ is connected. Thus, if $G/N$ is decomposable, then $G$ is decomposable.

(b) Assume next that $G$ is semisimple. The structure theory of compact connected semisimple groups says that there is a family of simple compact connected Lie groups $\{G_j\colon j \in J\}$ and a totally disconnected central subgroup $D$ such that $G \cong (\prod_{j\in J} G_j)/D$ (see, for example, [**49**, Chap. 9]). Identify $G$ with this factor group. If $G$ is not degenerate, pick an $i \in J$. Then $N \stackrel{def}{=} \prod_{j\in J-\{i\}} G_jD/D$ is a compact connected normal subgroup of $G$ and $G/N \cong G_i/(N \cap G_i)$ is a simple connected Lie group, and thus is a compact manifold of dimension at least 3. Hence it is decomposable. From part (a) of the proof we derive that nondegenerate compact connected semisimple groups are decomposable.

(c) The commutator subgroup $G'$ of a compact connected group $G$ is closed and semisimple [**49**]. Each compact connected group is a semidirect product of its (closed!) commutator subgroup $G'$ and an abelian subgroup $A$ which is isomorphic to $G/G'$ [**172**]. Thus if $G$ is not abelian it has a decomposable direct topological factor by part (b). We may therefore assume for the remainder of the proof that $G$ is abelian.

(d) Assume that $G$ is a compact connected abelian group. Then $\widehat{G}$ is a torsion free group whose rank is at least 2. Hence we find a pure subgroup $P$ of rank 2. (Indeed let $F$ be a free subgroup of maximal rank. Write it in the form $F_1 \oplus F_2$ with $\operatorname{rank} F_1 = 2$ and set $P = \{\chi \in \widehat{G}\colon n \cdot \chi \in F_1 \quad \text{for some} \quad n \in \mathbb{N}\}$.) Let $C = P^{\perp}$, the annihilator of $P$ in $G$. Then $\widehat{C} \cong \widehat{G}/P$ and $\widehat{G/C} \cong P$. Since $P$ is pure, $\widehat{G}/P$ and thus $\widehat{C}$ is torsion free. Hence $C$ is connected. In view of part (a) of the proof, $G$ is decomposable if $G/C$ is decomposable. Furthermore, $\operatorname{rank} P = 2$ implies $\dim G/C = 2$. In order to complete the proof it therefore suffices to prove that a 2-dimensional compact connected abelian group is decomposable. By Theorem 8.6.13(iv) we may assume that $G$ is torus free since $G$ is clearly decomposable if $G$ contains a circle group as a factor.

Using Proposition 8.6.14, we obtain a compact zero-dimensional subgroup $\Delta$ and a closed $\epsilon$-ball neighborhood $B$ of the origin in $\mathbb{R}^2$ such that $\phi$ maps $\Delta \times B$ homeomorphically onto an identity neighborhood $W$ of $G$. As in Proposition 8.6.2 and its proof we set $D = \exp^{-1}(\Delta)$ and note that $\exp X = \phi(d, Y) = \delta \exp Y$ with $\delta \in \Delta$ means $\delta = \exp(X - Y)$ and thus $X = y + d$ with $d \in D$. Thus, identifying $L(G)$ with $\mathbb{R}^2$ we see that $\exp\colon \mathbb{R}^2 \to G$ maps $D + B$ bijectively and continuously onto $W \cap \exp L(G)$. Let $U = \operatorname{Int} B$, the manifold interior of $B$. By Proposition 8.6.2(ii), the closed subgroup $D$ of $\mathbb{R}^2$ is a discrete lattice and therefore is countable. Thus $D + U$ is a countable disjoint union of open disks in the plane. Therefore its complement $E \stackrel{def}{=} \mathbb{R}^2 - \{D + U\}$ is connected. By Proposition 8.6.2(iii), the set $\exp L(G)$ is dense in $G$ and the complement $\exp L(G) - W = \exp E$ is dense in the complement $A \stackrel{def}{=} G - \phi(\Delta \times U)$. Note that $A$ is a compact subset of $G$ since $\phi$ is an open map and that $\Delta A = A$. Also, $A$, being the closure of a connected set, is connected. Every arc component of $W$ is of the form $\delta \exp B$, $\delta \in \Delta$, and its intersection with $A$ is $\delta \exp \partial B$.

Let $K_1$ be a proper compact open subgroup of $\Delta$ and set $K_2 = K - K_1$. Set $C_1 = A \cup \phi(K_1 \times B)$ and $C_2 = A \cup \phi(K_2 \times B)$. Since each arc component of the compact set $\phi(K_j \times B)$, $j = 1, 2$ intersects the continuum $A$, each of the sets $C_1$ and $C_2$ is a continuum. Therefore, since $C_1$, $C_2$ are proper subcontinua of $G$ and $G = G_1 \cup G_2$, the space $G$ is decomposable as claimed. □

**8.6.2. Compact Metrizable Groups and CIPH.** A *flow* on a space $X$ is a continuous function $\varphi\colon X \times \mathbb{R} \to X$ such that $\varphi(x, 0) = x$ and $\varphi(\varphi(x, s), t) = \varphi(x, s + t)$ for all $s, t \in \mathbb{R}$ and $x \in X$. The map $\varphi_t\colon X \to X$ defined by $\varphi_t(x) = \varphi(x, t)$ is a homeomorphism since $\varphi_{-t} = \varphi_t^{-1}$. A point $p \in X$ is an *invariant point* of $\varphi$ if $\varphi_t(p) = p$ for all $t \in \mathbb{R}$, and the set $\cap\{\operatorname{fix}(\varphi_t)\colon t \in \mathbb{R}\}$, called the *invariant set* of the flow, is a closed subset of $X$ if $X$ is a Hausdorff space.

DEFINITION 8.6.19. *A flow $\varphi\colon M \times \mathbb{R} \to M$ on a metric space $(M, d)$ is called* uniform *if it satisfies the following conditions:*

(i) *$d(x, \varphi(x, t)) \le C|t|$ for some positive $C$ and all $x \in M$, $t \in \mathbb{R}$.*

(ii) *For all $x \in M$, $t \in \mathbb{R}$ either $\varphi(x, t) = x$ iff $t = 0$ or there exists a positive real number $p$ such that $\varphi(x, t) = x$ iff $t \in p\mathbb{Z}$.*

PROPOSITION 8.6.20. *Let $(M, d)$ be a compact metric space with a uniform flow $\varphi$. Then every nonempty closed subset of $M$ is the fixed point set of an orbit-preserving autohomeomorphism of $M$. In particular, $M$ has the CIPH.*

PROOF. In the case where $p > 0$ it suffices to consider $p = 1$. The formulation here is a matter of technical convenience for later proofs. If 8.6.19(ii) is satisfied with $p > 0$, define a new metric $D$ and a new flow $\psi$ by $D(x, y) = (pC)^{-1}(d(x, y)$

and $\psi(x,t) = \varphi(x,pt)$. Then

$$D(x,\psi(x,t)) = p^{-1}d(x,\varphi(x,pt)) \leq (pC)^{-1}\cdot C|pt| = |t|$$

and $\psi$ satisfies 8.6.19(ii). When $p = 1$ there is a free circle action on $M$ and the proposition follows from [**216**, 2.3].

Now suppose we have the case corresponding to $p = 0$ and $A$ is a nonempty closed subset of $M$. Let $r(x) = \frac{1}{2C}d(x,A)$ for $x \in M$ and define a mapping $h\colon M \to M$ by $h(x) = \varphi(x,r(x))$ for $x \in M$. Clearly, $\operatorname{fix}(h) = A$.

To see that $h$ is one-to-one, suppose that $h(x) = h(y)$. Then $x$ and $y$ must lie in the same orbit and for some real number $t$, $y = \varphi(x,t)$. Then

$$\varphi(x,r(x)) = \varphi(\varphi(x,t),r(y)) = \varphi(x,t+r(y)).$$

Thus $t = r(x) - r(y)$. The triangle inequality applied to $x$, $y$, $A$ implies

$$d(x,y) \geq |d(x,A) - d(y,A)|.$$

Since $C|t| \geq d(x,y)$, we have $C|t| \geq 2C|r(x) - r(y)|$. Therefore $t = 0$ and $r(x) = r(y)$. Then we have

$$\varphi(\varphi(x,r(x)),-r(x)) = \varphi(\varphi(y,r(y)),-r(y)).$$

Consequently, $x = \varphi(x,0) = \varphi(y,0) = y$ and $h$ is one-to-one.

To complete the proof that $M$ has the CIPH, it suffices to show that $h$ is onto. To see this, let $y \in M$. For some $b \in \mathbb{R}$, the metric $d$ is bounded by $b$. Let $x$ denote the unique point in the orbit $\varphi(\{y\}\times\mathbb{R})$ such that $\varphi(x,b) = y$ and let $[x,y]$ denote the arc in $\varphi(\{y\}\times\mathbb{R})$ with endpoints $x$ and $y$. Since $h|[x,y]$ is an order-preserving homeomorphism from $[x,y]$ onto $h([x,y])$ such that $h(x) < y$ and $h(y) \geq y$, it follows that $y \in h([x,y])$. Consequently, $h$ is onto as required. □

LEMMA 8.6.21. *If $(X_j,d_j)$, $j = 1,2$, are metric spaces and $\varphi$ is a uniform flow on $(X_2,d_2)$, then $X = X_1\times X_2$ is a metric space with respect to the metric $D$ defined by*

$$D((x_1,x_2),(x_1',x_2')) = \max\{d_1(x_1,x_1'),d_2(x_2,x_2')\},$$

*and the flow $\Phi$ on $X$ given by $\Phi((x_1,x_2),t) = (x_1,\varphi(x_2,t))$ is uniform on $(X,D)$.*

PROOF. Note

$$D((x_1,x_2),\Phi((x_1,x_2),t) = D((x_1,x_2),(x_1,\varphi(x_2,t)) = d_2(x_2,\varphi(x_2\ t)).$$

Condition 8.6.19(i) follows readily and 8.6.19(ii) is straightforward. □

PROPOSITION 8.6.22. *If $(M,d)$ is a compact metric space with a uniform flow and $X$ is a metrizable space, then $M\times X$ has CIPH.*

PROOF. By Lemma 8.6.21 there is a metric $\rho$ for $M \times X$ and a flow $\varphi$ which is uniform on $(M \times X, \rho)$. The proof of Proposition 8.6.20 can be applied to $M \times X$. Since $M \times X$ need not be compact, the one adjustment required is to use the compactness of $M$ to show that the mapping $h$ is a closed mapping and hence an autohomeomorphism of $M \times X$. □

Recall each compact metrizable group and each abelian metrizable group $G$ has metrics $d$ satisfying $d(gxh, gyh) = d(x, y)$. Such metrics will be called *invariant*. They are completely characterized by the function $g \mapsto ||g||: G \to \mathbb{R}$ defined by $||g|| = d(1, g)$ from which $d$ is recoverable by $d(g, h) = ||g^{-1}h||$. The function $||\cdot||$ satisfies $||x^{-1}|| = ||x||$, $||gh|| \leq ||g|| + ||h||$, $||hgh^{-1}|| = ||h||$ and $||g|| = 0$ iff $g = 1$. We call a function $||\cdot||$ an invariant norm on $G$ and remember that the invariant norms are in bijective correspondence with the invariant metrics on a group. (See e.g. [**50**, Chap.IX, pp. TG IX.24] where this is discussed for abelian groups; the arguments work for invariant metrics quite generally.)

LEMMA 8.6.23. *Assume that $G$ is a group with an invariant norm $||\cdot||$ and there is a one-parameter subgroup $\alpha: \mathbb{R} \to G$ such that $||\alpha(t)|| \leq C|t|$ for some positive number $C$. Suppose that $N$ is a closed normal subgroup of $G$ not containing $\alpha(\mathbb{R})$. Then there is a metric $D$ on $G/N$ such that the flow $(Ng, t) \mapsto Ng\alpha(t)$ is uniform on $(G/N, D)$.*

PROOF. We define $||Ng||_N = \inf\{||ng||: n \in N\}$. Then $||\cdot||_N$ is an invariant norm on $G/N$. The topology defined by the metric $D$ associated with this norm is the quotient topology.

We must verify (i) and (ii) of Definition 8.6.19 for the flow $\varphi$ on $G/N$ defined by $\varphi(Ng, t) = Ng\alpha(t)$. Regarding (i) we compute

$$\begin{aligned} D(Ng, \varphi(Ng, t)) = D(Ng, Ng\alpha(t)) &= ||N\alpha(t)||_N \\ &= \inf\{||n\alpha(t)||: n \in N\} \leq ||\alpha(t)|| \leq C(t) \end{aligned}$$

by hypothesis on $\alpha$. Thus 8.6.19(i) is satisfied.

Regarding (ii) we note that $\varphi(Ng, t) = Ng$ means $gN\alpha(t) = Ng\alpha(t) = Ng = gN$, i.e., $\alpha(t) \in N$. Since $\alpha(\mathbb{R})$ is not contained in $N$, the inverse image $\alpha^{-1}(N)$ is a closed proper subgroup of $\mathbb{R}$ and thus is of the form $p\mathbb{Z}$ with a nonnegative $p \in \mathbb{R}$. If $p = 0$ then $\varphi(Ng, t) = Hg$ holds iff $t = 0$. If $p$ is positive, then $\varphi(Ng, t) = Hg$ holds iff $t \in p\mathbb{Z}$. Thus 8.6.19(ii) holds and the proof is complete. □

PROPOSITION 8.6.24. *The Cantor set and the space of irrationals have CIPH.*

PROOF. Note that the space of irrationals is homeomorphic to $\mathbb{Z}^\omega$ ([**8**]) and the Cantor set is homeomorphic to $\mathbb{D}^\omega$ where $\mathbb{D}$ denotes the two-point discrete group ([**56**]). We shall only deal here with the Cantor set.

Fix a metric $d$ for $\mathbb{D}^\omega$ and let $A$ be a nonempty closed subset of $\mathbb{D}^\omega$. Then the complement $\mathbb{D}^\omega - A$, as an open subspace of $\mathbb{D}^\omega$, can be written as a countable (finite or infinite) disjoint union $C_1 \cup C_2 \cup \ldots$ of closed and open subspaces of $\mathbb{D}^\omega$

each of which is homeomorphic to $\mathbb{D}^\omega$. Since translation by a nonzero element in an abelian group determines an autohomeomorphism of the group, it follows that there is a sequence of fixed point free autohomeomorphisms $h_1, h_2, \ldots$ of $C_1, C_2, \ldots$ respectively. Moreover, we can choose each $h_i$ so that $d(h_i(x), x) < \frac{1}{i}$ for $x \in C_i$. It only remains to observe that the function $h\colon \mathbb{D}^\omega \to \mathbb{D}^\omega$ defined by

$$h(x) = \begin{cases} x & \text{if } x \in A \\ h_i(x) & \text{if } x \in C_i \end{cases}$$

is an autohomeomorphism of $\mathbb{D}^\omega$ with $\operatorname{fix}(h) = A$. □

LEMMA 8.6.25. *Every compact metrizable totally disconnected infinite group is homeomorphic to the Cantor set and hence has CIPH.*

PROOF. The proof follows from [**56**] and Proposition 8.6.24. □

THEOREM 8.6.26. *In every infinite compact metrizable group $G$ each nonempty closed subset is the fixed point set of some autohomeomorphism. If $G$ is not totally disconnected, then $G$ admits a metric $d$ and a uniform flow on $(G, d)$ defined by the action of some one-parameter group of $G$ on the right. In particular, every nonempty closed subset of $G$ is the fixed point set of an orbit-preserving autohomeomorphism of $G$.*

PROOF. By Mostert's Cross Section Theorem [**173**, 1.14], every compact group $G$ possesses a zero-dimensional compact subset $Z$ homeomorphic to $G/G_0$ such that

$$(z, g) \longmapsto zg\colon Z \times G_0 \to G$$

is a homeomorphism. Since $G$ is infinite then at least one of $G/G_0$ or $G_0$ is infinite. If $G_0$ is not degenerate, in view of Lemma 8.6.21 it suffices to prove the theorem for $G_0$. If $G_0$ is singleton, then $G$ is totally disconnected and the assertion follows from Lemma 8.6.25.

Thus the proof will be completed if we establish the claim for $G$ connected which we shall assume for the remainder of the proof. Since every compact group is the projective limit of Lie groups, there is a closed normal subgroup $M$ such that $G/M$ is a Lie group. (See e.g. [**49**, Chap. IX, p. LIE IX.99, Corollaire 1].) The identity component $K$ of $M$ is a compact connected normal subgroup of $G$. There exists a compact connected normal subgroup $H$ of $G$ such that $K \cap H$ is zero-dimensional and central (see e.g. [**173**, p.299, 2.5]). Since $H/(K \cap H) \cong G/M$, the group $H$ is finite-dimensional. The map $q\colon K \times H \to G$, $q(k, h) = kh$ is a surjective continuous morphism. Thus $G \cong (K \times H)/N$ with $N = \ker q$. In view of Lemma 8.6.23 it suffices to prove the required data on $K \times H$. Then by Lemma 8.6.21 it suffices again to prove the claim for $H$. Thus we may assume now that $G$ is compact connected and finite-dimensional. Now $G$ is the semidirect product $G' \times A$ of the closed normal commutator subgroup

$G'$ and an abelian subgroup $A$ isomorphic to $G/G'$ (see [**172**]). Since $G'$ is finite-dimensional and semisimple, it is a Lie group. By Lemma 8.6.21 it suffices to prove the assertion for each of the two factors separately.
*Case 1:* $G$ is a semisimple Lie group. We claim that the theorem holds for every compact connected Lie group $G$. Every such $G$ contains a maximal torus (see [**49**, p.LIE IX.8]) and has an invariant Riemannian metric $d$ which induces on $T$ a Riemannian invariant metric and thus induces on each circle group in $T$ an invariant Riemannian metric. However, an invariant Riemannian metric on $\mathbb{R}/\mathbb{Z}$ is a positive multiple of the metric given by $d(s+\mathbb{Z}, t+\mathbb{Z}) = |s-t|$. Thus we find a one-parameter subgroup $\alpha\colon \mathbb{R} \to G$ such that $||\alpha(t)|| \leq C|t|$ with the invariant norm associated with $d$. Then Lemma 8.6.23 yields a proof of the assertion in this case with $N = \{1\}$.
*Case 2:* $G$ is a finite-dimensional abelian group. Then $G = T \times A$ with a unique maximal torus and a torus free factor by Theorem 8.6.13(iv). By Lemma 8.6.21 the proof is accomplished if it can be done for each of the factors separately. In the proof of Case 1 we took care of the case of a Lie group. Thus there remains only the case that $G$ is a torus free compact connected abelian group.

Now let $\beta\colon \mathbb{R} \to G$ denote any nondegenerate one-parameter subgroup. One such exists since $G$ is assumed to be nondegenerate. Since $G$ is torus free we have

$$\beta(t) = 0 \text{ iff } t = 0. \tag{8.6.1}$$

In particular, $\beta(1) \neq 0$. The characters of $G$ separate the points, hence there is a character $\chi\colon G \to \mathbb{T} = \mathbb{R}/\mathbb{Z}$ such that $\chi(\beta(1)) \neq 0$. The morphism $\chi \circ \beta\colon \mathbb{R} \to \mathbb{R}/\mathbb{Z}$ lifts to a morphism $\lambda\colon \mathbb{R} \to \mathbb{R}$ such that $\chi \circ \beta = p \circ \lambda$ with the quotient (covering) morphism $p\colon \mathbb{R} \to \mathbb{R}/\mathbb{Z}$. (This is elementary. This also follows from a topological fact since $\mathbb{R}$ is simply connected and $p$ is a covering map.) Every continuous endomorphism $\lambda$ of $\mathbb{R}$ is of the form $\lambda(t) = rt$ for some $r \in \mathbb{R}$. Since $p(r) = \chi \circ \beta(1) \neq 0$, we certainly have $r \neq 0$. We then have $\chi(\beta(t)) = rt + \mathbb{Z}$ and define $\alpha\colon \mathbb{R} \to G$ by $\alpha(t) = \beta(\frac{t}{r})$. Then (8.6.1) implies

$$\alpha(t) = 0 \text{ iff } t = 0 \tag{8.6.2}$$

and we have

$$\chi \circ \alpha = p\colon \mathbb{R} \to \mathbb{T}. \tag{8.6.3}$$

Now we set $K = \ker \chi$ and define

$$\Psi\colon K \times \mathbb{R} \to G, \quad \Psi(k,t) = k + \alpha(t). \tag{8.6.4}$$

*Claim: $\Psi$ is surjective.* For a proof let $g \in G$. Pick $t \in \mathbb{R}$ such that $\chi(g) = t+\mathbb{Z}$ and set $k = g - \alpha(t)$. Then $\chi(g - \alpha(t)) = \chi(g) - \chi(\alpha(t)) = \chi(g) - (t+\mathbb{Z}) = 0$ in view of (8.6.3). Hence $k \in K$ and thus $g = k + \alpha(t) = \Psi(k,t)$. This proves the claim.

Now Lemma 8.6.23 shows that it is sufficient to prove the claim for $K \times \mathbb{R}$. Then by Lemma 8.6.21 it suffices to prove the claim for $\mathbb{R}$ for which it is obvious. □

COROLLARY 8.6.27. *If $G$ is a positive-dimensional compact metrizable group and $X$ is a metrizable space, then $G \times X$ has CIPH.*

PROOF. Apply Theorem 8.6.26 and Proposition 8.6.22. □

COROLLARY 8.6.28. *A compact metrizable group has the CIPH iff it is either degenerate or infinite.*

PROOF. An autohomeomorphism of a nondegenerate finite discrete group cannot fix all points but one. □

COROLLARY 8.6.29. *Every compact metrizable group has the CIP.*

PROOF. Every nonempty subset of a finite discrete group is the fixed point set of a retraction. □

In [**235**, p.134] a space $Y$ is defined to have *property $W$(strong)* if it admits a homotopy $H: Y \times [0,1] \to Y$ such that $H(y,t) = y$ iff $t = 0$. We note that a metric space has property $W$(strong) if it admits a uniform flow since then we can set $H(x,t) = \varphi(x,t)$ (or $\varphi\left(x, \frac{t}{2p}\right)$ for $p > 0$). It is shown in [**218**, 3.14] that if $M$ is a connected metrizable space with property $W$(strong) and $X$ is a zero-dimensional compact (Hausdorff) space, then $M \times X \times X$ has CIP iff $X$ is metrizable. In the next result we use these facts and the concept of $CIPH$ to give a characterization of the compact zero-dimensional spaces which are metrizable

PROPOSITION 8.6.30. *If $(M,d)$ is a compact connected metric space admitting a uniform flow and $X$ is a compact zero-dimensional space, then $M \times X \times X$ has CIPH iff $X$ is metrizable.*

PROOF. *Sufficiency.* If $X$ is metrizable, then $X \times X$ is metrizable and $M \times X \times X$ has CIPH by Proposition 8.6.22.

*Necessity.* Since a space having the CIPH has the CIP, the necessity follows from the remarks preceding the proposition. □

We note that if $X$ is a nondegenerate compact zero-dimensional metrizable space, then $X^\lambda \times X^\lambda$ is metrizable iff $X^\lambda$ is metrizable iff $\lambda$ is a countable cardinal. Moreover, when $\lambda$ is uncountable we have $X^\lambda \times X^\lambda = X^\lambda$. Consequently, we can apply Proposition 8.6.30 to obtain the following result.

COROLLARY 8.6.31. *If $(M,d)$ is a compact connected metric space admitting a uniform flow and $X$ is a nondegenerate compact zero-dimensional metrizable space, then $M \times X^\lambda$ has the CIPH iff $\lambda$ is a countable cardinal.*

We note that if a Hausdorff space $D \times M$ has the CIPH, where $D$ is a totally disconnected space and $M$ is a connected space, then $M$ has the CIPH since, given a nonempty closed subset $A$ of $M$, an autohomeomorphism $h$ of $D \times M$ with $\operatorname{fix}(h) = D \times A$ must preserve components and thereby determine an autohomeomorphism of $M$ whose fixed point set is $A$. Since an $n$-manifold having a compact boundary component with a nonzero Euler characteristic does not have the CIPH ([**216**, 3.1]), it follows that $\mathbb{D}^\omega \times I^\lambda$, $I = [0,1]$, does not have the CIPH if $\lambda$ is an odd integer. If $G$ is a compact metrizable group which is not totally disconnected, then Corollary 8.6.27 shows that $G \times I^\lambda$ has the CIPH if $\lambda$ is a countable cardinal. However, no nonmetrizable $ANR$-compactum has the CIP ([**107**, 3.4]). Since a compact Lie group is an $ANR$-space, its product with a Tychonov cube is a nonmetrizable $ANR$-compactum and we obtain the following result.

PROPOSITION 8.6.32. *If $G$ is a compact Lie group, then $G \times I^\lambda$ has the CIPH iff $\lambda$ is a countable cardinal.*

*Remark* 8.6.33. We note that Corollary 8.6.31 and Proposition 8.6.32 provide examples which contrast the metrizable and nonmetrizable cases regarding the CIPH. In what follows, let $\lambda$ be an uncountable cardinal.

(i) Let $G$ be a compact Lie group and consider the Hilbert cube $Q = I^\omega$. By Proposition 8.6.32, $G \times I^\lambda = G \times Q^\lambda$ does not have the CIPH. However, since $G$ has the CIPH by Theorem 8.6.26 and $Q$ has the CIPH by [**216**, 12.4], $G \times Q^\lambda$ is a product of homogeneous spaces each of which has the CIPH. This is in contrast to the metric case where it is not known if every homogeneous locally contractible space has the CIPH.

(ii) Let $G$ be a compact connected metrizable group and let $C = \mathbb{D}^\omega$. Then $G \times \mathbb{D}^\lambda = G \times C^\lambda$ does not have the CIPH by Theorem 8.6.26 and Corollary 8.6.31. By Theorem 8.6.26, $G \times C^\lambda$ is a product of compact groups each of which has the CIPH. This is in contrast to the metric case since, by Corollary 8.6.28, a metrizable product of compact groups has the CIPH iff it is either degenerate or infinite.

We remark also that the hypothesis of compactness in Proposition 8.6.20 can be relaxed so that the conclusion of the proposition is valid for free $\mathbb{R}$-actions. However, one cannot conclude that connected metrizable groups must have CIP since there exists a connected subgroup of the plane $\mathbb{R}^2$ which fails to have CIP [**218**, 2.2].

Finally, we end this section by noting that any metrizable space $M$ admitting a flow satisfying condition (ii) of Definition 8.6.19 admits distinct flow structures corresponding to the closed subsets of $M$.

PROPOSITION 8.6.34. *Let $M$ be a metrizable space admitting a flow satisfying condition (ii) of Definition* 8.6.19. *If $A$ is a closed subset of $M$, then there is a flow on $M$ whose invariant set is $A$.*

PROOF. In [28] it is shown that every closed subset of a metrizable space $X$ is the invariant set of a flow on $X$ iff $X$ admits a flow having an empty invariant set. The proof follows since a flow $\varphi$ on $M$ satisfying condition (ii) of Definition 8.6.19 has no invariant points. □

By Lemma 8.6.21 a product of two metrizable spaces will admit a uniform flow if one the factor spaces does and thus we obtain the following result.

COROLLARY 8.6.35. *Let $M$ be a metrizable space and let $G$ be a compact metrizable positive-dimensional group. Then every closed subset of $G \times M$ is the invariant set of some flow on $G \times M$.*

**8.6.3. Compact Non-metrizable Groups and CIPH.** Unlike the metrizable case there are nonmetrizable compact groups which do not have CIP. The product $\mathbb{D}^{\omega_1} \times \mathbb{T}^{\omega}$ serves as a corresponding example (see Corollary 8.6.42 ). In this section we show that $G^{\tau}$ has CIPH for each nondegenerate metrizable compact group and each $\tau > \omega$.

Under Axiom of Choice every infinite cardinal $\tau$ is an aleph i.e. $\tau = \omega_{\alpha}$ for some ordinal $\alpha$. Recall that $\omega_{\beta} < \omega_{\alpha}$ whenever $\beta < \alpha$ and that $\omega_{\alpha+1} = \omega_{\alpha}^{+}$, where $\kappa^{+}$ denotes the smallest cardinal larger than $\kappa$. Also, if $\alpha$ is a limit ordinal, then $\omega_{\alpha} = \sup\{\omega_{\beta} \colon \beta < \alpha\}$. The *cofinality* of $\tau$, denoted by $\mathrm{cf}(\tau)$, is the smallest cardinal $\kappa$ such that $\tau$ has a cofinal subset of cardinality $\kappa$. Clearly $\mathrm{cf}(\omega_{\alpha+1}) = \omega_{\alpha+1}$ and in general $\mathrm{cf}(\tau) \leq \tau$.

Let $\tau$ be an uncountable cardinal. If $\tau = \omega_{\alpha+1}$ (i.e. if $\tau$ is a regular cardinal), then $\mathbb{D}^{\tau}$ can be represented as the limit space of the continuous inverse spectrum

$$\{(\mathbb{D}^{\omega_{\alpha}})^{\beta}, \pi_{\beta}^{\beta+1}, \beta < \mathrm{cf}(\tau)\}$$

where $\pi_{\beta}^{\beta+1} \colon (\mathbb{D}^{\omega_{\alpha}})^{\beta+1} \to (\mathbb{D}^{\omega_{\alpha}})^{\beta}$ denotes the projection $(\mathbb{D}^{\omega_{\alpha}})^{\beta} \times \mathbb{D}^{\omega_{\alpha}} \to (\mathbb{D}^{\omega_{\alpha}})^{\beta}$.

If $\tau$ is a singular carrdinal, then $\tau = \sup\{\omega_{\alpha_{\beta}} \colon \beta < \mathrm{cf}(\tau)\}$ where $\omega \leq \omega_{\alpha_{\beta}} < \tau$ for each $\beta < \mathrm{cf}(\tau)$. Consequently, in this case, $\mathbb{D}^{\tau}$ can be represented as the limit space of the continuous inverse spectrum

$$\lim\{\mathbb{D}^{\omega_{\alpha_{\beta}}}, \pi_{\beta}^{\beta+1}, \beta < \mathrm{cf}(\tau)\}$$

where $\pi_{\beta}^{\beta+1} \colon \mathbb{D}^{\omega_{\alpha_{\beta+1}}} \to \mathbb{D}^{\omega_{\alpha_{\beta}}}$ denotes the projection onto the corresponding subproduct.

Combining these to cases we see that $\mathbb{D}^{\tau} = \lim \mathcal{S}$ where

$$\mathcal{S} = \{Y_{\beta}, \pi_{\beta}^{\beta+1}, \beta < \mathrm{cf}(\tau)\}.$$

Here

$$Y_{\beta} = \begin{cases} (\mathbb{D}^{\omega_{\alpha}})^{\beta} & \text{if } \tau = \omega_{\alpha+1} \\ \mathbb{D}^{\omega_{\alpha_{\beta}}} & \text{if } \tau = \sup\{\omega_{\alpha_{\beta}} \colon \beta < \mathrm{cf}(\tau)\} \end{cases}$$

By $\pi_{\beta} \colon \mathbb{D}^{\tau} \to Y_{\beta}$ we denote the $\beta$-th limit projection of the spectrum $\mathcal{S}$.

LEMMA 8.6.36. *Let* $\tau > \omega$ *and* $X$ *be a zero-dimensional compact space of weight* $\leq \tau$. *Then* $X$ *is homeomorphic to the limit space of a continuous inverse spectrum*

$$\mathcal{S}_X = \{X_\beta, p_\beta^{\beta+1}, \beta < \mathrm{cf}(\tau)\}$$

*such that* $\dim X_\beta = 0$ *and*

$$w(X_\beta) \leq \begin{cases} \omega_\alpha & \text{if } \tau = \omega_{\alpha+1} \\ \omega_{\alpha_\beta} & \text{if } \tau = \sup\{\omega_{\alpha_\beta} \colon \beta < \mathrm{cf}(\tau)\} \end{cases}$$

*for each* $\beta < \mathrm{cf}(\tau)$.

PROOF. Since $\dim X = 0$ and $w(X) \leq \tau$ we may assume that $X \subseteq \mathbb{D}^\tau$. Let $\mathcal{S}$ be a spectrum defined above.

We let $X_\beta = \pi_\beta(X)$ and $p_\beta^{\beta+1} = \pi_\beta^{\beta+1}/X_{\beta+1}$. Then the spectrum

$$\mathcal{S}_X = \{X_\beta, p_\beta^{\beta+1}, \beta < \mathrm{cf}(\tau)\}$$

has all the required properties. □

We start with the following statement which in the case when $\tau = \omega$ can be viewed as a parametric version of Proposition 8.6.24.

THEOREM 8.6.37. *Let* $\tau \geq \omega$, $X$ *be a zero-dimensional compactum of weight* $\leq \tau$ *and* $F$ *be a nonempty closed subset of the product* $X \times \mathbb{D}^\tau$. *Then there exists a homeomorphism* $h \colon X \times \mathbb{D}^\tau \to X \times \mathbb{D}^\tau$ *such that* $\pi_1 h = \pi_1$ *and* $\mathrm{fix}(h) = F$, *where* $\pi_1 \colon X \times \mathbb{D}^\tau \to X$ *denotes the projection onto the first coordinate.*

PROOF. We shall carry out the transfinite induction with respect to $w(X)$.

**Case 1** ($\tau = \omega$). If $F$ is open and closed in $X \times \mathbb{D}^\omega$, then the statement is trivial. Therefore we may assume that the complement $X \times \mathbb{D}^\omega - F$ can be written as a countable disjoint union $C_1 \cup C_2 \cup \cdots \cup C_n \cup \ldots$ of closed and open subsets of $X \times \mathbb{D}^\omega$. The restriction $\pi_1/C_n \colon C_n \to \pi_1(C_n)$, $n \in \omega$, is an open map between zero-dimensional metrizable compacta whose fibers

$$(\pi_1/C_n)^{-1}(x) = \pi_1^{-1}(x) \cap C_n = (\{x\} \times \mathbb{D}^\omega) \cap C_n, x \in \pi_1(C_n),$$

being nonempty closed and open subsets of $\mathbb{D}^\omega$, are all homeomorphic to $\mathbb{D}^\omega$. Consequently, by Lemma 8.1.7, the restriction $\pi_1/C_n \colon C_n \to \pi_1(C_n)$ is a trivial bundle with fiber $\mathbb{D}^\omega$ for each $n \in \omega$. This means that for each $n \in \omega$ there is a homeomorphism $t_n \colon \pi_1(C_n) \times \mathbb{D}^\omega \to C_n$ such that $\pi_1 t_n = \pi_n$, where $\pi_n \colon \pi_1(C_n) \times \mathbb{D}^\omega \to \pi_1(C_n)$ denotes the projection onto the first coordinate. Here is the corresponding diagram

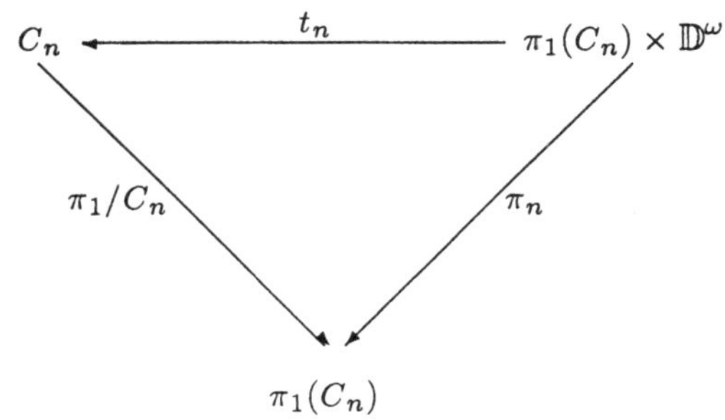


Let $d$ be a metric on $X \times \mathbb{D}^\omega$. Choose a fixed point free homeomorphism $f_n \colon \mathbb{D}^\omega \to \mathbb{D}^\omega$ so close to the identity that the homeomorphism $g_n \colon C_n \to C_n$ defined as the composition $g_n = t_n(\mathrm{id}_{\pi_1(C_n)} \times f_n)t_n^{-1}$ satisfies the condition $d(g_n(x), x) < \frac{1}{n}$ for each $x \in C_n$. Note that $g_n$ preserves the first coordinate, i.e. $\pi_1 g_n = \pi_1/C_n$ for each $n \in \omega$. Consider a homeomorphism $h_n \colon X \times \mathbb{D}^\omega \to X \times \mathbb{D}^\omega$, $n \in \omega$, defined as follows:

$$h(x) = \begin{cases} x & \text{if } x \in F \\ g_n(x) & \text{if } x \in C_n. \end{cases}$$

Clearly, $h$ preserves the first coordinate (i.e. $\pi_1 h = \pi_1$) and $\mathrm{fix}(h) = F$. This finishes the proof of Theorem in the case when $\tau = \omega$.

**Case 2** ($\tau > \omega$). Suppose that Theorem has been proved in the cases when $w(X) \leq \kappa$ where $\omega \leq \kappa < \tau$ and consider the case $w(X) = \tau$.

Represent $\mathbb{D}^\tau$ as the limit space of the continuos spectrum

$$\mathcal{S} = \{Y_\beta, \pi_\beta^{\beta+1}, \beta < \mathrm{cf}(\tau)\}$$

which was defined above.

By Lemma 8.6.36, $X = \lim \mathcal{S}_X$, where

$$\mathcal{S}_X = \{X_\beta, p_\beta^{\beta+1}, \beta < \mathrm{cf}(\tau)\}$$

is the spectrum from that lemma.

Observe that the limit space of the spectrum

$$\mathcal{S}_X \times \mathcal{S} = \{X_\beta \times Y_\beta, p_\beta^{\beta+1} \times \pi_\beta^{\beta+1}, \beta < \mathrm{cf}(\tau)\}$$

coincides with the product $X \times \mathbb{D}^\tau$. Let $F_\beta = (p_\beta \times \pi_\beta)(F)$ for each $\beta < \mathrm{cf}(\tau)$.

During the construction the projection $X_\beta \times Y_\beta \to X_\beta$ will be denoted by $\pi_1^\beta$.

Since $w(X_0) \leq w(Y_0) < \tau$ and $\dim X_0 = 0$, by the inductive hypotheses, there exists a homeomorphism $h_0 \colon X_0 \times Y_0 \to X_0 \times Y_0$ such that $\pi_1^0 h_0 = \pi_1^0$ and $\mathrm{fix}(h_0) = F_0$.

Suppose that for each $\beta$ with $\beta < \gamma$, where $\gamma < \mathrm{cf}(\tau)$, we have already constructed homeomorphisms $h_\beta \colon X_\beta \times Y_\beta \to X_\beta \times Y_\beta$ such that

$(i)_\beta$ $\pi_1^\beta h_\beta = \pi_1^\beta$ for each $\beta$ with $\beta < \gamma$.

$(ii)_\beta$ $\text{fix}(h_\beta) = F_\beta$ for each $\beta$ with $\beta < \gamma$.

$(iii)_\beta$ $(p_\delta^\beta \times \pi_\delta^\beta)h_\beta = h_\delta(p_\delta^\beta \times \pi_\delta^\beta)$ for each $\delta$, $\beta$ with $\delta \leq \beta < \gamma$.

$(iv)_\beta$ $h_\beta = \lim\{h_\delta \colon \gamma < \beta\}$ for each limit ordinal $\beta$ with $\beta < \gamma$.

Let us define the homeomorphism $h_\gamma \colon X_\gamma \times Y_\gamma \to X_\gamma \times Y_\gamma$.

If $\gamma$ is a limit ordinal, then let $h_\gamma = \lim\{h_\beta \colon \beta < \gamma\}$. Properties $(i)_\beta - (iv)_\beta$, $\beta < \gamma$, imply that $h_\gamma$ is a homeomorphism such that

$(i)_\gamma$ $\pi_1^\gamma h_\gamma = \pi_1^\gamma$.

$(ii)_\gamma$ $\text{fix}(h_\gamma) = F_\gamma$.

$(iii)_\gamma$ $(p_\delta^\gamma \times \pi_\delta^\gamma)h_\gamma = h_\delta(p_\delta^\gamma \times \pi_\delta^\gamma)$ for each $\delta$ with $\delta < \gamma$.

If $\gamma = \beta + 1$, then we proceed as follows. Consider the fibered product $Z_\beta$ of spaces $X_\beta \times Y_\beta$ and $X_{\beta+1}$ with respect to the maps $\pi_1^\beta \colon X_\beta \times Y_\beta \to X_\beta$ and $p_\beta^{\beta+1} \colon X_{\beta+1} \to X_\beta$. By $\varphi \colon Z_\beta \to X_\beta \times Y_\beta$ and $\psi \colon Z_\beta \to X_{\beta+1}$ we denote the corresponding projections of the above fibered product. Here is the induced diagram

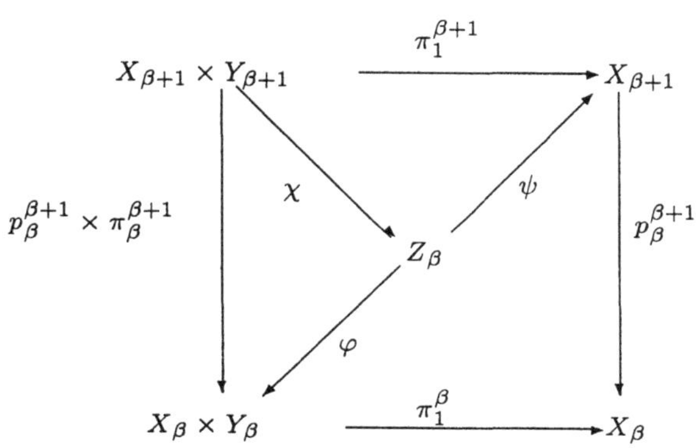


Since $\pi_1^\beta$ is a trivial bundle with fiber $Y_\beta$ and $\psi$ is the canonical projection of the fibered product, we see that $\psi$ also is a trivial bundle with fiber $Y_\beta$. Moreover, since $\pi_1^{\beta+1}$ is a trivial bundle with fiber $Y_{\beta+1}$, we conclude that in the described situation the map $\chi = (p_\beta^{\beta+1} \times \pi_\beta^{\beta+1}) \triangle \pi_1^{\beta+1}$ is a trivial bundle. Fiber of this bundle is $\mathbb{D}^\kappa$ where

$$\kappa = \begin{cases} \omega_\alpha & \text{if } \tau = \omega_{\alpha+1} \\ \omega_{\alpha_{\beta+1}} - \omega_{\alpha_\beta} & \text{if } \tau = \sup\{\omega_{\alpha_\beta} \colon \beta < \text{cf}(\tau)\}. \end{cases}$$

Clearly, $w(Z_\beta) \leq w(X_\beta \times Y_\beta \times X_{\beta+1})$. Therefore

$$w(Z_\beta) \leq \begin{cases} \omega_\alpha & \text{if } \tau = \omega_{\alpha+1} \\ \omega_{\alpha_{\beta+1}} & \text{if } \tau = \sup\{\omega_{\alpha_\beta} : \beta < \operatorname{cf}(\tau)\}. \end{cases}$$

Since

$$|\omega_{\alpha_{\beta+1}} - \omega_{\alpha_\beta}| = \omega_{\alpha_{\beta+1}},$$

there exists, by the inductive hypothesis, a homeomorphism

$$f_{\beta+1} \colon X_{\beta+1} \times Y_{\beta+1} \to X_{\beta+1} \times Y_{\beta+1}$$

such that $f_{\beta+1}\chi = \chi$ and $\operatorname{fix}(f_{\beta+1}) = F_{\beta+1}$. Consider the homeomorphism $g_\beta \colon Z_\beta \to Z_\beta$ defined by letting $g_\beta = h_\beta\varphi \triangle \psi$. Finally, let $h_{\beta+1} = f_{\beta+1}(g_\beta \times id_{\mathbb{D}^{(\omega_{\alpha_{\beta+1}} - \omega_{\alpha_\beta})}})$.

A straightforward verification shows that

- $\pi_1^{\beta+1} h_{\beta+1} = \pi_1^{\beta+1}$.
- $(p_\beta^{\beta+1} \times \pi_\beta^{\beta+1}) h_{\beta+1} = h_\beta (p_\beta^{\beta+1} \times \pi_\beta^{\beta+1})$.
- $\operatorname{fix}(h_{\beta+1}) = F_{\beta+1}$

This completes the construction of homeomorphisms $h_\beta$ for each $\beta < \operatorname{cf}(\tau)$. Then the homeomorphism

$$h = \lim\{h_\beta \colon \beta < \operatorname{cf}(\tau)\} \colon X \times \mathbb{D}^\tau \to X \times \mathbb{D}^\tau$$

preserves the first coordinate and $\operatorname{fix}(h) = F$. Proof is complete. □

Obvious changes of the above proof, coupled with the results of Subsection 8.6.2 allow us to prove the following.

THEOREM 8.6.38. *Let $\tau \geq \omega$, $X$ be a compactum of weight $\leq \tau$ and $F$ be a nonempty closed subset of the product $X \times G^\tau$ where $G$ is a positive-dimensional compact metrizable group. Then there exists a homeomorphism $h \colon X \times G^\tau \to X \times G^\tau$ such that $\pi_1 h = \pi_1$ and $\operatorname{fix}(h) = F$, where $\pi_1 \colon X \times G^\tau \to X$ denotes the projection onto the first coordinate.*

Theorems 8.6.37 and 8.6.38 imply the following.

COROLLARY 8.6.39. *Let $\tau \geq \omega$ and $G$ be a nondegenerate compact metrizable group. Then $G^\tau$ has the CIPH.*

Slightly modifying the proofs of Theorems 8.6.37 and 8.6.38 we obtain the following result.

THEOREM 8.6.40. *Let $X$ and $Y$ be compact spaces and $f \colon X \to Y$ be a principal $G^\tau$-bundle, $\tau \geq \omega$, where $G$ is a nondegenerate compact metrizable group. Let $w(Y) \leq \tau$ and assume that one of the following two conditions is satisfied:*

(a) $\dim X = \dim Y = \dim G = 0$, *or*

(b) $\dim G > 0$.

*Then $X$ has the CIPH. Moreover, each non-empty closed subset of $X$ can be fixed by a fiber-preserving autohomeomorphism of $X$.*

Theorem 8.6.40 tells us, in particular, that for any compactum $X$ of weight $\tau \geq \omega$ the product $X \times \mathbb{T}^\tau$ has the CIPH. Can we have less than $\tau$ copies of $\mathbb{T}$ and maintain the same conclusion? [**218**, Proof of Theorem 3.12] shows that if $\tau > \omega$ and $\kappa < \omega_1$, then the product $\mathbb{D}^\tau \times \mathbb{T}^\kappa$ does not even have the CIP. Our next result shows that the cardinality of the set of $\mathbb{T}$ factors in Theorem 8.6.40 cannot be reduced.

LEMMA 8.6.41. *Let $\omega \leq \kappa < \tau$ and $X$ and $Y$ be compact spaces such that $w(X) = \tau$ and $w(Y) = \kappa$. Let $F$ be a closed subset of the product $X \times Y$. Consider the following two conditions:*

(i) *There exists a map $h\colon X \times Y \to X \times Y$ such that*
  (a) *$fix(h) = F$, and*
  (b) *$\pi_X \cdot h = \pi_X$, where $\pi_X : X \times Y \to X$ denotes the natural projection onto the first factor.*
(ii) *$F$ is a $G_\kappa$-subset of the product $X \times Y$.*

*Then (i) implies (ii).*

PROOF. Represent $X$ as the limit space of some $\kappa$-spectrum $\mathcal{S} = \{X_\alpha, p_\alpha^\beta, A\}$. Obviously, the spectrum $\mathcal{S}' = \{X_\alpha \times Y, p_\alpha^\beta \times id, A\}$ is also a $\kappa$-spectrum with $\lim \mathcal{S}' = X \times Y$. By the spectral theorem for $\kappa$-spectra, there exists a $\kappa$-closed and cofinal subset $A'$ in $A$ such that for each $\alpha \in A'$, there is a map $h_\alpha\colon X_\alpha \times Y \to X_\alpha \times Y$ satisfying the following two conditions:

(a)$_\alpha$ $h_\alpha \cdot (p_\alpha \times id) = (p_\alpha \times id) \cdot h$.
(b)$_\alpha$ $\pi_{X_\alpha} \cdot h_\alpha = \pi_{X_\alpha}$, where $\pi_{X_\alpha} : X_\alpha \times Y \to X_\alpha$ denotes the natural projection onto the first factor.

Let $\alpha \in A'$ and $x_\alpha \in (p_\alpha \times id)(F)$. Fix a point $x \in F$ such that $(p_\alpha \times id)(x) = x_\alpha$. By (a), $x = h(x)$. Consequently, by (a)$_\alpha$, $x_\alpha = (p_\alpha \times id)(x) = (p_\alpha \times id)h(x) = h_\alpha \cdot (p_\alpha \times id)(x) = h_\alpha(x_\alpha)$. This shows that $(p_\alpha \times id)(F) \subseteq fix(h_\alpha)$. Let $Z = (p_\alpha \times id)^{-1}((p_\alpha \times id)(F))$. Clearly, $F \subseteq Z$. Suppose that $Z - F \neq \emptyset$ and consider a point $y \in Z - F$. Let $y = (x,s) \in X \times Y$ and $h(y) = (x', s') \in X \times Y$. By (b), $x = x'$. Since $(p_\alpha \times id)(F) \subseteq fix(h_\alpha)$, we see that $(p_\alpha(x'), s') = (p_\alpha \times id)(x', s') = (p_\alpha \times id) \cdot h(x,s) = h_\alpha \cdot (p_\alpha \times id)(x,s) = h_\alpha(p_\alpha(x), s) = (p_\alpha(x), s)$. Therefore, $s = s'$ and $y = h(y)$ which contradicts condition (a). Consequently $F = Z$, and since any closed subset of the product $X_\alpha \times Y$ has character $\leq \kappa$ (because $w(X_\alpha \times Y) = \kappa$), we see that $F$ is a $G_\kappa$-subset of the product $X \times Y$. $\square$

COROLLARY 8.6.42. *Let $\tau > \omega$, $\kappa \geq \omega$ and let $G$ be a nondegenerate compact connected metrizable group. If $\kappa \geq \tau$, then the product $\mathbb{D}^\tau \times G^\kappa$ has the CIPH. If $\tau > \kappa$, then the product $\mathbb{D}^\tau \times G^\kappa$ does not have the CIP.*

PROOF. The first part of our statement follows from Theorem 8.6.40.

Suppose now that $\tau > \kappa$. Let $F$ be a closed subset of the product $\mathbb{D}^\tau \times G^\kappa$ such that $\pi_{\mathbb{D}^\tau}(F) = \mathbb{D}^\tau$ and $F$ is not a $G_\kappa$-subset of that product. Suppose that $F = fix(h)$ for some map $h\colon \mathbb{D}^\tau \times G^\kappa \to \mathbb{D}^\tau \times G^\kappa$. By Lemma 8.6.41, there is a point $(x,s) \in \mathbb{D}^\tau \times G^\kappa - F$ such that $x \neq x'$, where $(x', s') = h(x,s)$.

Consider the connected set $S' = \pi_{\mathbb{D}^\tau}(h(\{x\} \times G^\kappa))$. Clearly, $x' \in S'$. On the other hand, since $\pi_{\mathbb{D}^\tau}(F) = \mathbb{D}^\tau$ we see that $x \in S'$. Therefore, $S'$ is a connected (as a continuous image of $G^\kappa$) and zero-dimensional (as a subspace of $\mathbb{D}^\tau$) set containing two distinct points. This contradiction shows that there is no map $h$ with $F = fix(h)$. Therefore the product $\mathbb{D}^\tau \times G^\kappa$ does not have CIP. □

*Historical and bibliographical notes* 8.6. The results of Subsections 8.6.1 and 8.6.2 are taken from [**103**].

## 8.7. Group actions

In this Section we investigate actions of nonmetrizable groups. Of course, each principal $G$-bundle generates a canonical free $G$-action (and vice versa for $G$ a Lie group). Consequently, one might ask:

- What sort of properties of a space $X$, in the presence of a free action of a "nice" group $G$ on $X$, guarantee that $X$ has the CIPH?

This question remains open even for particularly chosen groups. On the other hand, if $G = \mathbb{T}$ (or $G = \mathbb{R}$) the question is answered for any metrizable space [**216**]. The latter fact suggests that the difficulties involved in extending our result, concerning principal $G$-bundles (see Theorem 8.6.40), to free $G$-actions are perhaps of a set-theoretical nature. For instance, we show (Corollary 8.7.3) that there is no non-metrizable (locally) compact group $G$, acting freely on any locally compact $ANR$-space (note that the total space of any principal $\mathbb{T}^\tau$-bundle is not an $ANR$). This answers a non-metrizable version of R.D.Edwards's question[3] [**140**] asking whether there is a compact Hilbert cube manifold admitting a free action of the group $A_p$?

### 8.7.1. Spectral representations of group actions.

By an *action* of a topological group $G$ on a space $X$ we mean a continuous map $\lambda\colon G \times X \to X$ satisfying the following two conditions:

- $\lambda(g, \lambda(h, x)) = \lambda(gh, x)$ whenever $g, h \in G$ and $x \in X$.
- $\lambda(e, x) = x$ for each $x \in X$, where $e$ denotes the unit element of the group $G$.

We prefer to use the language of diagrams. If $\mu\colon G \times G \to G$ is a continuous multiplication operation on $G$, then the validity of the first condition is equivalent to the commutativity of the following square diagram:

[3]To the best of our knowledge, Edwards's question is still open. Obviously it is a weaker version of the well-known Hilbert-Smith conjecture which asks whether a locally compact group acting effectively on an $n$-manifold must be a Lie group.

$$\begin{array}{ccc} G \times G \times X & \xrightarrow{\ id_G \times \lambda\ } & G \times X \\ \Big\downarrow{\scriptstyle \mu \times id_X} & & \Big\downarrow{\scriptstyle \lambda} \\ G \times X & \xrightarrow{\ \ \lambda\ \ } & X \end{array}$$

The second condition is just the equality

$$\lambda/(\{e\} \times X) = \pi_X/(\{e\} \times X)$$

where $\pi_X : G \times X \to X$ denotes the projection onto the second coordinate.

THEOREM 8.7.1. *Let $\tau \geq \omega$ and $\lambda : G \times X \to X$ be an action. Suppose that $\mathcal{S}_X = \{X_\alpha, p_\alpha^\beta, A\}$ is a $\tau$-spectrum, the limit space of which is homeomorphic to $X$, and $\mathcal{S}_G = \{G_\alpha, s_\alpha^\beta, A\}$ is a $\tau$-spectrum consisting of topological groups and continuous homomorphisms, the limit of which is isomorphic to $G$. If $X$ and $G$ are both locally compact and Lindelöf, or if both are $AE(0)$-spaces (topologically), then $\lambda$ is induced by a cofinal and $\tau$-closed morphism of "level actions"*

$$\{\mu_\alpha : G_\alpha \times X_\alpha \to X_\alpha; B\} : \mathcal{S}_G/B \times \mathcal{S}_X/B \to \mathcal{S}_X/B.$$

*Moreover, if $X$ and $G$ are both locally compact and Lindelöf and $\lambda$ is an effective*[4] *action, we may assume that all actions $\lambda_\alpha$, $\alpha \in B$, also are effective.*

PROOF. The conditions imposed on $G$ and $X$ are needed to ensure that not only the given spectra $\mathcal{S}_G$ and $\mathcal{S}_X$, but also the spectrum

$$\mathcal{S}_G \times \mathcal{S}_X = \{G_\alpha \times X_\alpha, s_\alpha^\beta \times p_\alpha^\beta, A\}$$

is factorizing. Applying Theorem 1.3.6 to the map $\lambda$ (which maps the limit space of the spectrum $\mathcal{S}_G \times \mathcal{S}_X$ into the limit space of the spectrum $\mathcal{S}_X$), we conclude that there exists a cofinal and $\tau$-closed subset $B$ of the indexing set $A$ and maps

$$\lambda_\alpha : G_\alpha \times X_\alpha \to X_\alpha, \quad \alpha \in B$$

such that $\lambda = \lim\{\lambda_\alpha\}$.

Since $\lambda$ is an action, we easily see that each map $\lambda_\alpha$, $\alpha \in B$, is an action of the group $G_\alpha$ to the space $X_\alpha$. Indeed, let $e_\alpha$ denote the unit element of $G_\alpha$ and $x_\alpha \in X_\alpha$. Take a point $x \in X$ such that $p_\alpha(x) = x_\alpha$. Then

$$\lambda_\alpha(e_\alpha, x_\alpha) = \lambda_\alpha(s_\alpha(e), p_\alpha(x)) = \lambda_\alpha((s_\alpha \times p_\alpha)(e, x)) = p_\alpha\lambda(e, x) = p_\alpha(x) = x_\alpha.$$

[4]Recall that an action $\lambda : G \times X \to X$ is effective if for each $g \in G$, $g \neq e$, there exists a point $x \in X$ such that $g(x) \neq x$.

This proves the commutativity of the first diagram from the definition of an action. The second diagram from that definition appears as the bottom of the following cubic diagram:

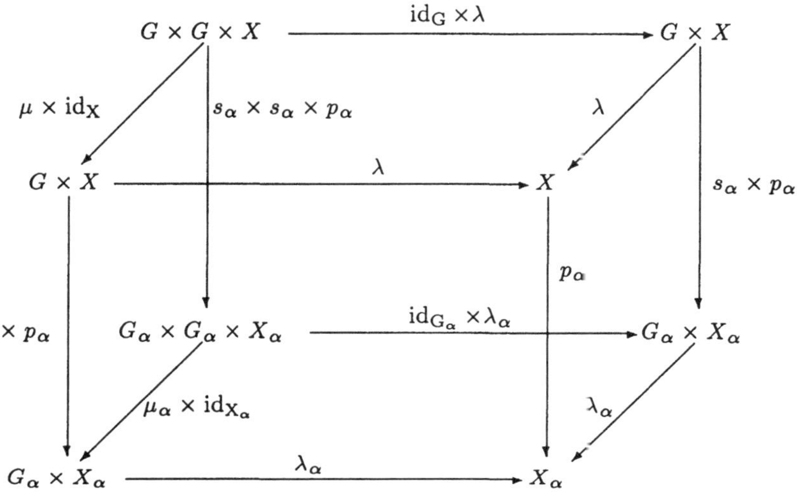


The commutativity of the top ($\lambda$ is an action) and of all the sides ($\{\lambda_\alpha\}$ is a morphism) of the above cube allow us to conclude the commutativity of the bottom. Indeed, let $(g_\alpha, g'_\alpha, x_\alpha) \in G_\alpha \times G_\alpha \times X_\alpha$. Choose a point $(g, g', x) \in G \times G \times X$ such that $s_\alpha(g) = g_\alpha$, $s_\alpha(g') = g'_\alpha$ and $p_\alpha(x) = x_\alpha$. Then

$$\begin{aligned}
\lambda_\alpha((\mathrm{id}_{G_\alpha} \times \lambda_\alpha)(g_\alpha, g'_\alpha, x_\alpha)) = \lambda_\alpha((\mathrm{id}_{G_\alpha} \times \lambda_\alpha)((s_\alpha \times s_\alpha \times p_\alpha)(g, g', x))) = \\
\lambda_\alpha((s_\alpha \times p_\alpha)((\mathrm{id}_G \times \lambda)(g, g', x))) = p_\alpha(\lambda((\mathrm{id}_G \times \lambda)(g, g', x))) = \\
p_\alpha(\lambda((\mu \times \mathrm{id}_X)(g, g', x))) = \lambda_\alpha((s_\alpha \times p_\alpha)((\mu \times \mathrm{id}_X)(g, g', x))) = \\
\lambda_\alpha((\mu_\alpha \times \mathrm{id}_{X_\alpha})((s_\alpha \times s_\alpha \times p_\alpha)(g, g', x))) = \lambda_\alpha((\mu_\alpha \times \mathrm{id}_{X_\alpha})(g_\alpha, g'_\alpha, x_\alpha)).
\end{aligned}$$

Therefore, $\lambda_\alpha$ is an action for each $\alpha \in B$. This proves the first part of theorem.

In order to prove the second part of theorem consider the following relation $L \subseteq B^2$:

$$\begin{aligned}
L = \; & \{(\alpha, \beta) \in B^2 \colon \alpha \leq \beta \text{ and for each } g \in G_\beta - \ker(s^\beta_\alpha) \\
& \text{there exists a point } x \in X_\beta \text{ such that } \lambda_\beta(g, x) \neq x\}.
\end{aligned}$$

Let us verify conditions of Proposition 1.1.29 for the above relation $L$.

*Existance.* Let $\alpha \in B$ and $g \in G - \ker s_\alpha$. Since $\lambda$ is an effective action, there exists a point $x_g \in X$ such that $\lambda(g, x_g) \neq x_g$. Let $O^g_1$ and $O^g_2$ be disjoint open neighborhoods in $X$ of the points $\lambda(g, x_g)$ and $x_g$ respectively. The continuity of $\lambda$ implies that the set $\lambda^{-1}(O^g_1)$ is open in the product $G \times X$. Consequently there exist open subsets $U_g \subseteq G$ and $V_g \subseteq O_2$ such that $(g, x_g) \in U_g \times V_g \subseteq \lambda^{-1}(O^g_1)$.

Clearly $\lambda(U_g \times V_g) \cap V_g = \emptyset$. This shows that $\lambda(g', x') \neq x'$ for each $(g', x') \in U_g \times V_g$. Observe that the open cover $\{U_g : g \in G - \ker s_\alpha\}$ contains an open refinement $\{U_t : t \in T\}$ of cardinality at most $\tau$, i.e. $|T| \leq \tau$ (to see this note that the restriction $s_\alpha/(G - \ker s_\alpha): (G - \ker s_\alpha) \to (G_\alpha - \{e_\alpha\})$ is a proper map and that the space $G_\alpha - \{e_\alpha\}$ can be written as an union of at most $\tau$ compact subsets of the group $G_\alpha$). If $t \in T$, then $U_t \subseteq U_g$ for some $g \in G - \ker s_\alpha$. By $V_t$ we denote the set $V_g$ constructed above. Without loss of generality we may assume that for each $t \in T$ there exist an index $\beta_t \in B$ and open sets $U_{\beta_t} \subseteq G_{\beta_t}$ and $V_{\beta_t} \subseteq X_{\beta_t}$ such that $U_t = s_{\beta_t}^{-1}(U_{\beta_t})$ and $V_t = p_{\beta_t}^{-1}(V_{\beta_t})$. There exists an index $\beta \in B$ such that $\beta \geq \alpha$ and $\beta \geq \beta_t$ for each $t \in T$. Let $U_t^\beta = (s_{\beta_t}^\beta)^{-1}(U_{\beta_t})$ and $V_t^\beta = (p_{\beta_t}^\beta)^{-1}(V_{\beta_t})$ for each $t \in T$. Then we see that $s_\beta^{-1}(U_t^\beta) = U_t$ and $p_\beta^{-1}(V_t^\beta) = V_t$ for each $t \in T$. Since $\lambda_\beta(s_\beta \times p_\beta) = p_\beta \lambda$ we conclude that $\lambda_\beta(U_t^\beta \times V_t^\beta) \cap V_t^\beta = \emptyset$. Finally observe that $G_\beta - \ker(s_\alpha^\beta) = \bigcup\{U_t^\beta : t \in T\}$. It only remains to note that in this situation for each $g \in G_\beta - \ker(s_\alpha^\beta)$ there is a point $x \in X_\beta$ such that $\lambda_\beta(g, x) \neq x$. This proves that $(\alpha, \beta) \in L$.

*Majorantness.* Let $(\alpha, \beta) \in L$ and $\gamma \geq \beta$ for some $\gamma \in B$. consider an element $g \in G_\gamma - \ker(s_\alpha^\gamma)$. Clearly $s_\beta^\gamma(g) \in G_\beta - \ker(s_\alpha^\beta)$. Since $(\alpha, \beta) \in L$, there exists a point $x \in X_\beta$ such that $\lambda_\beta(s_\beta^\gamma(g), x) \neq x$. Consider a point $y \in X_\gamma$ such that $p_\beta^\gamma(y) = x$. Then $\lambda_\gamma(g, y) \neq y$. Thus $(\alpha, \gamma) \in L$.

*$\tau$-closedness.* Let $P = \{\alpha_\gamma : \gamma < \tau\}$ be a chain of element of $B$ of cardinality at most $\tau$. Suppose also that $(\alpha_\gamma, \beta) \in L$ for each $\gamma < \tau$ and for some element $\beta \in B$. We need to show that $(\alpha, \beta) \in L$, where $\alpha = \sup P$. Let $g \in G_\beta - \ker(s_\alpha^\beta)$. Since the spectrum $\mathcal{S}_G$ is a $\tau$-spectrum, the space $G_\alpha$ is canonically homeomorphic (even isomorphic in our situation) to the limit space of the spectrum $\mathcal{S}_G/P$. Therefore $\{e_\alpha\} = \bigcap\{\ker(s_{\alpha_\gamma}^\alpha) : \gamma < \tau\}$. This allows us to find an ordinal $\gamma < \tau$ such that $g \in G_\beta - \ker(s_{\alpha_\gamma}^\beta)$. Since $(\alpha_\gamma, \beta) \in L$, there exists a point $x \in X_\beta$ such that $\lambda_\beta(g, x) \neq x$

Thus the relation $L$ satisfies all three conditions of Proposition 1.1.29. Consequently, by that Proposition, the set of all $L$-reflexive indexes, denote it by $C$, is cofinal and $\tau$-closed in $B$. It only remains to note that $L$-reflexibility of an index $\alpha$ means that the action $\lambda_\alpha$ is effective. This completes the proof of theorem. □

Recall that for a given action $\lambda: G \times X \to X$ and a point $x \in X$ the closed subgroup $G_x = \{g \in G : g(x) = x\}$ is called a *stabilizer* of $x$. An action $\lambda$ is said to be *semi-free* if the stabilizer of each point of $X$ coincides with $G$ or is trivial. If only the second possibility occurs for every point $x \in X$, then $\lambda$ is said to be *free*.

COROLLARY 8.7.2. *If an $AE(0)$-group $G$ acts semi-freely on a non-metrizable (locally) compact $ANR$-space $X$, then*

(a) *$\lambda$ is the trivial action*

*or*

(b) $w(G) < w(X)$.

PROOF. Let $w(X) = \tau > \omega$ and suppose that $w(G) \geq \tau$.

First consider the case $w(G) > \tau$. By Lemma 8.2.1, $G$ is isomorphic to the limit space of a factorizing $\tau$-spectrum $\mathcal{S}_G = \{G_\alpha, s_\alpha^\beta, A\}$ consisting of $AE(0)$-groups of weight $\tau$ and 0-soft limit homomorphisms. Applying Theorem 8.7.1 to this situation, we find an index $\alpha \in A$ and an action

$$\lambda_\alpha \colon G_\alpha \times X \to X$$

such that

$$\lambda = \lambda_\alpha(s_\alpha \times id_X). \tag{8.7.1}$$

Since $w(G) > \tau = w(G_\alpha)$, we see that $\ker s_\alpha$ contains at least two elements (otherwise $s_\alpha \colon G \to G_\alpha$ is an isomorphism). Let $g \in \ker s_\alpha$ be a non-trivial element of $G$. Then, by (1),

$$\lambda(g,x) = \lambda_\alpha(s_\alpha \times id_X)(g,x) = \lambda_\alpha(s_\alpha(g),x) = \lambda_\alpha(e_\alpha,x) = x$$

for each point $x \in X$ (here $e_\alpha$ denotes the trivial element of the group $G_\alpha$). This shows that $\lambda$ is the trivial action.

Next consider the case $w(G) = \tau$. The action $\lambda$ is trivial if (and only if) the set $X^G = \{x \in X : G_x = G\}$ coincides with $X$. Obviously, $X^G$ is closed in $X$. Therefore, we only need to show that $X^G$ is dense in $X$. By Lemma 8.2.1, $G$ is isomorphic to the limit space of a factorizing $\omega$-spectrum $\mathcal{S}_G = \{G_\alpha, s_\alpha^\beta, A\}$ consisting of Polish groups $G_\alpha$ and 0-soft limit homomorphisms $s_\alpha$. Similarly, by Proposition 6.3.5 and Remark 6.3.6, $X$ is the limit space of a factorizing $\omega$-spectrum $\mathcal{S}_X = \{X_\alpha, p_\alpha^\beta, A\}$ consisting of locally compact $ANR$-spaces (with countable bases) $X_\alpha$ and proper and soft limit projections $p_\alpha$. By Theorem 8.7.1, we may assume, without loss of generality, that $\lambda$ is the limit of "level actions" $\lambda_\alpha \colon G_\alpha \times X_\alpha \to X_\alpha$, $\alpha \in A$.

Let $U$ be a non-empty open $F_\sigma$-subset of $X$. Since the spectrum $\mathcal{S}_X$ is an $\omega$-spectrum, there is an index $\alpha \in A$ such that $U = p_\alpha^{-1}(p_\alpha(U))$. As above, there is a non-trivial element $g \in \ker s_\alpha$. Let $x_\alpha \in p_\alpha(U)$. We are going to show that the map $\lambda_g \colon X \to X$ (defined by letting $\lambda_g(x) = \lambda(g,x)$) preserves the fiber $p_\alpha^{-1}(x_\alpha)$ of the limit projection $p_\alpha$. Indeed, take a point $x \in p_\alpha^{-1}(x_\alpha)$. Then

$$p_\alpha\lambda(g,x) = \lambda_\alpha(s_\alpha \times p_\alpha)(g,x) = \lambda_\alpha(s_\alpha(g),p_\alpha(x)) = \lambda(e_\alpha,x_\alpha) = x_\alpha.$$

Therefore, $\lambda_g(p_\alpha^{-1}(x_\alpha)) \subseteq p_\alpha^{-1}(x_\alpha)$. The softness and properness of the projection $p_\alpha$ guarantees that the fiber $p_\alpha^{-1}(x_\alpha)$ is an $AR$-compactum and, hence, has the fixed point property. Therefore there is a point $z \in p_\alpha^{-1}(x_\alpha)$ such that $g(z) = z$. Since $\lambda$ is a semi-free action, the latter means that $z \in X^G$. It only remains to note that, by the choice of the index $\alpha$, $z \in U$. Therefore, $U \cap X^G \neq \emptyset$. $\square$

COROLLARY 8.7.3. *There is no non-metrizable $ANR$-compactum admitting a non-trivial semi-free action of any non-metrizable compact group.*

PROOF. Let $\lambda \colon G \times X \to X$ be a non-trivial semi-free action of a compact group $G$ of weight $\tau > \omega$ onto a non-metrizable compactum $X$ of weight $\kappa > \omega$. Since $\lambda$ is non-trivial, we have, by Corollary 8.7.2, $\tau < \kappa$. Represent $X$ as the limit space of a factorizing $\tau$-spectrum $\mathcal{S}_X = \{X_\alpha, p_\alpha^\beta, A\}$ consisting of $ANR$-compact spaces of weight $\tau$ and soft limit projections. By Theorem 8.7.1, we may assume that $\lambda$ is the limit of a morphism

$$\{\lambda_\alpha \colon G \times X_\alpha \to X_\alpha; A\}$$

of "level actions". Since $\lambda$ is non-trivial action, there is a pair $(g, x) \in G \times X$ such that $\lambda(g, x) \neq x$. Take an index $\alpha \in A$ such that

$$p_\alpha(\lambda(g, x)) \neq p_\alpha(x). \tag{8.7.2}$$

We claim that the action $\lambda_\alpha \colon G \times X_\alpha \to X_\alpha$ is also non-trivial and semi-free. Indeed, since $p_\alpha \lambda = \lambda_\alpha(id_G \times p_\alpha)$, we see that, by (2),

$$\lambda_\alpha(g, p_\alpha(x)) = \lambda_\alpha(id_G \times p_\alpha)(g, x) = p_\alpha \lambda(g, x) \neq p_\alpha(x).$$

This shows that the element $g \in G$ moves the point $p_\alpha(x)$ and, hence, $\lambda_\alpha$ is a non-trivial action.

Next we show that $\lambda_\alpha$ is semi-free. Take any point $x_\alpha \in X_\alpha$, and suppose that for some non-trivial element $g \in G$ we have $\lambda_\alpha(g, x_\alpha) = x_\alpha$. Take any point $x \in p_\alpha^{-1}(x_\alpha)$. Then we have

$$p_\alpha \lambda(g, x) = \lambda_\alpha(id_G \times p_\alpha)(g, x) = \lambda_\alpha(g, x_\alpha) = x_\alpha.$$

This shows that the map $\lambda_g \colon X \to X$ (defined, as above, by letting $\lambda_g(x) = \lambda(g, x)$) maps the fiber $p_\alpha^{-1}(x_\alpha)$ over the point $x_\alpha$ into itself. The softness of the limit projection $p_\alpha$ of the spectrum $\mathcal{S}_X$ guarantees that this fiber is an $AR$-compactum. Therefore, there is a point $y \in p_\alpha^{-1}(x_\alpha)$ such that $y = \lambda_g(y) = \lambda(g, y)$. Since $g$ is a non-trivial element of $G$, the stabilizer $G_y$ of the point $y \in X$ is non-trivial. Since $\lambda$ is a semi-free action, this implies that $G_y = G$, i.e. $\lambda(g', y) = y$ for each element $g' \in G$. But then

$$\lambda_\alpha(g', x_\alpha) = \lambda_\alpha(id_G \times p_\alpha)(g', y) = p_\alpha \lambda(g', y) = p_\alpha(y) = x_\alpha.$$

This shows that the stabilizer $G_{x_\alpha}$ of any point $x_\alpha \in X_\alpha$ coincides with $G$ whenever this stabilizer contains at least one non-trivial element. This means that $\lambda_\alpha$ is a semi-free action.

It only remains to observe that the equality $w(G) = \tau = w(X_\alpha)$, coupled with Corollary 8.7.2 (applied to the non-trivial semi-free action $\lambda_\alpha \colon G \times X_\alpha \to X_\alpha$), leads us to a contradiction. □

*Remark* 8.7.4. Note that the assumption of non-metrizability in Corollary 8.7.2 is essential. Indeed, the group $A_p$ acts on the Hilbert cube $I^\omega$, fixing a single point and acting freely off that point [**140**]. Hence $A_p$ acts freely on the non-compact Hilbert cube manifold $I^\omega - \{pt\} \approx I^\omega \times [0,1)$. We also remark that the product $M \times [0,1)$ has the CIPH for each $I^\omega$-manifold $M$ [**216**].

### 8.7.2. *Examples.*

PROPOSITION 8.7.5. *Let $f\colon X \to Y$ be a principal $R$-bundle between metrizable spaces. If $F$ is a nonempty closed subset of $X$, then there is a fiber preserving autohomeomorphism of $X$ whose fixed point set is $F$ (in particular, $X$ has the CIPH).*

PROOF. Let $|\cdot|$ denote the usual norm metric on the real line $R$ and let $d$ be a metric for $X$ such that $d(x,y) = |x-y|$ whenever $x, y \in X$ and $f(x) = f(y)$ (see [**295**] or [**300**]). Given a nonempty closed subset $F$ of $X$, set $r(x) = \frac{1}{2}d(x,F)$.

Consider (we maintain the notation in the proof of Theorem 8.6.40) a collection $\{U^i\colon i \in I\}$ of open subsets of $Y$, a collection

$$h^i\colon U^i \times R \to f^{-1}(U^i), \quad i \in A,$$

of fiber-preserving homeomorphisms and a collection

$$g^{i,j}\colon U^i \cap U^j \to R, \quad i,j \in A$$

of continuous maps such that

$$h^i(x,t) = h^j(x, t + g^{i,j}(x)) \text{ for each } (x,t) \in (U^i \cap U^j) \times R \text{ and } i,j \in A.$$

Here the plus sign represents the usual addition operation in the additive group $R$ and $A$ stands for an indexing set.

Let

$$H^i(x) = h^i(\pi_1^i((h^i)^{-1}(x)), \pi_2^i((h^i)^{-1}(x)) + r(x)) \text{ whenever } x \in f^{-1}(U^i),\ i \in A.$$

Here $\pi_1^i$ and $\pi_2^i$ denote the natural projections of the product $U^i \times R$ onto the first and the second coordinates respectively.

Note that if $x \in f^{-1}(U^i \cap U^j)$, $i,j \in A$, then

$$H^i/f^{-1}(U^i \cap U^j) = H^j/f^{-1}(U^i \cap U^j).$$

Therefore, the map $H\colon X \to X$ defined by letting $H(x) = H^i(x)$ for each $x \in f^{-1}(U^i)$, $i \in A$, is well defined.

Obviously, $fH = f$ and the set of all fixed points of $H$ coincides with $F$.

A straightforward verification (using a slight modification of the argument presented in the proof of [**216**, Theorem 2.2]) shows that $H$ is a homeomorphism. □

Another important class of spaces with the *CIPH* is the class of normed linear spaces.

PROPOSITION 8.7.6. *Any normed linear space has the CIPH.*

PROOF. Let $(X, \|\cdot\|)$ be a normed linear space. By $d$ we denote the metric generated by the norm, i.e. $d(x,y) = \|x-y\|$ for each $x, y \in X$.

Let $A$ be a closed subset of $X$. Without loss of generality we may assume that $0 \in A$. Define a map $h\colon X \to X$ as follows:

$$h(x) = \begin{cases} x + \frac{1}{2}d(x,A)\cdot\frac{x}{\|x\|} & \text{for } x \neq 0,\\ 0 & \text{for } x = 0.\end{cases}$$

Obviously, $fix(h) = A$. Let us show that $h$ is a homeomorphism.

Let $x, y \in X$ and $h(x) = h(y)$. Then

$$x + \frac{1}{2}d(x,A)\cdot\frac{x}{\|x\|} = y + \frac{1}{2}d(y,A)\cdot\frac{y}{\|y\|}.$$

Consequently,

$$(1 + \frac{1}{2\|x\|}d(x,A))\|x\| = (1 + \frac{1}{2\|y\|)}d(y,A))\|y\|.$$

Then

$$\|x\| + \frac{1}{2}d(x,A) = \|y\| + \frac{1}{2}d(y,A)$$

and

$$\|y-x\| = |\|x\| - \|y\|| = |\frac{1}{2}(d(x,A) - d(y,A)|.$$

On the other hand, the triangle inequality applied to $x, y$ and $A$ implies

$$\|y-x\| \geq |d(x,A) - d(y,A)|.$$

Therefore, $|d(x,A) - d(y,A)| = 0$ and $x = y$. This shows that $h$ is an injection.

We now show that for each point $x \in X$, there is a point $y \in X$ such that $h(y) = x$ (i.e. $h$ is a surjection). Since, by definition, $h(0) = 0$ we may assume that $x \neq 0$.

Since $0 \in A$ by our assumption, we conclude that $\|x\| \geq d(x,A)$. Choose a real number $t$ such that $t \leq \|x\| - d(x,A)$. Then $d(x,A) \leq \|x\| - t$ (in particular, $0 \leq t \leq \|x\|$). The triangle inequality applied to $\frac{tx}{\|x\|}$, $x$ and $A$ implies that

$$d(\frac{tx}{\|x\|}, A) \leq \|x - \frac{tx}{\|x\|}\| + d(x,A) = \|x\| - \|\frac{tx}{\|x\|}\| + d(x,A) = \|x\| - t + d(x,A) \leq$$
$$2(\|x\| - t).$$

Therefore

$$t + \frac{1}{2}d(\frac{tx}{\|x\|}, A) \leq \|x\|.$$

On the other hand,

$$\|h(\frac{tx}{\|x\|})\| = \|\frac{tx}{\|x\|} + \frac{1}{2}d(\frac{tx}{\|x\|}, A)\cdot\frac{\frac{tx}{\|x\|}}{\|\frac{tx}{\|x\|}\|}\| = \frac{t}{\|x\|}\|x\| + \frac{1}{2}d(\frac{tx}{\|x\|}, A)\cdot\frac{\frac{t}{\|x\|}\|x\|}{\|\frac{tx}{\|x\|}\|} =$$

$$t + \frac{1}{2}d(\frac{tx}{\|x\|}, A).$$

Therefore $\|h(\frac{tx}{\|x\|})\| \leq \|x\| \leq \|h(x)\|$. In this situation, the connected set $h(\{rx : \frac{t}{\|x\|} \leq r \leq 1\})$ (i.e. the image of the segment with end points $\frac{tx}{\|x\|}$ and $x$) is contained in $\{rx : r \in R\}$ and, as was shown above, contains the point $x$. Therefore there is a point $y \in [\frac{tx}{\|x\|}, x]$ such that $h(y) = x$ . This shows that $h$ is a surjection and hence a continuous bijection.

It only remains to show that the inverse function $h^{-1}\colon X \to X$ is also continuous. Let $\{x_i\}_{i=1}^{i=\infty}$ be a sequence of points of $X$ such that the sequence $\{h(x_i)\}_{i=1}^{i=\infty}$ converges in $X$ to the point $y$. In order to show that $\{x_i\}_{i=1}^{i=\infty}$ also converges, we need the following observation.

*Claim.* If $x, y \in X$, then $\frac{1}{2}\|x - y\| \leq \|h(x) - h(y)\|$.

*Proof of Claim.* If at least one of the given points coincides with 0, then the statement follows from the definition of $h$. Hence we may assume that $x \neq 0 \neq y$. The triangle inequality, applied to the triples $(x, y, h(y))$ and $(x, h(x), h(y))$, implies

$$\|x - y\| \leq \|h(y) - y\| + \|x - h(y)\| \leq \|h(y) - y\| + \|h(x) - x\| + \|h(x) - h(y)\| =$$

$$\frac{1}{2}(d(y, A) + d(x, A)) + \|h(x) - h(y)\| \leq \frac{1}{2}(\|y\| + \|x\|) \leq \frac{1}{2}\|y - x\| + \|h(x) - h(y)\|.$$

Therefore $\frac{1}{2}\|x - y\| \leq \|h(x) - h(y)\|$ as required. The claim is established.

Using the above claim we see that $\|h^{-1}(y) - x_i\| \leq 2\|y - h(x_i)\|$. Consequently, $\{x_i\}_{i=1}^{i=\infty}$ converges to the point $h^{-1}(y)$. □

*Historical and bibliographical notes* 8.7. The results of this Section are due to the author.

## 8.8. Baire isomorphisms

In this Section we show that the spectral theorem is valid not only for continuous maps but for Baire maps as well. While it is possible to cover the general case of arbitrary realcompact spaces, in order to simplify the situation, we consider below only spectrally complete spaces, i.e. spaces which can be represented as the limit spaces of factorizing $\omega$-spectra consisting of Polish spaces and surjective projections (see Section 6.2).

First of all, recall some definitions. Let $X$ be a space and $C(X)$ the set of continuous real-valued functions on $X$. Let $B_0(X) = C(X)$, and inductively define $B_\alpha(X)$ for each ordinal $\alpha \leq \omega_1$ to be the set of pointwise limits of sequences of functions in $\cup\{B_\beta(X)\colon \beta < \alpha\}$. Let $B^*_\alpha(X)$ be the set of bounded functions in $B_\alpha(X)$. The functions in

$$B_{\omega_1}(X) = \cup\{B_\alpha(X)\colon \alpha < \omega_1\}$$

are called *Baire functions* (or *Borel functions* if $X$ is perfectly normal).

The *Baire sets* (or *Borel sets* if $X$ is perfectly normal) of $X$ of *multiplicative* class $\alpha$, denoted by $\mathcal{Z}_\alpha(X)$, are defined to be the zero sets of functions in $B^*_\alpha(X)$. Those of *additive* class $\alpha$, denoted by $\mathcal{CZ}_\alpha(X)$, are defined as the complements of sets in $\mathcal{Z}_\alpha(X)$. Clearly, for each $\alpha < \omega_1$, $\mathcal{Z}_\alpha(X)$ is closed under countable intersections and finite unions, and $\mathcal{CZ}_\alpha(X)$ is closed under countable unions and finite intersections. It is well-known that

- Members of $\mathcal{Z}_{\alpha+1}(X)$ are countable intersections of members of $\mathcal{CZ}_\alpha(X)$.
- Members of $\mathcal{CZ}_{\alpha+1}(X)$ are countable unions of members of $\mathcal{Z}_\alpha(X)$.
- If $\alpha$ is a limit ordinal, then members of $\mathcal{Z}_\alpha(X)$ are countable intersections of countable unions of members of

$$\left(\bigcup_{\beta<\alpha} \mathcal{CZ}_\beta(X)\right).$$

- If $\alpha$ is a limit ordinal, then members of $\mathcal{CZ}_\alpha(X)$ are countable unions of countable intersections of members of

$$\left(\bigcup_{\beta<\alpha} \mathcal{Z}_\beta(X)\right).$$

Let $\alpha \in \omega_1$. A map $f\colon X \to Y$ is called a *Baire map of class* $\alpha$ (or *Borel map of class* $\alpha$ if $X$ and $Y$ are perfectly normal) if

$$f^{-1}(\mathcal{Z}_0(Y) \subseteq \mathcal{Z}_\alpha(X).$$

We say that $f\colon X \to Y$ is a *Baire map* (or *Borel map* if $X$ and $Y$ are perfectly normal) if

$$f^{-1}(\mathcal{Z}_0(Y) \subseteq \mathcal{Z}_{\omega_1}(X).$$

Obviously Baire maps are Baire maps of class $\omega_1$ (and conversely).

A bijection $f\colon X \to Y$ is called a *Baire isomorphism of class* $(\alpha, \beta)$ if $f$ is a Baire map of class $\alpha$ and $f^{-1}\colon Y \to X$ is a Baire map of class $\beta$. Baire isomorphisms of the class $(\omega_1, \omega_1)$ are, for simplicity, called *Baire isomorphisms.*

Finally a bijection $f\colon X \to Y$ is called an *$\alpha$-th level Baire isomorphism,* $0 \le \alpha \le \omega_1$, if $f(\mathcal{Z}_\alpha(X)) = \mathcal{Z}_\alpha(Y)$.

The proof of the following statement follows the proof of Theorem 1.3.4 (or Theorem 1.3.6).

THEOREM 8.8.1. *Let $0 \le \alpha, \beta \le \omega_1$. Each Baire map $f\colon \lim \mathcal{S}_X \to \lim \mathcal{S}_Y$ of class $\alpha$ between the limit spaces of factorizing $\omega$-spectra $\mathcal{S}_X = \{X_\alpha, p^\beta_\alpha, A\}$ and $\mathcal{S}_Y = \{Y_\alpha, q^\beta_\alpha, A\}$ , consisting of Polish spaces, is induced by a morphism (consisting of Baire maps of class $\alpha$) of cofinal and $\omega$-closed subspectra. If $f$ is a Baire isomorphism of class $(\alpha, \beta)$ (respectively, an $\alpha$-th level isomorphism), we may assume that the above morphism consists of Baire isomorphism of class $(\alpha, \beta)$ (respectively, of $\alpha$-th level Baire isomorphisms).*

Below we give several applications of Theorem 8.8.1 .

PROPOSITION 8.8.2. *If $X$ and $Y$ are first level Baire isomorphic compacta, then* $\dim X = \dim Y$.

PROOF. Our statement is true for metrizable compacta [**259**]. Consequently, without loss of generality we may assume that $w(X) = w(Y) > \omega$. Let $f: X \to Y$ be a first level Baire isomorphism and $\dim X \leq n$. Represent $X$ and $Y$ as the limit spaces of (factorizing) $\omega$-spectra $\mathcal{S}_X = \{X_\alpha, p_\alpha^\beta, A\}$ and $\mathcal{S}_Y = \{Y_\alpha, q_\alpha^\beta, A\}$ respectively. By Theorem 1.3.10, we may additionally assume that $\dim X_\alpha \leq n$ for each $\alpha \in A$. By Theorem 8.8.1, $f$ is the limit of the morphism

$$\{f_\alpha : X_\alpha \to Y_\alpha; A\} : \mathcal{S}_X \to \mathcal{S}_Y$$

consisting of first level Baire isomorphisms. By the above cited result from [**259**], $\dim Y_\alpha \leq n$ for each $\alpha \in A$. Therefore, by Theorem 1.3.10, $\dim Y \leq n$ as desired. □

Let us recall (see, for example, [**146**]) that the *transfinite dimensions* ind and Ind are the ordinal valued functions obtained by extending the usual concept of small and large inductive dimensions. The transfinite dimensions of separable metrizable spaces are always countable ordinals. It is also known (W. Hurewicz) that the small transfinite dimension $\operatorname{ind} X$ of a Polish space $X$ is defined if and only if $X$ is *countable dimensional* (i.e. $X$ can be represented as the union of countable many zero-dimensional subspaces). The last property is an invariant of first level Baire isomorphisms in the class of metrizable compacta [**259**].

Let $\mathcal{K}$ denote the class of compacta admitting zero-dimensional maps onto metrizable compacta.

PROPOSITION 8.8.3. *Let $X$ and $Y$ be first level Baire isomorphic compacta. If $X \in \mathcal{K}$ and the transfinite dimension* $\operatorname{ind} X$ *is defined, then $Y \in \mathcal{K}$ and the transfinite dimension* $\operatorname{ind} Y$ *is defined.*

PROOF. Let $f: X \to Y$ be a first level Baire isomorphism. Without loss of generality we may assume that $w(X) = w(Y) > \omega$. Represent $X$ and $Y$ as the limit spaces of (factorizing) $\omega$-spectra $\mathcal{S}_X = \{X_\alpha, p_\alpha^\beta, A\}$ and $\mathcal{S}_Y = \{Y_\alpha, q_\alpha^\beta, A\}$ respectively. By Theorem 8.8.1, $f$ is the limit map of the morphism

$$\{f_\alpha : X_\alpha \to Y_\alpha; A\} : \mathcal{S}_X \to \mathcal{S}_Y$$

consisting of first level Baire isomorphisms.

Since $X \in \mathcal{K}$, by [**76**, Theorem 5], there is a zero-dimensional map $g: X \to K$ onto a countable-dimensional metrizable compactum $K$. Represent $K$ as the union of countably many zero-dimensional $G_\delta$-subspaces $K_i$. Clearly

$$X = \cup\{g^{-1}(K_i) : i \in \omega\}$$

and for each $i \in \omega$ the subspace $g^{-1}(K_i)$, denoted by $X_i$, is zero-dimensional, Lindelöf and Čech-complete. Applying Theorem 1.3.10 to each of the subspaces

$X_i$, we may assume without loss of generality that $X_\alpha$ is countable-dimensional for each $\alpha \in A$. Note also that (again, without loss of generality) each limit projection $p_\alpha$ of the spectrum $\mathcal{S}_X$ is zero-dimensional. To see this observe that the map $g: X \to K$ can be factored through some index $\alpha \in A$, i.e. there is a map $g_\alpha: X_\alpha \to K$ such that $g = g_\alpha p_\alpha$. Therefore, the fibers of $p_\alpha$ are subsets of the corresponding fibers of $g$. This shows the zero-dimensionality of $p_\alpha$.

By **[259]**, each $Y_\alpha$ is countable dimensional. Next observe that the fibers of the limit projection $q_\alpha$ of the spectrum $\mathcal{S}_Y$ are first level Baire isomorphic to the corresponding fibers of the limit projection $p_\alpha$ of the spectrum $\mathcal{S}_X$ (this follows from the existence of the above indicated morphism). By Proposition 8.8.2, the fibers of the limit projection $q_\alpha$ are zero-dimensional. Consequently, $Y \in \mathcal{K}$ and, by **[76]**, the transfinite dimension of $Y$ is defined. □

PROPOSITION 8.8.4. *Let $\tau > 2^\omega$. Then there is no Baire isomorphism between $D^{\tau^+}$ and $expD^{\tau^+}$.*

PROOF. Consider the Cantor cube $D^{\tau^+}$ of weight $\tau^+$. Clearly $D^{\tau^+}$ is the limit space of the $\tau$-spectrum

$$\mathcal{S} = \{D^A, p_A^B, exp_\tau \tau^+\}$$

where $exp_\tau\tau^+$ denotes the collection of all subsets of $\tau^+$ of cardinality $\leq \tau$ and $p_A^B$ is the natural projection of $D^B$ onto $D^A$. Observe that all limit projections $p_A$ of $\mathcal{S}$ are homeomorphic to the natural projection of $D^{\tau^+}$ onto $D^\tau$. Consider also the $\tau$-spectrum

$$exp\mathcal{S} = \{expD^A, exp(p_A^B), exp_\tau\tau^+\}$$

whose limit space is homeomorphic to $expD^{\tau^+}$.

Suppose now that there exists a Baire isomorphism

$$f: D^{\tau^+} \to expD^{\tau^+}.$$

By Theorem 8.8.1, $f$ is the limit map of a Baire morphism

$$\{f_A: D^A \to expD^A; exp_\tau\tau^+\}: \mathcal{S} \to exp\mathcal{S}.$$

This simply means that $f_A p_A = exp(p_A)f$ for each $A \in exp_\tau\tau^+$. Consequently, for each point $F$ of $expD^A$, the fiber $exp(p_A)^{-1}(F)$ is Baire isomorphic to the corresponding fiber $p_A^{-1}(f_A^{-1}(F))$, which in turn is homeomorphic to $D^{\tau^+}$. Let $T$ be a subspace of $D^A$ such that $T$ is discrete in the relative topology and $|T| = \tau$. Let $F$ denote the closure of $T$ in $D^A$. Obviously, $F$ is a point of $expD^A$ and there exists a pairwise disjoint collection of cardinality $\tau$ of open subsets of $F$. One can easily check that the fiber $exp(p_A)^{-1}(F)$ also contains a pairwise disjoint collection of cardinality $\tau$ of open subsets. Then we can conclude that there exists a pairwise disjoint collection of cardinality $\tau$ of Baire sets in $p_A^{-1}(f_A^{-1}(F)) = D^{\tau^+}$. Since each Baire set is the union of functionally closed sets, we see that there exists a pairwise disjoint collection of cardinality

$\tau$ of functionally closed subsets of $D^{\tau^+}$. This is impossible by the inequality $\tau > 2^\omega$. □

COROLLARY 8.8.5. *Assume* $\omega_1 = 2^\omega$. *Then there is no Baire isomorphism between* $D^{\omega_3}$ *and* $expD^{\omega_3}$.

We conclude this Section by proving the following two statements.

PROPOSITION 8.8.6. *Let* $X$ *be a compact space and* $Y$ *a Baire subset of the Stone-Čech compactification* $\beta Y$. *If* $X$ *and* $Y$ *are first level Baire isomorphic, then* $Y$ *is* $\sigma$*-compact.*

PROPOSITION 8.8.7. *Let* $X$ *be a Baire subset of its Stone-Čech compactification* $\beta X$ *and* $Y$ *a Suslin subset of* $\beta Y$ *(generated by the collection of* $\mathcal{Z}_0(\beta Y)$*). If* $X$ *and* $Y$ *are Baire isomorphic, then* $Y$ *is a Baire subset of* $\beta Y$.

Both statements are known to be true for metrizable $X$ and $Y$ (see, for example, [**260**]). Our goal is to illustrate how Theorem 8.8.1 can be applied in order to get the general statements. The proof of Propositions 8.8.6 and 8.8.7 are similar. For this reason, below we sketch them simultaneously.

*Sketch of proof of Propositions* 8.8.6. It can be shown, using the standard argument (see [**260**, Theorem 5.8.7]), that $Y$ embeds as a closed subspace of the product $I^T \times B$ of the Tychonov cube $I^T$ and a Baire subset $B$ of some Polish space. Consider the images of $Y$ under the natural projections of this product onto the subproducts of the form $I^K \times B$, where $K$ is a countable subset of $T$. Obviously, each of these images is closed in the corresponding subproduct, and the restrictions of the indicated projections onto $Y$ are proper. Therefore, we can conclude that $Y$ is homeomorphic to the limit space of the inverse spectrum $\mathcal{S}_Y = \{Y_\alpha, q_\alpha^\beta, A\}$ consisting of the indicated images and restrictions of natural projections. It is also easy to see that this spectrum is an $\omega$-spectrum. Furthermore, the closedness of $Y$ in the normal space $I^T \times B$ guarantees that the spectrum $\mathcal{S}_Y$ is factorizing. Without loss of generality we may also assume that the compactum $X$ is the limit space of a factorizing $\omega$-spectrum $\mathcal{S}_X = \{X_\alpha, p_\alpha^\beta, A\}$ with the same indexing set. Let $f\colon X \to Y$ denote a Baire isomorphism of the first level. Applying Theorem 8.8.1 to our situation, we see that the set $A$ contains a cofinal and $\omega$-closed subset consisting of indexes $\alpha$ such that $f_\alpha p_\alpha = q_\alpha f$, where $f_\alpha\colon X_\alpha \to Y_\alpha$ is a Baire isomorphism of the first level. Since the spaces $X_\alpha$ and $Y_\alpha$ are separable and metrizable, we conclude that $Y_\alpha$ is $\sigma$-compact. Since the limit projection $q_\alpha\colon Y \to Y_\alpha$ is proper we see that $Y$ is also $\sigma$-compact. This completes the proof of Proposition 8.8.6.

*Sketch of proof of Proposition* 8.8.7. According to **260**, Theorem 5.8.9], $Y$ can be identified with a closed subspace of the product $I^T \times B$ of a Tychonov cube $I^T$ and an analytic subspace $B$ of a Polish space. Repeating the above argument, we easily conclude that $Y$ is the limit space of a factorizing $\omega$-spectrum $\mathcal{S}_Y = \{Y_\alpha, q_\alpha^\beta, A\}$ consisting of analytic subspaces of Polish spaces and proper limit projections. Similarly, $X$ is the limit space of a factorizing $\omega$-spectrum $\mathcal{S}_X = \{X_\alpha, p_\alpha^\beta, A\}$ consisting of Baire subspaces of Polish spaces and proper limit projections. By Theorem 8.8.1, there exists at least one index $\alpha \in A$ such that the spaces $X_\alpha$ and $Y_\alpha$ are Baire isomorphic. This obviously implies that $Y_\alpha$ is also a Baire subset of a Polish space. Then $Y$ is homeomorphic to a closed subspace of the product of a compactum and a Baire subspace of a Polish space (here we have used properness of the limit projection $q_\alpha$). This is enough (by Theorem 5.8.7 of [**260**]) to conclude that $Y$ is a Baire subset of its own Stone-Čech compactification. This finishes the proof of Proposition 8.8.7.

*Historical and bibliographical notes* 8.8. Theorem 8.8.1 and all other results of this Section were proved by the author [**87**], [**82**], [**95**]. Corollary 8.8.5, without assuming the Continuum Hypothesis, and also a related result (stating that there is no Baire isomorphism between $D^{\omega_2}$ and $expD^{\omega_2}$) was independently obtained by L. B. Shapiro. Propositions 8.8.6 and 8.8.7 solve the corresponding questions from [**260**].

## 8.9. Double spectra

**8.9.1. Direct spectra.** We consider direct spectra $\mathcal{S}_X = \{X_\alpha, i_\alpha^\beta, A\}$ all projections $i_\alpha^\beta \colon X_\alpha \to X_\beta$, $\alpha \leq \beta$, of which are closed embeddings. By $\varinjlim \mathcal{S}_X$ and by $i_\alpha \colon X_\alpha \to \varinjlim \mathcal{S}_X$ , we denote the direct limit and $\alpha$-th limit projection of the direct spectrum $\mathcal{S}_X$, respectively. We say that $\mathcal{S}_X = \{X_\alpha, i_\alpha^\beta, A\}$ is $\tau$-*continuous* if

- $i_\beta(X_\beta) = cl(\cup\{i_\alpha(X_\alpha) \colon \alpha \in B\})$ for each chain $B$ in $A$ with $\mid B \mid < \tau$ and $\sup B = \beta$.

Further we call a direct spectrum $\mathcal{S}_X = \{X_\alpha, i_\alpha^\beta, A\}$ a $\tau$-*spectrum* if the following conditions are satisfied:

$(i)_\rightarrow$ $d(X_\alpha) \leq \tau$ for each $\alpha \in A$.

$(ii)_\rightarrow$ The spectrum $\mathcal{S}_X$ is $\tau$-continuous.

$(iii)_\rightarrow$ The indexing set $A$ of the spectrum $\mathcal{S}_X$ is $\tau$-complete.

By a morphism between two direct spectra $\mathcal{S}_X = \{X_\alpha, i_\alpha^\beta, A\}$ and $\mathcal{S}_Y = \{Y_\alpha, j_\alpha^\beta, A\}$ we understand a collection $\mathcal{F} = \{f_\alpha \colon \alpha \in A\}$ of maps $f_\alpha \colon X_\alpha \to Y_\alpha$, $\alpha \in A$, such that $f_\beta i_\alpha^\beta = j_\alpha^\beta f_\alpha$ whenever $\alpha \leq \beta$, $\alpha, \beta \in A$, i.e. the diagrams of the following type commute

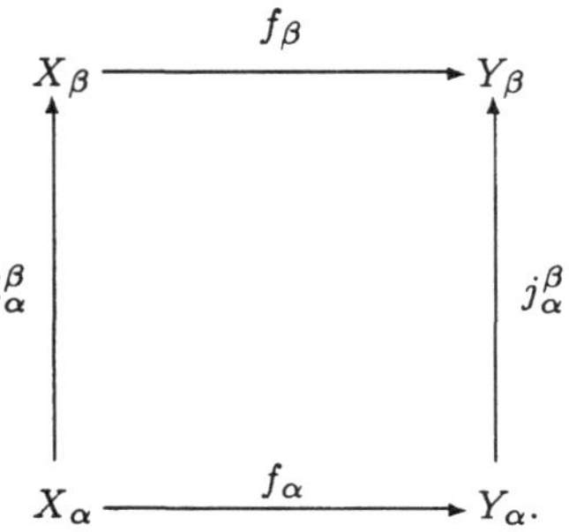


By the (direct) limit map of such a morphism we understand the map

$$\varinjlim \mathcal{F}\colon \varinjlim \mathcal{S}_X \to \varinjlim \mathcal{S}_Y$$

such that $\varinjlim \mathcal{F} \cdot i_\alpha = j_\alpha \cdot f_\alpha$ for each $\alpha \in A$.

Next we show that continuous maps between the limit spaces of (direct) $\tau$-spectra can be realized as the limit maps of morphisms in the above sense.

PROPOSITION 8.9.1. *Let $f\colon \varinjlim \mathcal{S}_X \to \varinjlim \mathcal{S}_Y$ be a map between the limit spaces of two direct $\tau$-spectra $\mathcal{S}_X = \{X_\alpha, i_\alpha^\beta, A\}$ and $\mathcal{S}_Y = \{Y_\alpha, j_\alpha^\beta, A\}$ . Then the collection $A' = \{\alpha \in A\colon f i_\alpha = j_\alpha f_\alpha\}$ is cofinal and $\omega$-closed in $A$.*

PROOF. Without loss of generality we may identify the spaces $X_\alpha$ and $Y_\alpha$ with their images $i_\alpha(X_\alpha)$ and $j_\alpha(Y_\alpha)$ in $\varinjlim \mathcal{S}_X$ and $\varinjlim \mathcal{S}_Y$ respectively. Consequently,

$$X_\alpha \subseteq \varinjlim \mathcal{S}_X \text{ and } Y_\alpha \subseteq \varinjlim \mathcal{S}_Y \text{ for each } \alpha \in A.$$

Consider the following relation on the set$A$:

$$L = \{(\alpha, \beta) \in A^2\colon \alpha \leq \beta \text{ and } f_\alpha(X_\alpha) \subseteq Y_\beta\}.$$

We are going to verify the conditions of Proposition 1.1.29.

*Existence condition.* Let $\alpha \in A$. Since $d(X_\alpha) \leq \tau$, $X_\alpha$ contains a dense subset $B_\alpha = \{b_t\colon t \in T\}$ of cardinality $\tau$, i.e. $|T| = \tau$. For each $t \in T$, there is an index $\alpha_t \in A$ such that $f(b_t) \in Y_{\alpha_t}$. By Corollary 1.1.28, there exists an index $\beta \in A$ such that $\beta \geq \alpha_t$, $t \in T$. We may assume without loss of generality that $\beta \geq \alpha$. Obviously,

$$f(X_\alpha) = f(\mathrm{cl}_{\varinjlim \mathcal{S}_X} B_\alpha) \subseteq \mathrm{cl}_{\varinjlim \mathcal{S}_Y} (f(B_\alpha)) \subseteq \mathrm{cl}_{\varinjlim \mathcal{S}_Y} Y_\beta = Y_\beta.$$

This shows that $(\alpha, \beta) \in L$.

*The majorantness and $\tau$-closedness conditions* are trivial.

Therefore, by Proposition 1.1.29, the set of all $L$-reflexive elements is cofinal and $\tau$-closed in $A$. It only remains to note that the index $\alpha \in A$ is $L$-reflexive (i.e. $(\alpha, \alpha) \in L$) if and only if $f_\alpha(X_\alpha) \subseteq Y_\alpha$. □

**8.9.2. $\Sigma$-products.** Let $P$ be the Hilbert space (or any other Polish $AR$-space) and $q^* \in P$. Let $T$ be an uncountable indexing set with $|T| = \tau > \omega$. For each $t \in T$, let $P_t = P$ and $q_t^* = q^*$. We denote the product $\prod\{P_t : t \in T\}$ by $P^T$. For a point $x = \{x_t : t \in T\} \in P^T$ let us set

$$supp(x) = \{t \in T : x_t \neq q^*\}.$$

The subspace

$$\Sigma(\tau, q^*) = \{x \in P^T : |supp(x)| \leq \omega\} \subseteq P^T$$

is called the $\Sigma$-*product* (of cardinality $\tau$) of $P$ with the base point $q^*$.

If $A \subseteq B \subseteq T$, then $\pi_A : P^T \to P^A$ and $\pi_A^B : P^B \to P^A$ denote the natural projections onto the corresponding subproducts. These retraction have naturally defined sections $i_A : P^A \to P^T$ and $i_A^B : P^A \to P^B$, where

$$i_A(\{x_t : t \in A\}) = \{x_t : t \in A\} \times \{q_t^* : t \in T - A\}$$

and

$$i_A^B(\{x_t : t \in A\}) = \{x_t : t \in A\} \times \{q_t^* : t \in B - A\}.$$

For each $A \subseteq B \subseteq T$ we use the following notation:

$$\Sigma_A(\tau, q^*) = \Sigma(\tau, q^*) \cap (\prod\{P_t : t \in A\} \times \{q_t^* : t \in T - A\}).$$

We also define the natural retractions

$$r_A : \Sigma(\tau, q^*) \to \Sigma_A(\tau, q^*)$$

and

$$r_A^B : \Sigma_B(\tau, q^*) \to \Sigma_A(\tau, q^*)$$

as follows:

$$r_A(\{x_t : t \in A\}, \{x_t : t \in T - A\}) = (\{x_t : t \in A\}, \{q_t^* : t \in T - A\})$$

and

$$r_A^B(\{x_t : t \in A\}, \{x_t : t \in B - A\}) = (\{x_t : t \in A\}, \{q_t^* : t \in B - A\}).$$

If $\{t_\alpha : \alpha < \tau\}$ is a well-ordering of the set $T$, then we use slightly different notations. Namely, for each $\alpha < \tau$ we let

$$\Sigma_\alpha(\tau, q^*) = \Sigma(\tau, q^*) \cap (\prod\{P_{t_\beta} : \beta \leq \alpha\} \times \{q_{t_\beta}^* : \beta > \alpha\}).$$

Definitions of the corresponding retractions

$$r_\alpha : \Sigma(\tau, q^*) \to \Sigma_\alpha(\tau, q^*) \text{ and } r_\alpha^\beta : \Sigma_\beta(\tau, q^*) \to \Sigma_\alpha(\tau, q^*)$$

are as follows:

$$r_\alpha(\{x_{t_\delta} : \delta < \tau\}) = (\{x_{t_\delta} : \delta \leq \alpha\}, \{q_{t_\delta}^* : \delta > \alpha\})$$

and

$$r_\alpha^\beta(\{x_{t_\delta} : \delta \leq \beta\}, \{q_{t_\delta}^* : \delta > \beta\}) = (\{x_{t_\delta} : \delta \leq \alpha\}, \{q_{t_\delta}^* : \delta > \alpha\}).$$

Straightforward verification shows that the following conditions are satisfied:

$(a)_\Sigma$ $\Sigma(\tau, q^*) = \cup\{\Sigma_\alpha(\tau, q^*) \colon \alpha < \tau\}$.

$(b)_\Sigma$ $\Sigma_\alpha(\tau, q^*) \subseteq \Sigma_\beta(\tau, q^*)$, whenever $\alpha \leq \beta < \tau$.

$(c)_\Sigma$ $\Sigma_\beta(\tau, q^*) = cl_{\Sigma(\tau,q^*)}(\cup\{\Sigma_\alpha(\tau, q^*) \colon \alpha < \beta\})$, whenever $\beta < \tau$ is a limit ordinal.

$(d)_\Sigma$ $\Sigma_0(\tau, q^*)$ is a copy of the Hilbert space $R^\omega$.

$(e)_\Sigma$ $r_\alpha^\gamma = r_\alpha^\beta r_\beta^\gamma$, whenever $\alpha < \beta < \gamma < \tau$.

$(f)_\Sigma$ If $\beta < \tau$ is a limit ordinal, then the diagonal product $\triangle\{r_\alpha^\beta \colon \alpha < \beta\}$ is an embedding of $\Sigma_\beta(\tau, q^*)$ into the limit space of the spectrum $\mathcal{S}_\beta(\Sigma) = \{\Sigma_\alpha(\tau, q^*), r_\alpha^{\alpha+1}, \alpha < \beta\}$.

$(g)_\Sigma$ $\Sigma_{\alpha+1}(\tau, q^*) = \Sigma_\alpha(\tau, q^*) \times P_{t_{\alpha+1}}$ and the map $r_\alpha^{\alpha+1} \colon \Sigma_{\alpha+1}(\tau, q^*) \to \Sigma_\alpha(\tau, q^*)$ is a trivial bundle with fiber $P_{t_{\alpha+1}}$, $\alpha < \tau$.

$(h)_\Sigma$ $\Sigma_\alpha(\tau, q^*)$ is a fibered $Z$-set in $\Sigma_{\alpha+1}(\tau, q^*)$ with respect to the projection $r_\alpha^{\alpha+1}$, whenever $\alpha < \tau$.

**8.9.3. Limits of double spectra and their properties.** In this Subsection we consider *double spectra* $\mathcal{S}_X = \{X_\alpha, p_\alpha^\beta, i_\alpha^\beta, A\}$ , i.e. systems that are inverse spectra and direct spectra simultaneously. This means, by definition, that a double spectrum consists of an indexing set $A$, spaces $X_\alpha$, $\alpha \in A$, retractions $p_\alpha^\beta \colon X_\beta \to X_\alpha$, $\alpha \leq \beta$, and their sections $i_\alpha^\beta \colon X_\alpha \to X_\beta$, $\alpha \leq \beta$. Further, retractions $p_\alpha^\beta$ and their sections $i_\alpha^\beta$ are chosen so that separately the systems $\mathcal{S}_X = \{X_\alpha, p_\alpha^\beta, A\}$ and $\mathcal{S}_X = \{X_\alpha, i_\alpha^\beta, A\}$ form inverse and direct spectra respectively. Therefore each double spectrum has two limits: the inverse limit $\varprojlim \mathcal{S}_X$ and the direct limit $\varinjlim \mathcal{S}_X$ .

All sensible notions for direct and/or inverse spectra will be automatically adapted to double spectra as well.

EXAMPLE 8.9.2. *As in Subsection* 8.9.2, *let* $q^* \in P$ *and* $T$ *be a set with* $|T| = \tau > \omega$. *Consider the system*

$$\mathcal{S}_\Sigma = \{P^A, \pi_A^B, i_A^B, exp_\omega T\},$$

*where the retractions* $\pi_A^B$ *and their section* $i_A^B$ *are defined as in Subsection* 8.9.2. *Obviously,* $\mathcal{S}_\Sigma$ *is a double* $\omega$*-spectrum. Its inverse limit coincides with the product* $P^T$ *and its direct limit coincides with the* $\Sigma$*-product* $\Sigma(\tau, q^*)$. *We suggest that the reader keep this basic example in mind, and analyze the situations described in various statements below in light of this example.*

First of all we show that for a double spectrum $\mathcal{S}_X = \{X_\alpha, p_\alpha^\beta, i_\alpha^\beta, A\}$ the direct limit $\varinjlim \mathcal{S}_X$ can be identified with a certain canonically defined subspace of its inverse limit $\varprojlim \mathcal{S}_X$ .

As noted above, for each $\alpha \leq \beta$ the map $i_\alpha^\beta \colon X_\alpha \to X_\beta$ is a section of the retraction $p_\alpha^\beta \colon X_\beta \to X_\alpha$. In other words, $p_\alpha^\beta \cdot i_\alpha^\beta = id_{X_\alpha}$ whenever $\alpha \leq \beta$. Let

$i_\alpha = \varprojlim\{i_\alpha^\beta \colon \beta \geq \alpha\}$. Obviously, $i_\alpha \colon X_\alpha \to \varprojlim \mathcal{S}_X$ is a section of the limit projection $p_\alpha \colon \varprojlim \mathcal{S}_X \to X_\alpha$. Thus $i_\alpha(X_\alpha)$ is a closed subspace (even a retract) of $\varprojlim \mathcal{S}_X$. Let $X' = \cup\{i_\alpha(X_\alpha) \colon \alpha \in A\}$.

LEMMA 8.9.3. *Let $\mathcal{S}_X = \{X_\alpha, p_\alpha^\beta, i_\alpha^\beta, A\}$ be a double $\tau$-spectrum. Suppose that $F$ is a subspace of $X' = \cup\{i_\alpha(X_\alpha) \colon \alpha \in A\}$ such that each of the intersections $F \cap i_\alpha(X_\alpha)$, $\alpha \in A$, is closed in $i_\alpha(X_\alpha)$. Then the collection*

$$A_F = \{\alpha \in A \colon p_\alpha(F) = p_\alpha(F \cap i_\alpha(X_\alpha))\}$$

*is cofinal and $\tau$-closed in $A$.*

PROOF. First let us show that $A_F$ is a cofinal subset of $A$. Let $\alpha \in A$. We need to find $\beta \in A_F$ such that $\beta \geq \alpha$. Denote $\alpha$ by $\alpha_0$ and consider an open base $\mathcal{U}_0$ of cardinality $\leq \tau$ of the space $X_{\alpha_0}$. Let $\mathcal{V}_0 = \{U \in \mathcal{U}_0 \colon U \cap p_{\alpha_0}(F) \neq \emptyset\}$. For each $U \in \mathcal{V}_0$, pick an index $\alpha_U \geq \alpha_0$ such that $p_{\alpha_0}^{-1}(U) \cap F \cap i_{\alpha_U}(X_{\alpha_U}) \neq \emptyset$. Thus we have a collection $\{\alpha_U \in A \colon U \in \mathcal{V}_0\}$ of cardinality $\leq \tau$ of indexes in $A$. By Corollary 1.1.28, there is an index $\alpha_1 \in A$ such that $\alpha_1 \geq \alpha_U$ for each $U \in \mathcal{V}_0$. If an index $\alpha_k \in A$ has already been constructed, the next index $\alpha_{k+1} \geq \alpha_k$ can be constructed as above. Namely, take an open base $\mathcal{U}_k$ of cardinality $\leq \tau$ of the space $X_{\alpha_k}$ and let $\mathcal{V}_k = \{U \in \mathcal{U}_k \colon U \cap p_{\alpha_k}(F) \neq \emptyset\}$. For each $U \in \mathcal{V}_k$, pick an index $\alpha_U \geq \alpha_k$ such that $p_{\alpha_k}^{-1}(U) \cap F \cap i_{\alpha_U}(X_{\alpha_U}) \neq \emptyset$. Finally, let $\alpha_{k+1} \in A$ be an index in $A$ such that $\alpha_{k+1} \geq \alpha_U$ for each $U \in \mathcal{V}_k$.

Now consider the countable chain $\{\alpha_k \colon k \in \omega\}$ and let $\beta = \sup\{\alpha_k \colon k \in \omega\}$. We claim that $\beta \in A_F$. Indeed, if this is not the case, then there is a point $x \in p_\beta(F) - p_\beta(F \cap i_\beta(X_\beta))$. Recall that the intersection $F \cap i_\beta(X_\beta)$ is closed in $i_\beta(X_\beta)$. Therefore, there is an open neighborhood $G$ of $x$ in $X_\beta$ such that $G \cap p_\beta(F \cap i_\beta(X_\beta)) = \emptyset$. Since $\mathcal{S}_X$ is $\tau$-continuous and $\tau \geq \omega$, the space $X_\beta$ is canonically homeomorphic to the (inverse) limit space $Y$ of the inverse sequence $\{X_{\alpha_k}, p_{\alpha_k}^{\alpha_{k+1}}, \omega\}$. Consequently, for some $\alpha_k$ (constructed above) and for some $U \in \mathcal{U}_k$ we have $x \in (p_{\alpha_k}^\beta)^{-1}(U) \subseteq G$. But, by the choice of the point $x$, $p_\beta^{-1}(x) \cap F \neq \emptyset$. Therefore, $p_{\alpha_k}^{-1}(U) = p_\beta^{-1}((p_{\alpha_k}^\beta)^{-1}(U)) \supseteq p_\beta^{-1}(x)$ and $p_{\alpha_k}^{-1}(U) \cap F \neq \emptyset$. In other words, $U \in \mathcal{V}_k$. This in turn implies (see the construction of $\alpha_{k+1}$) that there is a point $y \in p_{\alpha_k}^{-1}(U) \cap F \cap i_{\alpha_{k+1}}(X_{\alpha_{k+1}})$. Since $\alpha_{k+1} < \beta$, we have $p_\beta(y) \in p_\beta(F \cap i_\beta(X_\beta))$. Also $p_\beta(y) = i_{\alpha_k}^\beta p_{\alpha_k}(y) \in (i_{\alpha_k}^\beta)^{-1}(U) \subseteq G$. Thus, $p_\beta(y) \in G \cap p_\beta(F \cap i_\beta(X_\beta))$. This contradicts the choice of the neighborhood $G$. Cofinality of $A_F$ is proven.

Suppose now that $\{\alpha_\gamma \colon \gamma < \tau\}$ is a chain of cardinality $\leq \tau$ in $A$ consisting of elements of $A_F$. Let $\alpha = \sup\{\alpha_\gamma \colon \gamma < \tau\}$. By $\tau$-completeness of $A$, $\alpha \in A$. Let us show that $\alpha \in A_F$. The $\tau$-continuity of the inverse spectrum $\mathcal{S}_X = \{X_\alpha, p_\alpha^\beta, A\}$ guarantees that the space $X_\alpha$ is canonically embedded into the limit space $Y$ of the inverse spectrum $\{X_{\alpha_\gamma}, p_{\alpha_\gamma}^{\alpha_{\gamma+1}}, \tau\}$. Assuming that $\alpha \notin A_F$ we can find a point $x \in p_\alpha(F) - p_\alpha(F \cap i_\alpha(X_\alpha))$. Therefore, there exists an ordinal

$\gamma < \tau$ such that

$$p^{\alpha}_{\alpha_\gamma}(x) \notin p^{\alpha}_{\alpha_\gamma}(p_\alpha(F \cap i_\alpha(X_\alpha)) = p_{\alpha_\gamma}(F \cap i_\alpha(X_\alpha)).$$

Next observe that since $\alpha_\gamma \in A_F$ and since

$$p_{\alpha_\gamma}(F) = p_{\alpha_\gamma}(F \cap i_{\alpha_\gamma}(X_{\alpha_\gamma}) \subseteq p_{\alpha_\gamma}(F \cap i_\alpha(X_\alpha)),$$

we may conclude that $p^{\alpha}_{\alpha_\gamma}(x) \notin p_{\alpha_\gamma}(F)$. This contradicts the choice of $x$. Therefore $A_F$ is $\tau$-closed in $A$. □

COROLLARY 8.9.4. *Let $\mathcal{S}_X = \{X_\alpha, p^\beta_\alpha, i^\beta_\alpha, A\}$ be a double $\tau$-spectrum. Then the direct limit $\underrightarrow{\lim}\, \mathcal{S}_X$ coincides with $X' = \cup\{i_\alpha(X_\alpha) \colon \alpha \in A\}$.*

PROOF. Obviously $\underrightarrow{\lim}\, \mathcal{S}_X$ and $X'$ coincide as sets. Therefore we have only to show that the topology induced on $X'$ from $\underleftarrow{\lim}\, \mathcal{S}_X$ is the weak topology with respect to the collection $\{i_\alpha(X_\alpha) \colon \alpha \in A\}$. Consider a subset $F \subseteq X'$ such that $F \cap i_\alpha(X_\alpha)$ is closed in $i_\alpha(X_\alpha)$ for each $\alpha \in A$. We need to show that $F$ itself is closed in $X'$. Consider the closure $cl(F)$ (taken in $\underleftarrow{\lim}\, \mathcal{S}_X$ ) of the set $F$ and suppose that there is a point $x \in cl(F) \cap X'$. Since $x \in X'$, there exists an index $\alpha \in A$ such that $x \in i_\alpha(X_\alpha)$. Since $x \in cl(F)$ we see that $p_\gamma(x) \in p_\gamma(cl(F)) \subseteq cl_{X_\gamma}(p_\gamma(F))$ for each $\gamma \geq \alpha$. By Lemma 8.9.3, there is an index $\beta \geq \alpha$ such that $cl_{X_\beta}(p_\beta(F)) = cl_{X_\beta}(p_\beta(F \cap i_\beta(X_\beta)) = p_\beta(F \cap i_\beta(X_\beta))$. Therefore $x \in F$ as desired. □

COROLLARY 8.9.5. *The direct limit $\underrightarrow{\lim}\, \mathcal{S}_X$ of a double $\tau$-spectrum $\mathcal{S}_X = \{X_\alpha, p^\beta_\alpha, i^\beta_\alpha, A\}$ consisting of normal spaces is normal.*

PROOF. Let $F_1$ and $F_2$ be two disjoint closed subsets of $\underrightarrow{\lim}\, \mathcal{S}_X$ . By Lemma 8.9.3, the sets

$$A_{F_1} = \{\alpha \in A \colon p_\alpha(F_1) = p_\alpha(F_1 \cap i_\alpha(X_\alpha))\}$$

and

$$A_{F_2} = \{\alpha \in A \colon p_\alpha(F_2) = p_\alpha(F_2 \cap i_\alpha(X_\alpha))\}$$

are both cofinal and $\tau$-closed in $A$. Therefore, by Proposition 1.1.27, the intersection $A_{F_1} \cap A_{F_2}$ is cofinal and $\tau$-closed in $A$. In particular, $A_{F_1} \cap A_{F_2} \neq \emptyset$. Take an index $\alpha$ from this intersection. Then, as is easy to see, the sets $p_\alpha(F_1) = p_\alpha(F_1 \cap i_\alpha(X_\alpha))$ and $p_\alpha(F_2) = p_\alpha(F_2 \cap i_\alpha(X_\alpha))$ are disjoint closed subsets of the normal space $X_\alpha$. Let $G_1$ and $G_2$ be disjoint open subsets of $X_\alpha$ such that $p_\alpha(F_k) \subseteq G_k$, $k = 1, 2$. Obviously the sets $V_k = p_\alpha^{-1}(G_k) \cap \underrightarrow{\lim}\, \mathcal{S}_X$ , $k = 1, 2$, are disjoint and open in $\underrightarrow{\lim}\, \mathcal{S}_X$ . It only remains to note that $F_k \subseteq V_k$, $k = 1, 2$ □

COROLLARY 8.9.6. *Let* $\mathcal{S}_X = \{X_\alpha, p_\alpha^\beta, i_\alpha^\beta, A\}$ *be a double* $\tau$*-spectrum and* $F$ *be a closed subspace of* $\varinjlim \mathcal{S}_X$. *Then*

$$\mathcal{S}_F = \{p_\alpha(F), p_\alpha^\beta, i_\alpha^\beta, A_F\}$$

*is a double* $\tau$*-spectrum such that*

$$\varprojlim \mathcal{S}_F = cl_{\varprojlim \mathcal{S}_X}(F) \text{ and } \varinjlim \mathcal{S}_F = F.$$

PROOF. Apply Lemma 8.9.3 □

We call a double $\tau$-spectrum $\mathcal{S}_X = \{X_\alpha, p_\alpha^\beta, i_\alpha^\beta, A\}$ *factorizing* if the corresponding inverse spectrum $\mathcal{S}_X = \{X_\alpha, p_\alpha^\beta, A\}$ is factorizing in the sense of subsection 1.3.1. By a *Polish double spectrum* we understand a double $\omega$-spectrum $\mathcal{S}_X = \{X_\alpha, p_\alpha^\beta, i_\alpha^\beta, A\}$ consisting of Polish spaces $X_\alpha$, $\alpha \in A$, and open retractions $p_\alpha^\beta \colon X_\beta \to X_\alpha$, $\alpha \leq \beta$.

LEMMA 8.9.7. *Let* $\mathcal{S}_X = \{X_\alpha, p_\alpha^\beta, i_\alpha^\beta, A\}$ *be a Polish double spectrum. Then* $\varprojlim \mathcal{S}_X$ *is the Hewitt realcompactification of* $\varinjlim \mathcal{S}_X$.

PROOF. Inverse limits of $\omega$-spectra consisting of Polish spaces and open (surjective) bonding maps are perfectly $\kappa$-normal. This guarantees that every dense subspace of $\varprojlim \mathcal{S}_X$ is $z$-embedded (Proposition 1.1.21). $\omega$-completeness of the indexing set $A$, coupled with the fact that all limit projections $p_\alpha \colon \varprojlim \mathcal{S}_X \to X_\alpha$ are surjective, shows that $\varinjlim \mathcal{S}_X$ is $G_\delta$-dense in $\varprojlim \mathcal{S}_X$. Consequently, every continuous real- valued function $\varphi \colon \varinjlim \mathcal{S}_X \to R$ has a continuous extension $\bar{\varphi} \colon \varprojlim \mathcal{S}_X \to R$. □

THEOREM 8.9.8. *Let* $f \colon \varinjlim \mathcal{S}_X \to \varinjlim \mathcal{S}_Y$ *be a map between the direct limits of two double Polish spectra* $\mathcal{S}_X = \{X_\alpha, p_\alpha^\beta, i_\alpha^\beta, A\}$ *and* $\mathcal{S}_Y = \{Y_\alpha, q_\alpha^\beta, j_\alpha^\beta, A\}$. *Then the set*

$$A_f = \{\alpha \in A \colon \text{there is a map } f_\alpha \colon X_\alpha \to Y_\alpha \text{ with } f_\alpha p_\alpha = q_\alpha f \text{ and } j_\alpha f_\alpha = f i_\alpha\}$$

*is cofinal and* $\omega$*-closed in* $A$.

PROOF. By Lemma 8.9.7, $\varprojlim \mathcal{S}_X$ and $\varprojlim \mathcal{S}_Y$ are the Hewitt realcompactifications of $\varinjlim \mathcal{S}_X$ and $\varinjlim \mathcal{S}_Y$ respectively. Therefore, the map $f \colon \varinjlim \mathcal{S}_X \to \varinjlim \mathcal{S}_Y$ can be extended to the map $\nu f \colon \varprojlim \mathcal{S}_X \to \varprojlim \mathcal{S}_Y$. Now apply Proposition 8.9.1 and Theorem 1.3.6 to the maps $f$ and $\nu f$ respectively. Finally, use Proposition 1.1.27. □

**8.9.4. Closed subspaces of $\Sigma$-products.** Using properties $(a)_\Sigma - (h)_\Sigma$ as a guide, we give the following definition.

DEFINITION 8.9.9. *We say that a space $X$ is the limit space of a well-ordered $\Sigma$-system $\mathcal{S} = \{X_\alpha, p_\alpha^\beta, \tau\}$ consisting of closed subspaces $X_\alpha$ of $X$ and retractions $p_\alpha^\beta \colon X_\beta \to X_\alpha$, defined whenever $\alpha < \beta < \tau$, if the following conditions are satisfied:*

$(a)_X$ $X = \cup\{X_\alpha \colon \alpha < \tau\}$.

$(b)_X$ $X_\alpha \subseteq X_\beta$ *whenever* $\alpha \leq \beta < \tau$.

$(c)_X$ $X_\beta = cl_X(\cup\{X_\alpha \colon \alpha < \beta\}$, *whenever* $\beta < \tau$ *is a limit ordinal.*

$(d)_X$ $X_0$ *is a Polish space.*

$(e)_X$ $p_\alpha^\gamma = p_\alpha^\beta p_\beta^\gamma$ *whenever* $\alpha < \beta < \gamma < \tau$.

$(f)_X$ *If* $\beta < \tau$ *is a limit ordinal, then the diagonal product* $\triangle\{p_\alpha^\beta \colon \alpha < \beta\}$ *is an embedding of* $X_\beta$ *into the limit space of the spectrum* $\mathcal{S}_\beta(X) = \{X_\alpha, p_\alpha^{\alpha+1}, \alpha < \beta\}$.

$(g)_X$ *Retraction* $r_\alpha^{\alpha+1} \colon X_{\alpha+1} \to X_\alpha$ *has a Polish kernel whenever* $\alpha < \tau$.

Obviously the collection $\mathcal{S}_{\Sigma(\tau,q^*)} = \{\Sigma_\alpha(\tau, q^*), r_\alpha^\beta, \tau\}$ is a well-ordered $\Sigma$-system whose limit space coincides with the $\Sigma$-product $\Sigma(\tau, q^*)$.

Our next result characterizes the closed subspaces of $R^\tau$ ($\approx P^T$ with $|T| = \tau$) embeddable into $\Sigma(\tau, q^*)$ as the limit spaces of well-ordered $\Sigma$-systems (of length $\leq \tau$). We recall that the compact spaces embeddable into $\Sigma(\tau, q^*)$ for uncountable $\tau$ are known in the literature as *Corson compacta*. In [**227**, Theorem 6.1] these compacta were characterized as those which have point-countable, separating collections of open $F_\sigma$-subsets. Our result covers the non-compact case and, moreover, is given in totally different terms, and this allows us to use the spectral technique.

THEOREM 8.9.10. *Let $\tau > \omega$. The following conditions are equivalent for any space $X$:*

(i) *$X$ is homeomorphic to a closed subspace of $R^\tau$ embeddable into the $\Sigma$-product $\Sigma(\tau, q^*)$.*

(ii) *$X$ can be represented as the limit space of a well-ordered $\Sigma$-system of length $\tau$.*

PROOF. During the proof we will use the following notations: $\Sigma = \Sigma(\tau, q^*)$ and $\Sigma_\alpha = \Sigma_\alpha(\tau, q^*)$ for each $\alpha < \tau$.

**(ii) $\Rightarrow$ (i).** Suppose that $X$ is the limit space of a well-ordered $\Sigma$-system $\mathcal{S}_X = \{X_\alpha, p_\alpha^\beta, \tau\}$.

For each ordinal $\alpha < \tau$ we are going to construct a closed embedding $f_\alpha \colon X_\alpha \to \Sigma_\alpha$ so that the following conditions are satisfied:

(i) $f_\alpha p_\alpha^\beta = r_\alpha^\beta f_\beta$ for each $\alpha < \tau$, whenever $\alpha < \beta < \tau$.

(ii) $f_\beta = \lim\{f_\alpha \colon \alpha < \beta\}$ whenever $\beta < \tau$ is a limit ordinal.

(iii) $f_\alpha / X_\alpha = f_\beta / X_\alpha$ whenever $\alpha < \beta < \tau$.

Since $\Sigma_0$ is the Hilbert space $R^\omega$, there exists a closed embedding $f_0 \colon X_0 \to \Sigma_0$. Suppose now that the closed embeddings $f_\alpha \colon X_\alpha \to \Sigma_\alpha$, satisfying the conditions (i)-(iii) for appropriate ordinals, have already been constructed for each $\alpha < \beta$, where $\beta < \tau$. We now construct a closed embedding $f_\beta \colon X_\beta \to \Sigma_\beta$.

**Case 1** ($\beta = \alpha + 1$). By $(g)_X$, there is a Polish space $K$ such that $X_{\alpha+1} \subseteq X_\alpha \times K$ and $r_\alpha^{\alpha+1} = \pi_1/X_\alpha$, where $\pi_1 \colon X_\alpha \times K \to X_\alpha$ denotes the first projection. Since $K$ is a Polish space, there exists a closed embedding $g \colon K \to P_{t_{\alpha+1}}$. Clearly the map $f_\alpha \times g \colon X_\alpha \times K \to \Sigma_\alpha \times P_{t_{\alpha+1}}$ is a closed embedding. Therefore its restriction $f_{\alpha+1} = (f_\alpha \times g)/X_{\alpha+1} \colon X_{\alpha+1} \to \Sigma_{\alpha+1}$ also is a closed embedding. By the construction, $r_\alpha^{\alpha+1} f_{\alpha+1} = f_\alpha p_\alpha^{\alpha+1}$. Changing $f_{\alpha+1}$ if necessary (and preserving the first coordinate) we may additionaly assume, by $(h)_\Sigma$, that $f_{\alpha+1}/X_\alpha = f_\alpha$.

**Case 2** (**$\beta$ is a limit ordinal**). In this case, simply let $f = \lim\{f_\alpha \colon \alpha < \beta\}$. Condition (i) implies that $f$ maps the limit space of the spectrum $\{X_\alpha, p_\alpha^{\alpha+1}, \beta\}$ into the limit space of the spectrum $\{\Sigma_\alpha, r_\alpha^{\alpha+1}, \beta\}$. Since, by our constructions, all $f_\alpha$'s, $\alpha < \beta$, are closed embeddings, we conclude that $f$ is also a closed embedding. By $(g)_X$, $f(X_\beta) \subseteq \lim \mathcal{S}_\beta(\Sigma)$. Recall also that $\Sigma_\beta \subseteq \lim \mathcal{S}_\beta(\Sigma)$. Further, suppose that $x \in \cup\{X_\alpha \colon \alpha < \beta\}$ and choose an index $\alpha < \beta$ such that $x \in X_\alpha$. By condition (*iii*), one can easily conclude that $f(x) \in \Sigma_\alpha$. Therefore, $\cup\{X_\alpha \colon \alpha < \beta\} \subseteq \cup\{\Sigma_\alpha \colon \alpha < \beta\}$. By $(c)_X$ and $(c)_\Sigma$, we have

$$f(X_\beta) = f(cl_{X_\beta}(\cup\{X_\alpha \colon \alpha < \beta\})) \subseteq cl_{\Sigma_\beta}(f(\cup\{X_\alpha \colon \alpha < \beta\}) =$$

$$cl_{\Sigma_\beta}(\cup\{f(X_\alpha) \colon \alpha < \beta\}) \subseteq cl_{\Sigma_\beta}(\cup\{\Sigma_\alpha \colon \alpha < \beta\}) = \Sigma_\beta.$$

Therefore the map $f_\beta = f/X_\beta$ satisfies all the required conditions.

This finishes the construction of the maps $f_\alpha$ for each $\alpha < \tau$.

Now consider the inverse spectra $\mathcal{S}_X = \{X_\alpha, p_\alpha^\beta, \tau\}$ and $\mathcal{S}_\Sigma = \{\Sigma_\alpha, r_\alpha^\beta, \tau\}$. By condition (i), the limit map $g = \lim\{f_\alpha \colon \alpha < \tau\}$ maps $\lim \mathcal{S}_X$ into $\lim \mathcal{S}_\Sigma$. Also note that $r_\alpha g = f_\alpha p_\alpha$ for each $\alpha < \tau$ (here $p_\alpha \colon \lim \mathcal{S}_X \to X_\alpha$ and $r_\alpha \colon \lim \mathcal{S}_\Sigma \to \Sigma_\alpha$ denote the $\alpha$-th limit projections of these spectra). It can easily be seen that

$$X \subseteq \lim \mathcal{S}_X \text{ and } \Sigma \subseteq \lim \mathcal{S}_\Sigma.$$

It also follows from the construction that $g(X) \subseteq \Sigma$. It only remains to note that the map $f = g/X \colon X \to \Sigma$ is the desired closed embedding.

**(i) $\Rightarrow$ (ii)**. Suppose now that $X$ is a closed subspace $R^\tau$ contained in the $\Sigma$-product $\Sigma$. We are going to define closed subspaces $X_\alpha$, $\alpha < \tau$, and retractions $r_\alpha^\beta \colon X_\beta \to X_\alpha$ satisfying conditions $(a)_X - (g)_X$.

Let us call a subset $A$ of $T$ *admissible* if

$$r_A(X) = r_A(X \cap \Sigma_A).$$

Applying Lemma 8.9.3 to the double spectrum $\mathcal{S}_\Sigma$ and its closed subspace $X$, we see that the collection of countable admissible subsets of $T$ is $\omega$-closed and cofinal in $exp_\omega T$.

Now we proceed as in the final part of the proof of Theorem 6.3.1.

Since $|T| = \tau$, we can write $T = \{t_\alpha : \alpha < \tau\}$. Since the collection of countable admissible subsets of $T$ is cofinal in $exp_\omega T$, each point $t_\alpha \in T$ is contained in a countable admissible subset $B_\alpha \subseteq T$. Let $A_\alpha = \cup\{B_\beta : \beta \leq \alpha\}$. Finally, let

$$X_\alpha = X \cap \Sigma_{A_\alpha} \text{ and } p_\alpha^{\alpha+1} = r_{A_\alpha}^{A_{\alpha+1}}.$$

It follows from the properties $(a)_\Sigma - (h)_\Sigma$ and the definition of admissible subsets that the spectrum

$$\mathcal{S}_X = \{X_\alpha, p_\alpha^\beta, \tau\}$$

has all the required properties. This completes the proof of the Theorem. □

In particular we have the following.

PROPOSITION 8.9.11. *Let $\tau \geq \omega$. Then the following conditions are equivalent:*

(i) *$X$ is a Corson compactum of weight $\leq \tau$.*
(ii) *There is a $\Sigma$-system $\mathcal{S}_X = \{X_\alpha, p_\alpha^\beta, \tau\}$ consisting of closed subspaces $X_\alpha$ of $X$, $\alpha < \tau$, and retractions $p_\alpha^\beta : X_\beta \to X_\alpha$, defined whenever $\alpha < \beta < \tau$, with the following properties:*
    (a) $X = \cup\{X_\alpha : \alpha < \tau\}$.
    (b) *$X_\alpha \subseteq X_\beta$ whenever $\alpha \leq \beta < \tau$.*
    (c) *$X_\beta = \mathrm{cl}_X(\cup\{X_\alpha : \alpha < \beta\})$ whenever $\beta$ is a limit ordinal.*
    (d) *$X_0$ is a metrizable compactum.*
    (e) *$p_\alpha^\gamma = p_\alpha^\beta p_\beta^\gamma$ whenever $\alpha < \beta < \gamma < \tau$.*
    (f) *If $\beta \leq \tau$ is a limit ordinal, then the diagonal product $\triangle\{p_\alpha^\beta : \alpha < \beta\}$ is a homeomorphism of $X_\beta$ onto the limit space of the inverse spectrum $\mathcal{S}_\beta = \{X_\alpha, p_\alpha^{\alpha+1}, \alpha < \beta\}$.*
    (g) *Retraction $p_\alpha^{\alpha+1} : X_{\alpha+1} \to X_\alpha$ has a metrizable kernel whenever $\alpha < \tau$.*

Proposition 8.9.11 has several consequences. Some of them are mentioned below.

COROLLARY 8.9.12. *Every separable subspace of a Corson compactum has a countable base.*

COROLLARY 8.9.13. *A subset $F$ of a Corson compactum $X$ is open in $X$ if and only if its intersection $F \cap K$ is closed in $K$ for every metrizable compactum $K \subseteq X$.*

COROLLARY 8.9.14. *Let $\tau > \omega$. Then the Tychonov cube $I^\tau$ is not a Corson compactum.*

The following statement follows from the proof of Theorem 8.9.10.

COROLLARY 8.9.15. *If $X$ is the limit space of a $\Sigma$-system of uncountable length (in particular, if $X$ is a Corson compactum), then $d(X) = w(X)$.*

PROOF. Obviously always $d(X) \leq w(X)$. Assume that $d(X) < w(X)$ and let $Y$ be a dense subset (in $X$) of cardinality $\lambda < \tau = w(X)$. It follows from the properties of admissible subsets that $Y \subseteq X_\alpha$ for some $\alpha < \tau$. Therefore

$$X = \operatorname{cl} Y \subseteq \operatorname{cl} X_\alpha = X_\alpha.$$

But, again from the construction, it follows that $w(X_\alpha) < \tau$. This is a contradiction. □

*Historical and bibliographical notes* 8.9. The concept of double spectra was first presented in [**101**]. $\Sigma$-products have been extensively studied from different points of view. Variations of Lemma 8.9.3 appear in [**30**], [**160**]. They arise, in particular, in functional analysis. The topological characterization of Corson compacta was obtained in [**227**]. Corollary 8.9.15 is well known (see, for example, [**241**] where many other results in this direction are also presented). It is known [**9**] that for any Corson compactum $X$, the space $C_p(X)$ of all continuous real-valued functions on $X$ with the topology of pointwise convergence is Lindelöf. Measures on Corson compacta are investigated in [**201**]).

## 8.10. Skeletoids in Tychonov cubes

We have seen above that several important model spaces have naturally defined pseudo-boundaries (and pseudo-interiors). In most cases they have been constructed as absorbers (skeletoids). Moreover, in some cases these pseudo-boundaries have even been characterized topologically. What is the situation in the non metrizable case? For instance, do we have a satisfactory understanding of what exactly (if any) should be called a skeletoid in the Tychonov cube? Below we consider one of the possible approaches.

As a guide we take the property (∗) from Theorem 2.2.22. Namely, recall that a countable collection $\{A_i \colon i \in N\}$ of $Z$-sets in the Hilbert cube $I^\omega$ is a $Z$-skeleton (and the union $\cup\{A_i \colon i \in N\}$ is a $Z$-skeletoid) if for every $Z$-set $K \subseteq I^\omega$, every index $i \in N$ and every $\epsilon > 0$, there exist an index $j > i$ and an embedding $f \colon K \to I^\omega$ such that

$$(*) \quad d(f, id_K) < \epsilon,\ f/(K \cap A_i) = id \text{ and } f(K) \subseteq A_j.$$

Of course, here $d$ stands for a metric on $I^\omega$ (more formally, for a metric on $C(I^\omega, I^\omega)$) and we measure closeness of functions in the compact-open topology.

This topology works perfectly well when we deal with the Hilbert cube, but not with the Tychonov cube. In the non-metrizable case we need a finer topology on the space of functions (compare with Theorem 7.2.8). In Subsection 6.5.1 we have already introduced a topology on function spaces which depends on a cardinal $\tau$ and is finer for larger cardinals. It turns out that this topology works well in the present situation as well.

Let us, for simplicity, consider only the Tychonov cube

$$I^{\omega_1} = (I^{\omega})^{\omega_1}$$

of weight $\omega_1$. Every metrizable compactum $K \subseteq I^{\omega_1}$ is a $Z_\omega$-set in $I^{\omega_1}$ (see Definition 8.5.1 and Corollary 8.5.7) with respect to the topology $C_\omega(I^{\omega_1}, I^{\omega_1})$. Consider the $\Sigma$-product $\Sigma$ associated with $I^{\omega_1}$, i.e. the collection of those points of $I^{\omega_1}$ only countably many coordinates of which differ from $0 = (0,\dots,0,\dots)$. Obviously we can represent $\Sigma$ as the union

$$\Sigma = \{\Sigma_\alpha \colon \alpha < \omega_1\}$$

where each $\Sigma_\alpha$ is a copy of the Hilbert cube (we keep the notation of Subsection 8.9.2). In particular, the sets $\Sigma_\alpha$ satisfy properties $(a)_\Sigma - (j)_\Sigma$ from that Subsection. Let us show that the collection $\{\Sigma_\alpha \colon \alpha < \omega_1\}$ has a property similar, in some sense, to the property $(*)$ stated above.

Consider a metrizable compactum $A \subseteq I^{\omega_1}$, an ordinal $\alpha < \omega_1$ and a neighborhood $U$ of the inclusion map $i \colon A \subseteq I^{\omega_1}$ in the space $C_\omega(A, I^{\omega_1})$. Translating into the spectral language we may, without loss of generality, assume that for some $\gamma < \omega_1$ we have

$$U = \{f \in C_\omega(A, I^{\omega_1}) \colon f\pi_\gamma = i\pi_\gamma\}$$

where

$$\pi_\gamma \colon I^{\omega_1} = (I^{\omega})^{\omega_1} \to (I^{\omega})^{\gamma}$$

denotes the projection onto the corresponding subproduct.

We are going to show that there exist an index $\beta \geq \alpha$ and an embedding $f \colon A \to \Sigma_\beta$ such that

$$f/(A \cap \Sigma_\alpha) = id \text{ and } f \in U.$$

Since $A$ is a metrizable compactum, there exists an index $\delta < \omega_1$ such that the restriction

$$\pi_\delta/A \colon A \to \pi_\delta(A)$$

is a homeomorphism. Let $\beta = \max\{\alpha, \gamma, \delta\}$. Let

$$i_\beta \colon (I^{\omega})^{\beta} \to I^{\omega_1}$$

denote the section of the projection $\pi_\beta$ determined by the point $0 = (0,\dots,0,\dots)$ (see Subsection 8.9.2). Observe that the restriction

$$\pi_\beta/A \colon A \to \pi_\beta(A)$$

is also a homeomorphism. Finally let

$$f = i_\beta \pi_\beta / A\colon A \to \Sigma_\beta.$$

By the choice of the ordinal $\beta$, $f$ is an embedding. Since $\beta \geq \alpha$ we see that $f/(A \cap \Sigma_\alpha) = id$. Since $\beta \geq \gamma$ one can easily observe that $f \in U$.

Thus we have the following statement.

PROPOSITION 8.10.1. *The collection $\{\Sigma_\alpha\colon \alpha < \omega_1\}$ has the following property:*

$(*)_{(\omega,\omega_1)}$ *For each compactum $A \subseteq I^{\omega_1}$ of weight $\leq \omega$, for each ordinal $\alpha < \omega_1$ and for each neighborhood $U$ of the inclusion $A \hookrightarrow I^{\omega_1}$ in the space $C_\omega(A, I^{\omega_1})$, there exist an ordinal $\beta > \alpha$ and an embedding $f\colon A \to I^{\omega_1}$ such that*

$$f/(A \cap \Sigma_\alpha) = id \text{ and } f \in U.$$

It is easy to see that $\Sigma$ is a countably compact space. Therefore $\Sigma$ is not realcompact and, consequently, by Proposition 6.1.7, cannot be an $ANR$-space. Nevertheless the following statement shows that $\Sigma$ is an absolute extensor with respect to the class of metrizable compacta.

PROPOSITION 8.10.2. *Let $B$ be a metrizable compactum and let a map $f\colon A \to \Sigma$ be defined on a closed subspace $A$ of $B$. Then $f$ can be extended to a map $\bar{f}\colon B \to \Sigma$.*

PROOF. The image $f(A)$, being a metrizable compactum, is contained in $\Sigma_\alpha$ for some countable ordinal $\alpha$. But $\Sigma_\alpha$ is the copy of the Hilbert cube $I^\omega$. Therefore there exists an extension

$$\bar{f}\colon B \to \Sigma_\alpha \subseteq \Sigma$$

of $f$. □

The next statement shows that $\Sigma$ has some sort of strong universality property with respect to the class of metrizable compacta.

PROPOSITION 8.10.3. *Let $A$ be a closed subspace of a metrizable compactum $B$ and $f\colon B \to \Sigma$ be a map such that the restriction $f/A\colon A \to \Sigma$ is an embedding. Let $U$ be a neighborhood of $f$ in the space $C_\omega(B, \Sigma)$. Then there exists an embedding $g\colon B \to \Sigma$ such that*

$$g/A = f/A \text{ and } g \in U.$$

PROOF. Since $\Sigma$ is pseudocompact and a dense subspace of the Tychonov cube $I^{\omega_1}$, it follows easily that $I^{\omega_1}$ serves as the Stone-Čech compactification of $\Sigma$, i.e. $I^{\omega_1} = \beta\Sigma$. This in turn guarantees that any neighborhood $U$ of $f$ in the space $C_\omega(B, \Sigma)$ uniquely determines a neighborhood $\bar{U}$ of the same map $f$ in the space $C_\omega(B, I^{\omega_1})$. Further, we may assume without loss of generality that the neighborhood $\bar{U}$ has the form

$$\bar{U} = \{\varphi \in C_\omega(B, I^{\omega_1}) \colon \varphi\pi_\gamma = f\pi_\gamma\}.$$

As above we can find a countable ordinal $\beta$ such that $f(B) \subseteq \Sigma_\beta$. Let $\alpha = \max\{\beta, \gamma\}$. Since the projection

$$r_\alpha^{\alpha+1} \colon \Sigma_{\alpha+1} \to \Sigma_\alpha$$

is the trivial bundle with fiber $I^\omega$ (see the property $(i)_\Sigma$ and $\Sigma_\alpha$ is a fibered $Z$-set in $\Sigma_{\alpha+1}$ with respect to $r_\alpha^{\alpha+1}$ (see the property $(j)_\Sigma$) we immediately conclude that there is a map

$$g \colon B \to \Sigma_{\alpha+1}$$

such that

$$g/A = f/A \text{ and } r_\alpha^{\alpha+1} g = r_\alpha^{\alpha+1} f.$$

It only remains to observe that $g \in U$ (which is guaranteed by the inequality $\alpha \geq \gamma$ and the construction of $g$). □

It is not hard to see that above statements[5] remain valid for $\Sigma$-products associated with the representation $I^\tau = (I^\omega)^\tau$. Do these properties characterize $\Sigma$-products? At least as subspaces of the corresponding Tychonov cubes?

There is a stronger version of this question: Is there a corresponding theory of skeletoids in Tychonov cubes? More precisely, we ask:

PROBLEM 8.10.4. *Are any two skeletoids of $I^\tau$ equivalent as subspaces of $I^\tau$, i.e. if $A$ and $B$ are skeletoids in $I^\tau$, is there a homeomorphism $f \colon I^\tau \to I^\tau$ (as close[6] to the identity as we wish) such that $f(A) = B$?*

PROBLEM 8.10.5. *Let $K$ be a metrizable compactum in $I^\tau$ and $U$ be a neighborhood of $id_{I^\tau}$ in $C_\tau(I^\tau, I^\tau)$. Does there exist a homeomorphism $h \colon I^\tau \to I^\tau$ such that*

$$h \text{ is } U-\text{close to } id_{I^\tau} \text{ and } h(\Sigma \cup K) = \Sigma.$$

In particular, we ask the following.

PROBLEM 8.10.6. *In the notation of Problem 8.10.5, are $\Sigma \cup K$ and $\Sigma$ homeomorphic?*

[5] Recall that everywhere in this Section $\tau$ denotes an *uncountable* cardinal.

[6] Of course, closeness of $f$ and $id$ must be measured in the space $C_\tau(I^\tau, I^\tau)$.

PROBLEM 8.10.7. *Let $U$ be a neighborhood of $id_{I^\tau}$ in the space $C_\tau(I^\tau, I^\tau)$. Does there exist an embedding $f: I^\tau \to I^\tau$ which is $U$-close to $id_{I^\tau}$ and such that $f(I^\tau) \subseteq I^\tau - \Sigma$?*

All of these questions remain open at this point.

The next possible step in this direction is the investigation of various types of skeletoids of Tychonov cubes. For example, are there $n$-dimensional skeletoids in $I^\tau$? In particular, what about the easiest (perhaps) case $n = 0$?

Of course, it is interesting to investigate the complements of the above skeletoids, i.e. versions of pseudo-interiors of Tychonov cubes. The situation is not trivial at all. For instance, let us ask the following question: is the complement $I^\tau - \Sigma$ homeomorphic to $R^\tau$? The answer is negative. Moreover, the complement $I^\tau - \Sigma$ is not even realcompact. Indeed, assuming that $I^\tau - \Sigma$ is realcompact, we conclude, by Proposition 1.1.24, that $\Sigma$ contains a functionally closed subset of $I^\tau$. Since every functionally closed subset of $I^\tau$ contains, in turn, a copy of the Tychonov cube $I^\tau$, we see that a copy of the Tychonov cube must be contained in $\Sigma$. This implies that $I^\tau$ is a Corson compactum, which is an obvious contradiction (see Corollary 8.9.14). Thus $I^\tau - \Sigma \not\approx R^\tau$. This suggests the following problem.

PROBLEM 8.10.8. *Find a topological characterization of $I^\tau - \Sigma$.*

Of course, the positive solution of Problem 8.10.4 would imply the topological homogeneity of the space $I^\tau - \Sigma$. But at this point the following question remains unanswered.

PROBLEM 8.10.9. *Is the space $I^\tau - \Sigma$ topologically homogeneous?*

A positive solution of Problem 8.10.7 would guarantee that the space $I^\tau - \Sigma$ contains a copy of the Tychonov cube $I^\tau$ and, consequently, that $I^\tau - \Sigma$ is universal with respect to the class of spaces of weight $\leq \tau$. A more delicate problem is:

PROBLEM 8.10.10. *Characterize the closed subspaces of $I^\tau - \Sigma$.*

*Historical and bibliographical notes* 8.10. This concluding Section consists of questions suggested for the reader. Of course, there are several others closely related to those presented in the text. We hope that the interested reader will formulate (and solve!) them independently.

# Bibliography


1. **S. M. Ageev**, *Classifying spaces for free actions and the Hilbert-Smith conjecture*, Russian Acad. Sci. Sbornik **75** (1993), 137–144.
2. **P. S. Alexandrov**, *Sur les ensembles de la premiere et les ensembles abstraits*, C. R. Acad. Paris **178** (1924), 185–187.
3. ———, *Untersuchungen Über Gestalt und Lage abgeschlossener Mengen beliebiger Dimension*, Ann. Math. **30** (1929), 101–187.
4. ———, *Dimensiontheorie, ein Beitrag zur Geometrie der abgeschlossen Mengen*, Math. Ann. **106** (1932), 161–238.
5. ———, *Eine Problemstellungen in der mengentheoretischen Topologie*, Math. Sb. **1** (1936), 619–634.
6. ———, *Introduction to homological dimension theory and general combinatorial topology*, (Russian), "Nauka", Moscow, 1975.
7. **P. S. Alexandrov and B. A. Pasynkov**, *Introduction to dimension theory*, (Russian), "Nauka", Moscow, 1973.
8. **P. S. Alexandrov and P. S. Urysohn**, *Über nulldimensionale Mengen*, Math. Ann. **98** (1928), 89–106.
9. **K. Alster and R. Pol**, *On function spaces of compact subspaces of Σ-products of the real line*, Fund. Math. **107** (1980), 135–143.
10. **R. A. Alo and H. L. Shapiro**, *Normal topological spaces*, Cambridge Univ. Press, Cambridge, 1974.
11. **R. D. Anderson**, *Monotone interior dimension-raising mappings*, Duke Math. J. **19** (1952), 359–366.
12. ———, *Open mappings of compact continua*, Proc. Nat. Acad. Sci. No. 6, **42** (1956), 247–349.
13. ———, *Zero-dimensional compact groups of homeomorphisms*, Pacific J. Math. **7** (1957), 797–810.
14. ———, *A characterization of the universal curve and a proof of its homogeneity*, Ann. Math. **67** (1958), 313–324.
15. ———, *One-dimensional continuous curves and a homogeneity theorem*, Ann. Math. **68** (1958), 1–16.
16. ———, *The algebraic simplicity of certain groups of homeomorphisms*, Amer. J. Math. **80** (1958), 955–963.

17. ———, *On homeomorphisms as products of conjugates of a given homeomorphism and its inverse*, Topology of 3-manifolds and related topics (M. K. Fort, ed.), Prentice-Hall, 1961, 231–234.
18. ———, *Hilbert space is homeomorphic to the countable infinite product of lines*, Bull. Amer. Math. Soc. **72** (1966), 515–519.
19. ———, *On topological infinite deficiency*, Michigan Math. J. **126** (1967), 365–383.
20. ———, *Topologival properties of the Hilbert cube and the infinite product of open intervals*, Trans. Amer. Math. Soc. **126** (1967), 200–216.
21. **R. D. Anderson and R. H. Bing**, *A completely elementary proof that Hilbert space is homeomorphic to the countable infinite product of lines*, Bull. Amer. Math. Soc. **74** (1968), 771–792.
22. **R. D. Anderson and D. W. Curtis**, *Infinite-dimensional topology and AR-ANR theory*. Preprint, 1987.
23. **R. D. Anderson and T. A. Chapman**, *Extending homeomorphisms to Hilbert cube manifolds*, Pacific J. Math. **38** (1971), 281–293.
24. **R. D. Anderson and J. D. McCharen**, *On extending homeomorphisms to Frechet manifolds*, Proc. Amer. Math. Soc. **25** (1970), 283–289.
25. **R. D. Anderson and R. M. Schori**, *Factors of infinite-dimensional manifolds*, Trans Amer. Math. Soc. **142** (1969), 315–330.
26. **W. Barit**, *Small extensions of small homeomorphisms*, Notices Amer. Math. Soc. **16** (1969), p. 295.
27. **R. G. Bartle and L. M. Graves**, *Mappings between function spaces*, Trans. Amer. Math. Soc. **72** (1952), 400–413.
28. **A. Beck**, *On invariant sets*, Ann. of Math. **67** (1958), 99–103.
29. **M. Bell and A. Chigogidze**, *Topological groups homeomorphic to products of discrete spaces*, Topology Appl., **53** (1993), 67–73.
30. **Y. Benyamini**, *Continuous images of Eberlein compacta*, preprint.
31. **C. Bessaga and A. Pelczyński**, *The estimated extension theorem, homogenous collections and skeletons and their applications to the topological classification of linear metric spaces and convex sets*, Fund. Math. **69** (1970), 153–190.
32. ———, *Selected topics in infinite-dimensional topology* PWN, Warszawa, 1975.
33. **M. Bestvina**, *Characterizing k-dimensional universal Menger compacta*, Mem. Amer. Math. Soc. No. 380, **71** (1988).
34. **M. Bestvina, P. Bowers, J. Mogilski and J. Walsh**, *Characterizinf Hilbert space manifolds revisited*, Topology Appl. **24** (1986), 53–69.
35. **M. Bestvina and J. Mogilski**, *Characterizing certain incomplete infinite-dimensional absolute retracts*, Michigan Math. J. **33** (1986), 291–313.
36. **R. H. Bing**, *Partitioning a set*, Bull. Amer. Math. Soc. **55** (1949), 1101–1110.

37. ———, *A decomposition of $E^3$ into points and tame arcs such that the decomposition space is topologically different from $E^3$*, Ann. Math. **65** (1957), 484–500.
38. ———, *The cartesian product of a certain manifold and a line is $E^4$*, Ann. Math. **70** (1959), 399–412.
39. **G. Birkhoff**, *Moyennes de fonctions bornees.Algebre et theorie des nombres*, Colloques Internationaux du Centre National de la Recherche Scientifique. No. 24, (1950).
40. **R. L. Blair**, *Spaces in which special sets are z-embedded*, Canad. J. Math. **28** (1976), 673–690.
41. ———, *Filter characterizations of z-, $C^*$- and C-embeddings*, Fund. Math. **90** (1976), 285–300.
42. **R. L. Blair and A. W. Hager**, *Extensions of zero-sets and real-valued functions*, Math. Zeit. **136** (1974), 41–52.
43. **M. F. Bokshtein**, *On homological invariants of topological spaces I*, (Russian) Trudy Moskov. Mat. Obshch. **5** (1965), 3–80.
44. **V. G. Boltyanskii**, *On the dimensional full valuedness of compacta*, (Russian) Dokl. Akad. Nauk SSSR **67** (1949), 773–776.
45. **K. Borsuk**, *Über Isomorphie der Funktionalräume*, Bull. Int. Acad. Polon. Sci. et lettres, (1933), 1–10.
46. ———, *Theory of retracts*, PWN, Warszawa, 1967.
47. ———, *Theory of shape*, PWN, Warszawa, 1975.
48. **H. G. Bothe**, *Eine Einbettung m-dimensionalen Mengen in einen (m+1)-dimensionalen Absoluten Retrakt*, Fund. Math. **52** (1963), 209–224.
49. **N. Bourbaki**, *Groupes et algébres de Lie*, Masson, Paris, 1982.
50. **N. Bourbaki**, *Topologie générale*, Herman, Parix, 1974.
51. **P.Bowers**, *General position properties satisfied by finite products of dendrites*, Trans. Amer. Math. Soc. **288** (1985), 739–753.
52. ———, *Dense embeddings of nowhere locally compact separable metric spaces*, Topology Appl. **26** (1987), 1–12.
53. ———, *Limitation topologies for function spaces*, Trans. Amer. Math. Soc. **314** (1989), 421–431.
54. **B. Brechner**, *On the dimensions of certain spaces of homeomorphisms*, Trans. Amer. Math. Soc. **121** (1966), 516–548.
55. **G. E. Bredon**, *Introduction to compact transformation groups*, Academic Press, New York, 1972.
56. **L. E. J. Brouwer**, *On the structure of perfect sets of points*, Proc. Akad. Amsterdam **12** (1910), 785–794.
57. **W. Browder, J. Levine and G. R. Livesay**, *Finfing a boundary for an open manifold*, Amer. J. Math. **87** (1965), 1017–1028.
58. **M. Brown**, *Some applications of an approximation theorem for inverse limits*, Proc. Amer. Math. Soc. **11** (1960), 478–481.

59. **V. M. Bukshtaber and A. S. Mischenko**, *K-theory on the category of infinite cell complexes*, (Russian) Izv. AN SSSR **32** (1968), 560–604.
60. **J. W. Canon**, *The recognition problem: what is topological manifold?*, Bull. Amer. Math. Soc. **84** (1978), 832–866.
61. ———, *Shrinking cell-like decompositions of manifolds. Codimension three*, Ann. Math. **110** (1979), 83–112.
62. **R. Cauty**, *Un espace métrique linéaire qui n'est pas un rétracte absolu*, Fund. Math. **146** (1994), 85–99.
63. ———, *Une caractérisation des rétractes absolus de voisinage*, Fund. Math. **144** (1994), 11–22.
64. **T. A. Chapman**, *Dense sigma-compact subsets of infinite-dimensional manifolds*, Trans. Amer. Math. Soc. **154** (1971), 399–426.
65. ———, *On the structure of Hilbert cube manifolds*, Compos. Math. **24** (1972), 329–353.
66. ———, *Compact Hilbert cube manifolds and invariance of Whitehead torsion*, Bull. Amer. Math. Soc. **79** (1973), 52–56.
67. ———, *Locally trivial bundles and microbundles with infinite-dimensional fibers*, Proc. Amer. Math. Soc. **37** (1973),595–602.
68. ———, *Topological invariance of Whitehead torsion*, Amer J. Math. **96** (1974), 488–497.
69. ———, *Lectures on Hilbert cube manifolds*, Regional Series in Math. No. 28, Amer. Math. Soc., 1976.
70. **T. A. Chapman, L. C. Siebenmann**, *Finding a boundary for a Hilbert cube manifolds*, Acta Mathematica **137** (1976), 171–208.
71. **A. V. Chernavski**, *Generalization of L. V. keldysh's construction of monotone mappings of the cube onto a higher dimensional cube*, (Russian) Uspekhi Mat. Nauk **40** (1985), 209–211.
72. **A. Chigogidze**, *On the relative dimensions of completely regular spaces*, (Russian) Bull. Georgian Acad. Sci. **85** (1977), 45–48.
73. ———, *Inductive dimensions for completely regular spaces*, Comm. Math. Univ. Carolinae **18** (1977), 623–637.
74. ———, *On the dimensions of increments of Tychonoff spaces*, Fund. Math. **111** (1981), 25–36.
75. ———, *On the spectral theorem of E. V. Shchepin*, (Russian) Bull. Georgian Acad. Sci. **104** (1981), 25–28.
76. ———, *On infinite-dimensional bicompacta*, Siberian Math. J. **23** (1982), 157–164.
77. ———, *On a generalization of perfectly normal spaces*, Topology Appl. **13** (1982), 15–20.
78. ———, *Zero-dimensional open mappings which increase dimension*, Comm. Math. Univ. Carolinae **24** (1983), 571–579.
79. ———, *Uncountable powers of the line and the natural series and n-soft mappings*, Soviet Math. Dokl. **30** (1984), 342–345.

80. ______, *On relative dimensions*, (Russian) in: General Topology. Spaces of functions and dimension, Moscow, Moscow Univ. Press, 1985, pp. 67–117.
81. ______, *Characterization of the dimension* dim *of R-compacta and universal spaces of the given weight and of the given dimension*, (Russian) Uspekhi Mat. Nauk **40** (1985), 265–266.
82. ______, *On Baire isomorphisms of non-metrizable compacta*, Comm. Math. Univ. Carolinae **26** (1985), 811–820.
83. ______, *Trivial bundles with Tychonov cube fiber*, (Russian) Mat. Notes **39** (1986), 747–756.
84. ______, *Noncompact absolute extensors in dimension n, n-soft mappings and their applications*, Math. USSR Izv. **28** (1987), 151–174.
85. ______, *On the structure of non-metrizable $AE(0)$-spaces*, (Russian) Mat. Notes **41** (1987), 406–411.
86. ______, *Characterization of Polish $AE(n)$-spaces*, (Russian) Vestnik Moscow Univ. **5** (1987), 32–35.
87. ______, *Compacta lying in the n-dimensional universal Menger compactum and having homeomorphic complements in it*, Math. USSR Sb. **61** (1988), 471–484.
88. ______, *n-soft mappings of n-dimensional spaces*, (Russian) Mat. Notes **46** (1989), 88–95.
89. ______, *On $UV^n$-equivalent compacta*, (Russian) Vestnik Moscow Univ. **3** (1989), 33–35.
90. ______, *The theory of n-shapes*, Russian Math. Surveys **44** (1989), 145–174.
91. ______, *Trivial bundles and near homeomorphisms*, Fund. Math. **132** (1989), 89–98.
92. ______, *n-shapes and n-cohomotopy groups of compacta*, Math. USSR Sb. **66** (1990), 329–342.
93. ______, *Locally convex linear topological spaces that are homeomorphic to the powers of the real line*, Proc. Amer. Math. Soc. **113** (1991), 599–609.
94. ______, *Autohomeomorphisms of the universal Menger compacta are stable*, Bull. Georgian Acad. Sci. **142** (1991), 477–479.
95. ______, *On Baire isomorphisms*, (Russian) Vestnik Moscow Univ. **5** (1991), 67–69.
96. ______, *A note on cohomological dimension*, preprint.
97. ______, *$UV^n$-equivalence and n-equivalence*, Topology Appl. **45** (1992), 283–2
98. ______, *Classification theorem for Menger manifolds*, Proc. Amer. Math. Soc. **116** (1992), 825–832.
99. ______, *Shape properties of non-metrizable spaces*, Topology Appl. **53** (1993), 259–269.
100. ______, *Finding a boundary for a Menger manifold*, Proc. Amer. Math. Soc. **121** (1994), 631–640.

101. ———, *Double spectra*, Abstracts of Annual Spring Topology and Southeast Dynamics Conference, 1995, p.13.
102. **A. Chigogidze and V. V. Fedorchuk**, *Absolute retracts and infinite-dimensional manifolds*, "Nauka", Moscow, 1992 (Russian).
103. **A. Chigogidze, K. H. Hofmann and J. R. Martin**, *Compact groups and fixed point sets*, preprint.
104. **A. Chigogidze, K. Kawamura, R. B. Sher**, *Finiteness results in $n$-homotopy theory*, preprint.
105. **A. Chigogidze, K. Kawamura, E. D. Tymchatyn**, *Menger manifolds*, in: Continua with the Houston Problem Book, Lecture Notes in Pure and Appl. Math., **170** (1995), 37–88.
106. ———, *Nöbeling spaces and pseudo-interiors od Menger compacta*, Topology Appl., to appear.
107. **A. Chigogidze and J. R. Martin**, *Fixed point sets of Tychonov cubes*, Topology Appl. **64** (1995), 201–218.
108. **A. Chigogidze and V. Valov**, *Universal maps and surjective characterizations of completely metrizable $LC^n$-spaces*, Proc. Amer. Math. Soc. **109** (1990), 1125–1133.
109. ———, *Set-valued maps and $AE(0)$-spaces*, Topology Appl. **55** (1994), 1–15.
110. **H. Cohen**, *A cohomological definition of dimension for locally compact Hausdorff spaces*, Duke Math. J. **21** (1954), 209–224.
111. **D. W. Curtis**, *Property Z for function graphs and finite-dimensional sets in $I^\infty$ and $s$*, Compos. Math. **22** (1970), 19–22.
112. ———, *Some theorems and examples on local equiconnectedness and its applications*, Fund. Math. **72** (1971), 101–113.
113. ———, *Hyperspaces homeomorphic to Hilbert space*, Proc. Amer. Math. Soc. **75** (1979), 126–130.
114. **D. W. Curtis and J. van Mill**, *Zero-dimensional countable dense unions of $Z$-sets in the Hilbert cube*, Fund. Math. **108** (1983), 103–108.
115. **D. W. Curtis and R. M. Schori**, *$2^X$ and $C(X)$ are homeomorphic to the Hilbert cube*, Bull. Amer. Math. Soc. **80** (1974), 927–931.
116. ———, *Hyperspaces which characterize simple homotopy types*, Gen. Topol. Appl. **6** (1976), 153–165.
117. ———, *Hyperspaces of Peano continua are Hilbert cubes*, Fund. Math. **101** (1978), 19–38.
118. **R. J. Daverman and J. J. Walsh**, *An example of cell-like mapping which is not shape equivalence*, Michigan Math. J. **30** (1983), 17–30.
119. **J. J. Dijkstra**, *$k$-dimensional skeletoids in $R^n$ and the Hilbert cube*, Topology Appl. **19** (1985), 13–28.
120. ———, *Fake topological Hilbert spaces and characterization of dimension in terms of negligibility*, vol. CWI Tract 2, Mathematisch Centrum, Amsterdam, 1984.

121. **J. J. Dijkstra, J. van Mill, J. Mogilski**, *Classification of finite-dimensional pseudo-boundaries and pseudo-interiors*, Trans. Amer. Math. Soc. **332**, (1992), 693–709.
122. **J. Dixmier**, *Quelques propriétés des groupes abéliens localement compacts*, Bull. Sci. Math. (2) **81** (1957), 38–48.
123. **T. Dobrowolski and J. Mogilski**, *Problems on topological classification of incomplete metric spaces*, Open problems in topology (J. van Mill and G. M. Reed, eds.), North-Holland, New York, 1990, 409–429.
124. **T. Dobrowolski and H. Toruńczyk**, *On metric linear spaces homeomorphic to $l_2$ and compact convex sets homeomorphic to $Q$*, Bull. Acad. Polon. Sci. **27** (1979), 883–887.
125. **A. N. Dranishnikov**, *Absolute extensors in dimension n and n-soft mappings raising dimension*, (Russian) Uspekhi Mat. Nauk **39** (1984), 55–95.
126. ———, *Alexandrov diameters and cell-like mappings*, (Russian) Dokl Akad. Nauk SSSR **285** (1985), 550–555.
127. ———, *Universal Menger compacta and universal mappings*, (Russian) Mat.Sb. **129** (1986), 121–139.
128. ———, *On homological dimension modulo p*, (Russian) Mat. Sb. **132** (1987), 420–432.
129. ———, *On resolutions of $LC^n$-compacta*, Lecture Notes Math. **1283** (1087), 48–59.
130. ———, *Homological dimension theory*, (Russian) Uspekhi Mat. Nauk **43** (1988), 11–55.
131. ———, *On free actions of zero-dimensional compact groups*, Izv. Akad. Nauk USSR (in Russian), **32** (1988), 217–232.
132. **A. N. Dranishnikov and E. V. Ščepin**, *Cell-like mappings*, (Russian) Uspekhi Mat. Nauk **41** (1986), 49–90.
133. **J. Dugundji**, *An extension of Tietze's theorem*, Pacific J. Math. **1** (1951), 353–367.
134. **J. Dydak and J.Mogilski**, *Universal cell-like maps*, Proc. Amer. Math. Soc. **122** (1994), 943–948.
135. **J. Dydak and J. Segal**, *Shape theory*, Lect. Notes Math. **688**, 1978.
136. **J. Dydak and J. Walsh**, *Infinite-dimensional compacta having cohomological dimension two: an application of the Sullivan conjecture*, Topology **32** (1993), 93–104.
137. ———, *Spaces without cohomological dimension preserving compactifications*, Proc. Amer. Math. Soc. **113** (1991), 1155–1162.
138. **R. D. Edwards**, *A theorem and a question related to cohomological dimension and cell-like maps*, Notices Amer. Math. Soc. **25** (1978).
139. ———, *Characterizing infinite-dimensional manifolds topologically (after Henryk Toruńczyk)*, Lect.Notes Math. **540** (1999), 278–302.
140. ———, *Problems: 1987 western workshop in geometric topology*, Oregon State Univ. Press, 1987, 46–48.

141. **R. D. Edwards, R. C. Kirby**, *Deformations of spaces of embeddings*, Ann. Math. **93** (1971), 63–88.
142. **J. Eels and N. H. Kuiper**, *Homotopy negligible subsets of infinite-dimensional manifolds*, Compos. Math. **21** (1969), 155–161.
143. **S. Eilenberg and N. Steenrod**, *Foundations of algebraic topology*, Princeton Univ. Press, Princeton, 1952.
144. **R. Engelking**, *On functions defined on cartesian products*, Fund. Math. **59**, (1966), 221–231.
145. ———, *General topology*, PWN, Warszawa, 1977.
146. ———, *Dimension theory*, PWN, Warszawa, 1978.
147. **V. V. Fedorchuk**, *Completely closed mappings and compatibility of some theorems in general topology with the axioms of set theory*, (Russian) Mat. Sb. **99** (1976), 3–33.
148. ———, *On open mappings*, (Russian) Uspekhi Mat. Nauk **37** (1982), 187–188.
149. **S. Ferry**, *The homeomorphism group of a compact Hilbert cube manifold is an $ANR$*, Ann. Math. **106** (1977), 101–119.
150. ———, *$UV^k$-equivalent compacta*, Lecture Notes Math. **1283** (1987), 88–114.
151. **G. Fisher**, *On the group of all homeomorphisms of a manifold*, Trans. Amer. Math. Soc. **97** (1960), 193–212.
152. **H. Freudenthal**, *Entwicklungen von Räumen und ihren Gruppen*, Compos. Math. **4** (1937), 145–234.
153. **A. B. Forge**, *Dimensiona preserving compactifications allowing extensions of continuous functions*, Duke Math. J. **28** (1961), 625–627.
154. **R. H. Fox**, *On the Lusternik-Schnirelman category* Ann. of Math. **42** (1941), 333–370.
155. **D. Garity, J. P. Henderson, D. G. Wright**, *Menger spaces and inverse limits*, Pacific J. Math. **131** (1988), 249–259.
156. **R. Geoghegan**, *Open problems in infinite-dimensional topology*, Topology Proc. **4** (1979), 287–338.
157. **R. Geoghegan and R. C. Lacher**, *Compacta with the shape of finite complexes*, Fund. Math. **92** (1976), 25–27.
158. **R. Geoghegan and R. Summerhill**, *Pseudo-boundaries and pseudo-interiors in Euclidean space and topological manifolds*, Trans. Amer. Math. Soc. **194** (1974), 141–165.
159. **L. Gillman and M. Jerison**, *Rings of continuous functions*, Van Nostrand, New York, 1960.
160. **S. P. Gul'lko**, *On the structurw of spaces of continuous functions and their complete paracompactness*, Russian Math. Surveys **34** (1979), 36–44.
161. **O. Hanner**, *Some theorems on absolute neighborhood retracts*, Arkiv Mat. **1** (1951), 389–408.

162. ______, *Retractions and extensions of mappings of metric and non-metric spaces*, Arkiv Mat. **2** (1952), 315–360.

163. **W. E. Haver**, *Mappings between ANR's that are fine homotopy equivalences*, Pacific J. Math. **58** (1975), 457–461.

164. **R. Haydon**, *On a problem of Pelczyński: Milutin spaces, Dugundji spaces and $AE(0-\text{dim})$*, Studia Math. **52** (1974), 23–31.

165. **D. W. Henderson**, *Open subsets of Hilbert space*, Compos. Math. **21** (1969), 312–318.

166. ______, *Infinite-dimensional manifolds are open subsets of Hilbert space*, Topology **9** (1970), 25–34.

167. ______, *Z-sets in ANR's*, Trans. Amer. Math. Soc. **213** (1975), 205–216.

168. **D. W. Henderson and R. M. Schori**, *Topological classification of infinite-dimensional manifolds by homotopy type*, Bull. Amer. Math. Soc. **76** (1970), 121–124.

169. **L. J. Hernandez, T. Porter**, *An embedding theorem for proper n-types*, Topology Appl. **48** (1992), 215–233.

170. **E. Hewitt**, *Rings of real-valued continuous functions I*, Trans. Amer. Math. Soc. **64** (1948), 45–99.

171. **B. Hoffman**, *A surjective characterization of Dugundji spaces*, Proc. Amer. Math. Soc. **76** (1979), 151–156.

172. **K. H. Hofmann**, *Sur la décomposition semidirecte de groupes compactes connexes*, Symposia Mathematica **26** (1975), 471–476.

173. bf K. H. Hofmann and P. S. Mostert, *Elements of Compact Semigroups*, Charles E. Merill Books, Columbus,1966.

174. **W. Holstyński**, *An extension and axiomatic characterization of Borsuk's theory of shape*, Fund. Math. **70** (1971), 157–168.

175. ______, *Continuity of Borsuk's shape theory*, Bull. Acad. Polon Sci. **19** (1971), 1105–1108.

176. **S. T. Hu**, *Theory of retracts*, Wayne State Univ. Press, Detroit, 1965.

177. **W. Hurewicz**, *Über Abbildungen von endlich dimensionalen Räumen auf Teilmengen Cartesischer Räume*, Sitzungsberichte preuss. Akad. Wiss. **34** (1933), 754–765.

178. **D. M. Hyman**, *ANR divisors and absolute neighborhood contractibility*, Fund. Math. **62** (1968), 61–73.

179. **J. R. Isbell**, *Uniform neighborhood retracts*, Pacific J. Math. **11** (1961), 609–648.

180. ______, *Uniform spaces*, Amer. Math. Soc., Providence, 1964.

181. **L. N. Ivanovski**, *On a hypothesis of P.S.Aleksandrov*, Dokl. Akad. Nauk. USSR **123** (1958), 785–786.

182. **Y. Iwamoto**, *Menger manifolds homeomorphic to their n-homotopy kernels*, Proc. Amer. Math. Soc. **123** (1995), 945–953.

183. ______, *Fixed point sets of transformation groups of Menger manifolds, their pseudo-interiors and their pseudo-boundaries*, preprint.

184. **Y. Iwamoto, K. Sakai**, *A mapping theorem for Q-manifolds and $\mu^{n+1}$-manifolds*, preprint.
185. **M. Katětov**, *Measures in fully normal spaces*, Fund. Math. **38** (1951), 73–84.
186. **K. Kawamura**, *An Inverse system approach to Menger manifolds*, Topology Appl., to appear.
187. ———, *On n-homotopy equivalences between comapct ANR's*, preprint.
188. ———, *Characterizations of Menger manifolds and Hilbert cube manifolds in terms of partitions*, in: Continua with the Houston Problem Book, Lecture Notes in Pure and Appl. MAth. **170** (1995), 275–285.
189. **L. V. Keldysh**, *Monotone mappings of cube onto cube of larger dimension*, (Russian) Mat. Sb. **41** (1957), 129–158.
190. ———, *Transformation of monotone irreducible mappings in monotone open and monotone open mappings from cubes raising dimension*, (Russian) Mat. Sb. **43** (1957), 187–226.
191. ———, *Zero-dimensional open mappings*, (Russian) Izv. AN SSSR **23** (1959), 165–184.
192. **O. H. Keller**, *Die Homeomorphie der Kompakten Konvexen Mengen in Hilbertshen Räum*, Math. Ann. **105** (1931), 748–758.
193. **J. L. Kelley**, *General topology*, Van Nostrand, Princeton, 1955.
194. **R. C. Kirby**, *Stable homeomorphisms and annulus conjecture*, Ann. Math.**89** (1969), 575–582.
195. **V. L. Klee**, *Some topological properties of convex sets*, Trans. Amer. Math. Soc. **78** (1955), 30–45.
196. **A. N. Kolmogorov**, *Über offene Abbildeungen*, Ann. Math. **38** (1937), 36–38.
197. **P. Koszmider**, *On the complete invariance property in some uncountable products*, Canad. Math. Bull. **35** (1992), 221–229.
198. **G. Kozlowski and J. J. Walsh**, *Cell like mappings on 3-manifolds*, Topology **22** (1983), 147–153.
199. **I. M. Kozlowski**, *Absolute polyhedral expansions of metric spaces*, Trans. Moscow Math. Soc. **2** (1981), 79–114.
200. ———, *Dimension-raising open mappings of compacta onto polyhedra as the spectral limit of inessential mappings of polyhedra*, (Russian) Abstracts of Leningrad international topological Conference, "Nauka", 1982, p.85.
201. **K. Kunen and J. van Mill**, *measures on Corson compact spaces*, preprint.
202. **K. Kuratowski**, *Topology*, vol. 1, Academic Press, New York, 1966.
203. ———, *Topology*, vol. 2, Academic Press, New York, 1968.
204. **V. I. Kuzminov**, *On a hypothesis of P.S.Aleksandrov in the theory of topological groups*, Dokl. Akad. Nauk USSR **125** (1959), 727–729.
205. ———, *On cohomological dimensions of compacta*, (Russian) Siber. Math. J. **5** (1964), 1282–1304.

206. ———, *Homological dimension theory*, (Russian) Uspekhi Mat. Nauk **23** (1968), 3–49.

207. **R. C. Lacher**, *Cell-like mappings and their generalizations*, Bull. Amer. Math. Soc. **83** (1977), 495–552.

208. **M. A. Lavrentieff**, *Contribution a la theorie des ensembles homeomorphes*, Fund. Math. **6** (1924), 149–160.

209. **S. Lefshetz**, *On compact spaces*, Ann. Math. **32** (1931), 521–538.

210. ———, *Algebraic topology*, New York, 1942.

211. **O. V. Lokucievskii**, *An example of the open mapping of one-dimensional compactum onto Hilbert parallelepiped*, (Russian) Uch. zap. MGU **165** (1954), 118–130.

212. **S. Mac Lane**, *Categories for the working mathematician*, Springer-Verlag, 1971.

213. **S. Mardešić**, *On covering dimension and inverse limits of compact spaces*, Ill. J. Math. **4** (1960), 278–291.

214. ———, *Shapes for topological spaces*, Gen. Topol. Appl. **3** (1973), 265–282.

215. **S. Mardešić and J. Segal**, *Shape theory*, North-Holland, Amsterdam, 1982.

216. **J. R. Martin**, *Fixed point sets of homeomorphisms of metric products*, Proc. Amer. MAth. Soc. **103** (1988), 1293–1298.

217. **J. R. Martin, L. G. Oversteegen and E.D.Tymchatyn**, *Fixed point sets of products and cones*, Pacific J. Math. **101** (1982), 133-139.

218. **J. R. Martin and W. Weiss**, *Fixed point sets of metric and nonmetric spaces*, Trans. Amer. Math. Soc. **284** (1984), 337–353.

219. **J. C. Mayer, C. W. Stark**, *Group actions on Menger manifolds*, preprint.

220. **R. A. McCoy**, *Groups of homeomorphisms of normed linear spaces*, Pacific J. Math. **39** (1971), 735-743.

221. **K. Menger**, *Allgemeine Raume und Cartesische Raume Zweite Mitteilung: "Uber umfas - sendste n-dimensional Mengen"*, Proc. Akad. Amsterdam **29** (1926), 1125–1128.

222. ———, *Dimensionstheorie*, Verlag von B. G. Teubner, Leipzig, 1928.

223. **Y. Mibu**, *On Baire functions on infinite product spaces*, Proc. Imp. Acad. Tokyo **20** (1944), 661–663.

224. **E. Michael**, *Continuous selections I*, Ann. Math. **63** (1956), 361–382.

225. ———, *Continuous selections II*, Ann. Math. **64** (1956), 562–580.

226. ———, *Selected selection theorems*, Amer. Math. Monthly **58** (1965), 233–238.

227. **E. Michael and M. E. Rudin**, *A note on Eberlein compacta*, Pacific J. Math. **72** (1977), 487–495.

228. **J. van Mill**, *Infinite-dimensional topology*, North-Holland, Amsterdam, 1989.

229. **J. van Mill and J. Mogilski**, *Property C and fine homotopy equivalences*, Preprint, 1981.

230. **H. Miller**, *The Sullivan conjecture on maps from classifying spaces*, Ann. Math. **120** (1984), 39–87.

231. **R. T. Miller**, *Mapping cylinder neighborhoods of some ANR's*, Ann. Math. **103** (1976), 1–18.

232. **J. W. Milnor**, *On spaces having the homotopy type of a CW-complex*, Trans. Amer. Math. Soc. **90** (1959), 272–280.

233. **A. A. Milutin**, *On the space of continuous functions*, (Russian) Ph.D. Thesis, Moscow State Univ., 1952.

234. **J. Mogilski**, *Characterizing the topology of infinite dimensional $\sigma$-compact manifolds*, Proc. Amer. Math. Soc. **92** (1984), 111–118.

235. **J. C. Mayer, L. G. Oversteegen, E. D. Tymchatyn**, *The Menger curve: characterization and extension of homeomorphisms of non-locally separating closed subsets*, Diss. Math. **CCLII**, 1986.

236. **S. Mrówka**, *Some properties of Q-spaces*, Bull. Acad Polon. Sci. **5** (1957), 947–950.

237. ———, *On the unions of Q-spaces*, Bull. Acad. Polon. Sci. **6** (1958), 365–368.

238. ———, *Some comments on the author's example of a non-R-compact space*, Bull. Acad. Polon. Sci. **18** (1970), 443–448.

239. **L. Nachbin**, *On the continuity of positive linear transformations*, Proc. Intern. Congress of Mathematicians, Cambridge, Mass. 1950, vol. 1, Providence, 1952, 464–465.

240. **K. Nagami**, *Dimension of non-normal spaces*, Fund. Math. **109** (1980), 113–121.

241. **S. Negrepontis**, *Banach spaces and topology*, in: Handbook of Set-Theoretic Topology (K. Kunen and J. E. Vaughan eds), North-Holland, Amsterdam, 1984, 1045–1142.

242. **G. M. Nepomnyashchii**, *On the spectral decomposition of the multivalued absolute retracts*, (Russian) Uspekhi Mat. Nauk **36** (1981), 221–222.

243. **G. Nöbeling**, *Über eine n-dimensionale Universalmenge im $R^{2n+1}$*, Math. Ann. **104** (1931), 71–80.

244. **L. G. Oversteegen and E. D. Tymchatyn**, *On the dimension of certain totally disconnected spaces*, Proc. Amer. Math. Soc., to appear.

245. **B. A. Pasynkov**, *Zero-dimensional open mappings raising dimension*, (Russian) Uspekhi Mat. Nauk **18** (1963), 183–190.

246. ———, *On universal bicompactum of the given weight and of the given dimension*, (Russian) Dokl. Akad. Nauk SSSR **154** (1964), 1042–1043.

247. ———, *On the spectral decomposability of topological spaces*, (Russian) Mat. Sb. **66** (1965), 35–79.

248. ———, *Partial topological products*, (Russian) Trudy Moskov Mat. Obshch. **13** (1965), 136–245.

249. **A. Pelczyński**, *Linear extensions, linear averaging operators and their*

*applications to linear topological classification of spaces of continuous functions*, Dissert. Math. **99**, 1999.

250. **E. Pol**, *On the dimension of the product of metrizable spaces*, Bull. Acad. Polon. Sci. **26** (1978), 525–534.
251. **R. Pol and E. Puzio-Pol**, *Remarks on cartesian products*, Fund. Math. **93** (1976), 57–69.
252. **L. S. Pontrjagin**, *Sur une hypothèse fundamentale de la thèorie de la dimension*, Comp. Rend. Acad. Sci. **190** (1930), 1105–1107.
253. **L. S. Pontryagin**, *Topologische Gruppen I und II*, Teubner Leipzig 1954 (German translation of the second Russian edition).
254. ———, *Applicationcation of the combinatorial topology to compact metric spaces*, (Russian) Uspekhi Mat. Nauk **39** (1984), 131–164.
255. **L. S. Pontrjagin and G. Tolstowa**, *Beweis des Mengerschen Einbettungssatzes*, Math. Ann. **105** (1931), 734–747.
256. **F. Quinn**, *Resolutions of homology manifolds and the topological characterization of manifolds*, Invent. Math. **72** (1983), 267–284.
257. **F. Raymond, R. F. Williams**, *Examples of p-adic transformation groups*, Ann. Math. **78** (1963), 92–106.
258. **A. Robertson and W. Robertson**, *Topological vector spaces*, Cambridge tracts in Math. **53**, Cambridge Univ. Press, 1980.
259. **C. A. Rogers and J. E. Jayne**, *First level Baire functions and isomorphisms*, J. Math. Pure Appl. **61** (1982), 177–205.
260. **C. A. Rogers, J. E. Jayne at al**: editors, *Analytic sets*, Academic Press, London, 1980.
261. **K. A. Ross and A. H. Stone**, *Products of separable spaces*, Amer. Math. Monthly **71** (1964), 398–403.
262. **L. R. Rubin**, *Cell-like maps, dimension and cohomological dimension*, A Survey in Geometric and Algebraic Topology, PWN, Warszawa, 1986, 371–376.
263. **L. R. Rubin, R. M. Schori and J. J. Walsh**, *New dimension theory techniques for constructing infinite-dimensional examples*, Gen. Topol. Appl. **10** (1979), 93–102.
264. **W. Rudin**, *Functional analysis*, McGraw-Hill, New York, 1973.
265. **K. Sakai**, *Free actions of zero-dimensional compact groups on Menger manifolds*, preprint.
266. ———, *All autohomeomorphisms of connected Menger manifolds are stable*, preprint.
267. **R. M. Schori and J. E. West**, $2^I$ *is homeomorphic to the Hilbert cube*, Bull. Amer. Math. Soc. **78** (1972), 402–406.
268. **H. Schirmer**, *On fixed point sets of homeomorphisms of the n-ball*, Israel J. Math. **7** (1969), 46–50.
269. ———, *Fixed point sets of homeomorphisms of compact surfaces*, Israel J. Math. **10** (1971), 373–378.

270. **J. P. Serre**, *Groups d'homotopie at classes de groupes abeliens*, Ann. Math. **58** (1953), 258–294.

271. **L. B. Shapiro**, *The space of closed subsets of $D^{\omega_2}$ is not a dyadic bicompactum*, Soviet Math. Dokl. **17** (1976), 937–941.

272. **E. V. Ščepin**, *Topology of limit spaces of uncountable inverse spectra*, (Russian) Uspekhi Mat. Nauk **31** (1976), 191–226.

273. ———, *Real functions and canonical sets in Tychonov products and topological groups*, (Russian) Uspekhi Mat. Nauk **31** (1976), 17–27.

274. ———, *On topological products, groups and new class of spaces more general than metric spaces*, (Russian) Dokl. Akad. Nauk SSSR **226** (1976), 527–529.

275. ———, *Finite-dimensional bicompact absolute neighborhood retracts are metrizable*, (Russian) Dokl. Akad. Nauk SSSR **233** (1977), 304–307.

276. ———, *On κ-metrizable spaces*, (Russian) Izv. AN SSSR **43** (1979), 442–478.

277. ———, *On Tychonov manifolds*, (Russian) Dokl. Akad. Nauk SSSR **246** (1979), 551–554.

278. ———, *Functors and uncountable powers of compacta*, (Russian) Uspekhi Mat. Nauk **36** (1981), 3–62.

279. ———, *Soft maps of manifolds*, (Russian) Uspekhi Mat. Nauk **39** (1984), 207–224.

280. **R. B. Sher**, *A complement theorem in Menger manifolds*, Proc Amer. MAth. Soc. **121** (1994), 611–618.

281. **T. Shirota**, *A class of topological spaces*, Osaka Math J. **4** (1952), 23–40.

282. **L. C. Siebenmann**, *The obstruction to finding a boundary for an open manifold of dimension greater than five*, Doctoral Dissertation, Princeton Univ. Press, Princeton, 1965.

283. ———, *L'invariance topologique du type simple d'homotopie*, Seminaire Bourbaki **428** (1972/73).

284. **W. Sierpinski**, *Sur une courbe cantorienne qui contient une image biunivoque et continue de toute courbe donnee*, C.R.Acad. Paris **162** (1916), 629–632.

285. ———, *Sur les ensembles connexes et non connexes*, Fund. Math. **2** (1921), 81–95.

286. **S. Sirota**, *The spectral representation of spaces of closed subsets of bicompacta*, Soviet. Math. Dokl. **9** (1968), 997–1000.

287. **E. H. Spanier**, *Algebraic topology*, McGraw-Hill, New York, 1966.

288. **M. A. Štanko**, *The embedding of compacta into Euclidean space*, Math. USSR Sbornik **12** (1970), 234–254.

289. ———, *Solution of Menger's problem in the class of compacta*, Soviet Math. Dokl. **12** (1971), 1846–1849.

290. **N. Steenrod and S. Eilenberg**, *Foundations of Algebraic Topology*, Princeton University Press, Princeton, 1952.

291. **R. M. Switzer**, *Algebraic topology—homotopy and homology*, Springer-Verlag, Berlin, 1975.

292. **T. Terada**, *Note on z-, $C^*$- and C-embeddings*, Sci. Rep. Tokyo Kyoiku Daigaku **13** (1975), 129–132.

293. **H. Toruńczyk**, *Remarks on Anderson's paper "On topological infinite defeciency"*, Fund. Math. **66** (1970), 393–401.

294. ———, *Skeletonized sets in complete metric spaces and homeomorphisms of the Hilbert cube*, Bull. Acad. Polon. Sci. **18** (1970), 119–126.

295. ———, *$(G, K)$-absorbing sets in metric spaces*, Ph. D. Thesis, Institute of Mathematics, Polish Acad. Sci., Warszawa, 1970.

296. ———, *Absolute retracts as factors of normed linear spaces*, Fund. Math. **86** (1974), 53–67.

297. ———, *Concerning locally homotopy negligible sets and characterization of $l_2$-manifolds*, Fund. Math. **101** (1978), 93–110.

298. ———, *On CE-images of the Hilbert cube and characterization of Q-manifolds*, Fund. MAth. **106** (1980), 31–40.

299. ———, *Characterizing Hilbert space topology*, Fund. MAth. **111** (1981), 247–262.

300. **H. Toruńczyk and J. E. West**, *Fibrations and bundles with Hilbert cube manifold fibers*, Memoirs Amer. Math. Soc. No. 406, **80**, 1989.

301. **L. A. Tumarkin**, *Zur allegmeinen Dimensionstheorie*, Proc. Akad. Amsterdam **28** (1925), 994–996.

302. **P. S. Urysohn**, *Zum Metrizationsproblem*, Math. Ann. **94** (1925), 309–315.

303. **C. T. C. Wall**, *Finiteness conditions for CW-complexes*, Ann. of Math. **81** (1965), 56–69.

304. **J. J. Walsh**, *Monotone and open mappings on manifolds*, Trans. Amer. Math. Soc. **209** (1975), 419–432.

305. ———, *Dimension, cohomological dimension and cell-like mappings*, Lect. Notes Math. **870** (1981), 105–118.

306. ———, *Characterization of Hilbert cube manifolds: an alternate proof*, preprint.

307. ———, *The finite dimensionality of integral homology 3-manifolds*, Proc. Amer. Math. Soc. **88** (1983), 154–156.

308. **L. E. Ward, Jr.**, *Fixed point sets*, Pacific J. Math. **47** (1973), 553–565.

309. **A. Waśko**, *Spaces universal under closed embeddings for finite-dimensional complete metric spaces*, Bull. London Math. Soc. 218 (1986), 293–298.

310. **T. Watanabe**, *On spaces which have the shape of compact metric space*, Fund. Math. **104** (1979), 1–11.

311. **J. E. West**, *Fixed-point sets of transformation groups on infinite-product spaces*, Proc. Amer. Math. Soc. **21** (1969), 575–582.

312. ———, *Infinite products which are Hilbert cubes*, Trans. Amer. Math. Soc. **150**, (1970), 1–25.

313. ———, *The ambient homeomorphy of an incomplete subspace of infinite-dimensional Hilbert spaces*, Pacific J. Math. **34** (1970), 257–267.

314. ———, *Mapping Hilbert cube manifolds to ANR's. A solution of a conjecture of Borsuk*, Ann. Math. **106** (1977), 1–18.

315. ———, *Open problems in infinite dimensional topology*, Open problems in topology (J. van Mill and G. M. Reed, eds.), North-Holland, New York, 1990, 523–597.

316. **J. H. C. Whitehead**, *Combinatorial homotopy,I*, Bull. Amer. Math. Soc. **55** (1949), 213–245.

317. ———, *Simple homotopy types*, Amer. J. Math. **72** (1950), 1–57.

318. **G. T. Whyburn**, *Topological charac terization of Sierpinski curve*, Fund. Math. **45** (1958), 320–324.

319. **D. C. Wilson**, *Open mappings of the universal curve onto continuous curves*, Trans. Amer. Math. Soc. **168** (1972), 497–515.

320. ———, *Open mappings of manifolds and counterexample to the Whyburn conjecture*, Duke Math. J. **40** (1973), 705–716.

321. **M. Wojdyslawski**, *Retracts absolut et hyperspaces des continus*, Fund. Math. **32** (1939), 184–192.

322. **R. Y. T. Wong**, *A note on stable homeomorphisms of infinite-dimensional manifolds*, Proc. Amer. Math. Soc. **28** (1971), 271–272.

323. ———, *Periodic actions on (I-D) normed linear spaces*, Fund. Math. **80** (1973), 133–139.

324. **A. V. Zarelua**, *Universal bicompactum of the given weight and of the given dimension*, (Russian) Dokl. Akad. Nauk SSSR **154** (1964), 1015–1018.

325. **M. Zarichnyi**, *Universal map of $\sigma$ onto $\Sigma$ and absorbing sets in the classes of absolute Borelian and projective finite-dimensional spaces*, Topology Appl., to appear.

326. ———, *Universal spaces and absolute extensors for integral cohomological dimension*, preprint.

# Subject Index

www.ingramcontent.com/pod-product-compliance
Ingram Content Group UK Ltd.
Pitfield, Milton Keynes, MK11 3LW, UK
UKHW020229290926
15470UKWH00002B/438

9 780444 8222